Symbol	Descripti	
s	slip	
S	complex or apparent powe	
t	time	s
T	average torque	N · m
T_i	instantaneous torque	N · m
v	instantaneous voltage	V
V	electromotive force or electric potential	V
$\hat{V}$	voltage complexor	V
U	velocity	m/s
W	work, energy, heat	J
X	reactance	Ω
Y	admittance	s
Z	impedance	Ω
Greek Alphabet		
α	firing angle	deg, rad
α	temperature coefficient	1/°C
β	phase angle	deg, rad
γ	conduction angle	deg, rad
δ	torque angle	deg, rad
ε	permittivity	F/m
ε_0	permittivity of free space	F/m
ε_r	relative permittivity	—
η	efficiency	—
θ	phase angle	deg, rad
λ	flux linkage	Wb
μ	permeability	H/m
μ_0	permeability of free space	H/m
μ_r	relative permeability	—
ρ	resistivity	.m
Σ	summation	—
σ	real number	—
τ	time constant	s
φ	magnetic flux lines	Wb
ψ	electric flux lines	C/m^2
ω	angular speed	rad/s
Ω	resistance	ohms

Electric Machines

Principles, Applications, and Control Schematics

Electric Machines

Principles, Applications, and Control Schematics

Dino Zorbas
McGill University

WEST PUBLISHING COMPANY
St. Paul • New York • Los Angeles • San Francisco

Fortune favours the prepared mind.
Pasteur

Don't talk to me of your Archimedes lever. . . .
Give me the right word and the right
accent and I will move the world.
Joseph Conrad

Copy and art editing: Tage Publishing Service, Inc.
Design: Schneider & Company
Illustration: Scientific Illustrators
Composition: Polyglot Compositors
Cover design: Dakota Design

50 W. Kellogg Boulevard
P.O. Box 64526
St. Paul, MN 55164-1003

Printed in the United States of America

96 95 94 93 92 91 90 89 8 7 6 5 4 3 2 1 0

Library of Congress Cataloging-in-Publication Data

Zorbas, Dino.
Electric machines : principles, applications, and control schematics / Dino Zorbas.
p. cm.
Includes index.
ISBN 0-314-47011-5
1. Electric machinery. I. Title.
TK2000.Z67 1989
621.31'042—dc 19 88-28645
CIP

To
all those who believe that the
preparation of tomorrow's professionals
must include organized practical training.

CONTENTS

PREFACE

Electric machines are an essential part of almost all moving apparatus and of all manufacturing processes. Because electric motors consume about 60% of the electrical energy produced in industrialized nations, it is extremely important, for economic and social reasons, that their selection, operation, and maintenance be well understood. Choosing the best drive, and its associated control equipment, requires a very thorough knowledge of electric machinery.

This book provides the fundamentals of the knowledge required. It explains and demonstrates the governing principles, applications, and control schematics of electric machines—not only to meet the needs of the classroom, but also to assist those who practice in the field of electric machinery. As a result, each concept is presented with two goals in mind. The first is to help the engineering student analyze and use the concept to solve problems of an academic nature. The second is to demonstrate, whenever possible, how the concept may be applied to the everyday use of electric machines, for theoretical information can be best understood when illustrated by examples from the real world. Thus, this book *bridges* the gap between the academic work of the engineering student and the practical work of the specialist in the power option of electrical engineering. The concepts associated with the installation of power cables and instrument transformers, the starting of large ac machines, the advantages of power-factor improvement, the modern methods of speed control, and the entire chapter on control schematics—to name just a few—will challenge and motivate the student, and will also be extremely useful to those who select, design and/or use substations, transformers, and electric machines.

This book includes discussions (from an engineering perspective) of basic electromagnetic devices, transformers, generators, motors, and their control schematics. Each of the first six chapters is relatively self-sufficient, so this section of the book can be covered according to the priorities of a particular school's engineering program.

The first few sections of each chapter, which generally are devoted to basic principles, require a basic knowledge of calculus. The latter parts of each chapter develop more advanced concepts and applications and therefore also require a working knowledge of circuit analysis and control systems. On account of this, the first sections of each chapter are suitable for a general machinery course for nonelectrical engineering students (civil, mechanical, and so on), while the latter sections are suitable for electrical engineering students.

Each chapter also includes a summary of its basic concepts, review questions, and many solved and unsolved problems from the real world of engineering.

Traditionally, control schematics (Chapter 7) are not taught in the machinery courses. However, practical experience indicates that a knowledge of control schematics is absolutely necessary in order to effectively bridge the work of the engineering student and that of the practicing engineer.

Many engineering students (civil, mechanical, electrical, and so on) will some day find themselves in a technical environment, so an understanding of the language of the electrical department—that is, an understanding of electrical drawings, commonly known as control schematics—will help them to conscientiously fulfill their responsibilities.

If the chapter on control schematics cannot be covered during a semester's work, it can easily be mastered by the student himself when the need arises. Like the other six chapters, it is written in simple language and includes many realistic examples.

The appendixes include a detailed review of three-phase systems, the per-unit system of analysis, Laplace transforms, a description of the components of solid-state control devices, engineering tables, and a review of basic equations from engineering economics. The instructor's manual includes solutions to exercises and end of chapter problems.

Acknowledgments

The author wishes to acknowledge the many helpful suggestions offered by

J. C. Balda	Clemson University
H. Boettcher	University of Wisconsin, Milwaukee
Joseph D. Cole	University of Kentucky, Louisville
M. Giesselmann	Texas Tech University
G. Hespelt	University of Idaho
Herman Hill	Ohio University
S. Ratnajeevan H. Hoole	Harvey Mudd College
James Knickerbroker	University of Arizona
Hassan Nour	University of Southern California
Peter Sauer	University of Illinois, Urbana-Champaign
Alvin Day	Iowa State University

The author is indebted to all his teachers, students and authors (see Bibliography listing at the end of the book) who prepared the road for the realization of this undertaking.

The constructive comments and encouragement provided by my editor Michael Slaughter of West Publishing are greatly appreciated. Finally, I express my sincere thanks to Mrs. Collette Y. Julien for her great patience in typing the manuscript.

ONE

Basic Electromagnetic Concepts

1.0 Introduction

This chapter discusses the basic parameters and principles of electric and magnetic circuits and the development of force and torque. Electric circuit parameters are normally covered in other basic courses, but they are briefly reviewed here in order to facilitate the understanding of magnetic circuit parameters.

Each electric circuit concept is analogous to a corresponding magnetic circuit concept. As a result, mastering the theory of electric circuits will help one grasp the more abstract concepts of magnetism. In order to understand electric machines, one requires a very thorough knowledge of both magnetism and electric circuits.

1.1 Electric Circuit Concepts

This section deals first with the concepts of electric potential, resistance, capacitance, inductance, impedance, and power factor, then with the principles of electric circuits and the characteristics of the power drawn from a single-phase voltage source.

1.1.1 ELECTRIC POTENTIAL

Electric potential, electromotive force (emf), and voltage are synonymous terms. In the SI system of units, they are expressed or measured in volts.

When the voltage at a power outlet is 120 volts, an electric potential, an electric voltage, or an electromotive force of the said magnitude is available at that outlet.

All main and secondary power distribution networks are characterized by a voltage source. The voltage source may be of constant amplitude (a dc voltage source) or of alternating waveform (an ac voltage source). The various waveforms—whether they represent voltage, current, or power—can be described in terms of their instantaneous, effective, or average values.

Instantaneous Value

The instantaneous value* of a waveform is given by a general equation that describes the waveform as a function of time. For example, the instantaneous value of the voltage waveform that is available in a residential single-phase power outlet is

$$v = V_m \sin \omega t \tag{1.1}$$

where v is the instantaneous value of the voltage, V_m is its maximum value, and ω is its angular frequency of oscillation.

* Although this may appear as a misnomer, it has, however, some advantages especially when one deals with the concepts of torque, power, and so on.

The angular frequency of oscillation is given by

$$\omega = 2\pi f = \frac{2\pi}{T} \text{ rad/s} \tag{1.2}$$

where f and T are, respectively, the frequency and the period of oscillation of the voltage waveform. A sinusoidal voltage waveform is shown in Fig. 1-1.

The standard frequency of oscillation of the current and voltage waveforms of single-phase and three-phase power sources in North America is usually 60 Hz. In some other parts of the world it is 50 Hz.

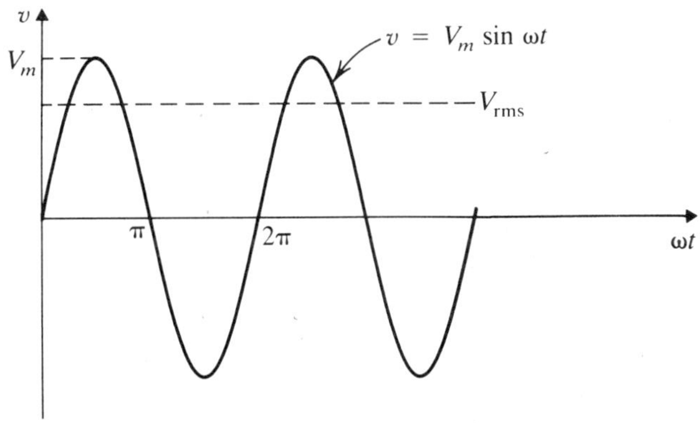

FIG. 1-1 A sinusoidal voltage waveform.

The instantaneous values of the various functions are in general represented by the lower-case letters with or without the time-function notation. Thus v (or $v(t)$) and i (or $i(t)$) represent, respectively, the instantaneous values of voltage and current.

Effective Value

The effective or root mean square (rms) value of a voltage waveform is given by

$$V_{\text{rms}}^2 = \frac{1}{T}\int_0^T v^2 \, dt \tag{1.3}$$

where v is the instantaneous value of the voltage waveform.

Root mean square values have been introduced because the heating effect of an ideal sinusoidal voltage waveform in a resistor is equal to that produced by a dc source that has the same voltage as the rms value of the given ac waveform. In designating rms values, the capital letter of the parameter is usually used without any subscript. Thus V and I correspond, respectively, to the rms values of the voltage and current. The voltage rating of all ac apparatus is given by the equivalent rms values. Thus, when one speaks of a voltage as being 208 V or 480 V, it is understood that the said magnitude is an rms value. The single-phase voltages delivered to a typical North American residence are 120 V (line-to-neutral) and 240 V (line-to-line).

The graphical interpretation of Eq. (1.3) is

$$V_{\text{rms}}^2 = \frac{\text{accumulation of areas under the voltage squared-time diagram over a complete cycle}}{\text{duration of one cycle}} \tag{1.4}$$

The graphical interpretation of Eq. (1.3), or of any other equation, helps one develop a better understanding of the physical significance of what the equation represents. One's feeling about a concept is often as important as the equation itself, because it is more easily retained and can be effectively used when calculus is forgotten, or when there is no time to review the equations, or when one deals with plant personnel who do not know much about integration.

Average Value

The average value (V_{av}) of a voltage waveform is defined as follows:

$$V_{\text{av}} = \frac{1}{T}\int_0^T v\, dt \tag{1.5a}$$

The graphical interpretation of Eq. (1.5a) is

$$V_{\text{av}} = \frac{\text{accumulation of areas under the } v - t \text{ diagram over a complete cycle}}{\text{duration of one cycle}} \tag{1.5b}$$

The average value of a symmetrical sinusoidal waveform is, as seen by inspection, equal to zero. The average value is also referred to as a dc value.

EXAMPLE 1–1 Determine the rms value of the sinusoidal voltage waveform shown in Fig. 1-1.

SOLUTION The instantaneous value of the voltage is

$$v = V_{\text{m}} \sin \omega t$$

Substituting into Eq. (1.3), we get

$$\begin{aligned} V_{\text{rms}}^2 &= \frac{1}{2\pi}\int_0^{2\pi} (V_m \sin \omega t)^2\, d\omega t \\ &= \frac{V_m^2}{2\pi}\int_0^{2\pi}\left(\frac{1 - \cos 2\omega t}{2}\right) d\omega t \\ &= \frac{V_m^2}{4\pi}\left[\omega t\Big|_0^{2\pi} - \frac{1}{2}\sin 2\omega t\Big|_0^{2\pi}\right] \\ &= \frac{V_m^2}{2} \end{aligned}$$

from which

$$V_{\text{rms}} = \frac{V_m}{\sqrt{2}}$$

The rms value of the voltage waveform is shown in Fig. 1-1.

Exercise 1–1 Find the rms and average values of the waveform shown in Fig. 1-2. Verify the answer graphically.

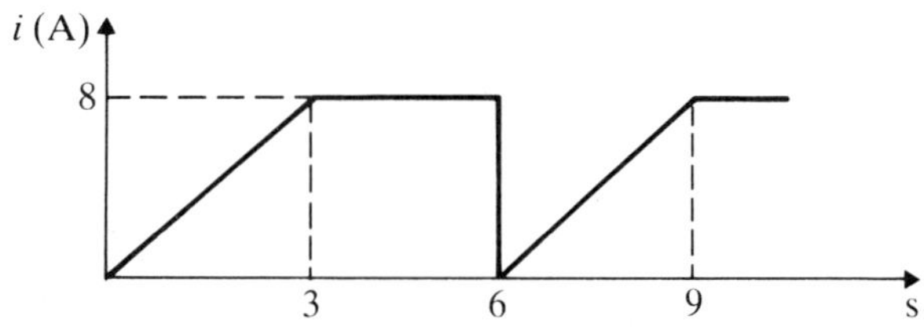

FIG. 1-2

Answer 6.53 A rms, 6A

Complexors and Phasors

A steady-state sinusoidal voltage of angular frequency of ω rad/s and of phase angle of β degrees can be of the form $V_m \cos(\omega t + \beta)$ or $V_m \sin(\omega t + \beta)$.

Recognizing that these two functions could be the real and imaginary component of a general voltage function, we have

$$\hat{V} = V_m[\cos(\omega t + \beta) + j \sin(\omega t + \beta)] \tag{1.6}$$

where $\hat{V}$ is called the voltage complexor.

A complexor is a time-variant function that can be represented by a line segment that rotates, on a fixed axis on the complex plane (Fig. 1-3(a)), with a velocity that is equal to the angular frequency of the voltage source (ω rad/s).

Using Maclaurin's series, the rectangular form of the complexor (Eq. (1.6)) can be written in its polar form as

$$\hat{V} = (V_m e^{j\beta})e^{j\omega t} \tag{1.7}$$

or

$$\hat{V} = V_m \underline{/\omega t + \beta} \tag{1.8}$$

The factor in the parenthesis of Eq. (1.7) is called the phasor (**V**) of the complexor, that is

$$\mathbf{V} = V_m e^{j\beta} \tag{1.9}$$

or

$$\mathbf{V} = V_m \underline{/\beta} \tag{1.10}$$

or

$$\mathbf{V} = \sqrt{2} V \underline{/\beta} \tag{1.11}$$

where V is the rms value of the voltage.

The phasor in general is a time-invariant parameter that can be represented by a line segment on the complex plane as shown in Fig. 1-3(b). The phasor of the input voltage to an ac machine or of the output voltage of a transformer is usually taken as reference ($\beta = 0$).

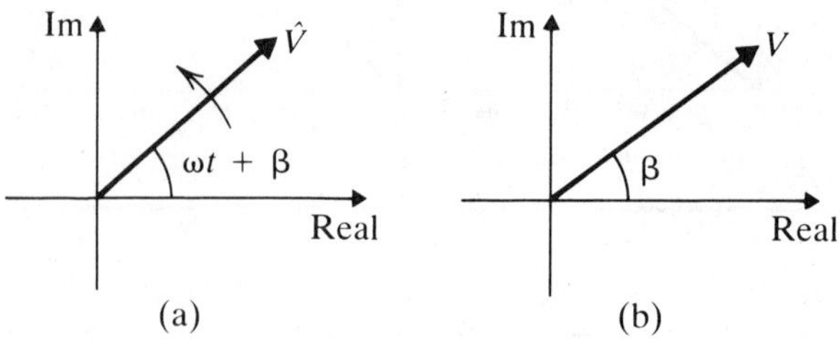

FIG. 1-3 General representation of a sinusoidal voltage on a complex plane: **(a)** its complexor, **(b)** its phasor.

Electric Field Intensity

Electric field intensity ($\vec{E}$) is related to electric potential (V) by

$$\vec{E} = -\nabla V \tag{1.12}$$

where the symbol ∇ is the del operator and ∇V is called the grad V. The grad (or gradient) of any scalar function is the space rate of change of that function. When V is described in the rectangular coordinate system, Eq. (1.12) is written as follows:

$$\vec{E} = -\left(\frac{\partial V}{\partial x}\tilde{a}_x + \frac{\partial V}{\partial y}\tilde{a}_y + \frac{\partial V}{\partial z}\tilde{a}_z\right) \tag{1.13}$$

where $\tilde{a}_x$, $\tilde{a}_y$, and $\tilde{a}_z$ are the unit vectors in the x, y, and z directions, respectively, and $\partial V/\partial x$, $\partial V/\partial y$, and $\partial V/\partial z$ are the distance rate of change of the potential in the x, y, and z directions, respectively.

From Eq. (1.13), when a scalar potential is given in a particular region, the electric field intensity in the same region is a vector quantity.

Electric field intensity is measured in volts/meter (V/m). The intensity of an electric field can have various effects. In the vicinity of radio or television stations, for example, the electric field intensity is relatively large. In such places, an incandescent lamp attached to a wire may—depending on the lamp's location and the length of the wire loop—produce a glow.

1.1.2 THE CONCEPT OF RESISTANCE

In a dc circuit, the opposition to the flow of current is called resistance. The resistance (R) of a conductor whose cross-sectional area is constant is given by

$$R = \rho \frac{l}{A} \tag{1.14}$$

where ρ is the resistivity of the conductor's material expressed in $\Omega \times$ m, l is its length expressed in m, and A is the conductor's cross-sectional area perpendicular to the flow of the current, expressed in m^2.

By inspecting Eq. (1.14), it is clear that conductors of larger cross-sectional area have lower resistance than conductors of smaller cross-sectional area. Thus, in order to minimize the resistance of a winding (wire forming one or more loops, often referred to as a coil or inductor), one has to use conductors of as large a cross-sectional area as can be physically accommodated.

The resistivity of any material depends on its atomic structure and on its temperature. At 20°C the resistivity of aluminum is 1.64 times greater than that of copper.

The resistance of a wire as a function of temperature is given by

$$R = R_0[1 + \alpha_{20}(T - 20)] \tag{1.15}$$

where R_0 is the resistance of the wire at 20°C, R is its resistance at T°C, and α_{20} is the temperature coefficient at 20°C. The temperature coefficient for copper is 0.00393/°C; for tungsten it is 0.0045/°C. Then, from Eq. (1.15), the temperatures at which copper and tungsten become superconductive (have resistance equal to zero) are −234.45°C and −202.2°C, respectively. Recently, a new class of ceramic material has been developed that becomes superconductive at much higher temperatures.

The power, or the rate of heat loss, in a winding depends on its resistance, which in turn depends on its operating temperature. When one speaks of the losses of an apparatus, then, one must specify the temperature to which these losses correspond.

1.1.3 THE CONCEPT OF CAPACITANCE

The capacitance of an electric circuit is a measure of the circuit's ability to store energy in its electric field. The higher the circuit's capacitance, the higher its ability to store energy.

Figure 1-4 shows two capacitor plates that are connected to a dc voltage supply of V volts. The upper plate is positively charged, the lower plate negatively charged.

The vector ψ between the plates represents the electric flux or the electric lines of force. The electric flux emerges from the positively charged plate and terminates in the negatively charged plate. According to Gauss's law, the electric flux is equal to the

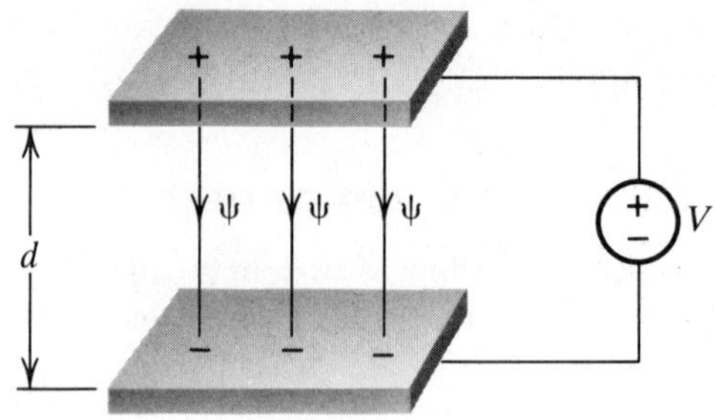

FIG. 1-4 Charged capacitor plates.

charge of one plate expressed in coulombs. In the SI system of units, capacitance is expressed or measured in farads (F).

In an ideal case, the capacitance of two parallel plates is given by

$$C = \varepsilon \frac{A}{d} \tag{1.16}$$

where

C is the capacitance of the two plates (F)

ε is the permittivity of the material between the capacitor plates (F/m)

A is the cross-sectional area of one plate (m^2)

d is the perpendicular distance between the plates (m).

The permittivity (ε) of a material is equal to the permittivity of free space (ε_0) times the relative permittivity (ε_r) of the material. That is,

$$\varepsilon = \varepsilon_r \varepsilon_0 \tag{1.17}$$

In the tuning of some radios, the parallel capacitor plates are moved in relation to each other and thus their effective area changes, resulting in a new capacitance. The new capacitance, according to the theory of resonant circuits, produces a new resonant frequency, and thus the radio can pick up a new station. A capacitor is sometimes referred to as a condenser.

The capacitance of the two plates is also given by

$$C = \frac{Q}{v_c} \tag{1.18}$$

where Q is the electric charge of one plate, and v_c is the voltage between the two plates.

The circuit representation of an ideal capacitor is shown in Fig. 1-5(a). The current (i_c) through the capacitor is, by definition, equal to the time-rate of change of the electric charge. That is,

$$i_c = \frac{dQ}{dt} \tag{1.19}$$

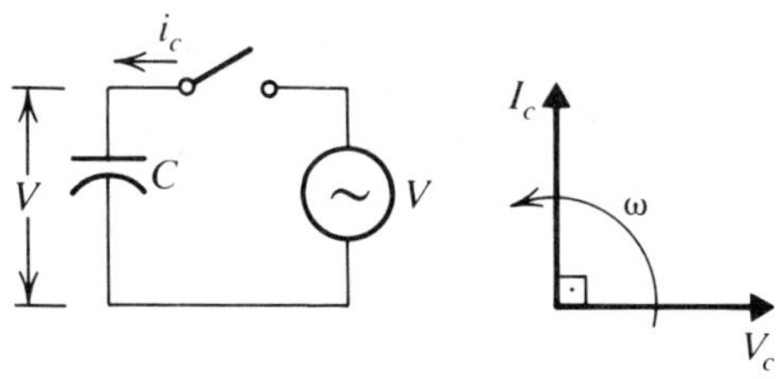

(a) (b)

FIG. 1-5 Capacitor: **(a)** circuit representation, **(b)** phasor diagram.

From Eqs. (1.18) and (1.19), we get

$$v_c = \frac{1}{C}\int i_c\,dt \tag{1.20}$$

Equation (1.20) is a general expression for the voltage across a capacitor. When the current is constant, the voltage across the capacitor increases linearly with time. When the current is sinusoidal, the rms steady-state value of the capacitor voltage (V_c) is

$$V_c = \frac{I_c}{j\omega C} \tag{1.21}$$

where I_c is the rms value of the current and ω is its angular frequency of oscillation expressed in rad/s. The operator j is the well-known factor of $\sqrt{-1}$.

From Eq. (1.21) we obtain

$$V_c = -j\frac{I_c}{\omega C} \tag{1.22}$$

or

$$V_c = -jX_cI_c \tag{1.23}$$

where X_c is called the capacitive reactance, given by

$$X_c = \frac{1}{\omega C} \text{ ohms} \tag{1.24}$$

or

$$X_c = \frac{1}{2\pi fC} \tag{1.25}$$

where f is the frequency of oscillation of the sinusoidal waveform.

From the complex plane notation we have

$$j = 1\underline{/90^\circ} \tag{1.26}$$

From the above and Eq. (1.23), we obtain

$$I_c = \frac{V_c\underline{/90^\circ}}{X_c} \tag{1.27}$$

That is, the current in a capacitor leads its voltage by 90 degrees. This is shown in Fig. 1-5(b).

The incremental energy (dW) gained by a capacitor in an infinitesimal amount of time dt is

$$dW = v_c i_c \, dt \tag{1.28}$$

From the above and Eq. (1.19) we obtain

$$dW = v_c \, dQ \tag{1.29}$$

Substituting for dQ its equivalent from Eq. (1.18), we get

$$dW = C v_c \, dv_c \tag{1.30}$$

Taking integrals and assuming that the voltage changes from zero up to V volts, we obtain the energy stored within the electric field of a capacitor:

$$W = C \int_0^V v_c \, dv_c = \frac{1}{2} C V^2 \tag{1.31}$$

where W is the energy stored in the electric field and V is the voltage across the capacitor.

When the switch is opened, the capacitor remains charged. Depending on the magnitude of its voltage, it may be hazardous to personnel. When the switch is closed, the supply voltage may aid the capacitor voltage. As a result, the voltage across the switch will be, instantaneously, twice that of the supply voltage. This again may be hazardous to personnel and may also carbonize the contacts of the switch, causing it to malfunction.

Carbonization is probably the most common cause of failure of equipment that is called upon to start and stop frequently: air conditioners, clothes washers, etc. Recently an international company that operates public clothes washing machines reported an annual failure rate of about 12 percent. The major cause, as determined by consultants, was the carbonization of the timer's contacts that resulted from the circuit capacitors.

To eliminate these problems the capacitors should be connected in parallel to properly sized resistors so that the stored energy may be safely dissipated as heat. For these reasons the electrical code requires that a capacitor in a power distribution network that is rated at less than 750 V should be equipped with a parallel-connected resistor so that the stored energy may be dissipated in less than 60 seconds.

Electric Flux Density

Electric flux density (D) is equal to the electric flux divided by the area (A) perpendicular to it. That is,

$$D = \frac{\psi}{A} \tag{1.32}$$

The unit of electric flux density is coulomb/m^2.

The electric flux density of any given material is related to the electric field intensity and the permittivity of the material by

$$\varepsilon = \frac{dD}{dE} \tag{1.33}$$

where ε is the permittivity of the material. Thus, when the flux density–field intensity curve of a material is given, its slope at the operating flux density is equal to the actual permittivity of the material.

1.1.4 THE CONCEPT OF SELF-INDUCTANCE

The self-inductance of a circuit is a measure of the circuit's ability to store energy in its magnetic field. The higher the inductance of a circuit, the higher its capacity to store energy. The circuit representation of an ideal inductor is shown in Fig. 1-6(a).

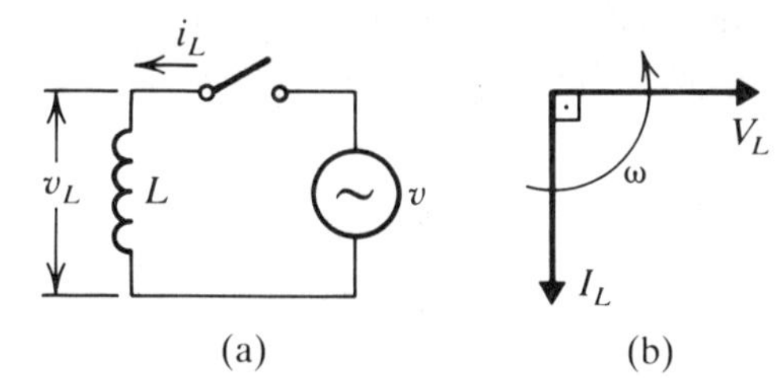

FIG. 1-6 Inductance: **(a)** circuit representation, **(b)** phasor diagram.

The voltage across an inductor (v_L) is equal to its self-inductance times the time-rate of change of its current. That is,

$$v_L = \frac{L\,di_L}{dt} \tag{1.34}$$

where L and i_L are, respectively, the self-inductance and the instantaneous current of the inductor. When the current through an inductor is constant, as in the steady-state operation of dc circuits, the voltage across the inductor is zero. In such cases the inductor is said to be short-circuited.

When the current is sinusoidal, the steady-state rms value of the inductor voltage (V_L) is

$$V_L = j\omega L I_L \tag{1.35}$$

where I_L is the rms value of the current and ω is its angular frequency of oscillation.

Equation (1.35) can also be written as

$$V_L = jX_L I_L \tag{1.36}$$

where X_L is called the inductive reactance, given by

$$X_L = \omega L \text{ ohms} \tag{1.37}$$

From the complex plane notation we have

$$j = 1 \angle 90^\circ \tag{1.38}$$

From the above and Eq. (1.36) we get

$$I_L = \frac{V_L \angle -90^\circ}{X_L} \tag{1.39}$$

That is, the current in an inductor lags its corresponding voltage by 90 degrees. The voltage–current phasors for an inductor are shown in Fig. 1-6(b). Resistors, capacitors, and inductors are referred to as "passive elements."

The incremental energy (dW_f) gained by an inductor in an infinitesimal amount of time dt is

$$dW_f = v_L i_L \, dt \tag{1.40}$$

From the above and Eq. (1.34) we obtain

$$dW_f = L i_L \, di_L \tag{1.41}$$

Taking integrals, and assuming a variation of the current from 0 to i_L A, we obtain, after integration:

$$W_f = \frac{1}{2} L i_L^2 \tag{1.42}$$

That is, the energy stored in a winding is proportional to its inductance and the square of the current. Equation (1.42) is also derived in Section 1.3.3 by using general magnetic circuit parameters.

The energy stored in an inductor (magnetic field) is released when the inductor is disconnected from its voltage source. The released energy is dissipated as heat. Depending on the circumstances, this heat can damage (carbonize) the contacts of the switches, relays, and so on through which the inductor's current is interrupted. When energy-storing elements are connected in parallel to properly sized resistors, however, the available energy is — following the current interruption — safely dissipated as heat in the resistor. Many field problems arise because this basic concept is misunderstood. The student must develop a feel for the quantitative and qualitative aspects of energy. For this reason, the entire *energy domain* is given in the appendix of this chapter (see Table 1-3).

As will be shown in Section 1.2.6, the self-inductance of a winding is proportional to its number of turns (N) squared. That is,

$$L \propto N^2 \qquad \textbf{(1.43)}$$

Owing to the actual number of winding turns of transformers and induction machines, it can be said that most loads in industrial plants are inductive in nature, or that the current lags the voltage. Capacitors and long transmission lines are, however, capacitive in nature. In contrast, synchronous motors can be adjusted to operate as either inductive or capacitive networks. The concept and physical significance of inductance will be illustrated with an analogous parameter from mechanical engineering.

The inductance multiplied by the current gives the so-called electrical momentum, which is analogous to mechanical momentum. In mathematical symbols,

$$LI \Rightarrow MU$$

$$\text{(inductance)(current)} \Rightarrow \text{(mass)(velocity)}$$

The flow of current through a circuit of high inductance cannot be interrupted *instantaneously*, just as a large moving object cannot be stopped instantaneously. Interrupting the flow of current in a highly inductive circuit (see Problem 1-7) is as destructive as attempting to stop a moving truck.

Fortunately, the hydraulic brakes of trucks and the circuit breakers of inductive loads are designed to withstand the forces that accompany the sudden stopping of trucks or the abrupt interruption of currents. In practice, in order to avoid fatal accidents, the operators of high-voltage "load break disconnects" must use a hook and, for a given voltage, must be a certain minimum distance away from the disconnects each time the current or the power is to be interrupted. The physical opening of an inductive circuit is followed by arcing until the magnetic field energy has been converted to heat.

The ratio of the inductance to the resistance (L/R) of a circuit defines the time constant of the circuit. The time constant indicates how fast the current increases or decreases following the closing or opening of a switch that connects the circuit to a dc source. For all practical purposes, it takes about four time constants for the current in an R-L circuit to reach its final value. Time constants are discussed extensively in Section 6.2.2.

1.1.5 IMPEDANCE

Physically, the impedance of a circuit represents the opposition of the circuit to the flow of current. The current may be of any waveform, but in this text only ac waveforms are considered. The impedance of various apparatus and machines varies widely. For example, the magnitude of the impedance of a TV set, as seen from the 120 V, 60 Hz voltage source, is several hundred ohms, while the impedance of an induction machine, as seen from two of its supply wires, is in many cases several ohms.

In general, impedance (Z) has a real and a positive or negative imaginary component. That is,

$$Z = R \pm jX \text{ ohms} \tag{1.44}$$

where R and X are, respectively, the resistance and reactance of the circuit.

The positive reactance ($X = \omega L$) is called inductive, while the negative reactance is called capacitive ($X = 1/\omega C$). The angular frequency of oscillation (ω) is expressed in rad/s, the inductance (L) in henrys, and the capacitance (C) in farads.

The inverse of the impedance is called admittance (Y), which is measured in siemens (S). The admittance has a real part and a negative or positive imaginary part. That is,

$$Y = G \mp jB = \frac{1}{Z} \text{ S} \tag{1.45}$$

The real part (G) of the admittance is called conductance, while its imaginary part (B) is called susceptance. The positive and negative signs in the equation represent capacitive and inductive susceptance, respectively. The real part of Y is not necessarily equal to the reciprocal of the real part of Z.

For example, if an impedance is equal to $3 + j4$ ohms, then its conductance is, as shown below $3/25\,S$.

$$Y = \frac{1}{Z} = \frac{1}{3 + j4} = \frac{3 - j4}{(3 + j4)(3 - j4)}$$

$$= \frac{3 - j4}{3^2 + 4^2} = \frac{3}{25} - j\frac{4}{25} \text{ S}$$

1.1.6 POWER FACTOR

The power factor of an electrical load is defined as the cosine of the phase angle between its voltage and current phasors. In other words, the power factor is the cosine of the phase angle of the circuit impedance.

Lagging power factor means that the current is lagging the voltage, and *leading* power factor means that the current is leading the voltage. The higher the power factor of a given ac load, the lower its upstream power losses and voltage drop. The advantages of higher-power-factor loads are qualitatively and quantitatively evaluated in Chapter 5 (see Example 5-4).

EXAMPLE 1–2 The *R-L* circuit shown in Fig. 1-7(a) is connected to a 120 V, 60 Hz voltage supply. Determine the circuit:

a) Impedance.
b) Power factor.

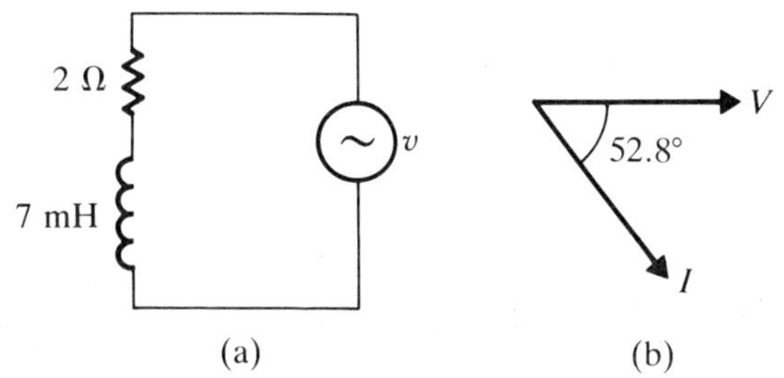

FIG. 1-7

SOLUTION

a) The impedance is

$$Z = R + jX_L = R + j2\pi fL = 2 + j2\pi(60)(7 \times 10^{-3}) = 2 + j2.64$$
$$= \underline{\underline{3.31 \angle 52.8^\circ \ \Omega}}$$

b) The power factor is

$$\cos\theta = \cos 52.8^\circ = \underline{\underline{.6 \text{ lagging}}}$$

The voltage and current phasors are shown in Fig. 1-7(b).

Exercise 1–2 The *R-C* circuit shown in Fig. 1-8 is connected to a 120 V, 60 Hz supply. When the power factor of the circuit is .8 leading, determine the circuit:

a) Capacitance.
b) Impedance.

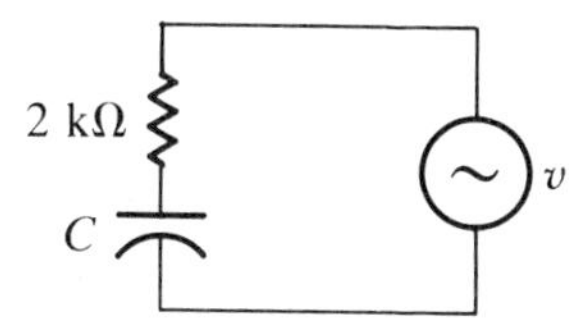

FIG. 1-8

Answer (a) 1.77 μF; (b) $2.5\angle -36.9^\circ$ kΩ

Exercise 1–3 Determine the resistance, reactance, impedance, admittance, conductance, and susceptance of the network shown in Fig. 1-9.

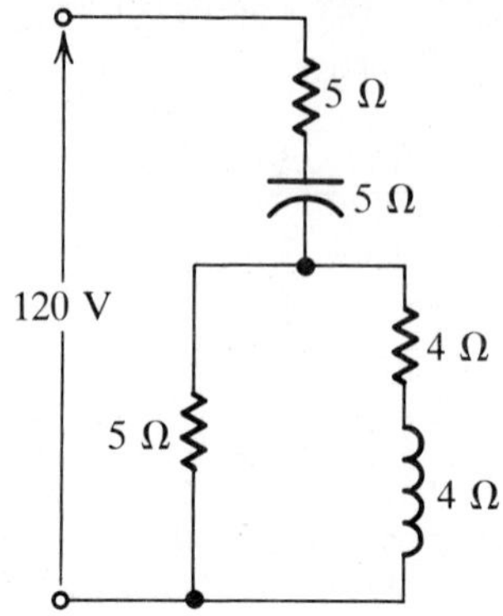

FIG. 1-9

Answer $Z = (7.68 - j3.97)\ \Omega$; $Y = (.10 + j.05)\ S$

1.1.7 PRINCIPLES OF ELECTRIC CIRCUITS

This section explains Ohm's and Kirchhoff's laws and their application in solving problems of electric circuits.

Ohm's Law

According to the generalized Ohm's* law, the current that will flow in an electric circuit is given by the ratio of the circuit potential divided by the circuit impedance. Referring to Fig. 1-10, we have

$$i = \frac{v}{Z} \tag{1.46}$$

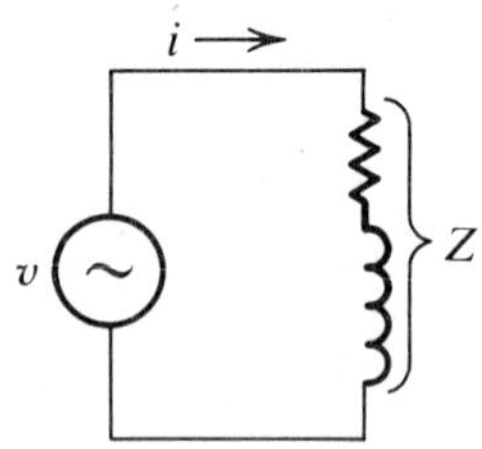

FIG. 1-10

* Ohm's experiments and conclusions dealt with dc circuits only.

where i, v, and Z are, respectively, the current, voltage, and impedance of the circuit in phasor form.

The impedance is a characteristic of the circuit and in electric machines — depending on the type of machine and the operating condition—may be a variable. The current through an electric circuit driven by a zero-impedance voltage source can be increased to infinity when the circuit impedance is shorted.

In the following example Ohm's law is used to derive general expressions for the voltage across two series resistors as a function of the supply voltage (voltage-divider concept) and the current through two parallel resistors as a function of the supply current (current-divider concept). The derived mathematical equations simplify the solution of many problems in electric and magnetic circuits.

EXAMPLE 1–3 Referring to the circuit of Fig. 1-11, prove the following:

a) $V_x = V \dfrac{R_b}{R_2}$

b) $I_1 = I \dfrac{R_2}{R_1 + R_2}, \quad I_2 = I \dfrac{R_1}{R_1 + R_2}$

The expression in (a) is the voltage-divider concept; those in (b) are current-divider concepts.

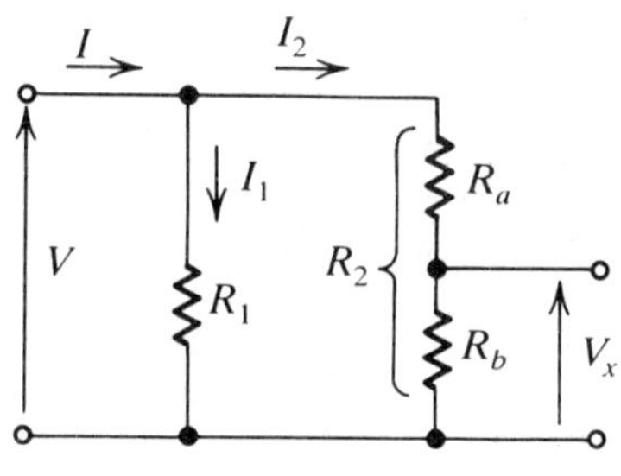

FIG. 1-11

SOLUTION The currents through the various resistors are identified as shown in Fig. 1-11.

a) Applying Ohm's law, we obtain:

$$V_x = I_2 R_b = \frac{R_b V}{R_a + R_b} = \frac{V}{R_2} R_b$$

b) Again from Ohm's law, we get the following relationships:

$$V = I(R_1 \mathbin{/\!/} R_2) = I \frac{R_1 R_2}{R_1 + R_2} \tag{I}$$

Also:

$$V = I_1 R_1 \tag{II}$$

and

$$V = I_2 R_2 \tag{III}$$

From (I), (II) and (III), we obtain:

$$I_1 = I \frac{R_2}{R_1 + R_2}$$

and

$$I_2 = I \frac{R_1}{R_1 + R_2}$$

Exercise 1–4 Find the current through the capacitor of Fig. 1-12.

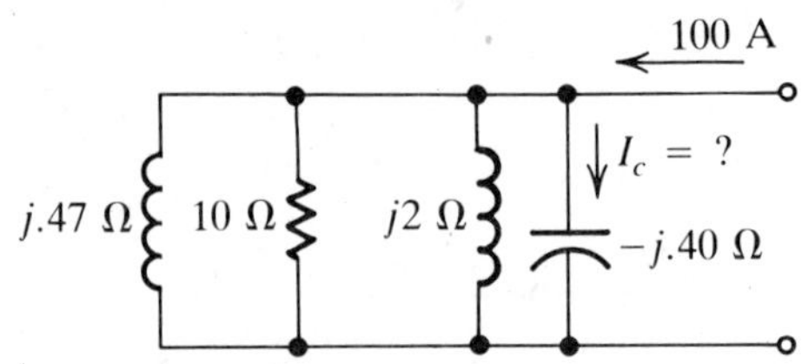

FIG. 1-12

Answer $1.54/142°$ kA

NOTE Figure 1.12 represents the per-phase equivalent circuit of a 20 MW plant to the flow of the 5th harmonic current at a particular load condition. The 100 A current is the 5th harmonic current produced by the solid state control devices of ac and dc machines. The .47 Ω reactance represents the reactance of the utility (source), the $10//j2\,\Omega$ impedance represents the load of the plant, and the .40 Ω reactance represents the plant's *incorrectly* sized capacitor banks, all at 4.16 kV level.

Kirchhoff's Laws

Voltage Law

According to Kirchhoff's voltage law (KVL), the algebraic sum of the electric voltages along a closed loop is equal to zero. That is,

$$\sum V_{\text{loop}} = 0 \tag{1.47}$$

In applying KVL, the following procedure is recommended.

a) Draw or assign loop currents (a loop current is a current that flows in a closed path) in whatever direction is preferable. That is, the clockwise (CW) or counterclockwise (CCW) current direction is arbitrary.

b) Assign positive (+) and negative (−) signs to the passive elements. The terminals where the current is entering should be marked positive and the terminals where the current is leaving, negative. In other words, the terminal of the passive element where the current emerges is at a relatively lower potential than the terminal where the current enters. A good analogy is that current, like free-running water, flows from a higher elevation (potential) to a lower one.

c) Identify the relative polarity of the sources (active elements). This is normally given. An arrow, placed between the terminals of a voltage source, indicates the direction of rising potential, and the tip of the arrow points toward the terminal of positive polarity.

d) Starting at any convenient point, proceed along a loop summing up all the voltages encountered. According to KVL, these voltages must be equal to zero. That is,

$$\sum V_{\text{loop}} = 0$$

It is absolutely necessary to sum up the voltages along a closed loop* even if there is no physical connection for the current to flow through. This condition is encountered in some equivalent circuits of electronic devices.

In proceeding along a loop, one does not have to follow the same direction as the loop current.

e) The sign of each voltage term in a KVL equation depends on the polarity of the element first seen as one proceeds along the loop. In going from the positive to the negative terminal of an element, the voltage across this element may be taken as negative. In contrast, in going from a negative to a positive terminal of an element, the voltage across this element must be taken as positive. However, the opposite sign convention could be selected.

f) When the correct solution of the loop equations yields a negative current, this means only that the assumed current direction is wrong.

EXAMPLE 1–4 Write the loop equations for the circuit shown in Fig. 1-13(a).

SOLUTION The circuit is redrawn in Fig. 1-13(b). Currents I_1 and I_2 flow in loop 1 and loop 2, respectively. (One may assign a loop current along the path $a-b-c-n-o-d-a$, but this is not included because it serves no purpose.) Clockwise direction is arbitrarily chosen for all loop currents. The sign, or the relative polarity across each element, is designated as shown.

Loop 1 $(a-b-c-d-a)$

. . .

* A closed loop, in this context, is simply the path one decides to follow in writing KVL; it may have nothing to do with the closed loop around which a loop current flows.

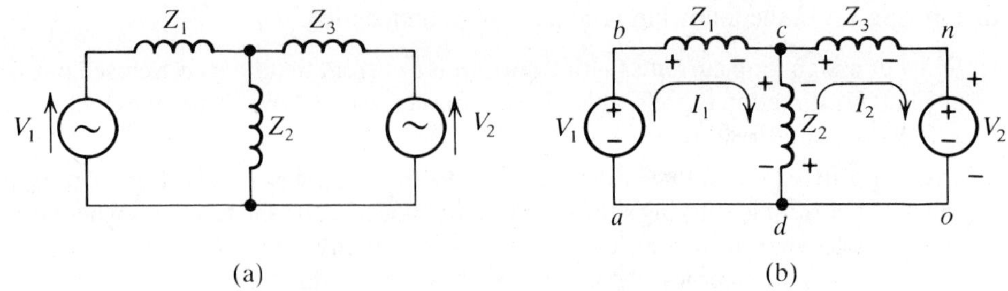

FIG. 1-13

Applying KVL, we obtain

$$V_1 - I_1Z_1 - I_1Z_2 + I_2Z_2 = 0$$

The voltage across Z_2 is due to the currents I_1 and I_2. These currents pass in opposite directions through the same impedance and will result in voltage drops of opposite polarity.

Loop 2 $(o-n-c-d-o)$

Applying KVL again, we get

$$V_2 + I_2Z_3 + I_2Z_2 - I_1Z_2 = 0$$

In proceeding along the loop, one encounters—owing to I_2—the same polarity across the passive elements as the given polarity on the voltage source V_2. This is why the first three terms in this loop equation have the same sign.

Current law

According to Kirchhoff's current law (KCL), the algebraic sum of the currents into a junction is equal to zero. That is,

$$\sum I_{\text{junction}} = 0 \tag{1.48}$$

In other words, the incoming currents to a junction must be equal to the outgoing currents.

In applying KCL, the following procedure is recommended:

a) Change the voltage sources to equivalent current sources by using Norton's theorem. According to this theorem, a voltage source (V) in series with an impedance (Z) is equivalent to a current source in parallel with the impedance (Z). The magnitude of the current source is V/Z, and its direction is the same as the current flow through the shorted series voltage-impedance network.

b) Identify and designate all the points on the circuit that are at different potentials. These points are referred to as the *nodes* of the circuit. For this reason Eq. (1.48) is often referred to as the "nodal equation."

c) Select a reference node — that is, a point of zero potential.

d) Apply KCL to each node. It is very important to note that, in a circuit with only current sources, *the node under consideration has the highest potential* in the circuit and thus the currents leave the node unless they are injected into it by a current source.

e) The currents that leave the node must be designated with a sign opposite to those approaching the node.

EXAMPLE 1–5 Write the nodal equations for the circuit shown in Fig. 1-14(a).

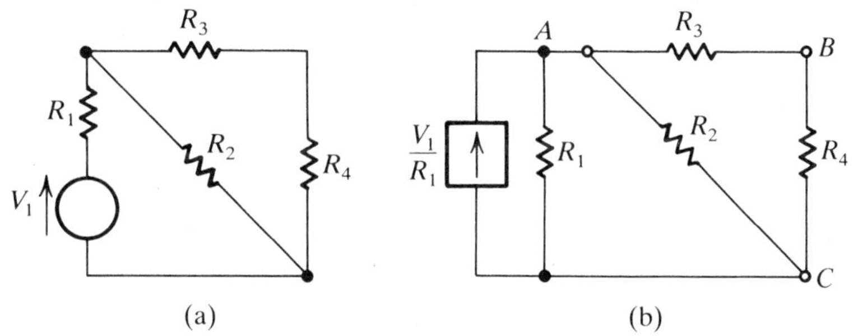

FIG. 1-14

SOLUTION The voltage source in series with the resistance R_1 is changed to an equivalent current source by using Norton's theorem. The resulting circuit is shown in Fig. 1-14(b). There are three nodes, identified with the letters A, B, and C. Node C is arbitrarily selected as reference.

Node A

This node is connected to four wires, and thus its nodal equation must have four terms.

From KCL we have

$$\frac{-V_1}{R_1} + \frac{V_A}{R_1} + \frac{V_A - V_B}{R_3} + \frac{V_A}{R_2} = 0 \qquad \text{(I)}$$

Notice that the current that flows from A to B is equal to the potential of A minus the potential of B divided by the resistance R_3. That is,

$$I_{AB} = \frac{V_A - V_B}{R_3}$$

This current flows from A to B because the node under consideration has the highest potential in the circuit.

Node B

Similarly, for node B we have

$$\frac{V_B}{R_4} + \frac{V_B - V_A}{R_3} = 0 \qquad \text{(II)}$$

From equations (I) and (II) the potential of points A and/or B can be easily determined, if required.

Ohm's and Kirchhoff's laws are the outstanding principles of electrical engineering. The design of all circuits must always satisfy these laws.

The high starting current in three-phase induction machines, the high short-circuit currents of ac machines, the sparking that accompanies the interruption of current through various coils, and the occasional fires that result from electric circuits are all caused by reactions of nature that attempt to satisfy these basic principles.

Exercise 1–5 The current through the resistor of the circuit in Fig. 1-15 is

$$i_r = 2 \sin\left(600t + \frac{\pi}{6}\right) \text{ A}$$

a) Determine the rms value of the total current.

b) Determine the power factor of the circuit.

c) Draw the phasor diagram.

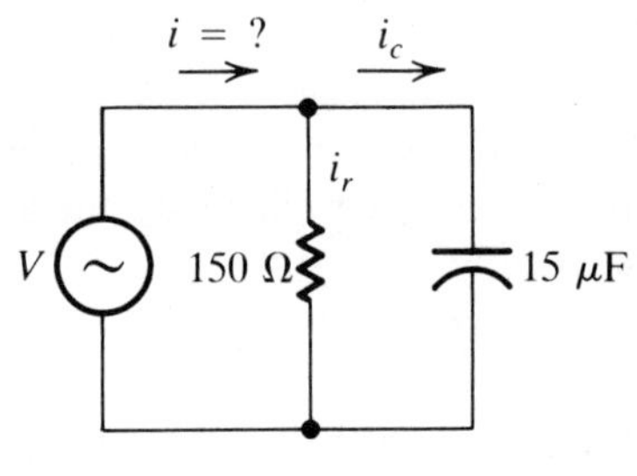

FIG. 1-15

Answer (a) 2.38 A; (b) .60 leading

1.1.8 POWER OF A SINGLE-PHASE SOURCE

Consider the circuit of Fig. 1-16(a). The input voltage waveform is shown in Fig. 1-16(b). The instantaneous power ($p(t)$) drawn by the load, in terms of the instantaneous values of the voltage and current, is given by

$$p(t) = vi \qquad \textbf{(1.49)}$$

where

$$v = V_m \sin \omega t \qquad \textbf{(1.50)}$$

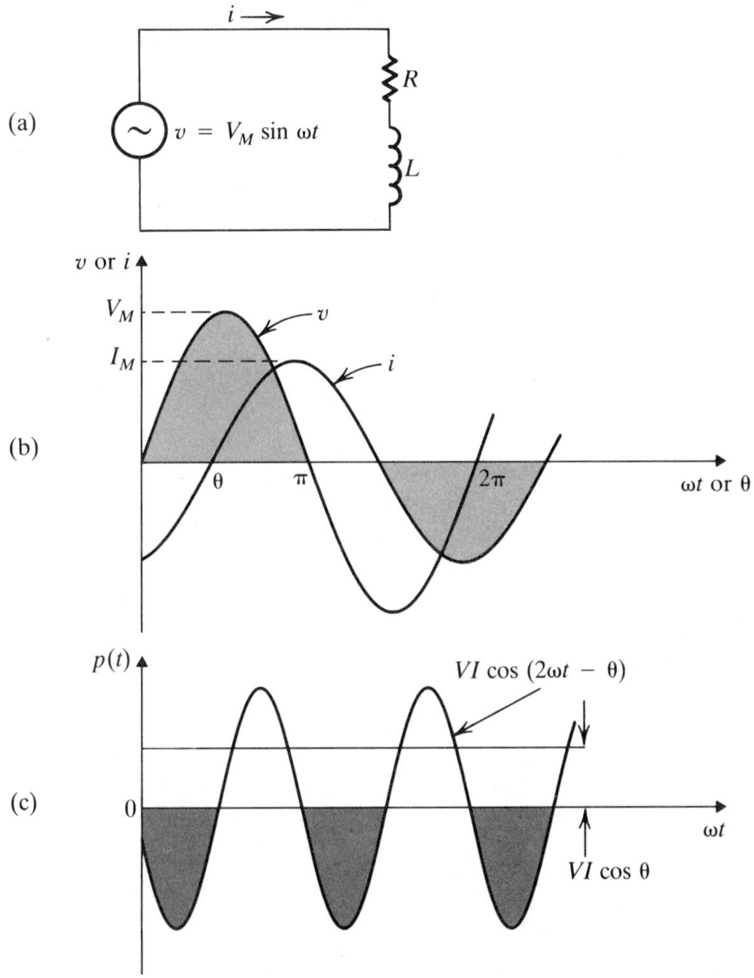

FIG. 1-16 **(a)** *R-L* circuit, **(b)** instantaneous supply voltage and current, **(c)** instantaneous power.

and

$$i = \frac{v}{Z} \qquad \textbf{(1.51)}$$

$$= \frac{V_m}{|Z|} \sin (\omega t - \theta) \qquad \textbf{(1.52)}$$

$$= I_m \sin (\omega t - \theta) \qquad \textbf{(1.53)}$$

where θ is the phase angle of the inductive impedance Z.

Substituting the instantaneous values of the voltage (Eq. (1.50)) and current

(Eq. (1.53)) into Eq. (1.49), we obtain:

$$p(t) = V_m I_m \sin \omega t \sin (\omega t - \theta) \tag{1.54}$$

Replacing the maximum values by their equivalent rms values, and using the identity

$$\cos (A \pm B) = \cos A \cos B \mp \sin A \sin B \tag{1.55}$$

we obtain

$$p(t) = VI \cos \theta - VI \cos (2\omega t - \theta) \tag{1.56}$$

Thus, the instantaneous power drawn from a single-phase network has a constant component ($VI \cos \theta$) and a pulsating component ($VI \cos (2\omega t - \theta)$) that oscillates at a frequency twice that of the supply voltage.

The variation of instantaneous power as a function of time is shown in Fig. 1-16(c). The shaded areas of the instantaneous power diagram are negative and thus indicate power returned to the source. The positive parts of the instantaneous power indicate power delivered to the load. The larger the phase angle of the load impedance, the larger the power returned to the source. When θ is 17°, the ratio of the power delivered to that of the power returned is about 4, while when θ is 71°, this ratio is about 2. When θ is 90°, the returned power is equal to that delivered and thus the load does not consume any power. In this case, power is absorbed during half of the voltage cycle, and the same amount of power is returned during the other half of the voltage cycle.

The *constant* component of instantaneous power is called the *average power.*

The average value (P) of the power drawn by an inductive load can be also derived as follows:

$$P = \frac{1}{T} \int_0^T p(t)\, dt \tag{1.57}$$

$$= \frac{1}{T} \int_0^T (V_m \sin \omega t)[I_m \sin (\omega t - \theta)]\, d\omega t \tag{1.58}$$

$$= \frac{V_m I_m}{4\pi} \int_0^{2\pi} [\cos \theta - \cos (2\omega t - \theta)]\, d\omega t \tag{1.59}$$

The second term in the integrand, being a sinusoidal function, has an average value equal to zero. Thus,

$$P = \left(\frac{V_m I_m}{4\pi} \cos \theta \right) \omega t \Big|_0^{2\pi} = \frac{V_m I_m}{2} \cos \theta \tag{1.60}$$

or

$$P = VI \cos \theta \tag{1.61}$$

As seen from the above, the average power consumed by a load is equal to its power factor times the product of the rms value of its voltage and current.

The complex power (S) drawn from a single-phase source is defined as

$$S = VI^* \tag{1.62}$$

where V is the rms value of the voltage and I^* is the complex conjugate of the current (if $I = 10\underline{/-30^\circ}$A, then $I^* = 10\underline{/30^\circ}$A).

For parallel-connected loads the total complex power (S_T) is

$$S_T = S_1\underline{/\theta_1} + S_2\underline{/\theta_2} + \cdots + S_n\underline{/\theta_n} \tag{1.63}$$

where $S_1\underline{/\theta_1}$, $S_2\underline{/\theta_2}$, $S_n\underline{/\theta_n}$ are the complex power phasors of the individual constitutent loads, each of which operates at a different lagging power factor.

The magnitude of the complex power is called the *apparent* power. Thus,

$$\text{apparent power} = |S| = VI \tag{1.64}$$

Apparent power is expressed in volt-amperes (VA) and it is used to designate the rating or nominal capacity of transformers and ac generators.

Refer to Fig. 1-17. The projection of the apparent power phasor on the real axis gives the real power. Thus, again,

$$P = VI \cos\theta \tag{1.65}$$

Real or average power is expressed in watts (W) and is used to designate the rating of motors and other electrical loads. Real power is designated as positive when it flows toward the load and as negative when it flows away from the load.

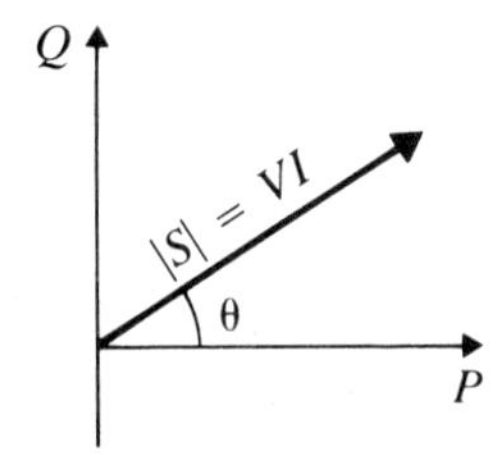

FIG. 1-17 Power phasors.

The projection of the apparent power on the imaginary axis gives the imaginary or reactive (Q) power. Thus,

$$Q = VI \sin\theta \tag{1.66}$$

Reactive power is expressed in volt-amperes reactive (VAR).

Reactive power is used to designate the rating of capacitors and inductors. Reactive power is positive when it flows toward the load and negative when it flows away from the load.

In the steady-state analysis of parallel-connected loads, it is convenient to use the power-triangle concept. The power triangles of the individual loads are drawn first, and then the overall power triangle is obtained by combining the individual power triangles. This is demonstrated by the next example.

EXAMPLE 1–6 A 230 V, 60 Hz voltage source supplies rated power to three loads whose characteristics are shown in Fig. 1-18(a). Assuming that each load's efficiency is 90%, determine:

a) The overall power factor of the system.

b) The magnitude of the current drawn from the voltage source.

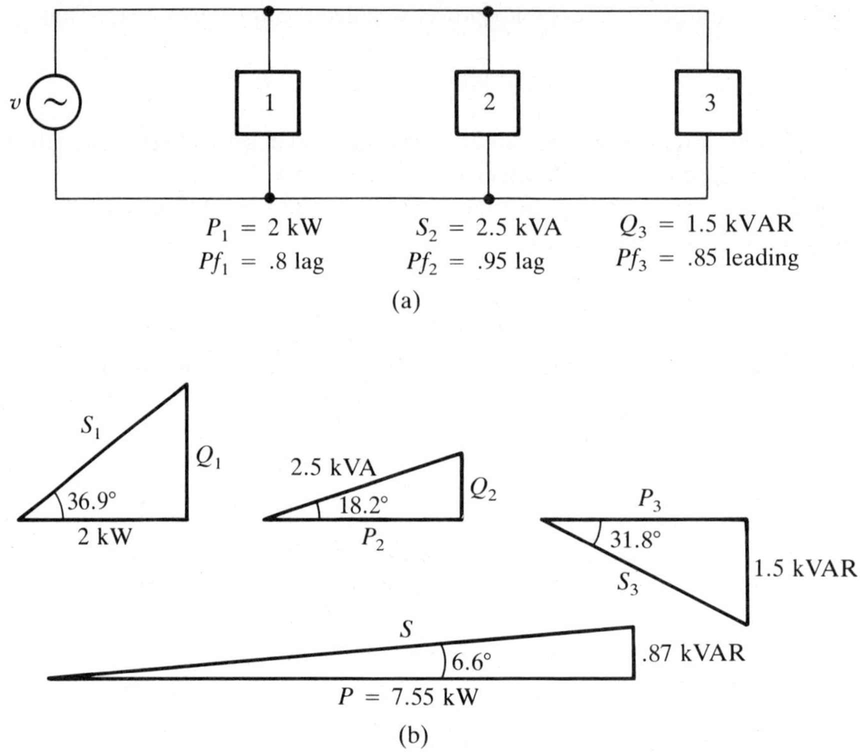

FIG. 1-18

SOLUTION

a) The individual power triangles are constructed from the given data. The input overall power triangle can be obtained graphically from the individual power triangles (see Fig. 1-18(b)) or mathematically as follows. The input real power is

$$P = \frac{1}{\eta}(P_1 + P_2 + P_3)$$

where η is the efficiency (output power/input power) of each load. Substituting the equivalent equations for P_2 and P_3, we obtain

$$\begin{aligned} P &= \frac{1}{\eta}\left(P_1 + S_2 \cos\theta_2 + \frac{Q_3}{\tan\theta_3}\right) \\ &= \frac{1}{.9}\left[2 + 2.5(.95) + \frac{1.5}{\tan 31.8^\circ}\right] \\ &= 7.55 \text{ kW} \end{aligned}$$

The input reactive power is

$$\begin{aligned} Q &= \frac{1}{\eta}(Q_1 + Q_2 - Q_3) \\ &= \frac{1}{.9}(P_1 \tan\theta_1 + S_2 \sin\theta_2 - Q_3) \end{aligned}$$

$$\begin{aligned} Q &= \frac{1}{.9}(2 \tan 36.9^\circ + 2.5 \sin 18.2^\circ - 1.5) \\ &= .87 \text{ kVAR} \end{aligned}$$

The complex power is

$$\begin{aligned} S &= P + jQ \\ &= 7.55 + j.87 = 7.6\underline{/6.6^\circ} \text{ kVA} \end{aligned}$$

The power factor is

$$\cos(6.6^\circ) = \underline{\underline{.99 \text{ lagging}}}$$

Alternatively, using Eq. (1.63) and Fig. 1-18(b), we obtain

$$S_T = S_1 + S_2 + S_3 = \frac{1}{\eta}\left(\frac{2\underline{/36.9^\circ}}{\cos 36.9^\circ} + 2.5\underline{/18.2^\circ} + \frac{1.5}{\sin 31.8}\underline{/-31.8^\circ}\right)$$

$= \frac{1}{.9}(6.8 + j.78) = 7.6\underline{/6.6^\circ}$ kVA, which agrees with the previous answer

b) The magnitude of the current is

$$I = \frac{|S|}{V} = \frac{7600}{230} = \underline{\underline{33.0 \text{ A}}}$$

Exercise 1–6 An impedance of $Z = 12 + j12$ ohms is connected to a 240 V, 60 Hz supply. Determine:

a) The voltage and current as trigonometric functions of time, taking the voltage as reference.

b) The instantaneous power as a function of time.
c) The maximum value of the instantaneous power.
d) The average power.
e) The volt-amperes and the VAR drawn from the 240 V supply.

Answer (a) 339.41 sin 377t V, 20 sin (377t – 45°) A
(b) 2.4 – 3.39 cos (754t – 45°) kVA
(c) 5.79 kVA
(d) 2.4 kW
(e) 3.39 kVA, 2.4 kVAR

Safety Considerations

Preceding sections have explained the basic elements and principles of electric circuits. Their physical significance can be further illuminated by considering the effects of voltage and of equipment grounding on the safety of personnel.

Effects of Voltage

A voltage higher than 50 V is considered dangerous to human beings. However, the resulting *current* through the human body determines the degree of danger, as the following data indicate:

Magnitude of Current	*Effects*
1 to 5 ma	Threshold of sensation
6 to 15 ma	Painful shock
16 to 20 ma	Will not let go of the stricken person
21 to 250 ma	Damage to blood vessels with mal-functioning of the heart
250 to 300 ma	Breathing stops

The magnitude of the current depends on the applied voltage and the resistance of the human body. The latter varies between individuals. Depending on the current path, the usual range of variation is 1.5 to 500 kΩ.

The degree of danger of any given current also depends on the current's duration. Hence, when one speaks of a circuit's current protective device, one must specify the *time* it takes to interrupt a specific amount of *current*.

Effects of Grounding

To ensure the safety of personnel and to protect equipment, the metallic frames of all electrical apparatus are grounded. To ground equipment, a properly sized copper wire is connected between the protected apparatus* and a point on the main grounding

* In practice, the grounding wire of each load is connected to a grounding terminal of the supply panel which in turn is connected to the grounded terminal of the transformer.

network. The main grounding network of a residence is the water supply conduit; in an industrial plant it is a specially designed bare copper wire loop commonly known as "grounding bus." The grounding network is connected to the neutral of the voltage source and is identified by the symbol ⏚.

Figures 1-19(a) and (b) show, respectively, a grounded and an ungrounded 240 V motor. When the live wire in the grounded motor accidentally makes contact with the frame of the motor (point A), the line current will substantially increase because the impedance of the current path is reduced. Depending on its magnitude, the increase in the current will be interrupted by the upstream current protective devices (not shown on the diagram), and thus the faulty apparatus will be removed from its electrical supply. If one touches the frame of the motor while the faulty condition exists, the voltage from the point of contact to ground is small and thus the danger to personnel is reduced.

In contrast, when the live wire accidentally makes contact with the frame of the ungrounded motor (Fig. 1-19(b)), the source current will not increase (the impedance of

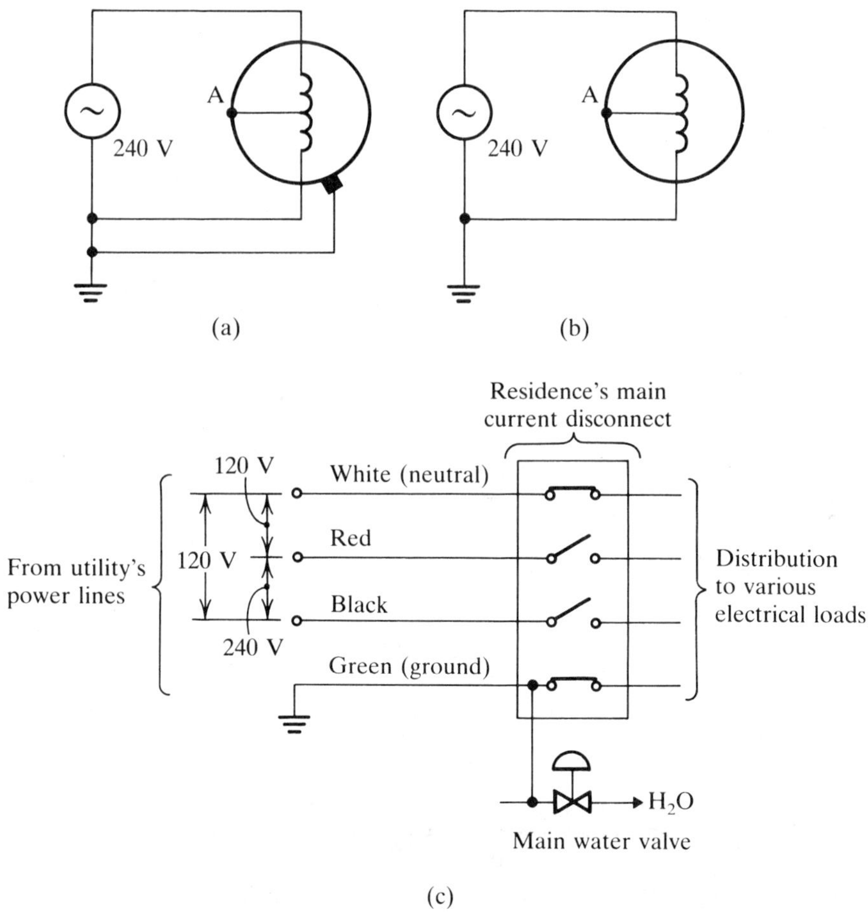

FIG. 1-19 The concept of "ground": **(a)** a grounded motor, **(b)** an ungrounded motor, **(c)** part of a grounded residential distribution system.

the current path remains substantially the same). Thus, the upstream current protective devices will not remove the motor from its voltage source. During this fault condition, the voltage from the motor's frame to ground is about equal to the source voltage. As a result, touching the metallic frame of the motor may result in electrocution.

Figure 1-19(c) shows 1-ϕ, 4-wire, 240/120 V power supply to a residence. The ground wire is connected upstream to the residence's main water valve and to the frames of all electrical apparatus.

Exercise 1–7 Figure 1-20 shows parts of a residence's power distribution diagram. The 120 V source supplies voltage to a three-prong and to a two-prong outlet.

a) When the live conductor makes contact with the frame of each of the electrical loads, what will be the level (high or low) of the current and the frame's voltage to ground?
b) If the outlets are in a very humid environment, what will happen when the loads are connected or disconnected?
c) Explain why a two-prong power outlet should not be interfaced to 1 – ϕ, 3-wire, 120 V load through a two-to-three-prong power outlet adaptor.

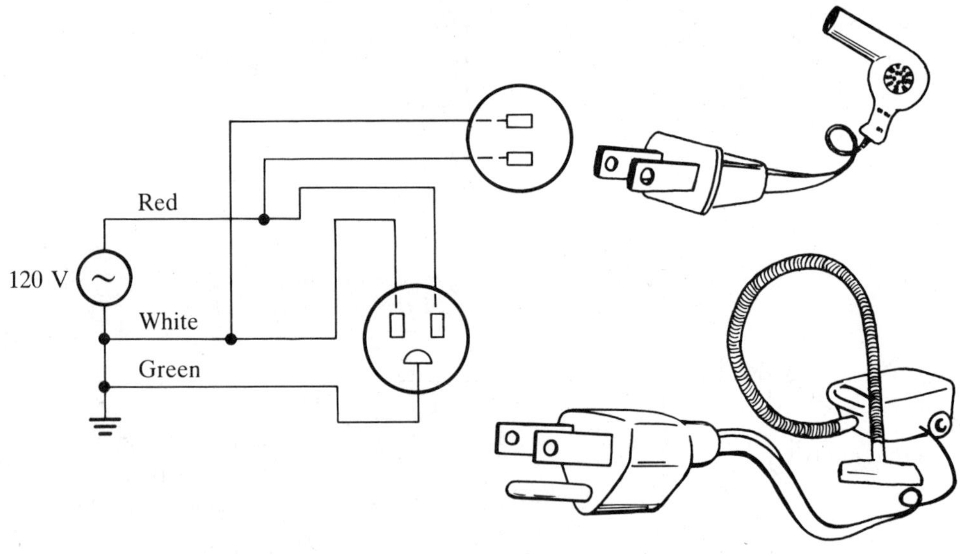

FIG. 1-20

1.2 Magnetic Circuit Concepts

This section discusses the fundamental concepts and principles of magnetic circuits. Knowledge of these elementary concepts will provide the basis for the understanding of the design and operation of electric machines.

1.2.1 MAGNETIC FLUX

Magnetic flux is a characteristic of magnetic fields, just as electric flux is a characteristic of electric fields. Magnetic flux may be thought of as representing the lines of force between a north magnetic pole and a south magnetic pole. Magnetic flux is represented by the Greek letter ϕ (phi), and in the SI system of units is measured in, or has the unit of, webers (Wb).

Refer to Fig. 1-21. The magnetic flux emerges from the north magnetic pole, goes through the south pole, and returns again to the north pole. In other words, the magnetic flux or the magnetic lines of force are continuous and thus they form a "*closed loop.*"

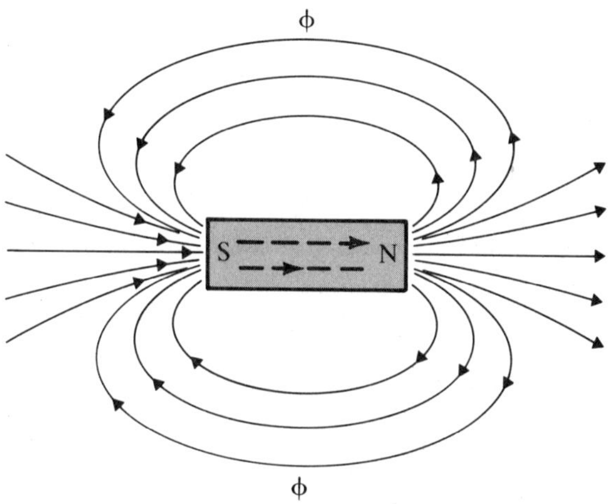

FIG. 1-21 A magnet and its magnetic flux.

The flux produced by a coil wound around a magnetic material depends on the properties of the magnetic material, the number of turns in the winding, and the current through the winding. The flux–current relationship is given by Ohm's law, which is discussed in detail in Section 1.2.10.

Since the current through a winding depends on the applied voltage, the flux can be also expressed as a function of the applied voltage. The flux–voltage relationship for sinusoidal voltages is derived in Section 1.2.8.

The direction of magnetic flux is obtained by using the so-called "right-hand" rule. According to this rule (see Fig. 1-22), when a conductor is held with the right hand and the thumb is pointing in the direction of current flow, then the other fingers curl in the direction of flux.

Consider coil 1, shown in Fig. 1-23. The flux generated by the coil's current has two components. Component ϕ_{12} links both coils; component ϕ_{11} links only coil 1. ϕ_{12} and ϕ_{11} are called the mutual and leakage flux, respectively. Thus, the total flux (ϕ_1) generated by coil 1 is

$$\phi_1 = \phi_{11} + \phi_{12} \tag{1.67}$$

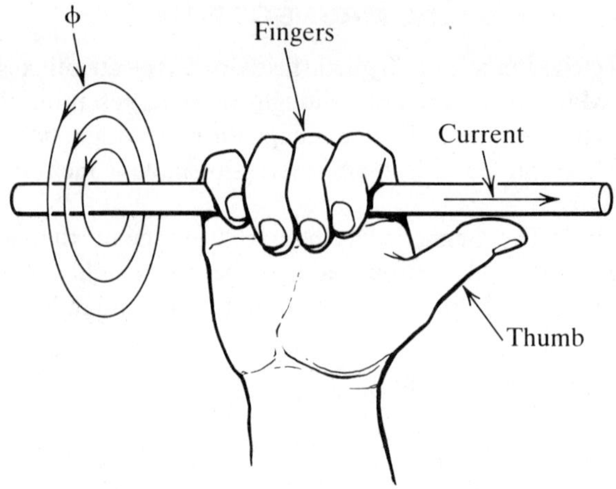

FIG. 1-22 Illustration of the right-hand rule.

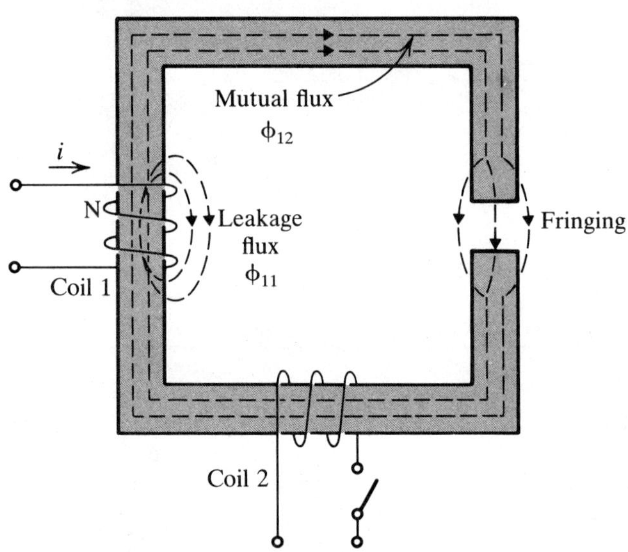

FIG. 1-23 An illustration of the concept of mutual flux, leakage flux, and fringing.

The spreading out of the mutual flux around the air gap is called fringing. This is further explained in Section 1.2.10.

The product of the winding turns and the flux that links them is called the "flux linkage." Flux linkage is generally represented by the Greek letter λ (lambda). Thus,

$$\lambda_{11} = N\phi_{11} \tag{1.68}$$

$$\lambda_{12} = N\phi_{12} \tag{1.69}$$

where λ_{11} and λ_{12} are the leakage and mutual flux linkages of winding 1, respectively.

The total flux linkage (λ_1) of coil 1 is then

$$\lambda_1 = \lambda_{11} + \lambda_{12} \tag{1.70}$$

The concept of flux linkage constitutes the basis for deriving general expressions for the inductance and voltage induced in a coil.

EXAMPLE 1–7 The total flux produced by the 100-turn coil in Fig. 1-24 is .005 Wb. Assuming that the leakage flux is 4% of the total flux, determine:

a) The flux linkage of each winding.
b) The relative polarity of the voltage source.

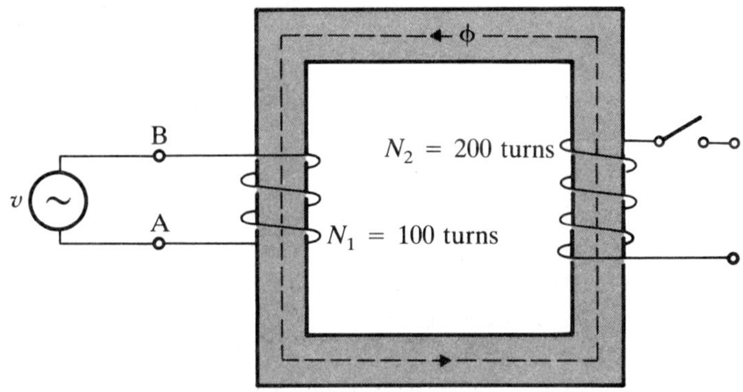

FIG. 1-24

SOLUTION

a) Designating with λ_1 and λ_2 the flux linkage of coil 1 and coil 2, we have

$$\lambda_1 = N_1(\phi_{11} + \phi_{12})$$
$$\lambda_2 = N_2\phi_{12}$$

where ϕ_{11} is the flux that links only coil 1, and ϕ_{12} is the flux that links both coils.
Substituting the given values, we obtain

$$\lambda_1 = 100(.005) = .5 \text{ Wb}$$

and

$$\lambda_2 = 200(.005)\left(1 - \frac{4}{100}\right) = .96 \text{ Wb}$$

b) The relative polarity of the voltage supply controls the direction of the current flow, which in turn controls the direction of the flux flow. Using the right-hand rule it is evident that the current must flow from terminal A to terminal B in a counterclockwise direction. Thus the potential of point A is higher than the potential of point B.

1.2.2 MAGNETIC FLUX DENSITY

Magnetic flux density is defined as the ratio of the magnetic flux divided by the area perpendicular to the flux. In mathematical symbols,

$$B = \frac{\phi}{A} \tag{1.71}$$

where B is the flux density.

The unit of flux density in the SI system of units is the tesla,* abbreviated by the letter T.

Considering incremental changes in the area, from Eq. (1.71) we get

$$\phi = \int B\, dA \tag{1.72}$$

The flux density of a material is a measure of its magnetization, which reveals the magnetic status of the material.

The nominal magnetization level of an apparatus' magnetic material depends on the design of its magnetic circuit. For example, the flux density of the coils used for communication purposes is maintained at about .001 T, while the flux density of power transformers, under normal operating conditions, is about .9 T.

The flux density in an air gap can be easily measured by using a Gauss meter. The design of such a meter may be based on the "Hall effect."

According to this principle, the voltage induced in a probe—usually made from a thin semiconductor material—that carries current (i) perpendicular to the magnetic field density (B) is given by

$$v = KiB \tag{1.73}$$

where K is a constant of proportionality that depends on the physical dimensions of the probe.

When the probe of a Gauss meter that carries a constant current is placed perpendicular to a magnetic field, the induced voltage will be indicated on the scale of the meter. The manufacturer of the meter gives the constants that relate the voltage to the magnetic flux and to the flux density.

. . .

* 1 tesla = 1 weber/m^2
= 10^4 gauss
1 gauss = 1 maxwell/cm^2
= 1 line/cm^2
10^8 lines = 1 weber

1.2.3 MAGNETOMOTIVE FORCE

The magnetomotive force (mmf), or the magnetic potential of a coil, is given by the product of the winding's current (i) times its number of turns (N). That is,

$$\text{mmf} = Ni \tag{1.74}$$

The mmf is also represented by the following symbols:

$$\text{mmf} = \mathcal{F} = U$$

The unit of the mmf in the SI system of units is ampere-turns (A).

When a coil is connected to an electric voltage, as shown in Fig. 1-25, a current will flow, which in turn will produce an mmf.

The mmf, like the current, has an instantaneous, average, and effective value. When the mmf is of sinusoidal waveform, it can be represented by a phasor having the same phase angle as that of the current that produces it.

The mmf may be thought of as the driving force of transformers and electric machines, just as the emf constitutes the driving force in incandescent lamps and electric heaters. The polarity of the mmf is as critical to magnetic circuits as the polarity of the emf is to electric circuits.

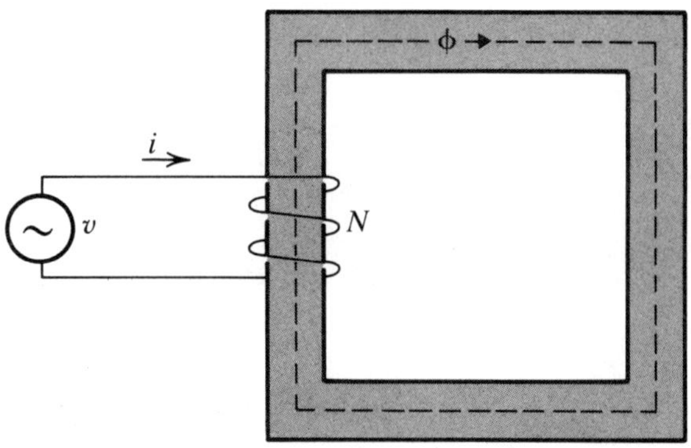

FIG. 1-25 Physical representation of a magnetic circuit (Leakage flux is neglected).

EXAMPLE 1–8 When a sinusoidal voltage source of 60 Hz and 120 V is connected to a 50-turn coil, a current of 15 A rms circulates. Assuming a linear magnetic circuit, determine the maximum value and the instantaneous value of the mmf.

SOLUTION The maximum value of the current is

$$\begin{aligned} I_m &= \sqrt{2}I_{\text{rms}} \\ &= \sqrt{2}(15) = 21.21 \text{ A} \end{aligned}$$

By definition, therefore, the maximum value of the coil's mmf is

$$\mathscr{F}_{\text{max}} = NI_{\text{max}} = 50(21.21) = \underline{\underline{1060.66\ \text{A}}}$$

Since the current is a sinusoidal function, the mmf will also be a sinusoidal function. Thus,

$$\begin{aligned}\mathscr{F} &= \mathscr{F}_{\text{m}} \sin \omega t \\ &= \underline{\underline{1060.66 \sin 377\,t\ \text{A}}}\end{aligned}$$

1.2.4 MAGNETIC FIELD INTENSITY

The magnetic field intensity (H) for a given magnetic circuit is given by the ratio of the mmf divided by the length of the mean magnetic path. Thus,

$$\text{magnetic field intensity} = \frac{\text{ampere-turns}}{\text{length of mean magnetic path}}$$

Using mathematical symbols,

$$H = \frac{Ni}{l} \tag{1.75}$$

Considering incremental length changes, we have

$$Ni = \int H\,dl \tag{1.76}$$

The unit of magnetic field intensity is

$$\frac{\text{ampere-turns}}{\text{meter}} \quad \left(\frac{\text{A}}{\text{m}}\right)$$

Magnetic field intensity is often referred to as magnetizing force.

Magnetic field intensity is related to flux density by

$$\mu = \frac{dB}{dH} \tag{1.77}$$

where μ represents the permeability of the material. This parameter is discussed extensively in Section 1.2.6.

Magnetic field intensity is used in the analysis of magnetic circuits. It is of unique importance in finding the voltage induced in a plane that is located at a given distance from a current-carrying conductor. The general expression for magnetic field intensity can be derived as follows. Consider Fig. 1-26. The conductor carries current of i amperes away from the reader and into the plane of the paper. The conductor's current produces a magnetic flux whose direction can be easily obtained from the right-hand rule. The mean magnetic paths of the flux at points P_1 and P_2 are given respectively by

$$l_{m_1} = 2\pi r_1 \tag{1.78}$$

and

$$l_{m_2} = 2\pi r_2 \tag{1.79}$$

From Eq. (1.75), the magnetic field intensity at the points P_1 and P_2 is

$$H_1 = \frac{i}{2\pi r_1} \tag{1.80}$$

$$H_2 = \frac{i}{2\pi r_2} \tag{1.81}$$

The direction of the vector H_1, as shown in the diagram, is perpendicular to the radius r_1 at the point P_1.

In general, the magnetic field intensity at a point P that is at a radius r from the center of the current-carrying conductor is

$$H = \frac{i}{2\pi r} \tag{1.82}$$

The voltage induced in a plane is due only to the perpendicular component of the H on the plane under consideration.

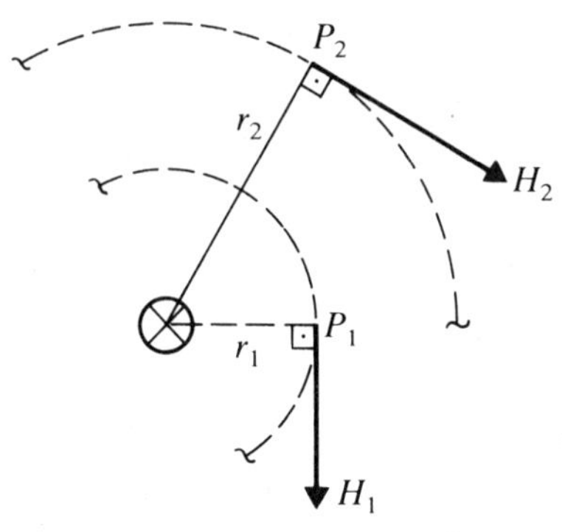

FIG. 1-26 The concept of field intensity.

EXAMPLE 1–9 The mean magnetic path of the material shown in Fig. 1-27 is .3 m, and its effective cross-sectional area is 5×10^{-4} m^2. The flux density is .008 T, and the dc resistance of the winding at the operating temperature is 2 ohms. Determine:

a) The current drawn from the 24 V dc supply.
b) The magnetic field intensity.
c) The flux.

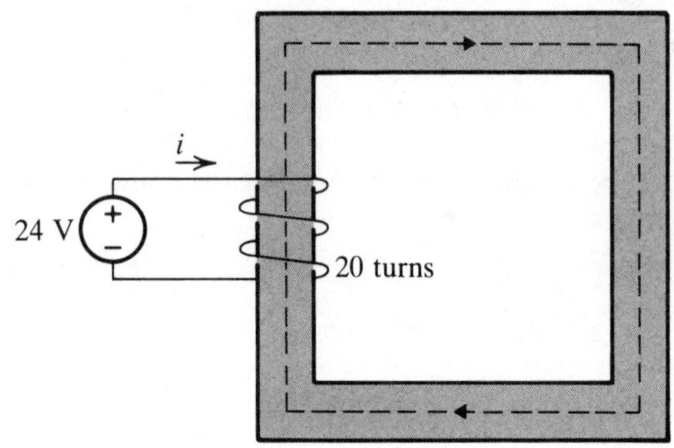

FIG. 1-27

SOLUTION

a) $i = \frac{V}{R} = \frac{24}{2} = \underline{\underline{12\ A}}$

b) $H = \frac{Ni}{l} = \frac{20(12)}{.3} = \underline{\underline{800\ A/m}}$

c) $\phi = AB = 5 \times 10^{-4}(.008) = \underline{\underline{4 \times 10^{-6}\ Wb}}$

Exercise 1–8 The flux through the toroidal magnetic material shown in Fig. 1-28 is 6 μWb, and the magnetic field intensity is 12 A/m. The length of the magnetic path is .5 m, and its effective cross-sectional area is 1.5×10^{-4} m^2. Determine:

a) The coil's mmf.
b) The flux density.

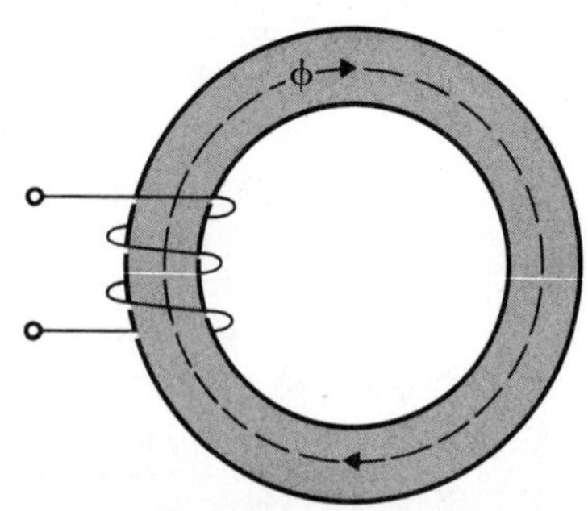

FIG. 1-28

Answer (a) 6.0 A; (b) .04 T

1.2.5 THE *B-H* CURVE AND MAGNETIC DOMAINS

The intrinsic properties of a magnetic material are, to a certain extent, revealed by the material's *B-H* curve. Typical *B-H* curves for cast iron and cast steel are shown in Fig. 1-29(a). The flux density (B) is the ordinate (y-axis) of the curves, and the field intensity (H) is their abscissa (x-axis).

Since the flux density is proportional to the voltage (v), and the field intensity is proportional to the current (i), then the *B-H* curve of a machine can be represented, as shown in Fig. 1-29(b), by its equivalent *v-i* characteristic curve. The *v-i* characteristic of a machine is referred to as its "open-circuit" or "magnetization" characteristic (MC).

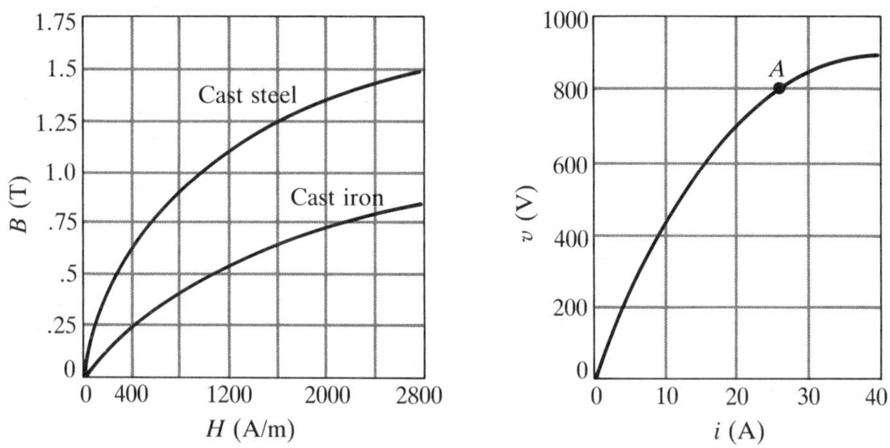

FIG. 1-29 Characteristics of magnetic materials: **(a)** *B-H* curve, **(b)** open-circuit characteristic.

Under normal operating conditions, a machine operates at a flux level that is very close to point A. Point A is referred to as the knee* of the magnetization characteristic. When the parameters of electric machines are measured at a flux density lower than that of point A, the resulting values are referred to as *unsaturated.* When they are measured at a flux density that is higher than that of point A, they are referred to as *saturated* values.

The intrinsic properties of magnetic materials, and thus their particular *B-H* characteristics, depend on the orientation of the materials' magnetic moments.

Magnetic moments are caused by the spinning and orbiting of an atom's electrons. Rotating electrons (electric charges) constitute a flow of current and thus produce flux and a corresponding magnetic moment whose direction can be determined by using the right-hand rule. The electrons' configuration depends on the temperature and on the chemical composition of the material under consideration.

At a very high temperature known as the "Curie point" magnetic moments cease to exist, and thus the magnetic material loses its magnetic properties. The Curie point for iron is about 775°C.

. . .

* The knee point voltage may be defined as the voltage at which a 10% increase in voltage will result in a 50% increase in current.

The strength of magnetic moments is also attenuated by time and ambient conditions. The degradation of a material's magnetic properties as a function of time and temperature is known as the "disaccommodation factor."

The smallest region in which a group of atoms has a common magnetic moment is called a "magnetic domain." Magnetic domains may be thought of as being the basic building blocks of magnetism.

The approximate volume and number of atoms within one domain are as follows:

Volume: 10^{-8} cm^3

Number of atoms: 10^9 to 10^{12}

Adjacent domains are separated by boundaries called domain walls. The application of an external magnetic field modifies the domain walls and orientates the magnetic moment vectors towards the direction of the applied field.

Figure 1-30(a) is an elementary representation of a material's random magnetization. For simplicity, only part of the magnetic material is considered. The orientation of the individual magnetic moments is such that no external field is produced. This is indicated by point A_1 in the flux density–field intensity characteristic of the material shown in Fig. 1-30(d).

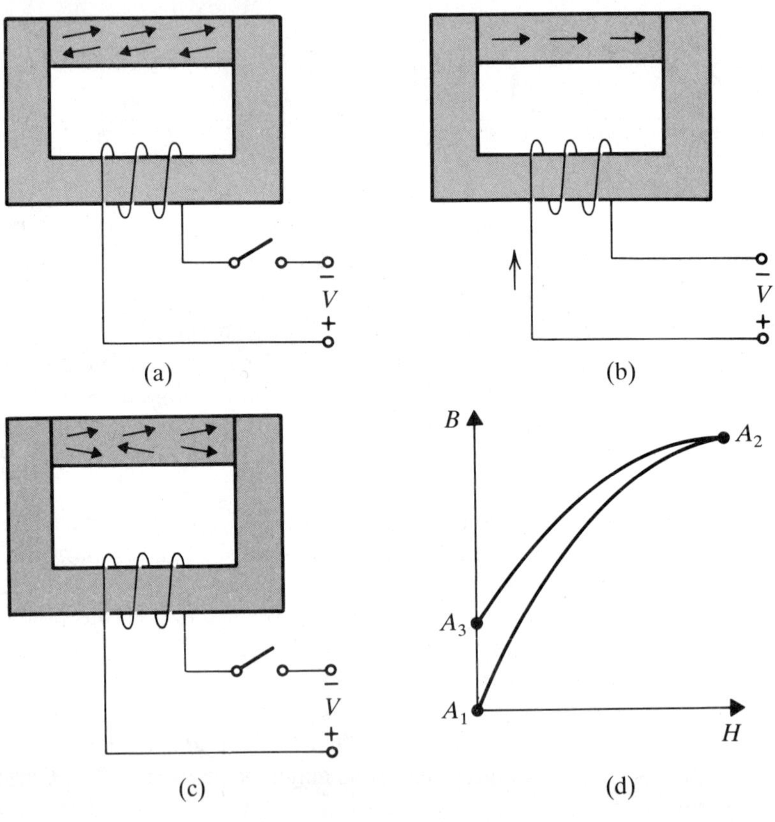

FIG. 1-30 Orientation of magnetic domains and the *B-H* curve: **(a)** opposite orientation, **(b)** parallel orientation toward the induced field, **(c)** partial orientation, **(d)** corresponding *B-H* curve.

In Fig. 1-30(b) the magnetic domains are completely oriented in the direction of the applied field, and the magnetization of the material is indicated by point A_2 in Fig. 1-30(d).

When the external field is removed, the magnetic domains remain partially oriented, as shown in Fig. 1-30(c). As a result, the material is said to have residual magnetism. This corresponds to point A_3 in the B-H curve of the material.

Hysteresis Loop

When a material is subjected to a cyclic variation of a magnetizing force, its B-H characteristic at steady state forms a closed loop, commonly referred to as the hysteresis loop. A typical hysteresis loop for cast iron is shown in Fig. 1-31.

The magnetizing force (H_c) which corresponds to zero flux density is referred to as the coercive force.

The product of the residual flux density (B_r) times the coercive force (H_c) is used to designate the quality of permanent magnet material. The higher the B_rH_c product of a material, the better the material is suited for permanent-magnet applications.

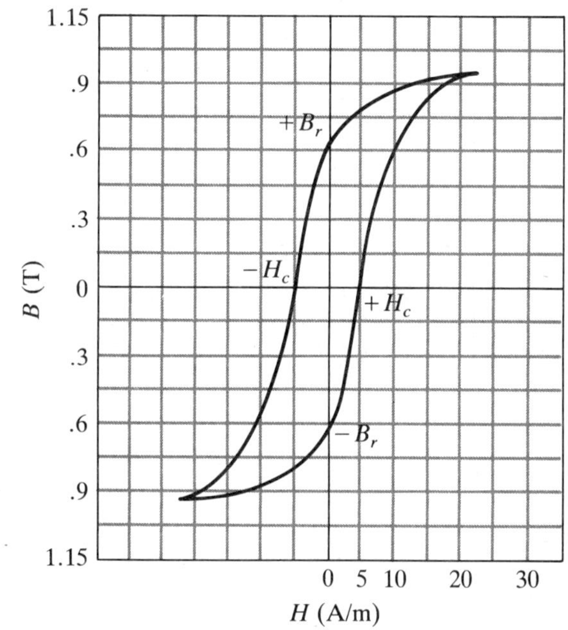

FIG. 1-31 A typical hysteresis loop.

1.2.6 PERMEABILITY

Permeability is a measure of how easily a material can be magnetized. The higher the permeability of a material, the easier it is to magnetize.

Magnetizing a material means aligning all microscopic magnetic forces or moments within the material in the direction of the externally applied field. The alignment of the tiny and random inner magnetic moments toward a particular direction results in the magnetization of the material or the development of a stronger external magnetic field.

In other words, then, permeability is a measure of how easily the magnetic domains within a material can be aligned in the direction of the magnetizing force.

Permeability may also be thought of as indicating how much flux density will be produced within a material by the amplication of a given magnetomotive force. The higher the permeability of a material, the greater the amount of flux density produced in a given value of field intensity.

Permeability is an intrinsic property of all materials and is designated by the Greek letter μ (mu). The unit of permeability in the SI system of units is the henry/meter (H/m).

The actual permeability (μ) of a magnetic material is given by the following equation:

$$\mu = \mu_0 \mu_r \tag{1.83}$$

where μ_r is the relative permeability of the material and μ_0 is the permeability of free space.

$$\mu_0 = 4\pi \times 10^{-7} \text{ H/m} \tag{1.84}$$

As shown in Table 1-1, the relative permeability of magnetic materials varies over a wide range.

TABLE 1-1 Relative permeability of some materials at low flux densities

Relative Permeability	*Material*			
	Air	*Powdered Iron*	*Typical Transformer Iron*	*Permalloy*
(μ_r)	1	100	3000	100,000

As can be seen from Eq. (1.77), the incremental permeability of magnetic materials is given by the slope of density versus intensity characteristic curve, That is,

$$\text{permeability} = \frac{\text{change in magnetic flux density}}{\text{corresponding change in magnetic field intensity}}$$

or

$$\mu = \frac{dB}{dH} \tag{1.85}$$

Permeability is not a constant parameter but depends on the flux density or on the applied mmf that is used to energize the magnetic circuit, as shown in Fig. 1-32. Thus,

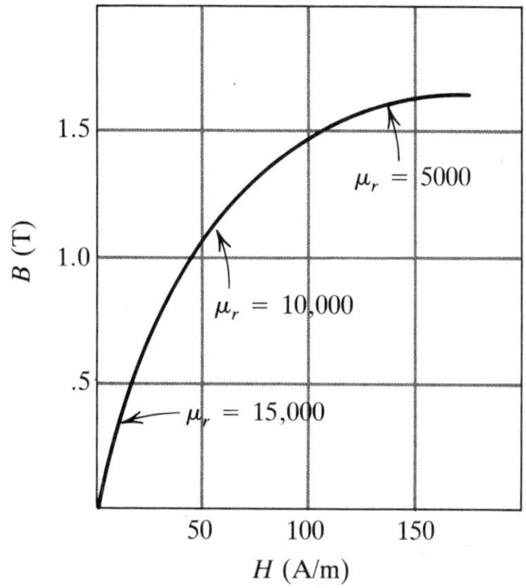

FIG. 1-32 Typical *B-H* characteristic.

when one speaks of the relative permeability of a material, one must always indicate at what flux level this permeability is measured.

The permeability of a material can also be expressed as a function of a coil's inductance. From the induction principle, we have

$$v = -N\frac{d\phi}{dt} = L\frac{di}{dt} \tag{1.86}$$

From the above, we obtain

$$N\phi = Li \tag{1.87}$$

where L and i are, respectively, the self-inductance and the instantaneous value of the coil's current. Solving for the flux, we get

$$\phi = \frac{Li}{N} \tag{1.88}$$

For operation along the linear section of the *B-H* curve,

$$\mu = \frac{B}{H} \tag{1.89}$$

From the above, we obtain

$$\mu = \frac{Li}{NA}\left(\frac{l}{Ni}\right) \tag{1.90}$$

or

$$\mu = \left(\frac{l}{N^2A}\right)L \tag{1.91}$$

or

$$\mu = KL \tag{1.92}$$

The constant of proportionality (K) can be calculated from the number of winding turns and the geometry of the magnetic core under consideration. If one knows the constant of proportionality and measures the inductance of a given coil, the permeability of the magnetic material can be easily established.

EXAMPLE 1–10 A coil of 50 turns is wound around the toroid shown in Fig. 1-33. When a certain voltage is applied across the coil, its inductance is found to be 200 μH. Determine the relative permeability of the magnetic material from which the toroid is made.

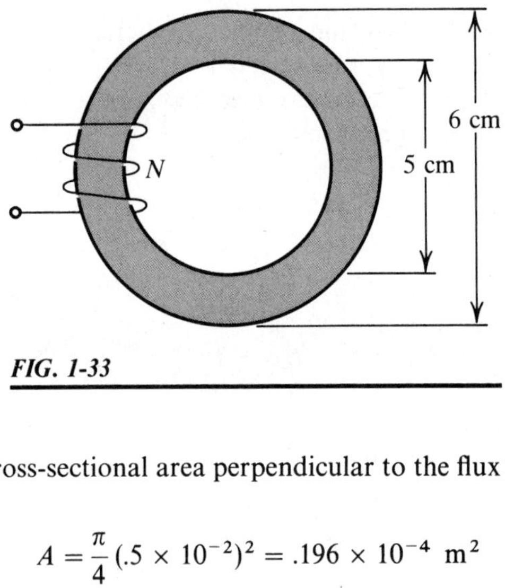

FIG. 1-33

SOLUTION The cross-sectional area perpendicular to the flux is

$$A = \frac{\pi}{4}(.5 \times 10^{-2})^2 = .196 \times 10^{-4} \text{ m}^2$$

The mean magnetic path is

$$l = 2\pi r_m = 2\pi(.0275) = .17 \text{ m}$$

From Eq. (1.91), the permeability is

$$\mu = \frac{Ll}{N^2A} = \frac{(200 \times 10^{-6})(.17)}{(50)^2(.196 \times 10^{-4})} = .70 \text{ mH/m}$$

Thus, the relative permeability is

$$\mu_r = \frac{\mu}{\mu_0} = \frac{.70 \times 10^{-3}}{4\pi \times 10^{-7}} = \underline{\underline{560.27}}$$

REMARKS The example would be more illuminating if the test voltage were specified. After all, the value of the permeability is meaningful if and only if the corresponding magnetization level is known.

Exercise 1–9 The coil shown in Fig. 1-34 has, at the operating temperature, a resistance of 4 ohms and draws 12 amperes from a 60 Hz, 120 V supply. The material's mean magnetic path is .75 m and its cross-sectional area is uniform at 80×10^{-4} m^2. Determine:

a) The relative permeability of the material.
b) The flux and its direction.
c) The power factor of the coil.

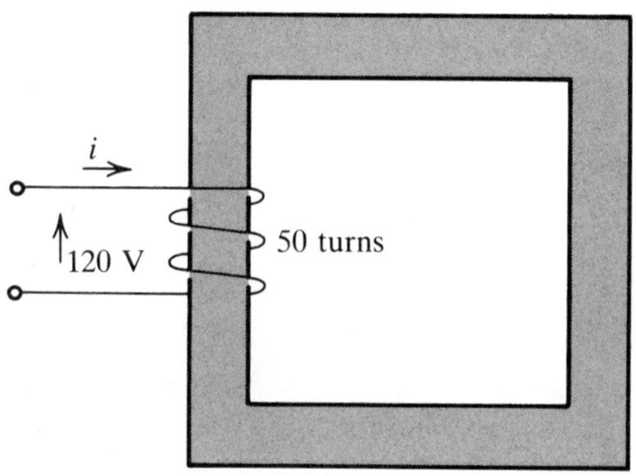

FIG. 1-34

Answer (a) 725.5; (b) 5.83 m Wb cw; (c) .4 lagging

1.2.7 RELUCTANCE

The opposition to the flux in a given magnetic circuit is termed reluctance ($\mathscr{R}$). As in the case of the resistance of electric circuits it is given by the ratio of the applied magnetic potential divided by the flux through the magnetic path under consideration. Mathematically,

$$\mathscr{R} = \frac{NI}{\phi} \left(\frac{\text{A}}{\text{Wb}}\right) \qquad \textbf{(1.93)}$$

The previous expression, as explained in Section 1.2.10, constitutes Ohm's law in magnetic circuits. The reluctance of a magnetic path, being analogous to the resistance of an electric circuit, is also given by the following relationship:

$$\mathcal{R} = \frac{l}{\mu A} \tag{1.94}$$

where l is the length of the mean magnetic path, μ is the permeability of the material, and A is the material's cross-sectional area perpendicular to the flux.

Reluctance, like electric resistance, varies with the impurity and temperature of the materials. Unsaturated magnetic materials have a relatively low reluctance, because their permeability is thousands of times greater than that of the air.

The equivalent series or parallel reluctances of a magnetic circuit are calculated in the same way as the equivalent series or parallel resistances of an electric circuit. For example, the total reluctance of the magnetic circuit shown in Fig. 1-35(b) is the sum of the reluctances of the magnetic material and of the air gap. That is,

$$\mathcal{R}_{\text{total}} = \mathcal{R}_{\substack{\text{magnetic}\\ \text{material}}} + \mathcal{R}_{\text{air gap}} = \mathcal{R}_{\text{m}} + \mathcal{R}_{\text{g}} \tag{1.95}$$

$$= \frac{l_m}{\mu A} + \frac{l_g}{\mu_0 A} \tag{1.96}$$

Thus,

$$\mathcal{R}_{\text{total}} = \frac{1}{\mu_0 A}\left(\frac{l_m}{\mu_r} + l_g\right) \tag{1.97}$$

As seen from the above, the relative effect of the air gap on the magnetic circuit depends on how the length of the air gap compares to the ratio l_m/μ_r of the magnetic material.

The closed-loop property of magnetic flux and the concept of reluctance will be demonstrated by the following practical illustrations.

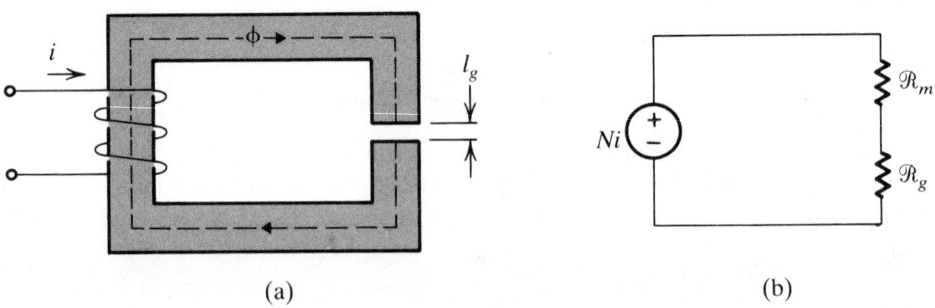

FIG. 1-35 Electromagnet: **(a)** physical representation, **(b)** equivalent magnetic circuit.

A Laboratory Case

In a laboratory, the arrangement shown in Fig. 1-36(a) was used to measure the magnetic properties of E-type laminations as a function of temperature. Although the applied voltage to the coil was gradually increased to maximum, and the temperature of laminations was varied over a wide range, no reading could be obtained in the measuring bridge.

The mistake was that the set-up used did not provide a loop of low resistance to allow for the existence of flux. Owing to its large reluctance, air presents very high opposition to the existence of magnetic flux.

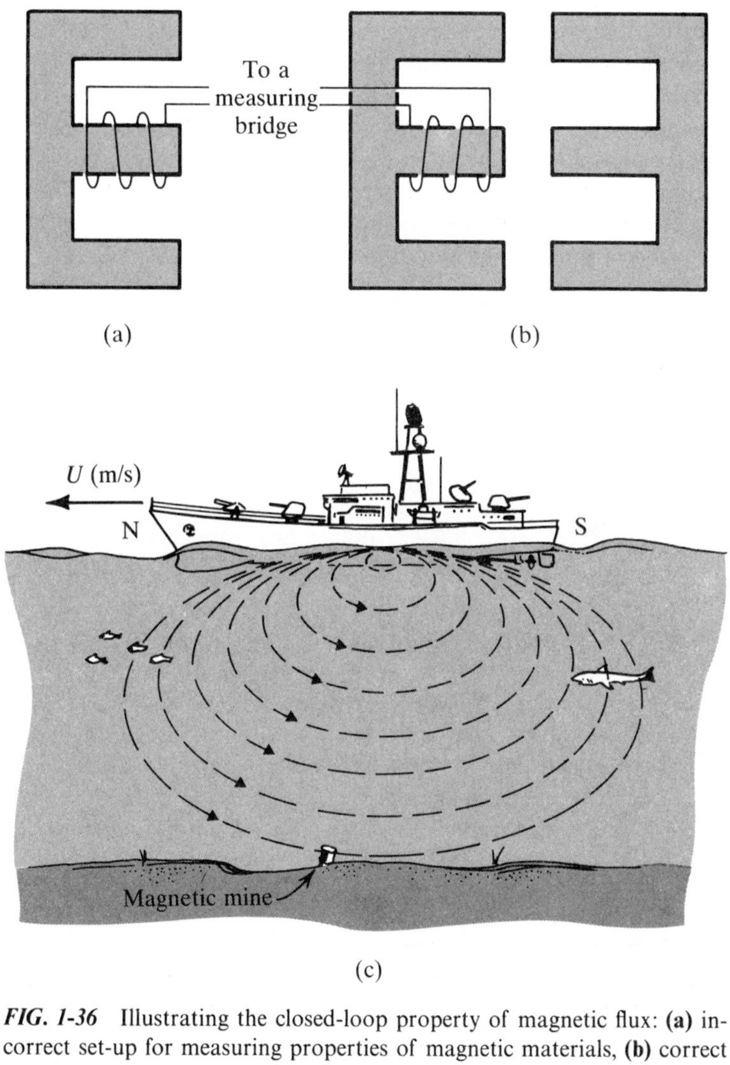

FIG. 1-36 Illustrating the closed-loop property of magnetic flux: **(a)** incorrect set-up for measuring properties of magnetic materials, **(b)** correct set-up, **(c)** triggering a magnetic mine.

The problem was solved by rearranging the laminations as shown in Fig. 1-36(b). In this case, magnetic materials were placed in the air path of the flux, and so the magnetic resistance was reduced a thousandfold. As a result, the quantity of flux through the coil increased. The measuring bridge then began to give readings that varied according to the ambient temperature and the applied voltage.

Demagnetizing the British Fleet

During the first phase of the Second World War, the German air force dropped magnetic mines in the navigated rivers and coastal waters of England. When a ship moved above a magnetic mine, the flux of its residual magnetism—a property of magnetic material—closed its loop by penetrating through the magnetic mine's structure, as shown in Fig. 1-36(c). The movement of the ship was accompanied by a change of the flux through the magnetic mine, and thus a voltage was induced that triggered an explosion. In all, 37 allied ships were destroyed.

The problem was solved by demagnetizing, or *degaussing*, the British Fleet. That is, flux of opposite direction and equal in strength to that produced by the ship's residual magnetism was passed through the ship's structure. As a result, the ship could not effectively emit any flux, and thus the magnetic mines remained untriggered. The neutralizing flux was generated by passing a specific amount of dc current through loops of wire properly wound around the ship.

EXAMPLE 1-11 Determine the length of the air gap (Fig. 1-35(a)) that will make its reluctance equal to that of the magnetic material. Assume that the cross-sectional areas are the same and that $l_m = .60$ m and $\mu_r = 1200$.

SOLUTION Rewriting the expression for the circuit reluctance, we have

$$\mathcal{R} = \frac{1}{\mu_0 A}\left(\frac{l_m}{\mu_r} + l_g\right)$$

In the above equation, the reluctance of the air gap will be equal to the reluctance of the magnetic material, provided that

$$\frac{l_m}{\mu_r} = l_g$$

Thus,

$$l_g = \frac{.6}{1200} = \underline{\underline{.5\ \text{mm}}}$$

Exercise 1–10 Understanding the "degaussing of the British Fleet" requires knowledge of four basic concepts or principles of magnetic circuits. Give the describing mathematical relationships and also the equivalent equations from the electric circuits.

1.2.8 THE CONCEPT OF GENERATED VOLTAGE

A voltage can be induced in a conductor either by a time-varying magnetic field or as a result of a conductor's motion in a stationary magnetic field. The first is referred to as *transformer voltage*, the second as *speed voltage*.

The instantaneous value of each of these voltages will be designated, for the purpose of minimizing the symbol usage, by the same letter (v).

Transformer Voltage

When a conductor is placed in a time-varying magnetic field, a voltage is induced in it. This voltage is commonly known as the transformer voltage. This voltage, first observed by Faraday, is given by

$$v = -N\frac{d\phi}{dt} \tag{1.98}$$

where v is the instantaneous value of the voltage induced in the conductor, N is the number of the conductor's turns, and $d\phi/dt$ is the time-rate of change of the magnetic flux. As per Lenz's law, the minus sign in the equation indicates that the voltage induced is of a polarity that will produce a current whose flux opposes the flux that induced the voltage. The polarity of the induced voltage is further explained in the section on magnetically coupled circuits (see Section 1.2.14).

Equation (1.98) can also be written as

$$v = -\frac{d\lambda}{dt} \tag{1.99}$$

where λ is the conductor's flux linkage.

Equations (1.98) and (1.99) both describe what is often referred to as the *induction principle.*

EXAMPLE 1–12

a) Derive the expression for the voltage induced per unit length in the telephone lines shown in Fig. 1-37(a). The single-conductor power line runs parallel to the telephone lines and carries a current given by

$$i = I_m \sin \omega t \text{ A}$$

b) Evaluate the induced voltage, given that

$$I_m = 1000 \text{ A}$$
$$\omega = 377 \text{ rad/s}$$
$$\mu_0 = 4\pi \times 10^{-7} \text{ H/m}$$
$$a = 3 \text{ m}$$
$$d = .5 \text{ m}$$

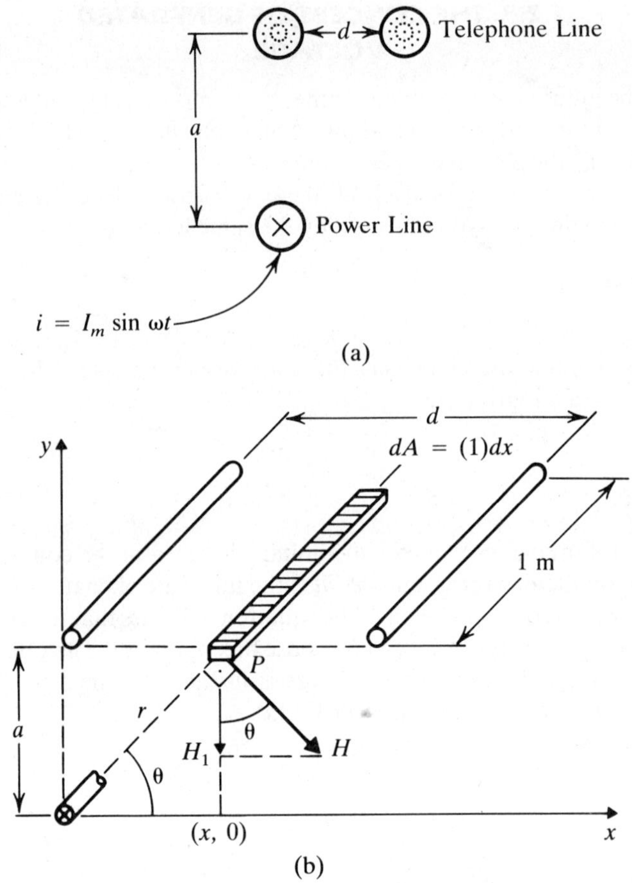

FIG. 1-37

SOLUTION

a) A voltage has meaning if it is specified between two points. In this example, the requirement is to find the voltage induced per unit length in the telephone lines. Rewriting the induction-principle equation, we have

$$v = -N\frac{d\phi}{dt} \qquad \text{(I)}$$

Thus the essential part of the problem is how to find the flux through the area between the telephone lines.

By definition,

$$\phi = \int B\,(dA) \qquad \text{(II)}$$

where dA is an increment of area perpendicular to the flux. (The induced voltage depends on the component of the magnetic field perpendicular to the area between the telephone wires.)

For linear variation of flux density, we have

$$B = \mu H \tag{III}$$

Also,

$$H = \frac{Ni}{l_m} \tag{IV}$$

The number of turns is equal to unity, and the mean magnetic path at point P, shown in Fig. 1-37(b), is

$$l_m = 2\pi r \tag{V}$$

Thus,

$$H = \frac{i}{2\pi r}$$

Its component (H_1), perpendicular to the area under consideration, is

$$H_1 = H\cos\theta = \left(\frac{i}{2\pi r}\right)\cos\theta \tag{VI}$$

From the geometry of Fig. 1-37(b) we obtain

$$\cos\theta = \frac{x}{r} = \frac{x}{\sqrt{a^2 + x^2}} \tag{VII}$$

From equations (VI) and (VII) we get

$$H_1 = \frac{i}{2\pi}\left(\frac{x}{a^2 + x^2}\right) \tag{VIII}$$

Then equations (III) and (VIII) give

$$B = \mu \frac{i}{2\pi}\left(\frac{x}{a^2 + x^2}\right) \tag{IX}$$

Considering the shaded strip, we have

$$dA = (1)\, dx \tag{X}$$

The corresponding incremental flux through this element of area is

$$d\phi = d(BA) = B\, dx \tag{XI}$$

Substituting equation (IX) into equation (XI), we obtain

$$d\phi = \mu\left[\frac{i}{2\pi}\left(\frac{x}{a^2 + x^2}\right)dx\right] \tag{XII}$$

Taking the integrals, we find:

$$\phi = \mu\frac{i}{2\pi}\int_0^d \frac{x\,dx}{a^2 + x^2}$$

The variation of x or the limits of integration are from $x = 0$ to $x = d$. After integrating, we obtain

$$\phi = \mu\frac{i}{2\pi}\left(\frac{1}{2}\right)\ln\,(a^2 + x^2)\Big|_0^d$$

$$\phi = \frac{\mu}{4\pi}\,i\ln\left(\frac{a^2 + d^2}{a^2}\right)$$

Substituting for the current, we get

$$\phi = \frac{\mu}{4\pi}\ln\left(\frac{a^2 + d^2}{a^2}\right)I_m \sin\omega t$$

Taking the derivative of the flux with respect to time, and substituting the result into equation (I), we obtain

$$v = -\frac{\mu}{4\pi}\ln\left(\frac{a^2 + d^2}{a^2}\right)I_m\omega\cos\omega t\ \text{V/m}$$

b) Substituting the given values, we have

$$v = -\frac{4\pi \times 10^{-7}}{4\pi}\ln\left(\frac{9 + .25}{9}\right)1000(377)\cos 377t$$

$$= \underline{\underline{-1.03\cos 377t\ \text{mV/m}}}$$

Speed Voltage

The voltage induced in a conductor that moves through a stationary magnetic field is referred to as the motional or speed voltage.

Speed voltage (v) is given by the following equation:

$$v = \int (U \times B)\,dl \tag{1.100}$$

where $U \times B$ is the cross product of the velocity and flux density vectors and dl is a length element along the conductor.

When U, B, and l are perpendicular to each other, then Eq. (1.100) is written as

$$v = BUl \tag{1.101}$$

where B, U, and l are the magnitudes of the flux density, velocity, and length of the conductor, respectively.

EXAMPLE 1–13 Figure 1-38(a) shows one coil turn of an ac generator that rotates at ω rad/s through a constant magnetic field of B tesla. Determine the voltage generated at the terminals of the generator if its coil has N turns connected in series.

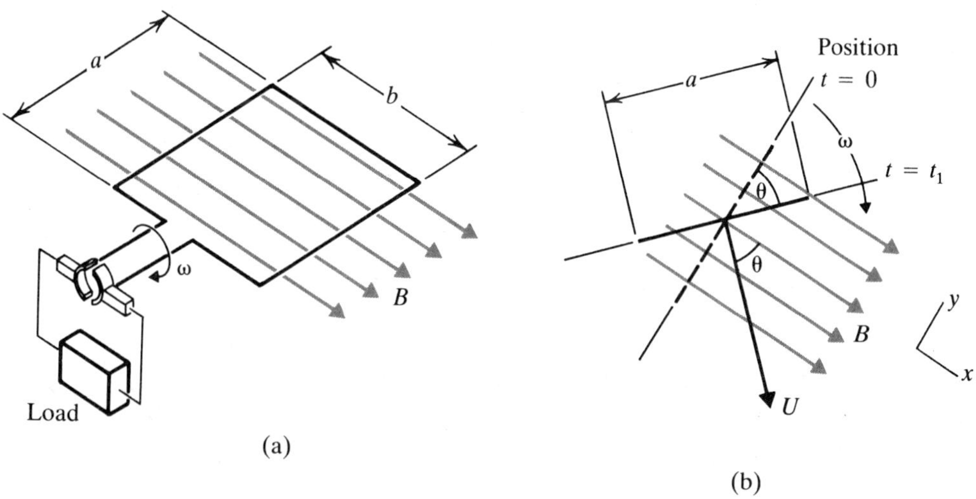

FIG. 1-38

SOLUTION Figure 1-38(b) shows the position of the conductor after it is rotated θ radians with respect to its position at $t = 0$. It should be noted that in the two sides of the conductor whose individual length is a meters a voltage is induced; while in the other two sides of the conductor whose individual length is b meters no voltage is induced. The area swept by each b-side of the conductor is parallel to the direction of the magnetic flux density; thus there is, in effect, no voltage induced in these two sides of the conductor.

The magnitude of the tangential component of the linear velocity is

$$U = \frac{b}{2}\omega \text{ m/s}$$

In vector form,

$$\vec{U} = \frac{b}{2}\,\omega(\cos\theta\tilde{a}_x - \sin\theta\tilde{a}_y)$$

where $\tilde{a}_x$ and $\tilde{a}_y$ are the unit vectors in the x and y direction, respectively.

The flux density vector is

$$\vec{B} = B\tilde{a}_x$$

The cross product of the velocity and flux density is

$$\vec{U} \times \vec{B} = \begin{vmatrix} \tilde{a}_x & \tilde{a}_y & \tilde{a}_z \\ \omega\frac{b}{2}\cos\theta & -\omega\frac{b}{2}\sin\theta & 0 \\ B & 0 & 0 \end{vmatrix}$$

$$= \frac{b}{2}\,\omega\tilde{a}_z \begin{vmatrix} \cos\theta & -\sin\theta \\ B & 0 \end{vmatrix}$$

$$= \frac{b}{2}\,\omega B\sin\theta\tilde{a}_z$$

Substituting the above into Eq. (1.100), we obtain

$$v = \frac{b}{2}\,\omega B\sin\theta \int dl$$

The effective length of the conductor is $2a$ m. Thus,

$$v = abB\omega\sin\theta$$

Since $\theta = \omega t$ and the generator has N turns, then the instantaneous value of the voltage generated is

$$v = \underline{\underline{abBN\omega\sin\omega t\ \text{V}}}$$

Exercise 1–11 For the ac generator of Example 1-13, determine the power furnished by the prime mover if the generator's efficiency is 90% and the resistance of the load at operating temperature is R ohms.

Answer $\dfrac{.56}{R}(abBN\omega)^2$

Voltage-flux Relationship

When the coil of a magnetic circuit is connected to an ac voltage source, a current circulates. As a result, flux is produced. The equation that describes the flux generated as a function of the coil's voltage is derived as follows.

Refer to the magnetic circuit in Fig. 1-39. If losses are neglected, the voltage induced is equal to the applied voltage (v_1). Assuming sinusoidal waveforms, we have

$$v_1 = V_m \sin \omega t \tag{1.102}$$

Equating the supply voltage and the voltage induced, we obtain

$$V_m \sin \omega t = -N \frac{d\phi}{dt} \tag{1.103}$$

from which

$$d\phi = -\frac{1}{N} V_m \sin \omega t \, dt \tag{1.104}$$

Integrating, we get

$$\phi = \frac{V_m}{\omega N} \cos \omega t \tag{1.105}$$

The flux, then, is a sinusoidal function whose maximum value is given by

$$\phi_m = \frac{V_m}{\omega N} \tag{1.106}$$

From the above,

$$V_m = 2\pi f N \phi_m \tag{1.107}$$

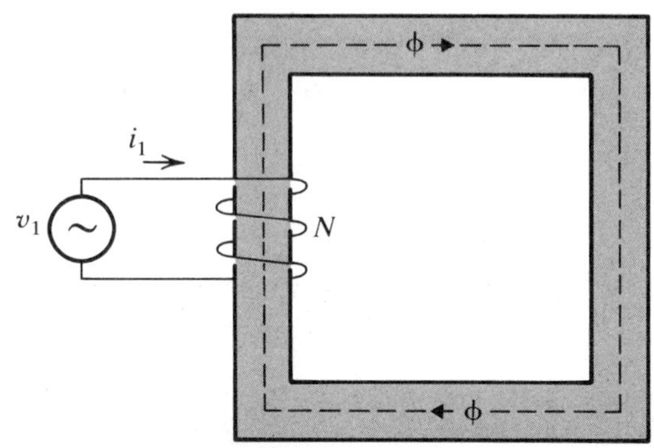

FIG. 1-39 Magnetic circuit.

or

$$\sqrt{2}V = 2\pi f N \phi_m \tag{1.108}$$

where V is the rms value of the applied voltage. From the last equation, we obtain

$$V = 4.44 N f \phi_m \tag{1.109}$$

Equation (1.109) relates magnetic flux to applied electric volts and is of fundamental importance to the understanding of transformers and electric machines. This equation is often used to calculate the flux within a toroidal magnetic core.

Exercise 1–12 When the 40-turn coil of Fig. 1-40 is connected to the 120 V, 60 Hz voltage supply, the voltages induced in the other coils are as shown. By neglecting leakage flux and winding losses, determine:

a) The number of turns of each coil.

b) The rms value of the mutual flux.

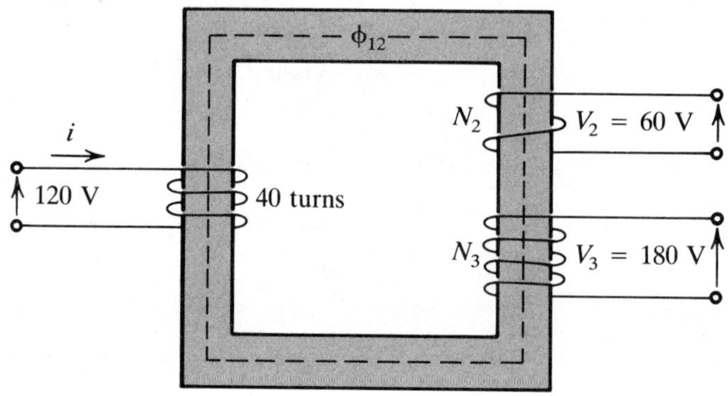

FIG. 1-40

Answer (a) 20 and 60 turns; (b) 7.96 m Wb

1.2.9 ENERGY CONTENT OF MAGNETIC MATERIALS

The energy content of a magenetic material depends on its level of magnetization, its physical size, and its intrinsic characteristics.

Refer to Fig. 1-41. The incremental change of its electrical input energy (dw) is

$$dw = vi\,dt \tag{1.110}$$

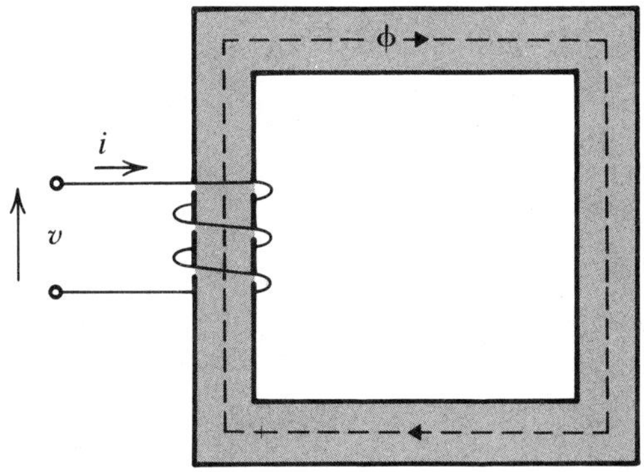

FIG. 1-41

where v is the voltage induced on its coil and i is the instantaneous value of the coil's current.

The magnitude of the voltage induced is

$$v = N\frac{d\phi}{dt} = NA\frac{dB}{dt} \tag{1.111}$$

and

$$H = \frac{Ni}{l} \tag{1.112}$$

or

$$i = H\frac{l}{N} \tag{1.113}$$

From Eqs. (1.110), (1.111), and (1.113) we obtain

$$dw = (Al)\, H\, dB \tag{1.114}$$

Recognizing that the product Al has the unit of volume, and assuming that the magnetic material is energized from zero up to B tesla, the energy stored per given volume of magnetic material is

$$\frac{W}{\text{volume}} = \int_0^B H\, dB \tag{1.115}$$

Assuming operation along the linear part of the magnetization characteristic, we obtain

$$\frac{W}{\text{volume}} = \frac{1}{\mu}\int_0^B B\, dB = \frac{1}{2\mu}B^2 \tag{1.116}$$

The energy stored in a magnetic material is

$$W = \frac{1}{2\mu} B^2 \,(\text{Vol}) \tag{1.117}$$

Thus, the energy stored in a magnetic system is directly proportional to its volume and to the square of its flux density, and inversely proportional to its permeability.

Consider Fig. 1-35(a). The total energy stored within the magnetic circuit is equal to the energy stored in the magnetic material plus the energy stored in its air gap. Mathematically,

$$W = \frac{B^2}{2\mu_0}\left(\frac{\text{Vol}_{\text{magn. mat.}}}{\mu_r} + \text{Vol}_{\text{air gap}}\right) \tag{1.118}$$

EXAMPLE 1–14 The relative permeability of the material shown in Fig. 1-35(a) is 1200. The mean length of the magnetic material is .75 m, and that of the air gap is .75 cm. The coil has 60 turns and draws 8 A from the 120 V ac supply. The volume of the magnetic material is 6×10^{-3} m^3 and the volume of the air gap is 60×10^{-6} m^3. Determine the energy stored:

a) In the magnetic material.

b) In the air gap.

SOLUTION The magnetic flux density is calculated as follows:

$$\begin{aligned} B &= \frac{\phi}{A} \\ &= \frac{Ni}{\mathcal{R}_T A} \\ &= \frac{Ni}{A\left(\dfrac{l_m}{\mu_m A_m} + \dfrac{l_g}{\mu_g A_g}\right)} \end{aligned}$$

Disregarding fringing, we obtain

$$B = \frac{Ni\mu_0}{\dfrac{l_m}{\mu_r} + l_g}$$

Substituting the given values, we get

$$\begin{aligned} B &= \frac{60(8)(4\pi \times 10^{-7})}{\dfrac{.75}{1200} + .0075} \\ &= 74.24 \times 10^{-3}\ \text{T} \end{aligned}$$

a) The energy stored in the magnetic material is

$$W_m = \frac{B^2}{2\mu} \text{Vol}_{(m)} = \frac{(74.24 \times 10^{-3})^2}{2 \times 1200(4\pi \times 10^{-7})}(6 \times 10^{-3})$$
$$= \underline{\underline{10.96 \text{ m J}}}$$

b) The energy stored in the air gap is

$$W_g = \frac{(74.24 \times 10^{-3})^2}{2(4\pi \times 10^{-7})}(60 \times 10^{-6})$$
$$= \underline{\underline{131.6 \text{ m J}}}$$

Comparing the results, it is clear that the energy stored in the air gap is about 12 times greater than the energy stored in the magnetic material.

1.2.10 PRINCIPLES OF MAGNETIC CIRCUITS

The basic principles of magnetic circuits are Ohm's law and Kirchhoff's laws. These laws are analogous to those of the electric circuits that are identified by the same name.

Ohm's Law

The flux of magnetic circuits is analogous to the current of electric circuits. The flux within a simple magnetic circuit is given by the ratio of the circuit's magnetic potential divided by the reluctance of the magnetic path. Mathematically,

$$\phi = \frac{Ni}{\mathcal{R}} \tag{1.119}$$

This equation is Ohm's law for magnetic circuits. In a complex electromagnetic system the numerator of the equation should include the algebraic sum of all mmf's in the particular loop under consideration, and the denominator should include the sum of the loop's reluctances.

An ideal magnetic circuit (Fig. 1-42(a)) can be represented in terms of the circuit mmf (Fig. 1-42(b)) or in terms of the circuit reluctance (Fig. 1-42(c)). The applied mmf is $N_1 i_1$ and that absorbed by the magnetic material is NI. The latter is, as per Ohm's law, equal to ϕR_m, where ϕ and R_m are, respectively, the flux and the reluctance of the magnetic circuit.

The magnetic circuit, unlike the electric circuit, cannot be easily shorted because, as the flux increases above the knee of the magnetization curve, the permeability of the magnetic material decreases and thus the reluctance of the magnetic path increases. Hence, there is a limit to the quantity of flux that can exist in a magnetic circuit.

The intrinsic properties of any given magnetic material are such as to limit the quantity of flux within its structure.

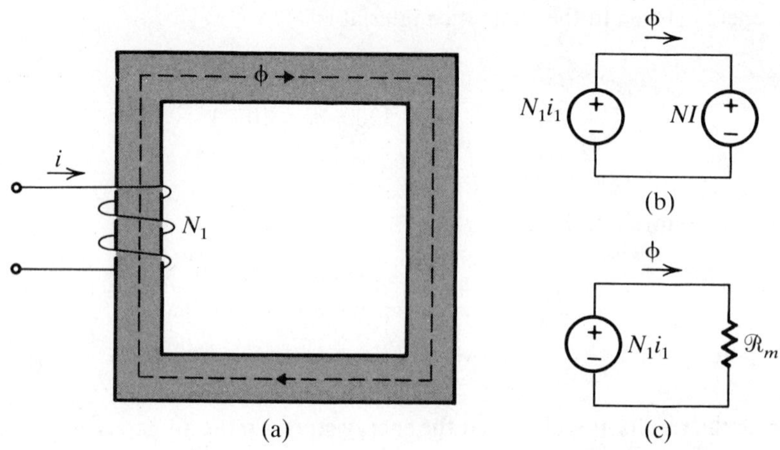

FIG. 1-42 Magnetic circuit: **(a)** elementary physical representation, **(b)** equivalent circuit in terms of mmf's, **(c)** equivalent circuit in terms of reluctance.

Exercise 1–13 The magnetic circuit shown in Fig. 1-42(a) has a mean magnetic path of .5 m, a relative permeability of 2000, and a uniform cross-sectional area equal to 5×10^{-4} m^2. The coil has 100 turns and, when connected to a 120 V, 60 Hz voltage source draws 5 A. Determine:

a) The inductance of the coil.

b) The maximum flux through the magnetic material.

Answer (a) 25.13 m H; (b) 1.78 m Wb

Kirchhoff's Laws

Voltage Law

The KVL of magnetic circuits is similar to that of electric circuits. That is, the sum of the mmf's along a closed loop is equal to zero. Mathematically,

$$\sum (Ni)_{\text{loop}} = 0 \tag{1.120}$$

This equation is analogous to Eq. (1.47), according to which the sum of the emf's along a closed loop in an electric circuit is equal to zero.

Current Law

According to Kirchhoff's current law, the sum of the magnetic flux in a junction is equal to zero. Mathematically,

$$\sum \phi_{\text{junction}} = 0 \tag{1.121}$$

That is, the sum of the flux entering a magnetic junction is equal to the sum of the flux leaving the junction.

For magnetic circuit calculations, one may take into consideration the leakage flux, the stacking factor, and the fringing of the flux through the circuit's air gaps.

Stacking Factor

The effective cross-sectional area of a magnetic core is less than the actual cross-sectional area because of the insulation on the surface of the laminations. The reduction in the actual area is accounted for by using what is called a stacking factor. The value of the stacking factor ranges between .85 to .95, depending on the thickness of the laminations and on the amount of their surface insulation.

Fringing

The spreading out of the flux around the edges of an air gap (Fig. 1-23) is called fringing, or diverging of the flux. In calculating the flux density through a short air gap, fringing is accounted for by increasing the given cross-sectional area of the air gap as shown below.

$$\text{Effective cross-sectional area} = (m + l_g)(n + l_g) \qquad \textbf{(1.122)}$$

where m and n are the dimensions of the magnetic core perpendicular to the flux lines and l_g is the length of the air gap.

Leakage Flux

The leakage flux is relatively small, owing to its large reluctance path. Thus, it is normally neglected in magnetic circuit calculations. In the analysis of transformers and electric machines, however, the effects of the leakage flux are included in their equivalent electric circuits.

EXAMPLE 1–15 In the circuit shown in Fig. 1-43(a), l_g and l_m represent, respectively, the length of the air gap and the total length of the mean magnetic path. Determine:

a) An expression for the total reluctance.

b) The length of the air gap that will absorb one-half of the applied mmf.

SOLUTION The equivalent circuit is shown in Fig. 1-43(b).

a) The total reluctance is the sum of the reluctance of the magnetic material and that of the air gap. Thus,

$$\mathcal{R}_T = R_m + R_{\text{air gap}} = \frac{l_m}{\mu A} + \frac{2l_g}{\mu_0 A}$$

$$= \frac{1}{\mu_0 A}\left(\frac{l_m}{\mu_r} + 2l_g\right)$$

where l_g is the length of each air gap.

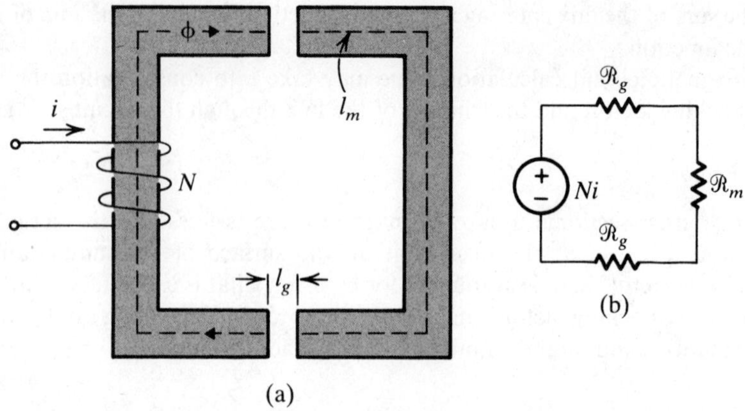

FIG. 1-43

b) From the equivalent circuit, the reluctances are in series. For one-half of the input mmf to be absorbed by the air gap, the reluctances of the magnetic material should be equal to that of the air gap. Thus,

$$\frac{l_m}{\mu} = \frac{2l_g}{\mu_0}$$

from which

$$l_g = \frac{l_m}{2\mu_r}$$

EXAMPLE 1–16 For the schematic shown in Fig. 1-44(a), draw the equivalent circuit in terms of:

a) MMf's

b) Reluctances.

Write the KVL and determine the flux through the various magnetic paths.

SOLUTION

a) The equivalent circuit, in terms of mmf's, is shown in Fig. 1-44(b). The mmf's in the air gap and in the other magnetic paths are identified with the same subscripts as the mean magnetic lengths.

Applying KVL, we have

$$Ni = (Ni)_a + (Ni)_c + (Ni)_g$$

or

$$Ni = (Ni)_a + (Ni)_b$$

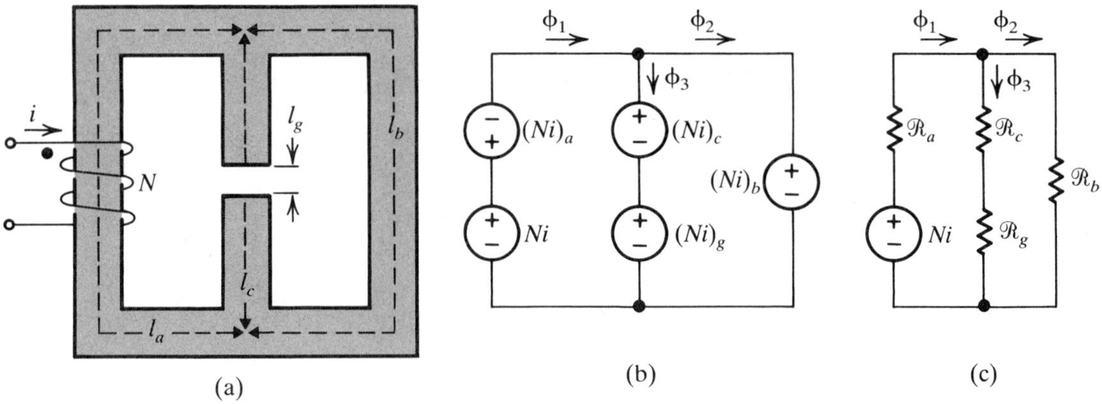

FIG. 1-44

It should be noted that, as in electric circuits, magnetic potentials across parallel branches of reluctances are equal.

b) The equivalent circuit with the various reluctances is shown in Fig. 1-44(c). Applying Ohm's law for magnetic circuits, we obtain

$$\phi_1 = \frac{Ni}{\mathcal{R}_a + \mathcal{R}_b \,/\!/\, (\mathcal{R}_c + \mathcal{R}_g)}$$

Applying the current-divider concept, we find

$$\phi_3 = \phi_1\left(\frac{\mathcal{R}_b}{\mathcal{R}_b + \mathcal{R}_c + \mathcal{R}_g}\right)$$

and

$$\phi_2 = \phi_1\left(\frac{\mathcal{R}_c + \mathcal{R}_g}{\mathcal{R}_b + \mathcal{R}_c + \mathcal{R}_g}\right)$$

Graphical Analysis of Electromagnetic Circuits

In calculations involving magnetic circuits that include air gaps, the magnetization and permeability of the materials are usually unknown. The level of magnetization and the corresponding permeability are factors of the basic magnetic circuit equations. In setting up these equations, often there are more unknowns than equations. This is similar to the analysis of electronic circuits (transistor or diode) where the voltage and current of electronic devices are unknown.

One way to solve these problems is by "trial and error," according to which an unknown parameter is assumed. If this assumed parameter satisfies the governing magnetic circuit equations, it is an acceptable solution. If it does not, then another value is assumed until a satisfactory solution is found.

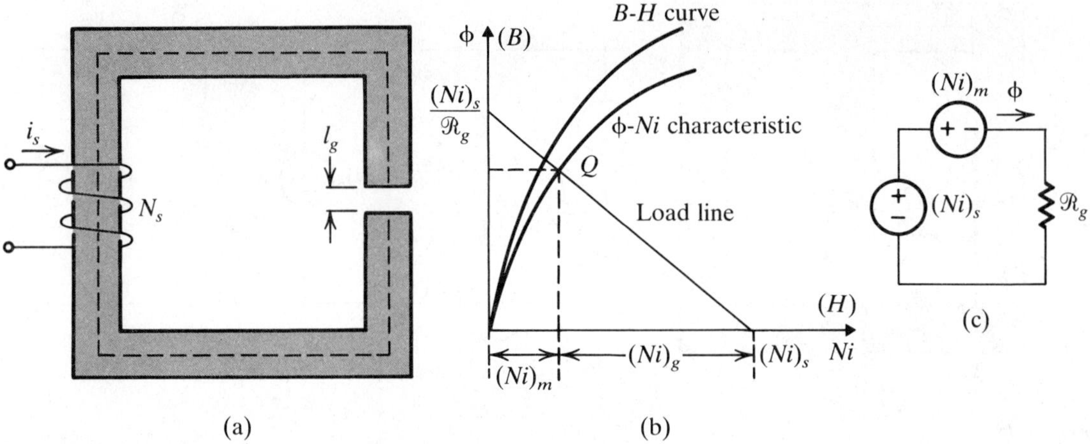

FIG. 1-45 Graphical solution of magnetic circuits: **(a)** simple magnetic circuit, **(b)** *B-H* and ϕ-*Ni* characteristic, **(c)** equivalent circuit.

A more effective method is to use a graphical solution. With this technique the given *B-H* characteristic is changed to its equivalent ϕ-*Ni* characteristic and the flux of the magnetic circuit is found as a *function* of the magnetomotive forces within a closed loop.

The resulting equation is drawn on the same coordinate system as the ϕ-*Ni* characteristic. The point of intersection of the two characteristics gives the actual flux and the magnetic potential drop within the magnetic material.

For a demonstration, consider the simple magnetic circuit of Fig. 1-45(a). Its *B-H* characteristic is shown in Fig. 1-45(b). The lengths of the magnetic material (l_m) and of the air gap (l_g) as well as their effective areas (A_m and A_g) are known. In order to find the flux through the magnetic material, one proceeds as follows:

Steps

1. *Draw the ϕ-$(Ni)_m$ characteristic.*
   Change the coordinates of the *B-H* characteristic to its equivalent ϕ-$(Ni)_m$ characteristic ($\phi = BA$ and $(Ni)_m = H_m \times l_m$).
2. *Write the governing equations* (Fig. 1-45(c)).

$$(Ni)_s = (Ni)_m + (Ni)_g \tag{1.123}$$

$$(Ni)_g = \phi \mathcal{R}_g \tag{1.124}$$

   where $(Ni)_s$, $(Ni)_m$, and $(Ni)_g$ represent the magnetic potentials of source, magnetic material, and air gap, respectively.
3. *Draw the load-line equation.*
   From Eqs. (1.123) and (1.124) we obtain

$$(Ni)_s = (Ni)_m + \phi \mathcal{R}_g \tag{1.125}$$

from which

$$\phi = -\frac{(Ni)_m}{\mathcal{R}_g} + \frac{(Ni)_s}{\mathcal{R}_g} \qquad \textbf{(1.126)}$$

Equation (1.126) is of the slope (m)–intercept (b) form ($y = mx + b$). Its intercepts are

$$y\text{-intercept} \quad \phi = \frac{(Ni)_s}{\mathcal{R}_g}, \qquad x\text{-intercept} = (Ni)_s$$

Since the intercepts are known, the equation is drawn as shown in Fig. 1-45(b). Equation (1.126) is known as the load-line equation. Its intersection with the ϕ-Ni characteristic gives the operating point, which is designated by the letter Q.

The operating point gives the actual flux, and the mmf drop within the magnetic material and through the air gap, as shown in Fig. 1-45(b).

Exercise 1–14 The coordinates of the B-H characteristic of the magnetic material shown in Fig. 1-46 are as follows:

B (T)	.10	.45	.78	.96	1.08
H (A/m)	50	150	350	800	1440

The mean length of the magnetic path is .75 m, and the effective area of the magnetic material and that of the air gap are 1.1×10^{-3} m^2 and 1.2×10^{-3} m^2, respectively. For an air-gap length of .60 mm, determine the flux density in the magnetic material.

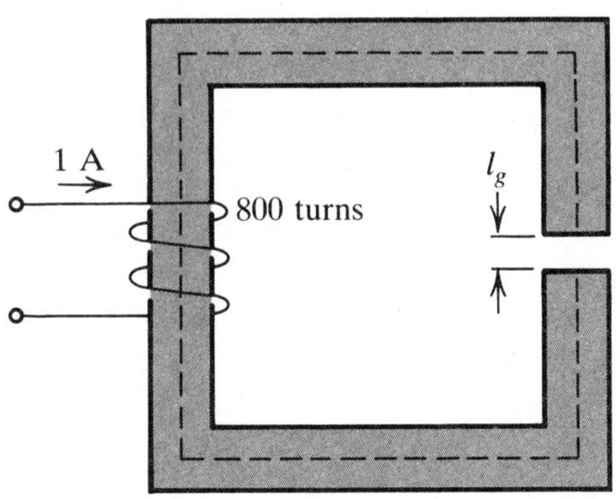

FIG. 1-46

Answer .90 T

Exercise 1–15 The *B-H* characteristic of the magnetic material shown in Fig. 1-47 is the same as that of Exercise 1-14. Determine the flux through the coil, given that

$$l_{bad} = .64 \text{ m}, \quad l_{bd} = .2 \text{ m}, \quad l_{bcd} = .4 \text{ m}$$
$$l_g = 1.0 \text{ mm}, \quad \text{stacking factor} = .95$$

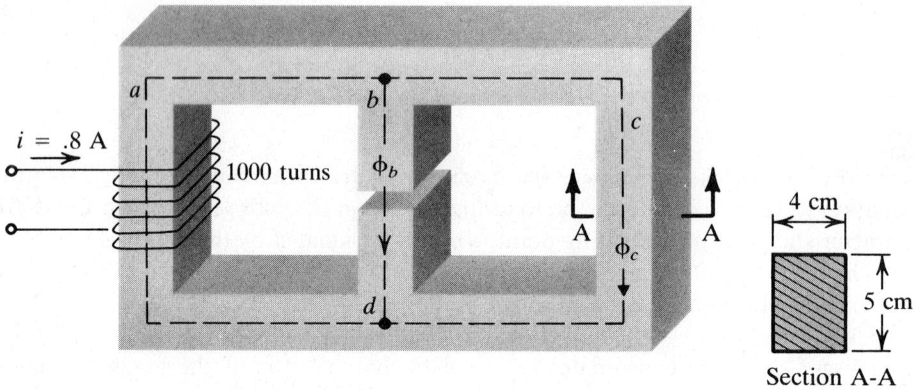

FIG. 1-47

Answer 1.94 m Wb

1.2.11 MAGNETIC LOSSES

Magnetic losses can be separated into eddy-current losses and hysteresis losses. Eddy currents are produced within the magnetic material because of the alternating nature of the flux, which in turn is produced by the winding's ac current. When the magnetic material has infinite resistance (zero conductivity), however, there will not be any eddy-current losses.

Hysteresis represents the friction among the walls of the magnetic domains. The magnetic domains are constantly being reoriented because of the alternating nature of the applied field. The total magnetic losses (P_m) of a coil are given by

$$P_m = P_e + P_h \tag{1.127}$$

where P_e and P_h represent, respectively, eddy-current and hysteresis losses.

Magnetic losses are also referred to as excitation losses or open-circuit losses. They are discussed in more detail in Section 2.1.3.

Eddy-Current Losses

The voltage applied to a coil wound around a toroidal magnetic material will be accompanied, as can be predicted from the induction principle, by a flux. Mathematically,

$$v = -N \frac{d\phi}{dt} \tag{1.128}$$

For operations along the linear section of the magnetization characteristic, the flux, according to Ohm's law for magnetic circuits, is proportional to the current. That is, for a linear change in the flux there will be a linear change in the current. Using mathematical symbols:

$$\phi \propto i \tag{1.129}$$

According to Lenz's law, the flux within the magnetic material will induce opposing voltages that produce circular currents (I_e, eddy currents). As shown in Fig. 1-48, these induced currents flow around the flux and produce eddy current loss.

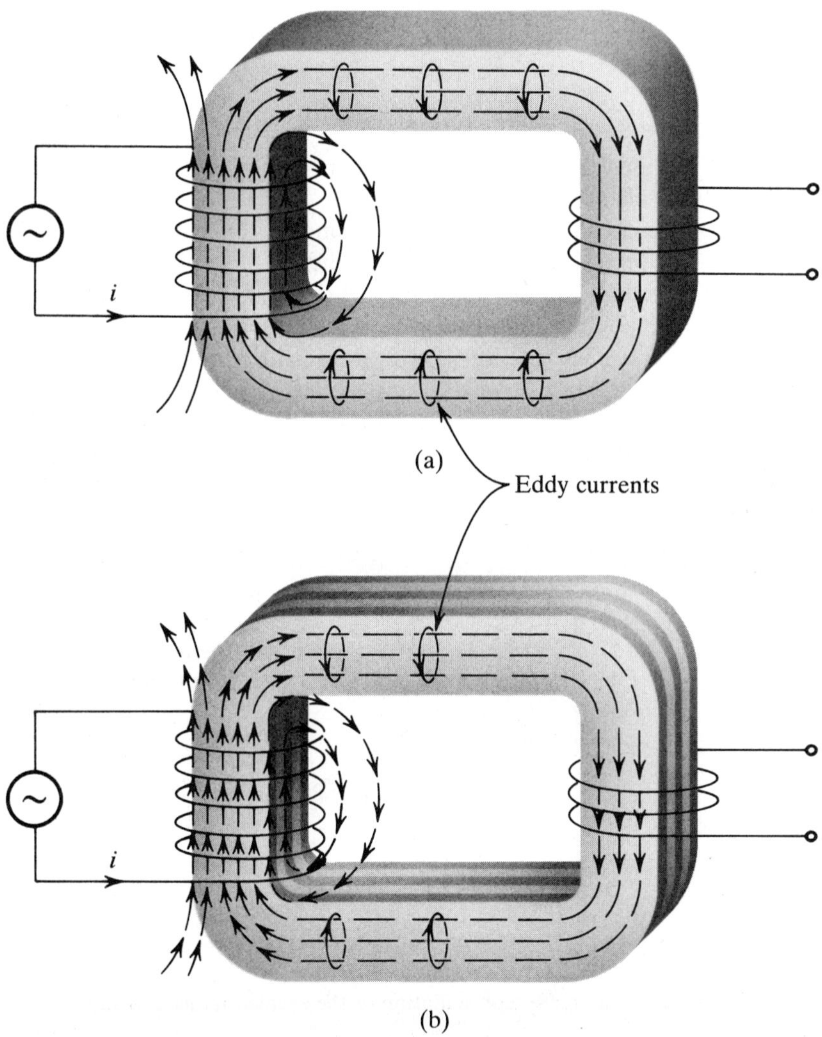

FIG. 1-48 An illustration of eddy currents: **(a)** nonlaminated core, **(b)** laminated core (notice the relative reduction in the eddy currents).

The eddy-current loss is proportional to eddy current squared. That is,

$$P_e \propto I_e^2 \tag{1.130}$$

In terms of the applied voltage, the last expression becomes

$$P_e = K_1 V^2 \tag{1.131}$$

where V is the effective value of the applied voltage and K_1 is a constant of proportionality. This constant is directly proportional to the volume of the material and inversely proportional to its resistivity.

Substituting for the voltage its equivalent expression from Eq. (1.109), we obtain

$$P_e = K_2(f\phi)^2 \tag{1.132}$$

The constant of proportionality K_2 is directly proportional to K_1 and to the square of the number of turns of the coil.

Eddy-current losses can be reduced by increasing the resistance to the flow of eddy currents and/or by decreasing the magnitude of the flux.

Resistance to the flow of eddy currents (R_e) can be increased by decreasing the thickness of the laminations and by increasing their surface insulation.

For a given coil voltage, the flux can be reduced by increasing the number of winding turns.

The laconic staircase representation shown in Fig. 1-49 is an alternative way to demonstrate the cause-and-effect relationship that leads to eddy-current losses.

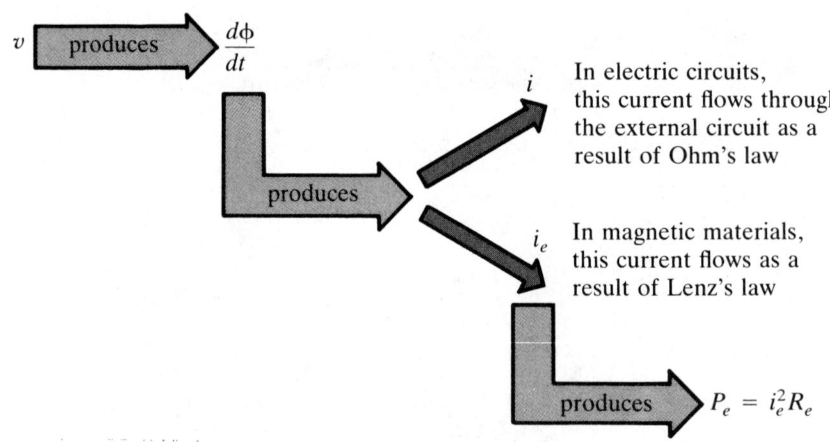

FIG. 1-49 A laconic staircase representation of the events that lead to eddy-current losses.

Hysteresis Losses

Hysteresis losses are heat losses due to the friction between magnetic domains. Each time the input voltage changes polarity, the magnetic field within the core of the coil also changes polarity. Thus the flux—or the lines of force—change direction. The consequent continuous alignment and realignment of the magnetic domains in the direction of the alternating flux produces friction, which manifests itself as heat within the material.

Hysteresis loss (P_h) is related to the flux of the core and to the frequency (f) of the voltage source by

$$P_h \propto f\phi^h \tag{1.133}$$

where h is a constant characteristic of the material. Its usual range is from 1.4 to 2.5.

From Eqs. (1.109) and (1.133) we obtain

$$P_h \propto f\left(\frac{V}{f}\right)^h \tag{1.134}$$

From the above,

$$P_h = K_3 \frac{V^h}{f^{h-1}} \tag{1.135}$$

The constant of proportionality K_3 is directly proportional to the volume of the magnetic material under consideration.

Etymologically, hysteresis is a Greek word which means lagging. In fact, owing to hysteresis the induced emf is not at 90° to the flux but lags it by $(90° + \theta_h)$, where θ_h is due to hysteresis phenomenon (see Fig. 1-50). This deviation in the phase angle is used to advantage in the production of torque in so-called hysteresis motors.

The hysteresis losses of a magnetic material can also be evaluated from the area of

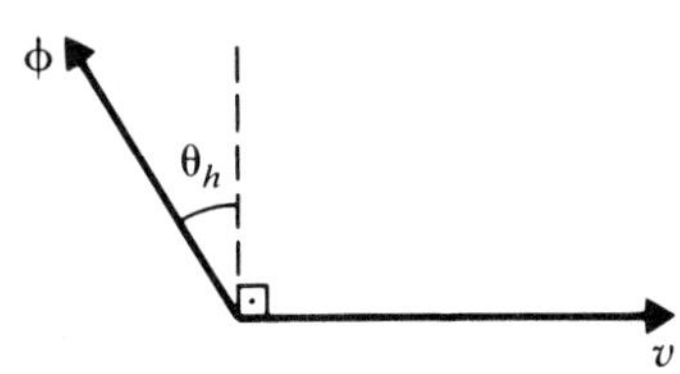

FIG. 1-50 Effect of hysteresis on the phase angle between flux and emf.

its hysteresis loop as follows (see Fig. 1-51):

$$\text{area of hysteresis loop} = \int H\, dB \tag{1.136}$$

Substituting the values

$$H = \frac{Ni}{l} \quad \text{and} \quad dB = \frac{d\phi}{A} = -\frac{v}{AN}\, dt$$

we obtain

$$\text{area of hysteresis loop} = -\frac{1}{Al}\int vi\, dt \tag{1.137}$$

As seen from the above, the area of a hysteresis loop is equal to the energy dissipated in the material per cycle per unit volume due to the hysteresis phenomenon.

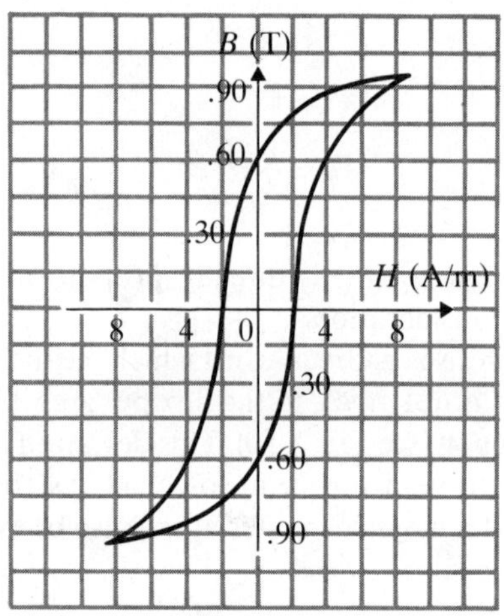

FIG. 1-51 Hysteresis loop.

An alternative method of evaluating the hysteresis losses of a material subjected to a cyclic variation of its magnetizing force is to measure the area of the hysteresis loop in proper units and to multiply that area by the period of the magnetizing force expressed in seconds.

For example, if the cyclic variation of the magnetizing force is 60 times per second and the area of the resulting hysteresis loop has 25 squares, the perpendicular sides of which are equal to .15 T and 2 A/m, respectively, the hysteresis losses are

$$25(.15)(2)(60) = 450 \ \text{W/m}^3$$

EXAMPLE 1–17

a) Figure 1-52 shows a section of an underground, three-phase, three-single-conductor, 4.16 kV power distribution network. The contractor mistakenly installed *steel* instead of *plastic* elbows as adaptors. As a result, the cables were damaged during the plant's operation, owing to induction currents. Explain the underlying principles and the nature of magnetic losses.

b) Prove that no induction currents are induced in steel conduits through which one three-conductor cable is installed.

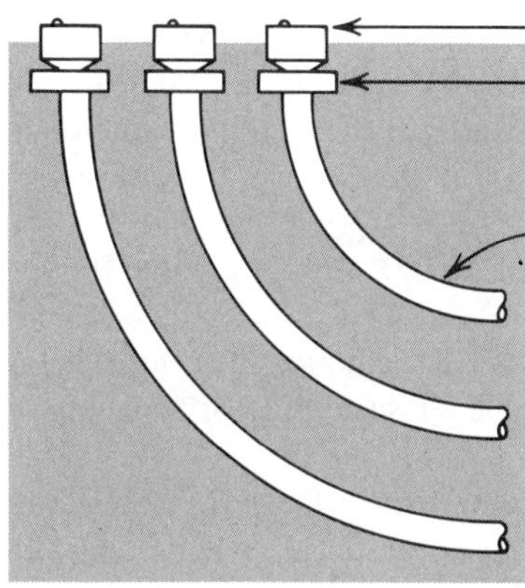

FIG. 1-52 Three-phase, single conductor, power distribution.

As per industrial practices the PVC conduit is used for exposed cable installations (such as going from the secondary transformer terminals to the underground cable distribution), while the PV duct is used only in concealed underground cable distribution. Both the conduit and the duct are made from plastic material, but the conduit is much more rugged than the duct.

SOLUTION

a) Owing to Ohm's law in magnetic circuits ($\phi = Ni/\mathscr{R}$), the conductor's current produced flux, which according to the "right-hand rule," encircled the area outside the conductor's current. As a result, flux was induced in the galvanized steel elbows.

The sinusoidal variation of the steel elbow's flux produced eddy currents and continuous realignment of the material's magnetic domains. According to Lenz's law, eddy currents represent the natural reaction of the material to oppose the induced flux. The realignment of the magnetic domains is a natural tendency of the basic building blocks of magnetism to follow the direction of the external field.

Eddy currents produced eddy current losses ($P_e \propto V^2$) and the alternating realignment of the magnetic domains produced hysteresis losses ($P_h \propto f\phi^{1.6}$). This resulted in overheating of the steel elbows, which led to the melting of the conductor's insulation, generation of sparks, and electrical shorting of the multiconductor distribution. Thus, owing to the induction principle, the cables were damaged and the plant's production was interrupted for the time necessary to identify the problem, remove the steel elbows, and purchase and install new cables.

b) Induction currents are not induced in a steel conduit through which a three-conductor cable (phases A, B, and C) passes, because the net induced flux is theoretically equal to zero. The proof of this is as follows:

The resultant flux (ϕ_r) induced in the steel conduit is due to the flux of three-phase currents. That is,

$$\phi_r = \phi_A + \phi_B + \phi_C$$

where ϕ_A, ϕ_B, and ϕ_C represent the flux produced by the individual phase currents.

From the theory of three-phase systems (see Appendix A) each current, or flux, is equal in magnitude to the others and is displaced 120 degrees from the others in phase. Thus, designating the rms value of the flux of the one-phase conductor with ϕ, we have

$$\phi_r = \phi(1\,\underline{/0^\circ} + 1\,\underline{/120^\circ} + 1\,\underline{/-120^\circ}) = 0$$

In actual cases, some flux (which has negligible heating effects) will be induced, owing to imperfect symmetry in the phase conductor's spacing relative to the steel conduit.

This industrial problem occurred in a factory in the Southwestern United States in August 1981. Damage was estimated at several hundred thousand dollars. It is astonishing that such a faulty installation took place, considering that (according to internal company memos) the consultants who designed the electrical distribution system, as well as the plant's electrical engineers, were very much aware of the induction problems associated with single-conductor distribution. Further, the construction manager who supervised the installation was an experienced mechanical engineer.

EXAMPLE 1–18 A firm that manufactures transformers requires magnetic materials that have a relative permeability equal to 100. A salesman of magnetic materials claims and demonstrates in the lab that his company's product has a permeability equal to 100. After testing the proposed material, however, the manufacturing firm's engineer rejects it. Give two possible technical reasons for his decision.

SOLUTION The permeability of a material has meaning only when the flux density at which it is measured is specified. As shown in Fig. 1-32, each magnetic material has a wide range of permeabilities. The engineer may have rejected the materials because, at the *level of flux density* he was interested in, the permeability was not 100.

A second reason may have been that the *losses* of the proposed materials were too high.

Additional reasons for the engineer's rejection of the proposed material may have been the unacceptable change in its permeability with *time* (disaccommodation factor) at high *temperatures*. The *cost* of the material is also of paramount importance.

This example demonstrates the various interrelated factors that one has to consider when a decision is to be made in selecting and/or designing a product. The phrasing of the specifications, or the description of the problem, is also important. In the above example, the manufacturer's requirements were not explicitly stated; an experienced salesman should have known the implied restrictions when dealing with properties and characteristics of magnetic materials.

1.2.12 EQUIVALENT CIRCUIT OF A COIL

A coil and its equivalent electric circuit are shown in Figs. 1-53(a) and (b), respectively. The impedance $R_1 + jX_1$ is referred to as the coil's leakage impedance, and the impedance $R_m + jX_m$ is the coil's magnetizing impedance. Leakage impedance is usually a small fraction of magnetizing impedance.

The resistance R_1 represents the dc resistance of the winding. As can be seen from Eq. (1.14), the winding resistance depends on the resistivity of the material, the length of the conductor, and its cross-sectional area. Resistivity depends on the conductor's chemical composition and temperature. The cross-sectional area of a conductor plays an important role in controlling the losses. There are, however, physical and economic restraints which determine how large a conductor's diameter can be.

The dc resistance per unit length of any gauge of wire is readily available in published tables.

The leakage reactance X_1 is due, as the name implies, to the coil's leakage flux.

The resistance R_m represents the equivalent electrical resistance of the magnetic material and depends on the material's eddy-current and hysteresis losses.

The reactance X_m is referred to as the magnetizing reactance of the coil. This reactance is due to the flux within the magnetic material, which is directly proportional to the applied voltage, as can be seen from Eq. (1.109).

The power consumed by R_1 constitutes the *electrical losses* of the coil; the power

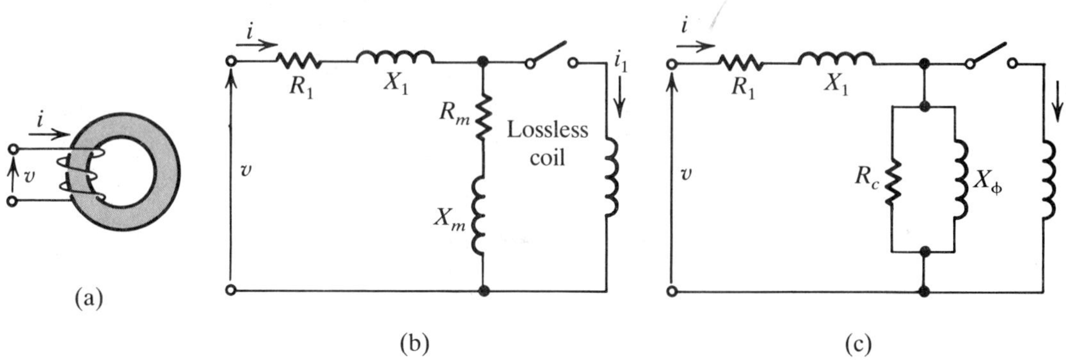

FIG. 1-53 Electromagnet: **(a)** elementary representation, **(b)** its equivalent electrical circuit, **(c)** equivalent circuit of (b).

consumed by R_m represents the *magnetic losses* of the iron core. Electrical losses are often referred to as copper losses, and magnetic losses as core, or iron, losses.

The coil's power consumption manifests itself as heat, which, if not promptly removed, may cause insulation breakdown and fire.

Winding losses are discussed further in Section 2.1.3.

The magnetizing impedance Z_m can also be represented by a resistance (R_c) connected in parallel to a reactance (X_ϕ), as shown in Fig. 1-53(c).

Equating the two equivalent impedances, we obtain

$$R_m + jX_m = \frac{jR_cX_\phi}{R_c + jX_\phi} \tag{1.138}$$

or

$$R_m + jX_m = \frac{jR_cX_\phi(R_c - jX_\phi)}{R_c^2 + X_\phi^2} \tag{1.139}$$

from which

$$R_c = \frac{R_m^2 + X_m^2}{R_m} \tag{1.140}$$

and

$$X_\phi = \frac{R_m^2 + X_m^2}{X_m} \tag{1.141}$$

The current through R_c provides the magnetic losses, while the current through X_ϕ magnetizes the magnetic material. The series representation of the magnetizing impedance of a coil is simpler to obtain than its parallel representation. The latter, however, reveals more about the magnetization process and for this reason is more common.

The lossless coil of Fig. 1-53(b) represents the capability of the winding to be magnetically coupled to other coils that may be connected to the system of Fig. 1-53(a). When no other coil is wound in the same core, the ideal coil is open-circuited and thus $i_1 = 0$.

Exercise 1–16 When connected to a 120 V, 60 Hz power supply, a coil draws 4.8 A and consumes 100 W of power. If the dc resistance of the coil is 2 ohms, determine

a) The magnetic losses.

b) The equivalent series electrical resistance of the magnetic material.

Answer (a) 53.92 W; (b) 2.34 Ω

1.2.13 MATHEMATICAL RELATIONSHIPS OF SELF- AND MUTUAL INDUCTANCES

The self-inductance of a single coil is given by the following relationships:

$$\text{self-inductance} = \frac{\text{flux linkage of the coil}}{\text{current through the coil}}$$
$$= \frac{\text{(number of turns)(quantity of flux)}}{\text{current through the coil}} \qquad \textbf{(1.142)}$$

Using mathematical symbols, we obtain

$$L = \frac{\lambda}{i} = \frac{N\phi}{i} \qquad \textbf{(1.143)}$$

For an electromagnetic system with two coils, such the one shown in Fig. 1-54(a), the flux linkage of each coil is defined as follows:

$\lambda_{11} = N_1\phi_{11}$ = flux linkage of coil 1 produced by its own current i_1
$\lambda_{12} = N_1\phi_{12}$ = flux linkage of coil 1 produced by the current i_2 of coil 2
$\lambda_{22} = N_2\phi_{22}$ = flux linkage of coil 2 produced by its own current i_2
$\lambda_{21} = N_2\phi_{21}$ = flux linkage of coil 2 produced by the current i_1 of coil 1

The total flux linkages of coil 1 (λ_1) and coil 2 (λ_2) are given by

$$\lambda_1 = N_1\phi_1 = N_1(\phi_{11} + \phi_{12}) \qquad \textbf{(1.144)}$$

$$\lambda_2 = N_2\phi_2 = N_2(\phi_{22} + \phi_{21}) \qquad \textbf{(1.145)}$$

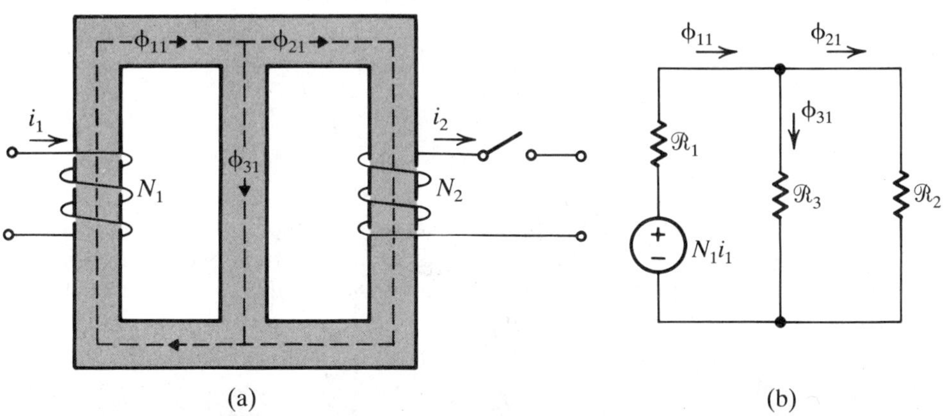

FIG. 1-54 Magnetic system with negligible leakage flux: **(a)** magnetically coupled coils ($i_2 = 0$), **(b)** its equivalent magnetic circuit.

The self- and mutual inductances are defined as follows:

$$L_{11} = \frac{\lambda_{11}}{i_1} = \frac{N_1\phi_{11}}{i_1} = \text{self-inductance of coil 1}$$

$$L_{12} = \frac{\lambda_{12}}{i_2} = \frac{N_1\phi_{12}}{i_2} = \begin{array}{l}\text{mutual inductance of coil 1,}\\ \text{due to current in coil 2}\end{array}$$

$$L_{22} = \frac{\lambda_{22}}{i_2} = \frac{N_2\phi_{22}}{i_2} = \text{self-inductance of coil 2}$$

$$L_{21} = \frac{\lambda_{21}}{i_1} = \frac{N_2\phi_{21}}{i_1} = \begin{array}{l}\text{mutual inductance of coil 2,}\\ \text{due to current in coil 1}\end{array}$$

Physically, the mutual inductance (L_{12}) of coil 1, with respect to coil 2, is the flux linkage that links coil 1 due to the current in coil 2.

In a two-subscript parameter, the first subscript identifies the affected coil while the second identifies the coil that produced this effect.

From the outlined mathematical relationships it is evident that, for the determination of self- and mutual inductances of an electromagnet, the equivalent magnetic circuit must be drawn and then basic principles must be used to obtain ϕ_{11}, ϕ_{12}, and ϕ_{21}.

Applying Ohm's law in Fig. 1-54(b), for example, we obtain

$$\phi_{11} = \frac{N_1 i_1}{\mathcal{R}_1 + \mathcal{R}_2 \,/\!/\, \mathcal{R}_3}$$

Using the current-divider technique,

$$\phi_{21} = \phi_{11} \frac{\mathcal{R}_3}{\mathcal{R}_2 + \mathcal{R}_3}$$

and

$$\phi_{31} = \phi_{11} \frac{\mathcal{R}_2}{\mathcal{R}_2 + \mathcal{R}_3}$$

In a multiple-coil system of constant permeability, the superposition theorem may be used. According to this theorem the effects of each mmf are found while the other mmf's are shorted, and then the results are summed vectorially.

The mathematical expression that relates the induced voltage to mutual inductance is similar to the equation that relates the induced voltage to the self-inductance. That is,

$$v_1 = \frac{d}{dt}(L_{12} i_2) \qquad \textbf{(1.146)}$$

For constant mutual inductance we have

$$v_1 = L_{12} \frac{d}{dt} i_2 \quad \textbf{(1.147)}$$

where v_1 is the voltage induced due to mutual inductance L_{12}.

Coefficient of Coupling

The coefficient of coupling is a factor that indicates the degree of magnetic coupling between two coils. It is always less than 1. The coefficient of coupling K is given by

$$K = \frac{L_{12}}{\sqrt{L_{11}L_{22}}} \quad \textbf{(1.148)}$$

In general, the higher the coefficient of coupling, the lower the leakage flux of the coils.

EXAMPLE 1–19 Using the data shown in Fig. 1-55, determine:

a) The flux linkage of coil 1.
b) The inductance, as seen from the terminals of coil 1.
c) The mutual inductance of coil 2, with respect to coil 1.

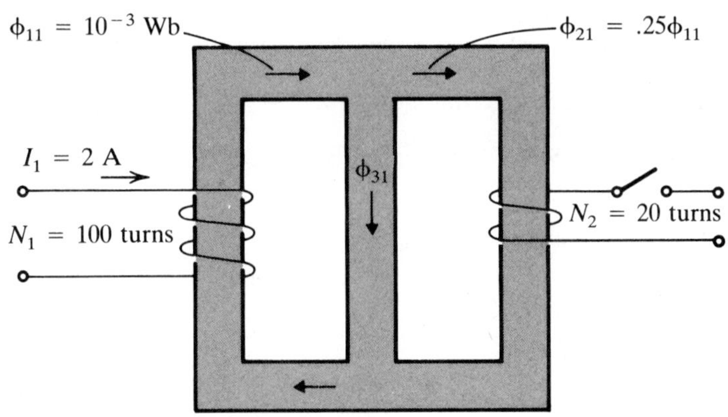

FIG. 1-55

SOLUTION

a) The flux linkage for coil 1 is

$$\lambda_{11} = N_1\phi_{11} = 100(10^{-3}) = \underline{\underline{.1 \text{ weber-turn}}}$$

b) The inductance of coil 1 is

$$L_{11} = \frac{\lambda_{11}}{I_1} = \frac{.1}{2} = \underline{\underline{50 \text{ mH}}}$$

c) The mutual inductance of coil 2 with respect to coil 1 is

$$L_{21} = N_2 \frac{\phi_{21}}{I_1} = \frac{20(.25 \times 10^{-3})}{2} = \underline{\underline{2.5 \text{ mH}}}$$

EXAMPLE 1–20 Laboratory measurements at 60 hertz for the two coils shown in Fig. 1-56 gave the following results.

Test	*Coil Condition*	*Voltage in Volts*	*Current in Amperes, rms*
1	Coil 1 connected to a voltage source	80	1.5
	Coil 2 open-circuited	30	0
2	Coil 2 connected to a voltage source	60	1
	Coil 1 open-circuited	20	0

Neglecting the resistance of the coils, determine:

a) The self-inductance of each coil.
b) The mutual inductances.
c) The coefficient of mutual coupling.

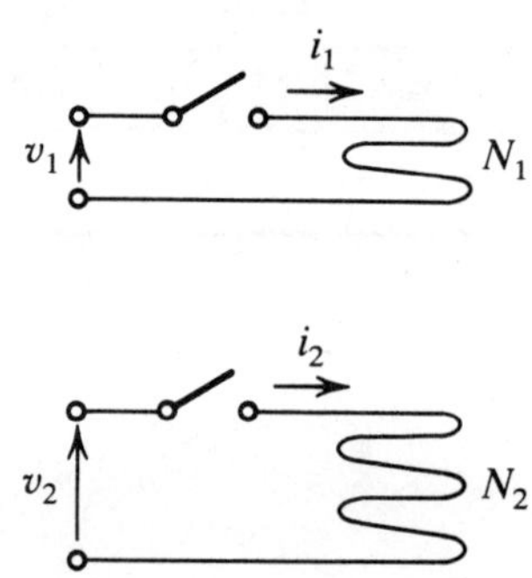

FIG. 1-56

<u>*SOLUTION*</u>

a) The self-inductance of coil 1 can be found from the following basic relationship:

$$v_1 = L_{11}\frac{di_i}{dt}$$

Using rms values and considering magnitudes only,

$$V_1 = L_{11}\omega I_1$$

Thus,

$$L_{11} = \frac{80}{2\pi(60)(1.5)} = \underline{\underline{.14\ \text{H}}}$$

Similarly,

$$L_{22} = \frac{60}{2\pi 60(1)} = \underline{\underline{.16\ \text{H}}}$$

Alternatively, from Eq. (1.143), using rms values, we obtain

$$L = \frac{N\phi}{I}$$

Since the current is given in rms, the flux must also be expressed in rms. Then, from the above and Eq. (1.109), the self-inductance of coil 1 is

$$L_{11} = \frac{N}{I}\left(\frac{\phi_m}{\sqrt{2}}\right) = \frac{N}{I\sqrt{2}}\left(\frac{V}{4.44Nf}\right)$$

Thus,

$$L_{11} = \frac{80}{1.5\sqrt{2}(4.44 \times 60)} = \underline{\underline{.14\ \text{H}}}$$

b) The mutual inductance of coil 1 with respect to coil 2 is found as follows:

$$v_1 = L_{12}\frac{d}{dt}i_2$$

Using rms values,

$$V_1 = L_{12}\omega I_2$$

From the above,

$$L_{12} = \frac{V_1}{\omega I_2} = \frac{20}{377(1)} = \underline{\underline{53.05\ \text{mH}}}$$

Similarly,

$$L_{21} = \frac{V_2}{\omega I_1} = \frac{30}{377(1.5)} = \underline{\underline{53.05 \text{ mH}}}$$

c) The coefficient of mutual coupling is

$$K = \frac{L_{12}}{\sqrt{L_{11}L_{22}}} = \frac{53.05 \times 10^{-3}}{\sqrt{.14 \times (.16)}} = \underline{\underline{.354}}$$

EXAMPLE 1–21 For coil 1 of the electromagnetic system shown in Fig. 1-57(a), neglect fringing, leakage flux, and the mmf within the magnetic material, then:

1. Determine:
   a) The flux linkage.
   b) The self-inductance.
   c) The mutual inductance.
2. If the current in coil 2 is reversed, repeat 1.

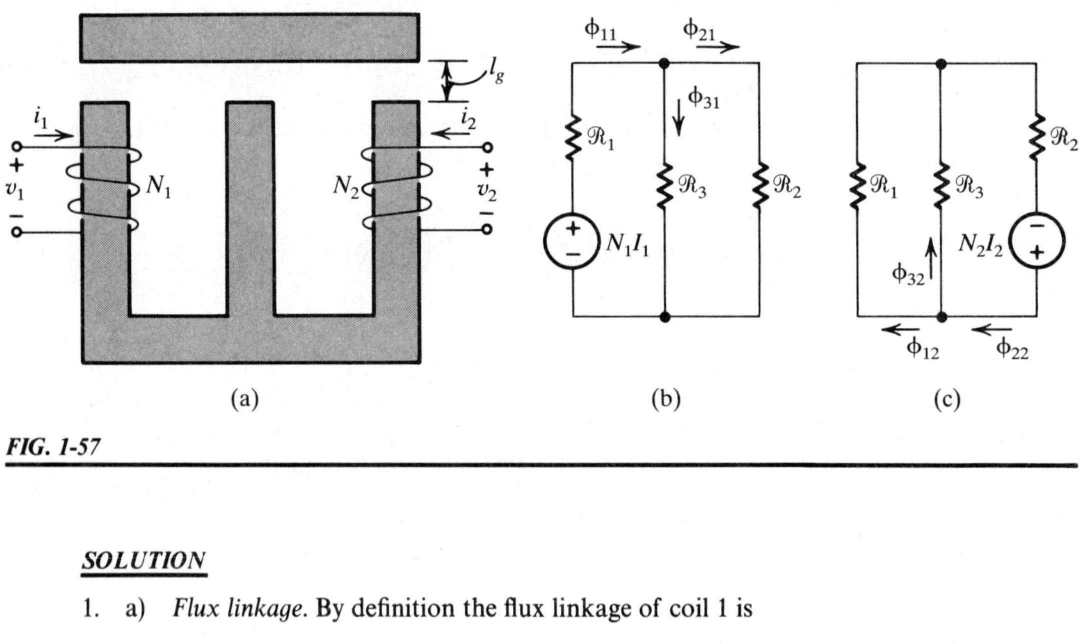

FIG. 1-57

SOLUTION

1. a) *Flux linkage.* By definition the flux linkage of coil 1 is

$$\begin{aligned}\lambda_1 &= \lambda_{11} + \lambda_{12}\\ &= N_1\phi_{11} + N_1\phi_{12}\end{aligned} \tag{I}$$

The term $N_1\phi_{11}$ represents the flux linkage of coil 1 due to its own current. To find $N_1\phi_{11}$, discard the current in coil 2 and draw the equivalent circuit as shown in Fig. 1-57(b). Thus,

$$\lambda_{11} = N_1\phi_{11} = N_1\left(\frac{N_1 I_1}{\mathscr{R}}\right) \tag{II}$$

The reluctance (as seen by the source of mmf $N_1 I_1$) is

$$\mathscr{R} = \mathscr{R}_1 + \mathscr{R}_2 \,/\!/\, \mathscr{R}_3$$

Since the air gaps have the same lengths, the reluctances $\mathscr{R}_1$, $\mathscr{R}_2$, and $\mathscr{R}_3$ are equal. Thus,

$$\mathscr{R} = \mathscr{R}_1 + \frac{\mathscr{R}_1}{2} = \frac{3}{2}\mathscr{R}_1 = \frac{3}{2}\frac{l_g}{\mu_0 A} \tag{III}$$

From equations (II) and (III) we obtain:

$$N_1\phi_{11} = N_1^2 I_1 \frac{2\mu_0 A}{3l_g} \tag{IV}$$

The term $N_1\phi_{12}$ of relationship (I) represents the flux linkage of coil 1 due to the current in coil 2. To find the effects of the second magnetic voltage source, apply the superposition principle and draw the equivalent circuit as shown in Fig. 1-57(c).

From the equivalent circuit,

$$N_1\phi_{12} = N_1\phi_{22}\frac{\mathscr{R}_3}{\mathscr{R}_1 + \mathscr{R}_3} = N_1\frac{\phi_{22}}{2}$$

$$N_1\phi_{12} = \left(\frac{N_1}{2}\right)\frac{N_2 I_2}{\mathscr{R}_2 + \mathscr{R}_1 \,/\!/\, \mathscr{R}_3} = \frac{N_1 N_2 I_2}{2\left(\frac{3}{2}\mathscr{R}_1\right)}$$

$$= N_1 N_2 I_2 \frac{\mu_0 A}{3l_g} \tag{V}$$

The total flux linkage of coil 1 is obtained by adding equations (IV) and (V). Thus,

$$\lambda_1 = \frac{\mu_0 A}{3l_g}(2N_1^2 I_1 + N_1 N_2 I_2) \tag{VI}$$

b) *Self-inductance.* By definition, the self-inductance of coil 1 is

$$L_{11} = \frac{\lambda_{11}}{I_1} = \frac{N_1\phi_{11}}{I_1}$$

From the last relationship and from equation (IV) we obtain

$$L_{11} = N_1^2 \frac{2}{3} \frac{\mu_0 A}{l_g} \tag{VII}$$

c) *Mutual inductance.* The mutal inductance of coil 1, with respect to coil 2, is

$$L_{12} = \frac{\lambda_{12}}{i_2} = \frac{N_1 \phi_{12}}{i_2}$$

Using rms values, from equation (V) and the last relationship, we obtain

$$L_{12} = N_1 N_2 \frac{\mu_0 A}{3l_g} \tag{VIII}$$

2. *Reversing current I_2*
   a) If the current in coil 2 changes direction, its flux will oppose the flux produced by the current in coil 1. Thus,

$$\lambda_1 = \frac{\mu_0 A}{3l_g}(2N_1^2 I_1 - N_1 N_2 I_2) \tag{IX}$$

b) and c) The self- and mutual inductances are scalar quantities, independent of current direction.

Exercise 1–17 Figure 1-58 shows an electromagnet and its equivalent magnetic circuit. Prove that

$$\phi_{12} = \frac{N_2 I_2}{\mathcal{R}_2 + \mathcal{R}_1 /\!/ \mathcal{R}_3}\left(\frac{\mathcal{R}_3}{\mathcal{R}_1 + \mathcal{R}_3}\right)$$

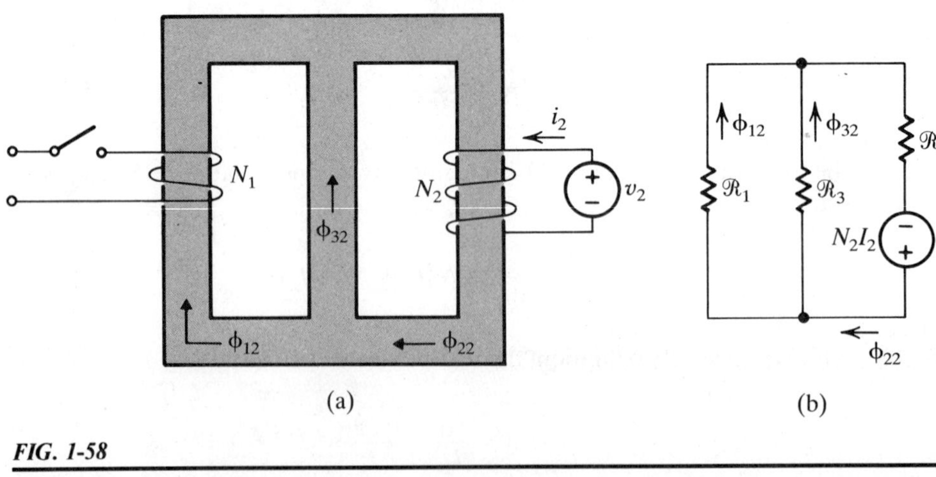

FIG. 1-58

Exercise 1–18 Figure 1-59(b) shows the idealized *B-H* characteristic of the magnetic material shown in Fig. 1-59(a). For a flux density of 1.4 *T*, determine:

a) The relative permeability of the magnetic material.
b) The inductance of the coil.

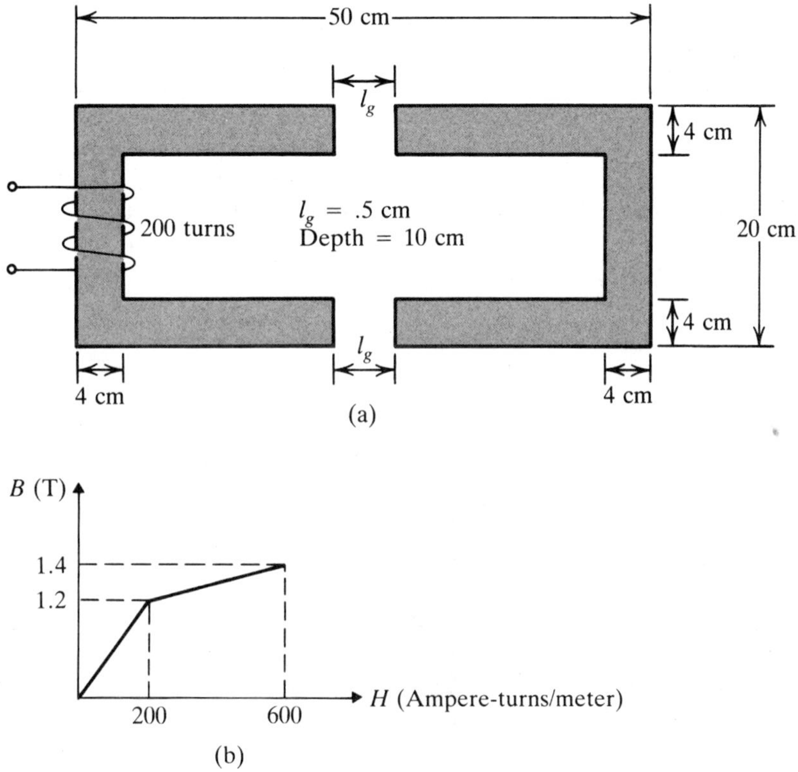

FIG. 1-59

Answer (a) 397.9; (b) 15.36 mH

1.2.14 POLARITY AND EQUIVALENT CIRCUITS OF MAGNETICALLY COUPLED COILS

Understanding magnetically coupled windings is very important, because it provides the basis for understanding the operation of transformers and electric machines. As will be explained in the chapters that follow, these devices have two or more magnetically coupled windings. The primary concern here is the polarity of the induced voltages and the equivalent circuit of each winding.

Polarity

Two magnetically coupled coils are shown in Fig. 1-60(a). The coil with N_1 winding turns, which is connected to the supply voltage, is referred to as the primary winding. The coil with N_2 winding turns, which supplies power to the load impedance Z, is the secondary winding.

The polarity of the voltage induced in a winding is indicated by a dot or by the positive (+) and negative (−) signs. A dotted terminal is at higher potential than an undotted terminal.

If the polarity of the induced voltage is known, the direction of current flow and that of the flux lines can easily be established.*

The significance of the dotted terminals is that the fluxes produced by the currents in the two coils are in opposition when the current in one coil enters the dotted terminal while the current in the other coil leaves the dotted terminal.

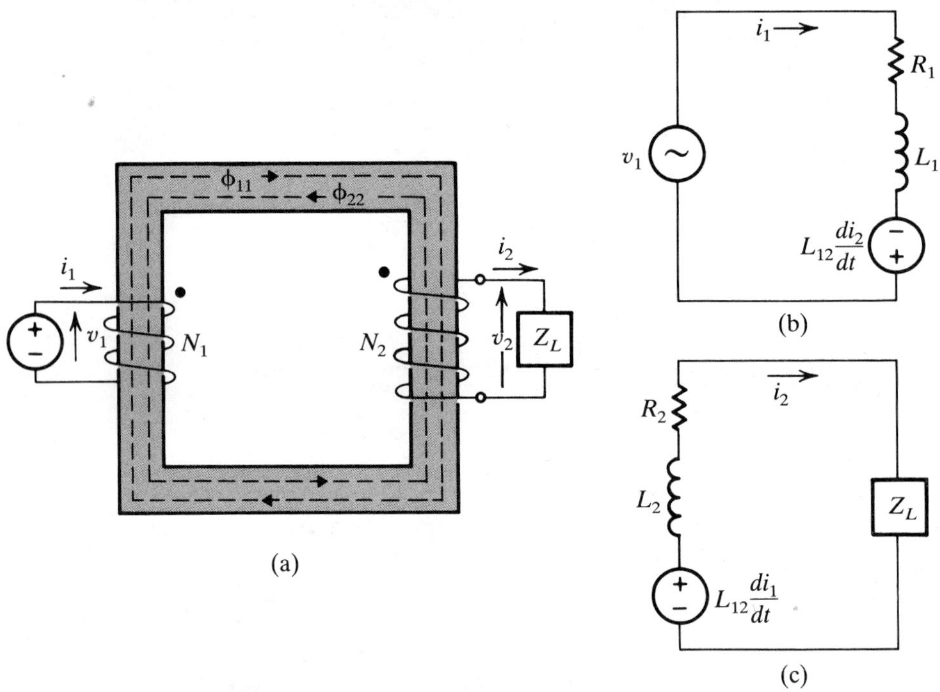

FIG. 1-60 A magnetically coupled circuit: **(a)** physical representation, **(b)** equivalent circuit of winding 1, **(c)** equivalent circuit of winding 2.

* When energy is drawn from a voltage source, the current comes out of its positive terminal. If a source is absorbing energy, current enters its positive terminal first.

In Fig. 1-60(a), notice that although the upper terminals of both windings are positive and the polarity of the voltage v_1, v_2 is the same, the currents flow in opposite directions. The current in the primary enters the dotted terminal because of the polarity of the input voltage, while the current in the secondary flows out of the dotted terminal because of the polarity of the induced voltage (v_2).

Equivalent Circuit

Assuming negligible magnetic losses, the equivalent circuits of coil 1 and coil 2 are shown in Fig. 1-60(b) and (c), respectively. The resistance and self-inductance of each coil are as shown.

From basic definitions, the voltages induced in the primary winding due to the flux of the secondary, and vice versa, are given by the following equations:

$$V_{12} = \frac{d}{dt}(L_{12}i_2)$$

$$= \frac{d}{dt}(N_1\phi_{12}) \tag{1.149}$$

and

$$V_{21} = \frac{d}{dt}(L_{21}i_1)$$

$$= \frac{d}{dt}(N_2\phi_{21}) \tag{1.150}$$

Assuming equal mutual inductances, we have

$$V_{12} = L_{12}\frac{d}{dt}i_2 \tag{1.151}$$

and

$$V_{21} = L_{12}\frac{d}{dt}i_1 \tag{1.152}$$

Using rms values, the above equations may be written in their phasor form as follows:

$$V_{12} = j\omega L_{12}I_2 \tag{1.153}$$

$$V_{21} = j\omega L_{12}I_1 \tag{1.154}$$

where ω is the angular frequency of the supply voltage.

As shown in the equivalent circuits, the polarities of these voltages are opposite to the polarities of the voltage drops across the leakage impedances because of Lenz's law. The equivalent circuits of magnetically coupled circuits whose core losses are not negligible, are discussed in Section 2.1.3.

EXAMPLE 1–22 Figure 1-61(a) shows two magnetically coupled coils (coil 1 and coil 2). When the coils are connected in series cumulatively (same sense):

a) Mark their polarity and show the direction of the flux.

b) Prove that the total equivalent inductance is given by the following relationship:

$$L_{\text{total}} = L_{11} + L_{22} + 2L_{12}$$

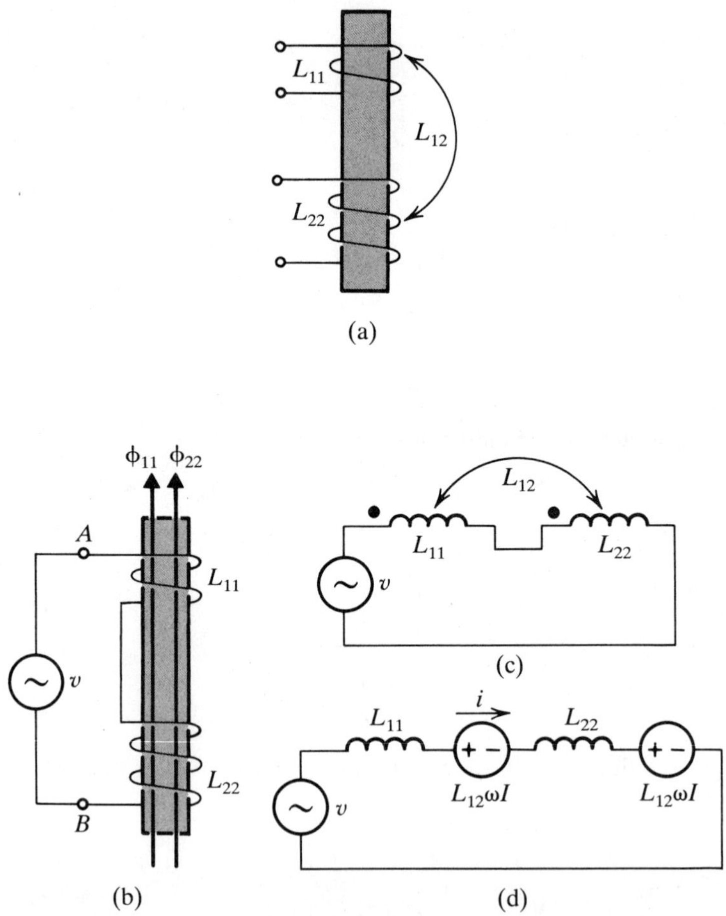

FIG. 1-61

SOLUTION

a) Cumulative connection of the coils means that each coil's flux is in the same direction as the other's. The coils are said to be of the same sense and are represented as shown in Fig. 1-61(b).

Since the flux of coil 2 aids, or is in the same direction as, the flux of coil 1, the schematic with the dotted terminals will be as shown in Fig. 1-61(c). The corresponding equivalent circuit is shown in Fig. 1-61(d).

b) Neglecting resistances from the equivalent circuit of Fig. 1-61(d), at steady state, we have

$$V = Ij\omega(L_{11} + L_{22} + 2L_{12})$$

or

$$V = I(j\omega L_T)$$

Thus,

$$L_T = L_{11} + L_{22} + 2L_{12}$$

Exercise 1–19 Refer to Fig. 1-61(a). When the coupled coils are connected in series differentially (opposite sense):

a) Mark their polarity and show the direction of flux.

b) Prove that their total equivalent inductance is given by the following equation:

$$L_{\text{total}} = L_{11} + L_{22} - 2L_{12}$$

1.3 Force, Energy, and Torque

In this section general expressions are derived for the following: the force exerted on a current-carrying conductor when it is placed in a magnetic field; the force exerted by an electromagnet; the energy stored in independently excited coils; and the torque developed by rotating transducers.

1.3.1 FORCE EXERTED ON A CURRENT-CARRYING CONDUCTOR WHEN IT IS PLACED IN A MAGNETIC FIELD

Ampere's experiments on the forces between current-carrying loops of thin wire led to the formulation of Eq. (1.155), which gives the force exerted on a current-carrying conductor when it is placed in a magnetic field.

$$\vec{f} = i\vec{l} \times B \tag{1.155}$$

where f is the force, in newtons, exerted on a conductor whose length is l meters and which carries a current of i amperes in a magnetic field whose flux density is B tesla. The factors il and B are vectors.

In the rectangular coordinate system these vectors may have components in the x, y, and z directions. That is,

$$\vec{il} = (il)_x\tilde{a}_x + (il)_y\tilde{a}_y + (il)_z\tilde{a}_z \tag{1.156}$$

and

$$\vec{B} = B_x\tilde{a}_x + B_y\tilde{a}_y + B_z\tilde{a}_z \tag{1.157}$$

where $\tilde{a}_x$, $\tilde{a}_y$, and $\tilde{a}_z$ are, respectively, the unit vectors in the x, y, and z directions; $(il)_x$, $(il)_y$, and $(il)_z$ are, respectively, the magnitudes of the current–length product of the conductor under consideration in the x, y, and z directions; and B_x, B_y, and B_z are the magnitudes of the components of the flux density in the x, y, and z directions, respectively.

Equation (1.155), being the cross product of two vectors, can be evaluated as follows:

$$\vec{f} = \begin{vmatrix} \tilde{a}_x & \tilde{a}_y & \tilde{a}_z \\ (il)_x & (il)_y & (il)_z \\ B_x & B_y & B_z \end{vmatrix} \tag{1.158}$$

From the above,

$$\vec{f} = \tilde{a}_x \begin{vmatrix} (il)_y & (il)_z \\ B_y & B_z \end{vmatrix} - \tilde{a}_y \begin{vmatrix} (il)_x & (il)_z \\ B_x & B_z \end{vmatrix} + \tilde{a}_z \begin{vmatrix} (il)_x & (il)_y \\ B_x & B_y \end{vmatrix} \tag{1.159}$$

$$= \tilde{a}_x(B_z(il)_y - B_y(il)_z) - \tilde{a}_y(B_z(il)_x - B_x(il)_z) + \tilde{a}_z(B_y(il)_x - B_x(il)_y) \tag{1.160}$$

Using the properties of the cross product of two vectors, Eq. (1.155) can also be written as follows:

$$\vec{f} = |il|\,|B| \sin \theta_{il,B} \tag{1.161}$$

where

il = the magnitude of the current–length vector

B = the magnitude of the flux density vector

$\theta_{il,B}$ = the smaller of the two angles between the vectors (il) and B.

When the flux density is perpendicular to the conductor's length, then, from Eq. (1.161) we obtain

$$\vec{f} = ilB \tag{1.162}$$

Equation (1.155) gives the magnitude and the direction of the force on a current-carrying conductor when it is placed in a magnetic field. The force is perpendicular to the current-carrying conductor and acts in the direction of the weaker magnetic field. This is illustrated in Fig. 1-62.

Refer to Fig. 1-62(a). The conductor's current flows into the page and perpendicular to the magnetic flux density. The direction of the conductor's field is determined by using the right-hand rule. To the right of the conductor, the field of its current opposes that of the existing magnetic field; as a result, the effective field in that region is weaker than the field to the left of the conductor. The force, as shown, always acts toward the region of the weaker field. Similar reasoning leads to the determination of the direction of the force when the current emerges out of the page (see Fig. 1-62(b)).

The torque (T) developed on a rotating conductor is equal to the cross product of the force and the perpendicular distance (r) from the conductor's axis of rotation to the point of application of the force. Mathematically,

$$T = \vec{r} \times \vec{f} \tag{1.163}$$

From Eq. (1.155) and the above we obtain

$$T = \vec{r} \times (\vec{il}) \times \vec{B} \tag{1.164}$$

Equation (1.164) gives the torque (motor action) produced by current-carrying conductors that rotate in a magnetic field. The opposite is also true; that is, when an external torque rotates a closed-loop conductor in a constant magnetic field, a voltage is induced in the conductor and thus a current is circulated (generator action). Thus Eq. (1.164) is the basis of motor and/or generator design. In a generator the input is a torque and the output is a voltage; in a motor the input is a voltage and the output is a torque.

In a generating system, the larger the torque on the generator's shaft, the larger is its output current. Thus, when an electric load such as student's desk lamp is switched on, the generator's output current increases in order to meet the new requirements; as a result, the generator's input torque must increase accordingly.

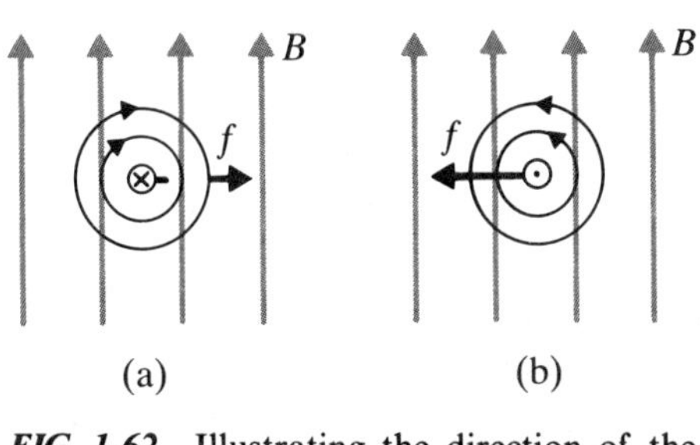

FIG. 1-62 Illustrating the direction of the force on a current-carrying conductor when placed in a magnetic field: **(a)** current into page, **(b)** current out of page.

In a motoring system, the higher the conductor's current (derived from an external source), the higher its output torque. When the shaft's mechanical load increases (increase in the requirements of torque), the current drawn by the motor's conductors must increase accordingly.

The motor's and generator's governing equations are also derived in Chapter 6 (Eqs. (6.13) and (6.22)).

EXAMPLE 1–23 Find the force exerted on the 6.5 m long, 4-segment conductor shown in Fig. 1-63.

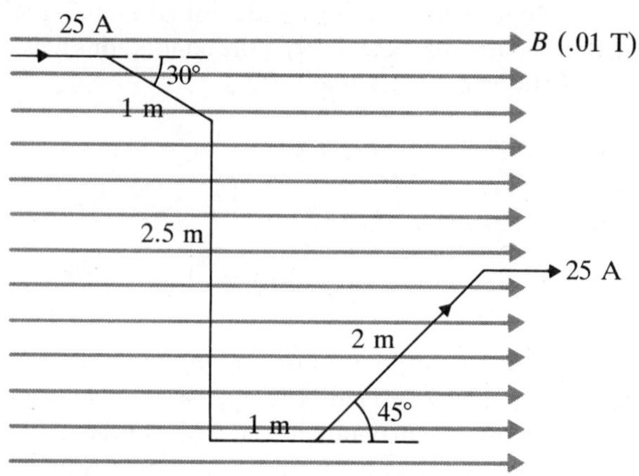

FIG. 1-63

SOLUTION From the given diagram, the current–length vector is

$$\begin{aligned}\vec{il} &= 25(\cos 30°\tilde{a}_x - \sin 30°\tilde{a}_y) - 25(2.5)\tilde{a}_y + 25\tilde{a}_x \\ &\quad + 50(\cos 45°\tilde{a}_x + \sin 45°\tilde{a}_y) \\ &= 82\tilde{a}_x - 39.65\tilde{a}_y\end{aligned}$$

The flux density factor is

$$\vec{B} = .01\tilde{a}_x$$

Substituting the above into equation Eq. (1.158), we obtain

$$\begin{aligned}\vec{f} &= \begin{vmatrix} \tilde{a}_x & \tilde{a}_y & \tilde{a}_z \\ 82 & -39.65 & 0 \\ .01 & 0 & 0 \end{vmatrix} = \tilde{a}_z \begin{vmatrix} 82 & -39.65 \\ .01 & 0 \end{vmatrix} \\ &= .396\tilde{a}_z \quad N\end{aligned}$$

Exercise 1–20 Figure 1-64 shows a 1.0 ohm conductor moving along the conducting rails at a constant speed of 70 m/s. Assuming that the flux density is constant, determine the force on the moving conductor and the current through the resistor.

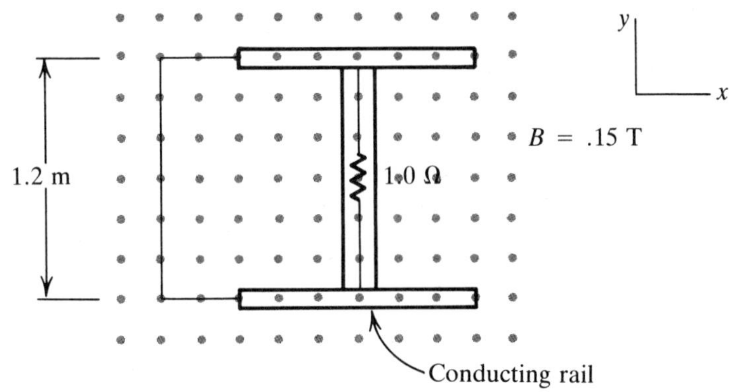

FIG. 1-64

Answer $-2.27\vec{a}_x$ N, 12.6 A

1.3.2 FORCE EXERTED BY AN ELECTROMAGNET

When a metallic body is placed in the air gap of an electromagnet, the latter exerts a force on the body and attempts to move it along the air gap in such a way as to *maximize* the coil's inductance. When this condition is satisfied, the electromagnet cannot be disturbed any further. Thus, a basic law of nature is fulfilled.

Referring to Fig. 1-65, from the principle of energy conservation we have

$$\begin{bmatrix}\text{energy supplied}\\ \text{by the source}\\ (W_s)\end{bmatrix} = \begin{bmatrix}\text{energy stored}\\ \text{in the field}\\ (W_f)\end{bmatrix} + \begin{bmatrix}\text{heat dissipated}\\ (I^2R)(\text{time})\end{bmatrix} + \begin{bmatrix}\text{mechanical}\\ \text{work done}\\ (W_m)\end{bmatrix} \qquad \textbf{(1.165)}$$

Using mathematical symbols, we obtain

$$W_s = W_f + I^2R(t) + W_m \qquad \textbf{(1.166)}$$

The resistance R represents the losses in the magnetic material and in the winding of the coil. For simplicity, these losses are lumped together and are represented by the resistance R, which is connected in parallel across the voltage source. In this section, however, the losses will be assumed to be negligible.

From Eq. (1.166), by considering incremental changes, the mechanical work is given by

$$dW_m = dW_s - dW_f \qquad \textbf{(1.167)}$$

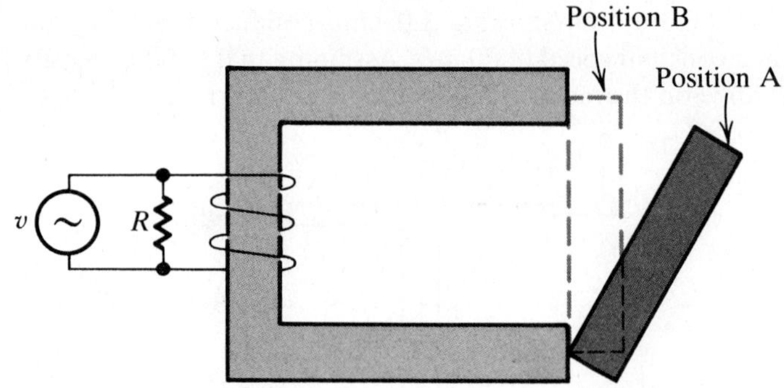

FIG. 1-65 Electromagnet.

For linear displacements, the force (f) developed is given by the partial derivative of the mechanical work with respect to the displacement in whose direction the force is acting. Mathematically,

$$\vec{f} = \frac{\partial}{\partial x} W_m \tag{1.168}$$

Equation (1.168) can be used to find the force developed by a linear transducer, or the weight an electromagnet can lift.

DC Excitation

When the coil of an electromagnet is connected to a dc voltage source, the current drawn is, for all practical purposes, constant. Thus from Eq. (1.167) we obtain the following:

$$\vec{f} = -\frac{\partial}{\partial x} W_f \tag{1.169}$$

The electromagnetic force, when it is applied to a moving body, produces a linear acceleration which is related to the mass M of the body by Newton's equation:

$$\sum F_n = Ma_n \tag{1.170}$$

where $\sum F_n$ represents the vectorial sum of all forces acting in the n direction, and a_n is the body's acceleration in the same direction.

Knowledge of the acceleration and the velocity of a body subjected to electromagnetic forces is a prerequisite for optimum design of weight-lifting electromagnets.

EXAMPLE 1–24 The electromagnet shown in Fig. 1-66(a) is used to control the position of the cylindrical plunger. The coil has N turns and carries a constant current of i amperes. Neglecting fringing, the mmf within the material and air gaps l_{g2}, determine general expressions for the following as a function of the length of the air gap l_{g_1}:

a) The self-inductance of the coil.
b) The energy stored in the magnetic field.
c) The electromagnetic force acting on the plunger.

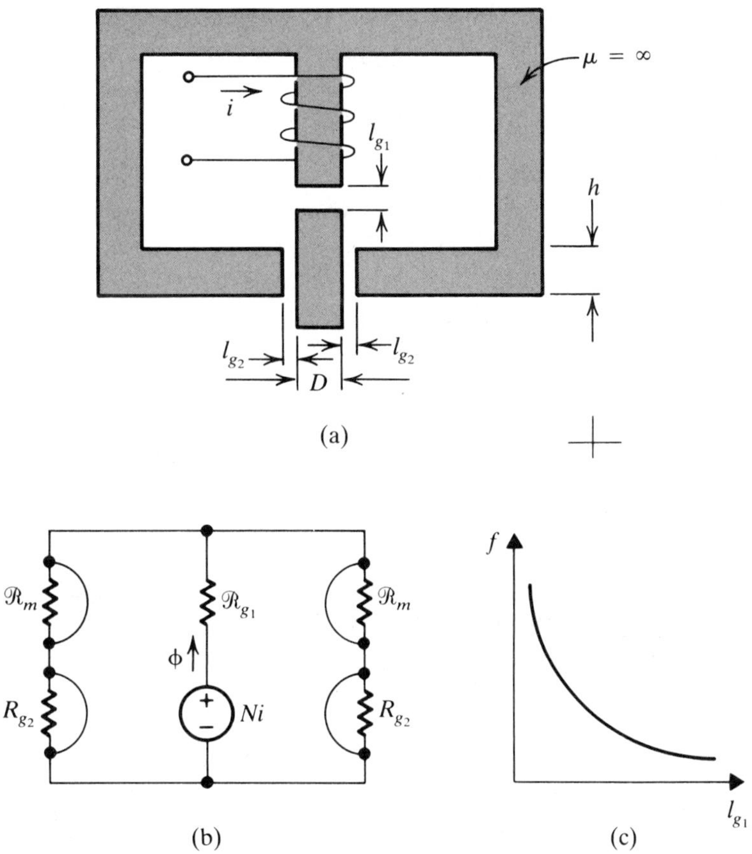

FIG. 1-66

SOLUTION

a) The equivalent magnetic circuit of the electromagnet is shown in Fig. 1-66(b). The reluctance of the magnetic material and that of the air gaps l_{g_2} are shorted due to the given assumption. By definition, the reluctance $\mathcal{R}_{g_1}$ is

$$\mathcal{R}_{g_1} = \frac{l_{g_1}}{\mu_0 A}$$

The area A perpendicular to flux is given by

$$A = \pi r^2$$

where r is the radius of the cylindrical plunger. From the above,

$$\mathscr{R}_{g_1} = \frac{l_{g_1}}{\pi \mu_0 r^2}$$

Then from Eq. (1.143) and Ohm's law in magnetic circuits, we obtain

$$L = \frac{N\phi}{i} = \frac{N}{i}\frac{Ni}{\mathscr{R}}$$

$$= \frac{N^2 \pi \mu_0 r^2}{l_{g_1}}$$

b) The energy stored in the magnetic field is

$$W_f = \frac{1}{2} L i^2$$

Substituting for the inductance, we obtain

$$W_f = \frac{1}{2} i^2 N^2 \pi \mu_0 \frac{r^2}{l_{g_1}}$$

c) From Eq. (1.169) we have

$$\vec{f} = -\frac{\partial W_f}{\partial l_{g_1}} = \frac{1}{2} i^2 \frac{N^2 \pi \mu_0 r^2}{l_{g_1}^2}$$

The force as a function of the air gap l_{g_1} is sketched in Fig. 1-66(c). The force acts in a way that reduces the length of the air gap and thus increases the inductance of the coil.

REMARKS

1. Determination of the mmf within the air gaps l_{g_2} requires an estimate of the area through which the flux perpendicularly enters the moving cylindrical plunger. A reasonable approximation is obtained by assuming that this area is equal to 60% of the outside area of a cylindrical surface whose height and diameter are, respectively, h and D meters (see Fig. 1-66(a)).
2. The stated assumptions simplify the solution. They yield, however, an unrealistic answer, for when $l_{g_1} = 0$, the force is equal to infinity.

1.3.3 ENERGY STORED IN THE MAGNETIC FIELDS OF COILS

The general expressions for the energy stored in the magnetic field of single and double excited coils are derived as follows.

Single Coil

The general expression for the energy stored in a single coil is derived in Section 1.1.4 (see Eq. (1.42)). Here the same equation will be derived using the concept of flux linkage.

Refer to Fig. 1-67(a). After the switch is closed, the instantaneous *power* $p(t)$ supplied by the electrical source to the coil is

$$p(t) = vi \tag{1.171}$$

and

$$p(t) = i\frac{d\lambda_{11}}{dt} \tag{1.172}$$

Neglecting losses, the *energy* supplied to the field of the coil (W_f) is equal to the energy drawn from the source (W_s). Considering incremental magnitudes, we have

$$dW_f = dW_s \tag{1.173}$$

$$dW_f = p(t)\,dt = \left(i\frac{d\lambda_{11}}{dt}\right)dt \tag{1.174}$$

$$= i\,d\lambda_{11} \tag{1.175}$$

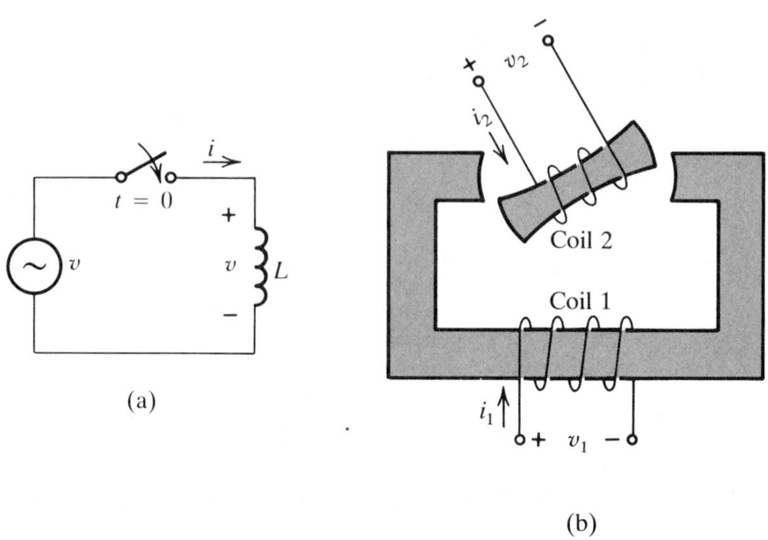

FIG. 1-67 Electromechanical systems: **(a)** single coil, **(b)** two coils.

For a linear system, while the current increases from zero to i amperes, the energy stored in the field of the coil is

$$W_f = \int_0^i i\, d\lambda_{11} = L_{11} \int_0^i i\, di = \frac{L_{11} i^2}{2} \tag{1.176}$$

where L_{11} is the self-inductance of the coil.

Two-Coil System

Similarly, for the two-coil electromagnetic system shown in Fig. 1-67(b), the energy supplied by the electrical source is equal to the energy stored in the field. Considering incremental magnitudes and neglecting losses, we have

$$dW_f = dW_s \tag{1.177}$$

Thus,

$$dW_f = v_1 i_1\, dt + v_2 i_2\, dt \tag{1.178}$$

and

$$dW_f = i_1\, d\lambda_1 + i_2\, d\lambda_2 \tag{1.179}$$

The flux linkages of coil 1 and coil 2 are, by definition,

$$\lambda_1 = \lambda_{11} + \lambda_{12} = L_{11} i_1 + L_{12} i_2 \tag{1.180}$$

$$\lambda_2 = \lambda_{22} + \lambda_{21} = L_{22} i_2 + L_{21} i_1 \tag{1.181}$$

For constant self- and mutual inductances, from Eqs. (1.179), (1.180), and (1.181) we obtain

$$dW_f = i_1(L_{11}\, di_1 + L_{12}\, di_2) + i_2(L_{22}\, di_2 + L_{21}\, di_1) \tag{1.182}$$

For $L_{12} = L_{21}$,

$$dW_f = L_{11} i_1\, di_1 + L_{12}(i_1\, di_2 + i_2\, di_1) + L_{22} i_2\, di_2 \tag{1.183}$$

and

$$dW_f = L_{11} i_1\, di_1 + L_{12}\, d(i_1 i_2) + L_{22} i_2\, di_2 \tag{1.184}$$

Taking integrals, we obtain

$$W_f = L_{11} \int_0^{i_1} i_1\, di_1 + L_{22} \int_0^{i_2} i_2\, di_2 + L_{12} \int_0^{i_1 i_2} d(i_1 i_2) \tag{1.185}$$

From the above,

$$W_f = \frac{1}{2} L_{11} i_1^2 + \frac{1}{2} L_{22} i_2^2 + L_{12} i_1 i_2 \tag{1.186}$$

When the coil inductances change while the currents remain constant, then from Eq. (1.186) the incremental change in the energy stored in the field of the two coils will be

$$dW_f = \frac{i_1^2}{2} dL_{11} + \frac{i_2^2}{2} dL_{22} + i_1 i_2 \, dL_{12} \tag{1.187}$$

1.3.4 TORQUE DEVELOPED BY ROTATING TRANSDUCERS

The instantaneous value of the torque developed by a body that rotates in a magnetic field is

$$T_i = \frac{\partial}{\partial \beta} (W_f) \tag{1.188}$$

where T_i is the instantaneous value of the torque, W_f is the energy stored in the field, and β is the angular displacement of the rotating body with respect to a reference axis. Field energy is a function of this angle.

From basic definitions, the average value of the torque (T) is

$$T = \frac{1}{2\pi} \int_0^{2\pi} T_i \, dt \tag{1.189}$$

In order to find the average value of the torque produced on a rotating body, one must first determine the energy stored in the magnetic system as a function of the body's angular position, then find its derivative and, finally, use Eq. (1.189).

---

*EXAMPLE 1–25** The coil of the transducer shown in Fig. 1-68 draws an rms current of I amperes at a phase angle θ from a sinusoidal voltage source whose angular frequency of oscillation is ω rad/s. The inductance of the coil is

$$L = L_1 + L_m \cos 2\beta$$

a) Show that the instantaneous value of the torque is given by

$$T = -2I^2 L_m \sin 2\beta \sin^2 (\omega t - \theta)$$

. . .

* Used by permission from O'Kelly and Simmons, *Introduction to Generalized Electrical Machine Theory* (New York: McGraw-Hill, 1968).

b) If the speed of the rotor is ω rad/s, show that the average value of the torque is

$$T = \frac{I^2 L_m \sin(2\theta)}{2}$$

c) Show that the average power produced is maximum when the resistance of the coil is equal to its reactance.

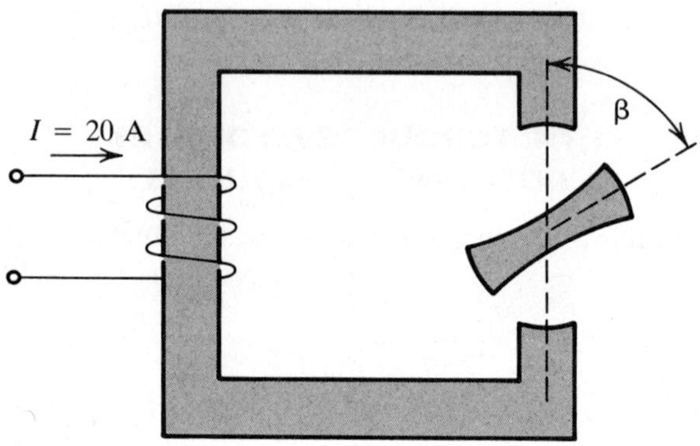

FIG. 1-68

SOLUTION

a) From basic considerations, the instantaneous value of the current in an inductive circuit is

$$i = I_m \sin(\omega t - \theta)$$

where

$$I_m = \sqrt{2} I$$

From the above and Eq. (1.176), the energy stored in the magnetic field is

$$W_f = \frac{1}{2}(L_1 + L_m \cos 2\beta)[\sqrt{2} I \sin(\omega t - \theta)]^2$$

Then, from Eq. (1.188), the instantaneous value of the torque is

$$\begin{aligned} T_i &= \frac{\partial}{\partial \beta}\left[\frac{1}{2}(L_1 + L_m \cos 2\beta)(\sqrt{2} I \sin(\omega t - \theta))^2\right] \\ &= -2I^2 L_m \sin 2\beta \sin^2(\omega t - \theta) \end{aligned}$$

b) The instantaneous value of the torque for $\omega t = \beta$ is

$$T_i = -2I^2 L_m \sin 2\omega t \sin^2(\omega t - \theta)$$

Using Eq. (1.189), the average value of the torque is

$$T = \frac{-2I^2 L_m}{2\pi} \int_0^{2\pi} \sin^2(\omega t - \theta) \sin 2\omega t \, d\omega t$$

Using the cosine double-angle identity, we obtain

$$T = -\frac{I^2 L_m}{\pi} \int_0^{2\pi} \left(\frac{1 - \cos(2\omega t - 2\theta)}{2} \right) \sin 2\omega t \, d\omega t$$

$$= -\frac{I^2 L_m}{2\pi} \left[\int_0^{2\pi} \sin 2\omega t \, d\omega t - \int_0^{2\pi} \cos(2\omega t - 2\theta) \sin 2\omega t \, d\omega t \right]$$

When integrated over a complete cycle, the first term, being a sinusoidal waveform, is equal to zero. The second term is expanded by using the following trigonometric identity:

$$\sin A \cos B = \frac{\sin(A + B) - \sin(A - B)}{2}$$

Substituting, we obtain

$$T = -\frac{I^2 L_m}{4\pi} \int_0^{2\pi} [-\sin(4\omega t - 2\theta) - \sin(2\theta)] \, d\omega t$$

When evaluated, the first term is again equal to zero. Then,

$$T = \frac{I^2 L_m}{4\pi} \sin(2\theta) \omega t \Big|_0^{2\pi}$$

from which

$$\underline{\underline{T = \frac{I^2 L_m}{2} \sin(2\theta) \text{ N.m}}}$$

c) The power is equal to the product of the torque times the angular speed of the rotor. That is,

$$P = T\omega$$

At constant speed, the average power is maximum when the average torque is maximum. From the expression derived in (b) the torque is maximum when

$$\theta = \frac{\pi}{4} \text{ radians}$$

The phase angle θ of the current is given by

$$\theta = \arctan \frac{X}{R}$$

where X and R are, respectively, the coil's reactance and resistance. From the last two equations,

$$\underline{\underline{X = R}}$$

Exercise 1–21 Figure 1-69 shows an elementary representation of a reluctance motor. The coil is connected to a 60 Hz voltage source and draws a sinusoidal current of 20 A rms at a power factor of .6 lagging. The approximate self-inductance of the coil is $L = .1 + .01 \cos 2\beta$ H. Determine an expression for:

a) The instantaneous value of the torque.
b) The speed at which the motor will develop average torque.
c) The average torque.

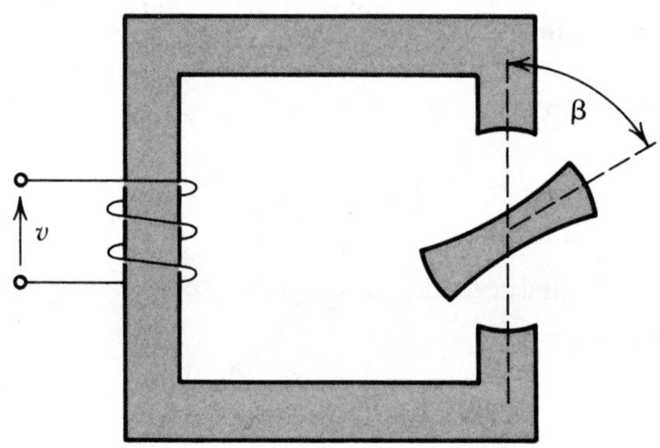

FIG. 1-69

Answer (c) 1.92 N · m

1.4 Summary

Each magnetic circuit parameter is analogous to a corresponding electric circuit's parameter. For example, the inductance, mmf, and reluctance of magnetic circuits correspond to the capacitance, emf, and resistance of electric circuits. The equivalent magnetic circuit of a particular system provides an in-depth view of its operation and limitations, while its equivalent circuit supplements and often simplifies the conclusions derived from the magnetic circuit.

The three fundamental laws of electric and magnetic circuits, upon which the operation of electric machines is based, are Ohm's law, Kirchhoff's voltage law (KVL), and Kirchhoff's current law (KCL). According to *Ohm's law*, in the absence of saturation, the flux in magnetic circuits is proportional to the mmf and inversely proportional to the reluctance of the circuit. The reluctance, in turn, is proportional to the length of the magnetic path and inversely proportional to the cross-sectional area and the permeability of the magnetic material. The permeability is mathematically given by the slope of the flux density–field intensity (B-H) curve of the material. This curve is an intrinsic characteristic of the material, and its slope varies from very high at low levels of magnetization to very low at high levels of magnetization (saturation levels). When one speaks of a certain permeability, then one must specify to which flux density it corresponds.

Alternating flux produces hysteresis and eddy-current losses. These magnetic losses are a limitation of coils used for communication purposes and constitute a potential danger to steel elbows and adjacent conducting plates in all monoconductor three-phase power distribution networks. The closed-loop property of the flux is basic to the design of transformers, electric motors, and all other types of electromagnetic transducers.

Kirchhoff's voltage law states that in magnetic circuits the algebraic sum of the magnetic potentials or ampere-turns in any loop is equal to zero. Although the magnetic potential absorbed within the magnetic material is relatively small, it plays an important role in transformers and electric machines. *Kirchhoff's current law* states that in magnetic circuits the algebraic sum of the flux in a junction must be equal to zero.

The *inductance* of a coil is analogous to the mass of an object or to the moment of inertia of a rotating body. As such, it represents the natural tendency of a coil to oppose those external forces that try to change its status. For this reason, precautions must be taken (use of properly sized switches, fuses, and circuit breakers) when the current is an inductive load is to be instantaneously interrupted.

The *energy stored* in the magnetic field of a coil is proportional to the coil's inductance. The *partial derivative* of the energy stored, with respect to an associated linear or angular displacement, gives the instantaneous value of the force, or torque, developed. The force, or torque, acts in a direction that increases the system's inductance. Thus the concept of inductance is of fundamental importance to the understanding of all force-, or torque-, developing mechanisms, such as electric machines and other electromechanical transducers.

Alternatively, the electromagnetic force can be derived from the experimental observations of Ampere. According to these experiments, the force produced on one current-carrying conductor by another is equal to the cross product of the current–length vector of one conductor and the flux density produced by the other. In other words, there is a force produced on a current-carrying conductor when that conductor is properly placed in a magnetic field.

A voltage is generated in a given coil either when an alternating flux cuts through the stationary coil (transformer voltage) or when the coil is properly rotated through a stationary magnetic field (speed voltage).

Tables 1-2 through 1-5 illustrate the concept of power and energy and summarize the various magnetic and electric circuit parameters, their units, and their interrelationships. The highlights of power and energy (Tables 1-2 and 1-3) will help the reader gain a broader view of these fundamental parameters, which are essential to the understanding of many engineering applications.

TABLE 1-2 Some highlights about power*

Source	Per-unit cost† of purchasing an electrical power source
Diesel generator	1
Large hydropower station	5
Wind power system	6
Atomic reactor	7
Uninterruptible power supply	15
Photovoltaic cells	60

Power	Description
100 W	Manual power capacity of an average human adult
(100–400) W/m^2	Power available from sun during sunlight hours (Chicago)
25 W/m^2	Lighting requirements of an office building
80 W/m^2	Requirements of average home for heating, cooling, cooking, lighting, etc. (Chicago)
To produce‡ a ton per year	Cement: 40 W Paper: 200 W Aluminum: 500 W

* *The given power requirements and costs are approximate.*
† *The per-unit cost in 1988 was about \$200/kW.*
‡ *Usual rating of plants.*

TABLE 1-3 Some highlights about energy*

Conversion factors

1 kWh = 3412 Btu
= 860 kcal
= 860 food calories
= 3.6 MJ

Some aspects of energy production are not fully understood. For example, we still do not know how a night's sleep refills the body's energy reservoir.

Residential rates (1988)

6¢/kWh, Canada
12¢/kWh, California
20¢/kWh, Japan
20¢/kWh, FRG

3 kWh	Energy required per adult per day (≈ 2500 food calories)
3 kWh	Energy used up when one walks 15 km
3 kWh	Energy required per shower or bath
8 kWh	Energy burned when one loses 1 kg of weight
$\frac{1}{6}$ kWh	Energy content of a bottle of beer
15 kWh	Energy content of a domestic hot water tank
100 kWh	Energy required per home per day in North America (includes heating, cooling, cooking, etc.)
6.4 kWh†	Energy produced per liter of gasoline or per cubic meter of natural gas
5.3 kWh	Energy produced per kg of wood
1000 kWh†	Energy produced per barrel of oil
900 kWh	Energy required to produce one ton of cement
1500 kWh	Energy required to produce one ton of paper
14,000 kWh	Energy required to produce one ton of aluminum

* *Cost and statements about energy requirements are approximate.*
† *Assumed efficiency of combustion is 62%.*

TABLE 1-4 Summary of electric and magnetic field parameters and their units

	Electric		*Magnetic*	
No	*Description*	*Symbol*	*Symbol*	*Description*
1	Electric flux (coulombs)	ψ	ϕ	Magnetic flux (weber, Wb = 10^8 lines = 10^8 maxwells)
2	Actual permittivity (farads/meter, F/m)	$\varepsilon = \varepsilon_0 \varepsilon_r$	$\mu = \mu_0 \mu_r$	Actual permeability (henrys/meter, H/m)
3	Relative permittivity	ε_r	μ_r	Relative permeability
4	Electric volts, emf (volts, V)	emf = V	mmf = $U = \mathscr{F} = NI$	Magnetic volts, mmf (ampere-turns, A)
5	Electric flux density (coulombs/m^2)	D	B	Magnetic flux density (tesla, Wb/m^2, T)
6	Electric field intensity (volts/meter, V/m)	E	H	Magnetic field intensity (ampere-turns/meter, A/m)
7	Resistance (ohms, Ω)	R	$\mathscr{R}$	Reluctance (ampere-turns/Weber, A/Wb)
8	Capacitance (farads, F)	C	L	Inductance (henry, H)

TABLE 1-5 Basic concepts in mathematical form

No	*Electric*	*Magnetic*	*Remarks*
1	$\varepsilon = \dfrac{dD}{dE}$	$\mu = \dfrac{dB}{dH}$	Physical constants
2	$D = \dfrac{\psi}{A}$	$B = \dfrac{\phi}{A}$	Flux densities
3	$E = -\dfrac{dV}{dl}$	$H = \dfrac{NI}{l}$	Field intensities
4	$R = \rho \dfrac{l}{A}$	$\mathscr{R} = \dfrac{l}{\mu A}$	Opposition to flow
5	$I = \dfrac{V}{R}$	$\phi = \dfrac{NI}{\mathscr{R}}$	Ohm's law
6	$\sum V_{\text{loop}} = 0$	$\sum (NI)_{\text{loop}} = 0$	KVL
7	$\sum I_{\text{junction}} = 0$	$\sum \phi_{\text{junction}} = 0$	KCL
8	$V = 4.44 N f \phi_{\max}$		Relates the electric and magnetic field parameters

1.5 REVIEW QUESTIONS

1. "A formula or an equation represents — in mathematical symbols — a physical or a natural law." Explain this statement and give two examples to demonstrate its significance.
2. Why were the effective or rms values of voltages and currents introduced?
3. What are the two essential differences between electric and magnetic flux?
4. What is the difference between emf and mmf?
5. In a laboratory, how would you measure the magnetic flux that crosses a given air gap of a magnetic circuit? How could you measure the mmf within a toroidal electromagnet?
6. Why do the resistance and the inductance of a material or a circuit change as a function of temperature? What is the Curie temperature of a material?
7. How would you short-circuit an electric and a magnetic circuit?
8. How would you explain inductance or permeability to a mechanical engineering student?
9. Define permeability and give five factors that should be considered when magnetic materials are to be purchased.
10. How would you increase the permeability of a material?
11. Is the energy stored within the components of a magnetic circuit inversely proportional to their permeabilities?
12. Differentiate between transformer and speed voltage.

1.6 PROBLEMS

1-1 **a)** One of the most expensive mishaps ($\$2.5 \times 10^8$) in the American space exploration program occurred in 1970 because of an error in the basic design of Apollo 13. The dc supply to the resistive heat element of the liquid oxygen tank was 48 volts while the resistor was rated 24 V, 5 kW,* with a maximum power capacity of 300%.* As the Apollo astronauts were approaching the gravitational field of the moon, they switched on the 48 V supply, and the oxygen tank was blown away. This is how the famous "drama in the skies" started. Determine by what percentage the maximum power capability of the heating element was exceeded when the 48 V dc supply was switched on.

b) One of the worst fatal mishaps in the Russian space exploration program took place during the flight of Salyut I in 1971. The astronauts' cabin was equipped with a valve, which was to open automatically on landing, thereby equalizing the cabin and the ambient pressures. After several weeks of record-breaking events in space, the astronauts were returning home. The valve opened accidentally on reentry, and the breathing atmosphere started to escape. The three astronauts attempted to close the valve but failed because, according to official statements, closing the valve manually required two minutes, while the cabin's atmosphere was emptied in one minute. If the valve had been electrically operated, what should the time constant (L/R) of its solenoid have been for a safety factor of 200%?

. . .

* Assumed

1-2 For the voltage waveform shown in Fig. P1-2, determine:

a) Its rms value.

b) Its average value.

c) The average power dissipated when this voltage is applied across a 4 ohm resistor.

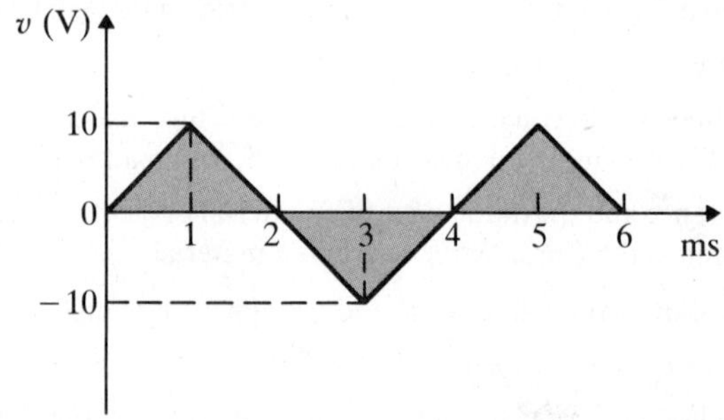

FIG. P1-2

1-3 **a)** For the circuit shown in Fig. P1-3, determine *graphically* the voltage across the 5 ohm resistor. Calculations with complex numbers must not be used. [*Hint:* In an *RL* or *RC* circuit, across a voltage *V*, the current is given by the intersection of two semicircles whose diameters are V/R and V/X.]

b) Check your answer *mathematically.*

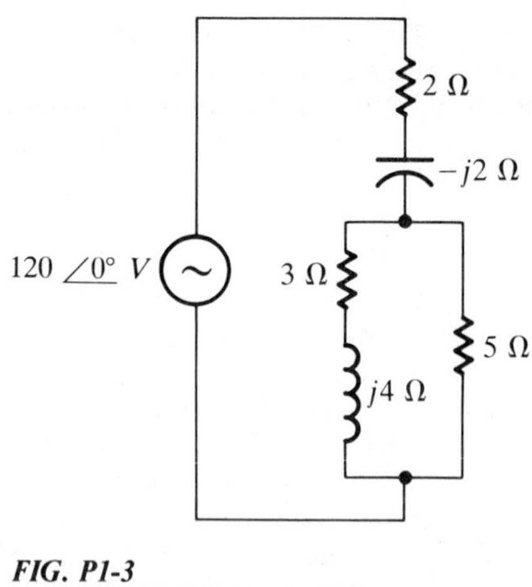

FIG. P1-3

1-4 For the circuit shown in Fig. P1-4, determine the currents I_1, I_2, and I_3 by using loop equations (KVL).

1-5 For the circuit in Fig. P1-5, determine the voltage V_{AB} by using nodal equations (KCL). Check your answer by using loop equations (KVL).

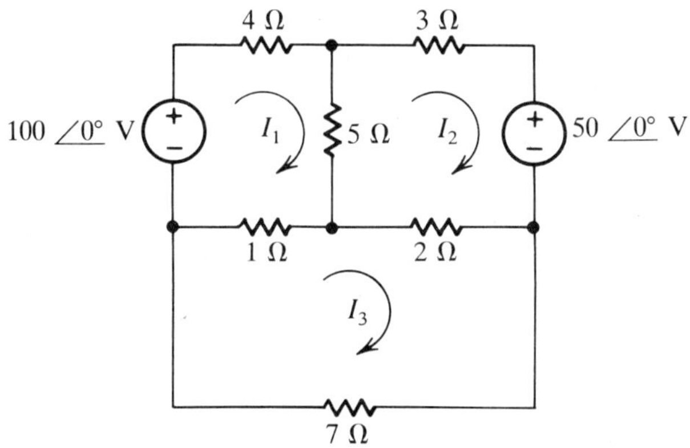

FIG. P1-4

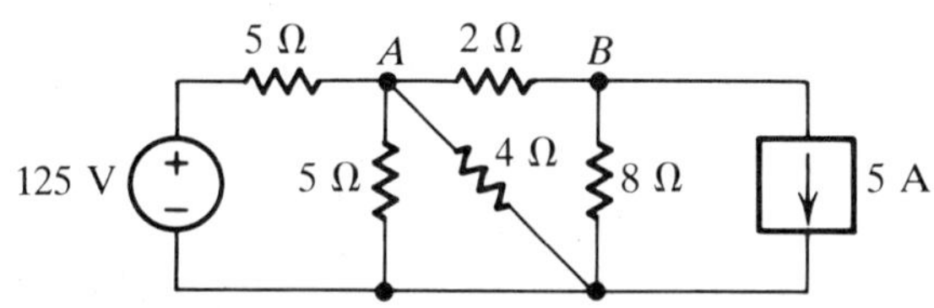

FIG. P1-5

1-6 Figure P1-6 shows (a) a properly and (b) an improperly grounded 120 V motor. The 20 Ω resistance represents the resistance of the grounding wire and of the earth's return path.

a) Compare the advantages and disadvantages of the two grounding schemes. Consider current, voltage to ground, and reaction of upstream protective devices when the winding of the motor is accidentally grounded through the motor's metallic frame.

b) Repeat (a), assuming that the 120 V source is not grounded.

c) Theoretically the resistance of the earth is zero but in practice it is always several ohms. Explain why.

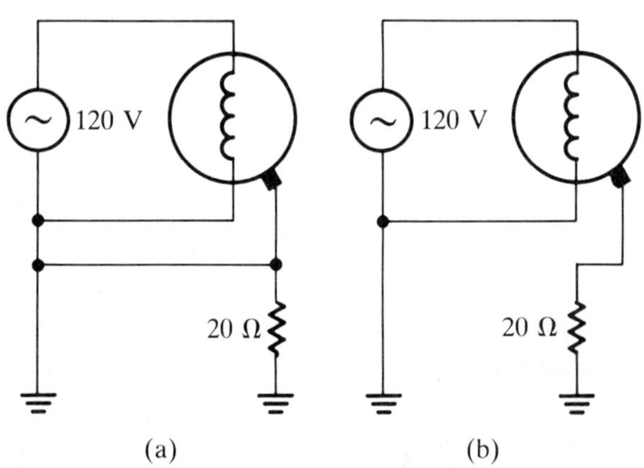

FIG. P1-6

1-7 A 4160 volt, 2000 kW, .8 power factor (Pf) leading synchronous motor (Fig. P1-7) was protected through a draw-out type of air circuit breaker. In general, these circuit breakers cannot be physically removed while the power is on because of standard electrical and mechanical safety interlocks. However, an electrician who mistakenly attempted to withdraw or roll out the circuit breaker from its "on" position — the circuit broke at Y-Z and not at X — was engulfed by a powerful flush-over and was electrocuted. The accident occurred because the safety interlocks malfunctioned (St-Constant Quebec, Canada).

a) Describe in simple statements — perhaps using equivalent concepts from mechanical engineering — the nature of the destructive forces that accompanied the interruption of the motor's current.

b) Estimate the maximum electric field intensity across Y-Z at the instant of the accident.

c) What was the effect of the flush-over on the other grounded equipment in the same electrical room?

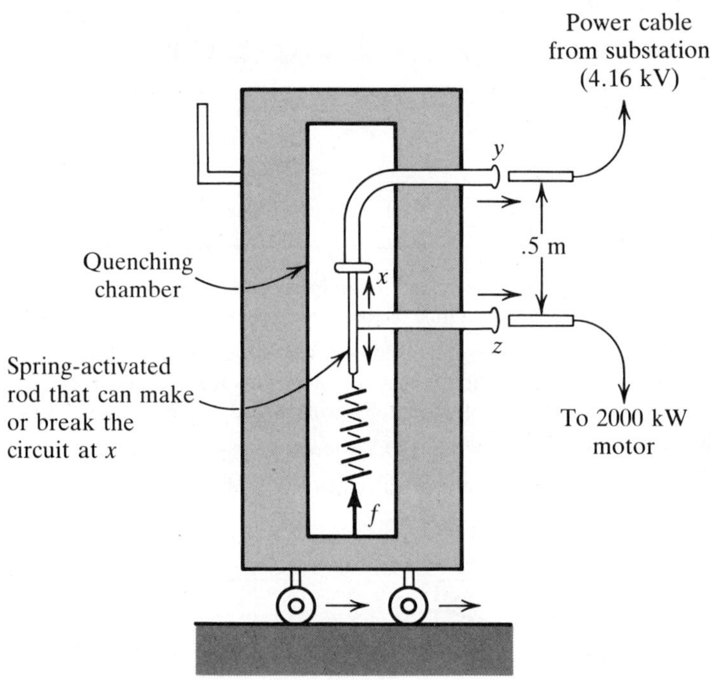

FIG. P1-7

1-8 The telephone line shown in Fig. P1-8 runs parallel to a balanced three-phase power line.

a) Show that the induced voltage per meter of the telephone line, due to current in phase "a" only, is given by

$$v = I_m \omega \frac{\mu_0}{4\pi} \ln\left[\frac{d_1^2 + (d_2 + d_3)^2}{d_1^2 + d_2^2}\right] \cos \omega t \ \text{V/m}$$

b) Determine the rms value of the net voltage induced in the telephone line per unit length, assuming

$$i_a = 1000 \sin 377t \text{ A}$$
$$d = 10 \text{ m}, \quad d_1 = 5 \text{ m}, \quad d_2 = 15 \text{ m}, \quad d_3 = 1 \text{ m}$$

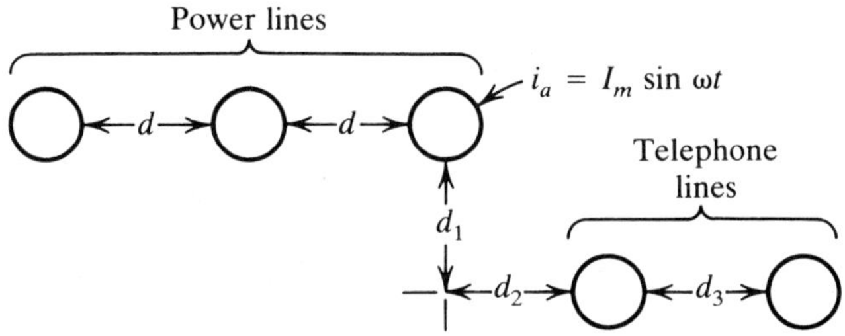

FIG. P1-8

1-9 The conductor shown in Fig. P1-9 moves at a velocity of U m/s in a uniform magnetic field. The magnitude of the magnetic flux density is B tesla, and its direction is as shown. Determine the voltage induced in the conductor.

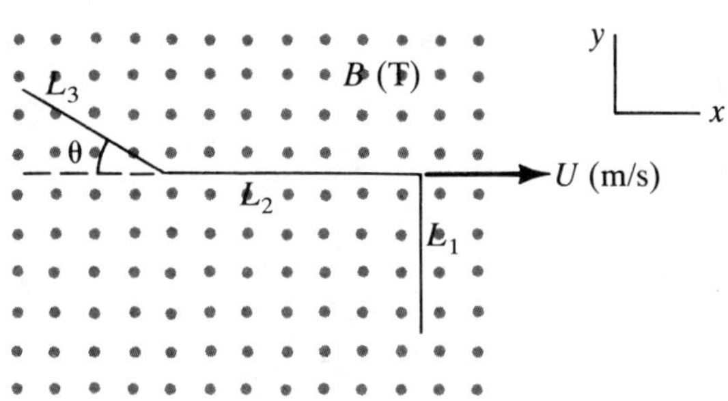

FIG. P1-9

1-10 The toroidal coil shown in Fig. P1-10 has a circular cross-sectional area and an inductance of .5 H. Neglecting leakage flux and losses, determine:

a) The relative permeability of the magnetic material.

b) The magnetic flux.

c) The magnetic flux density.

d) The magnetic flux intensity.

e) The magnetic potential drop within the magnetic material.

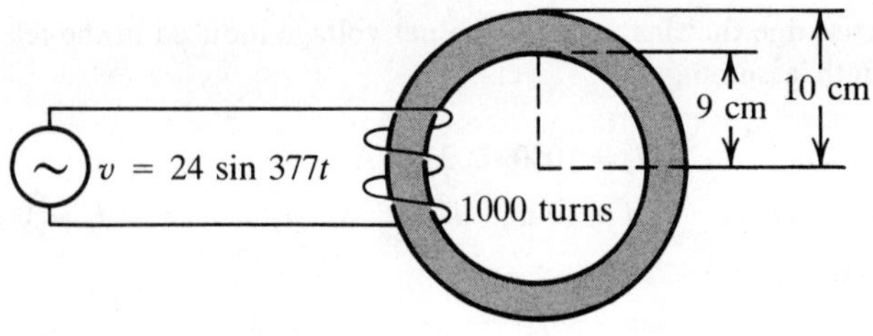

FIG. P1-10

1-11 In the magnetic circuit of Fig. P1-11,

$$l_{bad} = .5 \text{ m}, \quad l_{bd} = .15 \text{ m}, \quad l_{bcd} = .60 \text{ m}, \quad l_g = .8 \text{ mm}$$

The coordinates of the *B-H* characteristic of the magnetic material are as follows.

B(T)	.12	.54	.94	1.15
H (A/m)	100	300	700	1600

Assume uniform cross-sectional area. Determine the flux through the path *bcd* if the stacking factor is .95.

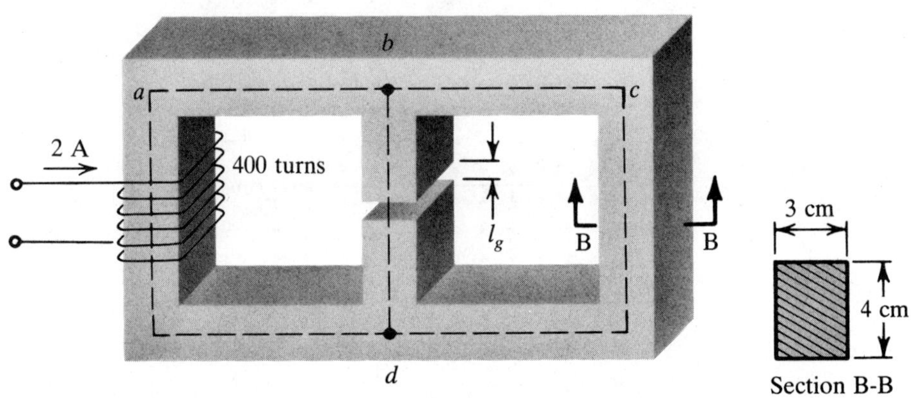

FIG. P1-11

1-12 In the magnetic circuit of Fig. P1-12,

$$l_{bad} = l_{bcd} = .25 \text{ m}, \quad l_{bd} = .10 \text{ m}$$
$$l_{g_1} = .25 \text{ mm}, \quad l_{g_2} = .35 \text{ mm}$$

The cross-sectional area of the magnetic material is uniform and equal to 6.5×10^{-3} m^2. The effective areas of the air gaps are

$$A_{g_1} = A_{g_2} = 6.8 \times 10^{-3} \text{ m}^2$$

The coordinates of the material's *B*-*H* characteristic are as follows.

B(T)	.11	.5	.95
H (A/m)	80	400	800

Determine the inductance of the coil.

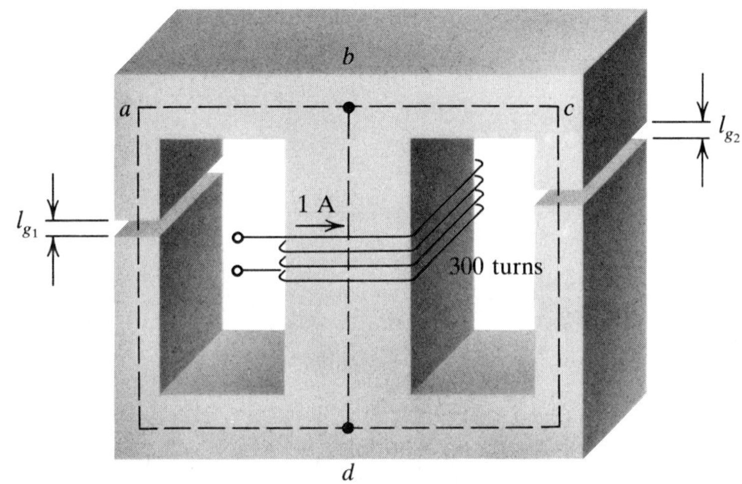

FIG. P1-12

1-13 Determine the force between the two coplanar conductors shown in Fig. P1-13.

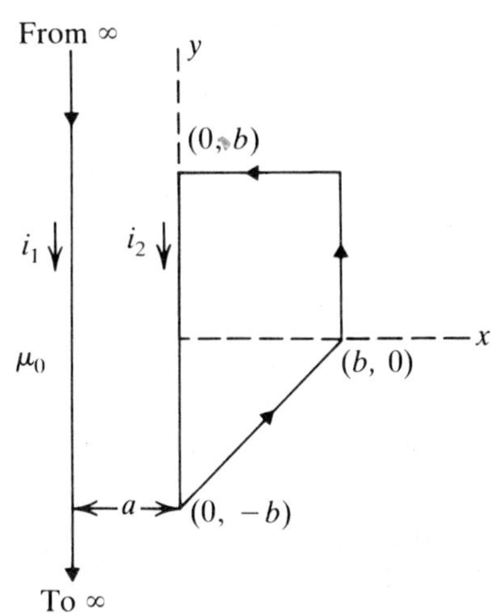

FIG. P1-13

1-14 Derive the general expression for the speed voltage (Eq. (1.101)) generated in a moving conductor by using the principle of energy conservation and Ampere's law (Eq. (1.155)).

1-15 The magnetic device shown in Fig. P1-15 has infinite permeability, and the rectangular plunger is allowed to move only in the vertical direction. Neglecting fringing.

a) Derive general expressions for the following:

1) The inductance of the coil.

2) The energy stored in the magnetic field.

3) The force acting on the rectangular plunger.

4) The flux density through the center leg of the electromagnet.

b) Evaluate the derived expressions using the following data:

$$l_{g_1} = 10 \text{ mm}, \quad l_{g_2} = 2 \text{ mm}$$
$$D = 50 \text{ mm}, \quad W = 60 \text{ mm}, \quad h = 40 \text{ mm}$$
$$N = 1000 \text{ turns}, \quad I = 2 \text{ A}$$

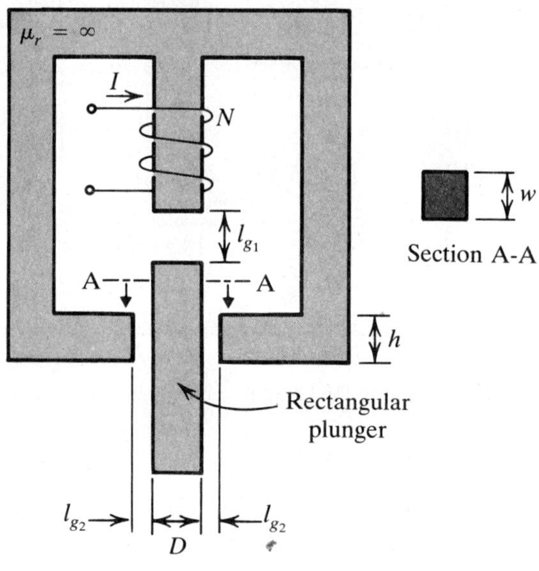

FIG. P1-15

1-16 The coil shown in Fig. P1-16 is energized from a dc voltage source. The mmf-ϕ characteristic of the magnetic material is given by

$$\mathscr{F} = 6.5 \times 10^4 \phi \text{ ampere-turns}$$

For an air gap of 10 mm determine:

a) The inductance of the coil.

b) The energy stored in the magnetic field.

c) The mechanical work done, if the air gap is slowly reduced to 1 mm. Sketch the variation of the force as a function of the air-gap length.

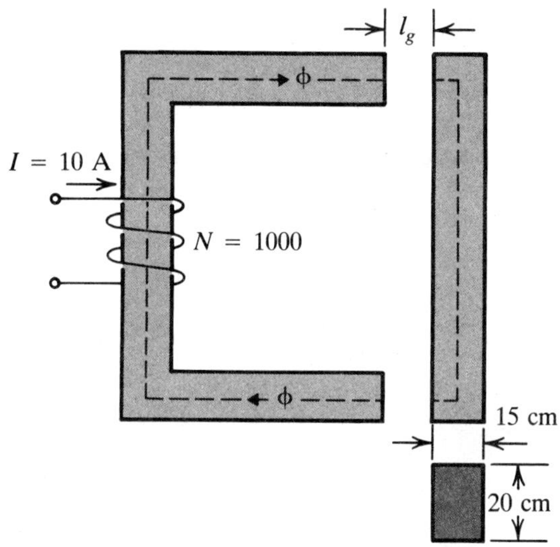

FIG. P1-16

1-17 Two coils, one mounted on the stator and the other on the rotor of a two-winding machine, have the following self- and mutual inductances:

$$L_s = L_1$$
$$L_r = L_2$$
$$L_{sr} = L_0 + L_{12} \cos 2\beta$$

where β is the angle between the axis of the two coils. The coils are connected in series and their current is $i = I_m \sin(\omega t - \theta)$. Determine:

a) The instantaneous torque.

b) The average torque.

TWO

Transformers

2.0 Introduction

This chapter covers the principles of operation, equivalent circuits, losses, and applications of the various types of transformers.

Single-phase transformers are covered in the greatest detail, because every other type of transformer is either a slightly modified single-phase transformer or a combination of single-phase transformers.

Two-winding, three-phase transformers are also discussed because they are of primary importance to industry.

Autotransformers and transformers in parallel operation have limited applications, but they are extensively analyzed because serious field problems may result from misunderstanding of their operation.

Instrument transformers are an essential part of any distribution network. Although covered in texts on electrical measurements, they are briefly discussed here in order to point out some of the common problems that may arise during their installation and operation, and to emphasize the effects of magnetic saturation.

The analysis of transformers requires a knowledge of *per-unit values* and of *three-phase networks*, which are common to all types of ac machines. Both these topics are briefly discussed in this chapter and are extensively covered in the Appendix of the book.

The end of this chapter includes some practical highlights of transformer applications, such as name-plate data (technical data inscribed on a metal plate fastened on an easily seen part of the transformer) and typical manufacturers' test results.

2.1 Single-Phase Transformers

As shown in Fig. 2-1, single-phase transformers usually have one input and one output winding, often referred to as the transformer's primary and secondary windings. These windings are not electrically connected, but are magnetically coupled. The primary winding draws energy from a voltage source, while the secondary winding delivers energy to a load.

Transformers, as the name implies, transform or change, from one level to another, the current and voltage that are applied to their input windings. An increase, or step up, in the voltage across one winding is accompanied by an equal decrease, or step down, of the current in the same winding.

Transformers are not power amplifiers. For all practical purposes, the apparent input power ($|S|$) to a transformer's primary winding is equal to the apparent power delivered to the load by its secondary winding. In other words, the volt-amperes of the primary winding ($V_1 I_1$) are approximately equal to the volt-amperes of the secondary

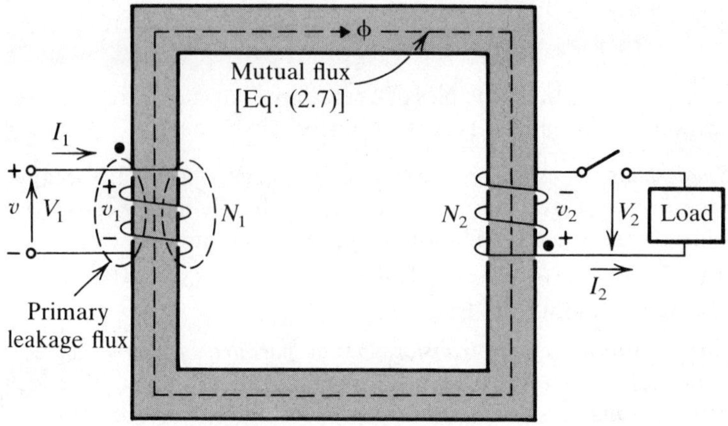

FIG. 2-1 Single-phase transformer.

winding ($V_2 I_2$). Mathematically,

$$|S| = V_1 I_1 \approx V_2 I_2 \tag{2.1}$$

In actual cases—because of transformer losses—the volt-amperes delivered to a load by a transformer are slightly less than the volt-amperes drawn from the voltage source.

2.1.1 PRINCIPLE OF OPERATION

The operation of single-phase transformers (and all other transformers) is based on the principle of induction. According to this principle, a voltage is induced in a winding when the flux linkages (λ) enclosed by the winding change as a function of time. The instantaneous value of the flux linkages is defined as

$$\lambda = N\phi \tag{2.2}$$

where N and ϕ are, respectively, the coil's number of turns and the instantaneous value of the flux per turn. Alternatively,

$$\lambda = Li \tag{2.3}$$

where L and i are, respectively, the coil's inductance and instantaneous current.

According to Faraday's law, the instantaneous value of the induced voltage (v) is given by

$$v = -\frac{d\lambda}{dt} \tag{2.4}$$

or

$$v = -\frac{d}{dt}(N\phi) \tag{2.5}$$

From Eq. (2.5):

$$v = -N\frac{d\phi}{dt} \tag{2.6}$$

The negative sign in the equations signifies that the polarity of the induced voltage opposes the change that produced it. This is a natural reaction of magnetically coupled coils, commonly known as Lenz's law.

As shown in Fig. 2-1, the polarity of the induced voltages is usually identified by the dot symbol. The physical significance of the dot symbol is explained in detail in Chapter 1 (Section 1.2.14).

Equation (2.6) is a basic law of nature and one of the most important mathematical relationships of electrical engineering. It governs the operation of motors and generators, and it demonstrates the coexistence of, and the quantitative relationship between, electric and magnetic fields.

The flux within the structure of the transformer changes because it is produced by the alternating voltage that is supplied to the transformer's input winding.

As Eq. (2.6) makes clear, for sinusoidal input voltages the flux is at a maximum when the voltage is at the zero point of its cycle. The flux and the voltage phasors must be 90° out of phase with each other. Typical voltage and flux waveforms for a transformer are shown in Fig. 2-2.

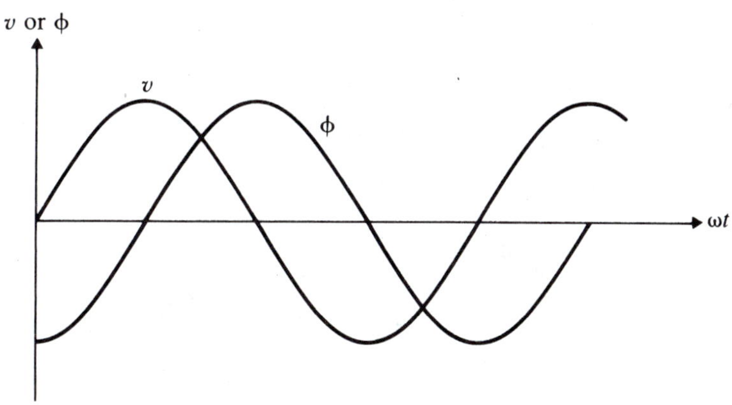

FIG. 2-2 Voltage and flux waveforms in a single-phase transformer.

The equation that relates the induced voltage to the flux produced is derived in Chapter 1 [Eq. (1.109)]. For convenience, it is rewritten here:

$$V = 4.44Nf\phi_m \tag{2.7}$$

where V is the rms value of the input voltage, f its frequency of oscillation in hertz, and ϕ_m the maximum value of the flux within the magnetic material in webers.

Thus, the rms value of the voltage induced in the secondary winding of the transformer, or in any other coil wound on the same core, will be given by Eq. (2.7). In each case the appropriate number of turns must be used.

Equation (2.7) is called the fundamental transformer equation, and it is often used in laboratories to calculate the flux level within a toroid or other type of magnetic core. This equation gives accurate results only if the leakage impedance of the coil is negligible.

The flux produced by the primary winding is divided into two parts: leakage flux and mutual flux. Leakage flux links only the windings of the primary coil and is associated with the transformer leakage impedance. Mutual flux links the windings of the primary and secondary coils and is associated with the magnetizing impedance of the transformer.

2.1.2 IDEAL TRANSFORMER RELATIONSHIPS

In this section we derive general equations that relate the parameters of the primary winding to those of the secondary winding. Only ideal transformers will be considered. This will provide the basis for the analysis of nonideal transformers, which will be introduced in later sections.

An ideal transformer has zero core loss, no leakage flux, and negligible winding resistances. Refer to Fig. 2-3. Using the induction principle, we have

$$v_1 = -N_1 \frac{d\phi}{dt} \tag{2.8}$$

$$v_2 = -N_2 \frac{d\phi}{dt} \tag{2.9}$$

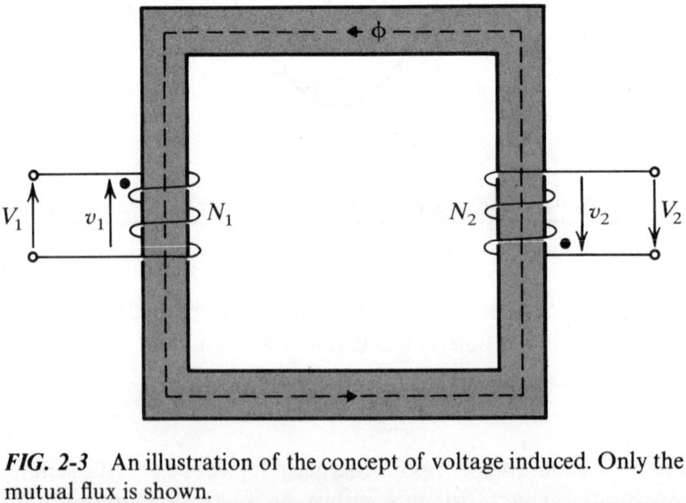

FIG. 2-3 An illustration of the concept of voltage induced. Only the mutual flux is shown.

From Eqs. (2.8) and (2.9) we obtain:

$$v_2 = v_1 \frac{N_2}{N_1} \tag{2.10}$$

Under the assumed ideal conditions, the induced voltages (v_1, v_2) are equal to their corresponding terminal voltages. Then, using rms values, we obtain

$$V_2 = V_1 \frac{N_2}{N_1} \tag{2.11}$$

That is, voltages of different magnitudes can be obtained by winding coils with different numbers of turns around a magnetic circuit. The effective flux (ϕ) within the magnetic material is dependent on the applied voltage and is essentially independent of the flux of the output current. Thus, under ideal conditions, the voltage induced in the output winding is independent of the load current.

Figure 2-4 shows a schematic of a two-winding transformer, its magnetic equivalent circuit, and its ideal electrical equivalent circuit.

Refer to Fig. 2-4(b). From KVL for magnetic circuits, we have

$$\sum (NI)_{\text{loop}} = 0 \tag{2.12}$$

That is, the sum of the magnetic potentials or ampere-turns within a closed magnetic circuit is equal to zero. In mathematical symbols,

$$(N_1 I_1)_{\text{input}} - (N_c I_c)_{\text{core}} - (N_2 I_2)_{\text{output}} = 0 \tag{2.13}$$

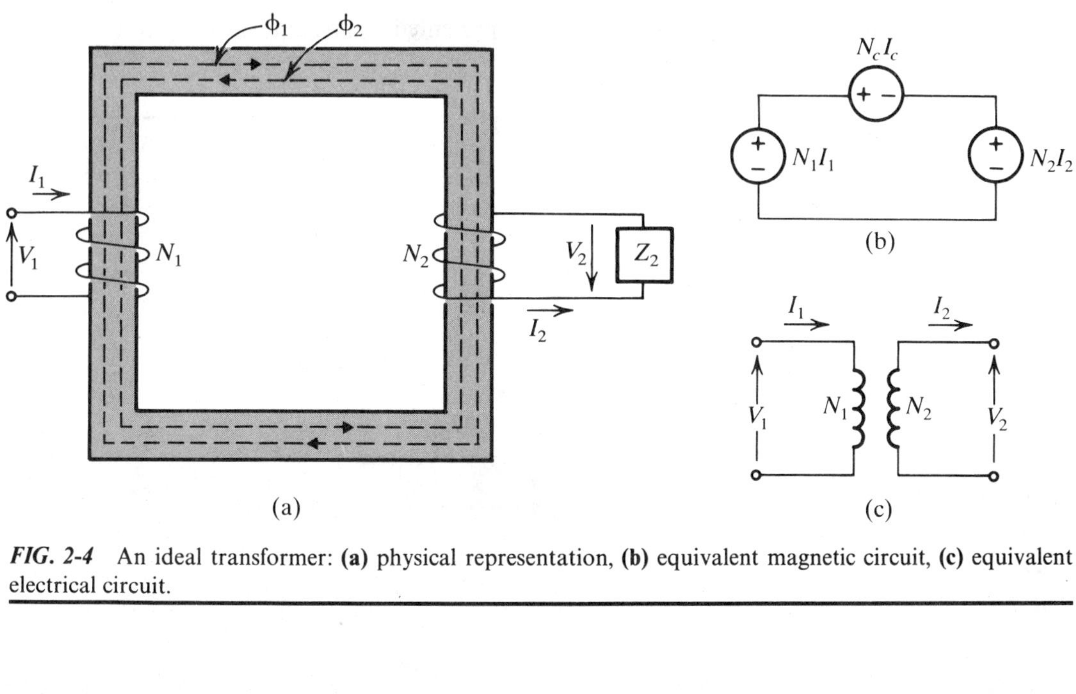

FIG. 2-4 An ideal transformer: **(a)** physical representation, **(b)** equivalent magnetic circuit, **(c)** equivalent electrical circuit.

The magnetic potential $N_c I_c$ is the amount of mmf required to magnetize the core of the transformer. For an ideal transformer ($\mu = \infty$) this mmf is equal to zero. That is, $(N_c I_c)_{core} = 0$.

Thus, from Eq. (2.13) we have

$$I_2 = I_1 \frac{N_1}{N_2} \tag{2.14}$$

From Ohm's law we obtain

$$Z_1 = \frac{V_1}{I_1} \tag{2.15}$$

and

$$Z_2 = \frac{V_2}{I_2} \tag{2.16}$$

where Z_1 and Z_2 are not the winding impedances, but the impedances as seen from the terminals of winding 1 and winding 2, respectively.

From the last two relationships and from Eqs. (2.11) and (2.14) we obtain the impedance transformation property of transformers. That is,

$$Z_1 = Z_2 \left(\frac{N_1}{N_2}\right)^2 \tag{2.17}$$

It should be emphasized that Eqs. (2.11), (2.14), and (2.17) are applicable only to ideal transformers. An ideal transformer is represented by the equivalent circuit shown in Fig. 2-4(c).

EXAMPLE 2–1 The 5 kVA, 480–120 V, single-phase transformer shown in Fig. 2-5 delivers rated current to a 120 volt load. Neglecting losses, determine the transformer currents and the supply voltage.

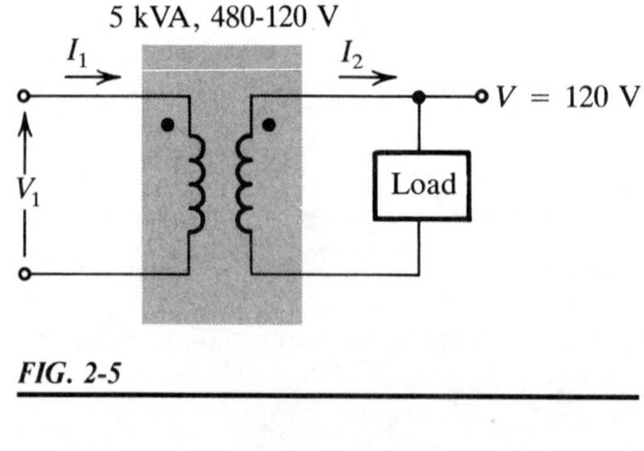

FIG. 2-5

SOLUTION The magnitude of the current through the secondary winding is

$$I_2 = \frac{|S|}{V} = \frac{5000}{120} = \underline{\underline{41.67\ A}}$$

The windings turns ratio is given by the ratio of the windings voltages:

$$\frac{N_2}{N_1} = \frac{V_2}{V_1} = \frac{120}{480} = \frac{1}{4}$$

From Eq. (2.14) the magnitude of the current through the primary winding is

$$I_1 = \frac{1}{4}(41.67) = \underline{\underline{10.42\ A}}$$

The primary voltage is given by the rating of the transformer. That is,

$$\underline{\underline{V_1 = 480 \text{ volts}}}$$

In an actual transformer the primary voltage and current would be slightly higher than calculated here because of the effect of the transformer's impedances.

Exercise 2–1 A 5 kVA, 240–120 volt, single-phase transformer supplies rated current to a load at 120 volts. Determine the magnitude of the load impedance as seen from the input terminals of the transformer.

Answer 11.52 Ω

2.1.3 DERIVATION OF THE EQUIVALENT CIRCUIT

The analysis of transformers is greatly simplified by using a model, called the *equivalent circuit.* A transformer's approximate equivalent circuit is shown in Fig. 2-6(b).

The resistance R_1 and the reactance X_1 represent the copper losses and the leakage flux of the primary winding, respectively. Similarly, the resistance R_2 and the reactance X_2 represent the copper losses and the leakage flux of the secondary winding, respectively. Then $R_1 + jX_1$ and $R_2 + jX_2$ represent the impedances of the primary and secondary windings, respectively. These impedances are often referred to as leakage impedances.

The winding with the higher voltage rating has higher leakage impedance than the winding with the lower voltage rating. R_m represents the core-loss equivalent resistance, and X_m represents the magnetizing reactance of the transformer. Usually, the magnetizing impedance is represented by a core resistance (R_c) in parallel with a magnetizing reactance X_ϕ [see Fig. 2-8(c)].

In Fig. 2-6(c), the leakage impedance of the secondary winding is transferred to the primary.

In Fig. 2-6(d), the magnetizing impedance is relocated to the input terminals of the transformer. This equivalent circuit, although approximate, simplifies the calculations.

In Fig. 2-6(e), all impedances shown in Fig. 2-6(d) are transferred to the secondary winding.

The leakage impedance is calculated from the data of the "short-circuit test" and the magnetizing impedance is calculated from the data of the "open-circuit test."

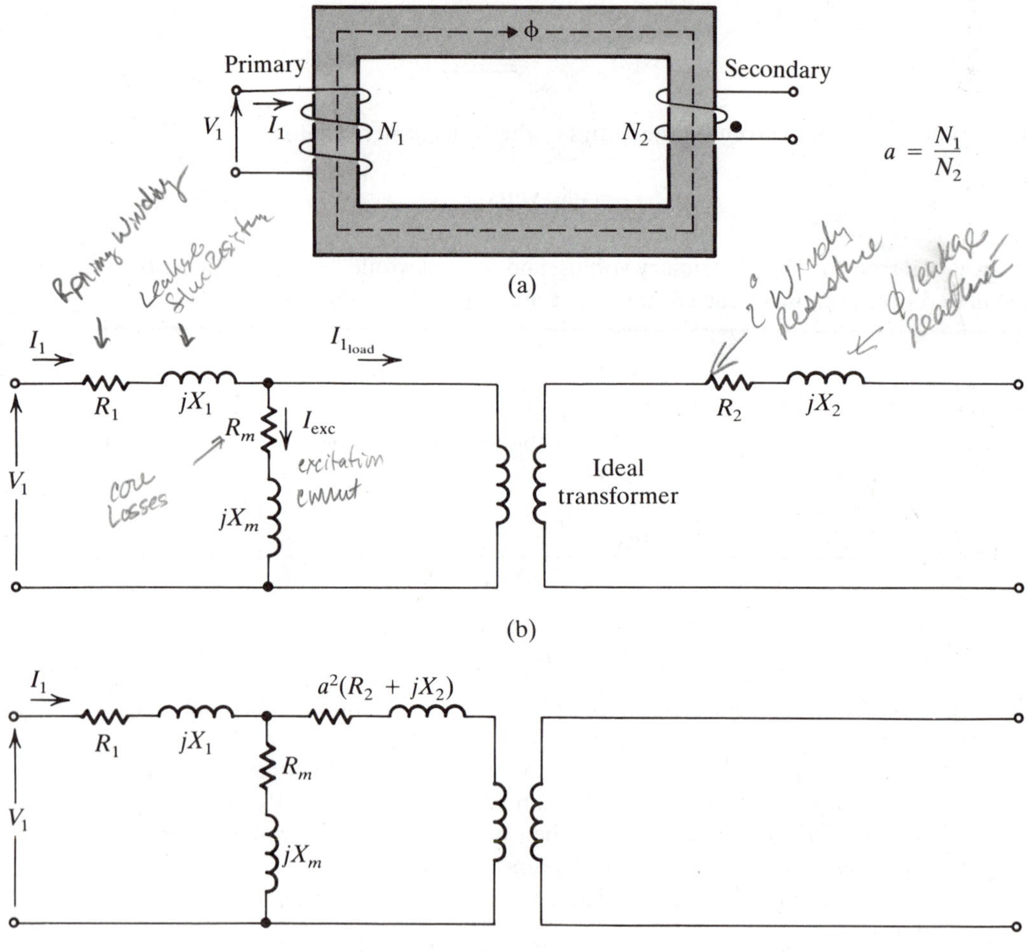

FIG. 2-6 Various forms of the equivalent circuit of a single-phase transformer: **(a)** elementary schematic, **(b)** individual winding impedances and the magnetizing impedance, **(c)** all impedances transferred to the primary, **(d)** an approximation of part **(b)**, **(e)** all impedances referred or transferred to the secondary.

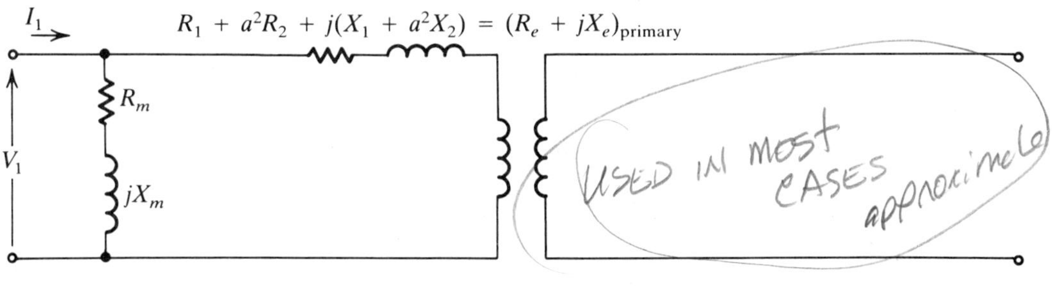

(d)

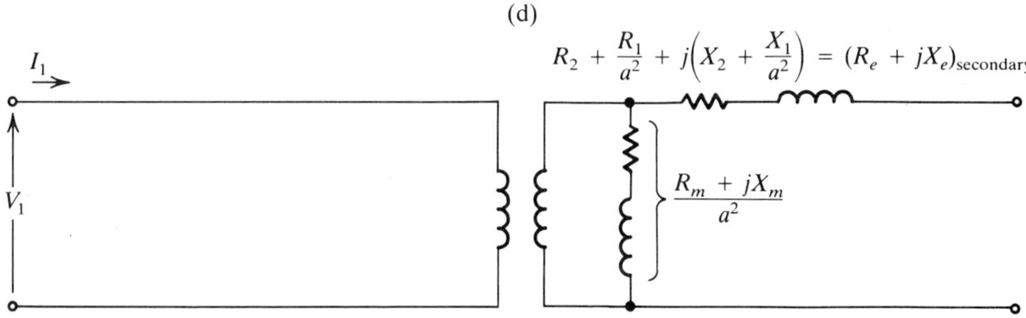

(e)

***FIG. 2-6** (Continued)*

Short-Circuit Test

The short-circuit test (also referred to as the impedance or copper-loss test) can be done on either side of the transformer. The input power (P_z), the applied voltage (V_z), and the input current (I_z) are measured in one winding, while the other winding, as shown in Fig. 2-7, is short-circuited. The objective of this test is to find the power loss in the windings of the transformer, and the equivalent winding impedances under rated conditions. The winding power losses affect the efficiency of the transformer, and the leakage impedance affects the short-circuit current and the output voltage of the transformer. For this reason, rated current is used in this test. Thus,

$$I_z = I_{\text{rated}}$$

In practice, a near rated value is used and then proper adjustments are made to related calculations so that the parameters obtained represent the transformer at nominal operating condition.

The voltage V_z is only a small percentage of the rated voltage and is sufficient to circulate rated current in the windings of the transformer. Usually,

$$V_z = (2\% \longrightarrow 12\%)V_{\text{rated}}$$

On account of this the transformer is magnetized at a relatively low flux density, as

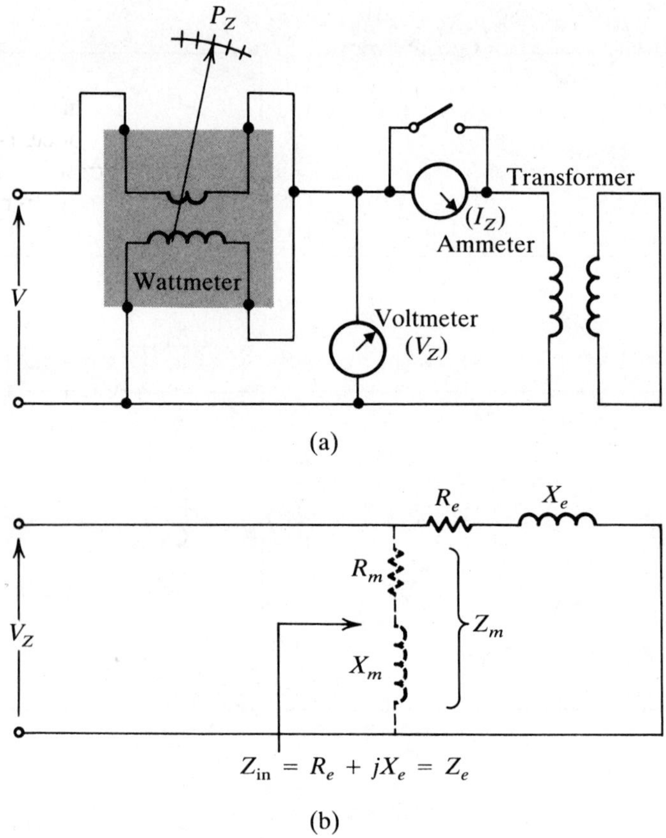

FIG. 2-7 Short-circuit test: **(a)** laboratory connection schematic, **(b)** equivalent circuit. (Notice that the magnetizing impedance Z_m is considered infinite.)

shown in Fig. 2-9. In performing this test the magnetizing impedance is assumed open because, in relative terms, it is very large.

Neglecting the losses of the measuring instruments, we have

$$R_e = \frac{P_z}{I_z^2} \text{ ohms} \tag{2.18}$$

$$|Z_e| = \frac{V_z}{I_z} \text{ ohms} \tag{2.19}$$

and

$$X_e = \sqrt{Z_e^2 - R_e^2} \text{ ohms} \tag{2.20}$$

It should be pointed out that this test does not give the individual winding impedances, but the combined impedances of primary and secondary windings. The subscript e is used to indicate that the short-circuit test gives the equivalent transformer leakage impedance, or the combination of the winding impedances as seen from one of the transformer windings.

The equivalent impedance, as calculated from the above relationships, is said to be referred to the side of the transformer where the instruments were placed. However, one can easily transfer this impedance to the other side of the transformer by multiplying it by the turns ratio squared. The turns ratio used must be the one seen from the side of the transformer to which this impedance is to be referred.

Open-Circuit Test

The open-circuit test (also referred to as the core-loss test, the magnetization test, the excitation test, the iron-loss test, or the no-load test) furnishes the core loss and the magnetizing impedance under rated conditions. It is usually done on the side of the transformer which has the lower rated voltage.

In doing this test the equivalent leakage impedance of the transformer is considered negligible because it is, relatively speaking, very small. The excitation power (P_{exc}), the excitation current (I_{exc}), and the excitation voltage (V_{exc}) are measured in one winding, as shown in Fig. 2-8(a), while the other winding is open-circuited. Neglecting the losses of the measuring meters, we have

$$R_m = \frac{P_{exc}}{I_{exc}^2} \text{ ohms} \tag{2.21}$$

$$|Z_m| = \frac{V_{exc}}{I_{exc}} \text{ ohms} \tag{2.22}$$

and

$$X_m = \sqrt{Z_m^2 - R_m^2} \text{ ohms} \tag{2.23}$$

The excitation current is a small percentage of the transformer's nominal current. Usually,

$$I_{exc} = (3\% \longrightarrow 10\%)I_{rated}$$

In order to obtain the core loss that corresponds to rated conditions, the flux on open-circuit test must be equal to the flux within the transformer when the latter delivers rated current. This is insured, as can be seen from Eq. (2.7), when the excitation voltage is equal to the rated voltage. That is,

$$V_{exc} = V_{rated}$$

Thus, as shown in Fig. 2-9, on open-circuit test the transformer is energized at approximately the same flux level as under normal operating conditions.

The magnetizing impedance calculated from the above relationships is as seen from the side of the transformer where the instruments were placed. It can be easily

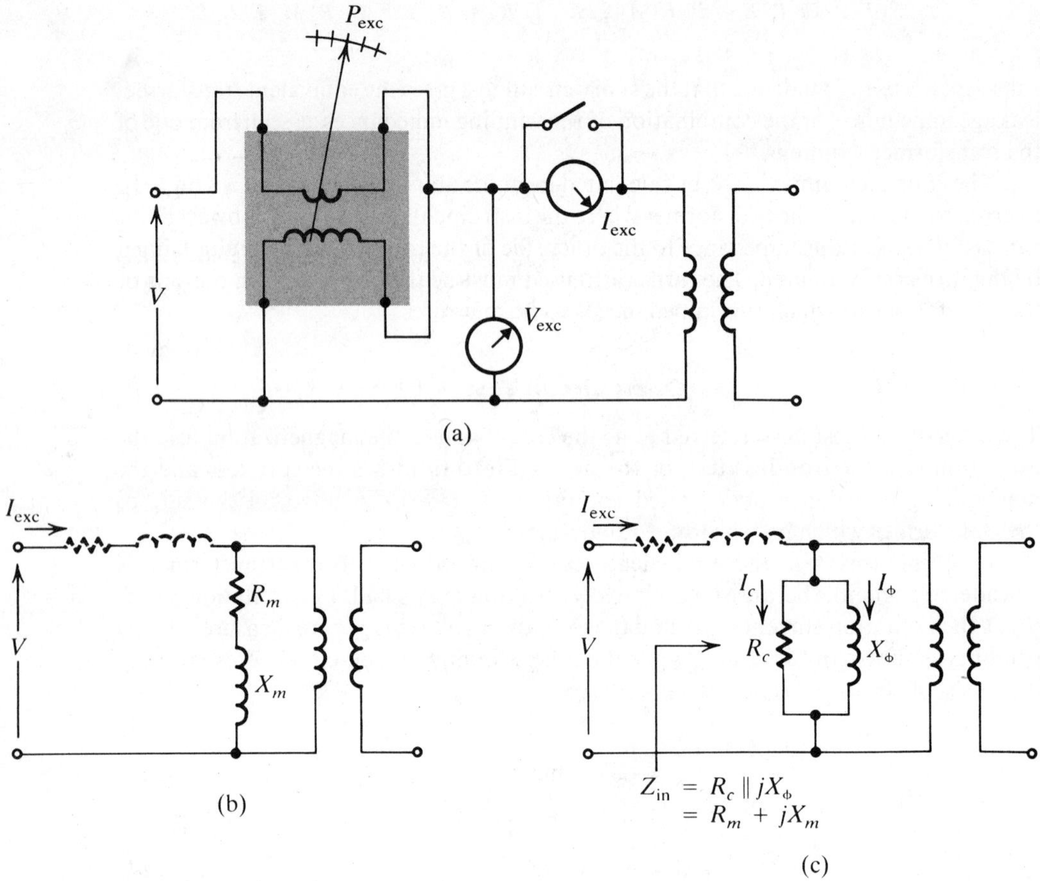

FIG. 2-8 Open-circuit test: **(a)** laboratory connection schematic, **(b)** series representation of the magnetizing impedance (notice that $R_1 + jX_1$ are considered negligible), **(c)** parallel representation of the magnetizing impedance.

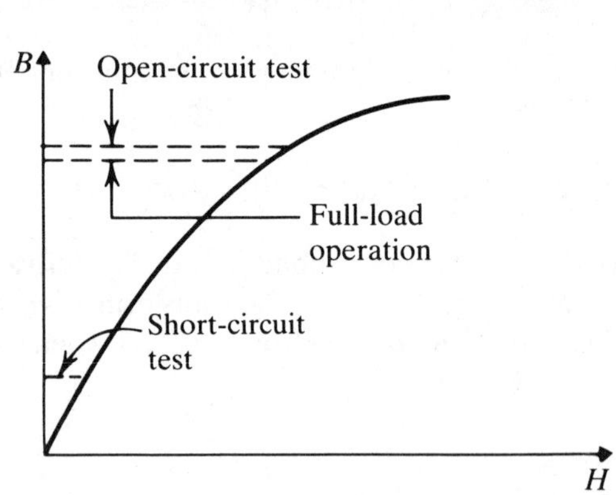

FIG. 2-9 *B-H* curve, showing the relative magnetization levels for a transformer during short-circuit test, open-circuit test, and under full-load operating conditions.

transferred or referred to the other side by using the impedance transformation property of ideal transformers.

When the core loss is measured at other than operating voltage, then the actual copper losses can be calculated by considering them as being proportional to the square of the applied voltage.

In the open-circuit test, both primary and secondary voltages are customarily recorded. This gives the effective turns ratio of the transformer, which might be slightly different than the ratio specified on its name-plate.

The series representation of the magnetizing impedance can be represented by its equivalent parallel impedance, as shown in Fig. 2-8(c). The equations that relate the components of the series equivalent impedance to those of the equivalent parallel impedance, Eqs. (1.140) and (1.141), are repeated here for convenience:

$$R_c = \frac{R_m^2 + X_m^2}{R_m} \tag{2.24}$$

$$X_\phi = \frac{R_m^2 + X_m^2}{X_m} \tag{2.25}$$

The series impedance representation tends to simplify the calculations, but the parallel representation reveals more about the magnetization process and thus is more common. For example, an inspection of the parallel impedance representation gives the core losses ($P_{exc} = V^2/R_c$) and the components of the exciting current ($I_\phi = V/jX_\phi$, $I_c = V/R_c$). The core or open-circuit losses are caused by eddy currents and hysteresis losses. These losses are further discussed in the sections that follow. Typical winding and iron losses for a 5 kVA single-phase transformer as a function of load current are shown in Fig. 2-10.

In concluding this section on transformer tests, it must be emphasized that Z_e is the input impedance of the transformer when its secondary is shorted, and Y_{exc} is the input

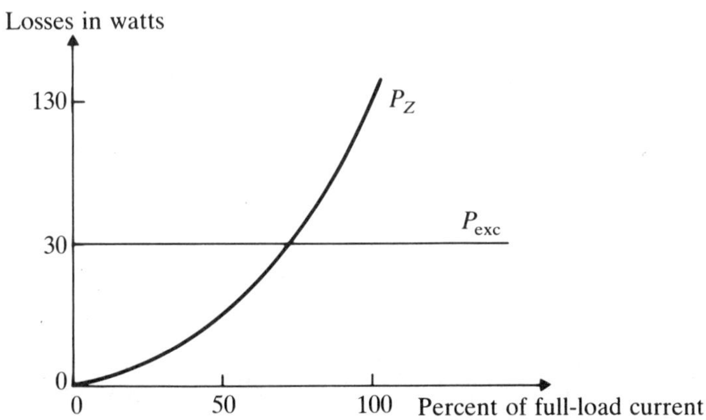

FIG. 2-10 Typical winding and iron losses as function of load current for a single-phase transformer (5 kVA, 480–120 volts).

admittance of the transformer when its secondary is open-circuited. These statements are, of course, approximate because of the underlying assumptions.

EXAMPLE 2–2 The following results were obtained from testing a 10 kVA, 480–120 V, single-phase transformer.

Test	*Voltage in Volts*	*Current in Amperes*	*Power in Watts*
Open-circuit	120	2.5	60
Short-circuit	26	20.83	200

Determine:

a) The equivalent leakage impedance of the transformer windings referred to the high-voltage (HV) and low-voltage (LV) winding.

b) The series and parallel components of the magnetizing branch referred to the LV winding.

c) The equivalent circuit of the transformer referred to the HV winding.

SOLUTION

a) The rated current of the transformer through the primary and secondary windings is:

$$I_H = \frac{|S|}{V} = \frac{10{,}000}{480} = 20.83 \text{ A}$$

$$I_L = \frac{10{,}000}{120} = 83.33 \text{ A}$$

Comparing the above with the given test data, it is clear that the short-circuit test was done on the HV winding.
Thus:

$$R_{eH} = \frac{P_z}{I_z^2} = \frac{200}{(20.83)^2} = .46 \ \Omega$$

and

$$|Z_{eH}| = \frac{V_z}{I_z} = \frac{26}{20.83} = 1.25 \ \Omega$$

$$X_{eH} = \sqrt{Z_{eH}^2 - R_{eH}^2} = \sqrt{(1.25)^2 - (.46)^2}$$
$$= 1.16 \ \Omega$$

Thus,

$$Z_{e_H} = \underline{\underline{.46 + j1.16\ \Omega}}$$

Using the impedance transformation property of the transformers, we obtain the equivalent impedance referred to the low-voltage side:

$$a = \frac{N_2}{N_1} = \frac{120}{480}$$

Thus,

$$Z_{e_L} = \left(\frac{120}{480}\right)^2 (.46 + j1.16) = \underline{\underline{.029 + j.073\ \Omega}}$$

b) The equivalent series elements of the magnetizing impedance are

$$|Z_{m_L}| = \frac{V_{\text{exc}}}{I_{\text{exc}}} = \frac{120}{2.5} = \underline{\underline{48\ \Omega}}$$

$$R_{m_L} = \frac{P_{\text{exc}}}{I^2_{\text{exc}}} = \frac{60}{(2.5)^2} = 9.6\ \Omega$$

and

$$X_{m_L} = \sqrt{Z^2_{m_L} - R^2_{m_L}} = \sqrt{48^2 - 9.6^2}$$
$$= \underline{\underline{47.03\ \Omega}}$$

The parallel components of the magnetizing impedance, as seen from the LV winding, are calculated from the series equivalent components by using Eqs. (2.24) and (2.25):

$$R_{c_L} = \frac{R^2_m + X^2_m}{R_m} = \frac{9.6^2 + 47.03^2}{9.6} = 240\ \Omega$$

and

$$X_{\phi L} = \frac{R^2_m + X^2_m}{X_m} = \frac{9.6^2 + 47.03^2}{47.03} = 48.99\ \Omega$$

c) Referring the resistance and the reactance of the magnetizing branch to the HV winding, we obtain

$$R_{c_H} + jX_{\phi H} = \left(\frac{480}{120}\right)^2 (240 + j48.99)$$
$$= (3840 + j783.84)\ \Omega$$

The equivalent circuit referred to the HV winding is shown in Fig. 2-11.

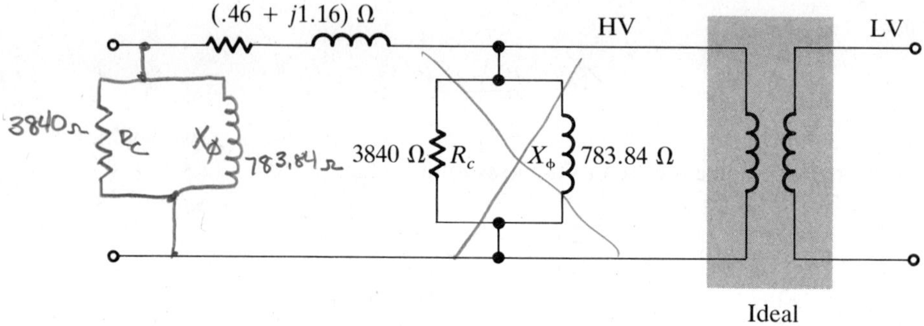

FIG. 2-11

2.1.4 WAVEFORM OF EXCITATION CURRENT

When one of the transformer's windings is open-circuited while the other is connected to rated voltage, the resulting current represents a small percentage of the transformer's rated current. This is due to the transformer's high magnetization impedance and to the infinite impedance seen by one winding while the other is open-circuited. This current is referred to as the excitation current (I_{exc}). In practice,

$$I_{exc} = (3\% \longrightarrow 10\%)I_{rated} \tag{2.26}$$

The theoretical derivation of the exciting current's waveform is obtained from the transformer's *B-H* characteristic as follows. Draw the waveform of the sinusoidal flux on an ωt-axis. Then, on the same coordinate system, draw the waveform of the exciting current. Refer to Fig. 2-12(a). The waveform of the flux can easily be obtained, because, it will be noted, the points $(a_1, 0)$, (d_1, d_2), and $(g_1, 0)$ are $\pi/2$ radians apart and correspond, respectively, to zero, maximum, and zero values of the flux. In other words, the critical points of the flux are known during half the period of its sinusoidal function. The resulting flux waveform is sketched in Fig. 2-12(b).

The magnitude of the exciting current at each of the coordinate points $[(a_1, 0)$, (b_1, b_2), (d_1, d_2), etc.$]$ of the *B-H* curve is given by the abscissa of these points. This is shown in Fig. 2-12(b). The transformer's voltage, flux, and exciting current are shown in Fig. 2-12(c).

The exciting current is not sinusoidal because of the nonlinearities of the *B-H* curve, and so a conventional phasor diagram of the exciting current cannot be drawn. However, its harmonics content, or its equivalent sinusoidal functions, can be found by using the Fourier series analysis. The Fourier series method is a mathematical tool used to determine the harmonic content of nonsinusoidal waveforms.

It can be shown that the exciting current is made up of "odd harmonics";* that

. . .

* According to Fourier series analysis, any function $f(t)$ of period T has only *odd* harmonics if $f(t)$ satisfies the following relationship: $f(t) = -f[t + T/2]$. The function which has such a characteristic is said to have half-wave *odd* symmetry.

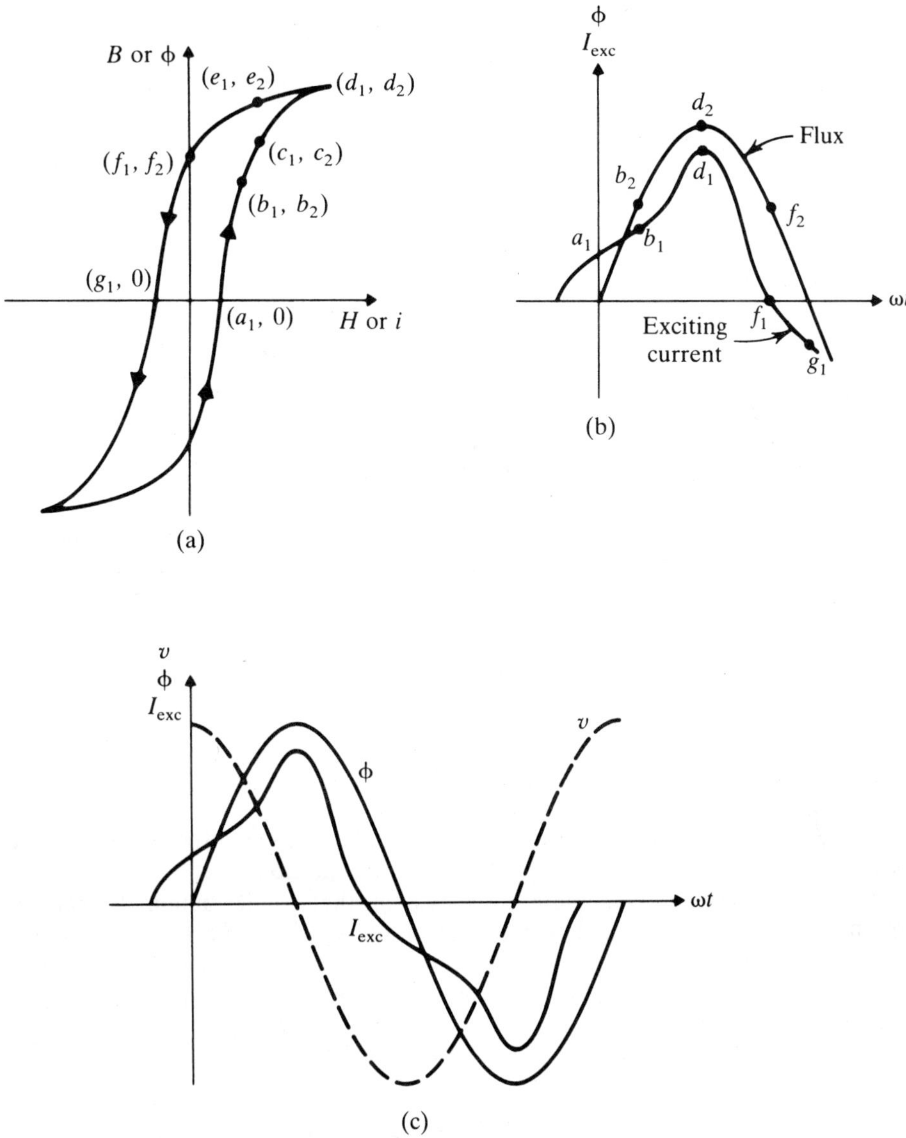

FIG. 2-12 Typical magnetic characteristics of transformers: **(a)** *B-H* curve, **(b)** instantaneous values of flux and exciting current, **(c)** instantaneous values of induced voltage, flux, and exciting current.

is, a fundamental term, a third harmonic, a fifth harmonic, and so on. The third harmonic can be as high as 40% of the fundamental term.

The exciting current is very important in the operation of all transformers, particularly three-phase transformers. It is discussed in greater detail in Section 2.2.4.

2.1.5 COMPONENTS OF PRIMARY CURRENT AND CORRESPONDING FLUXES

When a transformer is connected to its primary supply voltage while its secondary is open-circuited, it draws (as stated previously) a small percentage of its rated current (I_{exc}). Although this current is small, it produces rated flux within the core of the transformer.

As soon as an impedance is connected to the transformer's secondary (see Fig. 2-13(a)), a load current will flow. This current will produce the flux (ϕ_2) that opposes the flux of the primary current. This cannot be tolerated, because it would be accompanied by a reduction of the voltage induced in the primary [$v_i = -N(d\phi/dt)$], which would result in violation of KVL ($V_1 = v_i$). To overcome this opposition, the current drawn by the primary winding *increases* in such a way as to completely cancel the magnetic opposition of the secondary current.

The quantity of primary current needed to produce flux sufficient to completely neutralize the opposition of the secondary current is called the load component of the primary current. It is designated by I_{1_L}. Alternatively, the mmf of the secondary winding ($N_2 I_2$) is equal and opposite to the increase in the mmf of the primary winding ($N_1 I_{1_L}$). As a result, the effective mmf of the transformer at negligible leakage flux is $N_1 I_{exc}$.

The increase in the primary current that accompanies the flow of current in the secondary could also be justified by considering the following principle of energy conservation:

$$\begin{pmatrix}\text{energy drawn by} \\ \text{the transformer}\end{pmatrix} = \begin{pmatrix}\text{energy delivered} \\ \text{to the load}\end{pmatrix} + \begin{pmatrix}\text{energy consumed within} \\ \text{the transformer}\end{pmatrix}$$

The energy demand of the load is met by an increase in the magnitude of the primary current and its power factor relative to the no-load condition.

From the previous discussion it is evident that, under nominal operating conditions, the primary current is made up of two components. One, the exciting current, is required to magnetize the transformer; the other, its load component (L_{1_L}), is needed to cancel the opposition of the secondary current. In mathematical form, and as shown in Fig. 2-13(b),

$$I_1 = I_{exc} + I_{1_L} \tag{2.27}$$

The primary current I_1 is almost sinusoidal because its nonsinusoidal component, the exciting current, is negligible in comparison to its sinusoidal load component.

The flux of the primary current is shown in Fig. 2-13(a). Its mathematical representation is

$$\phi_1 = \phi_{exc} + \phi_{1_L} \tag{2.28}$$

where ϕ_{1_L} is the load component of the primary flux.

Phasor Diagram

Refer to Fig. 2-13(b). The transformer delivers power to a lagging-power-factor load. The windings turns ratio and the load voltage V_L normally are known. The load current

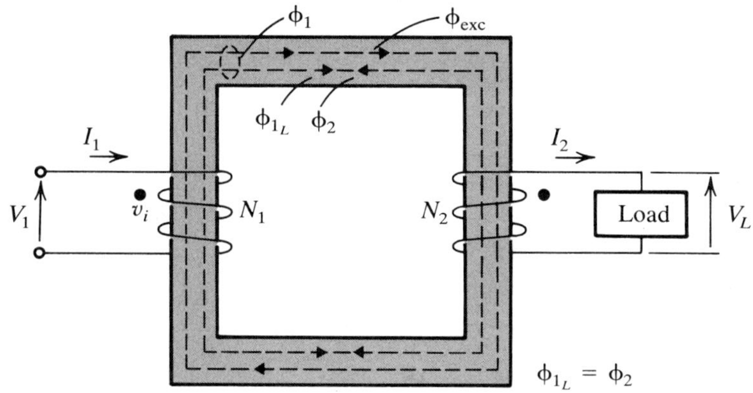

(a)

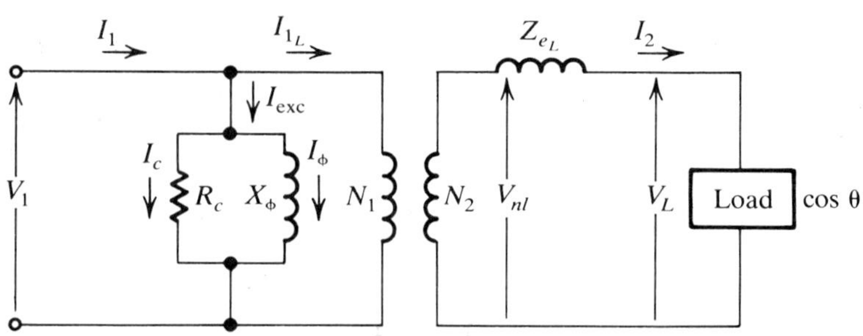

(b)

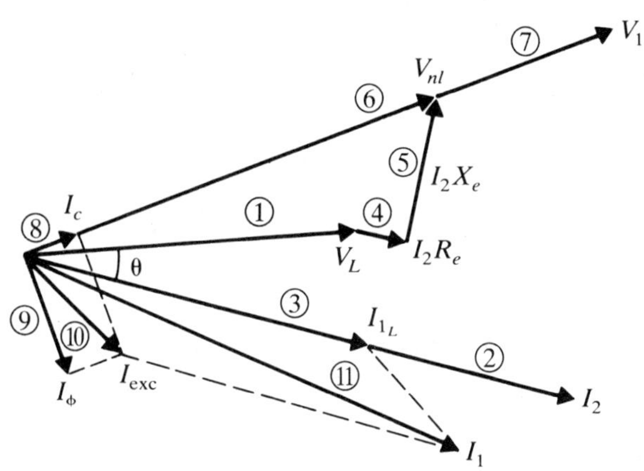

(c)

FIG. 2-13 Transformer: **(a)** polarity and component fluxes, **(b)** equivalent circuit, **(c)** phasor diagram.

I_2 is obtained from the load characteristics. The basic circuit equations are

$$V_1 N_2 = V_{nl} N_1 \tag{2.29}$$

and

$$V_{nl} = V_L + I_2 Z_{e_L} \tag{2.30}$$

These equations, together with Eq. (2.27), are used to construct the phasor diagram shown in Fig. 2-13(c). The encircled numbers indicate the sequence of steps that one may follow in order to simplify its construction.

2.1.6 TRANSFORMER CHARACTERISTICS

Inrush Current

When switching ON a transformer, the input or the inrush current is often many times larger than its rated current. This occurs when the switching takes place at the instant the input voltage is at the zero point on its time cycle. This results in a maximum flux that corresponds to a near saturation point on the *B-H* curve. At this point, the permeability of the magnetic material, the corresponding inductance, and the magnetizing impedance are at a minimum. Thus, the resulting current is at a maximum.

The magnitude and the waveform of the inrush current depend also on the leakage impedance of the transformer, on its magnetization characteristic, on the residual flux, and on the magnitude of the applied voltage at the instant of switching.

The magnitude of the inrush current is comparable to the current that results from external transformer shorts. However, the short-circuit current is of the fundamental frequency, while the magnetizing current has a large content (up to 40%) of third harmonic. The magnitude of the inrush current, as per standard industrial practice, is taken as being between 8 to 12 times the rated current of the transformer, and its duration is about 100 milliseconds.

Efficiency

By definition, the efficiency (η) of a transformer is given by:

$$\eta = \frac{P_{\text{out}}}{P_{\text{in}}} = \frac{P_{\text{in}} - P_{\text{loss}}}{P_{\text{in}}} = 1 - \frac{P_{\text{loss}}}{P_{\text{in}}} \tag{2.31}$$

where P_{in} and P_{out} correspond, respectively, to the input and output power of the transformer expressed in watts. The power loss (P_{loss}) is the sum of the core and copper losses of the transformer at the operating conditions under consideration. The copper loss depends on the kVA delivered. When the kVA drawn by the load varies over a 24-hour period, then one may calculate the so-called "24-hour transformer efficiency" by

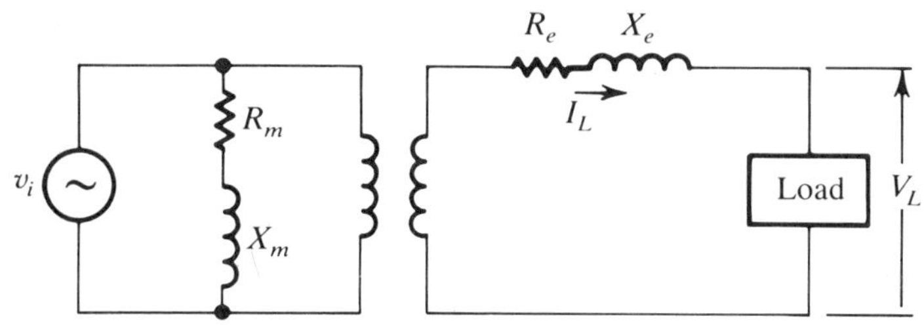

FIG. 2-14 Transformer's equivalent circuit and secondary load.

using energy instead of power. Generally speaking, the efficiency of transformers increases with their rating.

Industry continuously tries to minimize core loss by developing higher-quality magnetic materials from which the transformers can be economically manufactured. Where economics permit, copper windings are used instead of aluminum, because the latter give higher winding loss (I^2R).* The efficiency of the transformer is of primary importance, because the cost of the energy loss within the transformer over its lifetime is usually greater than that of the transformer itself. Typical transformer losses are shown in Tables 2-5 to 2-8. In the absence of winding condensation, idle transformers should be disconnected from their supply lines; otherwise, their energy consumption, due to core loss, is wasted.

Refer to Fig. 2-14. From basic definitions, we have

$$\text{efficiency} = \frac{\text{output power}}{\text{output power} + \text{losses}} \quad \textbf{(2.32)}$$

or

$$\eta = \frac{V_L I_L \cos\theta}{V_L I_L \cos\theta + P_{\text{exc}} + R_e I_L^2} \quad \textbf{(2.33)}$$

and

$$\eta = \frac{K I_L}{K I_L + P_{\text{exc}} + R_e I_L^2} \quad \textbf{(2.34)}$$

where

$$K = V_L \cos\theta \quad \textbf{(2.35)}$$

. . .

* The resistance of an aluminum conductor with the same current capacity as that of a copper conductor is about 10% higher.

Taking the derivative of the efficiency with respect to the variable parameter I_L, we obtain

$$\frac{d\eta}{dI_L} = \frac{(KI_L + P_{exc} + R_e I_L^2)K - KI_L(K + 0 + 2R_e I_L)}{(KI_L + P_{exc} + R_e I_L^2)^2} \tag{2.36}$$

Setting the last equation equal to zero gives

$$P_{exc} = R_e I_L^2 \tag{2.37}$$

or

$$P_{exc} = P_Z \tag{2.38}$$

Thus, the efficiency of a transformer at a constant load power factor is maximum when the iron loss is equal to its winding loss.

A transformer's efficiency depends on the load current and its power factor. The first condition is obvious, while the second becomes evident when one considers that the primary current, being the vector sum of the exciting current and the component of the load current (see Fig. 2-13(b)), depends on the power factor of the load. The magnitude of the primary current, in turn, controls the primary winding's copper losses. When one speaks of a transformer's efficiency, then, one must always identify the corresponding current and the power factor.

The variation of efficiency versus load current, as a function of the latter's power factor, is shown in Fig. 2-15.

The variation of iron loss, winding loss, and efficiency versus load current for a 25 kVA transformer is shown in Fig. 2-16.

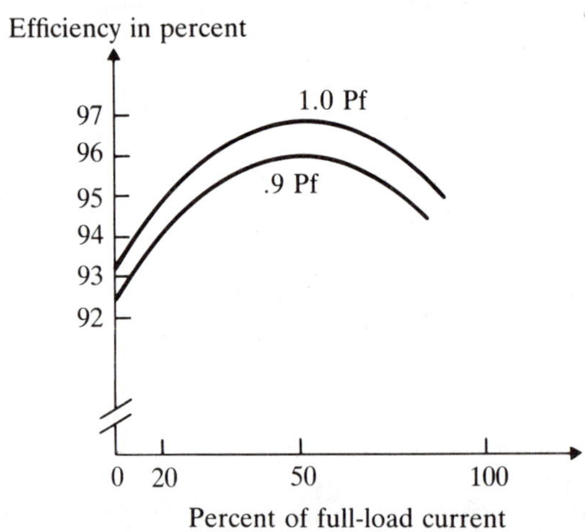

FIG. 2-15 Efficiency as a function of load current and power factor.

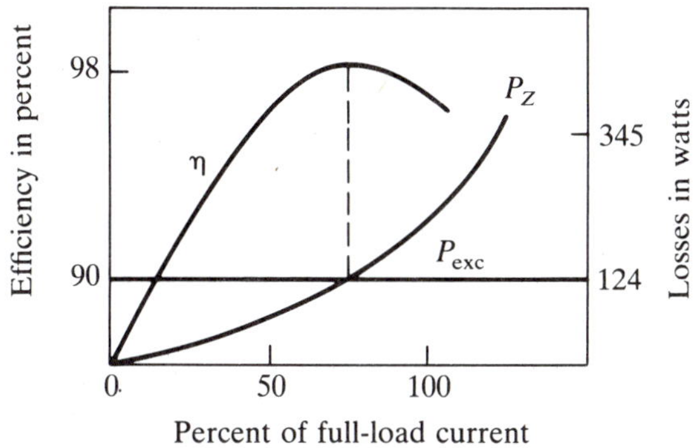

FIG. 2-16 Efficiency, iron loss, and winding loss as a function of load current for a 25 kVA, 480–120 V, single-phase transformer.

Regulation

Voltage regulation is a measure of the change in the magnitude of the output voltage while the load current varies from zero up to its rated value. In mathematical form, regulation is given by

$$\text{regulation}\ \% = \frac{|V_{nl}| - |V_{fl}|}{|V_{fl}|}(100) \tag{2.39}$$

where V_{nl} and V_{fl} are, respectively, the magnitudes of the voltages at the output terminals of the transformer at no-load and at full-load. Full-load voltage is also referred to as name-plate voltage, rated voltage, or nominal transformer voltage. No-load voltage can be obtained from the input voltage and the nominal voltage ratio of the transformer. Equation (2.39) can be used for any operating condition. In such cases, the full-load voltage is replaced by its actual value.

Regulation depends on the leakage impedance of the transformer and on the power factor of the load. At a particular value of load power factor and leakage impedance, voltage regulation becomes ideal or zero. Depending on the power factor of the load, the regulation could be positive or negative. Usually, under steady-state operating conditions it is less than plus or minus 5 percent ($\pm 5\%$). Regulation as a function of power factor is shown in Fig. 2-17.

At leading power factors, regulation is usually negative; that is, the voltage at the secondary terminals of a transformer is larger at full-load than it is when the load is disconnected. In such cases, the equipment connected to a transformer's secondary may be subjected to higher than rated voltages. This may occur when the power-factor-correction capacitor banks remain on the network while the plant operates at a reduced load.

The voltage across the load is of primary importance, because many operating characteristics of various pieces of equipment depend on it. For example, a 5% reduction of the rated voltage of an incandescent lamp noticeably reduces its output, and the light appears dimmer. For this reason, most power distribution transformers,

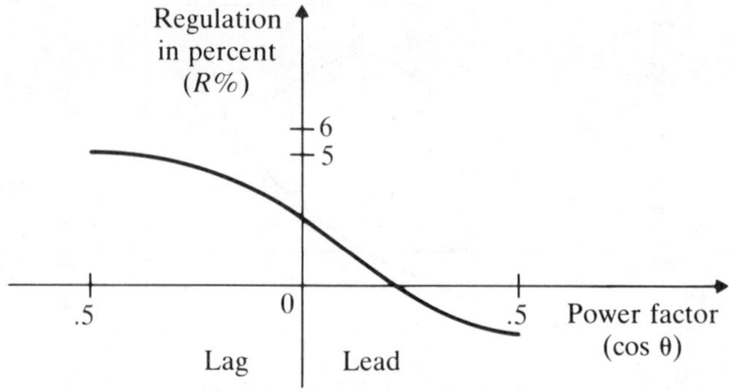

FIG. 2-17 Regulation as a function of load power factor for constant rated current.

as per manufacturers' standards, are equipped with "off-load" voltage tap changers through which the voltage can be changed by $\pm 2.5\%$ or by $\pm 5\%$ in relation to their nominal voltage. When the voltage is to be changed, the load of the transformer is disconnected and the number of turns of the primary winding is changed according to the requirements.

kVA Rating

The apparent power, or the kVA rating, of a transformer is always inscribed on its name-plate. It indicates the transformer's transformation capacity in terms of the volt-amperes a transformer is designed to deliver.

The capability of a transformer is limited by heating in the windings (hence there is a maximum permissible value of sustained current) and by excessively high exciting current and excessive core loss (hence there is a maximum permissible value for sustained voltage). Rated volt-amperes is the product of rated voltage and nominal current. In normal operation, the input voltage is close to the rated value.

As long as a transformer delivers rated or reduced kVA it will operate without being overheated. However, when it is cooled, it can dissipate more than nominal heat and, as a result, it can safely deliver higher than rated kVA. For this reason, most substation transformers are equipped with a set of ventilating fans, which enable the transformer's kVA capacity to be increased proportionally to the ventilation furnished.

A transformer of 1000 kVA supplied with fans, for example, can deliver 1333 kVA without being overheated. Such a transformer is identified as 1000/1333 kVA. Transformers with higher kVA ratings can transform higher magnitudes of voltage and current; that is, their windings have, on a relative basis, larger diameters and more insulation. As a result, the *physical* size and the *cost* of a transformer depend on its kVA rating.

The importance of this parameter is further explained in Section 2.6, which provides information about all the important parameters that appear on the name-plate of a transformer. A familiarity with name-plate data facilitates the solution of practical and theoretical transformer problems.

2.1.7 PER-UNIT VALUES

When engineering parameters such as power, torque, speed, voltage, current, and impedance are expressed in their corresponding per-unit values, many engineering concepts and calculations are simplified. For this reason, almost all design and problem solving in the power-distribution field is implemented by expressing all pertinent parameters in their equivalent per-unit values.

Actual engineering parameters are changed to their equivalent per-unit values as follows. The per-unit value of a parameter K is equal to its actual value divided by the base value. That is,

$$K = \frac{\text{actual value of } K}{\text{base value of } K} \text{ pu} \tag{2.40}$$

Base parameters are normally obtained from the name-plate data of the transformer, as follows:

$$\begin{aligned} \text{base power } (S_b) &= \text{volt-ampere rating of the transformer} \\ \text{base voltage } (V_b) &= \text{rated voltage of the winding under consideration} \\ \text{base current } (I_b) &= \text{rated current of the winding under consideration} \\ &= \frac{S_b}{V_b} \\ \text{base impedance } (Z_b) &= \frac{\text{base value of voltage}}{\text{base value of current}} \end{aligned}$$

From the above,

$$Z_b = \frac{V_b^2}{S_b} \tag{2.41}$$

The base power is the same in both the high- and the low-voltage windings ($S_{bH} = S_{bL}$). In contrast, the base voltage, base current, and base impedance are different for each winding of the transformer, and thus the subscript H or L should be included in the above expressions to represent parameters at the high- and low-voltage windings. The base value is the value of the parameter expressed in standard engineering units.

When the ohmic value of an impedance is $15\,\underline{/40^\circ}$ ohms, and the value of the base impedance is 300 ohms, then the per-unit value of the impedance is

$$Z = \frac{15\,\underline{/40^\circ}}{300} = .05\,\underline{/40^\circ} \text{ pu}$$

Similarly, when the actual value of a voltage is 250 volts, and the base voltage is 200 volts, then the per-unit value of the voltage is

$$V = \frac{250}{200} = 1.25 \text{ pu}$$

The per-unit value of a parameter, multiplied by 100, gives the value of the parameter in percent. Thus, when the per-unit value of a parameter is .04, its percentage value is 4%.

One advantage of the per-unit system is that when the voltage, current, and impedance of a transformer are expressed in per-unit, they have the same values regardless of the side of the transformer to which they refer. This greatly simplifies the understanding and solution of various transformer problems.

An additional advantage is that when the parameters of a machine are expressed in per-unit, they all fall within a known range, regardless of the rating of the machine. The per-unit values of the equivalent leakage impedance of most single-phase transformers, for example, are within the range of .035 – .055 p-u. Knowing this is of tremendous importance, greatly simplifying the solution of problems and the conceptual understanding of various transformer operating conditions.

It can be shown that the per-unit value of the transformer's leakage impedance is equal to the voltage required on short-circuit test, expressed in per-unit of the rated value. That is,

$$Z_{e_{pu}} = V_{z_{pu}} \tag{2.42}$$

This is left as student problem (see Exercise 2–3).

Because the per-unit system is used in most of the chapters of this text, it is described in detail in the Appendix.

One-Line Diagrams

Transformers, cables, and electric machines are conventionally identified in electrical drawings by their equivalent one-line representations. As the name implies, one-line diagrams represent electrical loads as being supplied through one wire, regardless of how many wires are used in the actual set-up.

The main parameters of a system—per-unit impedances, kVA rating of transformers, kW rating of motors, size of cables, protective devices, the short-circuit MVA of the supply voltage source, etc.—are written on the one-line diagrams.

For a single-phase system, the short-circuit apparent power ($|S|$) in volt-amperes is

$$|S| = VI_{sc} \tag{2.43}$$

where V and I_{sc} are, respectively, the nominal voltage and the short-circuit current. When the voltage is expressed in kV and the short-circuit current in kA, then Eq. (2.43) gives the short-circuit apparent power in MVA.

From Eq. (2.43) and Ohm's law we obtain

$$|S| = V\frac{V}{Z} \tag{2.44}$$

where Z is the impedance of the supply network in ohms/phase. From the above,

$$Z = \frac{V^2}{|S|} \tag{2.45}$$

For a three-phase system the short-circuit MVA is

$$S = \sqrt{3}\, V_{L-L} I_{sc} \tag{2.46}$$

where V_{L-L} and I_{sc} are, respectively, the line-to-line nominal voltage expressed in kV and the short-circuit current expressed in kA.

Thus, the short-circuit MVA of a given voltage source can be used to find the internal impedance of the source and/or the short-circuit current.

Protective equipment, such as circuit breakers and fuses, must be capable of safely withstanding the forces produced by the short-circuit currents. Since the latter are given indirectly in terms of short-circuit MVA, it is customary in the industry to rate the short-circuit capacity of equipment in terms of short-circuit MVA.

Typical short-circuit MVA's for a 25 kV and 46 kV distribution system are as follows:

kV	MVA
25	500
46	1500

Under ideal conditions, the source impedance is equal to zero and its frequency is constant. Such a system is identified on the one-line diagram as an "infinite bus." In more general terms an infinite bus is a power source whose voltage and frequency remain constant independent of the active or reactive power being supplied.

Since all electrical drawings represent electrical loads by their one-line diagram, an attempt is made throughout this book to use the one-line diagram as often as possible.

Figure 2-18(a) shows a two-wire diagram of a single-phase load. The 120 V, 60 Hz voltage supply has an internal reactance of .1 ohm. The 15 A fuses protect the 1.2 kW load from undesirable high currents. The two-conductor (2c) cable is size #12 AWG

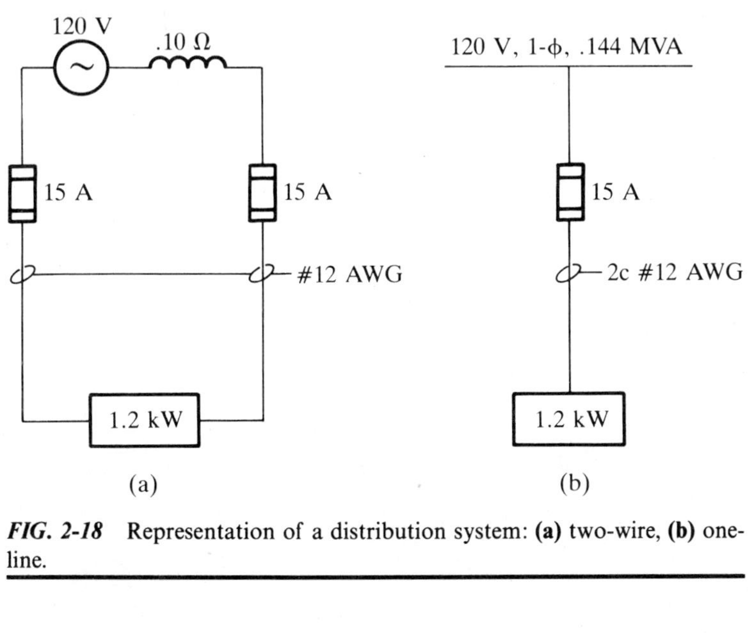

FIG. 2-18 Representation of a distribution system: **(a)** two-wire, **(b)** one-line.

(American Wire Gauge). When the load is shorted, the resulting short-circuit MVA is

$$|S| = V|I|$$
$$= 120\left(\frac{120}{.1}\right) \times 10^{-6}$$
$$= .144 \text{ MVA}$$

The one-line diagram of Fig. 2-18(a) is shown in Fig. 2-18(b).

EXAMPLE 2–3 A 5 kVA, 480–120 V single-phase transformer delivers rated current at .9 pf lagging. The transformer's equivalent leakage impedance referred to the high-voltage winding is $.92 + j1.84$ ohms. Find the per-unit values of the voltage, current, and leakage impedance on the HV and LV winding.

SOLUTION The approximate equivalent circuit of the transformer is shown in Fig. 2-19.

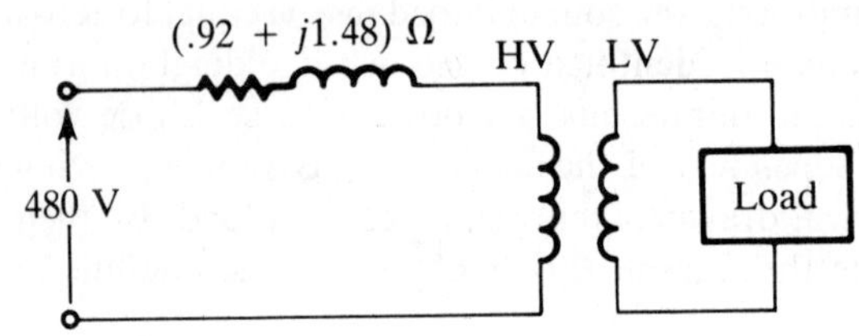

FIG. 2-19

HV Side

The base parameters are

$$S_b = 5000 \text{ VA}$$
$$V_{b_H} = 480 \text{ V}$$
$$I_{b_H} = \frac{5000}{480} = 10.42 \text{ A}$$
$$Z_{b_H} = \frac{(480)^2}{5000} = 46.08 \text{ ohms}$$

The per-unit values are

$$V_H = \frac{\text{actual value}}{\text{base value}} = \frac{480}{480} = \underline{\underline{1.0 \text{ pu}}}$$
$$I_H = \frac{10.42\angle{-25.8^\circ}}{10.42} = \underline{\underline{1.0\angle{-25.8^\circ} \text{ pu}}}$$
$$Z_H = \frac{.92 + j1.84}{46.08} = \underline{\underline{.02 + j.04 \text{ pu}}}$$

LV Side

The base parameters are

$$S_b = 5000 \text{ VA}$$
$$V_{b_L} = 120 \text{ V}$$
$$I_{b_L} = \frac{5000}{120} = 41.67 \text{ A}$$
$$Z_{b_L} = \frac{(120)^2}{5000} = 2.88 \text{ ohms}$$

The per-unit values are

$$V_L = \frac{120}{120} = \underline{\underline{1.0 \text{ pu}}}$$

$$I_L = \frac{41.67\underline{/-25.8^\circ}}{41.67} = 1.0\underline{/-25.8^\circ} \text{ pu}$$

The leakage impedance referred to the LV side is

$$Z_e = (.92 + j1.84)\left(\frac{120}{480}\right)^2 = .06 + j.12 \ \Omega$$
$$= \frac{.06 + j.12}{2.88} = \underline{\underline{.02 + j.04 \text{ pu}}}$$

EXAMPLE 2–4 The 100 kVA, 440–220 V, 60 Hz single-phase transformer shown in the one-line diagram representation (Fig. 2-20(a)) has negligible magnetizing current and a leakage impedance of .03 + j.040 pu. The transformer is connected to a 440 V source through a feeder whose impedance is .04 + j.08 ohms. Determine:

a) The voltage across the load when the transformer is connected to a 440 V source and delivers rated current to a load of .80 power factor lagging.
b) The voltage regulation.
c) The primary and secondary short-circuit MVA when the secondary of the transformer is shorted.
d) The rating of the capacitor in kVAR which, when connected across the load, will improve the power factor from .80 lagging to unity.

SOLUTION

a) The base impedance in the HV winding is

$$Z_{b_H} = \frac{(440)^2}{100{,}000} = 1.94 \ \Omega$$

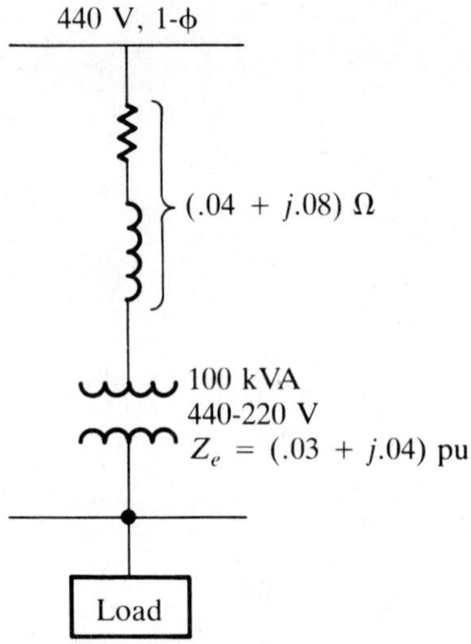

FIG. 2-20(a)

The per-unit value of the voltage is

$$V_{pu} = \frac{\text{actual value}}{\text{base value}} = \frac{440}{440} = 1.0 \text{ pu}$$

Similarly,

$$I_{pu} = 1.0\underline{/-36.9^\circ} \text{ pu}$$

The per-unit value of the feeder impedance is

$$Z_f = \frac{\text{actual value}}{\text{base value}}$$

$$Z_f = \frac{.04 + j.08}{1.94} = .02 + j.04 \text{ pu}$$

Applying KVL to the equivalent circuit of Fig. 2-20(b), and using per-unit values, we have

$$V_L = V_1 - IZ$$

or

$$V_L\underline{/0^\circ} = 1.0\underline{/\beta_1^\circ} - 1.0\underline{/-36.9^\circ}[.03 + .02 + j(.04 + .04)] \qquad \text{(I)}$$

The given power factor of the load establishes the reference for the phase angles. In

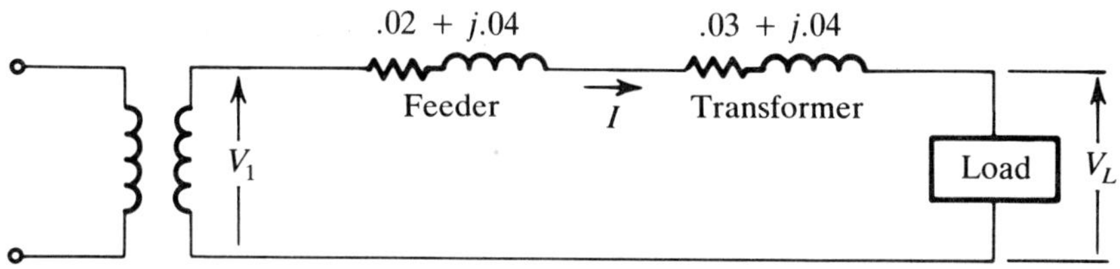

FIG. 2-20(b)

other words, the voltage across the load has a phase angle equal to zero, while the input voltage will be at an angle of β_1 degrees to the reference.

From equation (I), and after some algebraic calculations, we obtain

$$V_L\underline{/0^\circ} = 1.0\underline{/\beta_1^\circ} - .096\underline{/21.2^\circ}$$

Equating the imaginary components of the above equation, we get

$$\beta_1 = \arcsin(.096 \sin 21.2^\circ) = 2^\circ$$

Equating the real components, we obtain

$$\begin{aligned} V_L &= 1.0 \cos 2^\circ - .096 \cos 21.2^\circ \\ &= .91 \text{ pu} \\ &= .91(220) \\ &= \underline{\underline{200.21 \text{ V}}} \end{aligned}$$

Alternatively, *using ohmic values*, from KVL we have

$$Z_t = 1.94(.03 + j.04) = .06 + j.08 \ \Omega$$

The total impedance referred to the HV winding is

$$Z = (.06 + j.08) + (.04 + j.08) = .19\underline{/58^\circ} \ \Omega$$

The magnitude of the current in the HV is

$$\begin{aligned} |I| &= \frac{100{,}000}{440} \\ &= 227.27 \text{ A} \end{aligned}$$

Thus,

$$V_L = \frac{1}{2}(440\underline{/\beta_1^\circ} - IZ)$$

or

$$V_L\underline{/0^\circ} = \frac{1}{2}[440\underline{/\beta_1^\circ} - 227.27\underline{/-36.9^\circ}(.19\underline{/58^\circ})]$$

$$= \frac{1}{2}(440\underline{/\beta_1^\circ} - 42.156\underline{/21.2^\circ})$$

Equating the imaginary components of the last equation, we obtain

$$\beta_1 = \arcsin\frac{15.25}{440} = 2^\circ$$

Equating the real components, we obtain

$$V_L = \frac{1}{2}(440\cos 2^\circ - 39.3) = \underline{\underline{200.21\ \text{V}}}$$

b) Using Eq. (2.39), the regulation in percent is

$$\text{regulation}\ \% = \frac{220 - 200.21}{200.21}(100) = \underline{\underline{9.9\%}}$$

c) The short-circuit volt-amperes are given by

$$(\text{VA})_{sc} = V_{\text{rated}} I_{sc}$$

$$I_{sc} = \frac{V}{|Z|} = \frac{440}{.19}$$

$$= 2372\ \text{A}$$

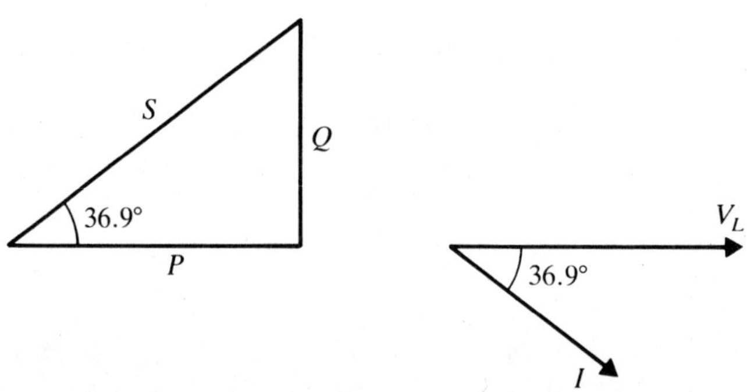

FIG. 2-20(c)

The short-circuit VA expressed in MVA is

$$(\text{MVA})_{sc} = .440(2.372) = \underline{\underline{1.044 \text{ MVA}}}$$

Owing to the transformation properties of the transformer, the short-circuit MVA is the same whether it is referred to the high- or low-voltage winding. The power triangle and the phasor diagram of the load are shown in Fig. 2-20(c).

d) The required reactive power (Q_c) of the capacitor is

$$Q_c = P \tan \theta$$

from which

$$Q_c = 100(.8)(\tan 36.9°) = \underline{\underline{60 \text{ kVAR}}}$$

Exercise 2–2 The approximate equivalent circuit of a single-phase 5 kVA, 480–240 V transformer is shown in Fig. 2-21.

a) Draw the equivalent circuit as seen from the high-voltage side.
b) Draw the equivalent circuit as seen from the low-voltage side.
c) Find the per-unit value of the transformer's leakage impedance.
d) Determine the primary voltage when the transformer delivers rated current to a load at 240 volts and .80 power factor lagging.

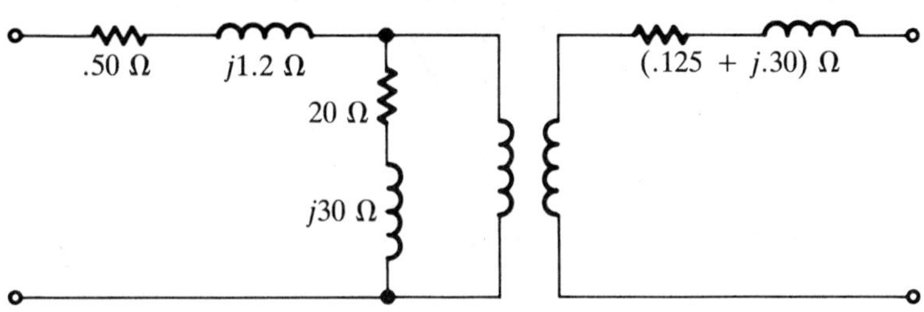

FIG. 2-21

Answer (a) $Z_{e_H} = 1 + j2.4\ \Omega$; (b) $Z_{e_L} = .25 + j.6\ \Omega$; (c) $Z_e = (.022 + j.052)$ pu; (d) $503.52\underline{/1.56°}$ V.

Exercise 2–3 Show that the per-unit value of a transformer's equivalent leakage impedance is equal to the per-unit value of the voltage employed on the short-circuit test.

2.2 Three-Phase, Two-Winding Transformers

2.2.1 INTRODUCTION

Three-phase, two-winding transformers are used to interconnect two distribution systems of different voltages, such as a 25 kV transmission line and a 480 V plant. The majority of industrial transformers are of this type. They are called two-winding transformers because for each primary phase winding there is only one secondary phase winding. As shown in Fig. 2-22, the transformers actually have six coils—three for the incoming three-phase power and three for the output three-phase power.

Three-phase transformers can be obtained either by interconnecting three separate single-phase transformers, as shown in Fig. 2-22, or by winding the six coils on a

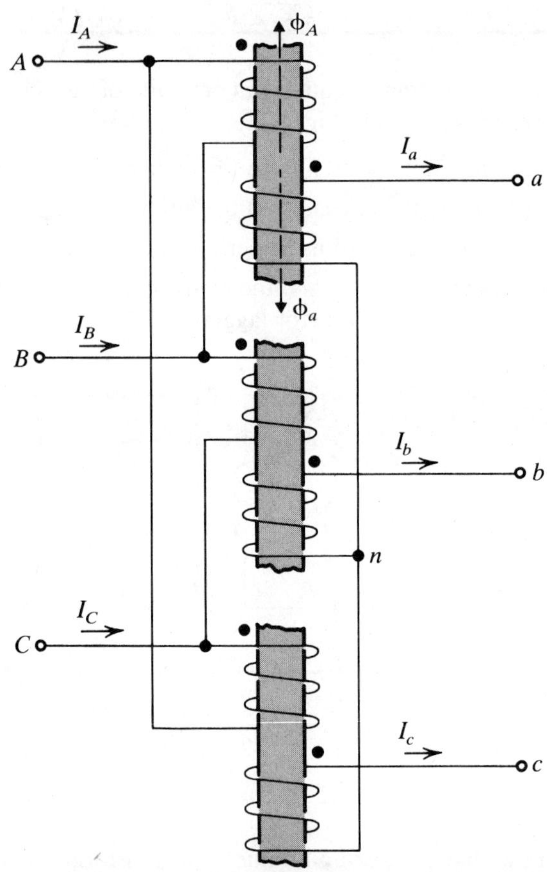

FIG. 2-22 Three single-phase transformers connected Δ-Y to form a two-winding, three-phase transformer bank.

common structure to form a core- or shell-type transformer, shown in Figs. 2-23 and 2-24, respectively.

Most transformers are of the core type. Their main advantage is that they prevent (ideally, if the phase core branches are identical and leakage is neglected) the presence of the third-harmonic flux, and hence avoid inducing third-harmonic voltages. They are also less expensive than shell-type transformers of equivalent rating.

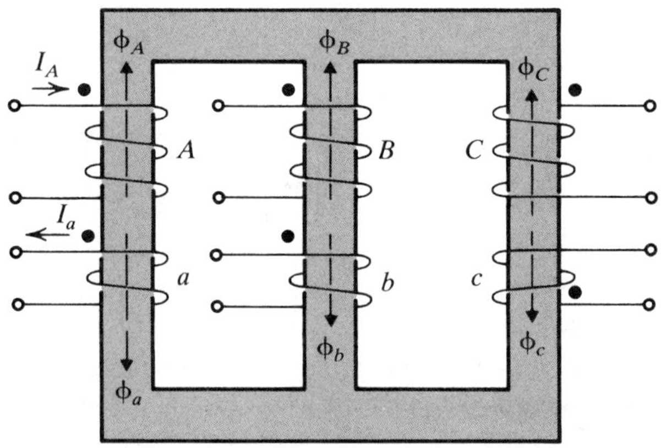

FIG. 2-23 Core-type 3-ϕ, two-winding transformer.

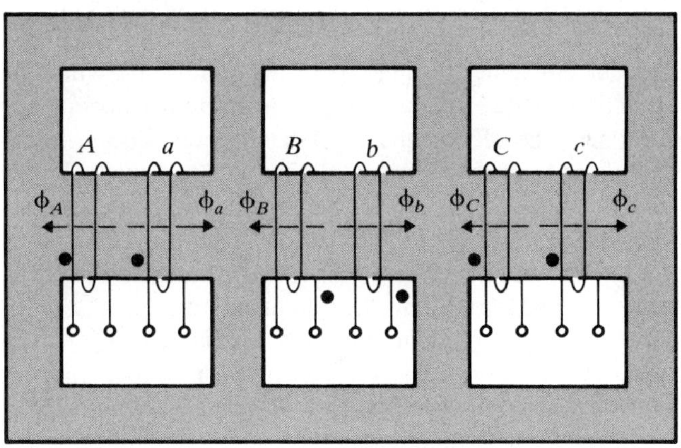

FIG. 2-24 Shell-type 3-ϕ, two-winding transformer.

2.2.2 ONE-LINE DIAGRAM

Three-phase transformer coils can be connected in either star (Y) or delta (Δ) configurations. Figure 2-25(a) shows the industry's standard terminal identification for two-winding, three-phase transformers. The letters H and X represent high and low voltage, respectively. The subscripts 1, 2, and 3 correspond to phases 1, 2, and 3, or

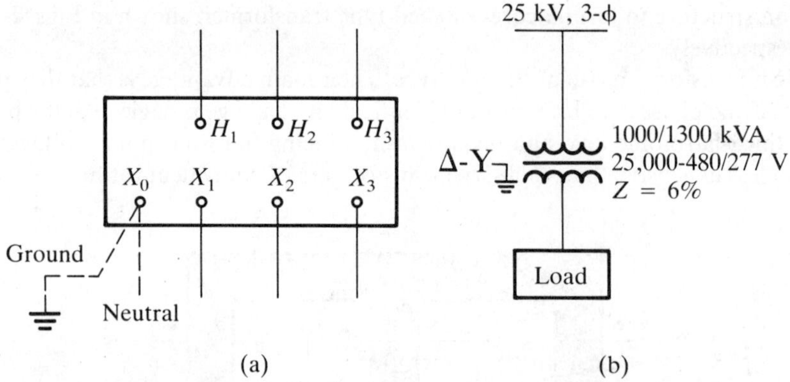

FIG. 2-25 Three-phase, two-winding transformers: **(a)** terminal identification, **(b)** one-line diagram representation of a 1000/1300 kVA, 25000–480/277 V.

phase rotation ABC.* The subscript 0 is used to designate the grounded terminal of a star-connected transformer. The neutral conductor of the system is also connected to this point when a four-wire distribution system is required. A star-star connected three-phase, two-winding transformer has, on the HV and LV windings, four terminals to which as many as five conductors can be connected.

The three conductors correspond to the 3-ϕ power cables, the fourth conductor constitutes the neutral of the system, and the fifth conductor is the ground of the transformer tank. The neutral is required when single-phase loads are to be supplied with line-to-neutral voltage. The ground and the neutral conductors are shown dotted in order to differentiate them from the other three main cables. The neutral is an insulated cable and carries current under normal operating conditions, while the ground conductor is usually not insulated and carries no current under normal operating conditions. However, when one of the phases of a grounded equipment is accidentally grounded, the resulting high short-circuit current will flow through the ground cable.

In the one-line diagram (Fig. 2-25(b)) the type of winding connections (Δ or Y) is shown. The first symbol indicates the primary or incoming terminal connection, the second the outgoing or secondary terminal coil connection. The kVA, voltage, and impedance ratings are also marked on the one-line diagram.

When the transformer is manufactured with special insulation, or equipped with fans, it can transform higher kVA without having its life expectancy reduced. Transformers equipped with one set of ventilating fans cost about 5% more but can deliver 33% more current at nominal voltages.

The per-unit impedance shown in the one-line diagram is calculated using the base kVA power of the transformer.

. . .

* In practice the phase rotation ABC is often designated as BRB. The latter expression identifies the black, red, and blue colors of the actual cables. When the ground wire is insulated, its color is green. The insulation of the neutral wire is normally white.

2.2.3 MATHEMATICAL ANALYSIS

To understand three-phase transformers, one needs a knowledge of the per-unit system, of the principles of operation of single-phase transformers, and of the three-phase systems. Although three-phase systems are extensively discussed in Appendix A, they are briefly reviewed here for convenience.

Power Considerations

The active (P), reactive (Q), and complex power (S) drawn by a balanced three-phase load are given by

$$P = \sqrt{3}\, V_{L-L} I_L \cos\theta$$

$$Q = \sqrt{3}\, V_{L-L} I_L \sin\theta$$

$$S = \sqrt{3}\, V_{L-L} I_L{}^*$$

where V_{L-L} is the rms value of the line-to-line voltage, I_L is the rms value of the line current, and $I_L{}^*$ is its conjugate in phasor form. θ is the phase angle of the load impedance.

When analyzing a three-phase network, it is convenient to draw its power triangle. The real power is taken as reference, and the reactive power is 90 degrees out of phase with it. When the current lags the voltage, the power triangle is drawn leading. In other words, a lagging-power-factor load (which is characteristic of most industrial loads) draws positive VAR, and a leading-power-factor load (capacitive) draws negative VAR (references occasionally differ from this convention).

Current and Voltage Considerations

Depending on the type (Δ, Y) of three-phase connection, the phase and line parameters of a three-phase load are, as shown below, interrelated.

Delta-Connected Load

In a delta-connected load, the line and phase voltages are equal, while the line current lags[†] the phase current by 30° and is larger by a factor of $\sqrt{3}$. Thus,

$$V_{L-L} = V_{p-p} \tag{2.47}$$

$$I_L = \sqrt{3}\, I_p \underline{/-30^\circ} \tag{2.48}$$

where the subscript p stands for the phase parameters.

· · ·

[†] The phase shift between the line and phase parameters depends on the order of rotation of the supply voltages (see Appendix A, Section A-2).

Star-Connected Load

In a star-connected load, the line and phase currents are the same, while the line voltage leads† the phase voltage by 30° and is larger by a factor of $\sqrt{3}$. Thus,

$$I_L = I_p \tag{2.49}$$

$$V_{L-L} = \sqrt{3}\, V_{L-N} \underline{/30^\circ} \tag{2.50}$$

Equations (2.48) and (2.50) are derived from a voltage phase sequence *ABC*.

The analysis of three-phase, two-winding transformers is simplified when their per-phase equivalent electric or magnetic circuits are used. The equivalent circuit of one phase is drawn, and the computations are identical to those used for single-phase transformers. Then the standard three-phase system relationships are employed to find the overall transformer parameters, such as currents, voltages, or power.

Phase Shift between Primary and Secondary Voltages

The phase shift, or the angular displacement, between the primary and the corresponding secondary voltage depends on the type of transformer connection. In a delta-delta or a star-star transformer, the angular displacement is zero; in a delta-star or a star-delta transformer, it is 30 degrees. According to the ASA,* the voltages on the HV side lead the corresponding voltages on the LV side by 30 degrees, regardless of which side is star or delta.

Magnetic Circuit Analysis

The magnetic circuit approach to the solution of three-phase transformer problems has the advantage of giving a physical insight into what is happening within the transformer. It can also be of great assistance in computations involving the unbalanced operation of transformers.

A per-phase magnetic equivalent circuit is shown in Fig. 2-26(c). Neglecting the effects of the exciting current, and applying KVL, we obtain

$$N_A I_A = N_a I_a \tag{2.51}$$

Also, from KCL, we get

$$\phi_A = \phi_a \tag{2.52}$$

As before, the upper-case subscripts represent the parameter of the primary side, while the lower-case subscripts represent the parameter of the secondary side.

. . .

† The phase shift between the line and phase parameters depends on the order of rotation of the supply voltages (see Appendix A, Section A-2)

* American Standards Association, C57.

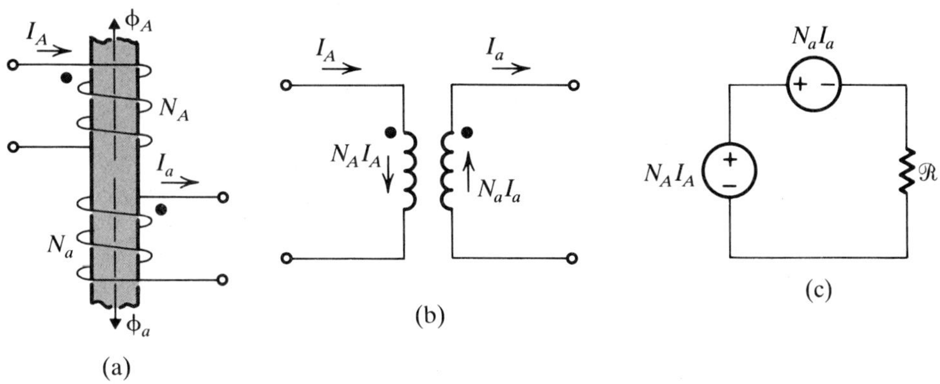

FIG. 2-26 Schematic for one phase of a two-winding, three-phase transformer: **(a)** polarity of terminals and direction of flux, **(b)** relative direction of mmf's, **(c)** magnetic equivalent circuit per-phase.

In analyzing problems, it is helpful if the coupled coils, with their actual magnetic volts, are drawn as shown in Fig. 2-26(b).

Electric Circuit Analysis

When electric circuit analysis is used to solve three-phase transformer problems, the following procedure is recommended:

1. Draw the per-phase electrical equivalent circuit.
2. Determine the turns ratio. This is given by the ratio of the phase to neutral voltages of the primary and secondary windings. This ratio is established by taking into consideration the delta- or star-type winding connection.
3. Draw the equivalent star-star connection of the given transformers for delta-star (Δ-Y) or delta-delta (Δ-Δ) type of winding connection. This is easily accomplished by using the standard delta-star equivalent impedances. That is,

$$Z_y = \frac{Z_\Delta}{3} \tag{2.53}$$

   where Z_Δ and Z_y represent, respectively, the equivalent impedances connected delta and star. The above relationship is applicable only when the per-phase winding impedances are identical. When the winding impedances are unequal, changing a delta-connected load to its equivalent star is accomplished by using Eq. (A.6), found in Appendix A.
4. Select the reference voltage from the given data of the problem. Recall that when the transformer is connected delta-star or star-delta, the HV line-to-line voltages lead the corresponding LV line-to-line voltages by 30 degrees.

EXAMPLE 2–5 For the three-phase transformer shown in the one-line diagram of Fig. 2-27(a), determine:

a) The ohmic value of its reactance referred to the low- and high-voltage windings.
b) The magnitude of the short-circuit current in the transformer windings when a three-phase short takes place across the transformer's secondary terminals.

SOLUTION

a) The base parameters for the per-unit calculations are obtained from the rating of the transformer:

$$S_b = 1000 \text{ kVA}$$

$$V_{b_H} = 25 \text{ kV}, \qquad V_{b_L} = 480 \text{ V}$$

$$Z_{b_H} = \frac{(25.0)^2}{1.0} = 625 \ \Omega/\text{phase}, \qquad Z_{b_L} = \frac{(.48)^2}{1.0} = .23 \ \Omega/\text{phase}$$

Z_{b_H} is the equivalent star-connected impedance in Ω/phase. The base currents are

$$I_{b_H} = \frac{1000}{\sqrt{3} \times 25} = 23.09 \text{ A}, \qquad I_{b_L} = \frac{1000}{\sqrt{3}(.48)} = 1202.81 \text{ A}$$

Then the ohmic values of the reactances are:

HV Winding:

$$X_H = .05(625) = \underline{\underline{31.25 \ \Omega/\text{phase, star equivalent}}}$$

LV Winding:

$$X_L = .05(.23) = \underline{\underline{.012 \ \Omega/\text{phase, star equivalent}}}$$

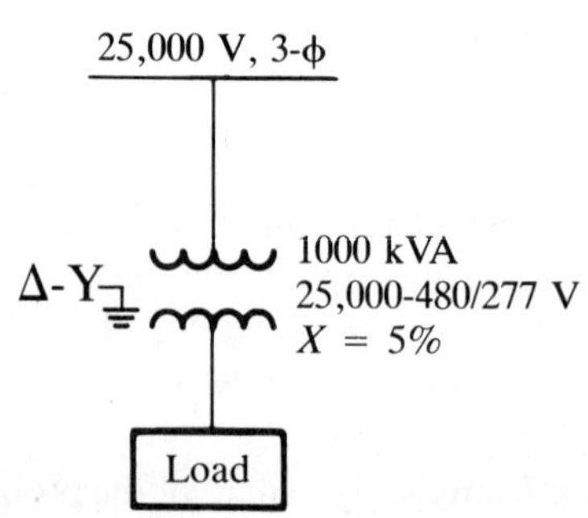

FIG. 2-27(a)

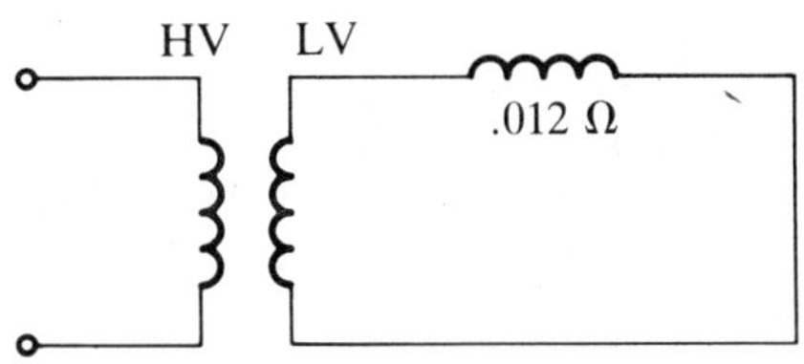

FIG. 2-27(b)

b) The short-circuit current can be found by using any of the following methods.

LV Winding:

(1) Applying Ohm's law in the circuit of Fig. 2-27(b), the magnitude of the current in the LV winding is

$$I = \frac{V}{X} = \frac{480/\sqrt{3}}{.012} = \underline{\underline{24{,}056.26 \ \text{A}}}$$

(2) Using per-unit values:

$$I = \frac{1.0}{.05} = 20 \ \text{pu}$$
$$= 20(1202.81) = \underline{\underline{24{,}056.26 \ \text{A}}}$$

H.V. Side

(1) Applying Ohm's law in the star equivalent circuit of Fig. 2-27(c), we obtain the magnitude of the short-circuit current in the HV windings:

$$I = \frac{V}{X} = \frac{25{,}000/\sqrt{3}}{31.25} = 461.88 \ \text{A line current}$$

The magnitude of the phase current is

$$I_p = 461.88/\sqrt{3} = \underline{\underline{266.67 \ \text{A}}}$$

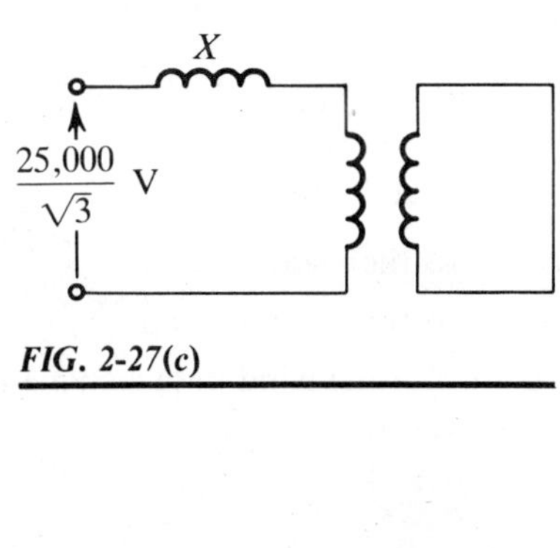

FIG. 2-27(c)

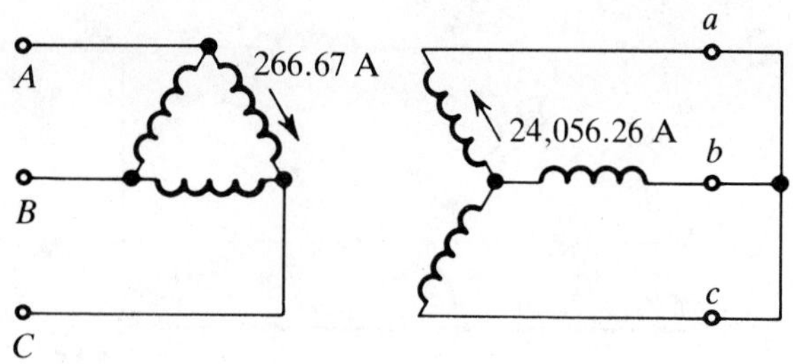

FIG. 2-27(d)

(2) Using Ohm's law, for the delta-connected equivalent circuit, we obtain

$$X_\Delta = 3X_y = 3(31.25) = 93.75\ \Omega$$

$$I_p = \frac{25{,}000}{93.75} = 266.67\ \text{A}$$

and

$$I_L = \sqrt{3}(266.67) = 461.88\ \text{A}$$

(3) Refer to Fig. 2-27(d). By using the current transformation property of transformers, we obtain

$$I_p = 24{,}056.26\,\frac{480/\sqrt{3}}{25{,}000} = \underline{\underline{266.67\ \text{A}}}$$

and

$$I_L = \sqrt{3}\,I_p = 461.88\ \text{A}$$

EXAMPLE 2–6 Three identical single-phase transformers, each of 500 kVA, 4160–480 V, 60 Hz, are connected Δ-Y to form a 3-ϕ transformer bank. Data for the short-circuit test on one of the single-phase transformers is as follows:

Low voltage winding shorted:

$$V_H = 250\ \text{V}, \qquad I_H = 120.19\ \text{A}, \qquad P = 10\ \text{kW},$$

The transformer, as shown in Fig. 2-28(a), is connected to an upstream substation of 4160 volts through a feeder whose impedance is (.2 + j.4) ohms per-phase. When the transformer delivers rated current at .9 power factor lagging, determine:

a) The line-to-line voltage across the load.
b) The power supplied by the upstream substation.
c) The current through the transformer windings when a 3-ϕ short takes place at the secondary terminals of the transformer.

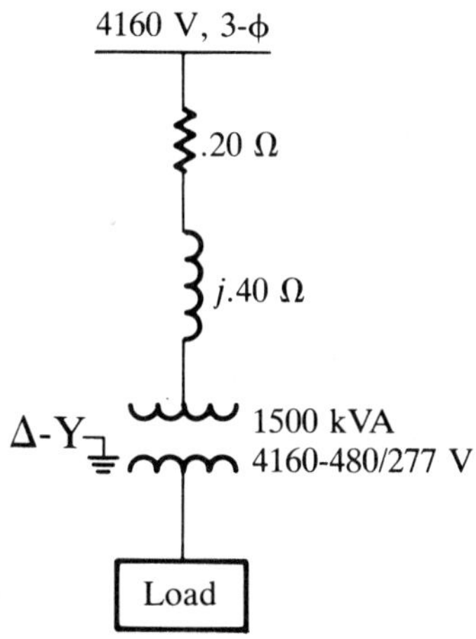

FIG. 2-28(a)

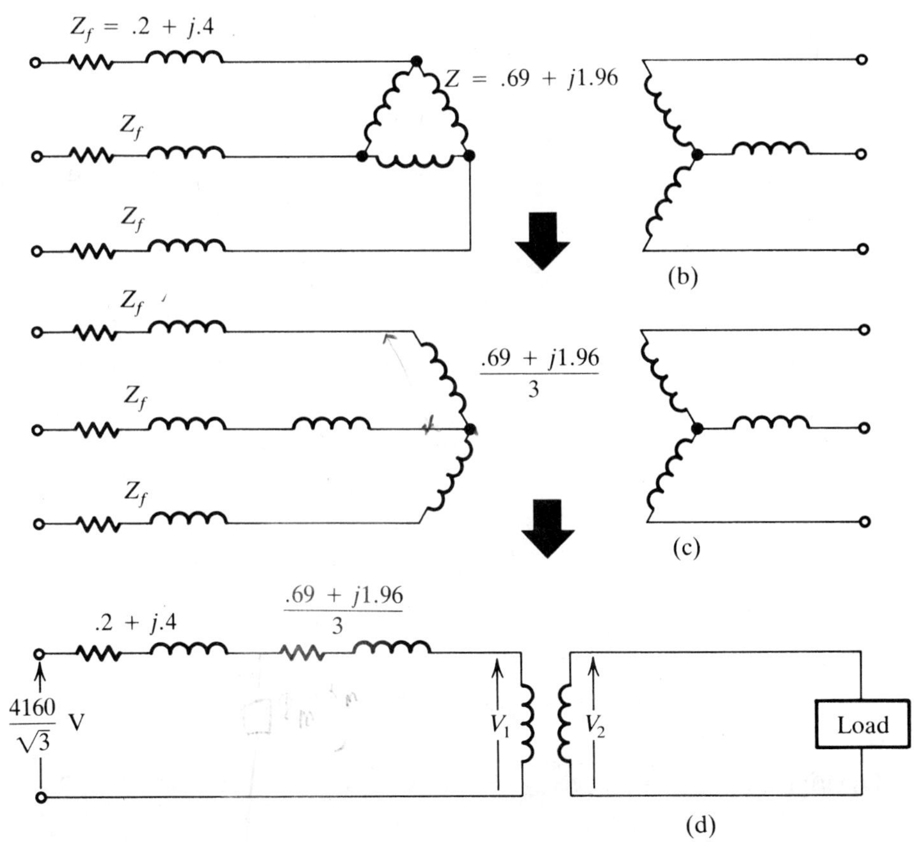

FIG. 2-28(b), (c), and (d)

SOLUTION

a) From the short-circuit test data, we get

$$|Z_{eH}| = \frac{250}{120.19} = 2.08 \ \Omega/\text{phase}$$

$$R_{eH} = \frac{10{,}000}{(120.19)^2} = .69 \ \Omega/\text{phase}$$

$$X_{eH} = \sqrt{2.08^2 - .69^2} = 1.96 \ \Omega/\text{phase}$$

Thus,

$$Z_{eH} = .69 + j1.96$$
$$= 2.08 \angle 70.6^\circ \ \Omega/\text{phase}$$

This is the per-phase impedance referred to the HV winding, which is delta-connected. In order to simplify the calculations the primary, as shown in Fig. 2-28(c), is changed to an equivalent star. Then the per-phase equivalent circuit, as seen from the HV winding, is shown in Fig. 2-28(d).

The magnitude of the line current in the primary of the Δ-Y transformer is

$$I = \frac{1500}{\sqrt{3}\,4.160} = 208.18 \text{ A}$$

Refer to Fig. 2-28(d). The total impedance in the primary circuit is

$$Z = .2 + j.4 + \frac{.69 + j1.96}{3} = 1.14 \angle 67.8^\circ \ \Omega/\text{phase}$$

Applying KVL to the per-phase equivalent circuit of Fig. 2-28(d), we obtain

$$V_1 \angle 0^\circ = \frac{4160}{\sqrt{3}} \angle \beta_1^\circ - 208.18 \angle -25.8^\circ \, 1.14 \angle 67.8^\circ$$
$$= \frac{4160}{\sqrt{3}} \angle \beta_1^\circ - 237 \angle 41.9^\circ$$

Equating the imaginary parts, we obtain

$$\beta_1 = \arcsin \frac{\sqrt{3}}{4160} (237 \sin 41.9^\circ) = 3.8^\circ$$

Equating the real parts, we get

$$V_1 = \frac{4160}{\sqrt{3}} \cos 3.8^\circ - 237(\cos 41.9^\circ) = 2220.21 \text{ V/phase}$$

The voltage across the secondary of the transformer is

$$V_2 = V_1\left(\frac{480}{4160}\right) = 2220.21\,\frac{480}{4160} = 256.18\ \text{V/phase}$$

$$= (256.18)\sqrt{3} = \underline{\underline{443.71\ \text{V, L–L}}}$$

b) The power supplied is

$$P = \sqrt{3}\,V_{L-L}I_L\cos\theta$$

$$= \sqrt{3}(4160)(208.18)\cos(3.8^\circ + 25.8^\circ)$$

$$= 1303.96\ \text{kW}$$

c) The magnitude of the short-circuit current through the feeder is

$$I_L = \left|\frac{V}{Z}\right| = \frac{4160}{\sqrt{3}(1.14)} = 2109.60\ \text{A}$$

The current in the Δ-connected windings is

$$I_P = \frac{I_L}{\sqrt{3}} = \frac{2109.60}{\sqrt{3}} = \underline{\underline{1218\ \text{A}}}$$

The current in the Y-connected windings is

$$I_s = (1218)\frac{4160}{480/\sqrt{3}} = \underline{\underline{18{,}283.20\ \text{A}}}$$

Alternatively, using per-unit values,

$$Z_{b_H} = \frac{(4.160)^2}{1.5} = 11.54\ \Omega/\text{phase}$$

Thus,

$$Z_{e_{pu}} = \frac{1.14}{11.54} = .099\ \text{pu}$$

$$I_{sc} = \left(\frac{1}{.099}\right) = 10.13\ \text{pu}$$

In the LV side:

$$I_{sc} = 10.13\frac{1500}{\sqrt{3}(.480)} = \underline{\underline{18283.20\ \text{A}}}$$

In the HV side:

$$I_{sc} = 10.13\frac{1500}{\sqrt{3}(4.16)} = \underline{\underline{2109.60\ \text{A}}}$$

Exercise 2–4 The three-phase, three-winding transformer shown in Fig. 2-29 delivers 25 A to a single-phase load. The tertiary winding (T) is shorted, and the coils have an equal number of turns.

a) Mark the polarity of each coil and show the direction of its flux.

b) Determine the current of each winding. (*Hint:* The net flux through each leg must be equal to zero.)

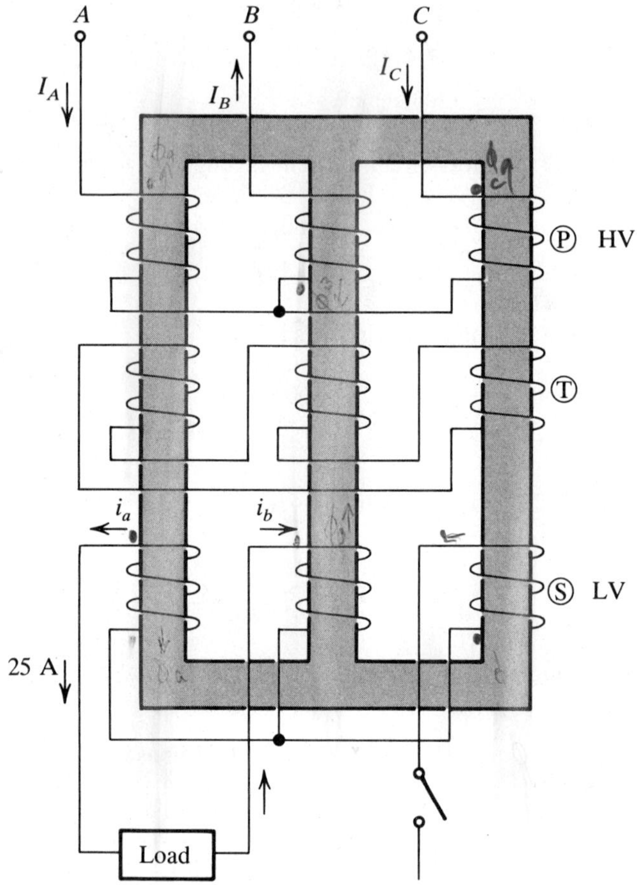

FIG. 2-29

Answer $I_T = 0$, $i_A = i_B = 25$ A, $i_C = 0$.

Exercise 2–5 The three-phase transformer shown in the one-line diagram of Fig. 2-30 is connected to a 4.16 kV source. The magnetizing current is negligible.

1. If the transformer delivers its base kVA at rated voltage, determine:
   a) The primary and secondary line and phase currents.

b) The ohmic values of the leakage reactance referred to the HV and LV windings.

c) The primary and secondary line and phase currents when the secondary terminals of the transformer are shorted.

2. Repeat (a), (b) and (c) when the transformer operates with the fans and delivers 2000 kVA at rated voltage.

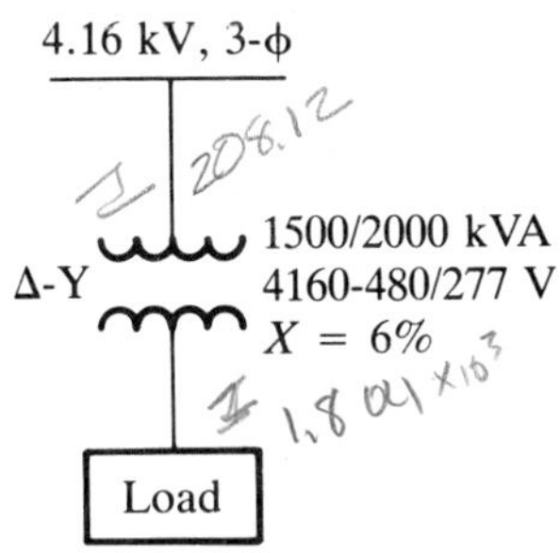

FIG. 2-30

Answer 1. (a) 1804.2 A, 208.18 A, 120.2 A
(b) 2.08 Ω, .0092 Ω/ph
(c) 30070.3 A, 2003.2 A, 3469.6 A
2. (a) 2405.6 A, 277.6 A, 160.3 A
(b) and (c) no change

2.2.4 HARMONICS OF THE EXCITING CURRENT AND WINDING CONNECTIONS

This section discusses the harmonics of the exciting current and the advantages and disadvantages of the various transformer winding connections.

Harmonics of the Exciting Current

Consider the exciting current of one phase only. Neglecting its fourth and higher-order harmonics, we have

$$\begin{aligned} i_{\text{exc}} &= i_1 + i_2 + i_3 \\ &= I_{m_1} \cos \omega t + I_{m_2} \cos 2\omega t + I_{m_3} \cos 3\omega t \end{aligned} \qquad \textbf{(2.54)}$$

where I_{m_1}, I_{m_2}, and I_{m_3} are the maximum values of the fundamental, second, and third harmonics.

Second Harmonics

Designating I_{m_2} as the maximum value of the second harmonic in one phase, the vectorial sum for all three phases is

$$i_2 = I_{m_2} \cos 2\omega t + I_{m_2} \cos 2(\omega t + 120°) + I_{m_2} \cos 2(\omega t + 240°) \qquad \textbf{(2.55)}$$

From the above:

$$i_2 = 0$$

Thus, the sum of the second, or even, harmonics of a three-phase system is equal to zero, regardless of the transformer winding connections.

Third Harmonics

The third-harmonic components of the exciting current in the lines of a three-phase system do not add up to zero, as in the case of the fundamental and second harmonics. Mathematically:

$$i_3 = I_{m_3} \cos 3\omega t + I_{m_3} \cos 3(\omega t + 120°) + I_{m_3} \cos 3(\omega t + 240°) \qquad \textbf{(2.56)}$$

From the above:

$$i_3 = 3I_{m_3} \cos 3\omega t \qquad \textbf{(2.57)}$$

When third harmonics are permitted to flow, the exciting current is nonsinusoidal in waveform. This waveform—as can be verified by the *B-H* characteristic of high-permeability magnetic materials—is accompanied by sinusoidal flux. Sinusoidal flux produces sinusoidal secondary voltages, and nonsinusoidal flux produces nonsinusoidal secondary voltages. The latter are undesirable becuse they may cause the electrical apparatus—motors, computers, etc.—to malfunction. Thus, it is desirable that the waveform of the exciting current be nonsinusoidal.

The nonsinusoidal waveform of the exciting current is ensured when the transformer connections permit the flow of third harmonics. Star-connected windings with a grounded neutral provide a physical link so the harmonics can flow from the system's lines down to ground, then eventually return to the generator's grounded neutral. Therefore, star-grounded windings permit the flow of nonsinusoidal excitation currents. Thus, their secondary phase voltages are sinusoidal. Delta-connected windings also produce sinusoidal fluxes and voltages, because they permit the third harmonics to circulate within their windings.

In contrast, ungrounded star-connected windings do not permit the flow of third harmonics of exciting currents, and thus the induced voltages are nonsinusoidal.

Three-Phase Winding Connections

Ungrounded Star-Star Transformer

An ungrounded Y-Y transformer is shown in Fig. 2-31. When the transformer's load is slightly unbalanced, the neutral points (*n*) may not be at a fixed potential with respect to ground. This is referred to as the "floating" neutral. The potentially unstable neutral is due mainly to unbalanced load currents.

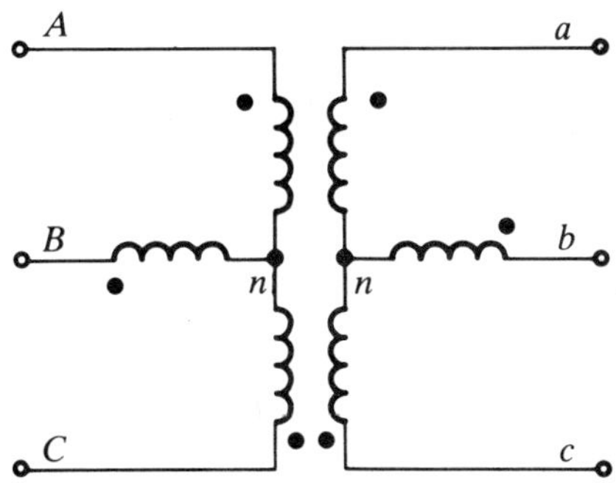

FIG. 2-31 Ungrounded star-star (Y-Y) connected transformer, supplying unbalanced loads.

Considering Fig. 2-31, from KVL in the primary side we have

$$V_{Cn} + V_{nA} + V_{AC} = 0 \tag{2.58}$$

or

$$V_{AC} = V_{nC} + V_{An} \tag{2.59}$$

The potential V_{AC} is constant, owing to the primary voltage source. The other voltages (V_{nC} and V_{An}) will vary in a way that satisfies the above relationship.

Owing to unequal load impedances, the secondary line currents will be different for each phase, and thus the corresponding primary phase currents and voltages will be different for each phase. In other words, the potentials V_{nC} and V_{An} will vary and thus the neutral point will be at unstable, or floating, potential with respect to ground. Since the safety of personnel and the normal operation of equipment require a neutral with a fixed potential, the ungrounded star-star winding connection has very few applications.

To stabilize the potential of the neutral point a delta-connected winding—the so-called tertiary—is wound on the core of the transformer. The transformer, in this case, becomes a three-phase, three-winding transformer. Such transformers have high MVA ratings and are normally used by the utilities.

Delta-Star Transformer

A Δ-Y transformer is shown in Fig. 2-32(a). This type of winding connection is very popular because of the following advantages:

1. A neutral cable can be connected to the grounded terminal, and thus a 3-ϕ, 4-wire system that satisfies the electrical requirements of most industries is established.
2. Relatively speaking, the Δ-Y transformer maintains sinusoidal line-to-line waveforms in the primary and secondary windings.

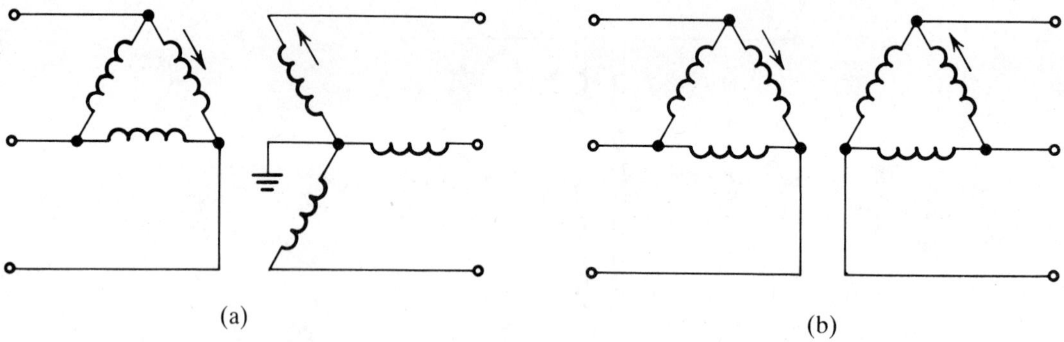

FIG. 2-32 Transformer winding connections: **(a)** delta-star, **(b)** delta-delta.

3. The Δ-connected winding provides a short circuit to the flow of third harmonics. The zero-sequence* current components produced by the unbalanced loads and the third harmonics of the nonsinusoidal magnetizing current are in phase for each winding and can circulate within the Δ-connected circuit.
4. The star-connected winding is subjected to $1/\sqrt{3}$ of the line-to-line voltage, and thus reduced insulation is required for the transformer windings.

One *disadvantage* of the Δ-Y transformer is that it cannot be paralleled with a Y-Y transformer. The latter's line-to-line voltages are in phase in the high- and low-voltage sides, while the corresponding Δ-Y line-to-line voltages between primary and secondary are 30° apart. This phase difference is undesirable for parallel operation. It results in large circulating currents, which reduce the kVA output capability of the paralleled transformers.

Delta-Delta Winding Connection

A Δ-Δ winding connection is shown in Fig. 2-32(b). This type of transformer has the following advantages:

1. When a line-to-ground fault—the most common type of fault—occurs in the secondary distribution of the transformer, the resulting current is very small and therefore the power flow is not interrupted. This is of tremendous importance to industries whose processes—such as melting of metals—cannot be interrupted.
2. This type of transformer connection maintains sinusoidal voltages at its output terminals. This is because the delta-delta winding connection provides a path for the flow of the third-harmonic component of the magnetizing current and for the circulation of the zero-sequence component of the unbalanced load currents.

• • •

* The zero-sequence currents are due to unbalanced supply voltages and to nonidentical load impedances. They are similar to third-harmonic currents. For more information see Reference 14.

3. When three single-phase transformers are connected in Δ-Δ to form a 3-ϕ transformer bank, one of the single-phase transformers can be removed and the resulting open-delta or VEE transformer bank can deliver about 58% of its original rating to a three-phase secondary load.

EXAMPLE 2–7 Show that, when a single-phase load is connected from one line to neutral in the secondary of an ungrounded Y-Y, 3-ϕ transformer, the voltage across the loaded phase is reduced while the other voltages from line-to-neutral will be increased.

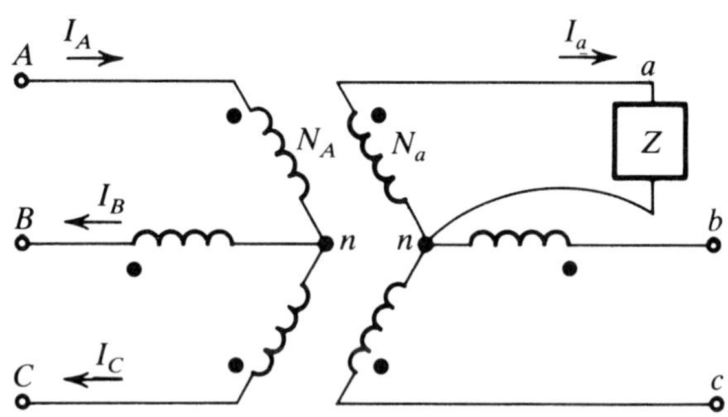

FIG. 2-33

SOLUTION Assume a load is connected from line "a" to neutral, as shown in Fig. 2-33. The load impedance will draw a current I_a because of the voltage induced in the secondary winding:

$$I_a = \frac{V_a}{Z}$$

Having a current on the secondary means there must also be a current I_A in the primary, because of the current-transformation property of transformers, namely:

$$N_A I_A = N_a I_a$$

Applying KCL, in the primary winding we have

$$I_A = I_B + I_C$$

I_A exists, and, since $I_b = I_c = 0$, then I_B and I_C must represent increases in the exciting currents in phases B and C, respectively. Thus their corresponding voltages V_{Bn} and V_{Cn} must also be *increased*. This leads to the reduction of V_{AB} ($V_{AB} = V_{nB} + V_{An}$) and V_{ab}.

Exercise 2–6 Compare and contrast the Δ-Δ and Δ-Y type of three-phase, two-winding transformers. Consider: insulation levels, exciting currents, and output voltage waveforms.

2.3 Autotransformers

Figure 2-34(a) shows the schematic representation of a 10 kVA, 200–100 V, single-phase transformer. Its autotransformer type connection can be obtained as shown in Figs. 2-34(b) and (c), by a special interconnection of its primary and secondary windings.

Autotransformers are smaller in size and more economical than two-winding transformers of the same rating. They are normally used for voltage-transformation ratios close to 1:1 and in variacs, which provide a variable secondary voltage.

Understanding and analyzing autotransformers is simplified by noting the following:

1. The ampere-turns of each coil are the same whether the transformer is connected as an autotransformer or as a two-winding transformer. That is,

$$(NI) = \text{constant} \tag{2.60}$$

   In other words, the magnitude of the current through each coil remains the same, regardless of whether the transformer operates as an autotransformer or as a conventional two-winding transformer.

   Referring to Fig. 2-34, the ampere-turns of each coil are the same in each of the three situations shown, but the complex power transferred from the supply to the load differs from one case to the next.

2. Since the current through each autotransformer winding is the same as in the conventional two-winding transformer, the winding loss remains the same. However, the efficiency of the autotransformer is increased if the output power is increased.

3. Neglecting losses, the complex power (kVA) at the input is equal to the complex power delivered to the output:

$$(\text{kVA})_{\text{input}} = (\text{kVA})_{\text{output}} \tag{2.61}$$

   The kVA transformation capability of the autotransformer is the same as that of the two-winding transformer. An autotransformer, however, delivers higher kVA than the conventional transformer because of the direct electrical connection between the primary and secondary windings. In other words, part of the output kVA is conducted from the primary to the secondary winding. The conducted kVA are referred to as untransformed kVA.

4. The autotransformer has a common terminal between the primary and the secondary windings, and when the secondary is shorted, the voltage applied to primary is much higher than its rated voltage. This results in higher short-circuit currents and thus in more expensive protective devices.

5. The two-winding conventional transformer has its primary and secondary circuits electrically isolated, while in the autotransformer electrical disturbances in the primary can be easily passed to the secondary through their direct electrical connection.

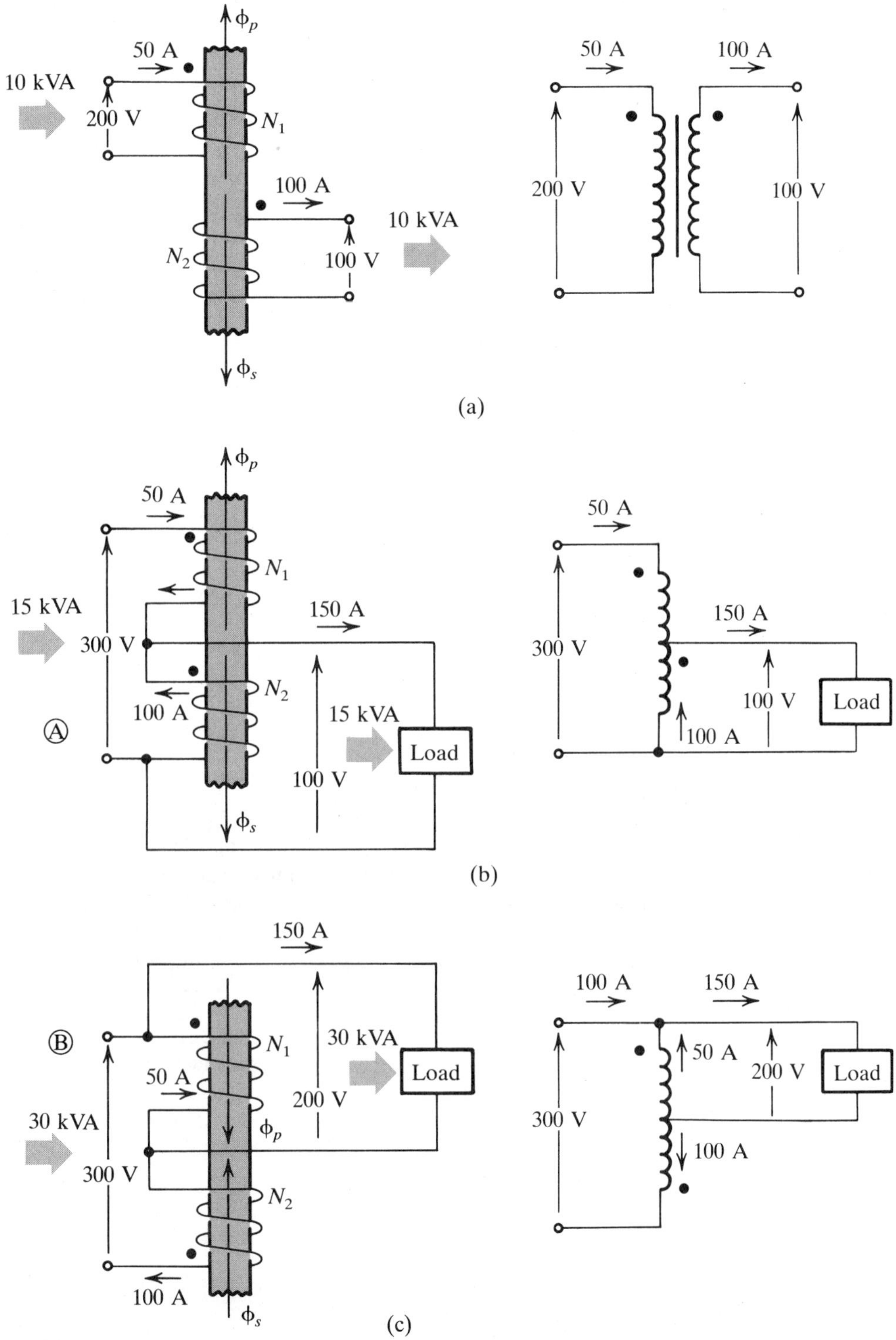

FIG. 2-34 Transformer representations and their schematics: **(a)** a two-winding transformer, **(b)** type A autotransformer connection, **(c)** type B autotransformer connection.

TABLE 2-1 Parameters of Two-Winding Transformers and Autotransformers

Parameter	*Conventional Two-Winding Transformer*	*Autotransformer Connection Type A: Load Across the Same Winding as in the Two-Winding Transformer*	*Autotransformer Connection Type B: Load Connected Across the Primary Winding of the Two-Winding Transformer*
Turns ratio	$a = \frac{N_1}{N_2}$	$a + 1$	$\frac{a+1}{a}$
Winding voltage	V	V	V
Winding current	I	I	I
Efficiency	High	Higher	Much higher
Input short-circuit current	I_{sc}	$\frac{(1+a)}{a} I_{sc}$	$a(a+1)I_{sc}$
kVA transformed	S	S	S
kVA untransformed (conducted)	0	$\frac{S}{a}$	aS
Total kVA output	S	$S\frac{(a+1)}{a}$	$S(a+1)$

A two-winding transformer, as shown in Fig. 2-34, can be connected in two ways to supply a load as an autotransformer. The two different connections are identified as A and B. For reasons of comparison, the highlights of type A and B connections are summarized in Table 2-1. In both cases, the output voltage is smaller than the input voltage. However, when one of the voltage windings is reversed, the output voltage will be larger than the input voltage.

EXAMPLE 2–8 A 480–120 V, 20 kVA, two-winding transformer with a leakage impedance of $(.02 + j.05)$ per-unit has an efficiency of .96 when it delivers rated current at .90 power factor lagging. This transformer is to be connected as an autotransformer to a 600 V source to supply a load at either 480 or 120 V. For each *type* of autotransformer connection determine:

a) The kVA rating of the autotransformer. What percent of this kVA passes through the autotransformer untransformed?
b) The efficiency at full-load and .90 power factor lagging.
c) The input current when the load is shorted. Compare the results with those of the two-winding operation.

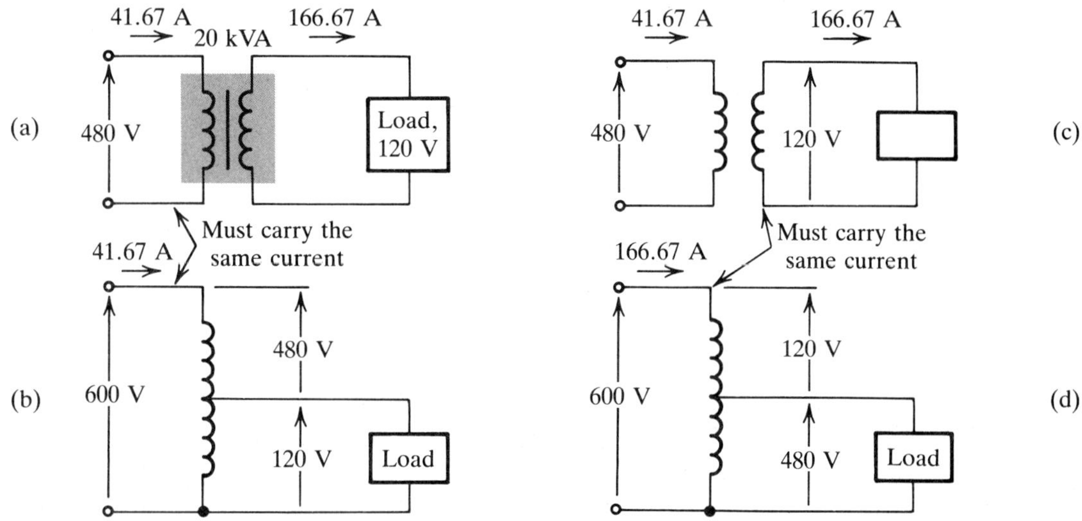

FIG. 2-35(a) and (b): Two-winding and autotransformer connection (type A).
(c) and (d): Two-winding and autotransformer connection (type B).

SOLUTION This problem compares the essential parameters of a two-winding transformer with those of its autotransformer connections.

Type (A) Autotransformer Connection (Load across the 120 V Winding)

a) Referring to Fig. 2-35(b), the magnitude of the current through the input will be the current rating of the 480 V winding. That is,

$$I = \frac{20{,}000}{480} = 41.67 \text{ A}$$

Therefore, the input apparent power to the autotransformer is

$$S = 41.67(600)$$
$$= \underline{\underline{25 \text{ kVA}}}$$

Of these 25 kVA, 5 kVA pass unaltered, or untransformed, from input to output; the remaining 20 kVA are transformed, as in the case of the two-winding transformer connection. Thus,

$$\text{percentage of untransformed kVA} = \frac{5}{25}(100)$$
$$= \underline{\underline{20\%}}$$

b) The losses of the two-winding transformer are

$$\text{losses} = \text{power input} - \text{power output} = \frac{P_{\text{out}}}{\eta} - P_{\text{out}}$$

$$= 20{,}000(.9)\left(\frac{1}{.96} - 1\right) = 750 \text{ W}$$

For the autotransformer connection, the losses remain constant. Thus, the efficiency is

$$\eta = \frac{25(.90)}{25(.90) + .750} = .97$$

c) For the two-winding transformer, the magnitude of the short-circuit current is

$$I_{sc} = \left|\frac{1}{.02 + j.05}\right|$$
$$= 18.57 \text{ pu}$$

For the HV winding, the current in amperes is

$$I_{sc} = 18.57(41.66)$$
$$= \underline{\underline{773.73 \text{ A}}}$$

The leakage impedance, in ohms, is

HV side:

$$Z_{b_H} = \frac{(480)^2}{20{,}000} = 11.52 \ \Omega$$
$$Z_H = (.02 + j.05)(11.52) = .62\angle 68.2^\circ \ \Omega$$

LV side

$$Z_{b_L} = \frac{(120)^2}{20{,}000} = .72 \ \Omega$$
$$Z_L = (.02 + j.05)(.72) = .04\angle 68.2^\circ \ \Omega$$

In the case of the autotransformer, when the 120 V winding is shorted, the source of 600 V is applied across the 480 V winding. Thus, the magnitude of the short-circuit current is

$$I_{sc} = \frac{600}{.62} = \underline{\underline{967.16 \text{ A}}}$$

or, by using the appropriate relationship from Table 2-1,

$$I_{sc} = 773.73\frac{(1+a)}{a} = 773.73\left(\frac{5}{4}\right)$$
$$= \underline{\underline{967.16\ \text{A}}}$$

Type (B) Autotransformer Connection (Load across the 480 V Winding)

a) Referring to Fig. 2-35(d), the current through the 600 V source will be the rated current of the 120 V winding. That is,

$$I = \frac{20 \times 10^3}{120} = 166.67\ \text{A}$$

This current must pass through the input terminals. Therefore, the apparent input power for this type of autotransformer connection is

$$S = 166.67(600)$$
$$= \underline{\underline{100}}\ \text{kVA}$$

Of these 100 kVA, 80 kVA pass unaltered from the input to the output, while the remaining 20 kVA are transformed as in the case of the two-winding connection.

Expressing the untransformed kVA in percentage of the total output kVA, we have

$$(kVA)_{\text{untr.}} = \frac{80}{100}(100) = \underline{\underline{80\%}}$$

b) Since the losses are constant at .75 kW, the efficiency is

$$\eta = \frac{100(.9)}{100(.9) + .75}$$
$$= \underline{\underline{.99}}$$

c) By shorting the load, the 600 V are applied across the 120 V winding. Thus, the magnitude of the short-circuit current is

$$I_{sc} = \frac{600}{.04} = \underline{\underline{15{,}474.6\ \text{A}}}$$

For comparison purposes, the results are summarized in Table 2-2. The student should check the results by using the applicable relationships given in Table 2-1.

TABLE 2-2 Summary of results of Example 2–8

Parameter	Two-Winding Transformer Connection	Autotransformer Connection: Type A. Load Across 120 V	Autotransformer Connection: Type B Load Across 480 V
Output kVA	20	25	100
Efficiency	.96	.97	.99
Short-circuit current through the source in A	773.7	967.16	15,474.6

Exercise 2–7 A single-phase, 1000 VA, 60 Hz autotransformer has four coils, as shown in Fig. 2-36. The voltage rating of the HV windings is 240 V/coil and that of the LV windings is 120/coil. Show the field connections, and determine the maximum VA that can be delivered to the load when:

a) The transformer supplies 120 V from a 240 V source.
b) The transformer supplies 240 V from a 480 V source.
c) Two of these transformers are connected in an open-delta configuration to supply a 3-ϕ, 480 load from a 3-ϕ, 600 V source.

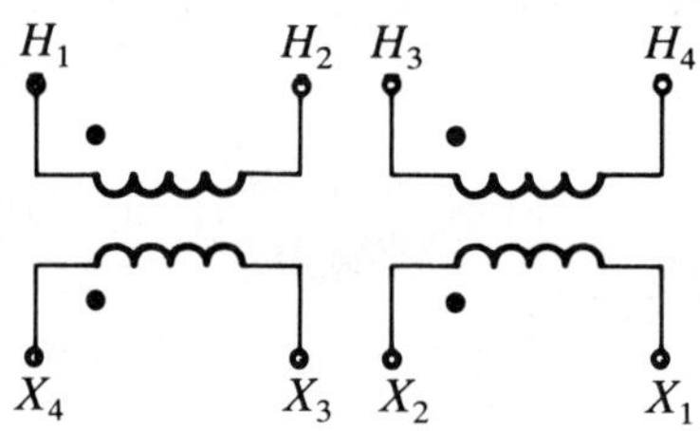

FIG. 2-36

Answer (a) 1000 VA; (b) 1000 VA; (c) 8660 VA

2.4 Parallel Operation of Transformers

Transformers are sometimes paralleled in order to meet the increased kVA demands of a particular load. Transformers that are to be paralleled satisfactorily should satisfy the following conditions:

a) Equal impedances.

b) Equal turns ratio.

c) Equal phase shift between the primary and secondary open-circuited voltages.

d) The same phase rotation.

When these conditions are not satisfied, the transformers—as it is demonstrated below—will be damaged.

The one-line representation of parallel transformers, as well as their corresponding equivalent circuits, are shown in Fig. 2-37. For the analysis of paralleled transformers, the per-phase equivalent circuits of Fig. 2-37(b) and Fig. 2-37(c) can be used.

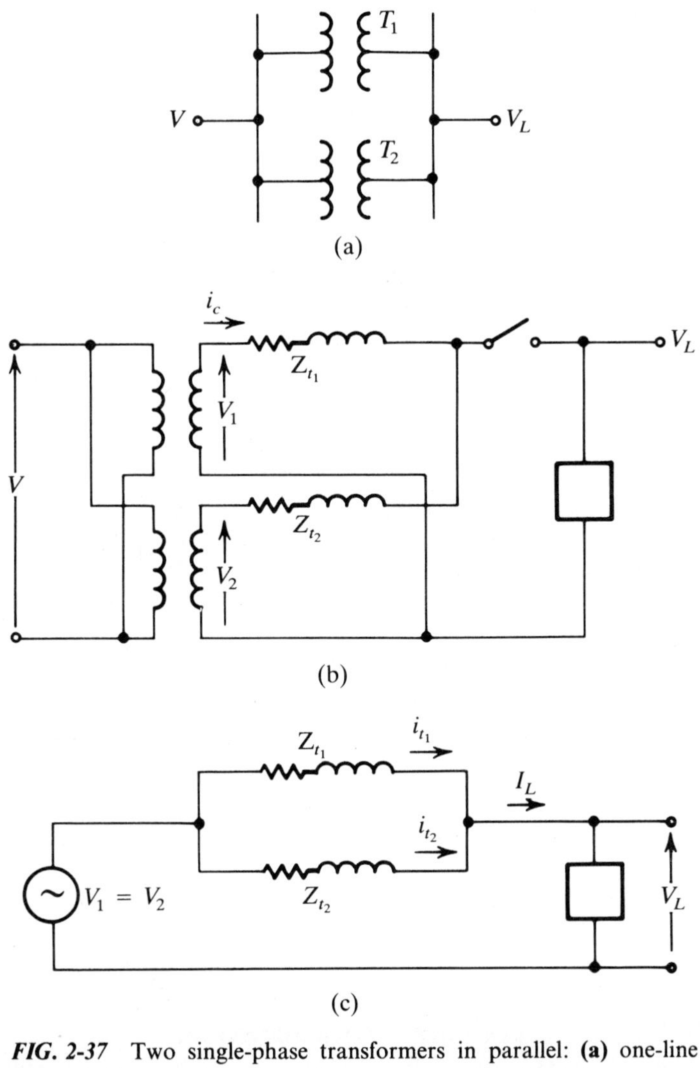

FIG. 2-37 Two single-phase transformers in parallel: **(a)** one-line diagram, **(b)** equivalent circuit for determining the circulating current, **(c)** equivalent circuit for determining the load on each transformer. (The turns ratio for all units are assumed to be equal.)

In Fig. 2-37(b) the load is disconnected from the common secondary transformer terminals, and thus this circuit can be used to find the transformer's common circulating current. When the transformers have an equal number of winding turns, then Fig. 2-37(c) may be used to find each transformer's load as a function of the power delivered to the common load.

Transformers Must Have Equal Leakage Impedances

Assuming equal transformer voltage ratios, then from Fig. 2-37(c), and by applying the current-divider concept, we obtain each transformer's current as a function of the load current I_L:

$$I_{t_1} = I_L \frac{Z_{t_2}}{Z_{t_1} + Z_{t_2}} \tag{2.62}$$

and

$$I_{t_2} = I_L \frac{Z_{t_1}}{Z_{t_1} + Z_{t_2}} \tag{2.63}$$

where Z_{t_1} and Z_{t_2} are the leakage impedances of transformers 1 and 2, respectively.

From the above, when the leakage impedances are the same, each transformer supplies half of the load current requirements. However, when the leakage impedances are unequal, the transformer currents are also unequal. As a result, the transformer that delivers the higher current may be overheated.

Transformers Must Have Equal Turns Ratios

When transformers are connected in parallel, their turns ratios must be equal. Otherwise large no-load circulating currents will flow between the transformers.

From Fig. 2-37(b), the circulating current is

$$I_c = \frac{V_1 - V_2}{Z_{t_1} + Z_{t_2}} \tag{2.64}$$

where V_1 and V_2 are the voltages induced on the secondaries of transformers 1 and 2, respectively.

To minimize the circulating current, the "off-load" voltage tap changers should, if possible, be properly adjusted to equalize the voltages induced in the secondaries of the transformers.

If the transformers supply a load when their turns ratios are different, ordinary circuit-analysis techniques can be used to find the individual transformer currents.

Equal Phase Shift Between the Voltages of Paralleled Units

Besides meeting the two previous conditions, three-phase units must also produce secondary line-to-line voltages in phase when they are paralleled. Otherwise, large circulating currents will flow.

For this reason, a Δ-Y transformer cannot be paralleled with a Y-Y transformer. As previously mentioned, the line-to-line voltages on the delta side of a Δ-Y transformer lead the corresponding line-to-line voltages on the star side by 30°, while the line-to-line voltages on the primary of a star-star (Y-Y) transformer are in phase with the line-to-line voltages on the secondary side.

The line-to-line voltages for one phase only of a Δ-Y and Y-Y three-phase transformer are shown in Figs. 2-38(a) and (b), respectively.

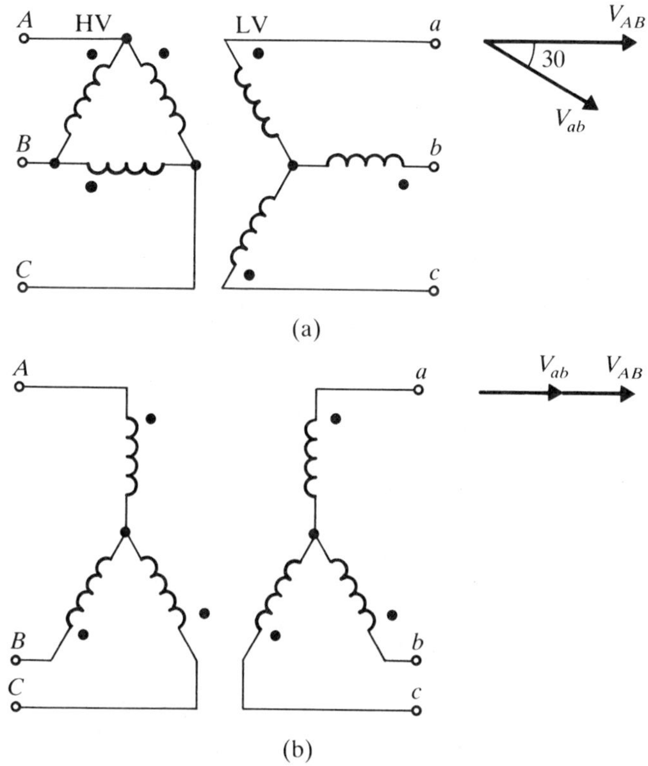

FIG. 2-38 Two-winding, three-phase transformers and a partial voltage phasor diagram: **(a)** delta-star, **(b)** star-star.

Same Phase Rotation

When two transformers are paralleled, the primary and secondary voltages must have the same phase rotation. This condition is easily verified in the field, and if necessary the connections of the two phases can be interchanged. This will result in the same phase rotation of the three-phase voltages. When the phase rotation is not the same, large circulating currents will flow through the transformer windings, and unbalanced voltages will be delivered to the load.

Paralleling of transformers, in general, results in higher rated and short-circuit currents through the common primary feeders. This, in turn, would require more expensive primary protective devices.

EXAMPLE 2–9 Two single-phase transformers are connected in parallel as shown in Fig. 2-39(a), and their primaries are connected to a 25 kV supply. The data on the nameplate of each transformer is as follows:

Transformer	*Rating in kVA*	*Voltage Ratio*	*Leakage impedances in Per-Unit*
1	600	25,000–600	$.02 + j.05$
2	500	25,000–610	$.02 + j.06$

Determine:

a) The no-load circulating current.

b) The power loss due to the no-load circulating current.

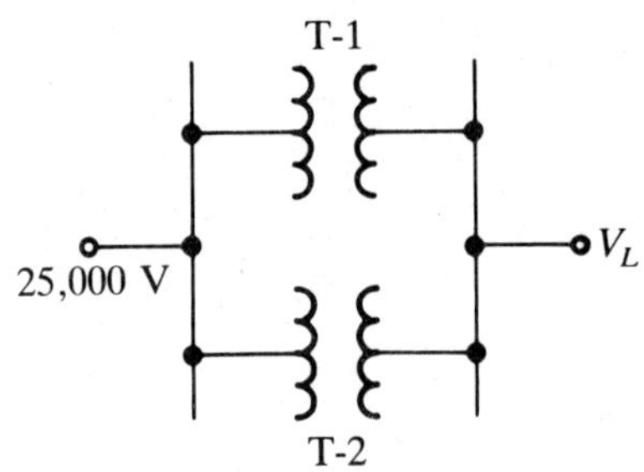

FIG. 2-39(a)

SOLUTION

a) Change the per-unit values of the transformer impedances to the equivalent ohmic values of the low-voltage winding.

For Transformer 1:

$$Z_{b_L} = \frac{(600)^2}{600 \times 10^3} = .6\ \Omega$$

and

$$Z_{t_1} = (.02 + j.05)(.6)\ \Omega$$
$$= .012 + j.03\ \Omega$$

For Transformer 2:

$$Z_{b_L} = \frac{(610)^2}{500 \times 10^3} = .74\ \Omega$$

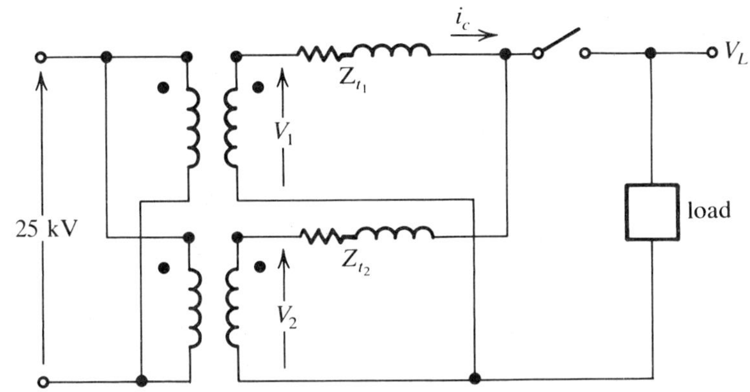

FIG. 2-39(b)

and

$$Z_{t_2} = (.02 + j.06)(.74)\ \Omega$$
$$= .015 + j.045\ \Omega$$

Thus, from the equivalent circuit of Fig. 2-39(b), we have

$$I_c = \frac{V}{Z} = \frac{610 - 600}{.012 + .015 + j(.03 + .045)} = \frac{10}{.027 + j.075}$$
$$= \underline{\underline{125.95\angle{-70.2^\circ}\ \text{A}}}$$

b) The power loss due to circulating current is

$$I^2R = (125.95)^2(.027) = \underline{\underline{426.73\ \text{W}}}$$

Exercise 2–8 Two single-phase transformers are connected in parallel, as shown in Fig. 2-40, to supply a load of 500 kVA at 480 volt and .90 power factor lagging. The data on the transformer's name-plate is as follows:

Transformer	*Rating in kVA*	*Voltage Ratio*	*Leakage impedance in Percent*
1	300	4160–480	$2 + j3.5$
2	250	4160–480	$1.8 + j4.0$

For each transformer, neglect the magnetizing impedances, and determine

a) The current.

b) The kVA loading.

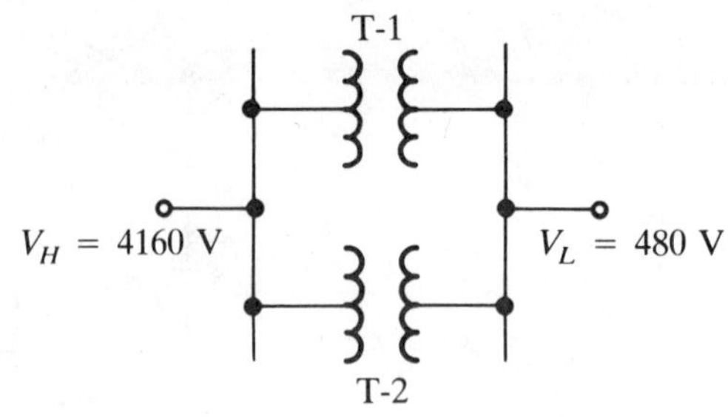

FIG. 2-40

Answer (a) $590.4\underline{/-23.4^\circ}$ A, $452.3\underline{/-29^\circ}$ A
(b) $283.4\underline{/23.4^\circ}$, $217.1\underline{/29^\circ}$ kVA

2.5 Instrument Transformers and Wiring Diagrams

In this section instrument transformers and wiring diagrams are discussed. Wiring diagrams illustrate how the instrument transformers are interconnected to the various measuring and indicating meters. These topics not only supplement the basic transformer concepts but also give an overview of the practicing engineer's work.

2.5.1 INSTRUMENT TRANSFORMERS

Instrument transformers are used to measure and control electrical parameters (voltage and currents). There are many different classes, each with a specific range of measuring accuracy.

Instrument transformers used for relaying and indicating purposes have a larger margin of error than those used for measuring the power and the energy upon which the billing demand of a particular plant is based.

There are two types of instrument transformers: potential transformers, used for measuring voltage, and current transformers, used for measuring current. The equipment connected to their secondaries constitutes the so-called load, or burden. In normal operation, the burden of the current transformer is of very low impedance, while the load on the potential transformer is of very high impedance. An elementary representation of instrument transformers and some of the meters connected to their secondaries is shown in Fig. 2-41.

The polarity and the grounding of the potential and current transformers are as shown. Correct polarity identification will lead to correct instrument connections; proper device grounding will ensure the safety of personnel and the protection of equipment.

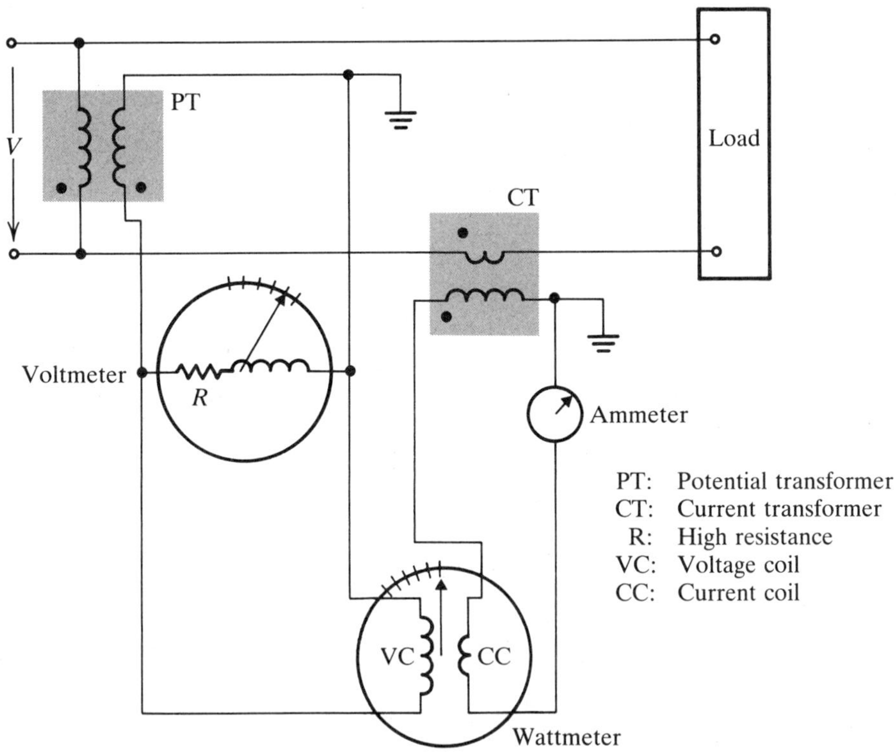

FIG. 2-41 An elementary representation of instrument transformers and measuring meters.

Potential Transformers

Potential transformers are used for stepping down voltages to nonhazardous and conventional voltage levels—normally 120 volts—for measuring and controlling purposes.

As shown in Fig. 2-42, their primary is connected in parallel to the high-voltage circuit whose voltage is to be measured. The voltage across the secondary winding is applied to potential coils of meters, relays, and other instruments, depending on what is desired.

Refer to Fig. 2-42(c). By applying KVL, we obtain

$$\text{input mmf} = N_c I_c + N_2 I_2 \tag{2.65}$$

where $N_c I_c$ is the mmf within the material and on whose magnitude the magnetization level of the PT depends. For a PT, the above expression becomes

$$I_p N_1 = N_c I_c + N_2 I_2 \tag{2.66}$$

where I_p is the current through the PT's primary winding. The magnitude of the

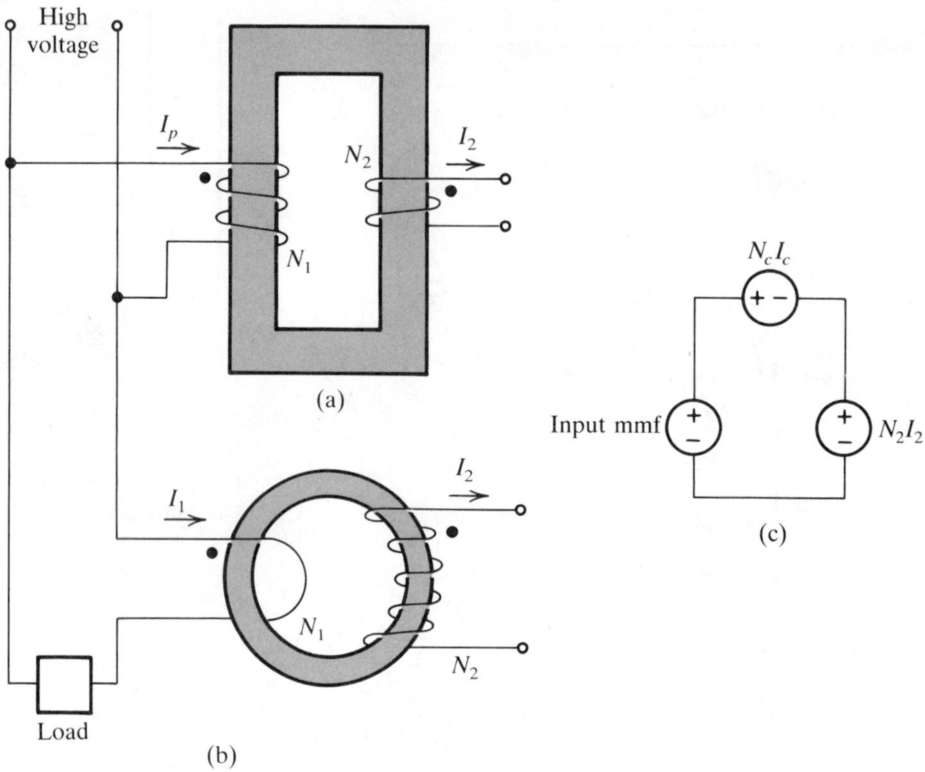

FIG. 2-42 Instrument transformers: **(a)** schematic diagram for a PT, **(b)** schematic diagram for a CT, **(c)** magnetic equivalent circuit for instrument transformers.

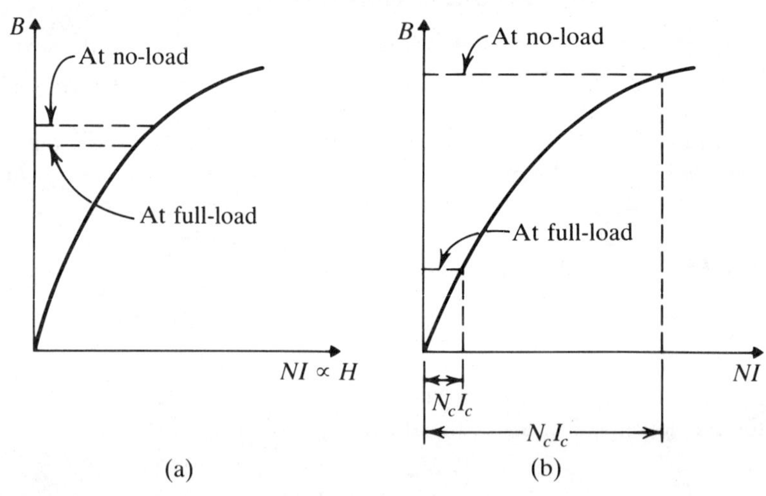

FIG. 2-43 Levels of magnetization from no-load to full-load for instrument transformers: **(a)** potential transformer, **(b)** current transformer.

primary voltage of a potential transformer is nearly constant, and therefore the magnitude of the flux within the magnetic material is also constant (see Eq. (2.7)). For all practical purposes, this results in a constant-magnitude mmf within the material, regardless of the magnitude of the load current.

The variation of the PT's magnetization level from no-load ($I_2 = 0$) to full-load operating condition is shown in Fig. 2-43(a). The small change in the levels of magnetization (from no-load to full-load operating condition) is due to leakage flux that saturates part of the magnetic material.

When a PT's secondary winding is shorted, the resulting high current will produce high winding losses. Consequently, if the transformer is not properly protected, it will be damaged and may become a fire hazard.

Current Transformers

Current transformers step down the current of the primary circuit to conventional standard levels (normally 5 amperes) for measuring and controlling purposes. As shown in Fig. 2-42(b), their primary is connected in series with the circuit whose current is to be measured. The current of the secondary winding is applied to current coils of meters and relays. Under normal operating conditions, the current through the load is constant, and thus the input mmf $N_1 I_1$ to a CT is also constant.

Rewriting KVL in the magnetic circuit of Fig. 2-42(c), we have

$$\text{input mmf} = N_c I_c + N_2 I_2 \tag{2.67}$$

The above expression for a CT becomes

$$N_1 I_1 = N_c I_c + N_2 I_2 \tag{2.68}$$

From the last equation it becomes evident that a CT's magnetizing mmf ($N_c I_c$) varies from its maximum value at zero secondary current (open secondary) to its minimum

TABLE 2-3 Main Characteristics of PT's and CT's

Description	*PT*	*CT*
Input winding connection, with respect to plant's load	Parallel	Series
Step down the magnitude of the voltage	Yes	
Step down the magnitude of the current		Yes
Levels of magnetization depends on	Supply line voltage	Current through its secondary
Damaged when its secondary is	Short-circuited	Open-circuited
Cause of damage	High copper loss	High core loss

value at maximum secondary current (shorted secondary). The variation of the CT's magnetization level from the no-load to full-load operating condition is shown in Fig. 2-43(b).

A low magnetic potential drop within the material corresponds to low levels of flux and core loss. For this reason, energized CT's have their secondaries either shorted or connected to low-impedance equipment. When an energized CT has its secondaries open-circuited, it will be damaged (a problem usually encountered in the commissioning of new plants) by excessive magnetic losses.

It should be emphasized that a CT hooked up to a live feeder should never have its secondary open-circuited. This would result in a higher level of magnetization, which in turn would induce a high potential across the secondary terminals. This will endanger working personnel and will destroy the CT (owing to excessive core loss).

For the purposes of comparison, the characteristics of CT's and PT's are summarized in Table 2-3.

Exercise 2–9 When the single-phase load of Fig. 2-42 is accidentally shorted, what would be the effect on the potential and current transformers?

2.5.2 WIRING DIAGRAMS

Wiring diagrams are electrical drawings that use conventional symbols to indicate the actual hookup of meters. Typical one-line representations of PT's and CT's and their corresponding internal wiring diagrams are shown in Figs. 2-44 and 2-45, respectively.

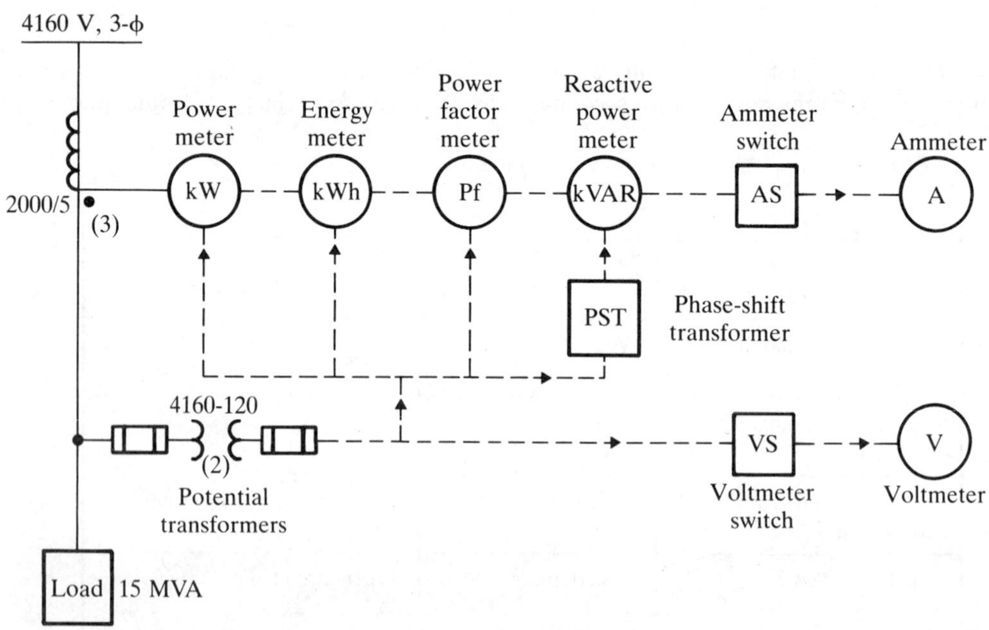

FIG. 2-44 Typical one-line representation of metering equipment.

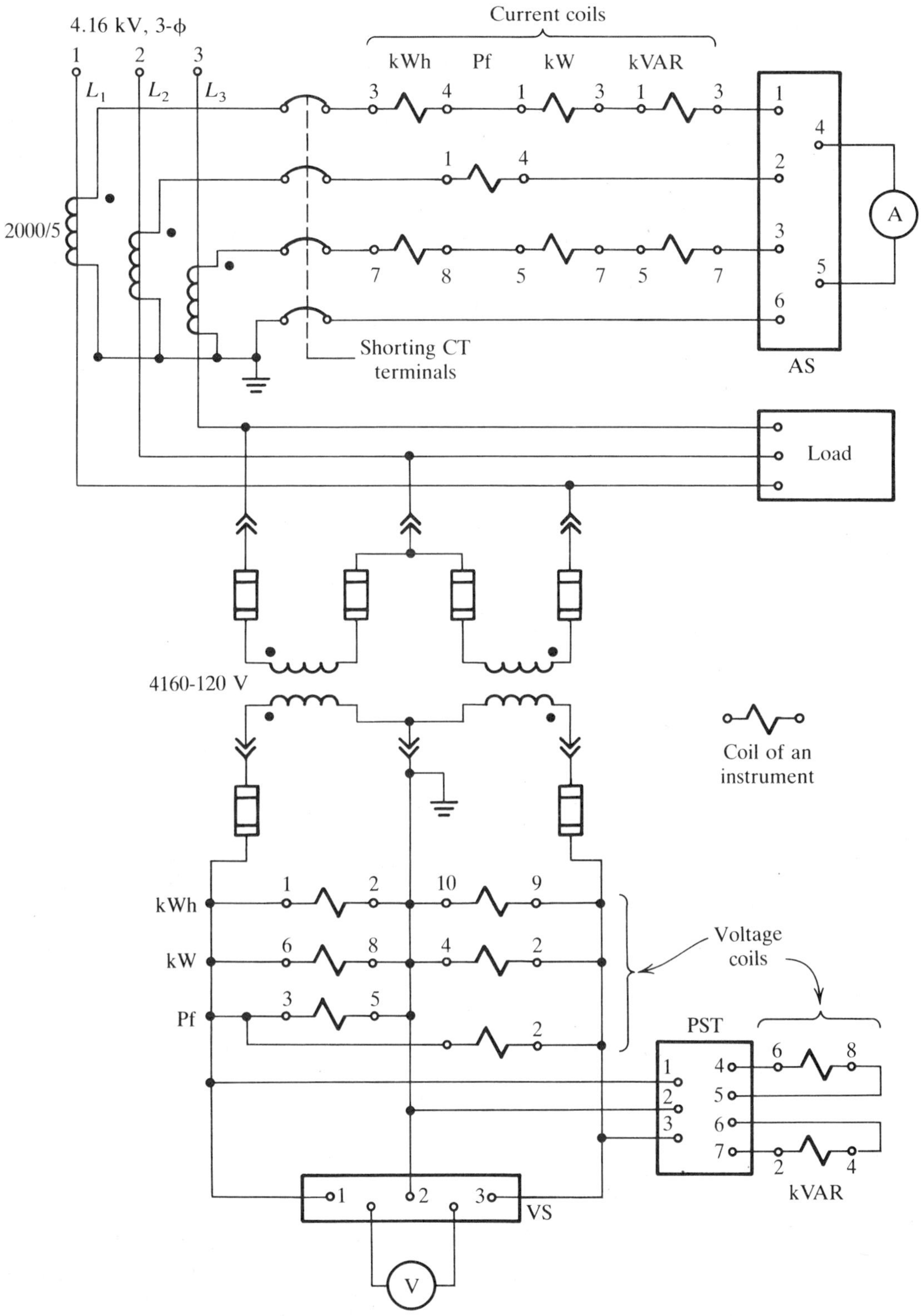

FIG. 2-45 Typical wiring diagram of the meters shown in Fig. 2-44. The coils of the various meters are represented as shown.

Referring to Fig. 2-45, notice that the potential coils are connected in parallel across the secondaries of the PT's, while the current coils are connected in series with the secondaries of the CT's. This is because the potential coils have a resistance of several thousand ohms, while the resistance of the current coils is a fraction of an ohm. For a three-phase system, two PT's are required and three CT's.

The numbered identification of the various coil terminals corresponds, as shown in Fig. 2-46, to the external identification of the meter's terminals. The identification of the coil terminals is not standardized among the various manufacturers, and it is not rare—during the first stages of a plant's start-up—to have problems with meter indications. A customary approach to these problems is to check the wiring diagrams and, perhaps, to reverse a coil connection.

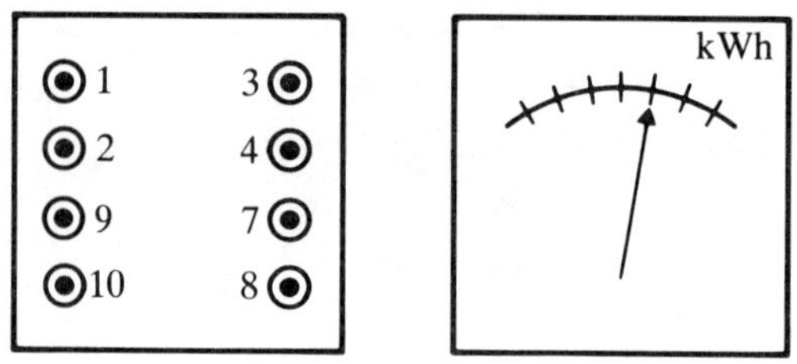

(a) (b)

FIG. 2-46 kWh Meter: **(a)** hookup terminal identification, **(b)** front view.

The ammeter is calibrated to indicate the actual load current, although the CT's secondary current is only a fraction of the load current. Referring to Fig. 2-44, the ammeter selector switch (AS) is externally positioned so that the ammeter can read the current through a selected line (L_1, L_2, L_3) or can be used to switch OFF the ammeter. In this case, the secondary terminals of the CT are, of course, shorted.

The voltmeter is calibrated to indicate the actual line-to-line voltages, even though its secondary voltage is a fraction of the primary voltage.

The voltmeter selector switch (VS) can be externally positioned so that the voltmeter measures the line-to-line (V_{1-2}, V_{2-3}, V_{1-3}) voltages. When the switch is in the OFF position, the voltmeter indicates zero and the secondaries of the PT's are open-circuited.

As the name implies, the phase-shift transformer (PST) introduces a phase shift between the measured voltages and currents, a necessary condition for the operation of the meter that measures reactive power.

The voltage and current coils for the kW, kWh, kVAR, and power-factor meters are shown in Fig. 2-45.

Nowdays microprocessor-type meters are available that combine the functions of the individual meters in one unit. Such units are equipped with a display and a set of push buttons. When a parameter is to be displayed, a corresponding push button is depressed. These meters can also send "measuring pulses" to computers, which in turn can print out the instantaneous variations of the system parameters.

EXAMPLE 2–11 For the single-phase transformer shown in the one-line diagram in Fig. 2-47, determine the ammeter reading for the following operating conditions:

a) The transformer delivers 80% of its base kVA at nominal voltage.
b) The transformer delivers 100% of its fan-rated kVA.
c) The transformer's secondary feeder is shorted.

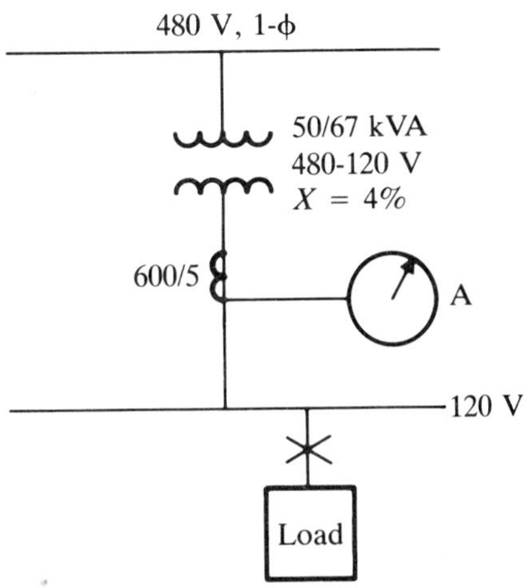

FIG. 2-47

SOLUTION

a) When the transformer delivers 80% of its rated kVA, the magnitude of the current in the line (or the ammeter reading) is

$$I = .8\left(\frac{50 \times 10^3}{120}\right) = 333.33 \text{ A}$$

The CT's secondary current will be

$$I_{CT} = I_{\text{actual}} \times \text{turns ratio}$$
$$= 333.33\,\frac{5}{600} = 2.78 \text{ A}$$

This is also the current through the ammeter, but the latter is always calibrated to give the actual current through the load.

b) Similarly, the ammeter reading will be

$$I = \frac{67 \times 10^3}{120} = 558.33 \text{ A}$$

The CT's secondary current will be

$$I_{CT} = 558.33\left(\frac{5}{600}\right) = 4.65 \text{ A}$$

c) The impedance of the transformer depends on its base kVA and not on its fan-rated kVA. The latter's significance is that if the transformer windings are externally cooled, they can carry higher currents without becoming overheated. Since the applied voltage and the transformer impedance remain constant, the short-circuit current is independent of the level of operating KVA. Thus, the ammeter reading on short-circuit conditions is

$$I = \left(\frac{1}{.04}\right)\frac{50 \times 10^3}{120} = 10.42 \text{ kA}$$

The current through the CT's secondary winding will be

$$I = 10.42 \times 10^3 \frac{5}{600} = 86.81 \text{ A}$$

EXAMPLE 2–12

a) Refer to Fig. 2-44. When the ammeter reads 1500 A and has an accuracy of 99%, what is the apparent power drawn by the inductive load?

b) What is the power factor of the load when the kW meter reads 2000 kW and the reactive power meter reads 1500 kVAR?

SOLUTION

a) The scales on switchboard-type meters are calibrated in terms of the line-side parameters. Thus, the current through the 4.16 kV lines is

$$I_L = \frac{1}{.99}(1500) = 1515.15 \text{ A}$$

By definition, the apparent power is

$$|S| = \sqrt{3}\, V_{L-L} I_L = \sqrt{3}(4.16)(1515.15) = \underline{\underline{10{,}917.17 \text{ kVA}}}$$

b) The real and reactive powers are given, respectively, by

$$P = \sqrt{3}\, V_{L-L} I_L \cos\theta$$

$$Q = \sqrt{3}\, V_{L-L} I_L \sin\theta$$

Thus,

$$\theta = \arctan \frac{Q}{P} = \arctan \frac{1500}{2000} = 36.9^\circ$$

The power factor is

$$\cos 36.9^\circ = \underline{\underline{.80 \text{ lagging}}}$$

Exercise 2–10 Refer to Fig. 2-44.

a) Explain why a voltmeter never measures the voltage and an ammeter never measures the current.

b) What is the function of the voltage switch? The current switch?

c) Why, when the ammeter and voltmeter switches are in the OFF position, is the PT open-circuited and the CT short-circuited?

d) What is the power triangle (P, Q, and S) of the load if the meters indicate

power factor:	.95 lagging
voltmeter:	600 V, line-to-line
ammeter:	750 A

Answer (d) 740.45 kW, 779.42 kVA, 243.37 kVAR

Exercise 2–11 Refer to Fig. 2-45.

a) Why is the secondary of a PT normally supplied with a fuse but not the secondary of the CT?

b) Why are the voltage coils of the kVAR meter connected to the output of the phase-shift transformer?

c) Draw the phasor diagram of the voltages on the HV and LV winding of the PT's.

2.6 Transformer's Name-Plate Data

Some of the essential markings on a transformer's name-plate are described below.

Apparent Power Rating in kVA

This parameter indicates the designed kVA base transformation capability of the transformer. As long as the transformer operates at or below its marked kVA, it will not overheat. This is also the base power for the per-unit calculations of the transformer impedances. When transformers are equipped with cooling fans, they can deliver higher kVA than their rated value without being overheated.

For example, a transformer designated as

$$1500/2000 \text{ kVA}$$

has a base transformation capacity of 1500 kVA. However, with the use of cooling fans, the transformer can deliver 2000 kVA (33% increase) without being overheated. When a second set of fans is used, the transformer can deliver an additional 25% kVA without being overheated. Also, when a special winding insulation is used, the transformer can safely deliver an additional 12% kVA. For example, a transformer with name-plate data of 12/16/20/22.4 MVA should be interpreted as follows:

12 MVA	is the base power
16 MVA	is its capacity with one set of fans (1.33)
20 MVA	is its capacity with two sets of fans (1.25)
22.4 MVA	is its capacity due to upgraded winding insulation (1.12)

The ohmic value of the transformer's impedance remains the same, regardless of whether the transformer delivers its base or its fan-rated kVA. The resistance of the windings depends on the temperature of the conductors and will change somewhat with increases or decreases in load current.

Impedance in Percent

This is the magnitude of the equivalent winding or leakage impedance. The transformer's magnetizing impedance is not included on the name-plate data. The winding impedance (expressed in percent) indirectly gives the equivalent transformer's impedance in ohms and also indicates what percent of the rated voltage is required to circulate rated current through the windings of the transformer when the secondary is shorted.

The leakage impedance varies with the size of the transformer, class voltage, type of winding (copper or aluminium), and frequency. For the standard voltages and in the range of 1000 to 2000 kVA, the transformer's impedance is about 6%.

Efficiency in Percent

This is the efficiency of the transformer when it delivers rated kVA at unity power factor. Generally, the higher the kVA capacity of the transformer, the higher its efficiency. In the range of 100 to 2000 kVA the efficiency is usually between 96 to 99%.

"OFF-Load" Voltage Tap Changers

Normally, transformers are equipped with four "OFF-load" voltage tap changers ($\pm 2.5\%$ and $\pm 5\%$). The positive tap settings are used to compensate for voltage drop through the transformer's upstream feeders, while the negative tap settings are used to compensate for generated voltages that are slightly above nominal values.

Winding Connections — Rated Voltages

The name-plates of three-phase transformers also indicate whether their windings have a delta (Δ) or a star (Y) connection and give their corresponding rated voltages. The latter may be used in establishing the line-to-line or the phase-to-neutral turns ratio.

Name-plate data also indicates whether provisions are made for grounding the neutral terminal of the star-connected windings.

2.7 Conclusion

Transformers change the voltage and the current from one level to another and, as such, constitute the main interconnecting link between two systems of different voltages.

The design, operation, and analysis of all types of transformers is based on the induction principle, which says that a voltage is induced in coils whose flux linkages ($N\phi$) change as a function of time. Under ideal conditions, the induced voltage and the resulting current in a secondary coil are related to the corresponding parameters of the source coil (primary coil) by the windings turns ratio.

Each type of transformer can be analyzed by using either the magnetic or the electric equivalent circuits. Although using the magnetic circuit is simpler, the trend throughout the industry is to employ the electric equivalent circuits.

The core and copper losses of transformers—even though they are a small fraction of the power-transformation capability of any transformer—often constitute the criteria for selecting transformers. This is because the purchasing cost of a transformer is much lower than the cost of the energy losses within the transformer over its operating life. (A kilowatt consumed continuously over a 20-year period has a present worth of \$7215 when the annual interest rate is 6% and the cost of energy is 5¢/kWh and increases 4% per year. In comparison, the cost of a 25-kVA transformer that has 1 kW of power loss is about \$1000.)

In the analysis of circuits that include transformers, it is much more convenient to use per-unit values, because the impedances, voltages, or currents, when expressed in per-unit, are the same in either side of the transformer.

Throughout the various sections of this chapter, the short-circuit MVA of transformers is calculated because of its importance in equipment selection (circuit breakers, fuses, etc.). When properly selected, this equipment safeguards personnel and downstream apparatus.

Single-phase power transformers have relatively small ratings, usually in the range of 1–100 kVA.

The coils of the transformer are magnetically coupled, and the relative polarity of the induced voltages is normally identified by a dot.

By convention, the dotted terminal is at higher potential than the undotted terminal. This leads to the determination of the direction of current flow and of flux. This is of extreme practical importance for proper field connections of measuring equipment and of protective relays. Misinterpretation of the dotted terminals will result in incorrect meter indication, a problem that usually arises during the commissioning of industrial plants.

The analysis of *three-phase*, two-winding transformers is simplified by using their electric or magnetic per-phase equivalent circuits. An additional simplification results when the per-unit values of the transformer parameters are used. The latter removes the difficulty of long calculations and the confusion that may arise from the various three-phase transformer connections.

Transformers of .5 to 20 MVA rating have a nominal impedance of 5% to 8% and an efficiency of .97 to .99.

The phase shift between the primary and secondary line-to-line voltages depends on the transformer's connections. For the delta-star connection, the voltages on the HV side lead their corresponding LV side by 30 degrees. The majority of three-phase transformers are of Δ-Y type. They tend to maintain sinusoidal secondary voltages, and the secondary star windings can be used to ground the distribution network and to connect line-to-neutral loads.

Three-phase transformers are normally equipped with a set of ventilating fans (the additional cost is 5% above nominal). As a result, the transformer's transformation capacity is increased by 33%.

Autotransformers are used in large power systems and in testing. Standard transformers are paralleled in order to increase the reliability of a substation (two smaller units are more reliable than a larger unit) and to provide additional transformed kVA that meets a plant's increased demand. In order to parallel two transformers they must have equal impedances, equal turns ratios, the same phase rotations, and equal phase shifts between the primary and secondary open-circuited voltages.

Instrument transformers step down the voltage and the current to more standard and less dangerous levels.

Usually, the secondary of a potential transformer is rated at 120 V while the secondary of a current transformer is rated at 5 A.

The burden, or the load (voltmeters, voltage coils of kW-meters, etc.), of potential transformers is of very high impedance; conversely the burden of the current transformers is of very low impedance.

Energized potential transformers are damaged—owing to high copper loss—when their secondary is shorted. Current transformers are damaged—owing to high core loss—when their secondary is open-circuited.

Wiring diagrams are electrical drawings that use conventional symbols to indicate the actual field connections of the measuring equipment. As such, wiring diagrams are of fundamental importance to power-distribution engineers, to those who participate in the commissioning of industrial plants, or to those whose responsibility it is to maintain the plant's electrical equipment.

The main concepts of this chapter are condensed in Table 2-4, and typical manufacturer's data are given in Tables 2-5 to 2-8.

TABLE 2-4 Summary of Main Transformer Concepts

Item	*Description*	*Remarks*
1	Principle of operation of transformers $$v = -N\frac{d\phi}{dt}$$	Eq. (2.6)
2	Fundamental transformer equation $$V = 4.44 N f \phi_m$$	Eq. (2.7)
	Ideal Transformer Relationships	
3	$$V_2 = V_1\frac{N_2}{N_1}$$	Eq. (2.11)
4	$$I_2 = I_1\frac{N_1}{N_2}$$	Eq. (2.14)
5	$$Z_1 = Z_2\left(\frac{N_1}{N_2}\right)^2$$	Eq. (2.17)
6	*Short-circuit test* $$I_z = I_{\text{rated}}$$ $$V_z = (2\% \rightarrow 12\%)V_{\text{rated}}$$ Measure I_z, V_z, and P_z Calculate $$R_e = \frac{P_z}{I_z^2},\quad \lvert Z_e\rvert = \frac{V_z}{I_z},\quad X_e = \sqrt{Z_e^2 - R_e^2}$$	Section 2.1.3
7	*Open-circuit test* $$I_{\text{exc}} = (3\% \rightarrow 10\%)I_{\text{rated}}$$ $$V_{\text{exc}} = V_{\text{rated}}$$ Measure: $$I_{\text{exc}},\quad V_{\text{exc}}, \text{ and } P_{\text{exc}}$$ Calculate $$R_m = \frac{P_{\text{exc}}}{I_{\text{exc}}^2}, \lvert Z_m\rvert = \frac{V_{\text{exc}}}{I_{\text{exc}}},\quad X_m = \sqrt{Z_m^2 - R_m^2}$$	Section 2.1.3
8	Transformer operates at maximum efficiency when $$P_z = P_{\text{exc}}$$	Eq. (2.38)
9	$$\text{Regulation } \% = \frac{\lvert V_{nl}\rvert - \lvert V_{fl}\rvert}{\lvert V_{fl}\rvert}(100)$$	Eq. (2.39)

(Continued)

TABLE 2-4 (Continued)

Item	*Description*	*Remarks*
10	$Z_{e_{pu}} = V_{z_{pu}}$	Eq. (2.42)
	3-ϕ Transformers	
11	$N_A I_A = N_a I_a$	Eq. (2.51)
12	Star-delta impedance $Z_y = \frac{Z_\Delta}{3}$	Eq. (2.53)
13	Third harmonic component of exciting current $i_3 = 3I_{m_3} \cos 3\omega t$	Eq. (2.57)
	Autotransformers	
14	The ampere-turns of each winding are constant, regardless of the type of transformer connection. That is, $(NI) = \text{constant}$	Eq. (2.60)
15	Neglecting losses, $(\text{kVA})_{\text{input}} = (\text{kVA})_{\text{output}}$	Eq. (2.61)
	Parallel Operation of Transformers	
16	To parallel two transformers, the following conditions must be satisfied: a) equal impedances b) equal turns ratios c) equal phase shift between the primary and the secondary open-circuited voltages d) the same phase rotations	Section 2.4
	PT's and CT's	
17	When short-circuited, energized potential transformers are damaged because of high copper loss	Section 2.5.1
18	When open-circuited, energized current transformers are damaged because of high core loss	Section 2.5.1

TABLE 2-5 Typical parameters of dry type transformers with copper windings at 60 Hz*

a) Single phase (480–240/120 V)

kVA	Losses in Watts at 170°C		Efficiency in Percent at 170°C		Impedance in Percent at 170°C
	Core	Copper	Full-Load	1/4 Full-Load	
10	70	275	96.7	96.6	3.6
25	110	825	96.4	97.5	4.1
50	150	2000	95.9	97.8	5.7
100	280	3350	98.5	98.1	4.7

b) Three-phase (480–208/120 V)

kVA	Core	Copper	Full-Load	1/4 Full-Load	Impedance
15	130	410	96.5	96.0	3.6
75	300	3100	95.7	97.4	5.7
150	540	5000	96.4	97.8	4.7
500	1120	11531	97.5	98.5	5.0

* *Courtesy of Westinghouse Canada Inc.*

TABLE 2-6 Parameters of 4160–480/277 V, dry-type transformers with copper windings*

kVA	Resistance in Ohms at 170°C		Exciting Current Percent	Losses in kW		Efficiency at Full-Load	Impedance in Percent
	HV	LV		Core	Copper		
500	2.9	.011	1.5	1.2	7.6	98.3	5.5
1000	1.09	.004	1.12	3.8	13.1	98.3	5.9
1500	.75	.003	0.70	4.0	20	98.4	6.1
2000	.48	.016	.65	4.8	23	98.6	6.5

**Courtesy of Westinghouse Canade Inc.*

TABLE 2-7 Parameters of a 15/20/22.4 MVA, 60–4.16/2.4kV, liquid-type transformer with copper windings*

Exciting Current in Percent	*Core Loss in kW*	*Copper Loss in kW*	*Impedance in Percent*
.41	15.0	65	7.0

Courtesy of Westinghouse Canada Inc.

TABLE 2-8 Parameters of 4160–480/277 V, 60 Hz, dry-type transformers with copper and/or aluminum windings*

kVA	*Cost in 1988*		*Core Loss in kW at 170°C*		*Copper Loss in kW at 170°C*		*Total Weight in kg*		*Estimated Impedance in Percent*
	Windings		*Windings*		*Windings*				
	Copper	*Aluminum*	*Copper*	*Aluminum*	*Copper*	*Aluminum*	*Copper*	*Aluminum*	
750	$22,800	$19,000	3.0	3.4	8.1	8.8	2100	2050	6
1000	26,400	22,000	3.8	4.0	13.1	14.2	2500	2425	6
1500	30,600	25,500	4.2	4.2	20	21.5	3300	3100	6

* *Courtesy of Westinghouse Canada Inc.*

2.8 REVIEW QUESTIONS

1. In general, what fraction of the transformer's output power is its iron and winding losses?
2. Show that 1 kW consumed continuously over a 20-year period has a present-worth value of \$7215. Assume the interest rate is 6% per year, the cost of energy is 5 cents/kWH, and the increase in the cost of energy is 4% per year.
3. Explain why the magnetizing impedance is not a constant parameter but depends on the level of magnetization.
4. How would you measure the magnetic potential drop ($N_c I_c$) in a magnetic material? How is this potential related to the permeability of the material?
5. Explain why the per-phase analysis of a three-phase balanced system is analogous to the per-pole analysis of a magnetic circuit.
6. For maximum efficiency, how is winding loss related to core loss?
7. Why does the selection of the transformer's primary protective devices depend on the time-current waveform of the inrush current?
8. Sinusoidal voltages produce sinusoidal fluxes, which in turn produce nonsinusoidal exciting currents. Explain.
9. What is the predominant harmonic of the exciting current and of the inrush current of a transformer?
10. Show that the per-unit value of the transformer's leakage impedance is the same whether it is referred to the HV or LV winding.

11. Draw the power triangle for a balanced three-phase inductive load and give the formulae for the real, imaginary, and complex power.

12. Nonsinusoidal exciting currents produce sinusoidal secondary voltages. What are the *two* essential requirements for the existence of nonsinusoidal exciting currents?

13. Explain why the most important characteristic of an autotransformer is that the current in its windings is the same as when the transformer operates as a conventional two-winding transformer.

14. What four conditions must be satisfied before any two transformers are paralleled?

15. What is the essential difference between a CT and a PT?

16. Explain the name-plate data marked on your laboratory's transformer.

2.9 PROBLEMS

2-1 A 50 kVA, 2300–230 V, 60 Hz, single-phase step-down transformer has primary and secondary leakage impedances of $(.5 + j2.6)$ and $(.005 + j.026)$ ohms, respectively. Neglecting the magnetizing impedance, determine:

a) The equivalent circuit, as seen from the high-voltage and the low-voltage windings.

b) The secondary voltage, when the transformer is connected to a 2300 V source and delivers rated current to a load at .9 Pf lagging.

2-2 Data from the open- and short-circuit test on a 10 kVA, 2400–240 V, 60 Hz step-down transformer are as follows:

Test	*Voltage in Volts*	*Current in Amperes*	*Power in Watts*
Open circuit	2400	.35	150
Short circuit	12	41.67	320

Determine:

a) The equivalent circuit, as seen from the high- and low-voltage winding.

b) The primary voltage, when the transformer delivers rated current at .9 power factor to a 240 V inductive load.

c) The voltage regulation and efficiency.

d) The per-unit value of the transformer's equivalent leakage impedance.

2-3 A single-phase 50 kVA, 2400–120 V, 60 Hz transformer has a leakage impedance of $(.023 + j.05)$ per-unit and a core loss of 600 watts at rated voltage. Determine:

a) The load current for maximum efficiency.

b) The efficiency for the condition in (a) and the efficiency when the transformer delivers rated current. In both cases, assume unity power factor.

c) The power factor of the load that will result in the best regulation (0%). Assume that the load current and the primary voltage remain constant at their rated values.

2-4 A 100 kVA, 480–120 V, 60 Hz, single-phase transformer has an efficiency of 95.75% at rated conditions and unity power factor. Its leakage impedance is (2.5 + j5.0) percent. The transformer's open-circuit test data are as follows:

Voltage in Volts	*Power in Watts*	*Frequency in Hertz*
100	1400	50

a) Determine the eddy-current and hysteresis losses at 60 Hz and 120 V.

b) Determine the rate at which heat is produced within the structure of the transformer at 100 V and 50 Hz due to the friction caused by the motion of the magnetic domains.

c) Can the transformer be used to deliver its rated power at rated voltage and 50 Hz? Explain.

2-5 A 15 kVA, 480-240/120 V, 60 Hz, single-phase transformer has equal primary and secondary leakage impedances of (.01 + j.02) per-unit. Determine:

a) The no-load voltage across the secondary terminals at rated flux.

b) The voltage regulation when the transformer is connected to a 480 V source and delivers rated current at a power factor of:

(1) Unity.
(2) .80 leading.
(3) .80 lagging.

2-6 A transformer has a leakage impedance of (.015 + j.06) per-unit. When the transformer delivers rated current at .9 Pf lagging, its efficiency is 97%. Determine:

a) The nominal core loss in per-unit and the voltage regulation.

b) The no-load voltage tap setting which will minimize the voltage regulation.

2-7 Figure P2-7 shows a core-type 150 kVA, 4160–480/277 V, three-phase transformer. Connect the primary in delta and the secondary in star. The neutral is to be solidly grounded. Show coil polarities and direction of flux lines and find the coil currents for the following load conditions:

a) A load drawing 60 A is connected from line "a" to neutral.

b) A load drawing 60 A is connected from line "a" to line "b".

c) A 3-ϕ delta-connected load of 100 kW, 480 V, .90 efficient, and .8 power factor lagging is connected across the secondary.

2-8 A three-phase, 2000 kVA, 25,000–480/277 V, Δ-Y transformer supplies two three-phase loads through feeder B as shown in Fig. P2-8. The Δ-connected high-voltage transformer windings are connected through cable A to a 3-ϕ, 60 Hz power source whose short-circuit MVA is 500. The ratings of the various equipment are as shown in the diagram. If the three-phase loads draw rated currents at rated voltage, determine:

a) The per-unit values of the impedances of cables A and B, using the ratings of the transformer as base values.

b) The total losses of the transmission system.

c) The voltage required at the source.

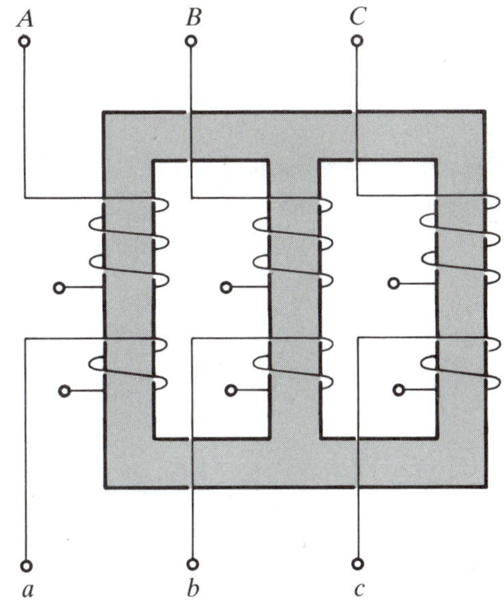

FIG. P2-7

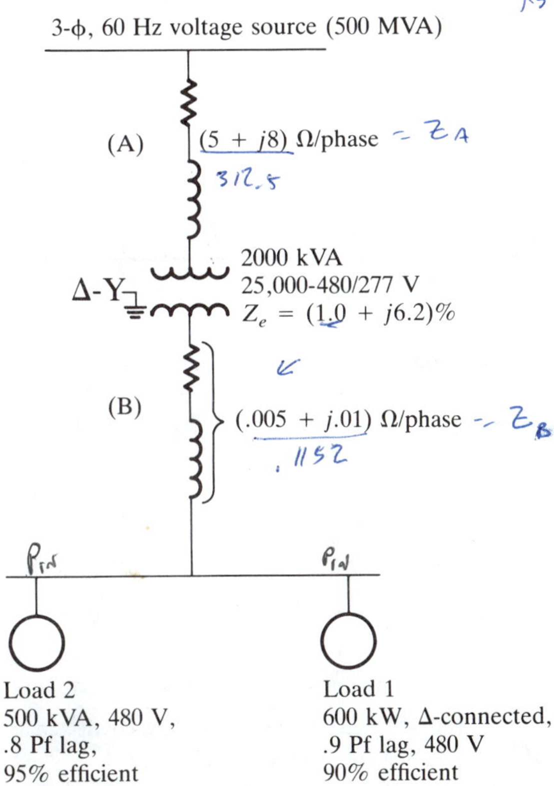

FIG. P2-8

2-9 Figure P2-9 shows a three-phase, five-wire (3-ϕ, 5W) distribution system. If the connected loads draw rated currents, determine:

a) The minimum required kVA capacity of the transformer.

b) The phase currents through each of the six transformer coils.

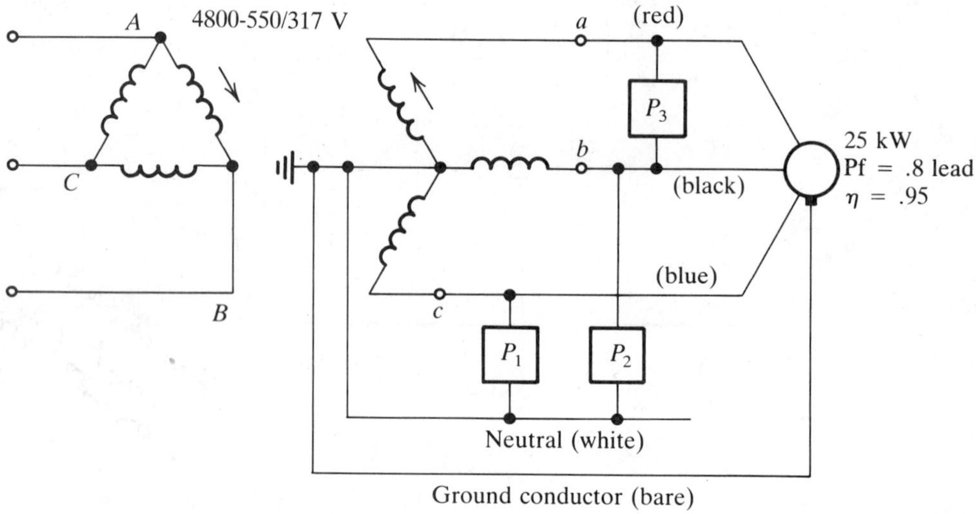

P_1 = 12 kW
Pf = .8 lag
η = .95

P_2 = 10 kW
Pf = .9 lag
η = .9

P_3 = 20 kW
Pf = 1.0

FIG. P2-9

2-10 Figure P2-10 shows a three-phase transformer whose neutral point is grounded through a 1.0 ohm resistor. The transformer delivers rated voltage to the three-phase heater. When the incoming live part of cable C (phase ABC) accidentally makes contact with the grounded metallic enclosure of the heater, estimate (assuming negligible transformer leakage impedance) the following:

a) The line and ground currents.
The voltage from the heater's enclosure to ground.
The reaction of the upstream protective devices.

b) Repeat (a), assuming that the enclosure is not grounded.

c) Repeat (a), assuming that the neutral of the transformer is solidly grounded.

d) Repeat (a), assuming that the neutral of the transformer is not grounded.

2-11 A 5 kVA, 480–120 V, two-winding transformer is to be connected as an autotransformer to supply a 480 V load from a 600 V supply. Determine:

a) The autotransformer kVA.

b) The untransformed kVA delivered to the load.

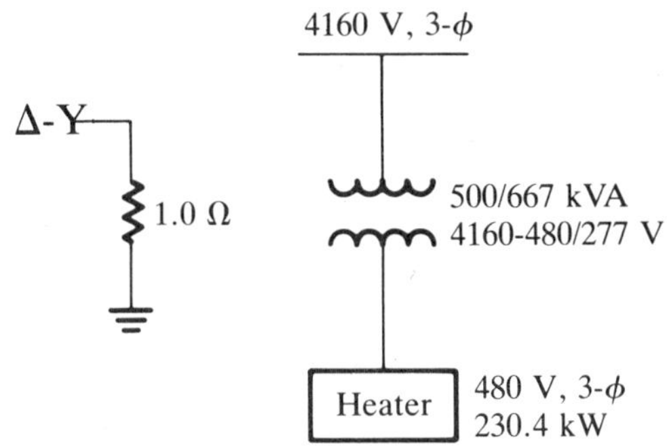

FIG. P2-10

2-12 A two-winding, 2400–1000 V, 200 kVA transformer with a leakage impedance of (.02 + j.06) per-unit has an efficiency of .96 when it delivers rated current at .85 power factor lagging. This transformer is to be connected as an autotransformer in order to supply a 2400 V load from a 3400 V supply. Determine:

a) The kVA rating of the autotransformer. What percentage of this kVA passes through the transformer unaltered?

b) The efficiency at full-load and at .85 power factor.

c) The primary current when the secondary of the transformer is shorted.

2-13 Two single-phase transformers are connected in parallel, as shown in Fig. P2-13, to supply rated voltage and current to a 1000 kVA, 480 V, .95 power-factor-lagging load. The name-plate information for each transformer is as follows:

Transformer	*Rating in kVA*	*Voltage Ratio*	*Leakage Impedances in Per-unit*
1	550	4160–480	(.02 + j.04)
2	550	4160–468	(.02 + j.05)

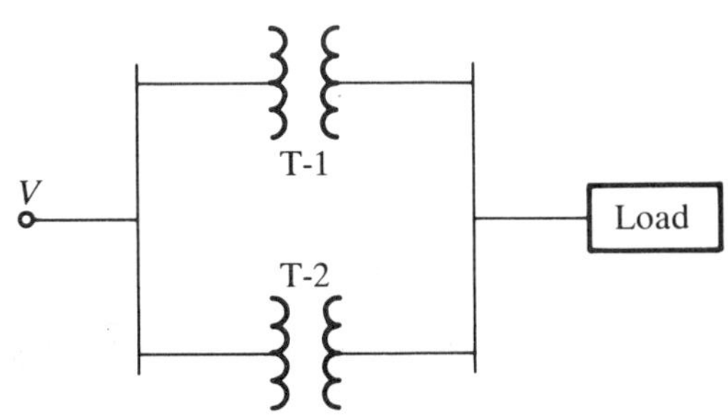

FIG. P2-13

Assuming constant primary voltage and negligible magnetizing current, determine:

a) The primary supply voltage.

b) The current through each transformer.

c) The output kVA of each transformer.

d) The no-load circulating current.

e) The current in the common primary feeder if the load is shorted.

2-14 A transformer is connected to a 25 kV substation through a cable, as shown in the one-line diagram in Fig. P2-14. The characteristics and parameters of the system's components are as shown.

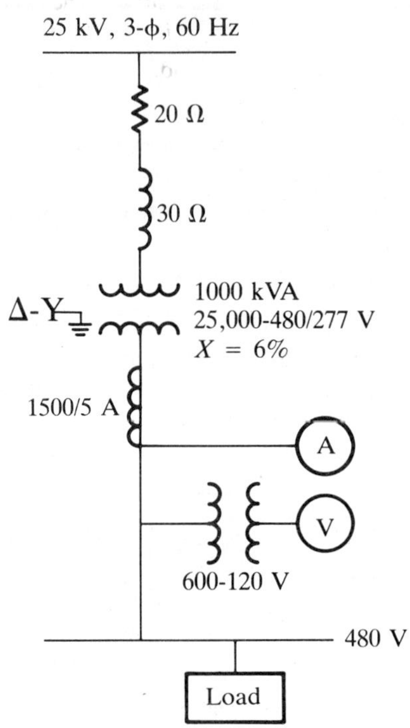

FIG. P2-14

a) Determine the voltmeter indication, the current through the secondary of the CT, and the ammeter indication for the following operating conditions:

(1) The transformer delivers rated current at .90 power factor lagging.

(2) A sustained three-phase short takes place at the 480 V terminals of the transformer.

(3) The transformer delivers rated current at .9 Pf leading. Assume that the short-circuit MVA of the source to be 500.

b) Explain why "a PT must not be energized with its secondaries short-circuited, and a CT must not be energized with its secondaries open-circuited."

2-15 Two manufacturers quoted the following on a request for the purchase of one (1) 1000 kVA, three-phase, 4160–480 V transformer:

Manufacturer	*Cost*	*Core Loss in kW*	*Copper Loss in kW*
A	$30,000	3	12
B	$35,000	2.5	9.5

Under the following operating conditions, estimate the present worth of each manufacturer's transformer over a five-year operating period.

a) Continuous operation at 75% of full-load.

b) Continuous operation at 50% of full-load.

c) Operation at full-load, 50% of the time.

Assume that the interest rate is 10%, the cost of energy is 6¢/kWh, and its annual increase is 8%.

THREE

Three-Phase Induction Machines

3.0 Introduction

Three-phase induction motors are by far the most widely used in industry: they constitute about 80% of the total number of motors used in the average plant.* In fact, there are many small and medium-sized industries whose motors above 3 kW are all three-phase induction motors. Three-phase induction motors are popular because they are more economical, last longer, and require less maintenance than other types of motors. Because of their importance, this chapter provides a detailed analysis of three-phase motors and a thorough discussion of their applications. The following subjects are specifically discussed: principles of operation, rotating magnetic fields, equivalent circuits, governing mathematical relationships, industrial considerations, voltage–frequency multipliers, induction generators, electromagnetic transients, and solid state control.

Manufacturers' data for three-phase induction machines are included in the appendix to this chapter. Their control schematics are extensively covered in the chapter on control schematics.

The physical distribution of the machine windings, and their unbalanced voltage operation, are covered in more specialized textbooks.

3.1 Three-Phase Induction Motors

3.1.1 STATOR AND ROTOR

Like all motors, three-phase induction motors transform electrical power into mechanical power by means of a stationary part called the stator and a rotating part called the rotor. Stator or *armature windings* are housed on the stator; *rotor windings* are installed on the rotor. An elementary representation of stator and rotor windings is shown in Fig. 3-1.

Stator windings can be either star- or delta-connected. Their main function is to receive three-phase ac power and to produce a *single rotating* magnetic field that has an approximately sinusoidal space distribution. This rotating field completes its magnetic path through the stator, two air gaps, and the rotor structure. Figure 3-2 shows a manufacturer's picture of an actual armature winding. The stator windings are brought out in a terminal box where they can be connected to a suitable three-phase voltage supply.

* However, only about 50% of the energy used by a plant is consumed by 3-ϕ induction motors. Roughly 35% is absorbed by synchronous and dc machines, and the remaining 15% is consumed by single-phase motors, heating, lighting, and miscellaneous.

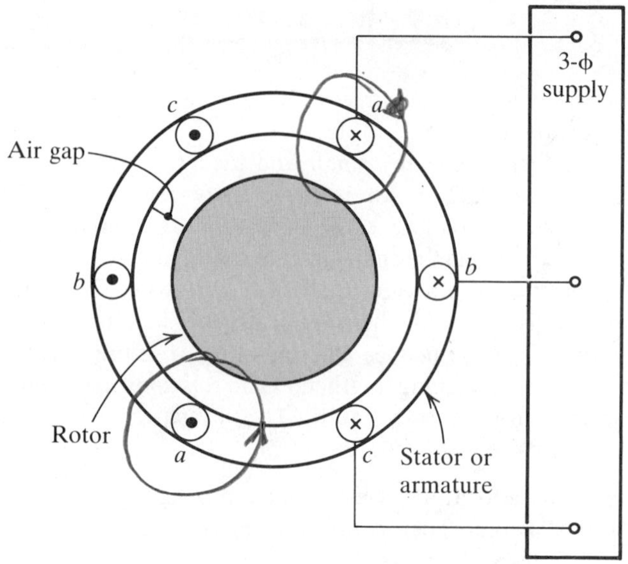

FIG. 3-1 An elementary representation of a three-phase induction motor. (In an actual machine, the armature and the field windings are placed in the stator and rotor slots, respectively.)

FIG. 3-2 Stator windings. Courtesy of General Electric Company.

There are two types of rotor windings: the squirrel-cage and the wound type. A manufacturer's pictures of squirrel-cage and wound-rotor windings are shown in Figs. 3-3 and 3-4, respectively.

A *squirrel-cage* rotor winding is generally made up of bare aluminum bars that are connected at their terminals to shorted end rings. In other words, the rotor windings

FIG. 3-3 Squirrel-cage rotor. Courtesy of General Electric Company.

FIG. 3-4 Wound rotor. Courtesy of General Electric Canada, Inc.

(a)

(b)

FIG. 3-5 Three-phase induction motors. **(a)** 200 kW, 460 V; **(b)** 6300 kW, 4160 V. Courtesy of General Electric Canada, Inc.

FIG. 3-6 Induction motor rotors. Courtesy of General Electric Canada, Inc.

are always short-circuited regardless of motor operating condition. The rotor bars are not parallel to the rotor axis but are set at a slight skew. This reduces mechanical vibrations, so the motor is less noisy.

Rotor windings cannot be electrically connected to a circuit outside the rotor. As a result, the rotor resistance is constant, and for a given stator voltage its torque–speed characteristic is fixed (see Section 3.1.6).

In some special designs (for a particular motor application, for instance) the rotor may have double squirrel-cage windings, each with a different resistance. This construction gives higher starting torque, lower starting current, and higher full-load power factor.

A very small percentage of induction machines have a wound rotor. *Wound-rotor windings* terminate at the slip rings on which the brushes rest. The brushes can then be connected to a three-phase variable resitor and the resistance of the rotor winding can be externally controlled. This variable resistor, as demonstrated in the sections that follow, controls the torque–speed characteristic of the motor. Figures 3-5 and 3-6 show, respectively, manufacturer's pictures of three-phase induction motors and rotors.

3.1.2 PRINCIPLES OF OPERATION

The operation of three-phase induction machines is based on the generation of a revolving field, the transformer action, and the alignment of the magnetic field axes.

When balanced three-phase currents are injected into the stator windings, they produce a *rotating magnetic field* which—unless the rotor is revolving at the same speed as the magnetic field—will *induce* voltages* in the rotor windings. This results in rotor current and therefore rotor flux. The magnetic fields of the stator and rotor try to *align their magnetic axes*—a natural phenomenon—and in so doing, a torque is developed.

Refer to Fig. 3-7. The flux of the synchronously rotating stator field (ϕ_s) and the flux of the rotor currents (ϕ_f) combine vectorially to produce the net or resultant flux (ϕ_R) within the structure of stator and rotor.

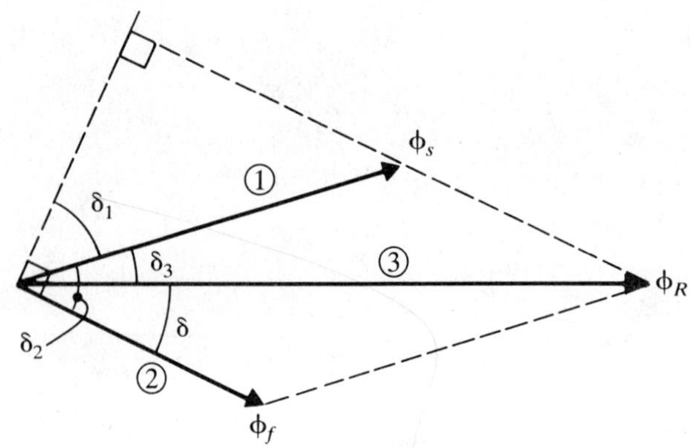

FIG. 3-7 Stator and rotor flux in a 3-ϕ induction machine at standstill.

From Ampere's law (Eq. (1.161)), the force and thus the torque (T) developed by two interacting fields is proportional to the product of the strength of the two fields times the sine of the smallest angle between the two fields. Mathematically,

$$T = K\phi_f\phi_s \sin \delta_2 \tag{3.1a}$$

Since

$$\delta_1 + \delta_2 = 90^\circ$$

Then,

$$T = K\phi_f\phi_s \cos \delta_1 \tag{3.1b}$$

where the constant of proportionality K is a function of the physical parameters of the machine, and the factor ($\phi_s \cos \delta_1$) is the quadrature component of the stator flux with respect to the rotor field.

. . .

* For these reasons, three-phase induction motors are sometimes referred to as rotating transformers.

From basic trigonometric concepts we have:

$$\phi_s \cos \delta_1 = \phi_R \cos (\delta_1 + \delta_3) \tag{3.2}$$

and

$$\phi_R \cos (\delta_1 + \delta_3) = \phi_R \sin \delta \tag{3.3}$$

where δ is the phase angle between the resultant field and the rotor field. This angle is referred to as the torque angle. From Eqs. (3.1b), (3.2), and (3.3) we obtain

$$T = K\phi_f \phi_R \sin \delta \tag{3.4}$$

Under ideal conditions, the torque produced in the rotor structure is delivered to the shaft load, while the opposing torque produced in the stator structure is transmitted to the motor's foundation.

3.1.3 ROTATING MAGNETIC FIELD

This section will analyze the generation of a rotating field by the stator windings of 3-ϕ induction machine and will derive formulae that give the strength and the speed of the rotating field.

Consider the elementary two-pole, three-phase machine shown in Fig. 3-8(a). The dots in the center of the three conductors indicate that the direction of current is

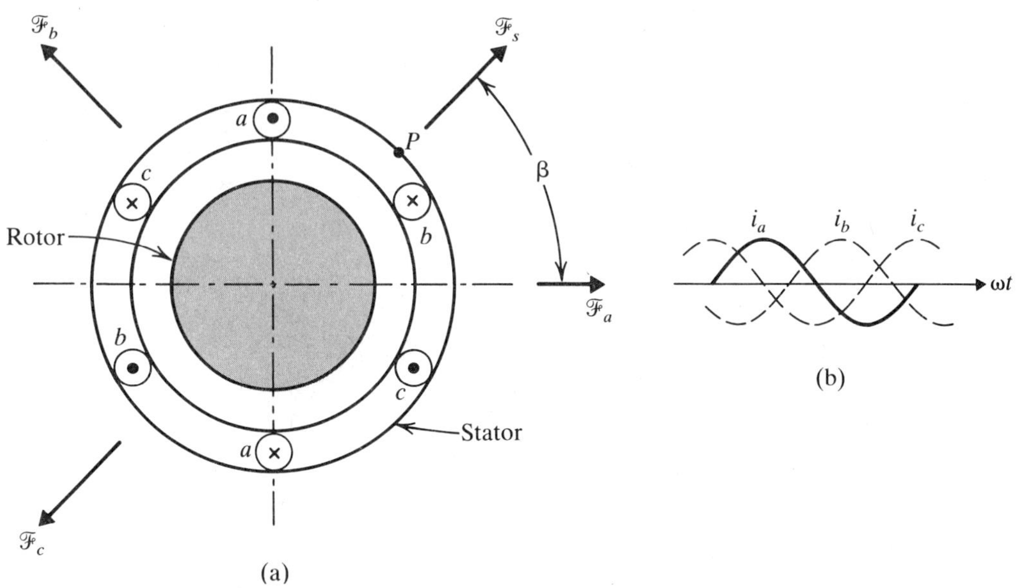

FIG. 3-8 Three-phase induction machine: **(a)** an elementary two-pole induction machine, **(b)** balanced three-phase stator currents of sequence ABC.

toward the reader, while the crosses indicate that the assumed positive direction of current is away from the reader. The stator windings carry balanced three-phase currents as shown in Fig. 3-8(b).

For a phase sequence ABC, the phase mmf's as functions of time are as follows:

$$\mathscr{F}_a = N_A I_{ma} \cos \omega t \tag{3.5}$$

$$\mathscr{F}_b = N_B I_{m_b} \cos (\omega t - 120°) \tag{3.6}$$

$$\mathscr{F}_c = N_C I_{m_c} \cos (\omega t + 120°) \tag{3.7}$$

where I_{m_a}, I_{m_b}, and I_{m_c} are the maximum values of the phase currents, and N_A, N_B, and N_C are the number of turns of the phase windings. The quantity ω is the angular frequency of oscillation of the stator currents, which by definition is

$$\omega = 2\pi f \text{ electrical radians per second} \tag{3.8}$$

where f is the frequency of the stator currents in hertz.

At the instant under consideration, the mmf of phase a coincides with the horizontal axis, as shown in Fig. 3-8(a). The directions of phase mmf's $\mathscr{F}_a$, $\mathscr{F}_b$, and $\mathscr{F}_c$ are obtained by using the right-hand rule.

The resultant stator mmf($\mathscr{F}_s$) along an axis at an angle β to the horizontal is found by summing up the projections of the phase mmf's along this line:

$$\mathscr{F}_s = \mathscr{F}_a \cos \beta + \mathscr{F}_b \cos (120° - \beta) + \mathscr{F}_c \cos (-120° - \beta) \tag{3.9}$$

The number of winding turns and the maximum value of the current for each phase are the same. Designating $\mathscr{F}_1$ as the maximum mmf of any one phase, we have

$$\mathscr{F}_1 = N_A I_{m_a} = N_B I_{m_b} = N_C I_{m_c} \tag{3.10}$$

Substituting Eqs. (3.5), (3.6), (3.7), and (3.10) into Eq. (3.9), we obtain

$$\begin{aligned}\mathscr{F}_s = \mathscr{F}_1[&\cos \omega t \cos \beta + \cos (\omega t - 120°) \cos (120° - \beta) \\ &+ \cos (\omega t + 120°) \cos (-120° - \beta)]\end{aligned} \tag{3.11}$$

By use of the identity

$$\cos x \cos y = \frac{1}{2} \cos (x + y) + \frac{1}{2} \cos (x - y)$$

we obtain

$$\mathcal{F}_s = \frac{\mathcal{F}_1}{2} [\cos(\omega t + \beta) + \cos(\omega t - \beta) + \cos(\omega t - 120^\circ + 120^\circ - \beta) + \cos(\omega t - 120^\circ - 120^\circ + \beta) + \cos(\omega t + 120^\circ - 120^\circ - \beta) + \cos(\omega t + 120^\circ + 120^\circ + \beta)] \quad \textbf{(3.12)}$$

Simplifying, we get

$$\mathcal{F}_s = \frac{\mathcal{F}_1}{2} [\underline{\cos(\omega t + \beta)} + \cos(\omega t - \beta) + \cos(\omega t - \beta) + \underline{\cos(\omega t + \beta - 240^\circ)} + \cos(\omega t - \beta) + \underline{\cos(\omega t + \beta + 240^\circ)}] \quad \textbf{(3.13)}$$

The sum of the three underlined terms is equal to zero because these phasors are displaced from each other by 120° and because their magnitudes are equal. Therefore Eq. (3.13) becomes

$$\mathcal{F}_s = \frac{3}{2} \mathcal{F}_1 \cos(\omega t - \beta) \quad \textbf{(3.14)}$$

Equation (3.14) describes a revolving field that rotates counterclockwise with an angular velocity of ω radians per second. The speed of the revolving field is normally designated by ω_s and is referred to as synchronous speed ($\omega_s = \omega$). The flux at a point P that is at β degrees to the horizontal will vary sinusoidally with the same frequency as that of the stator currents.

The revolving field may be visualized as being equivalent to the field generated by a permanent magnet which is rotated about an axis that coincides with the rotor of the machine.

The effective mmf through the stator structure is equal to 1.5 times the mmf produced by one phase alone. The resulting effective flux, at the absence of saturation and rotor current, is directly proportional to this mmf and inversely proportional to the reluctance of the path through which the flux completes its magnetic circuit.

In general, a rotating field of constant amplitude is produced by an m-phase winding wound $2\pi/m$ electrical radians apart and excited by balanced m-phase currents. The magnitude of the rotating field is $m/2$ times the field produced by any one phase, and its speed of rotation is given by Eq. (3.8). The direction of rotation of the field depends on the phase sequence of the applied currents. When the phase sequence of the supply voltages is reversed, the direction of rotation of the stator field and the speed of the motor are also reversed. The revolving fields of single and two-phase ac motors are developed in Chapter 4.

Mechanical and Electrical Radians

Consider the elementary four-pole machine shown in Fig. 3-9(a). Starting at point *A* and going counterclockwise (2π mechanical radians) around the periphery of the machine's stator, the following field polarities are encountered: N_1S_1-S_2N_2-N_3S_3-S_4N_4. This corresponds to two complete electrical cycles (4π electrical radians), as can be seen in Fig. 3-9(b).

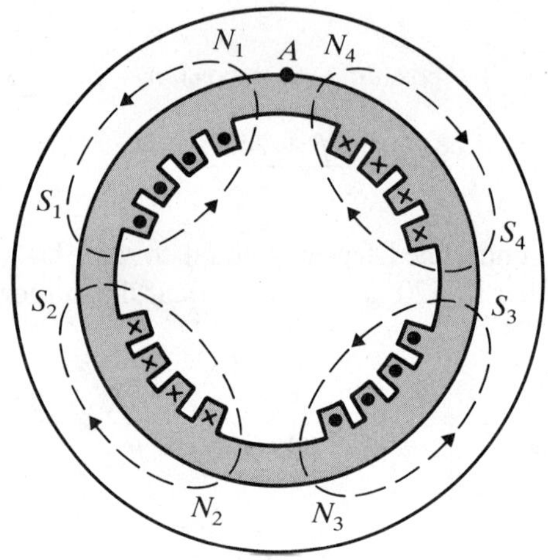

(a)

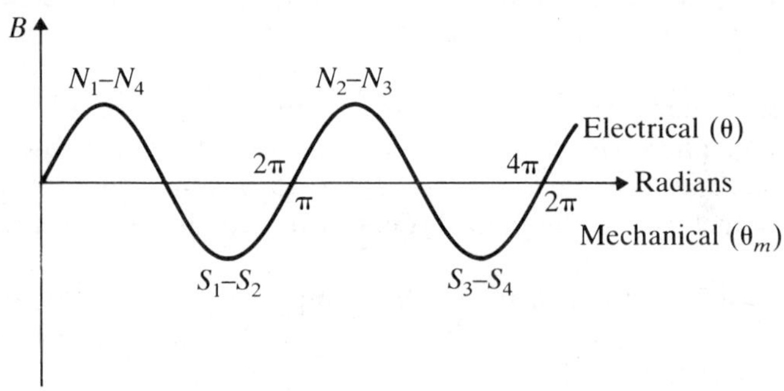

(b)

FIG. 3-9 **(a)** Elementary four-pole machine with stator slots and windings omitted, **(b)** space distribution of field illustrating the relationship between mechanical (θ_m) and electrical (θ) radians.

For a four-pole machine, then, one mechanical or physical revolution will correspond to two complete cycles of the field. In general, for a p-pole machine

$$\theta = \frac{p}{2}\,\theta_m \tag{3.15}$$

where θ and θ_m are the electrical and mechanical radians, respectively.

From Eqs. (3.8) and (3.15), we get the speed of the rotating field in mechanical radians per second:

$$\omega_m = \frac{2}{p}(2\pi f) = 4\pi\frac{f}{p} \quad \text{mechanical radians/s} \tag{3.16}$$

Since one revolution is equivalent to 2π mechanical radians, the speed in revolutions per second (n_s) is

$$n_s = 2\frac{f}{p} \quad \text{r/s} \tag{3.17}$$

The speed in revolutions per minute is

$$n_s = 120\frac{f}{p} \quad \text{r/min} \tag{3.18}$$

Thus, for a 60 Hz system, the synchronous speed of the stator field will be 3600, 1800, or 1200 r/min when the number of poles is 2, 4, or 6, respectively.

EXAMPLE 3–1 For a 480 V, 3-ϕ, 4-pole, 60 Hz induction motor determine the speed of the stator field in

a) Electrical radians per second.
b) Mechanical radians per second.
c) Revolutions per second.
d) Revolutions per minute.

SOLUTION

a) From Eq. (3.8)

$$\omega_s = 2\pi(60) = \underline{\underline{377 \text{ electrical rad/s}}}$$

b) From Eq. (3.16)

$$\omega_m = 377\left(\frac{2}{4}\right) = \underline{\underline{188.5 \text{ mechanical rad/s}}}$$

c) From Eq. (3.17)

$$n_s = 2\left(\frac{60}{4}\right) = \underline{\underline{30\ \text{rps}}}$$

d) From Eq. (3.18)

$$n_s = 120\left(\frac{60}{4}\right) = \underline{\underline{1800\ \text{r/min}}}$$

Exercise 3–1 Prove that the speed of rotation of the stator field is reversed when the order of rotation of the stator currents is reversed.

3.1.4 SLIP

Under normal operating conditions, the rotor rotates in the same direction as the magnetic field of the stator, but at a reduced speed. The difference between the synchronous speed of the stator field and the actual rotor speed (n_a) defines the slip (s) of the motor. The slip of a motor is an important parameter, used extensively in the design and analysis of induction machines. The per-unit value of the slip is given by

$$s = \frac{n_s - n_a}{n_s}\ \text{pu} \tag{3.19}$$

where n_s is the synchronous speed of the rotating field, given by Eq. (3.18).

To achieve higher efficiency, the majority of three-phase induction motors are designed to operate at a very small slip (usually less than 5%), when delivering rated power.

When three-phase induction machines operate as induction generators, their actual rotor speed is higher than their synchronous speed, and in the same direction as the synchronously rotating stator field. Thus, their slip is *negative.* Induction generators are used to connect small generating stations to large utility networks. A detailed analysis of induction generators is provided in Section 3.4.2.

When an induction machine is driven by another motor in such a way that its actual rotor speed is in the direction opposite to that of the synchronously rotating stator field, then n_a in Eq. (3.19) becomes negative, and thus the slip is *greater than unity.* Induction machines operating at a slip greater than unity are used as frequency and voltage multipliers. They are covered in Section 3.4.1.

EXAMPLE 3–2 A 75 kW, 60 Hz, 480 V, 1176 r/min three-phase induction motor has, at full-load, a power factor of .80 lagging and an efficiency of .90. Under rated operating conditions, determine:

a) The magnitude of the current drawn by the motor.

b) The real, reactive, and complex power drawn by the motor.

c) The output torque at full-load.

d) The operating slip.

SOLUTION

a) By definition:

$$\text{efficiency} = \frac{\text{output power}}{\text{input power}}$$

In mathematical symbols,

$$\eta = \frac{P_{\text{out}}}{P_{\text{in}}}$$

The output power is 75 kW. The input power is

$$P_{\text{in}} = \sqrt{3}\, V_{L-L} I_L \cos\theta$$

From the above, the magnitude of the line current is

$$I_L = \frac{P_{\text{out}}}{\eta(\sqrt{3}\, V_{L-L} \cos\theta)} = \frac{75{,}000}{.90(\sqrt{3}(480)(.80)} = \underline{\underline{125.29\text{ A}}}$$

The current phasor is shown in Fig. 3-10(a).

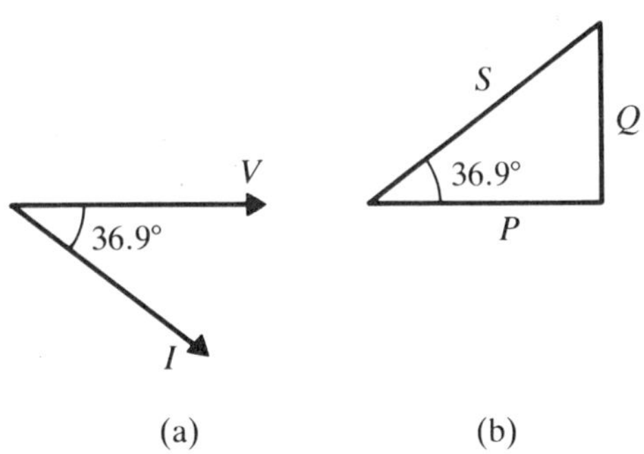

FIG. 3-10

b) The real or consumed power is

$$\begin{aligned} P_{\text{in}} &= \sqrt{3}\, V_{L-L} I_L \cos\theta \\ &= \sqrt{3}(480)(125.29)(.80) = \underline{\underline{83.33\text{ kW}}} \end{aligned}$$

or

$$P_{in} = \frac{P_{out}}{\eta} = \frac{75}{.9} = 83.33 \text{ kW}$$

The reactive power is

$$Q = \sqrt{3}\, V_{L-L} I_L \sin\theta$$
$$= \sqrt{3}(480)(125.29)(\sin 36.9^\circ) = \underline{\underline{62.5 \text{ kVAR}}}$$

or

$$Q = P \tan\theta = 83.33 \tan 36.9^\circ = \underline{\underline{62.5 \text{ kVAR}}}$$

The complex power is

$$S = \sqrt{3}\, V_{L-L} I^*$$
$$= \sqrt{3}(480)(125.29)\angle 36.9^\circ = \underline{\underline{104.17\angle 36.9^\circ \text{ kVA}}}$$

Alternatively,

$$S = P + jQ$$
$$= 83.33 + j62.5 = \underline{\underline{104.17\angle 36.9^\circ \text{ kVA}}}$$

The motor's power triangle is shown in Fig. 3-10(b).

c) The full-load output torque is

$$T = \frac{\text{power}}{\text{speed}} = \frac{75{,}000}{1176\dfrac{2\pi}{60}} = \underline{\underline{609 \text{ N}\cdot\text{m}}}$$

d) By inspection of the given data, it can be determined that the motor's synchronous speed is 1200 r/min. The operating slip is

$$s = \frac{n_s - n_a}{n_s} = \frac{1200 - 1176}{1200} = \underline{\underline{.02 \text{ pu}}}$$

Exercise 3–2 A 480 V, 50 kW, delta-connected three-phase, 60 Hz, 6-pole induction motor operates at a slip of 1.5% and has a power factor of 82% and an efficiency of 88%. Determine the following:

a) The magnitude of the line and phase currents.

b) The actual rotor speed.

c) The real, reactive, and complex power drawn by the motor.

Answer (a) 83.34 A, 48.12 A; (b) 1182 r/min;
(c) 56.82 kW, 39.66 kVAR, 69.29 $\underline{/34.9^\circ}$ kVA.

Exercise 3–3 In a laboratory a 4-pole, 60 Hz induction machine is operated as follows:

a) As an induction motor at a speed of 1760 r/min.

b) As an induction generator at a speed of 1850 r/min.

c) As a frequency multiplier at a speed of 900 r/min.

In each of the above cases, determine the operating slip and the direction of the rotor's rotation relative to the clockwise rotation of the stator field.

Answer (a) 2.22%, cw; (b) −2.78%, cw; (c) 150%, ccw.

3.1.5 EQUIVALENT CIRCUIT

The analysis of polyphase electric machines is simplified by using their per-phase equivalent circuits. The derivation of an equivalent circuit is based on the consideration of the electromagnetic coupling between the stator and rotor when the latter is stationary, and when it is running.

Rotor Stationary

When the rotor of the motor is stationary, the time-rate of change of the flux linkages between the stator and the rotor depends on the number of winding turns in the rotor and on the speed and magnitude of the rotating mmf of the stator. This is similar to the transformer action, where the alternating flux of the primary current sets up flux linkages between the primary and secondary windings. The voltage induced in the secondary winding of a transformer is proportional to the time-rate of change of its flux linkages. Similarly, the voltage induced in the rotor windings of a three-phase induction motor is proportional to the time-rate of change of the winding's flux linkages.

The frequency (f_r) of the voltage induced in the rotor windings is equal to that of the stator windings. That is,

$$f = f_r \tag{3.20}$$

where f is the frequency of the stator currents.

The per-phase equivalent circuit of a three-phase induction motor at standstill, then, is similar to that of a transformer. Refer to Fig. 3-11(a). The impedance $R_1 + jX_1$

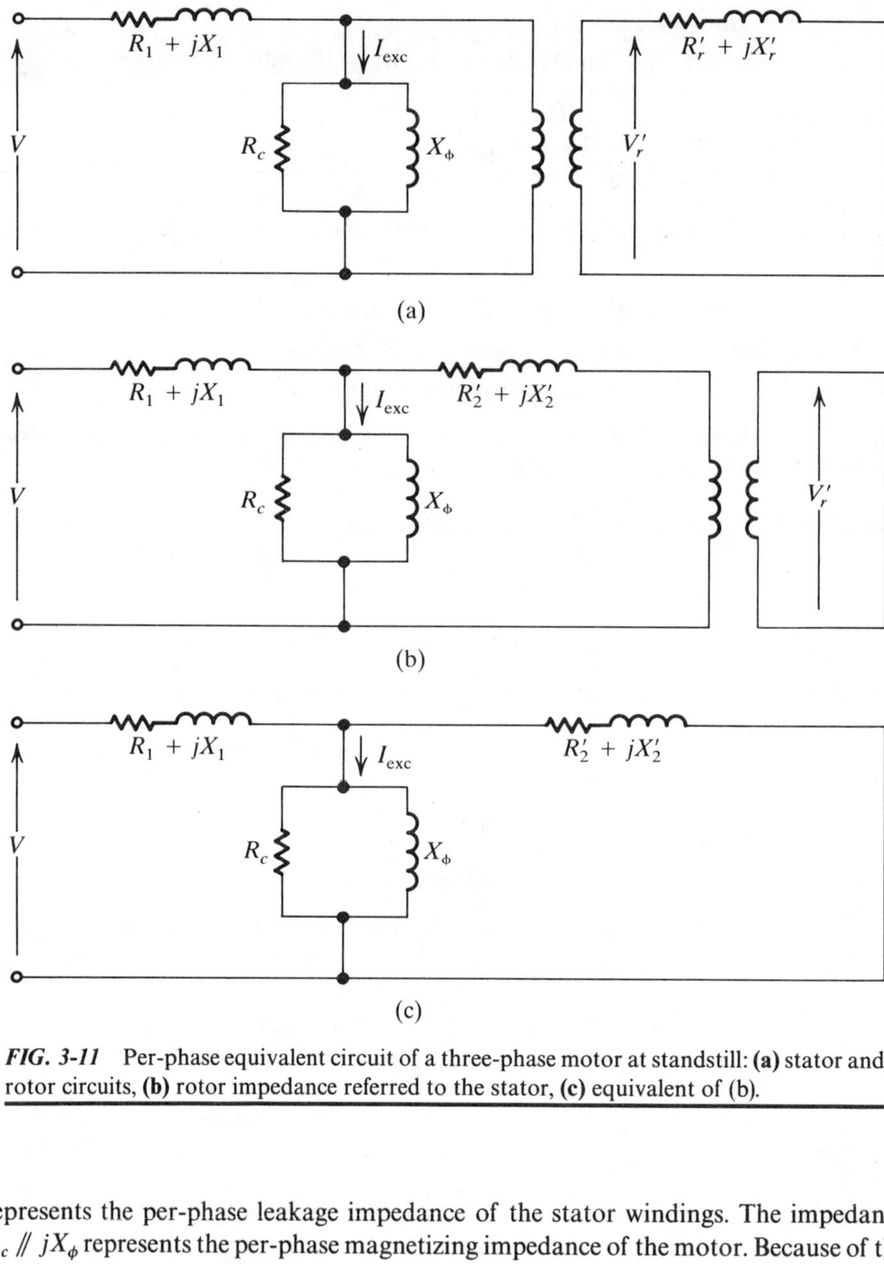

FIG. 3-11 Per-phase equivalent circuit of a three-phase motor at standstill: **(a)** stator and rotor circuits, **(b)** rotor impedance referred to the stator, **(c)** equivalent of (b).

represents the per-phase leakage impedance of the stator windings. The impedance $R_c \parallel jX_\phi$ represents the per-phase magnetizing impedance of the motor. Because of the air gap between the stator and rotor structures, this impedance is substantially larger than that of a static transformer of equivalent voltage and volt-ampere rating. The impedance $R_r' + jX_r'$ represents the per-phase leakage impedance of the rotor windings at standstill.

The rotor impedance as seen from the stator (Fig. 3-11(b)) is

$$Z_2' = a^2(R_r' + jX_r') \tag{3.21}$$

or

$$Z'_2 = R'_2 + jX'_2 \tag{3.22}$$

where Z'_2 is the rotor impedance at standstill as seen from the stator, and a represents the number of effective turns between the stator and rotor windings.

The rotor windings are shorted, and thus the ideal transformer shown in Fig. 3-11(b) is also shorted. This leads to the per-phase equivalent circuit shown in Fig. 3-11(c). The per-phase input voltage to the stator windings and the excitation current are represented, respectively, by V and I_{exc}.

Rotor Running

When the rotor is running, the time-rate of change of the flux linkage between the stator and rotor windings depends on the relative motion between the synchronously rotating stator mmf and the actual rotor speed. Since the relative motion between the stator mmf and the rotor speed is given by the slip, the voltage induced in the rotor windings (V_r) is given by

$$V_r = sV'_r \tag{3.23}$$

where s is the operating slip of the motor, and V'_r is the voltage induced in the rotor windings when the rotor is stationary.

The frequency (f_r) of the voltage induced in the rotor windings is related to the frequency (f) of the stator windings by

$$f_r = sf \tag{3.24}$$

Then the per-phase rotor impedance at a slip s is

$$Z_r = R_r + jsX_r \tag{3.25}$$

where R_r and X_r represent the rotor resistance and reactance, respectively. The per-phase equivalent circuit of the motor is shown in Fig. 3-12(a). The rotor current (I_r) is

$$I_r = \frac{V_r}{Z_r} \tag{3.26}$$

$$= \frac{sV'_r}{R_r + jsX_r} \tag{3.27}$$

From the above:

$$I_r = \frac{V'_r}{\dfrac{R_r}{s} + jX_r} \tag{3.28}$$

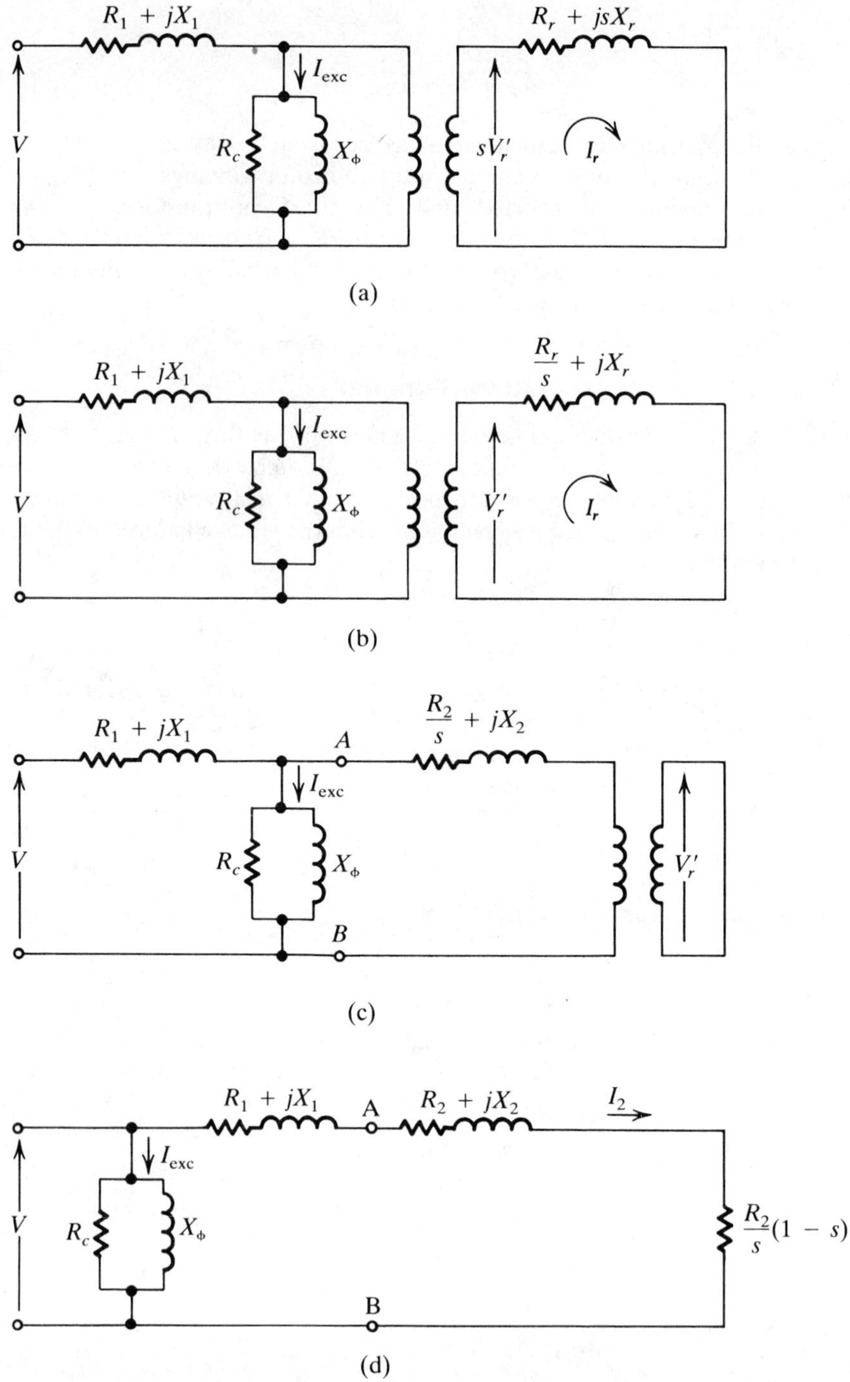

FIG. 3-12 Per-phase equivalent circuit of a three-phase induction motor: **(a)** stator and rotor circuits, **(b)** equivalent of (a), **(c)** equivalent of (b), **(d)** approximation of (c).

Then the per-phase equivalent circuit of the motor can be drawn as shown in Fig. 3-12(b).

The rotor impedance (Z_2) as seen from the stator is

$$Z_2 = a^2\left(\frac{R_r}{s} + jX_r\right) \tag{3.29}$$

or

$$Z_2 = \frac{R_2}{s} + jX_2 \tag{3.30}$$

The corresponding equivalent circuit is shown in Fig. 3-12(c).

The real component of the rotor impedance, as seen by the stator, can also be written as follows:

$$\frac{R_2}{s} = R_2 + R_2\frac{1-s}{s} \tag{3.31}$$

The second term on the right-hand side of this equation is referred to as the equivalent mechanical load (R_L) resistance. That is,

$$R_L = \frac{R_2(1-s)}{s} \tag{3.32}$$

By assuming that

$$R'_2 + jX'_2 = R_2 + jX_2 \tag{3.33}$$

then Fig. 3-12(c) represents the general per-phase equivalent circuit of a three-phase induction motor at any speed, as seen from the input terminals of the motor.

It should be pointed out that when the motor is running, the values of the per-phase rotor resistance (R_r) and inductance (L_r) are different than their corresponding values at standstill (R'_r, L'_r) because of higher temperatures, saturation, and the skin effect.

The skin effect (nonuniform current density within the conductors caused by magnetic flux which passes through them and which varies with time) is of particular importance because it causes the rotor's circuit resistance and inductance to vary with the frequency of the rotor currents. This variation is significant during starting, but negligible within the speed range of normal load operation. For accurate results, the actual values of the rotor parameters should be used at starting and running conditions.

In order to further simplify the calculations, the magnetizing branch is usually transferred to input terminals of the stator winding, as shown in Fig. 3-12(d). The magnetizing impedance could also be represented, as in the case of transformers, by its equivalent series-connected components ($R_m + jX_m$). The measurement of the parameters of the equivalent circuit is discussed in Section 3.3.

3.1.6 TORQUE AND POWER RELATIONSHIPS

Torque and power relationships will be derived by using the equivalent circuit shown in Fig. 3-12(d). In deriving the general expression for the torque of an induction motor, the effects of the magnetizing impedance will be neglected. Because this magnetizing impedance is relatively large, its effect on the torque developed by the motor at full-load is negligible.

From basic definitions, the per-phase torque developed is

$$T = \frac{\text{power developed}}{\text{actual speed}} = \frac{I_2^2 R_L}{\omega_a} \tag{3.34}$$

The equivalent mechanical load resistance (R_L) is

$$R_L = R_2 \frac{1-s}{s}$$

and the actual speed is

$$\omega_a = \omega_s(1-s)$$

Substituting the above into Eq. (3.34), we obtain

$$T = \frac{I_2^2 R_2 \dfrac{1-s}{s}}{\omega_s(1-s)} \tag{3.35}$$

$$= \frac{I_2^2 R_2}{\omega_s s} \ \text{N}\cdot\text{m/phase} \tag{3.36}$$

From the equivalent circuit, the magnitude of the current I_2 is

$$I_2 = \left|\frac{V}{Z}\right| = \frac{V}{\sqrt{\left(R_1 + \dfrac{R_2}{s}\right)^2 + X^2}} \tag{3.37}$$

where $X = X_1 + X_2$.
From the above, the torque developed in terms of the motor's parameters is

$$T = \frac{V^2}{\left[\left(R_1 + \dfrac{R_2}{s}\right)^2 + X^2\right]}\left(\frac{R_2}{\omega_s s}\right) \ \text{N}\cdot\text{m/phase} \tag{3.38}$$

where V is the phase-to-neutral voltage.

The variation of torque versus slip is shown in Fig. 3-13. The abscissa of this curve is the motor's speed or slip. The symbols T_{fl}, T_{st}, and T_m correspond to the full-load,

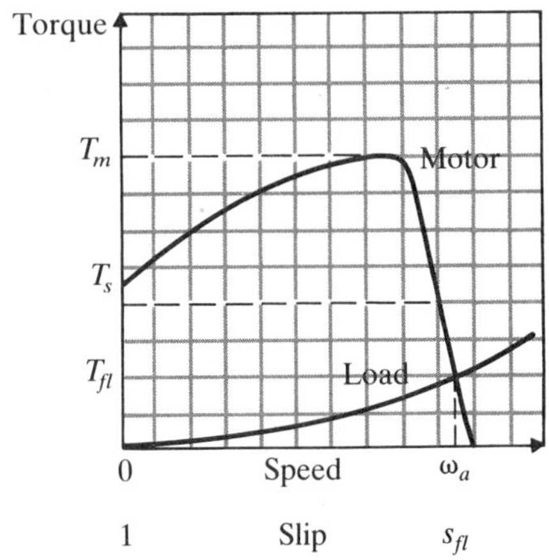

FIG. 3-13 Typical torque–speed characteristics.

starting, and maximum torque of the motor, respectively. The same diagram also shows the torque-speed characteristic of the mechanical load. The intersection of these characteristics gives the operating torque and speed of the motor. On account of this it is extremely important, before a suitable motor is selected, to accurately calculate the torque–speed characteristic of the driven load.

The motor and load torque–speed characteristics are also used, as explained in Section 3.5, to find the accelerating and decelerating times of the motor. These parameters are important in the transient analysis of machines and in the proper selection of the motor's protective devices.

From Eq. (3.38), it is evident that, for a constant slip, the torque is proportional to the applied voltage squared:

$$T \propto V^2 \tag{3.39}$$

In practice, one may have to evaluate the starting torque capability of a motor because its high starting currents—depending on the impedance of the supply network—reduce the voltage delivered to the stator, and as a result the starting torque is also reduced. When the Thevenin impedance ($R_{th} + jX_{th}$) of the supply network is known, R_1 in Eq. (3.38) should be replaced by $R_1 + R_{th}$, and X should be replaced by $X + X_{th}$.

From Eq. (3.38), after expanding and simplifying, we obtain

$$T = \frac{V^2 sR_2}{\omega_s[(sR_1)^2 + 2sR_1R_2 + R_2^2 + (sX)^2]} \quad \text{N} \cdot \text{m/phase} \tag{3.40}$$

For slips in the full-load range, the predominant term in the denominator of the above equation is R_2^2. Therefore,

$$T \approx \frac{V^2}{\omega_s} \frac{s}{R_2} \tag{3.41}$$

or

$$T \approx K \frac{s}{R_2} \tag{3.42}$$

where K is the constant of proportionality given by

$$K = \frac{V^2}{\omega_s} \tag{3.43}$$

From Eq. (3.42), it is evident that in the full-load operating range, the higher the actual speed of the motor, the lower its output torque. It can be also shown that for a certain range of rotor resistance, the starting torque (T_{st}) of the motor is proportional to its rotor resistance. Mathematically,

$$T_{st} \propto R_2 \tag{3.44}$$

The higher the rotor resistance, the higher the motor's starting torque, and the lower its inrush current. The derivation of Eq. (3.44) is left as a student problem (see Problem 3-2).

Because of the effects of rotor resistance on starting torque and starting current, the high-power motors supplied through weak power systems (i.e., a supply network with relatively high impedance) are of the wound-rotor type.

Maximum or Breakdown Torque

Refer again to the approximate equivalent circuit of Fig. 3-12(d). For maximum power transfer from stator to the rotor-load resistance, the following relationship must be satisfied:

$$|R_1 + jX| = \frac{R_2}{s} \tag{3.45}$$

Equating the magnitudes of this relationship, we get the equation for the slip of the motor at maximum torque (s_{mt}):

$$s_{mt} = \frac{R_2}{\sqrt{R_1^2 + X^2}} \tag{3.46}$$

where

$$X = X_1 + X_2$$

Substituting Eq. (3.46) into Eq. (3.38), after simplification we find the following expression for the maximum torque developed by the motor:

$$T_m = \frac{V^2}{[(R_1 + \sqrt{R_1^2 + X^2})^2 + X^2]} \times \frac{\sqrt{R_1^2 + X^2}}{\omega_s} \text{ N} \cdot \text{m/phase} \tag{3.47}$$

Thus, the maximum torque is independent of rotor resistance. The latter controls the slip or the speed where the maximum torque occurs. Typical torque–speed characteristics, as a function of rotor resistances, are shown in Fig. 3-14.

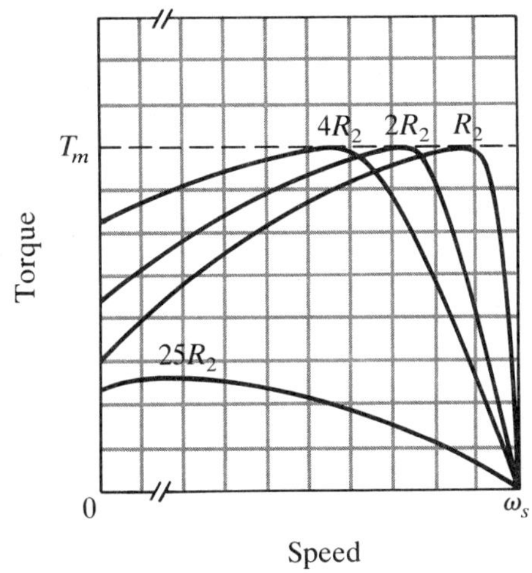

FIG. 3-14 Torque–speed characteristics as function of rotor resistance.

With negligible stator resistance ($R_1 = 0$), it can be shown that the torque (T) at any slip (s) is related to the maximum torque (T_m) and its corresponding slip (s_{mt}) by

$$\frac{T}{T_m} = \frac{2ss_{mt}}{s_{mt}^2 + s^2} \tag{3.48}$$

If the torque–speed characteristic of the motor is not available, then the above equation can be used, in conjunction with some of the motor parameters (T_{st}, T_m, T_{fl}, s_{fl}), to sketch it. The derivation of Eq. (3.48) is left as a student problem (see Problem 3-2).

The torque developed by a motor must be sufficient to provide the motor's rotational losses and the torque requirements of the mechanical load. The torque requirements of a mechanical load may include any or all of the following torque components: the torque component $[(J(d\omega_a/dt)]$ required to accelerate the rotating masses up to the steady-state speed; the torque component ($B\omega_a$) that represents the viscous friction of the rotating parts; and any other torque (T_L) requirements of the load. In mathematical form,

$$T = J\frac{d\omega_a}{dt} + B\omega_a + T_L \tag{3.49}$$

where

J = the polar moment of inertia of the rotating masses in Newton-meter-second squared

B = the coefficient of viscous friction in N · m/(rad/s)
T_L = a component of load torque in N · m
ω_a = the actual speed of the motor in rad/s.

Equation (3.49) is derived from Newton's second law of motion for rotating bodies. This law states that the sum of the torques ($\sum T$) about an axis of rotation is equal to the polar moment of inertia (J) times the angular acceleration (α) of the rotating masses. Mathematically,

$$\sum T = J\alpha \tag{3.50}$$

This law is of primary importance to the dynamic analysis of all machines, and its applications are discussed in various chapters of this book.

In developing the torque relationships, the magnetizing impedance of the motor was neglected in order to simplify the calculations. The magnetizing impedance, as already mentioned, does not affect the torque of the motor at full-load; however, at light loads it has significant effects. For accurate torque calculations, the magnetizing impedance should be taken into consideration. This can easily be done by replacing, in Eqs. (3.38) and (3.47), the voltage source (V) and stator impedance (Z_1) by the equivalent Thevenin's voltage and impedance. Referring to Fig. 3-12(c), and looking toward the input from terminals $A - B$, we obtain

$$V_{th} = \frac{V(R_c \mathbin{/\!/} jX_\phi)}{R_1 + jX_1 + R_c \mathbin{/\!/} jX_\phi} \tag{3.51}$$

and

$$Z_{th} = (R_1 + jX_1) \mathbin{/\!/} (R_c \mathbin{/\!/} jX_\phi) \tag{3.52}$$

From Eqs. (3.37) and (3.46), we find that the current I at a slip s is related to the current at maximum torque (I_{mt}) and its corresponding slip (S_{mt}) by

$$\left(\frac{I}{I_{mt}}\right)^2 = \frac{2s^2}{s_{mt}^2 + s^2} \tag{3.53}$$

It must be emphasized that the equations developed in this section are based on the approximate equivalent circuit of Fig. 3-12(d). They are *not* applicable to motors with deep-bar rotor construction or to double squirrel-cage rotors. In such cases the proper equivalent circuit should be drawn; then, from basic principles, the desired relationships can be derived.

Power Considerations

The power across the air gap (P_{ag}) provides the power developed and the rotor copper loss. That is,

$$P_{ag} = I_2^2 R_2 \frac{1 - s}{s} + I_2^2 R_2 \tag{3.54}$$

Thus:

$$P_{ag} = I_2^2 \frac{R_2}{s} \text{ W/phase} \tag{3.55}$$

As can be seen from Eq. (3.55), the rotor winding losses are directly proportional to the air-gap power and the operating motor slip. Thus, the smaller the slip, the lower the rotor copper losses. For this reason, the slip, at full-load, is usually kept below 5%.

The power developed (P_d) furnishes the output power plus the rotational losses. Mathematically,

$$P_d = I_2^2 R_2 \frac{1-s}{s} \text{ W/phase} \tag{3.56}$$

$$= P_{\substack{\text{rotational}\\ \text{losses}}} + P_{\text{output}}$$

The rotational losses include windage, friction, and stray load losses.

EXAMPLE 3–3 Figures 3-15(a) and (b) show, respectively, the one-line diagram of a three-phase induction motor and its per-phase equivalent circuit referred to the stator. Assuming constant rotor impedance, and neglecting the effects of the magnetizing current and rotational losses, determine:

a) The number of poles.
b) The synchronous speed of the stator field.
c) The per-unit slip.
d) The line current at starting.
e) The starting torque.
f) The full-load current, power factor, torque, and efficiency.
g) The slip and current at maximum torque.
h) The maximum torque.

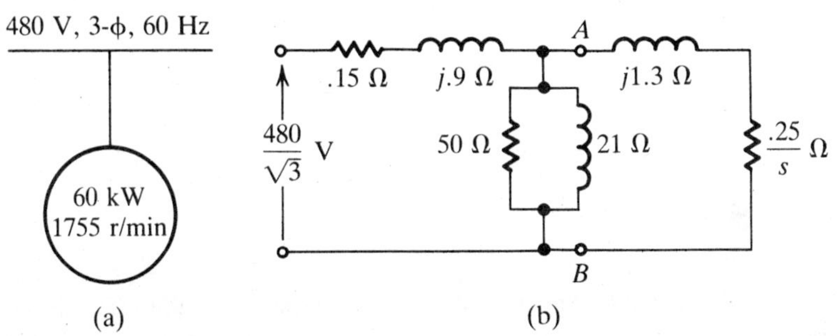

FIG. 3-15

SOLUTION

a) Under normal operating conditions, the slip of a three-phase induction motor is kept small because of its direct relationship to the copper losses. Thus, the actual speed of the motor should be very close to the speed of the synchronously rotating stator field. Since the actual speed is given as 1755 r/min, the closest synchronous speed is 1800 r/min. From Eq. (3.18), the number of poles is

$$p = \frac{120 \times 60}{1800} = \underline{\underline{4 \text{ poles}}}$$

b) The speed of the stator field is

$$n_s = \underline{\underline{1800 \text{ r/min}}}$$

or

$$\omega_s = 1800 \frac{2\pi}{60} = \underline{\underline{188.5 \text{ rad/s}}}$$

c) Using Eq. (3.19), the slip in per-unit is

$$s = \frac{1800 - 1755}{1800} = \underline{\underline{.025 \text{ pu}}}$$

d) The line current at starting is found by applying Ohm's law in the equivalent circuit of Fig. 3-15(c):

$$I_{st} = \frac{480/\sqrt{3}}{.15 + .25 + j2.2} = \underline{\underline{123.94\angle -79.7^\circ \text{ A}}}$$

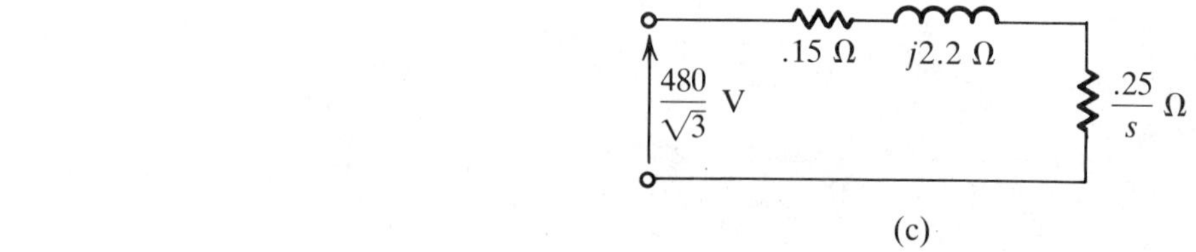

FIG. 3-15(c)

e) Since the motor is star-connected and the magnetizing impedance is neglected, the rotor current as seen from the motor terminals is equal to the stator or line current.

Substituting the known data into Eq. (3.36), the motor's starting torque is

$$T_{st} = 3(123.94)^2 \frac{(.25)}{188.5} = \underline{\underline{61.12 \ \text{N} \cdot \text{m}}}$$

f) The current at full-load is

$$I_{fL} = \frac{480/\sqrt{3}}{.15 + \dfrac{.25}{.025} + j2.2} = 26.68 \underline{/-12.2^\circ} \ \text{A}$$

By definition, the power factor is cos 12.2° = .977 lagging. The power factor is unrealistically high because the magnetizing impedance was assumed to be negligible.

The full-load torque is

$$T_{fL} = \frac{3(26.68)^2}{188.5}\left(\frac{.25}{.025}\right) = \underline{\underline{113.32 \ \text{N} \cdot \text{m}}}$$

The efficiency of the motor at full-load is

$$\eta = 1 - \frac{P_{\text{loss}}}{P_{\text{in}}} = 1 - \frac{3I^2R}{\sqrt{3}\, V_{L-L} I_L \cos\theta} = 1 - \frac{3(26.68)(.40)}{\sqrt{3}(480)(.977)}$$

$$= \underline{\underline{.96}}$$

g) From Eq. (3.46), the slip at maximum torque is

$$s_{mt} = \frac{.25}{\sqrt{.15^2 + 2.2^2}} = \underline{\underline{.11 \ \text{pu}}}$$

The corresponding motor's speed is

$$n_a = 1800(1 - .11) = 1596 \ \text{r/min}.$$

The current at maximum torque is

$$I = \frac{480/\sqrt{3}}{.15 + \dfrac{.25}{.11} + j2.2} = \underline{\underline{85.99 \underline{/-43^\circ} \ \text{A}}}$$

h) The maximum torque is

$$T_m = \frac{3(85.99)^2(.25)}{188.5(.11)} = 259.44 \ \text{N} \cdot \text{m}$$

For the purposes of comparison, the results are summarized in Table 3-1.

TABLE 3-1 Summary of the Results of Example 3-3

Operating Condition	*Speed in r/min*	*Current in A*	*Power Factor Lagging*	*Torque in N · m*
Starting ($s = 1$)	0	123.94	.18	61.12
Full-load ($s = .025$)	1755	26.68	.977	113.32
Maximum torque ($s_{mt} = .11$)	1596	85.99	.73	259.44

Exercise 3–4 The per-phase equivalent circuit of a 480 V, 60 Hz, 1176 r/min induction motor is shown in Fig. 3-16. Neglecting the magnetizing impedance, determine the starting, full-load, and maximum torque of the motor.

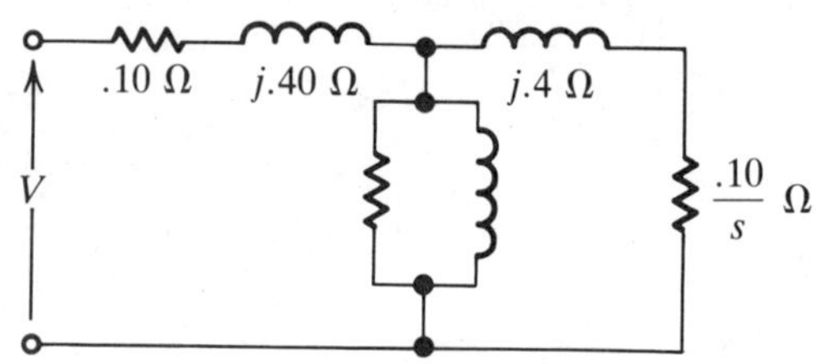

FIG. 3-16

Answer 269.63 N · m, 343.99 N · m, 1011.6 N · m.

3.2 Industrial Considerations

This section discusses the industrial classification of squirrel-cage induction motors and mechanical loads, the change in a motor's input parameters that accompany variations in the load requirements, a motor's efficiency, power factor, and the effects of the supply voltage on the torque–speed characteristic.

3.2.1 CLASSIFICATION OF INDUCTION MOTORS

The National Electrical Manufacturers Association (NEMA) has developed a code-letter system by which a letter (A, B, C, or D) designates a *particular class of motors* with specific characteristics. The various characteristics are mainly obtained by a unique rotor design. Table 3-2 shows the general characteristics, and the corresponding applications, of each class of motor. Typical torque–speed characteristics for the various code-letter designations of squirrel-cage induction motors are shown in Fig. 3-17. Before a motor is selected, the characteristics of the load must be known. The load characteristics include power requirements, starting and full-load torque, speed variation, acceleration characteristics, and the environment in which the motor is to operate.

TABLE 3-2 Characteristics and Applications of 3-ϕ Induction Motors

Design	*Starting Torque, in Per-Unit*	*Starting Current, in Per-Unit*	*Maximum Torque, in Per-Unit*	*Range of Full-Load Slip*	*Applications*
A	1.5 to 1.75	5 to 8	2 to 2.5	2% to 5%	General-purpose motors, fans, blowers, most machinery tools
B	1.5 to 1.75	4.5 to 5	2 to 3	3% to 6%	As Class A
C	2 to 2.5	3.5 to 5	1.9 to 2.25	4% to 8%	Used in conveyors, compressors
D	2.75 to 3	3 to 8	2.75	7% to 17%	Used for high-inertia loads; low efficiency

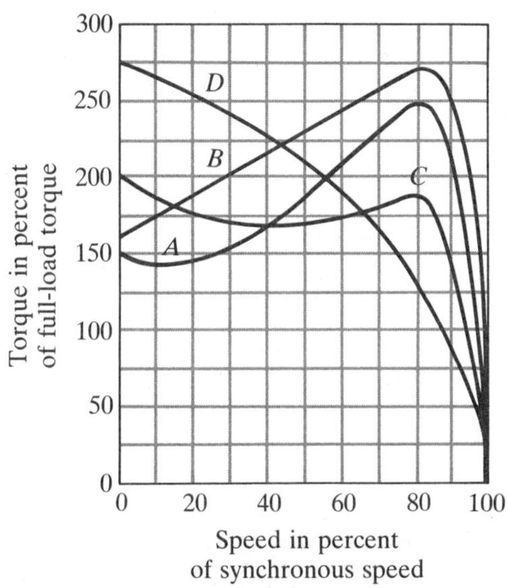

FIG. 3-17 Typical torque–speed characteristics of squirrel-cage induction motors.

To select a particular type of motor for a given mechanical load, the torque–speed characteristic of the motor must be compared to that of the mechanical load. The intersection of the motor–load characteristics must correspond not only to the speed and torque requirements of the load, but also to the highest efficiency for the motor.

The *starting torque* of the motor must be larger than the load requirement, or the motor will not be able to start rotating. As a result, it will draw its locked-rotor current. This may damage the motor.

The minimum value of the torque in the region between starting and maximum torque is called the *pull-in torque*. As can be seen from the general torque–speed characteristics (Fig. 3.17), only class A and C motors have a meaningful pull-in torque. The torque requirements of a load, at a speed that corresponds to the pull-in torque of the motor, must be lower than those of the motor itself; otherwise, the rotor will not accelerate smoothly.

The *maximum torque* developed by a motor indicates the capability of the machine to overcome high transient-load torques. A motor with a maximum torque that is relatively low may stall when a sudden load torque exceeds the motor's breakdown torque.

The *accelerating torque* of a motor can be obtained from its torque–speed characteristics, by comparing the differences in the areas under the torque–speed curves of the motor and the load. This is explained in more detail in Section 3.5.

A motor's accelerating time (the time it takes a motor, starting from rest, to reach its steady-state speed) is inversely proportional to its accelerating torque. For all practical purposes, during acceleration a motor draws locked-rotor current. The longer a motor's accelerating time, then, the greater the possibility of thermal damage to its windings. On account of this, the maximum number of permissible starts during a specific time interval should be carefully determined, and the motor's manufacturer should—as is usually the practice—be consulted.

For any given mechanical load, a class D motor is best equipped to reduce accelerating time (see Fig. 3-17). The cost of a motor, however, increases progressively from class A to class D.

3.2.2 MECHANICAL LOADS

The selection of a particular motor for a specific load requirement is based on the torque–speed characteristic of the mechanical load. Manufacturers of motors will provide the characteristics of a motor: its starting current and torque, maximum torque, starting and full-load power factors, etc. Similarly, manufacturers of fans, pumps, etc. can provide the torque–speed characteristics of their equipment.

When the torque–speed characteristic of a particular load is not available, it can be derived by drawing the free-body diagram of the system under consideration and by applying Newton's laws of motion.

Load Classification

Mechanical loads are classified according to their torque–speed characteristics. There are actually many such characteristics but all may be grouped into one of four general categories: constant torque; torque proportional to speed; torque inversely proportional to speed; and torque proportional to the square of the speed (see Fig. 3-18). The mathematical descriptions and the actual torque–speed characteristics of mechanical loads—usually evaluated by mechanical process engineers—can generally be found in standard mechanical engineering handbooks.

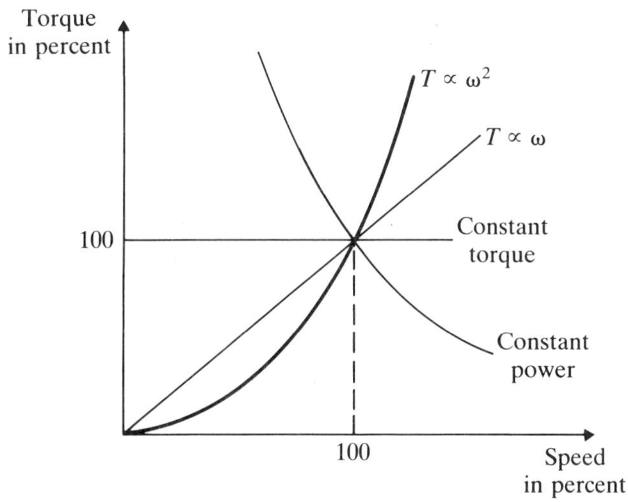

FIG. 3-18 General torque–speed characteristics of mechanical loads.

Constant Torque

A constant-torque load is a load whose torque remains constant within the operating speed range of the load. Examples of constant-torque loads are conveyors, grinding mills, and crane-hoist systems.

Torque Proportional to Speed

A load whose torque requirement is proportional to its speed is the calender—a machine used for pressing and smoothing cloth or paper between rollers. The smoothness of this page depends on the calender of the paper machine that produced it.

Torque Inversely Proportional to the Speed

A load whose torque is inversely proportional to its speed is also referred to as constant-power load. Examples of constant-power loads are circular saws and lathe drives.

Torque Proportional to the Square of the Speed

Loads whose torque requirements are proportional to the square of their speed (power is proportional to the third power of the speed) are centrifugal fans and blowers. All the air-handling fans used in heating, ventilating, and air conditioning systems (HVAC) are of this type. The opposition or the resistance to the flow of air in these systems is proportional to the square of the speed of the air flow.

Exercise 3–5 Prove the following:

a) The mechanical load on a motor that drives a crane-hoist is of the constant torque type. (*Hint:* Torque is equal to the force times its moment arm.)
b) The torque of a motor that drives a fan in an air-handling system is directly proportional to the square of the motor's speed. (*Hint:* From the energy equation of hydraulics, the frictional losses are proportional to the square of the speed of the air flow.)
c) The torque of a motor that drives rollers between which paper is pressed and smoothed is directly proportional to the speed of the rollers.
d) The torque required by a circular saw is inversely proportional to its speed.

3.2.3 MECHANICAL LOAD CHANGES AND THEIR EFFECTS ON A MOTOR'S PARAMETERS

This section details how the current, power factor, and efficiency of a motor are affected by changes in the load requirements. A load characteristic may be changed when the dampers of a ventilating fan are adjusted, for example, or when a pump's control valve is reset. Such adjustments may be necessary in order to satisfy a particular rate-of-flow condition, which in turn enables the successful completion of a process.

Figure 3-19 shows the torque–speed characteristics of a motor and of a variable mechanical load. Under *nominal* operating conditions the speed and the torque developed by the motor are given by the intersection of the motor and load characteristics, identified by the symbol Q_1. Under this load requirement, the motor operates at full-load condition. Its losses (copper and rotational) are minimum in comparison to the output power; its stator and rotor currents are of nominal

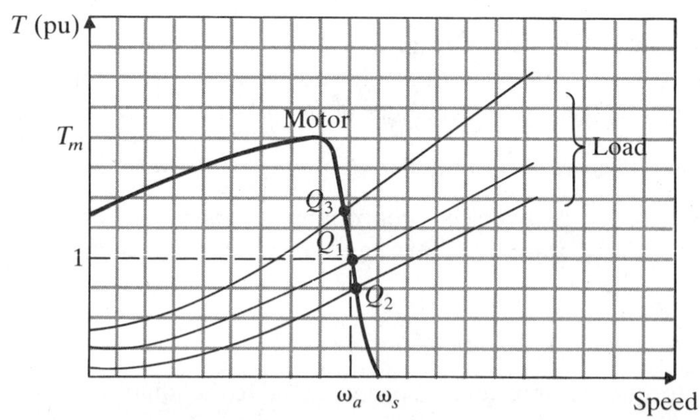

FIG. 3-19 Torque–speed characteristics of an induction machine and its mechanical load.

magnitude, and thus the motor—all other conditions being equal—will operate satisfactorily during its entire life. The motor's power factor, efficiency, torque, and speed under nominal operating conditions are all inscribed on its name-plate.

When the torque requirements of the load are *reduced*, the new operating point is given, as before, by the intersection of the torque–speed characteristics. This point is identified by Q_2. As shown, the speed of the motor will increase and its slip will be reduced. Then the impedance* of the motor, as can be seen from the motor's equivalent circuit, will also increase. As a result, the current—at constant terminal voltage—will be reduced.

The reduction in the stator current will reduce the copper losses of the motor but will not appreciably affect the rotational losses. Although the total losses decrease—as does the output power—the *efficiency* of the motor *decreases*. The power factor will increase because the reduction in slip increases the real component of motor impedance. The motor, as seen from its supply lines, appears more resistive.

When the torque requirements of the load *increase*, the motor's torque–speed characteristic is shifted upward. The new operating point of the motor–load system is given by Q_3. The speed of the motor is reduced, and its torque is increased in order to satisfy the new load requirements. The slip of the motor increases [see Eq. (3.19)], and as a result the impedance of the motor decreases and becomes more inductive. This increase in the operating slip is accompanied by an increase in the current drawn by the motor; consequently, the motor's copper losses increase. The rotational losses remain about constant and the efficiency of the motor normally decreases, although the output power of the motor increases.

Exercise 3–6 A 6-pole, 60 Hz, 3-ϕ induction motor drives a ventilating fan. The motor torque at full-load, as a function of slip, is

$$|T_{m_{fL}}| = 4000s \ \ \text{N} \cdot \text{m}$$

The torque of the load as a function of the speed is

$$T_L = 10 + 4.616 \times 10^{-3}\, \omega^2 \ \ \text{N} \cdot \text{m}$$

a) Sketch the torque–speed characteristics of the motor and load, and determine the torque and power of the motor.
b) Repeat (a), given that the dampers of the ventilating system are adjusted so that the load characteristic becomes

$$T_L = 10 + 16.484 \times 10^{-3}\, \omega^2 \ \ \text{N} \cdot \text{m}$$

Answer (a) 80 N · m, 9.85 kW; (b) 240 N · m, 28.35 kW.

---

* $Z = R_1 + (R_2/s) + jX$ ohms.

3.2.4 VOLTAGE, EFFICIENCY, AND POWER-FACTOR CONSIDERATIONS

Voltage Considerations

Induction motors whose nominal capacity is less than 200 kW are designed to operate at voltages less than 1 kV. These are referred to as low voltages (LV). In contrast, motors with a power rating of more than 200 kW are generally designed to operate at voltages within the 1 kV to 15 kV range. These are referred to as medium voltages (MV). The most popular MV levels are 2.3 kV, 4.16 kV, 6.6 kV, and 13.2 kV. The 4 MW grinding mills used in the cement industry, for example, are rated at 4.16 kV, while the 8 MW motors used in the pulp and paper industry are designed to operate at 13.2 kV. MV induction machines are certainly more expensive than LV motors of the same kW rating, but they draw reduced inrush and full-load currents, require less expensive cables, and reduce the losses throughout the upstream electrical network.

In general, the torque developed by a motor at a given speed is proportional to the square of the applied voltage, as demonstrated by Eq. (3.38). Refer to Fig. 3-20. For a given voltage, while all other parameters remain constant, the motor's torque–speed characteristic is fixed. Changes in the supply voltage will change the motor's characteristic accordingly. By reducing the voltage from V' to V volts, the motor's torque–speed characteristic is shifted downward. As a result, the motor's torque—at any speed—is reduced.

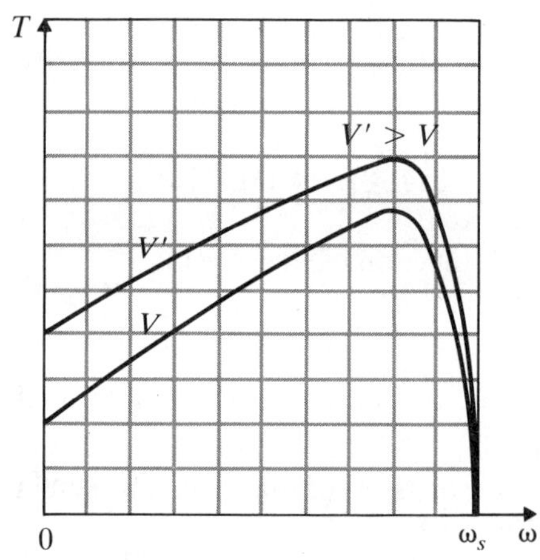

FIG. 3-20 Torque–speed characteristics as a function of voltage.

For small variations in the load requirements, and for a constant ac voltage supply, the motor's *speed variation is limited.* A moderate reduction in the speed will be accompanied by a drastic increase in the current drawn by the motor (see manufacturer's data, Fig. 3-23). If the current protective devices do not remove the motor from its supply voltage, it will overheat. For a given mechanical load, on the other hand, a substantial increase in the speed of the motor is not possible. Under normal conditions, the motor operates very close to synchronous speed, regardless of supply voltage. This is evident from the torque–speed characteristics of any 3-ϕ induction motor.

Thus, squirrel-cage induction machines supplied through a constant ac voltage source have a very small speed-variation range; consequently, they cannot be used for mechanical loads that require even a $\pm 2\%$ variation in speed. Another disadvantage of the constant ac voltage supply is that the inrush current of the motor—about six times its nominal value—may cause a significant voltage drop through the upstream impedances. As a result, the motor's *starting torque* is reduced, and thus the motor may not be able to start rotating its mechanically coupled load.

Furthermore, large inrush currents produce excessive heat through the rotor–stator structure, so successive starts within seconds or even within minutes may damage the motor's windings. For this reason, large motors that draw intolerably high starting currents are provided with suitable external winding connections through which their winding's voltage is reduced during the starting of the machine. The stator winding of these motors may be connected in star during starting and then switched to delta when nominal speed is approached. When a motor's winding is star-connected, it receives 58% ($1/\sqrt{3}$) of the supply voltage; when delta-connected, it receives 100% of the supply voltage. At starting, the motor's supply voltage may also be reduced by using an autotransformer, an inductor, or a current-limiting resistor.

The motor's supply voltage can also be easily reduced to desired levels by using solid state controls. These are discussed in Section 3.6.

Exercise 3–7 Under normal operating conditions, the windings of a 3-ϕ induction motor are connected in delta (Δ); at starting they are connected in star (Y). Determine the motor's torque and line current at starting in per-unit.

Answer $T = .33$ pu.

Efficiency Considerations

As illustrated in Fig. 3-21, the efficiency of the motor depends on its speed, its power rating, and on the driven load. These values are all normally selected by mechanical process engineers, and not by electrical engineers, who handle all other aspects of the motor—its specifications, purchasing, protection, etc.

In recent years—owing to the continuously increasing cost of energy—two types of induction machines have been made available: those with "standard efficiency" and those with "high efficiency." High-efficiency machines are more expensive than standard-efficiency machines, but in the long run—because of their lower energy losses—they are more economical. The efficiencies of these two types of motors, and their approximate costs, are shown in Table 3-3.

Efficiency measurements may differ from manufacturer to manufacturer; this is particularly true when motors are built overseas. In evaluating the advantages of higher-efficiency machines, then, one has to know how these efficiency ratings are established.

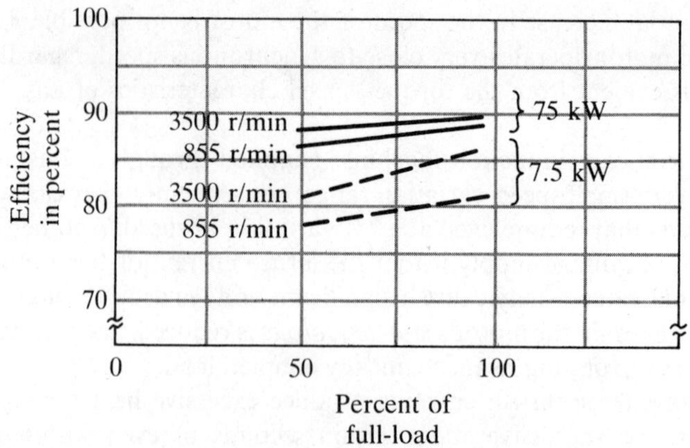

FIG. 3-21 Efficiency of 480 V, squirrel-cage induction motors as a function of load, speed, and nominal power.

TABLE 3-3 Squirrel-cage, 480 V, 3-ϕ, 4-pole induction machines: Typical "standard" and "high efficiencies" and their corresponding costs. (Courtesy of General Electric Canada Inc.)

Rating in kW	*Efficiency in Percent*		*Cost Per Unit**	
	Standard	*High*	*Standard Efficiency*	*High Efficiency*
7.5	85.9	90.2	1	1.4
50	91.3	95	8	9.8
75	93	95	10	12
100	95	95.8	16	19

* *The cost per unit, in 1988 dollars, is $350.*

Power-Factor Considerations

As illustrated in Fig. 3-22, the power factor of the motor depends on its speed, its kW rating, and the operating load. When the power factor of a plant is low, it can be improved by installing sufficient capacitor banks. Selecting a capacitor bank requires consideration and analysis of the following:

— Optimum kVAR and voltage ratings.

— Optimum connection within the power distribution network.

— Resonance conditions and the effects of harmonics.

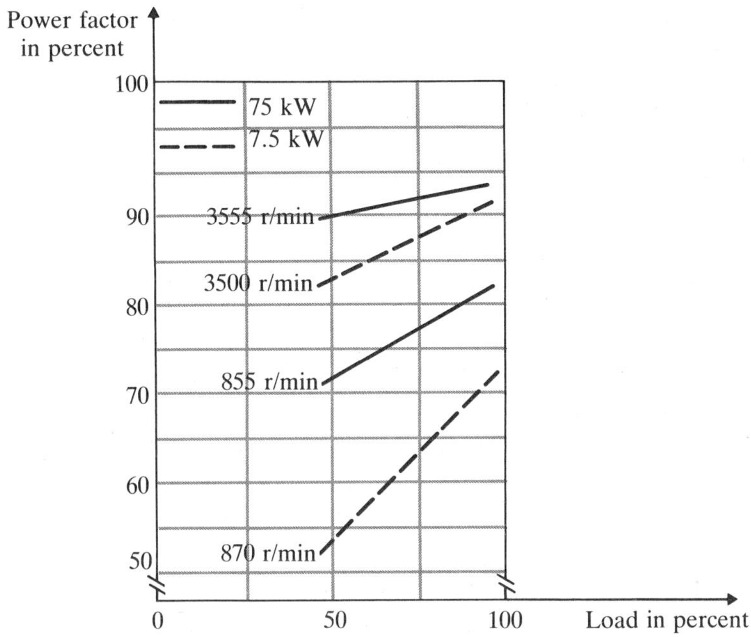

FIG. 3-22 Power factor of 480 V induction machines as a function of load, speed, and nominal power.

When a motor and its power-factor improvement capacitor are switched through a common disconnect device, the rating of the capacitor must be such as not to magnetize the motor each time the latter is switched OFF. In order to avoid this condition, the rated current of the capacitor must be less than the reactive no-load magnetizing current of the motor. If the capacitor bank is oversized, it may, during motor stopping, magnetize (source of excitation) the stator windings. Therefore, owing to the motor's decelerating rotor (source of rotation), the motor may operate as a generator. This may be accompanied by high transient voltages and torques, and as a result the normal life expectancy of the motor may be shortened.

Exercise 3–8 Explain the following:

a) The higher the nominal speed of a motor, the higher its power factor and efficiency.

b) The designing of a plant with a low power factor is due to oversized equipment.

Actual Motor Characteristics

Figure 3-23 shows the torque, current, and power factor versus speed characteristic for a particular 2400 kW, 2300 V, 3-ϕ, 60 Hz squirrel-cage induction machine. It should be noted that a small decrease in the operating speed will be accompanied by a large increase in the current. For this reason, the characteristics of the mechanical load

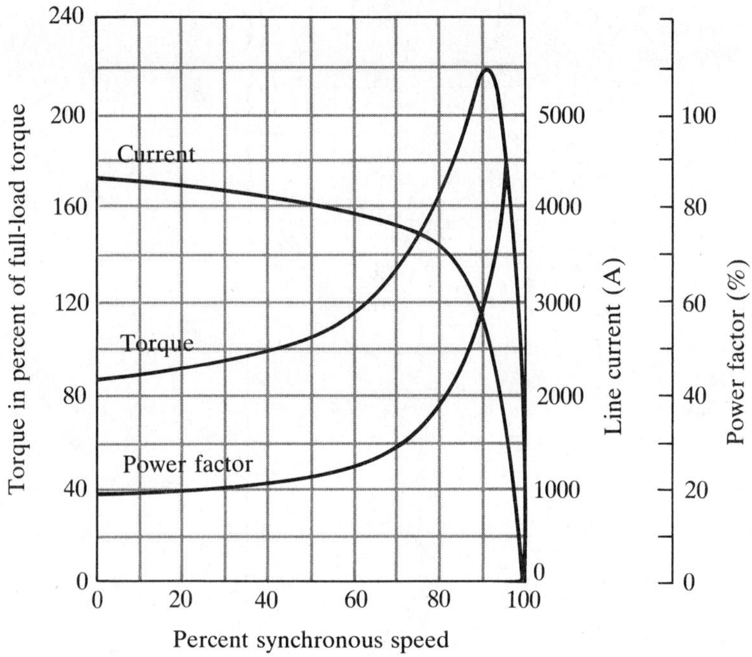

FIG. 3-23 Output torque, line current, and power factor versus speed for a 4-pole, 60 Hz, 2400 kW, 2300 V squirrel-cage induction motor. Rated speed: 1765 r/min; rated current : 705 A; power factor: 90%. (Courtesy of Westinghouse Canada Inc.)

should be accurately evaluated; otherwise, the motor will operate at larger slips and thus higher motor currents. In practice, all ac machines can withstand a continuous current overload of 10% without overheating.

EXAMPLE 3–4 The three-phase induction motor shown in the one-line diagram of Fig. 3-24(a) has the following starting characteristics:

$$T_{st} = 2.2 \text{ pu}, \qquad I_{st} = 6\underline{/-75^\circ} \text{ pu}$$

The motor receives its voltage supply through a transformer whose name-plate data are shown on the diagram. Neglecting the impedance of the cables and that of the 4.16 kV source, determine, at motor starting:

a) The voltage at the motor terminals.
b) The starting torque of the motor in per-unit.
c) The size of the capacitor bank that, when connected in parallel to the motor, will minimize the current through the upstream network.
d) The torque developed by the motor after the capacitor bank has been added.

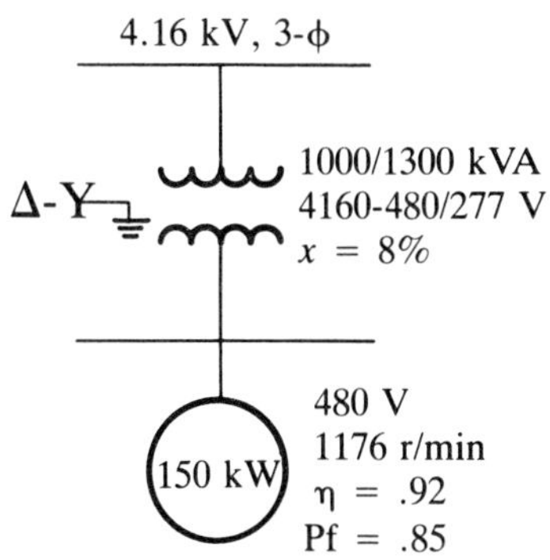

FIG. 3-24(a)

SOLUTION The magnitude of the rated current of the motor is

$$I = \frac{P}{\sqrt{3}\, V_{L-L} \cos\theta(\eta)} = \frac{150{,}000}{\sqrt{3}(480)(.92)(.85)} = 230.72 \text{ A}$$

The magnitude of the starting current of the motor is

$$I_{st} = 230.72(6) = 1384.31 \text{ A}$$

In phasor form:

$$I_{st} = 1384.31 \underline{/-75^\circ} \text{ A}$$

Taking the name-plate data of the transformer as base parameters, we have:

base impedance:

$$X_{b_L} = \frac{V_b^2}{S_b} = \frac{(.48)^2}{1.0} = .23 \ \Omega/\text{phase}$$

transformer reactance:

$$X_{a_L} = X_{b_L}(X_{pu}) = .23(.08) = .018 \ \Omega/\text{phase}$$

a) From KVL (Fig. 3-24(b)), the terminal voltage across the motor at starting is

$$V = V_s - IZ$$

$$V\underline{/0^\circ} = \frac{480}{\sqrt{3}}\underline{/\theta_1} - 1384.31\underline{/-75^\circ}(.018\underline{/90^\circ})$$

from which

$$\theta_1 = 1.4^\circ$$

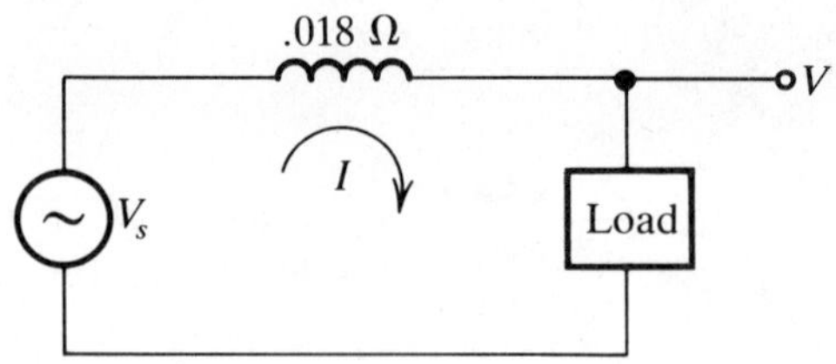

FIG. 3-24(b)

and

$$V = 252.40 \text{ V/phase}$$
$$= \underline{\underline{437.18 \text{ V, L-L}}}$$

Since the motor's terminal voltage is reduced from 480 V to 437.18 V, the starting current must also be proportionately reduced. By repeating the calculations with a new starting current, one will find that the motor's terminal voltage remains close to 437 V.

b) The torque developed by the motor is proportional to the square of the voltage. That is,

$$T \propto V^2$$

from which

$$\frac{T_2}{T_1} = \left(\frac{V_2}{V_1}\right)^2$$

where subscript 1 identifies the parameters at starting as given by the data of the motor, and subscript 2 identifies the corresponding motor parameters at the actual starting conditions. Substituting the known values, we obtain

$$T_{st} = 2.2\left(\frac{437.18}{480}\right)^2 = \underline{\underline{1.82 \text{ pu}}}$$

The starting-torque capability of the motor has been reduced from 2.2 pu to 1.82 pu, owing to the voltage drop on the upstream network. If the driven mechanical load requires a starting torque of 2.0 pu, for example, the motor will not be able to start rotating.

c) One method of reducing the voltage drop at starting is to connect a capacitor bank in parallel to the motor. The capacitor bank must be of a magnitude that will cancel the reactive component of the starting current. Thus, the capacitor's current (I_c) should be

$$I_c = 1384.31 \sin 75° = 1337.14 \text{ A}$$

The size of the capacitor bank is

$$\begin{aligned} Q &= \sqrt{3}\, V_{L-L} I \sin\theta \\ &= \sqrt{3}(480)(1337.14)(1) \\ &= 1111.68 \ \text{kVAR*} \end{aligned}$$

d) The upstream current is

$$\begin{aligned} I &= I_m + I_c \\ &= 1384.31\,\underline{/-75^\circ} + 1337.14\,\underline{/90^\circ} \\ &= 358.29\,\underline{/0^\circ}\ \text{A} \end{aligned}$$

The voltage across the motor is

$$V\underline{/0^\circ} = \frac{480}{\sqrt{3}}\,\underline{/\theta_2} - 358.29\,\underline{/0^\circ}(.018\,\underline{/90^\circ})$$

from which

$$\theta_2 = 1.4^\circ$$

and

$$\begin{aligned} V &= 277.05\ \text{V/phase} \\ &= 479.86\ \text{V, L–L} \end{aligned}$$

The torque developed at starting is

$$T = 2.2\left(\frac{479.86}{480}\right)^2 = \underline{\underline{2.2\ \text{pu}}}$$

The results are summarized in Table 3-4.

TABLE 3-4 Summary of the Results of Example 3–4

Operating Condition	*Starting Torque (pu)*	*Feeder Current at Starting (A)*
Nominal	2.2	1384.31
Without the capacitor	1.82	1384.31
With the capacitor	2.2	358.29

• • •

* The cost of this capacitor is about 40% of the cost of the motor.

Exercise 3–9 For the motor whose characteristics as a function of speed are shown in Fig. 3-23, the nominal speed is 1765 r/min. Estimate the current, power factor, and torque of the motor when its speed varies by $\pm 2\%$ of its nominal value. Tabulate the results.

3.3 Measurement of Equivalent-Circuit Parameters

The parameters of the approximate equivalent circuit* of a three-phase induction machine can be obtained from the equivalent standard transformer tests. The actual measurements are carried out on the stator of the machine. Consequently, the calculated parameters and the equivalent circuit are as seen from the stator.

Measurement of Stator Resistance

Per-phase stator resistance (R_1) is measured by any of the standard dc methods. This resistance is usually adjusted in order to compensate for the effects of higher operating motor temperatures. Usually $R_{ac} \approx 1.3R_{dc}$.

Locked-Rotor Test (s = 1)

The locked-rotor test on an induction machine—often referred to as the blocked-rotor test—corresponds to the short-circuit test on a transformer. Keeping the rotor stationary, the slip is equal to unity, and thus the equivalent mechanical load is shorted.

Refer to Fig. 3-25(b). The locked-rotor test gives the equivalent impedance of the

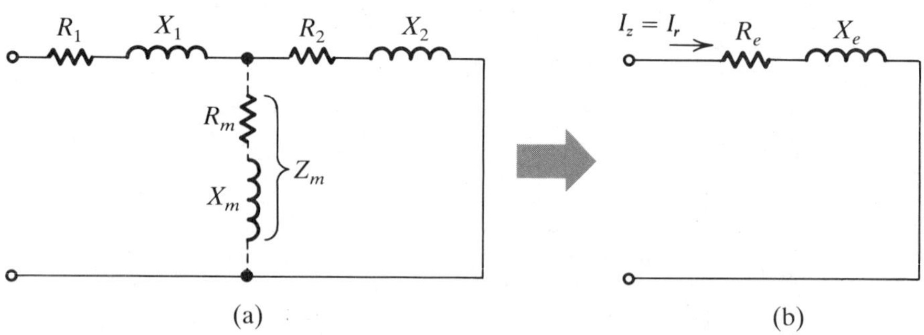

FIG. 3-25 Equivalent circuit corresponding to locked-rotor test ($s = 1$): **(a)** actual (Z_m large). **(b)** approximation of (a).

* For a more accurate and complete discussion of tests on induction machines, see IEEE Test Code for Polyphase Induction Motors and Generators No. 112A (New York: Institute of Electrical and Electronic Engineers 1964).

stator and rotor windings. That is,

$$Z_e = R_e + jX_e \tag{3.57}$$

These parameters are calculated by measuring the stator's 3-ϕ power (P_z), the line voltage (V_z), and the line current (I_z).

From basic considerations, we have

$$Z_e = \frac{V_z}{I_z} \tag{3.58}$$

$$R_e = \frac{P_z}{I_z^2} = R_1 + R_2 \tag{3.59}$$

and

$$X_e = \sqrt{Z_e^2 - R_e^2} = X_1 + X_2 \tag{3.60}$$

where V_z and P_z are the phase-to-neutral voltage and per-phase power, respectively.

The empirical relationships between the stator and rotor reactances are given in Table 3-5.

TABLE 3-5 Empirical Relationships for Induction Motor Leakage Reactances (From IEEE Test Code)

Type of Motor	*Class A*	*Class B*	*Class C*	*Class D*	*Wound-Rotor*
X_1	$.5X_2$	$.67X_2$	$.43X_2$	$.5X_2$	$.5X_2$

The current for the locked-rotor test (I_z) is usually the rated current of the motor:

$$I_z = I_{\text{rated}}$$

The voltage (V_z) required by this test is normally a small percentage of the rated voltage. Usually,

$$V_z \approx (10\% \longrightarrow 25\%)V_{\text{rated}}$$

The power measured is essentially the rate of heat loss (I^2R) in the stator and rotor windings under normal operating conditions. A small percentage of P_z is core loss, which is normally assumed to be negligible.

For class B and C 60 Hz motors the frequency recommended for the locked-rotor test is 15 Hz. This is necessary in order to simulate a rotor-heating effect equivalent to running conditions. The reactances measured at 15 Hz are only one-quarter of their actual values.

For class A motors, the recommended frequency is the same as that required under normal operating conditions.

No-Load Test (s = 0)

The no-load test of an induction motor corresponds to the open-circuit test on a transformer. The machine runs unloaded at a speed very close to synchronous while the excitation power (P_{exc}), the voltage (V_{exc}), and the current (I_{exc}) are measured.

The excitation voltage should be equal to the rated voltage of the motor so that rated flux conditions prevail. That is,

$$V_{exc} = V_{rated}$$

The exciting current is a large percentage of the rated current:

$$I_{exc} = (25\% \longrightarrow 40\%) I_{rated}$$

The excitation current is larger than that of a transformer with the same kVA and voltage ratings because of the larger air gap between the rotor and stator structures and because of the rotational losses that accompany the no-load test.

The data of the no-load test gives the so-called gross open-circuit motor impedance (Z'_m), as shown in Fig. 6.26(a),

$$Z'_m = R'_m + jX'_m \tag{3.61}$$

This impedance includes the stator and magnetization impedances, and also the rotational losses in their equivalent ohmic values. From basic definitions, we have

$$Z'_m = \frac{V_{exc}}{I_{exc}} \tag{3.62}$$

$$R'_m = \frac{P_{exc}}{I^2_{exc}} \tag{3.63}$$

and

$$X'_m = \sqrt{Z'^2_m - R'^2_m} \tag{3.64}$$

The no-load resistance (R'_m) is equal to the stator resistance (R_1) plus the equivalent resistances of the core loss (R_m) and the rotational losses. (see Fig. 3.26(b))

$$R'_m = R_1 + R_m + R_{\substack{\text{rotational}\\\text{losses}}} \tag{3.65}$$

The rotational losses* are normally incorporated in the mechanical load, and are considered part of the power consumed by the equivalent load resistance $R_2(1 - s)/s$.

. . .

* The rotational losses are made up of the stray-load losses and the mechanical losses. The latter can be accurately measured by plotting, at no-load, the power versus voltage squared. The intercept with zero voltage gives the mechanical losses.

Reasonable results are obtained by assuming the actual core-loss resistance to be two-thirds of the total no-load resistance. That is,

$$R_m = \frac{2}{3} R'_m \text{ ohms/phase} \tag{3.66}$$

The no-load reactance (X'_m), as shown in Fig. 3.26(c), is equal to the sum of the magnetizing and stator reactances. That is,

$$X'_m = X_1 + X_m \tag{3.67}$$

The stator reactance is relatively small, and for all practical purposes

$$X'_m \approx X_m \tag{3.68}$$

The series magnetizing impedance is often represented by an equivalent parallel impedance, as shown in Fig. 3-26(d).

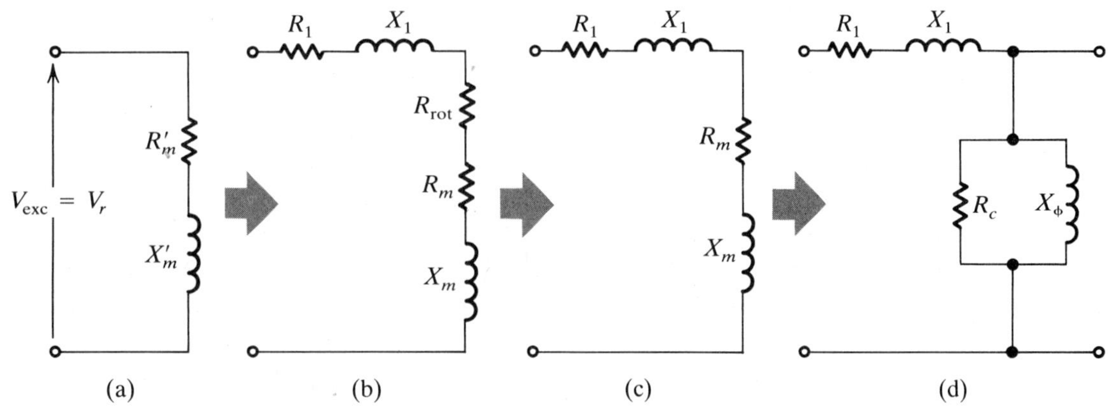

FIG. 3-26 Equivalent circuit obtained from the no-load test: **(a)** measured parameters ($Z'_m = R'_m + jX'_m$); **(b)** detailed representation of the measured parameters, $Z'_m = (R_1 + R_{rot} + R_m) + j(X_1 + X_m)$; **(c)** the resistance (R_{rot}) that represents the rotational losses; ***(d)*** parallel representation of the magnetizing impedance shown in (c).

The parallel representation of the magnetizing impedance is more common and can easily be found from its equivalent series representation, as discussed in Chapter 2 (Eqs. (2.24) and (2.25)).

The parameters of the motor's equivalent circuit are shown in Fig. 3-27. The most commonly used approximate equivalent circuit for the engineering analysis of induction machines is that of Fig. 3-27(c).

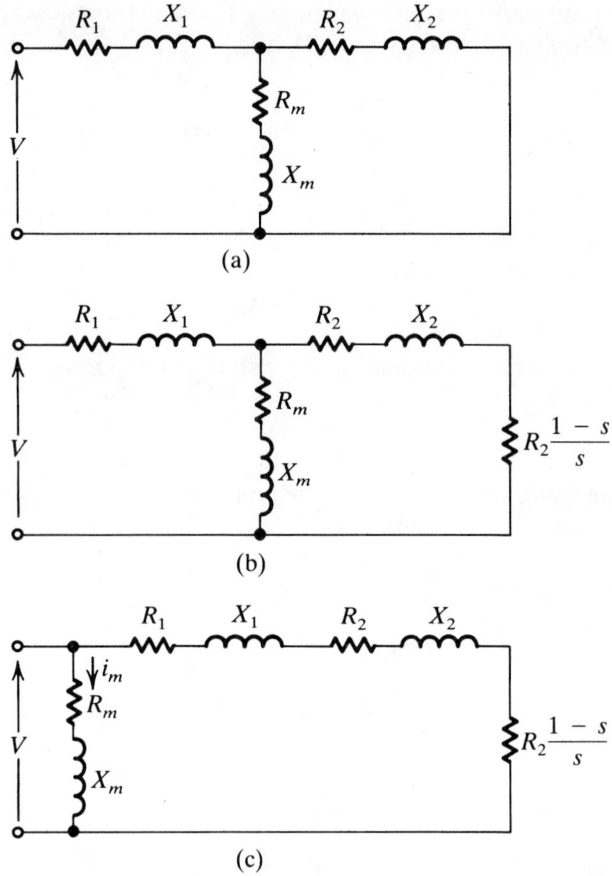

FIG. 3-27 Per-phase approximate equivalent circuits as seen from the stator: **(a)** rotor stationary, **(b)** rotor running, **(c)** a further approximation of (b); the magnetizing impedance has been relocated in order to simplify the calculations.

EXAMPLE 3–5 A 100 kW, three-phase, 60 Hz, 480 V, star-connected 1146 r/min class C induction motor yielded the following results when tested:

1. Average value of dc resistance between stator terminals: .20 ohms
2. No-load test:

$$V_{\text{exc}} = 480 \text{ V, L–L}$$
$$P_{\text{exc}} = 1920 \text{ W, three-phase}$$
$$I_{\text{exc}} = 40 \text{ A}$$

3. Locked-rotor test:

$$V_z = 32 \text{ V, L–L}$$
$$P_z = 2560 \text{ W, three-phase}$$
$$I_z = 65.6 \text{ A}$$
$$f = 15 \text{ Hz}$$

Determine the per-phase equivalent circuit of the motor.

SOLUTION

No-Load Test (Fig. 3-28)

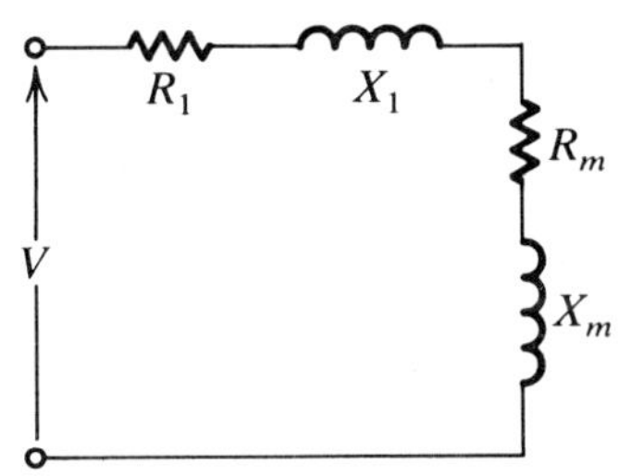

FIG. 3-28 Cirucit corresponding to no-load test.

Using Eq. (3.63):

$$R'_m = \frac{P_{\text{exc}}}{I^2_{\text{exc}}} = \frac{1920}{3(40)^2} = .40\ \Omega/\text{phase}$$

From Eq. (3.66):

$$R_m = \frac{2}{3}(.40) = .27\ \Omega/\text{phase}$$

The per-phase stator resistance is

$$R_1 = \frac{.2}{2} = .10\ \Omega/\text{phase}$$

The rotational loss equivalent resistance is found by using Eq. (3.65):

$$R_{r.l.} = .40 - .10 - .27 = .033\ \Omega/\text{phase}$$

This resistance is not explicitly shown in the equivalent circuit, but its power consumption will be part of the power developed by the motor.

From Eq. (3.62), the gross magnetizing impedance is

$$Z'_m = \frac{V_{\text{exc}}}{I_{\text{exc}}} = \frac{480/\sqrt{3}}{40} = 6.92 \ \Omega/\text{phase}$$

The gross magnetizing reactance is

$$X'_m = \sqrt{6.92^2 - .40^2} = 6.92 \ \Omega/\text{phase}$$

Locked-Rotor Test (Fig. 3-29)

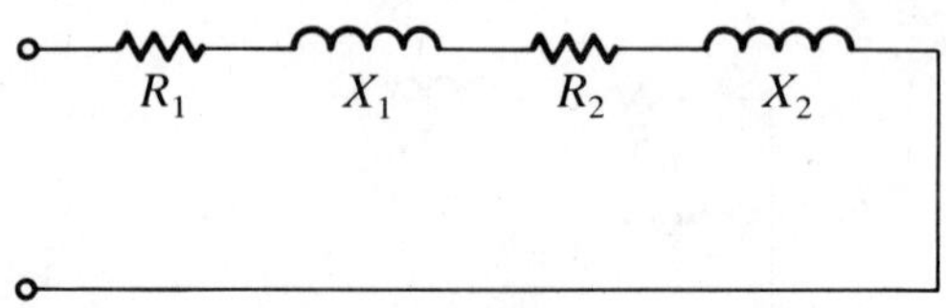

FIG. 3-29 Locked-rotor test.

From the data of the locked-rotor test and Eqs. (3.58) and (3.59), we obtain

$$Z_e = \frac{V_z}{I_z} = \frac{32/\sqrt{3}}{65.6} = .28 \ \Omega/\text{phase}$$

$$R_e = \frac{P_z}{I_z^2} = \frac{2560}{3(65.6)^2} = .2 \ \Omega/\text{phase}$$

The rotor resistance, as seen from the stator, is found by using Eq. (3.59):

$$\begin{aligned} R_2 &= R_e - R_1 \\ &= .20 - .10 \\ &= .10 \ \Omega/\text{phase} \end{aligned}$$

The equivalent motor reactance is found by using Eq. (3.60):

$$X_e = \sqrt{Z_e^2 - R_e^2} = \sqrt{.28^2 - .20^2} = .20 \ \Omega/\text{phase}$$

The stator and rotor reactances can be found by using Table 3-5 and Eq. (3.60), as follows:

$$X_e = X_1 + X_2$$

Substituting, we get

$$.2 = X_1 + \frac{X_1}{.43}$$

From the last two equations:

$$X_1 = .06 \ \Omega/\text{phase}$$

and

$$X_2 = .14 \ \Omega/\text{phase}$$

The above reactances were measured at a frequency of 15 Hz. Thus, at the rated frequency of 60 Hz:

$$X_1 = 4(.06) = .24 \ \Omega/\text{phase}$$

and

$$X_2 = 4(.14) = .56 \ \Omega/\text{phase}$$

The magnetizing reactance is found by using Eq. (3.67):

$$\begin{aligned} X_m &= X'_m - X_1 \\ &= 6.93 - .24 = 6.69 \ \Omega/\text{phase} \end{aligned}$$

The per-phase equivalent circuit, as obtained from the data of the given tests, is shown in Fig. 3-30.

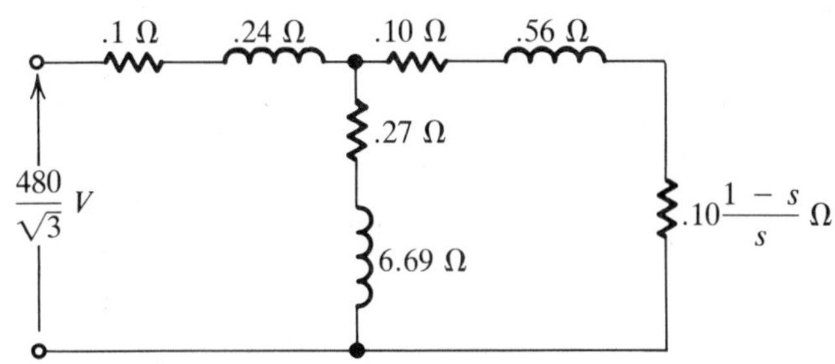

FIG. 3-30

3.4 Special Applications

This section discusses two special applications of induction machines: the use of wound-rotor machines for voltage–frequency multiplication, and the use of squirrel-cage machines as induction generators. Voltage–frequency multipliers have limited laboratory applications, while induction generators are often used to economically interconnect small power-generating stations (down to 20 kW) to utility distribution systems.

3.4.1 FREQUENCY AND VOLTAGE MULTIPLIERS

Three-phase wound-rotor induction machines can be used to furnish, at their slip-ring terminals, a 3-ϕ source of variable voltage and frequency. For this application, the induction machine is connected to a normal voltage and frequency supply and is driven by a variable-speed motor. The operating slip is positive and larger than unity because the induction machine is driven in a direction opposite to the rotation of its synchronously rotating magnetic field.

According to Eqs. (3.23) and (3.24), the voltage induced in the rotor windings, and its frequency, are directly proportional to the operating slip. Rewriting these equations, we have

$$V_r = sV'_r \tag{3.69}$$

and

$$f_r = sf \tag{3.70}$$

As seen from the above, for operating slips greater than unity, the voltage and its frequency across the rotor terminals will be larger than that of the stator. The slip rings of the wound rotor can thus become a power source of variable voltage and frequency.

Power Considerations

Figure 3-31(a) shows a variable-speed motor driving a three-phase induction motor. The induction motor is of the wound-rotor type and is driven in the direction opposite to that of its synchronously rotating field.

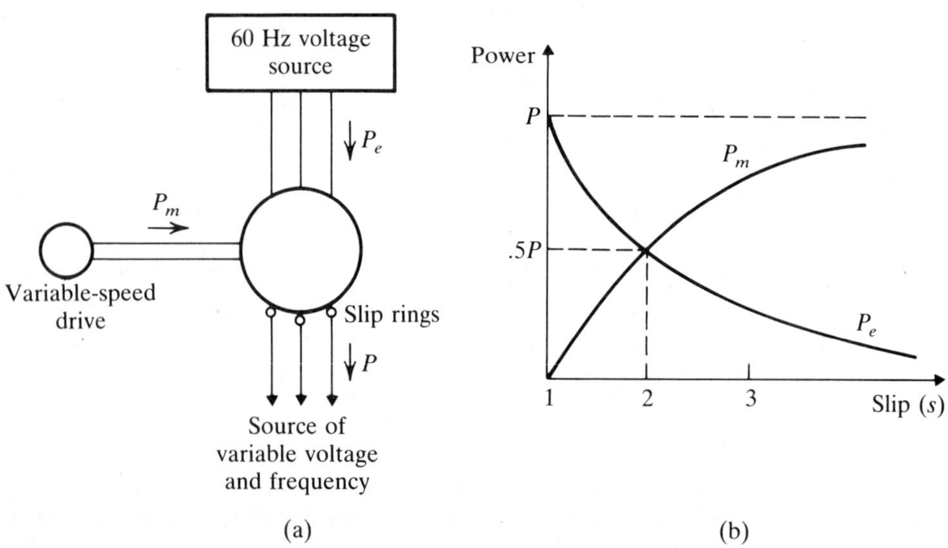

FIG. 3-31 Voltage and frequency multipliers: **(a)** power flow diagram, **(b)** graphs of various powers versus operating slip.

The available electrical output power (P) across the slip rings is the sum of the power input from the induction machine (P_e) and the power input (P_m) from the variable-speed drive. In mathematical form,

$$P = P_e + P_m \tag{3.71}$$

Considering the variable-speed drive, from basic definitions we have

$$\begin{aligned} P_m &= \text{(torque)(actual speed)} \\ &= T\omega_a \end{aligned} \tag{3.72}$$

Neglecting losses, the power contribution of the induction machine is equal to the air-gap power. That is,

$$\begin{aligned} P_e &= P_{\text{air gap}} \\ &= \text{(torque)(synchronous speed)} \\ &= T\omega_s \end{aligned} \tag{3.73}$$

For operating slips larger than unity, it can be shown from the previous equations that

$$P_m = P\frac{s-1}{s} \tag{3.74}$$

and

$$P_e = \frac{P}{s} \tag{3.75}$$

The derivations of Eqs. (3.74) and (3.75) are left as student exercises (see Exercise 3-10).

The power variations of the variable-speed motor and of the induction motor are sketched in Fig. 3-31(b).

Exercise 3–10 For a three-phase, wound-rotor type of induction motor:

a) Explain how a voltmeter can be used to determine the speed of the motor.
b) Show the intermediate steps in the derivation of Eqs. (3.74) and (3.75).

3.4.2 INDUCTION GENERATORS

When the rotor of a 3-ϕ induction machine is driven above synchronous speed and in the same direction as the synchronously rotating field, the slip—as can be seen from the general slip definition—becomes negative. The current and the voltage of the rotor windings are reversed in polarity, and the torque developed is in a direction opposite to the rotation. In this mode of operation, the machine delivers power to the stator terminals: that is, the motor acts as an induction generator.

An ordinary 3-ϕ induction motor, driven in the proper direction at hypersynchronous speeds, becomes an induction generator. The *excitation* of the generator is provided by the source of the stator voltage, and the *shaft's rotation* is derived from a water, wind, solar, steam, or gas-driven turbine.

Figures 3-32(a) and (b) show an elementary representation of an induction motor and an induction generator. In many respects, the induction generator is identical to the induction motor. The equivalent circuits are the same, but the energy flows in opposite directions. An induction machine that drives a hoist operates as an induction generator for descending loads and as a motor for ascending loads.

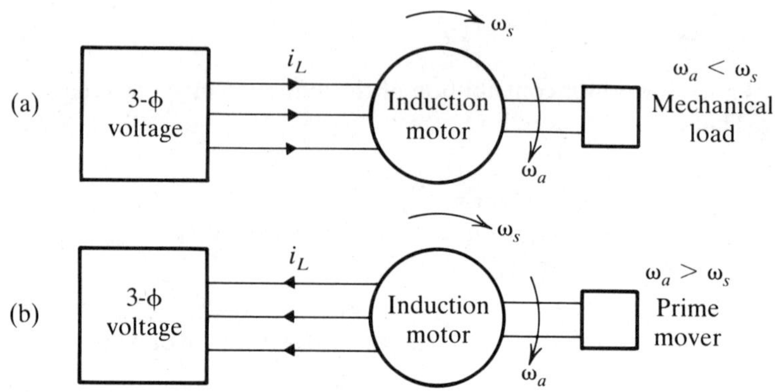

FIG. 3-32 Induction machines: **(a)** induction motor, **(b)** induction generator.

The induction generator finds increasing applications because it provides the most economical method of interconnecting a small power station to a large power-distribution network. Since the late 70's, small power-generating stations have been operating in many places, because utilities are obliged by law to purchase, at reasonable rates, the electricity produced by their own consumers.

The induction generator is not *self-excited*: it requires magnetizing kVAR. This is drawn from the power-distribution network to which the generator is connected. Capacitor banks connected in parallel to the generator can, if they are the right size, compensate for the magnetizing or reactive power drawn by the induction generator. However, the power factor of the network—as seen from the terminals of the generator—depends on the impedance of the other loads that are connected to the same power grid. The prime movers of induction generators are equipped with speed-control governors that limit the no-load speed of the machines to safe levels. This is necessary because, when the electrical load of the generator is suddenly disconnected, its rotor—owing to constant prime mover's power and diminished electrical counter-torque—will overspeed.

Overspeeds are undesirable because the resulting centripetal and centrifugal forces may damage the rotor if the latter is not designed to accommodate them. The level of overspeed depends on the ability of the control apparatus to instantly reduce the power developed by the prime mover. It is not always possible to instantly reduce the power, however. If a hydropower station were to attempt this, for example, destructive pressure rises would be produced within upstream water conduits (penstocks).

High overspeeds are also undesirable because they may induce high stator voltages. The magnitude of the induced voltage depends on the machine's electrical time constant (L/R) and on the prime mover's speed. For example, if an induction generator has an open-circuit time constant of 6 seconds, then its terminal voltage after 2 seconds—at a constant prime-mover speed—will be 72% ($e^{-2/6}$) of nominal voltage. If, however, the speed of the prime mover is increased at the same time by 200%, then the terminal stator voltage will be about 144%. This overvoltage has undesirable effects, not only on the generator but also on the auxiliary equipment (PTs, relays, etc.) that is part of the generator's control apparatus.

The power rating of installed induction generators is in the range of 20 to 2000 kW. If a generator's voltage rating is different from that of the distribution network, a transformer is used to interconnect the two voltage levels.

EXAMPLE 3–6 A three-phase induction motor drives a hoist. During lowering, the load accelerates the motor above synchronous speed, and thus the motor runs as a generator. The per-phase equivalent circuit of the motor under rated conditions, and its one-line diagram, are shown in Figs. 3-33(a) and (b), respectively. If the resulting speed is 1900 r/min, determine:

a) The line current.
b) The power returned to the supply.

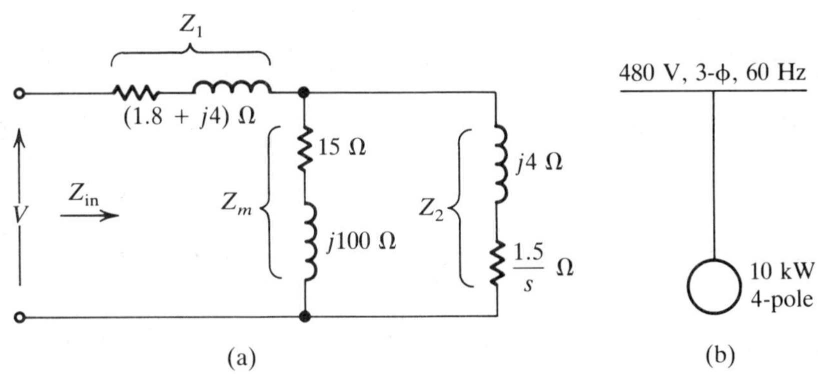

FIG. 3-33

SOLUTION

a) The synchronous speed is 1800 r/min and, at operating condition, the slip is

$$s = \frac{1800 - 1900}{1800} = -.056 \text{ pu}$$

The total impedance is

$$Z_{in} = Z_1 + Z_m /\!/ Z_2$$

$$Z_{in} = (1.8 + j4) + (15 + j100) \Big/\!\!\Big/ \left(\frac{1.5}{-.056} + j4\right)$$

$$= 1.8 + j4 - 24.2 + j10.53 = 26.67\underline{/147^\circ}\ \Omega/\text{phase}$$

The voltage at the terminals of the machine is assumed to be constant and independent of the mode of operation of the machine. Thus the line current is

$$I = \frac{480/\sqrt{3}}{26.67\underline{/147^\circ}} = \underline{\underline{10.39\underline{/-147^\circ}\ \text{A}}}$$

b) The power returned to the supply is

$$P = \sqrt{3}\, V_{L-L} I \cos\theta = \sqrt{3}(480)(10.39)\cos 147^\circ = \underline{\underline{-7.24\ \text{kW}}}$$

The negative sign indicates the reverse direction of power flow—that is, from the induction generator to the 480 V distribution network.

Exercise 3–11 The open-circuit time constant of an 480 V, three-phase induction generator is 4 s. Determine the stator voltage 2 s after the removal of the generator's load. Assume that the speed of the prime mover at this time interval:

a) Is constant.

b) Increases by 250% owing to the removal of the counteracting electrical load.

Answer (a) 291.1 V; (b) 727.84 V.

3.5 Electromechanical Transients

The objective of this section is to derive general equations that give the time that it takes a motor to go from one operating condition to another, and to derive general equations that determine how much energy is dissipated in the rotor's windings as a function of the rotor's kinetic energy. Knowing how much time it takes to start or reverse a motor is very important in determining energy consumption, in selecting a motor's protective devices, and in evaluating the suitability of a motor for a particular application. In general, a three-phase induction motor that takes more than 20 s to reach its nominal speed (starting from rest) will be dangerously overheated. This is because it draws close to locked-rotor current—about 5 to 6 times larger than its rated current—during its acceleration period.

As demonstrated below, the time expressions depend on the characteristics of the load and on the parameters of the motor.

3.5.1 ACCELERATING TIME

Analytical Method

The torque balance between the motor and the load is given by Eq. (3.49), which for convenience is rewritten below:

$$T = J\frac{d\omega_a}{dt} + B\omega_a + T_L \tag{3.76}$$

That is, the torque developed by a motor (T) is used to overcome the moment of inertia (J) of the rotating masses, the frictional opposition (B), and whatever other torque (T_L) is required by the load.

Using Eq. (3.38), the torque developed by a motor can also be written as follows:

$$T = \frac{V^2R_2s}{\omega_s[s^2(R_1^2 + X^2) + R_2^2 + 2sR_1R_2]} \tag{3.77}$$

In Eq. (3.77), and in the expressions that follows, the voltage V is the line-to-line voltage, and T represents the total torque developed by the motor.

Assuming that a motor drives a pure-inertia load, then from Eq. (3.76) the torque developed becomes

$$T = J\frac{d}{dt}(\omega_s(1 - s)) = -J\omega_s\frac{ds}{dt} \tag{3.78}$$

From the last two equations, we get

$$-J\omega_s\frac{ds}{dt} = V^2R_2\frac{s}{\omega_s(s^2(R_1^2 + X^2) + R_2^2 + 2sR_1R_2)} \tag{3.79}$$

Rearranging and taking integrals, we have

$$\int_0^t dt = -\frac{J\omega_s^2}{V^2R_2}\int_1^{s_1}\frac{s^2(R_1^2 + X^2) + R_2^2 + 2sR_1R_2}{s}\,ds \tag{3.80}$$

The upper and lower limits of the right-hand side integrand are selected to give the time a motor requires to go from zero speed ($s = 1$) up to a speed that corresponds to a slip s_1. Integrating the last relationship, and after some algebraic manipulations, we obtain

$$t = \frac{J\omega_s^2}{V^2R_2}\left[(1 - s_1^2)\frac{R_1^2 + X^2}{2} + 2R_1R_2(1 - s_1) + R_2^2\ln\frac{1}{s_1}\right] \tag{3.81}$$

Thus, the time it takes for a motor starting from rest to reach a particular speed is directly proportional to the moment of inertia and to the square of the synchronous speed, and inversely proportional to the rotor resistance and to the square of the stator voltage. For greater accuracy, the magnetizing impedance should be taken into consideration by using the Thevenin's equivalent. In that case the stator impedance and voltage in the previous relationships should be replaced by the Thevenin equivalent voltage and impedance, as discussed in Section 3.1.6.

Graphical Method

Very often, the torque–speed characteristics of a motor and its load are given by their corresponding curves, as shown in Fig. 3-34(a). In this case it is simpler to use a graph to find the time that it takes a motor to reach the common load–motor speed of ω_1 rps.

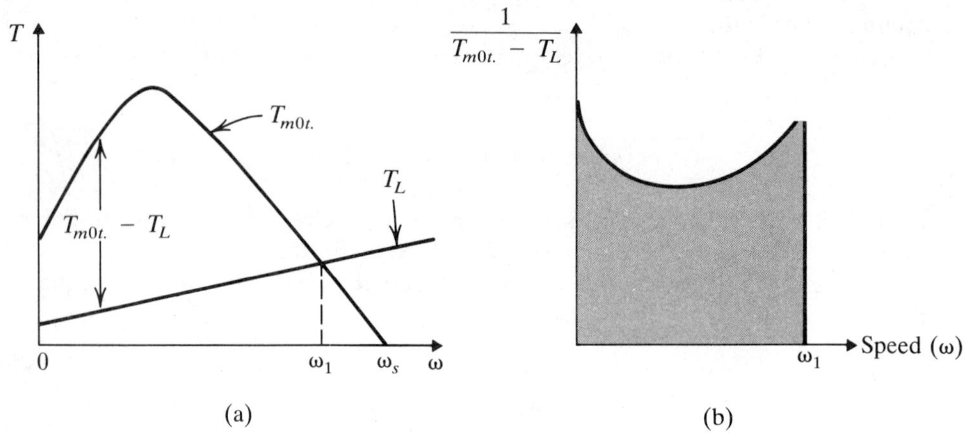

FIG. 3-34 **(a)** Torque–speed characteristics of motor and load, **(b)** typical inverse accelerating torque ($1/(T_{mot} - T_L)$) versus the speed of the motor.

Neglecting the frictional component of the torque, from Eq. (3.76) the accelerating torque (T_a) is

$$T_a = T_{mot} - T_L = J\frac{d\omega}{dt} \qquad \textbf{(3.82)}$$

From the above, the time it takes a motor starting from rest to reach steady-state speed of ω_1 rad/s is

$$t = J\int_0^{\omega_1} \frac{d\omega}{T_{mot} - T_L} \qquad \textbf{(3.83)}$$

By definition, the integral sign represents the accumulation of areas under the functions specified in its integrand and between the indicated limits. Therefore,

$$t = J\begin{pmatrix}\text{the area under the inverse} \\ \text{accelerating torque–speed curve}\end{pmatrix} \qquad \textbf{(3.84)}$$

This area is represented in Fig. 3-34(b) by the crosshatched section.

The area can be plotted on squared graph paper and then measured by counting the number of squares. When the accelerating torque is expressed in newton-meters, the speed in radians per second, and the polar moment of inertia in kg-m^2, then Eq. (3.84) gives the time in seconds. For example, suppose we are required to find the time that it takes a motor to reach operating speed. It is given that the number of squares under the accelerating torque-versus-speed diagram is 200 and the two perpendicular sides of each square are 2×10^{-3} $(\mathrm{N \cdot m})^{-1}$ and 5 rad/s, respectively. The polar moment of rotor-plus-load is 4 kg-m^2. Then,

$$\begin{aligned} t &= 4(2 \times 10^{-3})(5)(200) \\ &= \underline{\underline{8 \text{ s}}} \end{aligned}$$

3.5.2 SPEED AS A FUNCTION OF TIME

If an induction motor is running unloaded, and a mechanical load is suddenly applied, its speed will be reduced and its torque will be increased in order to satisfy the new load requirements. In the full-load region the torque of the motor is given by Eq. (3.41). From Eqs. (3.41) and (3.76) we obtain

$$\frac{V^2}{\omega_s} \frac{s}{R_2} = J \frac{d\omega_a}{dt} + B\omega_a + T_L \tag{3.85}$$

The left-hand side of this equation represents the torque developed by the motor; the right-hand side represents the torque requirements of the load. Substituting for the slip s the following value:

$$s = \frac{\omega_s - \omega_a}{\omega_s} \tag{3.86}$$

and after simplification and rearrangement, we obtain

$$\frac{V^2}{\omega_s R_2} - T_L = J \frac{d\omega_a}{dt} + B\omega_a + \frac{V^2}{\omega_s^2 R_2} \omega_a \tag{3.87}$$

For a constant load torque (T_L) component, by taking Laplace transforms we have

$$\left(\frac{V^2}{\omega_s R_2} - T_L\right)\frac{1}{S} = JS\omega_a(s) - J\omega_s + B\omega_a(s) + \frac{V^2}{\omega_s^2 R_2}\,\omega_a(s) \tag{3.88}$$

where the capital S is the Laplace operator, and ω_s is the initial speed of the motor just before the application of the load. After simplification, we obtain

$$\omega_a(s) = \frac{\omega_s}{\tau_1 + S} + \frac{A}{S(\tau_1 + S)} \tag{3.89}$$

where

$$\tau_1 = \frac{1}{J}\left(B + \left(\frac{V}{\omega_s}\right)^2 \frac{1}{R_2}\right) \; s^{-1} \tag{3.90}$$

and

$$A = \frac{V^2 - \omega_s R_2 T_L}{J\omega_s R_2} \text{ rad/s} \tag{3.91}$$

Taking the inverse Laplace of Eq. (3.89), we obtain the following expression for the speed of the motor as a function of time:

$$\omega_a(t) = \omega_s e^{-\tau_1 t} + \frac{A}{\tau_1}(1 - e^{-\tau_1 t}) \tag{3.92}$$

The graphical representation of Eq. (3.92) is shown in Fig. 3-35. This equation is approximate because of the underlying assumptions. A more practical and up-to-date approach would be to use a digital computer to perform the numerical calculations.

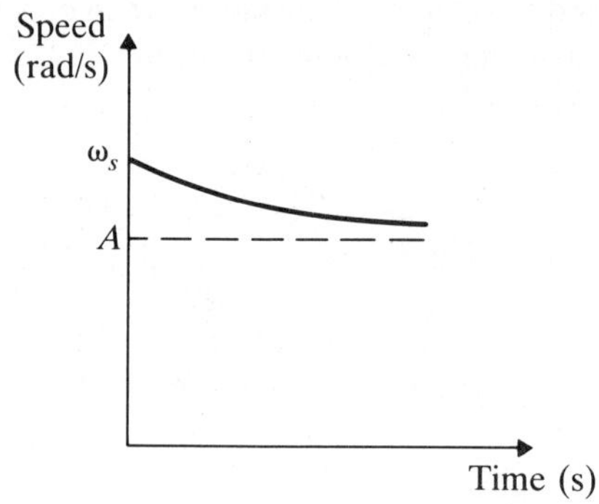

FIG. 3-35 Speed as a function of time following a sudden increase in the load requirements.

EXAMPLE 3–7 A 75 kW, Y-connected, 480 V, 60 Hz, 4-pole, 1746 r/min, three-phase induction motor has a per-phase equivalent circuit referred to the stator, as shown in Fig. 3-36. The combined motor–load moment of inertia is 20 kg · m^2. Neglecting the effects of the magnetizing current, determine the time it takes for the

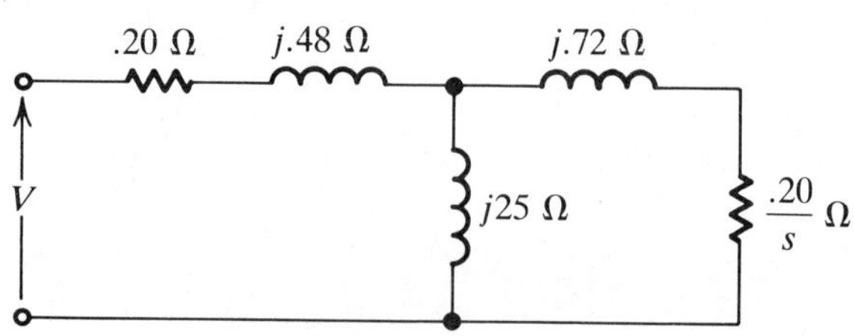

FIG. 3-36

motor to reach maximum torque and rated speed if:

a) The applied voltage is 480 V.
b) The applied voltage is reduced to 300 V.
c) The stator resistance is negligible and the applied voltage is 480 V.

SOLUTION

a) The speed of the stator field is:

$$n_s = 120\frac{f}{p} = 120\frac{60}{4} = 1800 \ \text{r/min}$$

and

$$\omega_s = 188.50 \ \text{rad/s}$$

The slip at full-load is

$$s_1 = \frac{1800 - 1746}{1800} = .03 \ \text{pu}$$

The time it takes for the motor to reach its rated speed of 1746 r/min is calculated by substituting numerical values in Eq. (3.81). Thus,

$$t = \frac{20(188.5)^2}{(480)^2(.2)}\left[(1 - .03^2)\left(\frac{.2^2 + 1.2^2}{2}\right) + 2(.2)(.2)(1 - .03) + .2^2 \ln\frac{1}{.03}\right]$$
$$= \underline{\underline{14.77 \ \text{s}}}$$

The slip at maximum torque is

$$s_{mt} = \frac{.20}{\sqrt{.2^2 + 1.2^2}} = .16 \ \text{pu}$$

Similarly, from Eq. (3.81), the time it takes for the motor to reach maximum torque is

$$t = \frac{20(188.5)^2}{(480)^2(.2)}\left[(1 - .16^2)\left(\frac{.2^2 + 1.2^2}{2}\right) + 2(.2)(.2)(1 - .16) + .2^2 \ln\frac{1}{.16}\right]$$
$$= \underline{\underline{13.25 \ \text{s}}}$$

b) When the voltage applied to the motor is reduced to 300 V, the time it takes to reach rated speed is found from

$$t \propto \frac{1}{V^2}$$

Thus,

$$t = 14.77\left(\frac{480}{300}\right)^2$$
$$= \underline{\underline{37.79 \ \text{s}}}$$

Similarly, the time it takes to reach the speed at maximum torque is

$$t = 13.25\left(\frac{480}{300}\right)^2$$

$$= \underline{\underline{33.91 \text{ s}}}$$

c) By neglecting the stator resistance, Eq. (3.81) yields the required times for the following operating conditions. The full-load slip is

$$s = .03$$

and the time to reach rated speed is calculated to be

$$t = \underline{\underline{13.26 \text{ s}}}$$

The slip at maximum torque is

$$s_{mt} = \frac{.2}{1.2} = .167 \text{ pu}$$

and the time to reach this torque is calculated to be

$$t = 11.90 \text{ s}$$

For the purposes of comparison the results are summarized in Table 3-6. By inspecting the tabulated results, it is clear that the stator resistance does not have any drastic effect on the time it takes for the motor to reach a particular speed. The applied voltage, however, has significant effects.

TABLE 3-6 Summary of Results of Example 3–7

Operating Condition	*Motor Voltage (V)*	*Stator Resistance, R_1 (ohms)*	*Time It Takes To Reach Operating Condition (s)*
Rated speed	480	.20	14.77
Maximum torque	480	.20	13.25
Rated speed	300	.20	37.79
Maximum torque	300	.20	33.91
Rated speed	480	0	13.26
Maximum torque	480	0	11.90

Exercise 3–12 A three-phase, 480 V, 60 Hz, 6-pole, squirrel-cage rotor induction motor has per-phase stator and rotor resistances of .10 and .15 ohms, respectively. The per-phase reactance of its windings is 1.2 ohms. The motor plus load moment of inertia is 15 kg · m^2. Determine the time it takes for the motor, starting from rest, to reach a speed of 1176 r/min.

Answer 5.77 s.

3.5.3 PLUGGING

A motor can be stopped by any of the following methods:

1. Removing the supply voltage.
2. Reversing the phase sequence of the applied voltages.
3. Applying a dc field to a stator phase winding while the ac voltage supply is removed.
4. Using a mechanical brake.

With each method, a specific amount of time is required to bring the motor to a complete stop.

This section will discuss reversing the phase sequence of the applied voltages, commonly known as plugging.

Reversing the Phase Sequence of the Applied Voltages

By interchanging two of the leads of a three-phase, three-wire distribution system, the phase sequence of the transmitted voltages is reversed. As a result, a backward-rotating field is generated. This field, of course, opposes the forward field that existed before the phase sequence of the applied voltages was reversed. On phase reversal, the motor will go through zero speed and then start in the reverse direction. When the motor approaches zero speed, however, the supply voltage is removed and the motor comes to a complete stop. The time that it takes for a motor to go from a speed ω_1 to a complete stop following a phase reversal of the applied voltages is derived as follows:

Assuming that a motor drives a pure-inertia load, from Eqs. (3.76) and (3.77) we obtain

$$J\frac{d\omega_a}{dt} = \frac{V^2R_2s}{\omega_s[s^2(R_1^2 + X^2) + R_2^2 + 2sR_1R_2]} \tag{3.93}$$

By substituting for the actual speed of the motor and rearranging, we have

$$dt = \frac{-J\omega_s^2}{V^2R_2}\frac{1}{s}[R_2^2 + 2sR_1R_2 + s^2(R_1^2 + X^2)]\,ds \tag{3.94}$$

from which

$$t = -\frac{J\omega_s^2}{V^2R_2}\int_{2-s_1}^{1}\frac{1}{s}[R_2^2 + 2sR_1R_2 + s^2(R_1^2 + X^2)]\,ds \tag{3.95}$$

The limits of integration are obtained by noticing that the negative or backward-rotating field goes from a slip of $(2 - s_1)$ to a slip of unity at zero speed. The slip (s_1) of the forward field is given by Eq. (3.19).

Integrating Eq. (3.95), and after some algebraic manipulations, we obtain

$$t = J\frac{\omega_s^2}{V^2R_2}\left[R_2^2 \ln(2 - s_1) + 2R_1R_2(1 - s_1) + \frac{1}{2}(R_1^2 + X^2)(3 - 4s_1 + s_1^2)\right] \quad \textbf{(3.96)}$$

The time required for a complete speed reversal is equal to the time that the motor takes to go from a forward speed of ω_1 rad/s to a complete stop (Eq. (3.96)) *plus* the additional time that the motor takes to go from zero speed to a new speed of ω_1 rad/s (Eq. 3.81) in the opposite direction.

The value of the rotor resistance that will result in minimum time while plugging a motor to stop can be obtained from Eq. (3.96) by using the calculus theory of "maximum and minimum." From this theory, it can be shown that

$$R_2 = \sqrt{\frac{(R_1^2 + X^2)(3 - 4s_1 + s_1^2)}{2\ln(2 - s_1)}} \text{ ohms/phase} \quad \textbf{(3.97)}$$

The resistance R_2 is the per-phase rotor resistance as seen from the stator. With a wound-rotor type of three-phase induction motor, this resistance represents the rotor winding resistance plus whatever external resistance is connected across the slip rings.

The intermediate steps in the derivation of Eqs. (3.96) and (3.97) are left as a student exercise (see Exercise 3–13).

Exercise 3–13 Derive Eqs. (3.96) and (3.97).

3.5.4 ENERGY CONSIDERATIONS AND ROTOR WINDING LOSSES

From the basic principles of dynamics, the kinetic energy (K.E.) stored in a body that has a mass of M kg and a linear velocity of U m/s is

$$\text{K.E.} = \frac{1}{2}MU^2 \text{ watt-seconds} \quad \textbf{(3.98)}$$

Similarly, the kinetic energy stored in the rotating shaft of an induction machine that runs at near synchronous speed is

$$\text{K.E.} = \frac{1}{2}J\omega_s^2 \quad \textbf{(3.99)}$$

where J is the polar moment of inertia of the rotating masses, and ω_s is the synchronous speed of the motor.

Rotor Winding Losses

The general expression for the power dissipated in the rotor windings ($I_2^2 R_2$) can be derived from Eq. (3.55) as follows:

$$\text{rotor power loss } (I_2^2 R_2) = sP_{\text{air gap}} \tag{3.100}$$

$$= s(\text{angular velocity} \times \text{torque}) \tag{3.101}$$

$$= s\omega_s \left(J \frac{d\omega_a}{dt} \right) \tag{3.102}$$

Thus

$$I_2^2 R_2 = -J\omega_s^2 s \frac{ds}{dt} \tag{3.103}$$

The energy dissipated in the rotor windings, then, is

$$I_2^2 R_2 \, dt = -J\omega_s^2 s \, ds \tag{3.104}$$

In driving a motor from rest up to synchronous speed, the energy dissipated in the rotor windings (W_{dr}) is

$$W_{\text{dr}} = \int_1^0 (-J\omega_s^2 s \, ds) \tag{3.105}$$

From the above,

$$W_{dr} = J \frac{\omega_s^2}{2} \tag{3.106}$$

Comparing Eqs. (3.99) and (3.106), it is evident that the energy dissipated in the rotor windings, when a motor is brought from rest up to synchronous speed, is equal to the kinetic energy stored in the rotating masses.

Rotor Energy Loss Due to Plugging

When a motor is running at synchronous speed and is brought to rest by interchanging two stator leads (phase reversal), the energy dissipated in the rotor windings $(W_{dr})_p$ is found by using Eq. (3.104). Using proper limits of integration, we obtain

$$(W_{dr})_p = -J\omega_s^2 \int_{s=2}^{s=1} s \, ds \tag{3.107}$$

From the above,

$$(W_{dr})_p = 3 \left(J \frac{\omega_s^2}{2} \right) \tag{3.108}$$

Using superposition, the energy dissipated in the rotor windings when the motor is subjected to a complete speed reversal $(W_{dr})_r$ is found by adding the results of Eqs. (3.106) and (3.108). Thus,

$$(W_{dr})_r = 4\left(\frac{1}{2} J\omega_s^2\right) \qquad \textbf{(3.109)}$$

Equation (3.109) can also be derived from Eq. (3.105) by using 0 and 2 as upper and lower integration limits. These integration limits are obtained, as before, from the definition of the per-unit slip. It should be noted that at synchronous speed the per-unit slip is zero, while at negative synchronous speed the per-unit slip is equal to 2.

Eqs. (3.106), (3.108), and (3.109) are represented graphically in Fig. 3-37.

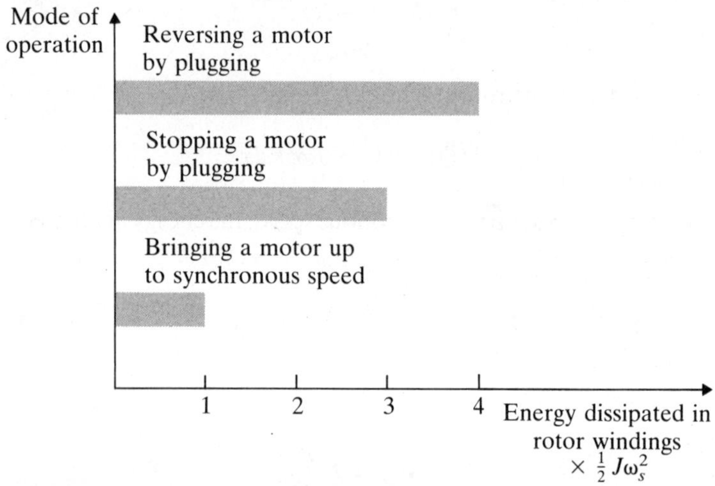

FIG. 3-37 Mode of operation versus the energy dissipated in the rotor windings.

A three-phase induction motor will be overheated if the speed reversals are numerous enough to generate more heat than the machine is designed to dissipate. The name-plate efficiency of the machine specifically indicates the amount of power losses, and thus the maximum allowable rate of heat dissipation, under continuous operation. For example, the maximum rate of heat a 100 kW motor with an efficiency of 94% can safely dissipate is

$$\text{losses} = P_{\text{out}}\left(\frac{1}{\eta} - 1\right) = 100\left(\frac{1}{.94} - 1\right) = 6.38 \text{ kW}$$

EXAMPLE 3–8 Figures 3-38(a) and (b) show, respectively, the one-line diagram representation and the per-phase equivalent circuit of a three-phase induction motor. The motor drives a pure-inertia load of 6.5 kg · m^2. Neglecting the effects of the mag-

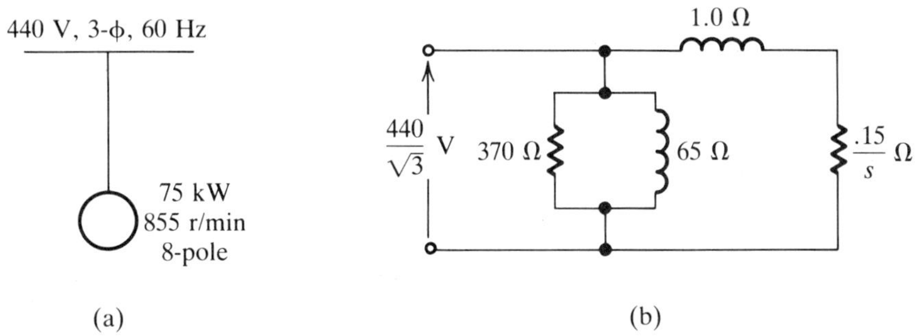

FIG. 3-38

netizing current and stator resistance, determine:

a) The energy stored in the rotating masses while the motor is running unloaded.
b) The time it takes to plug the motor to a complete stop and the associated energy dissipation in the rotor windings.
c) The time it takes to reverse the speed of the motor by plugging, and the associated energy dissipation in the rotor windings.

SOLUTION

a) From the name-plate data, the synchronous speed of the motor is

$$n_s = 900 \ \text{r/min}$$

or

$$\omega_s = 94.25 \ \text{rad/s}$$

At no-load, the energy stored in the rotating masses is

$$\text{K.E.} = \frac{1}{2} J\omega_s^2 = \frac{1}{2}(6.5)(94.25)^2 = \underline{\underline{28.87 \ \text{kW} \cdot s}}$$

b) The full-load operating slip is

$$s = \frac{900 - 855}{900} = .05 \ \text{pu}$$

The time it takes to stop the motor by plugging is found by substituting the various parameters into Eq. (3.96):

$$t_1 = \frac{6.5(94.25)^2}{(440)^2(.15)}\left(.15^2 \ln (2 - .05) + \frac{1}{2}(1.0)^2(3 - 4(.05) + .05^2)\right)$$
$$= \underline{\underline{2.82 \ \text{s}}}$$

The energy dissipated in the rotor windings when the motor is plugged to a complete stop is found by using Eq. (3.108):

$$(W_{dr})_p = 3\left(\frac{1}{2}J\omega_s^2\right)$$

$$= 3(28.87) = \underline{\underline{86.61\ \text{kW}\cdot s}}$$

c) The time (t) that it takes to reverse the motor's speed is equal to the time that it takes to bring the motor to a complete stop (t_1) plus the time that it takes to reach full-load speed in the reverse direction (t_2). The time t_1, as calculated previously, is 2.82 seconds, while the time t_2 is calculated by substituting the given data into Eq. (3.81):

$$t_2 = \frac{6.5(94.25)^2}{(440)^2(.15)}\left[(1 - .05^2)\frac{1.0}{2} + .15^2 \ln\frac{1}{.05}\right] = \underline{\underline{1.13\ \text{s}}}$$

Thus, the time that it takes to reverse the motor's speed by plugging is

$$t = 2.82 + 1.13 = \underline{\underline{3.94\ \text{s}}}$$

The energy dissipated in the rotor windings when the speed of the motor is reversed is found by using Eq. (3.109):

$$(W_{dr})_r = 4(28.87) = \underline{\underline{115.47\ \text{kW}\cdot\text{s}}}$$

Exercise 3–14 A three-phase, 60 Hz, 8-pole induction motor has a total moment of inertia of 20 kg · m^2. Determine the energy dissipated in the rotor windings when:

a) Starting from rest, the motor is brought up to its nominal speed.

b) The motor is stopped by plugging.

c) The motor is reversed by plugging.

Answer (a) 88.83 kJ; (b) 266.48 kJ; (c) 355.31 kJ.

3.6 Solid State Control

3.6.1 FREQUENCY CONVERTERS

The torque-speed characteristic of a three-phase, squirrel-cage induction motor can be controlled by a frequency converter. As shown in Fig. 3-39, frequency converters have three distinct sections: the rectifier, the inverter, and the regulator.

The rectifier, usually a six-pulse diode bridge, changes the incoming ac voltage to a dc waveform. The inverter changes the rectifier's output dc voltage waveform to a

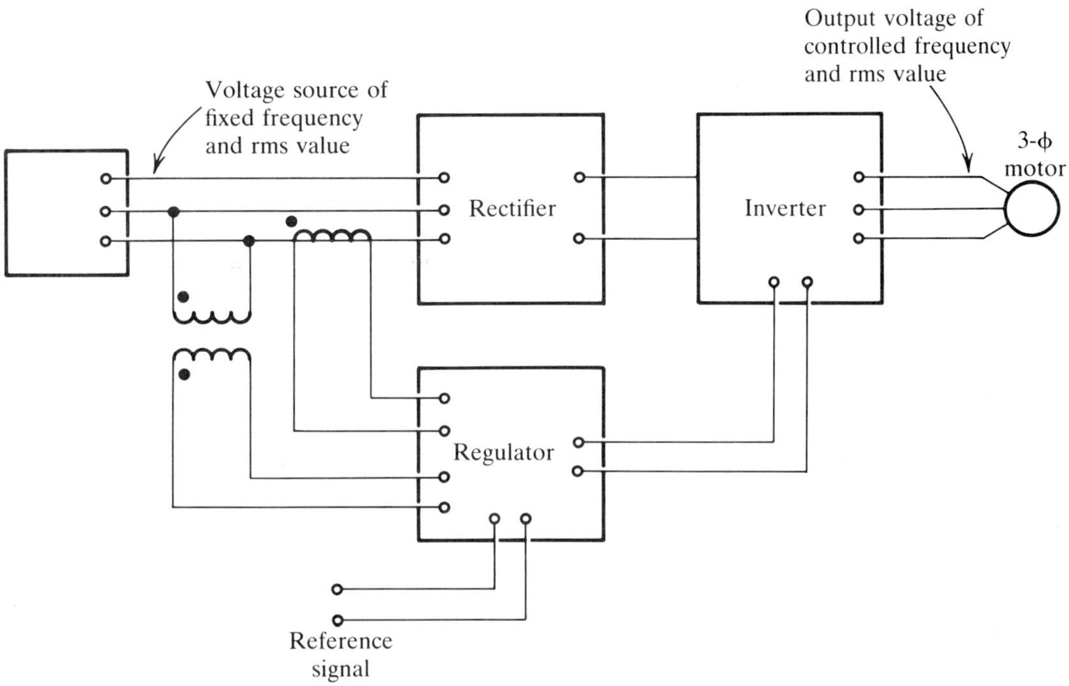

FIG. 3-39 A three-phase frequency converter.

three-phase ac signal of desired voltage and frequency. The regulator incorporates all the devices necessary for the protection and automatic control of the inverter's output. These components of frequency converters are discussed in some detail in the sections that follow.

Rectifiers

A rectifier receives a three-phase ac waveform at its input terminals and produces a dc voltage waveform at its output terminals. A typical six-pulse, diode-bridge rectifier is shown in Fig. 3-40(a). It is called a six-pulse rectifier because, during a complete cycle of the input voltage, its output is comprised of six voltage pulses. The diodes act as switches which are either open- or short-circuited, according to the relative polarity of the voltage from their anode (A) to cathode (K) (see diode D_1). When the diode's anode is at a higher potential than its cathode, the diode is shorted (forward-biased or conducting). When the diode's cathode is at higher potential than its anode, the diode is open-circuited (reverse-biased or nonconducting).

The ON–OFF condition of the diodes (such as D_1 and D_2) connected to the terminals a and b depends on the magnitude of the voltage between a and b (v_{ab}) relative to the other line-to-line voltages. At the particular time interval when the voltage v_{ab} is the highest, then diodes D_1 and D_2 conduct. Refer to Fig. 3-40(b). The conducting period of these two diodes is from $\pi/3$ to $2\pi/3$ radians.

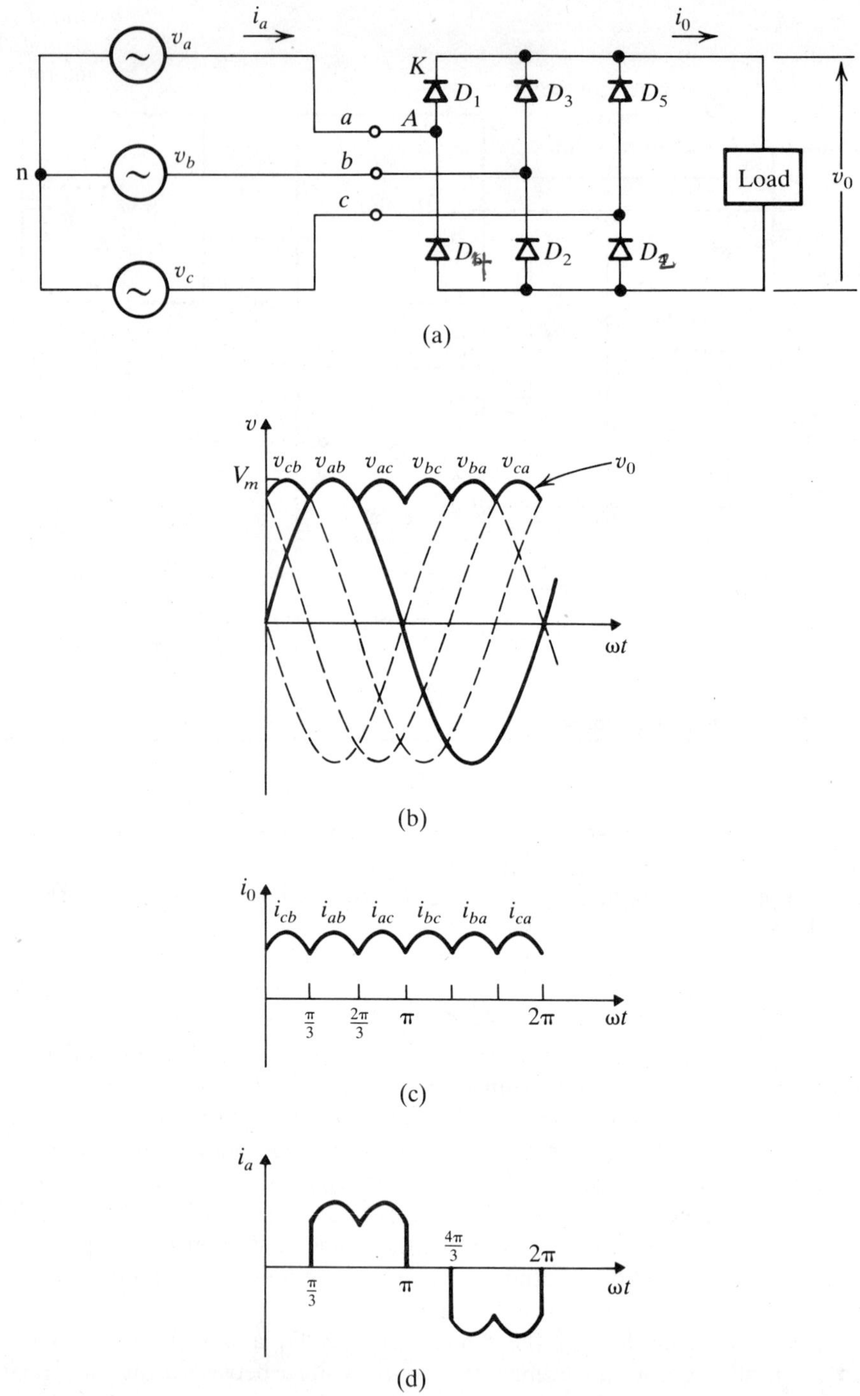

FIG. 3-40 Rectifier: **(a)** typical circuit, **(b)** input-output voltage waveforms, **(c)** output current, **(d)** line current (I_a).

The conducting period of each pair of diodes can be determined in this manner. The output current pulses for a resistive load are shown in Fig. 3-40(c). According to Ohm's law, these current pulses are similar to the output voltage pulses shown in Fig. 3-40(b).

Each ac supply line carries four of the six output current pulses in an alternating symmetry, and thus the line currents do not have a dc component. This is shown in Fig. 3-40(d).

By definition, the average value (V_{av}) of the rectifier's output voltage is

$$V_{av} = \frac{1}{T}\int_0^T v_0 \, dt \qquad \textbf{(3.110)}$$

where v_0 is the instantaneous value of the output voltage whose period is T seconds. Using symmetry, and considering only the output voltage pulse that is due to v_{ab}, we have

$$V_{av} = \frac{6}{2\pi}\int_{\pi/3}^{2\pi/3} V_m \sin \omega t \, d\omega t$$

$$= -\frac{3V_m}{\pi}(\cos 120^\circ - \cos 60^\circ) \qquad \textbf{(3.111)}$$

from which

$$V_{av} = \frac{3}{\pi} V_m \qquad \textbf{(3.112)}$$

or

$$V_{av} = .95 V_m \qquad \textbf{(3.113)}$$

where V_m is the maximum value of the supply line-to-line voltage.

When a capacitor of the proper size is placed across the output of the rectifier, the dc output voltage is equal to V_m. When a variable dc voltage is desirable, the diodes are replaced with control switches, such as power transistors or thyristors. The use of transistors as switches is explained in the section on inverters, and the use of thyristors is described in detail in the chapter on dc machines (see Section 6.3).

Exercise 3–15 Assuming that the output current remains constant,

a) Derive Eq. (3.112) by considering the output voltage pulse that is due to voltage V_{cb}.

b) Prove that the rms value of the supply line current (I) is related to the dc value of the rectifier's output current (I_{av}) by the following equation:

$$I = \sqrt{\frac{2}{3}} I_{av}$$

Regulators

The regulator is the brain of a frequency converter: not only does it provide a closed loop control for the motor, but it also monitors the motor's performance and protects it from abnormal operating conditions.

A regulator in block diagram form is shown in Fig. 3-41. The input section of the regulator receives the reference signal and the outputs from the potential and the current transformers.

Input section	Analog-to-digital converter (A/D)	Microprocessor	Digital-to-analog converter (D/A)

FIG. 3-41 Block-diagram representation of a digital regulator.

The output of the potential transformers (PT) is used to produce the dc voltages required to operate the various digital circuits. The output of the current transformers (CT) is directly proportional to the current drawn by the motor; when properly evaluated, it reveals the magnitudes of the motor's torque and speed, as can be seen from Fig. 3-23.

In some applications, a dc type of CT may be connected to the dc lines of the rectifier. The operation of such a CT is based on the Hall effect, not on the electromagnetic coupling of two coils. The incoming section of the regulator subtracts the CT's signal from the reference signal, and the resulting signal is converted to an equivalent digital signal. This digital signal is fed into the microprocessor, which compares it to a set of stored instructions and produces an output accordingly.

The output of the microprocessor is converted to an analogue signal which operates the switches (S) of the inverter (see Fig. 3-42). These switches, in turn, control the ON–OFF operation of the transistors. As a result, the motor's performance is electronically controlled.

Inverters

The function of an inverter is to produce a three-phase ac waveform of controlled frequency from a single dc voltage supply. A typical transistor three-phase inverter is shown in Fig. 3-42(a).

Depending on its base current, each transistor* in an inverter is either open-circuited or short-circuited; that is, the transistor operates as a switch and not as an amplifier. When the base current is sufficiently high, the transistor is short-circuited ($v_{CE} \approx 0$); when the base current is interrupted, the transistor is open-circuited ($I_c = 0$).

When a transistor is short-circuited, it is said to be operating in the saturation region. When a transistor is open-circuited, it is said to be operating in the cut-off region.

. . .

* Transistor operation is discussed in Appendix D.

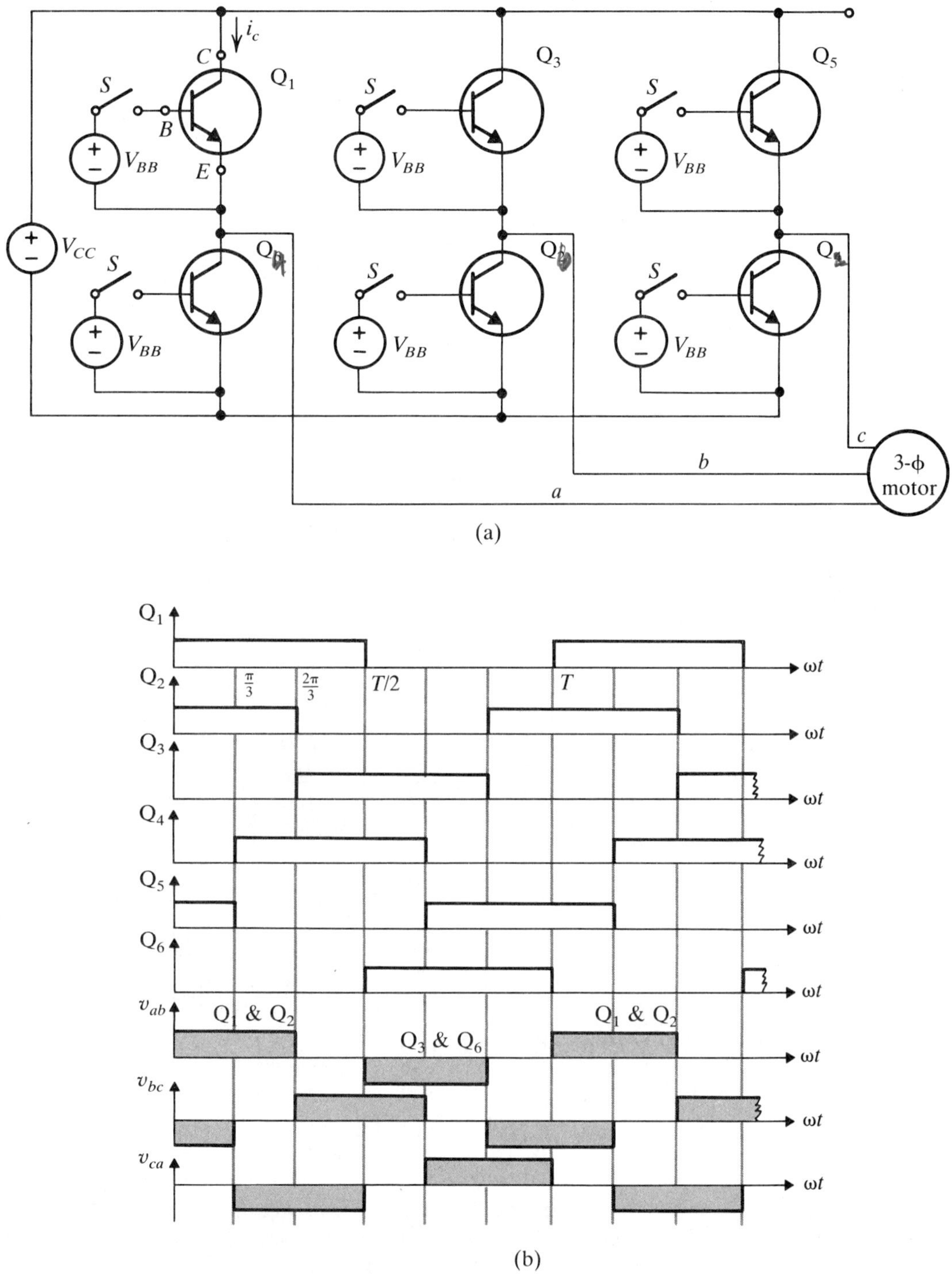

FIG. 3-42 3-ϕ Inverter: **(a)** circuit, **(b)** pertinent waveforms.

The base current depends on the magnitude of the dc supply (V_{BB}). This current is switched ON and OFF by the switch (S), whose status is in turn controlled by the microprocessor of the regulator.

In inverters of higher power rating, the voltage V_{BB} is obtained by amplifying a control signal (not shown in the diagram), and thus the switch S is located in series with the control signal. Refer to Fig. 3-42(b). When the base current flows, it is understood that it is of sufficient magnitude to saturate the transistor. The ON and OFF time for each transistor's base current is equal to half the period ($T/2$) of the output voltage.

The switching ON of a transistor is delayed $\pi/3$ radians relative to the transistor that was previously conducting. For the current to flow, two transistors must conduct simultaneously.

Consider the output voltage waveform v_{ab}. As shown in the diagram and as tabulated below, for each component of this output voltage a unique pair of transistors conduct.

Time Interval of v_{ab}	*Conducting Transistors*
$0 - 2\frac{\pi}{3}$	Q_1 and Q_2
$2\frac{\pi}{3} - \pi$	None
$\pi - 5\frac{\pi}{3}$	Q_3 and Q_6
$5\frac{\pi}{3} - 2\pi$	None

The pair of conducting transistors of the other line-to-line voltages can be determined in the same manner. The 3-ϕ voltages produced have a phase sequence *ABC*.

Normally, the operation of an inverter's switches produces an output voltage that, over a complete voltage cycle, is made up of a series of pulses whose width, as shown in Fig. 3-43(a), is variable. This is referred to as "pulse-width modulation" (PWM). Although the output voltage is made up of a series of pulses, the resulting motor current resembles a sine function, as shown in Fig. 3-43(b).

The output current, as expected in an induction machine, lags the voltage. Since this current is not a pure sinusoidal waveform, it will increase the motor's losses, slightly reduce its output torque, and may give rise to vibration problems. However, when the output of the inverter is passed through a properly designed filter (not often used, in practice), an almost pure sine wave is produced.

The width of the voltage pulses controls the rms value of the output voltage, and the number of pulses within a voltage cycle controls the frequency of the output voltage. Evaluating the output voltage's rms value and frequency that results from *n*

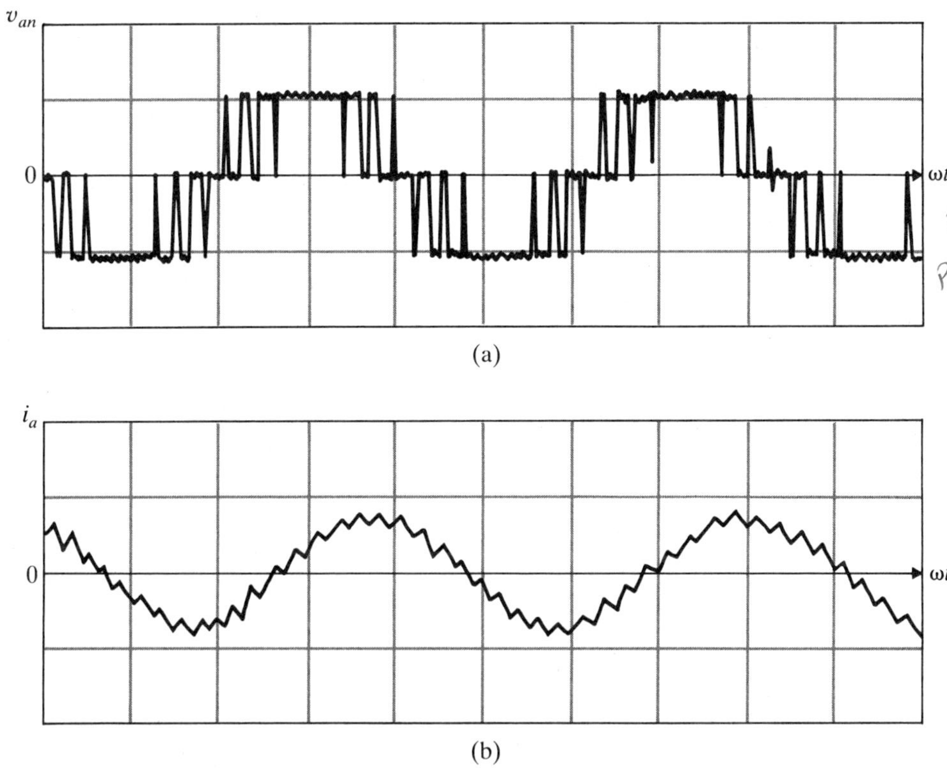

FIG. 3-43 Actual output waveforms of an inverter: **(a)** voltage, **(b)** current.

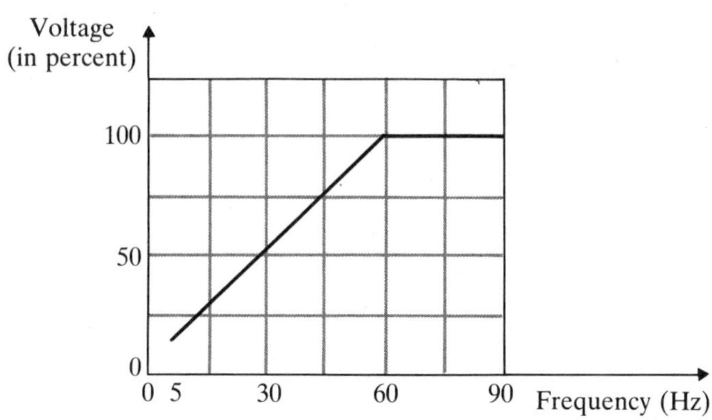

FIG. 3-44 Inverter's output voltage–frequency characteristic.

pulses per cycle—whose width forms an arithmetic progression—is left as a student exercise (see Problem 3-18).

The output voltage–frequency characteristic of a frequency converter is shown in Fig. 3-44. At frequencies above 60 Hz the voltage is kept constant at the motor's rated voltage. At operating frequencies between 5 Hz and 60 Hz the voltage is proportional to the frequency, so that nominal flux levels within the motor's magnetic material can be maintained. Thus,

$$V = Kf \tag{3.114}$$

where V and f are, respectively, the inverter's output voltage and frequency, and K is a constant of proportionality that depends on the motor's nominal flux level and winding turns. Under ideal conditions, this constant can be obtained from Eq. (1.109).

The inverters of motors with high power ratings incorporate thyristors instead of transistors. Thyristors can handle larger quantities of power than transistors, but they are more difficult to control and require circuits that are much more complex.

The torque–frequency characteristic of a 3-ϕ squirrel-cage induction motor is shown in Fig. 3-45.

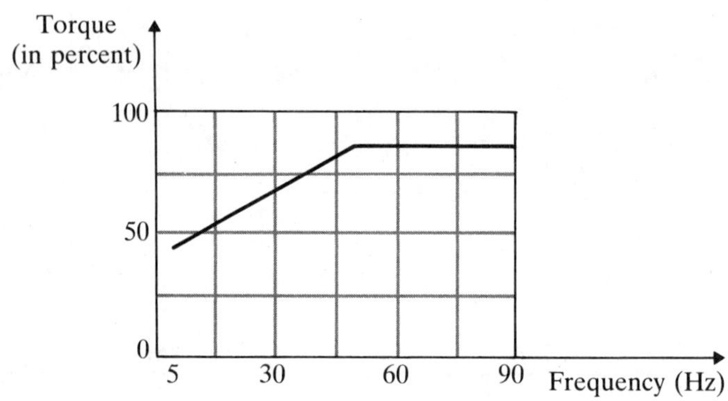

FIG. 3-45 Typical torque–frequency characteristic.

The approximate per-phase equivalent circuits of a motor at 60 Hz and at a reduced frequency of $60K_1$ Hz ($K_1 < 1$) are shown in Figs. 3-46(a) and 3-46(b), respectively.

Referring to Fig. 3-46(a), the stator current I_1 at 60 Hz and at a slip of s_1 is

$$I_1 = \frac{V}{R_1 + \dfrac{R_2}{s_1} + jX} \tag{3.115}$$

The stator current I_2 at a reduced frequency of $60K_1$ Hz is

$$I_2 = \frac{K_1 V}{R_1 + \dfrac{R_2}{s_2} + jK_1 X} \tag{3.116}$$

where s_2 is the per-unit slip of the motor at the new operating condition.

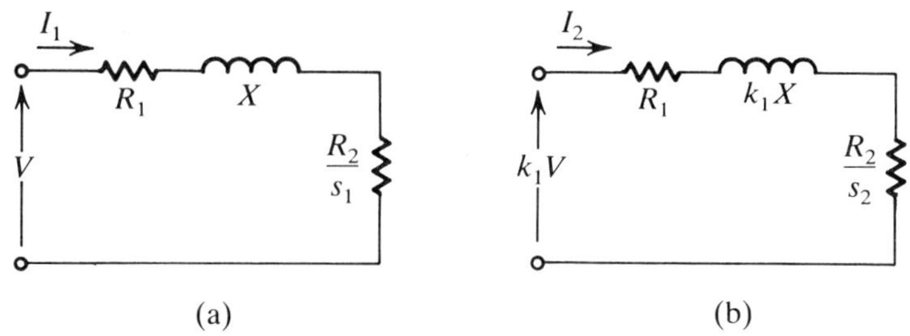

FIG. 3-46 Approximate per-phase equivalent circuit of a 3-ϕ induction machine: **(a)** at 60 Hz and at a slip of s_1, **(b)** at $60K_1$ Hz and at a slip of s_2.

When the motor is running, the predominant term in the denominators of the above expressions is the term that includes R_2. Thus, from Eqs. (3.115) and (3.116) we obtain

$$\frac{I_1}{I_2} \approx \frac{s_1}{K_1 s_2} \tag{3.117}$$

The output torque of the motor at a reduced frequency can be obtained by using the methods described in Section 3.1.6.

EXAMPLE 3–9 A 480 V, 50 A, 60 Hz, 1176 r/min, .8 Pf three-phase induction motor has negligible stator resistance. The motor is controlled by a frequency converter apparatus. When the supply frequency is changed to 30 Hz, the motor operates at a Pf of .9 lagging. Estimate:

a) The voltage at the motor terminals.
b) The per-unit slip.
c) The current drawn by the motor.

SOLUTION

a) From Eq. (3.114), it is clear that since the frequency is halved, the voltage must also be halved. Thus,

$$V = \frac{480}{2}$$
$$= 240 \text{ V, L–L}$$

and

$$K_1 = \frac{30}{60} = .5$$

b) The per-unit slip of the motor at 60 Hz is

$$s_1 = \frac{1200 - 1176}{1200} = .02 \text{ pu}$$

The phase angle θ_1 of the motor's impedance is

$$\theta_1 = \arccos(.8) = 36.9°$$

Also,

$$\tan \theta_1 = \frac{X}{R_2/s_1} \qquad \text{(I)}$$

where X and R_2 are the reactance and resistance of the motor's impedance, respectively. At 30 Hz:

$$\tan \theta_2 = \frac{K_1 X}{R_2/s_2} \qquad \text{(II)}$$

From relationships (I) and (II) we get

$$\frac{\tan \theta_1}{\tan \theta_2} = \frac{s_1}{K_1 s_2}$$

Substituting the given values, we obtain

$$s_2 = \frac{.02 \tan 25.8°}{.5 \tan 36.9°} = \underline{\underline{.026 \text{ pu}}}$$

c) Substituting into Eq. (3.117), we obtain

$$I_2 = \frac{50(.5)(.026)}{.02} = \underline{\underline{32.3 \text{ A}}}$$

In phasor form,

$$I_2 = 32.3\underline{/-25.8°} \text{ A}$$

EXAMPLE 3–10 A 480 V, 60 Hz, 6-pole, 1176 r/min, 3-ϕ, squirrel-cage induction motor has the approximate per-phase equivalent circuit shown in Fig. 3-47(a). The motor drives a ventilating fan whose torque–speed characteristic, with the dampers

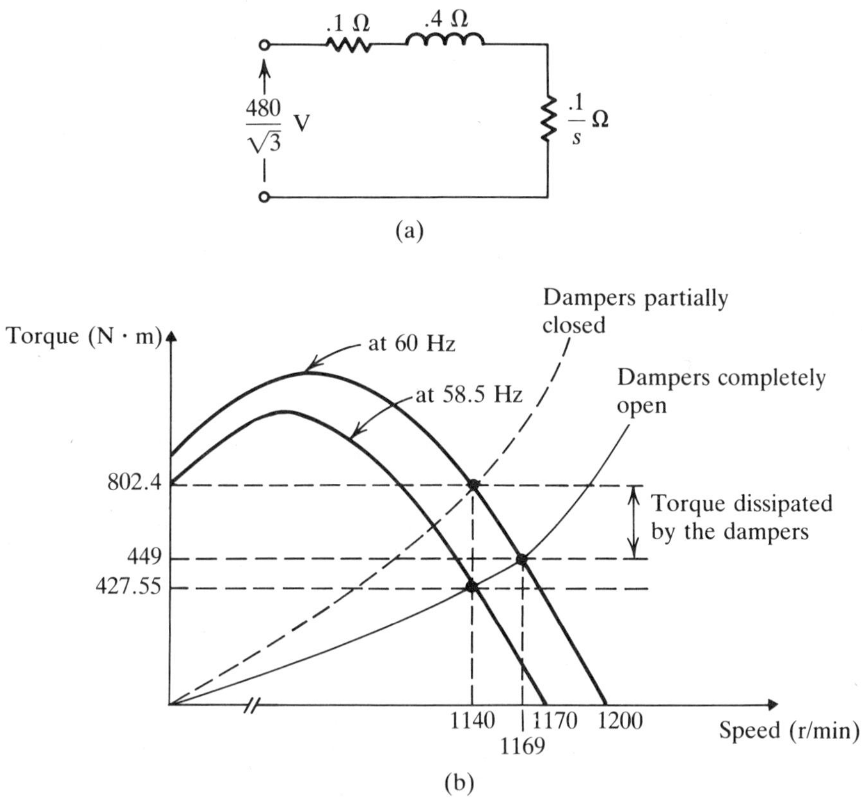

FIG. 3-47

completely open, is given by

$$T = .03\omega^2 \ \text{N} \cdot \text{m}$$

Determine the motor's current, torque, and copper losses for the following operating conditions:

a) The motor drives the fan while the dampers are completely open.
b) The dampers are partially closed and the motor operates at 1140 r/min.
c) The dampers are completely open, and a frequency converter is used to drive the motor's load at 1140 r/min.

SOLUTION

a) At 60 Hz, the motor's stator field rotates at

$$\omega_s = 120\left(\frac{60}{6}\right)\frac{2\pi}{60} = 125.66 \ \text{rad/s}$$

The motor's operating slip will be such as to produce a torque that is equal to the load requirements. Equating the torque of the motor to that of the load, we obtain

$$.03(125.66(1 - s))^2 = \frac{480^2}{125.66\left[\left(.1 + \frac{.1}{s}\right)^2 + .4^2\right]}\left(\frac{.1}{s}\right)$$

where s is the per-unit slip of the motor. From the above, by trial and error we obtain

$$s = .026 \text{ pu}$$

The corresponding speed, torque, current, and copper losses are

$$\begin{aligned} n_a &= 1200(1 - .026) \\ &= \underline{\underline{1169 \text{ r/min}}} \end{aligned}$$

$$\begin{aligned} T &= .03\left(1169\frac{2\pi}{60}\right)^2 \\ &= \underline{\underline{449.43 \text{ N}\cdot\text{m}}} \end{aligned}$$

$$\begin{aligned} |I| &= \frac{480/\sqrt{3}}{\sqrt{\left(.1 + \frac{.1}{.026}\right)^2 + (.4)^2}} \\ &= \underline{\underline{69.87 \text{ A}}} \end{aligned}$$

$$\begin{aligned} P_{cl} &= 3(69.87)^2(.2) \\ &= \underline{\underline{2.93 \text{ kW}}} \end{aligned}$$

b) At 1140 r/min the per-unit slip is

$$\begin{aligned} s &= \frac{1200 - 1140}{1200} \\ &= .05 \text{ pu} \end{aligned}$$

The magnitude of the current is

$$\begin{aligned} |I| &= \frac{480/\sqrt{3}}{\sqrt{\left(.1 + \frac{.1}{.05}\right)^2 + .4^2}} \\ &= \underline{\underline{129.64 \text{ A}}} \end{aligned}$$

The copper losses are

$$P_{cl} = 3(129.64)^2(.2) = \underline{\underline{10.1 \text{ kW}}}$$

The torque is

$$T = \frac{3(129.64)^2(.1)}{125.66(.05)} = \underline{\underline{802.39 \text{ N} \cdot \text{m}}}$$

c) Using a frequency converter to modify the motor's torque–speed characteristic instead of using the dampers to modify the fan's torque–speed characteristic, we have:

$$\text{actual speed } n_a = 1140 \text{ r/min}$$

$$\text{load torque } T = .03\left(1140\frac{2\pi}{60}\right)^2 = \underline{\underline{427.55 \text{ N} \cdot \text{m}}}$$

The output torque of the motor must equal that of the load, Thus,

$$427.55 = \frac{[K_1(480)]^2}{(125.66)K_1\left[\left(.1 + \frac{.1}{s}\right)^2 + (.4K_1)^2\right]}\left(\frac{.1}{s}\right) \qquad \text{(I)}$$

where K_1 is the factor by which the converter must reduce the nominal 60 Hz frequency. The operating slip at this frequency is

$$s = \frac{1200K_1 - 1140}{1200K_1} \qquad \text{(II)}$$

Substituting equation (II) into (I), by trial and error we find that

$$K_1 = .975$$

The corresponding slip of the motor is

$$s = \frac{1200(.975) - 1140}{1200(.975)} = .026 \text{ pu}$$

The converter's output frequency is

$$f = .975(60) = 58.5 \text{ Hz}$$

The magnitude of the motor's current is

$$|I| = \frac{.975(480/\sqrt{3})}{\sqrt{\left(.1 + \dfrac{.1}{.026}\right)^2 + (.4(.975))^2}}$$
$$= \underline{\underline{67.13\ \text{A}}}$$

The corresponding copper losses of the motor are

$$P_{cl} = 3(67.13)^2(.2)$$
$$= \underline{\underline{2.71\ \text{kW}}}$$

The torque–speed characteristics are shown in Fig. 3-47(b). The dotted characteristic represents the combination of the original load plus the load added to it by the dampers of the fan. The torque (T_d) contribution of the dampers is

$$T_d = 802.39 - 427.55 = 374.84\ \text{N} \cdot \text{m}$$

For purposes of comparison, the results are summarized in Table 3-7.

TABLE 3-7 Summary of Results of Example 3–10

Speed r/min	*Current (A)*	*Copper Losses (kW)*	*Output Torque (N · m)*	*Notes*
1169	69.87	2.93	449.43	
1140	129.64	10.1	802.39	This speed is obtained by using the fan's dampers
1140	67.13	2.71	427.55	This speed is obtained by using a frequency converter

Exercise 3–16 Referring to Fig. 3-42(a),

a) Why are transistors, and not switches such as those of the base circuit, used to directly switch the load current ON and OFF?
b) When the transistors are switched OFF, they are subjected to high voltages from emitter to collector due to the energy stored in the magnetic field of the motor. Design a simple circuit that will short-circuit these high transient voltages.

Exercise 3–17 A frequency converter is used to operate a 3-ϕ, 480 V, 60 Hz, 1710 r/min, squirrel-cage induction motor. The per-phase impedance of the motor at 60 Hz is

$$Z = .15 + \frac{.1}{s} + j.6$$

The motor operates at 1500 r/min and delivers 100 kW to a load. Neglecting the magnetizing current and the motor's rotational losses, estimate the output frequency of the converter.

Answer 52.5 Hz.

3.7 Conclusion

The operation of induction machines is based on the development of the rotating field, on the induction principle, and on the natural tendency of magnetic fields to align their axes.

When balanced 3-ϕ currents are circulated through properly distributed coils, a rotating field is produced. This rotating field completes its loop by cutting through the rotor and stator structures. As a result, a voltage is induced in the rotor coils. This voltage, in turn, causes current to flow through the rotor windings, and thus a rotor field is produced. The stator and rotor fields try to align their magnetic axes and as a result, the motor produces torque.

The rotor of an induction machine is of either the squirrel-cage or the wound-rotor type. Induction machines with squirrel-cage rotors are much more popular than those with wound rotors because they are less expensive to both purchase and operate. The National Electrical Manufacturers Association (NEMA) groups motors with squirrel-cage rotors into four categories—class A, B, C, or D—according to their torque–speed characteristics. Each class of motor is suited to a particular application. Class B motors, for example, are for general-purpose use and can be found in fans and blowers.

Induction machines with wound rotors are used for special applications, but because they are less efficient at higher slips, they are gradually being replaced by variable-speed drives. One may find wound-rotor induction machines used in the following special applications:

1. To operate variable torque–speed characteristics for loads that operate over a wide range of speed and torque.
2. To limit the starting current of a motor in a distribution system where the supply power system is weak (large upstream impedance).
3. To limit the starting current of motors that are required to start and restart frequently, such as those used in the crushing machines of the mining industry. The reduction in the starting current increases the number of permissible starts during any given time interval, so the motor's life is not reduced.
4. To give high torque at high slip.

A motor's slip indicates the relative motion between the speed of the synchronously rotating stator field and the speed of the shaft of the motor. The slip can be *positive* and less than unity (in an induction motor); *positive* and larger than unity (in voltage and frequency multipliers); or *negative* and less than unity (in induction generators).

For all practical purposes, the real part of the impedance of a motor as seen from its stator is inversely proportional to the slip. Thus, at starting ($s = 1$) the impedance is small, while under full-load conditions ($s \approx .015$) the impedance is very high Consequently, a motor's current at this operating condition—known as the rated condition—is about six times smaller than it is at starting. The large starting currents of a motor not only lower its starting-torque capability—owing to a high voltage drop in the upstream impedances—but also develop extremely high temperatures in the rotor and stator windings. These large currents may limit the number of motor starts permissible within a given time interval.

The motor's equivalent circuit is of paramount importance. It simplifies the understanding of the motor's operation and facilitates the analysis of a particular problem. In practice, however, the circuit parameters are often not known, and an application engineer has to make decisions based on other data—starting and full-load torque and currents, etc.—that are made readily available by the manufacturers.

Induction generators are used to generate electricity from either a constant-speed prime mover or a decelerating load. Induction generators are gaining popularity because they provide the most economical method of connecting a small power station to a distribution network. Small power stations driven by wind, solar, water, or process heat are increasingly common because of recent legislation requiring utilities to buy, at reasonable rates, electricity produced by their consumers.

The time it takes for a motor, starting from rest, to reach its rated speed is directly proportional to the moment of inertia of load plus motor and inversely proportional to the square of the voltage. Analytical expressions for the speed and the current of a motor as functions of time are approximate at best and are—owing to the nonlinearities of the parameters involved—complex to derive. Empirical methods, though slightly inaccurate, are simple to use.

Frequency converters electronically change the frequency and the voltage applied to a motor. As a result, a motor's torque–speed characteristic can be modified to meet diverse load requirements. Frequency converters, additionally, contribute more significantly to the reduction in energy consumption per-unit output than do the other methods of speed control.

The tables that follow summarize the basic concepts of this chapter in their mathematical form and also furnish typical manufacturers' data for three-phase induction machines.

3.8 Tables

TABLE 3-8 Summary of Important Equations

Item	*Description*	*Remarks*
1	Magnitude of rotating field $$\mathscr{F}_s = \frac{3}{2}\mathscr{F}_1 \cos(\omega t - \beta)$$	Eq. (3.14)
2	Mechanical and electrical degrees or radians $$\theta = \frac{P}{2}\theta_m$$	Eq. (3.15)

TABLE 3-8 (Continued)

Item	*Description*	*Remarks*
3	Speed of rotating field in r/min $$n_s = 120\frac{f}{P}$$	Eq. (3.18)
4	Per-unit slip $$s = \frac{n_s - n_a}{n_s}$$	Eq. (3.19)
5	Voltage induced in rotor windings $$V_r = sV'_r$$	Eq. (3.23)
6	Frequency of voltage induced in the rotor windings $$f_r = sf$$	Eq. (3.24)
7	Mechanical load in equivalent electrical resistance $$R_L = R_2\frac{1-s}{s}$$	Eq. (3.32)
8	Torque developed $$T = \left[\frac{V^2}{\left(R_1 + \frac{R_2}{s}\right)^2 + X^2}\right]\left(\frac{R_2}{\omega_s s}\right)$$	Eq. (3.38)
9	Torque in the full-load region $$T \approx K\frac{s}{R_2}$$	Eq. (3.42)
10	Slip at maximum torque $$s_{mt} = \frac{R_2}{\sqrt{R_1^2 + X^2}}$$	Eq. (3.46)
11	Motor's output torque $$T = J\frac{d\omega_a}{dt} + B\omega_a + T_L$$	Eq. (3.49)
12	Power developed $$P_d = I_2^2 R_2\frac{1-s}{s}$$	Eq. (3.56)
13	Starting current $$I_{st} = (5 \rightarrow 6)I_{fl}$$	
	Voltage and Frequency Multipliers	
14	Power input from the variable speed drive $$P_m = P\left(\frac{s-1}{s}\right)$$	Eq. (3.74)

(Continued)

TABLE 3-8 *(Continued)*

Item	*Description*	*Remarks*
15	Power input from the induction machine $$P_e = \frac{P}{s}$$	Eq. (3.75)
	Electromechanical Transients	
16	Accelerating time: *Mathematically* $$t = \frac{J\omega_s^2}{V^2R_2}\left[(1 - s_1^2)\left(\frac{R_1^2 + X^2}{2}\right) + 2R_1R_2(1 - s_1) + R_2^2 \ln \frac{1}{s_1}\right]$$	Eq. (3.81)
	Graphically $$t = J\int_0^{\omega_1} \frac{d\omega}{T_{\text{mot}} - T_L}$$	Eq. (3.83)
	Plugging	
17	Time required to stop the motor $$t = \frac{J\omega_s^2}{V^2R_2}\left[R_2^2 \ln(2 - s_1) + 2R_1R_2(1 - s_1) + \frac{1}{2}(R_1^2 + X^2)(3 - 4s_1 + s_1^2)\right]$$	Eq. (3.96)
	Energy and Rotor Winding Losses	
18	Energy dissipation in rotor windings $$W_{dr} = \frac{1}{2}J\omega_s^2$$	Eq. (3.106)
19	Rotor energy dissipation due to a complete speed reversal by plugging $$(W_{dr})_r = 4\left(\frac{1}{2}J\omega_s^2\right)$$	Eq. (3.109)
	Frequency Converters	
20	Output voltage of a six-pulse bridge rectifier $$V_{av} = \frac{3}{\pi}V_m$$	Eq. (3.112)
21	Characteristic of a frequency converter $$V = Kf$$	Eq. (3.114)

TABLE 3-9 Data for three-phase 60 Hz, 460 V, NEMA Design B, service factor 1.0, class insulation B, standard squirrel-cage induction motors*

Power in kW	Full-Load Speed in r/min	Approximate Amperes		Starting Torque in Percent	Maximum Torque in Percent	Efficiency in Percent		Power Factor in Percent	
		Full-Load	Locked-Rotor			Full-Load	50% Full-Load	Full-Load	50% Full-Load
.75	1730	1.64	15	275	300	78.5	72	73	51
	870	2.3	15	135	215	70.5	61	58.5	39.5
7.5	3500	12.23	81.25	135	200	85.5	82.5	90	82
	880	15.86	81.25	125	200	83	79.5	71.5	52.5
20	3540	32.2	182	130	200	86.5	83	90	82.5
	885	35.2	182	135	200	88.5	87.5	80.5	64.5
40	3545	62.3	362.5	125	250	89	87	90.5	83
	880	69.3	362.5	145	230	90.5	88.5	80	65
50	3560	75.4	542.5	115	250	90.5	88	92	85
	885	86.2	542.5	145	210	91	90	80	66.5
75	3555	112.4	725	110	200	90.5	88.5	92.5	88.5
	885	128.4	725	125	200	90.5	89.5	81	70
100	3565	149.9	1085	110	240	91	88	92	88.5
	1775	151.6	1085	125	210	92	90.5	90	86

**Courtesy of Siemens Electric Limited.*

TABLE 3-10 Data for three-phase, 60 Hz, 4000 V, totally enclosed, fan-cooled, squirrel-cage induction motors*

kW	Full-Load Speed in r/min	Full-Load Amperes	Locked-Rotor Amperes	Starting Torque in Percent	Maximum Torque in Percent	Efficiency in Percent		Power Factor in Percent	
						Full-Load	50% Load	Full-Load	50% Load
150	1764	28	162	100	200	91.8	88.7	84.1	71
	1178	27.1	162	120	200	91.8	88.7	87	75.5
200	589	32.4	162	60	175	90.8	88.6	73.7	53.5
	589	42.6	202	60	175	91.1	89	74.3	54.5
225	1180	40.4	242	100	175	92.3	89.2	87	76.4
	589	47.5	242	60	175	91.4	89.3	74.8	55.5
300	1182	53.8	323	60	175	92.7	89.6	87	76
400	1184	71.4	404	60	175	93	89.9	87	77.1
1500	1176	259	1510	140	210	95	94.5	88	77

**Courtesy of Siemens Electric Limited.*

TABLE 3-11 Test results and measurements of a 260 kW, 4000 V, .75 Pf, .88 efficient, 10-pole, squirrel-cage induction motor*

a) Tests

Test	*Frequency (Hz)*	*Voltage (V)*	*Current (A)*	*Power (kW)*
No-load test	60	4000	33.1	9.5
Locked-rotor test	60	975	57	27
Locked-rotor test	30	537.8	57	19.2

b) Calculation of Load Characteristics at Rated Voltage and Frequency

Load in percent	25	50	75	100	125
Line current in A	35	40.2	47.9	57.6	68.7
Efficiency in percent	86.6	91.6	92.7	92.7	92.2
Power factor in percent	31.1	51.2	63.7	70.7	74.4
Slip in percent	.32	.65	1.01	1.4	1.83

Starting current (locked rotor) in pu: 4.6
Starting torque in pu: 1.5
Maximum torque in pu: 2.2

c) Temperature rise

Hours run: 4
Volts: 4000
Load: 100%

	Frame	*Bearings* *Load Side*	*Bearings* *Opposite Side*
Temperature in °C	50	19	15

d) Winding resistance between terminals

At 20°C: 1.69 ohms

e) Air-gap measurement

1.13 mm, average

**Courtesy of Siemens Electric Limited.*

TABLE 3-12 Data for three-phase, 4000 V, 60 Hz, squirrel-cage rotor induction motors*

Rating in kW	Number of Poles	Polar Moment of Inertia Referred to Speed of Motor in kg-m²			Torque in Percent			Accelerating Time in Seconds	Maximum Permissible Stall Time in Seconds	
		Rotor	Load	Total	Starting	Pull-In	Maximum		Starting from Cold	Restarting Immediately after Operation (Hot-Start)
200	4	6	100	106	110	90	230	20	25	20
200	4	6	40	46	110	90	230	9	25	20
250	6	31	217	248	90	70	220	20	25	20
250	14	68	34	102	135	100	260	2	20	16
300	6	46	125	171	175	130	210	8	20	16
375	6	53	150	203	175	130	210	8	20	16

*Courtesy of Siemens Electric Limited.

3.9 REVIEW QUESTIONS

1. What are the essential differences between squirrel-cage and wound-rotor three-phase induction machines?
2. Why are three-phase induction machines often called "rotating transformers"?
3. Differentiate between a synchronous speed and an actual speed.
4. What is "slip" and under what operating conditions is it:
   a) Positive and smaller than unity.
   b) Positive and larger than unity.
   c) Negative and smaller than unity.
5. What are the frequencies of rotor currents at starting and at full-load? How does frequency affect the impedance of the rotor?
6. Why is rated voltage used in the "no-load test" and near full-load current used in the locked-rotor test?
7. Why, in general, is the starting torque for a three-phase squirrel-cage induction motor 1.5 to 2 times the full-load torque, and the starting current 5 to 6 times the full-load current?
8. For the A, B, C, and D classifications of motors, how does the rotor reactance compare with that of the stator?
9. *Explain*: The torque at the starting region is inversely proportional to the operating slip, while at the full-load region it is proportional to the slip.
10. Why is the time that it takes a motor, starting from rest, to reach its rated speed of paramount importance?

11. Describe the four methods by which a three-phase induction motor can be stopped.
12. How does the motor of a hoist change direction of rotation?
13. What provides the excitation power for an induction generator?
14. *Explain*: The maximum rate of heat dissipation of a machine is given by its efficiency.
15. What are the advantages and disadvantages of frequency converters?

3.10 PROBLEMS

3-1 A 480 V, 60 Hz, 1740 r/min, .90 Pf, 7.5 kW three-phase induction motor has a full-load efficiency of 85%. Determine:

a) The full-load slip.

b) The number of poles.

c) The current drawn by the motor.

3-2 For a three-phase induction motor, show that:

a) At starting, the motor's torque is directly proportional to its rotor resistance. That is,

$$T_{st} \propto R_2$$

b) The magnitude of the maximum torque is independent of rotor resistance.

c) By neglecting stator resistance, the torque (T) at slip (s) is related to the maximum torque (T_m) and its corresponding slip (s_{mt}) by the following equation:

$$\frac{T}{T_m} = \frac{2ss_{mt}}{s_{mt}^2 + s^2}$$

3-3 A 240 V, 3-ϕ, 1140 r/min, three-phase induction motor has the per-phase equivalent circuit shown in Fig. P3-3. Neglecting the rotational losses and the magnetizing impedance, determine:

a) **(1)** The synchronous speed of the stator field.
(2) The per-unit slip.
(3) The line current at starting.
(4) The full-load current, power factor, torque, and efficiency.
(5) The per-unit impedance of the stator and rotor circuits.
(6) The slip and current at maximum torque.
(7) The maximum torque.

b) Repeat (a3), taking the magnetizing impedance into consideration.

3-4 A 480 V, 6-pole, 1140 r/min three-phase induction motor has a starting current, at rated voltage and frequency, of 4.0 pu. Neglecting stator resistance, determine:

a) The full-load slip.

b) The slip at maximum torque.

c) The starting torque in pu.

d) The maximum torque in pu.

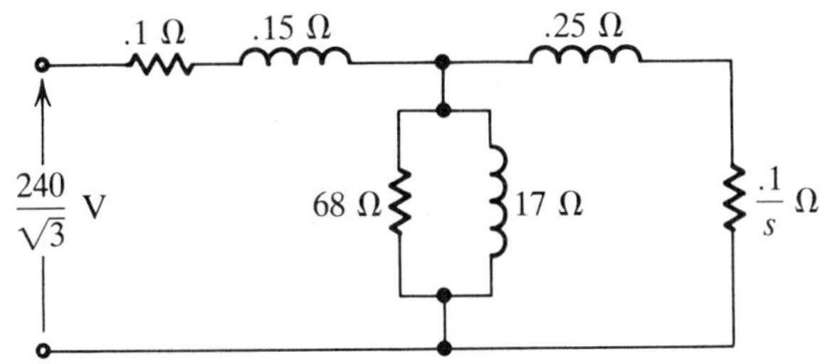

FIG. P3-3

3-5 A 440 V, 60 Hz, 1746 r/min, 60 kW, star-connected, class B, three-phase induction motor yielded the following results when tested:

1. Average value of dc resistance between stator terminals = .16 Ω
2. Locked-rotor test:

$$V_z = 28 \text{ V, line-to-neutral}$$
$$P_z = 2.24 \text{ kW, three-phase}$$
$$I_z = 60 \text{ A}$$

3. No-load test:

$$V_{exc} = 440 \text{ V, line-to-line}$$
$$P_{exc} = 1.6 \text{ kW, three-phase}$$
$$I_{exc} = 12 \text{ A}$$

Determine:

a) The per-phase approximate equivalent circuit.

b) The torque developed at full-load.

3-6 Figures P3-6(a) and (b) show, respectively, the one-line diagram and the per-phase equivalent circuit of a 3-ϕ wound-rotor induction motor. The motor drives a load whose

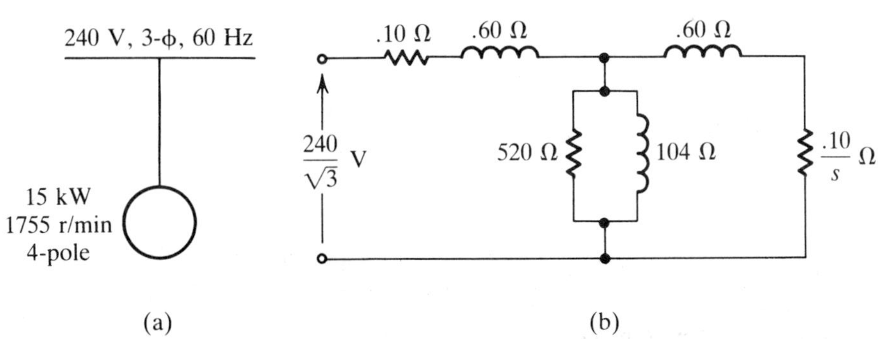

FIG. P3-6

torque–speed characteristic is given by

$$T_L = 50 + 4 \times 10^{-2} n_a \ \text{N}\cdot\text{m}$$

The actual speed (n_a) is expressed in r/min. Neglecting the effects of the magnetizing current, determine:

a) The load viscous damping factor (B) in N · m/rad/s.

b) The minimum required additional rotor resistance that must be inserted in each phase in order to start the motor.

c) The operating speed with the external resistance in the rotor circuit reduced to zero.

3-7 A ventilating fan whose torque is proportional to the square of its speed is driven by a squirrel-cage induction motor at nominal conditions.

a) Assuming that the torque requirements of the fan change by ±10% in comparison to the full-load torque, estimate qualitatively how this will affect (increase or decrease) the speed, efficiency, line current, and power factor of the motor.

b) Repeat (a) assuming that the motor is oversized by 10% above the nominal requirements.

c) Repeat (a) assuming that the motor is undersized by 10% below the nominal requirements.

3-8 The 208 V, 60 Hz, three-phase, 4-wire distribution system shown in Fig. P3-8 delivers power to a three-phase induction motor and to single-phase loads. The starting current of the three-phase induction motor is four times the full-load value and lags the applied voltage by 84°. Determine the currents through the transmission line and through the neutral wire of the distribution system for each of the following operating conditions:

a) The motor at starting.

b) The motor running at rated condition.

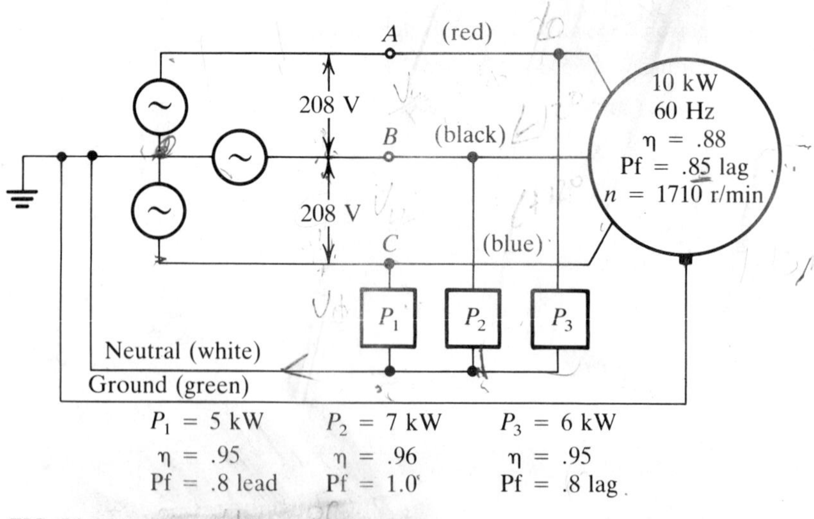

FIG. P3-8

3-9 Referring to Table 3-11(a):

a) Determine the maximum permissible size of a capacitor bank that can be used to correct the power factor of the 260 kW motor. Assume that the motor and capacitor bank are switched through the same disconnect device.

b) What is the power factor of the motor and capacitor, as seen from their common disconnect device?

3-10 The three-phase induction motor shown in Fig. P3-10 has a starting torque and current as follows:

$$T_{st} = 1.8 \text{ pu}, \qquad I_{st} = 6\angle -70^\circ \text{ pu}$$

The reactances of the cables are as shown. When the 500 kW load draws rated current and the 200 kW motor starts, determine:

a) The torque developed by the motor.

b) The size of the capacitor bank that, when connected in parallel with the motor, will minimize the starting current through the upstream network.

c) The torque developed at minimum starting current.

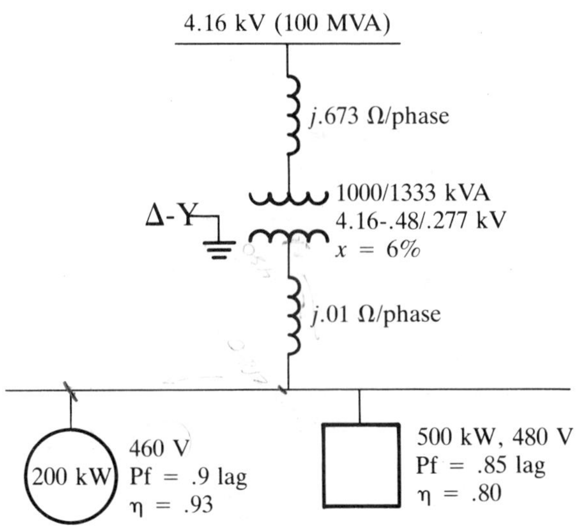

FIG. P3-10

3-11 A 480 V, 3-ϕ, 60 Hz, 6-pole induction motor has the per-phase equivalent circuit shown in Fig. P3-11. The motor is used to drive a hoist. When a hoist is lowered, the load accelerates the rotor above synchronous speed, and thus the motor runs as an induction generator. If the resulting speed is 1260 r/min, determine:

a) The line current.

b) The accelerating torque.

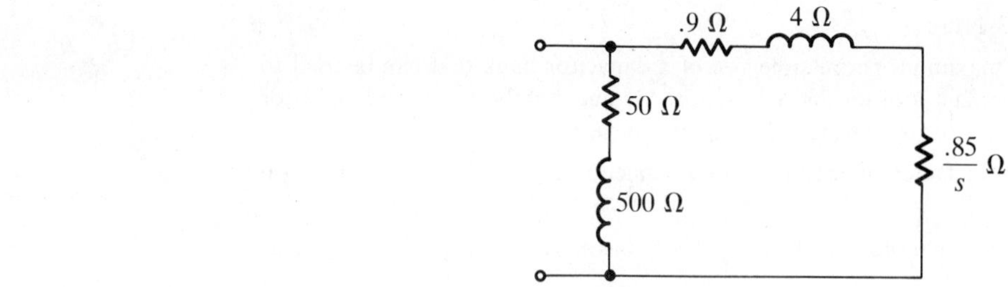

FIG. P3-11

3-12 A 3-ϕ, 480 V, three-phase induction generator delivers 375 kW when powered through a hydro turbine. The water's rate of flow and head at the turbine entrance are 4.70 m^3/s and 10 m, respectively. The open-circuit time constant of the induction generator is 4 ms, and at the sudden removal of the electrical load the speed of the turbine reaches 2.2 pu in 1 ms. Data from the open-circuit test of the generator are as follows:

V:	480 V, line-to-line
I:	332 A
P:	72 kW, 3-ϕ

Estimate:

a) The wire-to-water efficiency of the power station.

b) The maximum stator voltage following the removal of the electrical load.

c) The magnetizing kVAR drawn from the power distribution network to which the generator is connected.

3-13 A three-phase, 75 kW, 600 V, 60 Hz, 1160 r/min induction motor drives a pure-inertia load of 5 kg · m^2. The torque–slip characteristic of the load and the motor are as follows:

Motor torque (N · m)	200	390	455	375	250	0
Load torque (N · m)	0	60	90	120	135	150
Slip (pu)	1	.6	.40	.2	.1	0

Determine:

a) The speed of the motor at steady state.

b) The approximate time it takes the motor to reach this speed.

c) The approximate time it takes the motor to reach the speed of maximum torque.

3-14 A 240 V, 60 Hz, 6-pole, star-connected three-phase induction motor has a rotor impedance of .20 + j.6 ohms per phase referred to the stator. It runs unloaded at a speed very close to synchronous when a mechanical load is suddenly connected to the rotor. The parameters of the load are

$$B = 8 \text{ N} \cdot \text{m/rad/s}$$
$$J = 20 \text{ kg} \cdot \text{m}^2$$
$$T_L = 10 \text{ N} \cdot \text{m}$$

Making reasonable assumptions, estimate:

a) The time it takes to reach the steady-state speed.

b) The steady-state speed.

3-15 The motor in Fig. P3-15 drives a pure-inertia load and has a per-phase equivalent circuit as shown. Neglecting stator resistance and the effects of the magnetizing current, determine:

a) The energy stored in the rotating masses while the motor is running unloaded.

b) The time it takes to plug the motor to a complete stop, and the associated energy loss in the rotor windings.

c) The time it takes to reverse the speed of the motor by plugging, and the associated energy loss in the rotor windings.

d) The value of the rotor resistance that will furnish the minimum time for (b).

e) The number of maximum reversals per minute. Assume total motor loss to be 1.8 times the rotor copper loss.

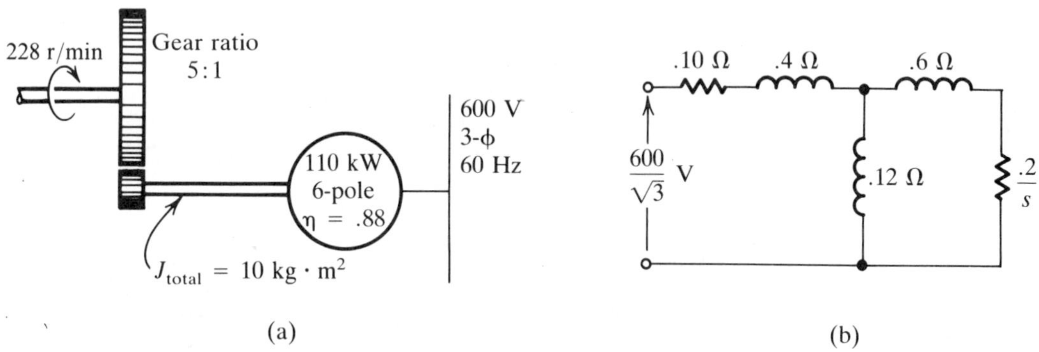

FIG. P3-15

3-16 A 6-pole, three-phase, star-connected, 60 Hz, 460 V induction motor has a stator leakage impedance of $.5 + j1.0$ ohms per phase, a rotor leakage impedance of $.6 + j1.2$ ohms per phase, and a magnetizing impedance of $4 + j40$ ohms per phase.

a) Draw the per-phase equivalent circuit at 60 Hz and at 15 Hz.

b) Estimate the motor torque when operating at 15 Hz and a slip of 10%. Neglect the magnetizing current.

3-17 The torque–speed characteristic of a 60 Hz, 3-ϕ squirrel-cage induction motor is controlled by a frequency converter. Making reasonable assumptions, prove the following:

a)

$$\frac{T_1}{T_2} = \frac{s_1}{K_1 s_2}$$

where T_1 is the torque at a slip s_1 at 60 Hz, and T_2 is the torque at slip s_2 at a frequency of $60K_1$ Hz ($K_1 < 1$)

b) For equal torques the corresponding winding currents are equal.

3-18 The output of a 3-ϕ inverter is obtained through the PWM technique. The number of voltage pulses per cycle, per phase, is equal to n. The width of the pulses forms an arithmetic progression whose first term is d seconds, and the common difference between two successive pulse widths is $2d$ seconds (see Fig. P3-18). The arithmetic progression ascends (increases) in the first quarter of the cycle, and descends in the second quarter of the cycle. Prove the following:

a) The rms value of the output line-to-neutral voltage is given by

$$V = \frac{V_{dc} n}{\sqrt{n^2 + 8n - 32}}$$

where V_{dc} is the dc voltage supply to the inverter.

b) The frequency of oscillation of the ac voltage is

$$f = \frac{4}{d(n^2 + 8n - 32)}$$

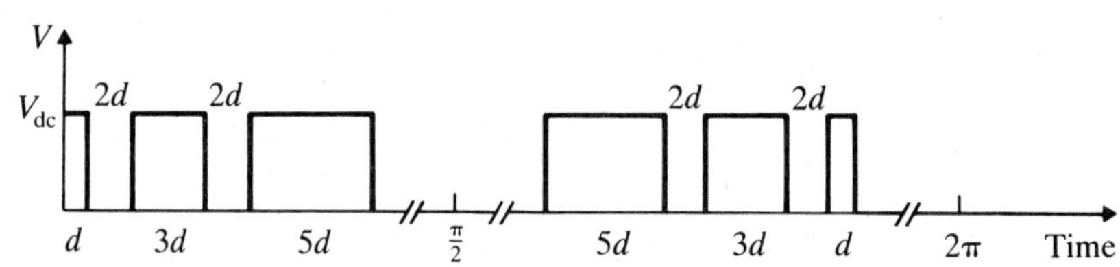

FIG. P3-18

3-19 Referring to the data in Table 3-3, compare the present cost of the energy losses of the 50 kW squirrel-cage induction motor for the two commercially available efficiencies. Assume that:

The motor operates at full-load 16 hours a day, 46 weeks a year, over a five-year period.

The energy cost is 10¢/kWh and increases by 8% per year.

The nominal interest rate is 12% and is compounded annually.

FOUR

Two-Phase and Single-Phase Motors

4.0 Introduction

This chapter discusses first two-phase and then single-phase induction machines. This sequence is necessary because the mathematical analysis of two-phase motors provides the basis for understanding single-phase motors. Two-phase motors are used for speed-control and measurement purposes, while single-phase motors are used in almost all home appliances. This chapter examines in detail the development of spacewise revolving fields, the derivation of equivalent circuits, and the motors' principles of operation and characteristics.

4.1 Two-Phase Induction Motors

In industry today two-phase motors are almost never used for continuous operation. However, two-phase motors have found important applications in control systems where intermittent operation or speed reversal is required.

A two-phase motor's principle of operation, equivalent circuit, and characteristics depend on whether the applied voltages are balanced (equal in magnitude and at 90° to each other) or unbalanced. The analysis of two-phase motors with *balanced* stator voltages is similar to that of three-phase induction motors. The analysis of two-phase motors with *unbalanced* stator voltages is similar that of single-phase motors.

4.1.1 BALANCE-VOLTAGE OPERATION

As shown in Fig. 4-1(a), two-phase induction motors have two windings in space quadrature. Under normal operating conditions, the voltages applied to the two windings are equal in magnitude and at 90° to each other, as shown in Fig. 4-1(b). Mathematically,

$$V_d = V_q \angle -90^\circ \tag{4.1}$$

where V_d and V_q are the rms values of the voltage across the direct and quadrature windings, respectively.

In this chapter, the direct and quadrature axis windings will be designated as main (m) and auxiliary (a) windings, respectively. The symbols and subscripts for two-phase motors will then be identical to those used for the analysis of single-phase induction motors. Therefore

$$V_m = V_a \angle -90^\circ \tag{4.2}$$

Two-phase balanced voltages produce a rotating field, which induces a rotor voltage. The interaction of the stator's and rotor's fields produces motor torque.

The stator's main and auxiliary windings are shown in Fig. 4-1(c). The resultant

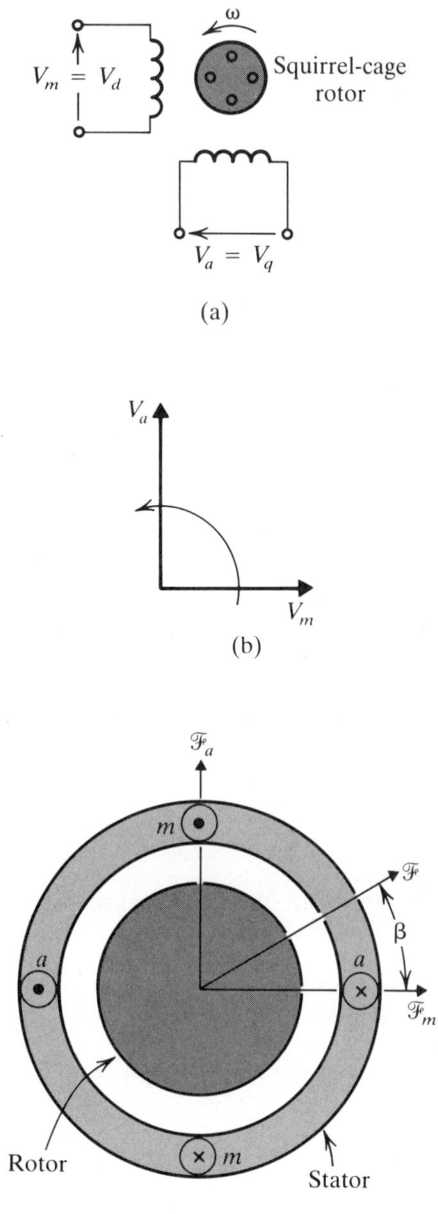

FIG. 4-1 Two-phase ac motors: **(a)** simplified two-phase, two-pole winding, **(b)** balanced applied voltages, **(c)** physical representation.

mmf ($\mathscr{F}$) is obtained by adding up the projections of the main and auxiliary winding's mmf's along the line drawn at an angle of β degrees with respect to the horizontal axis. The horizontal axis coincides with the direction of the mmf produced by the main winding. Thus,

$$\mathscr{F} = \mathscr{F}_m \cos \beta + \mathscr{F}_a \cos (90° - \beta) \tag{4.3}$$

The instantaneous values of mmf's F_m and F_a are

$$\mathscr{F}_m = \mathscr{F}_{\max} \cos (\omega t - \theta) \tag{4.4}$$

$$\mathscr{F}_a = \mathscr{F}_{\max} \cos (\omega t + 90° - \theta) \tag{4.5}$$

where $\mathscr{F}_{\max}$ is the maximum value of each winding's mmf and θ is the phase angle of the winding's impedance.

Substituting Eqs. (4.4) and (4.5) into Eq. (4.3), we obtain

$$\mathscr{F} = \mathscr{F}_{\max}[\cos \beta \cos (\omega t - \theta) + \cos (90° - \beta) \cos (\omega t + 90° - \theta)] \tag{4.6}$$

Expanding the trigonometric functions, and simplifying, we find

$$\mathscr{F} = \mathscr{F}_{\max} \cos (\omega t + \beta - \theta) \text{ ampere-turns} \tag{4.7}$$

The speed of rotation of the resultant mmf is given by

$$\omega = 2\pi f \text{ electrical rad/s} \tag{4.8}$$

or

$$\omega = 2\pi f \frac{2}{p} \text{ mechanical rad/s} \tag{4.9}$$

where p is the number of poles and f the frequency of the supply voltages. The angular speed of rotation of the resultant mmf is designated ω_s and is referred to as synchronous speed ($\omega_s = \omega$).

From the above, the speed of the synchronously rotating field is

$$n_s = 4\pi \frac{f}{p}\left(\frac{60}{2\pi}\right) \text{ r/min} \tag{4.10}$$

and

$$n_s = 120 \frac{f}{p} \text{ r/min} \tag{4.11}$$

Thus, when balanced voltages are applied to the stator windings of a two-phase motor, a single revolving field is produced. The speed of rotation of the stator field is directly proportional to the frequency of the stator voltages. The maximum value of the rotating mmf is equal to the maximum mmf produced by only one phase winding.

Equation (4.7) represents a spacewise synchronously rotating field. That is, a given point on the stator structure sees a sinusoidal mmf. As a result, at two successive times (t_1 and t_2) of the input-voltage cycle, the same point is subjected to two different levels of magnetization.

This rotating stator mmf completes its loop by passing through the two air gaps and the rotor–stator structures, inducing a voltage in the rotor windings. The mmf of the rotor currents attempts to align its magnetic axis with that of the stator field—a natural phenomenon—and as a result, a torque is developed.

A two-phase motor's per-phase equivalent circuit is shown in Fig. 4-2(a). The method used to derive it is similar to that used for a three-phase induction machine. Similarly, the relationships for torque and power, at various operating conditions, are obtained in the same way as for three-phase induction motors with balanced supply voltages. A typical torque–speed characteristic is shown in Fig. 4-2(b).

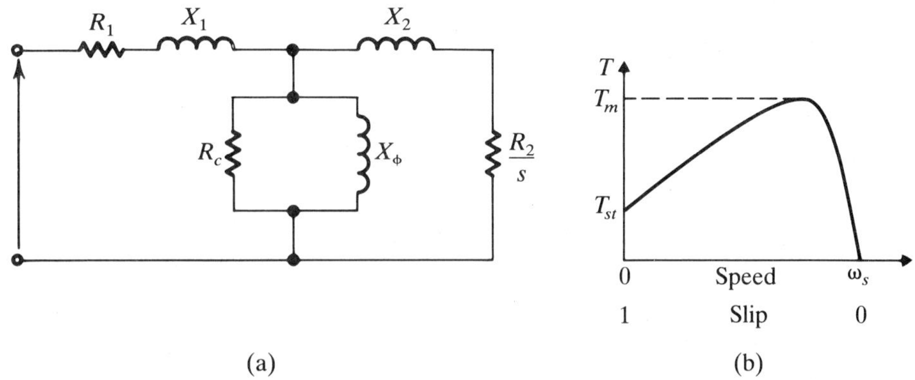

FIG. 4-2 Two-phase induction motor: **(a)** per-phase equivalent circuit for balanced applied voltages, **(b)** torque–speed characteristic.

4.1.2 UNBALANCE-VOLTAGE OPERATION

When unbalanced voltages are applied to a two-phase motor, their equivalent balanced-voltage components can be found using the theory of symmetrical components. According to this theory, any two unbalanced voltages can be resolved into two sets of balanced voltages, one called the "positive sequence" and the other the "negative sequence." The phase angle between the voltage components of any voltage sequence is 90 degrees. The positive-sequence voltages have the same phase rotation as the given unbalanced voltages, while the negative-sequence voltages have an opposite phase rotation. On account of this, the positive-sequence voltages produce a field that rotates in a direction opposite to that produced by the negative-sequence voltages.

The symmetrical components of any two phasors (voltages, current, mmf's, etc.) can be obtained from the standard mathematical relationships presented in the following sections.

Voltage Considerations

By definition, the voltage of the main winding (V_m) and its symmetrical components are related by

$$V_m = V_{m_1} + V_{m_2} \tag{4.12}$$

where V_{m_1} and V_{m_2} are, respectively, the positive- and negative-sequence components of the main winding's voltage. These symmetrical components are related to the unbalanced-voltage phasors by the following standard matrix equation:

$$\begin{bmatrix} V_{m_1} \\ V_{m_2} \end{bmatrix} = \frac{1}{2}\begin{bmatrix} 1 & -j \\ 1 & j \end{bmatrix}\begin{bmatrix} V_m \\ V_a \end{bmatrix} \tag{4.13}$$

where V_a is the voltage of the auxiliary winding. From Eq. (4.13), we obtain

$$V_{m_1} = \frac{1}{2}(V_m - jV_a) \tag{4.14}$$

and

$$V_{m_2} = \frac{1}{2}(V_m + jV_a) \tag{4.15}$$

Similarly, the voltage of the auxiliary winding, and its symmetrical components, are related by

$$V_a = V_{a_1} + V_{a_2} \tag{4.16}$$

Also by definition, the symmetrical components of the auxiliary winding's voltage are related to the symmetrical components of the main winding's voltage by

$$V_{a_1} = jV_{m_1} \tag{4.17}$$

$$V_{a_2} = -jV_{m_2} \tag{4.18}$$

From the above,

$$V_a = j(V_{m_1} - V_{m_2}) \tag{4.19}$$

where the subscripts 1 and 2 represent, as before, the positive- and negative-sequence voltage components.

Figure 4-3 shows two unbalanced voltages and their positive and negative sequence components.

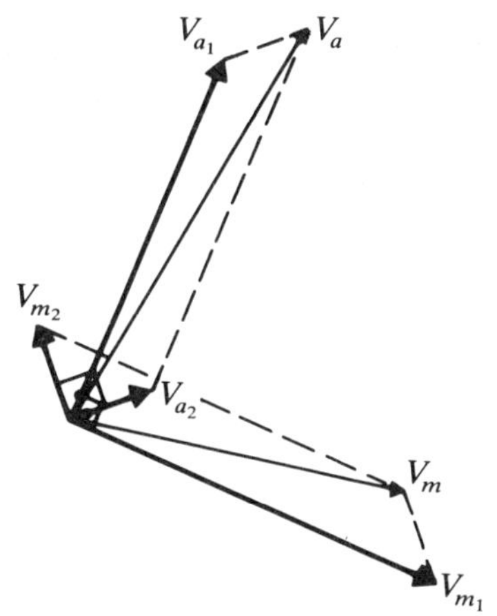

FIG. 4-3 Two-phase unbalanced voltages and their symmetrical components.

Advantages of Symmetrical Components

Resolving the two unbalanced voltages into their equivalent symmetrical or balanced-voltage components has the advantage of permitting a per-phase analysis and the use of known standard relationships and techniques. In other words, using symmetrical components simplifies the understanding and the analysis of machines that are connected to an unbalanced-voltage supply.

Current Considerations

The positive- and negative-sequence components of the current through the main winding can be obtained from the following matrix equation:

$$\begin{bmatrix} I_{m_1} \\ I_{m_2} \end{bmatrix} = \frac{1}{2}\begin{bmatrix} 1 & -j \\ 1 & j \end{bmatrix}\begin{bmatrix} I_m \\ I_a \end{bmatrix} \tag{4.20}$$

where

I_m, I_a = respectively, the currents through the main and auxiliary windings of the motor

I_{m_1}, I_{m_2} = respectively, the positive- and negative-sequence components of the current through the main winding.

MMF Considerations

Each winding's current produces a corresponding mmf. The mmf of the main winding ($\mathscr{F}_m$) and that of the auxiliary winding ($\mathscr{F}_a$) are related to the positive- ($\mathscr{F}_{m_1}$) and negative-sequence ($\mathscr{F}_{m_2}$) components of the main winding's mmf by the following matrix equation:

$$\begin{bmatrix} \mathscr{F}_{m_1} \\ \mathscr{F}_{m_2} \end{bmatrix} = \frac{1}{2}\begin{bmatrix} 1 & -j \\ 1 & j \end{bmatrix}\begin{bmatrix} \mathscr{F}_m \\ \mathscr{F}_a \end{bmatrix} \tag{4.21}$$

From the above,

$$\mathscr{F}_{m_1} = \frac{1}{2}(\mathscr{F}_m - j\mathscr{F}_a) \tag{4.22}$$

and

$$\mathscr{F}_{m_2} = \frac{1}{2}(\mathscr{F}_m + j\mathscr{F}_a) \tag{4.23}$$

The net positive-sequence field ($\mathscr{F}_1$) and the net negative-sequence field ($\mathscr{F}_2$) are given by

$$\mathscr{F}_1 = \mathscr{F}_{m_1} + \mathscr{F}_{a_1} \tag{4.24}$$

$$\mathscr{F}_2 = \mathscr{F}_{m_2} + \mathscr{F}_{a_2} \tag{4.25}$$

The positive and negative sequence mmf's rotate at synchronous speeds, but in opposite directions.

Equivalent Circuit

The positive-sequence voltages produce a so-called forward-rotating field, which normally rotates in the same direction as the shaft. The per-unit slip of a motor to this rotating field is the same as in the case of 3-ϕ induction motors. Thus, the positive-sequence slip is

$$s = \frac{n_s - n_a}{n_s} \text{ pu} \tag{4.26}$$

where n_s and n_a are the field's synchronous speed and the rotor's actual speed, respectively.

The negative-sequence voltage components produce a negative or backward-rotating field that rotates in a direction opposite to that produced by the positive-sequence voltages. The slip of a motor to this backward-rotating field is

$$s_b = \frac{n_s + n_a}{n_s} \tag{4.27}$$

or

$$s_b = \frac{n_s + n_s(1 - s)}{n_s} \tag{4.28}$$

and

$$s_b = (2 - s) \text{ pu} \tag{4.29}$$

where s_b is the negative-sequence slip or the slip of the backward-rotating field.

A motor's equivalent circuit, for positive- and negative-sequence voltages, is shown in Figs. 4-4(a) and (b), respectively. The actual phase currents are given by the vector sum of their components, which are furnished by positive- and negative-sequence equivalent circuits.

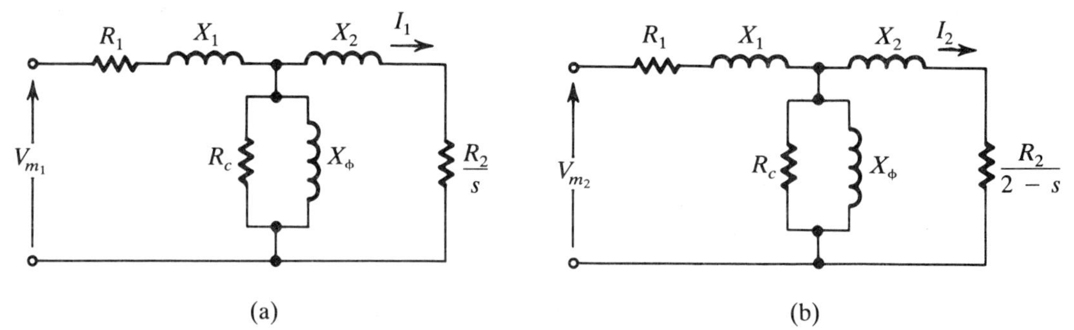

FIG. 4-4 Per-phase equivalent circuit of a two-phase motor: **(a)** positive-sequence voltages, **(b)** negative-sequence voltages.

Torque Developed

The mathematical relationships for positive and negative torques are derived in the same way as for the 3-ϕ induction motors, and are given by the following equations:

Positive-sequence torque:

$$T_1 = \frac{I_1^2 R_2}{\omega_s s} \text{ N} \cdot \text{m/phase} \tag{4.30}$$

Negative-sequence torque:

$$T_2 = \frac{I_2^2 R_2}{\omega_s (2 - s)} \text{ N} \cdot \text{m/phase} \tag{4.31}$$

where I_1 and I_2 are the positive- and negative-sequence components of the rotor current, respectively.

The net torque (T) developed by a motor is given by the difference of the torques developed by the positive and negative sequence fields. That is,

$$T = T_1 - T_2 \tag{4.32a}$$

Similarly, the net power developed by the motor is given by

$$P = R_2\left[I_1^2\frac{1-s}{s} - I_2^2\frac{s-1}{2-s}\right] \text{ watts/phase} \tag{4.32b}$$

EXAMPLE 4–1 The parameters of the per-phase equivalent circuit of a 220 V, 2-pole, 60 Hz, 2-phase motor are as follows:

$$r_1 = .5\ \Omega, \qquad r_2 = .8\ \Omega$$
$$X_1 = 2\ \Omega, \qquad X_2 = 2.5\ \Omega$$

The excitation current is negligible. The motor runs at a slip of 4%.

1. When balanced voltages of 220 V are applied to the motor, determine:
   a) The phase currents.
   b) The torque developed.
   c) The power developed.
2. Repeat 1 if the speed of the motor remains constant and the applied voltages are unbalanced, as follows:

$$V_m = 220\underline{/0^\circ}\ \text{V}$$
$$V_a = 200\underline{/60^\circ}\ \text{V}$$

SOLUTION

1. a) When the applied voltages are balanced, the motor's per-phase equivalent circuit is as shown in Fig. 4-5(a).
   The per-phase current is

$$I = \frac{V}{Z}$$
$$= \frac{220\underline{/0^\circ}}{.5 + \dfrac{.8}{.04} + j(2 + 2.5)}$$
$$= 10.48\underline{/-12.4^\circ}\ \text{A/phase}$$

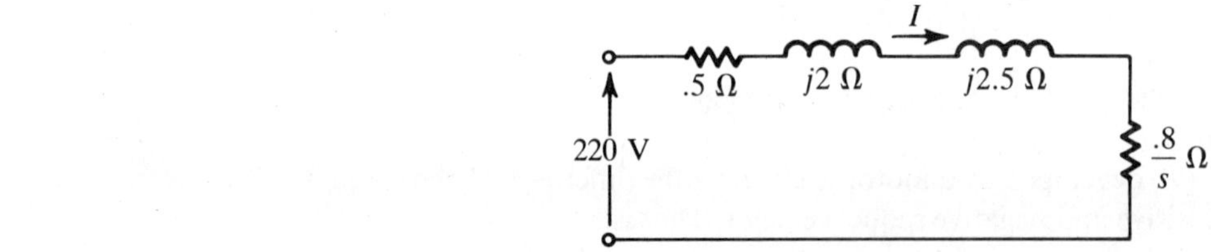

FIG. 4-5(a)

b) The synchronous speed of the rotating field is

$$\omega_s = 120\frac{f}{p}\left(\frac{2\pi}{60}\right) = 120\left(\frac{60}{2}\right)\frac{2\pi}{60}$$
$$= 376.99 \text{ rad/s}$$

The torque developed is

$$T = \frac{\text{power}}{\text{speed}}$$
$$= \frac{I^2 R_2}{\omega_s s} \text{ N}\cdot\text{m/phase}$$

Substituting, we obtain

$$T = \frac{(10.48)^2(.8)}{376.99(.04)} = 5.83 \text{ N}\cdot\text{m/phase}$$

Thus, the motor's torque is

$$T = \underline{\underline{11.66 \text{ N}\cdot\text{m}}}$$

c) The power developed is

$$P = 2I^2\frac{R_2}{s}(1-s) = 2(10.48)^2(.8)\left(\frac{1-.04}{.04}\right)$$
$$= \underline{\underline{4219.2 \text{ W}}}$$

2. a) The symmetrical components of the unbalanced voltages are obtained as follows. Using Eq. (4.14), the positive-sequence voltage of the main winding is

$$V_{m_1} = \frac{1}{2}(220\underline{/0^\circ} - j200\underline{/60^\circ})$$
$$= 202.86\underline{/-14.3^\circ} \text{ V}$$

Using Eq. (4.15), the negative-sequence voltage of the main winding is

$$V_{m_2} = \frac{1}{2}(220\underline{/0^\circ} + j200\underline{/60^\circ})$$
$$= 55.2\underline{/64.9^\circ} \text{ V}$$

The given voltages, and their corresponding positive- and negative-sequence components are shown in Fig. 4-5(b).

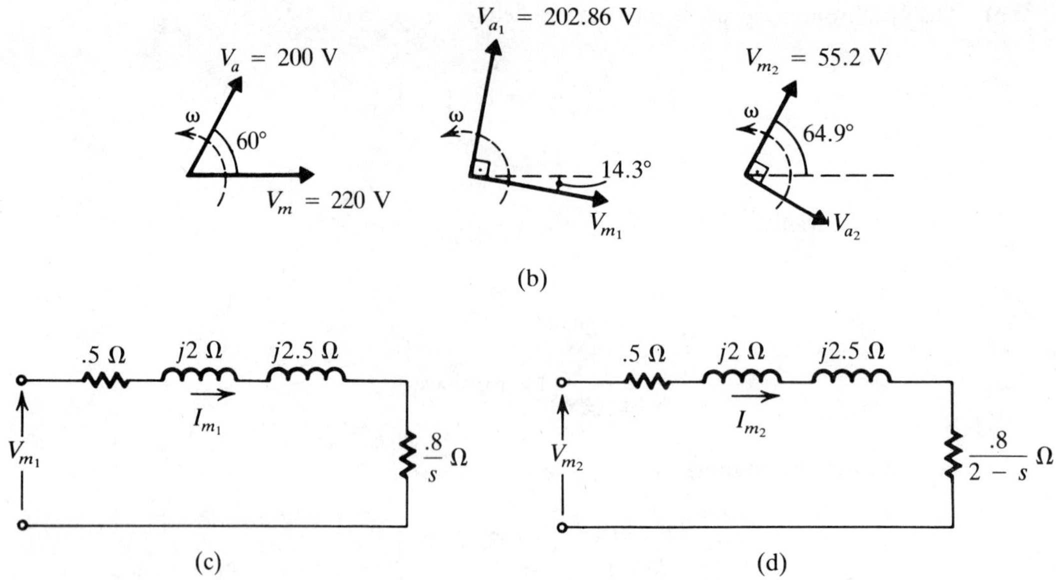

FIG. 4-5 (b), (c), (d)

From the equivalent circuit of Fig. 4-5(c), the per-phase positive-sequence current is

$$I_{m_1} = \frac{V_{m_1}}{Z_1} = \frac{202.86\,\underline{/-14.3^\circ}}{.5 + \dfrac{.8}{.04} + j(2 + 2.5)} = 9.67\,\underline{/-26.7^\circ}\text{ A}$$

and

$$I_{a_1} = jI_{m_1} = 9.67\,\underline{/63.3^\circ}\text{ A}$$

Similarly, from Fig. 4-5(d), the negative-sequence current is

$$I_{m_2} = \frac{V_{m_2}}{Z_2} = \frac{55.2\,\underline{/64.9^\circ}}{.5 + \dfrac{.8}{2 - .04} + j4.5} = 12.03\,\underline{/-13.7^\circ}\text{ A}$$

and

$$I_{a_2} = -jI_{m_2} = 12.03\,\underline{/-103.7^\circ}\text{ A}$$

The current in the main winding is

$$\begin{aligned} I_m &= I_{m_1} + I_{m_2} \\ &= 9.67\,\underline{/-26.7^\circ} + 12.03\,\underline{/-13.7^\circ} \\ &= 21.55\,\underline{/-19.5^\circ}\text{ A} \end{aligned}$$

The current in the auxiliary winding is

$$I_a = I_{a_1} + I_{a_2}$$
$$I_a = 9.67\,\underline{/63.3^\circ} + 12.03\,\underline{/-103.7^\circ} = 3.39\,\underline{/-63.9^\circ}\ \text{A}$$

When the rated current is compared with the current that will result from the application of the unbalanced voltages, it is obvious that the main winding is overrated by a factor of

$$\left(\frac{I_{m_{\text{unbalanced}}}}{I_{m_{\text{balanced}}}}\right)^2 = \left(\frac{21.55}{10.48}\right)^2 = \underline{\underline{4.23}}$$

b) The positive-sequence or forward-rotating torque is

$$T_1 = T_f = 2\frac{I_{m_1}^2}{\omega_s}\frac{R_2}{s}$$
$$= \frac{2(9.67)^2}{376.99}\left(\frac{.8}{.04}\right)$$
$$= 9.91\ \text{N}\cdot\text{m}$$

The negative-sequence or backward-rotating torque is

$$T_2 = T_b = 2\frac{I_{m_2}^2}{\omega_s}\left(\frac{R_2}{2-s}\right)$$
$$= \frac{2(12.03)^2}{376.99}\left(\frac{.8}{2-.04}\right)$$
$$= .31\ \text{N}\cdot\text{m}$$

Thus the net torque developed is

$$T = T_1 - T_2 = 9.91 - .31 = \underline{\underline{9.60\ \text{N}\cdot\text{m}}}$$

c) The power developed is

$$P = 2R_2\left[I_1^2\left(\frac{1-s}{s}\right) - I_2^2\left(\frac{s-1}{2-s}\right)\right]$$
$$= 2(.8)\left[9.67^2\left(\frac{.96}{.04}\right) - (12.03)^2\left(\frac{.96}{1.96}\right)\right]$$
$$= 3587.4 - 113.32$$
$$= 3474\ \text{W}$$

For purposes of comparison, the results of the balanced- and unbalanced-voltage operation are summarized in Table 4-1.

TABLE 4-1 Summary of Results of Example 4-1

Parameter	*Balanced Voltages*	*Unbalanced Voltages*
Main winding current in A (I_m)	10.48	21.55
Auxiliary winding current in A (I_a)	10.48	3.39
Positive-sequence torque in N · m (T_1)	11.66	9.91
Negative-sequence torque in N · m (T_2)	0	.31
Net torque developed in N · m (T)	11.66	9.60
Power developed in W	4219.2	3474

Exercise 4–1 For the motor of Example 4-1, determine the winding copper losses that result from the application of the given balanced- and unbalanced-voltage phasors.

Answer 285.67 W, 618.86 W.

Exercise 4–2 Show that the magnitudes of the net positive ($\mathscr{F}_1$) and the net negative ($\mathscr{F}_2$) mmf's of the stator fields are related to the positive- and negative-sequence mmf's of the main winding by the following equations:

$$\mathscr{F}_1 = \sqrt{2}\,\mathscr{F}_{m_1}$$
$$\mathscr{F}_2 = \sqrt{2}\,\mathscr{F}_{m_2}$$

4.1.3 TWO-PHASE CONTROL MOTORS

Two-phase control motors are used for automatic control purposes in "closed loop" systems. Their analysis requires a basic knowledge of Laplace transforms, which are reviewed in Appendix Section C.

A typical schematic representation of a two-phase control motor, and its idealized torque–speed characteristics, are shown in Figs. 4-6(a) and (b), respectively. The linear characteristics are due to the high rotor resistance. The amplifier controls the voltage to the auxiliary winding, which, as will be demonstrated below, controls the speed of the motor.

The voltage applied to the main winding has constant rms value, while the voltage applied to the auxiliary or so-called control winding has a variable magnitude. However, the frequency of both voltages must be identical, or no useful mmf will be produced.

The governing equations for the transient and steady-state analyses are derived as described below.

The general equation for the torque developed by the motor is derived from the given motor characteristics by using the standard slope–intercept equation of a

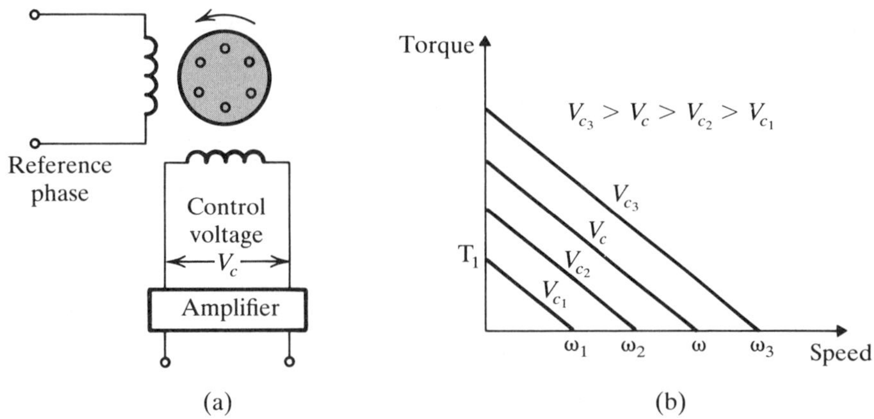

FIG. 4-6 Two-phase control motors: **(a)** schematic representation, **(b)** ideal torque-speed characteristic.

straight line. Thus, from Fig. 4-6(b) we obtain

$$T = -m\omega + KV_c \tag{4.33}$$

where m is the slope of the line and K is a constant that, depending on the control voltage V_c, establishes the y-intercept of the torque characteristic. This constant is given by

$$K = \frac{y\text{-intercept}}{\text{control voltage corresponding to the same } y\text{-intercept}} \tag{4.34}$$

When the motor operates at a control voltage of V_{c_1}, for example, the constant is

$$K = \frac{T_1}{V_{c_1}}$$

Taking the Laplace transform of Eq. (4.33), we obtain

$$T(s) = -m\omega(s) + KV_c(s) \tag{4.35}$$

where the subscript s stands for the Laplace operator.

The torque developed by the machine is used to overcome the moment of inertia (J) of the rotating masses, the viscous frictional force (B), and whatever other torque-load (T_L) requirements the motor has to satisfy. In mathematical form,

$$T = J\frac{d\omega}{dt} + B\omega + T_L \tag{4.36}$$

Taking the Laplace transform and neglecting the initial conditions, we obtain

$$T(s) - T_L(s) = (Js + B)\omega(s) \tag{4.37}$$

Equations (4.35) and (4.37) are represented in block diagram form in Fig. 4-7(a). Combining them, we obtain

$$KV_c(s) - T_L(s) = (Js + B + m)\omega(s) \tag{4.38}$$

which, in block diagram form, is shown in Fig. 4-7(b). From Eq. (4.38) the transfer function, relating the output speed and the input control voltage, can easily be found.

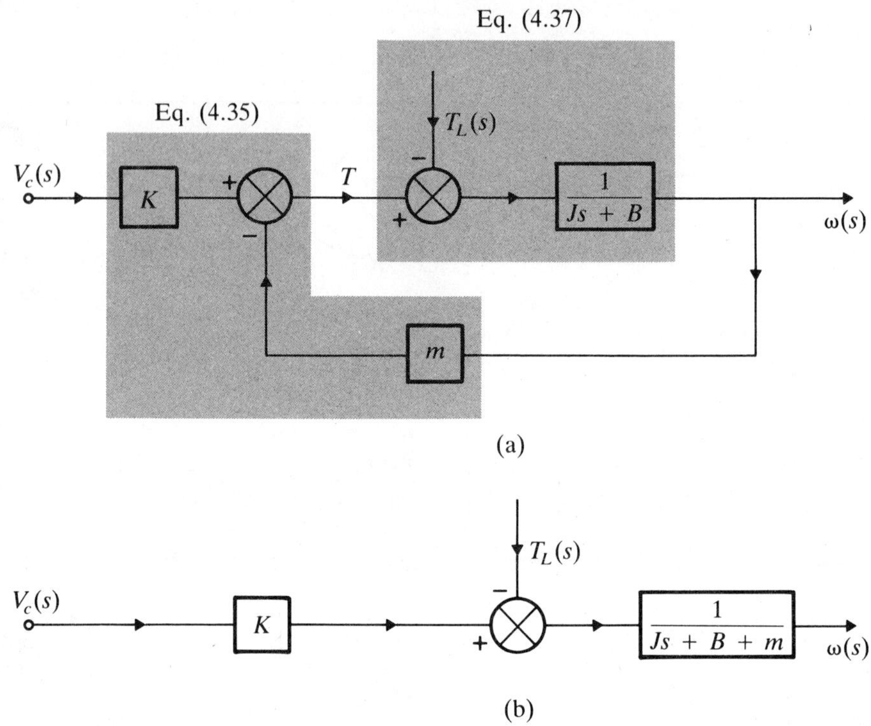

FIG. 4-7 Two-phase control motor: **(a)** block diagram based on Eqs. (4.35) and (4.37), **(b)** simplified block diagram based on Eq. (4.38).

EXAMPLE 4–2 A 60 Hz, 120 V, two-phase control motor has the torque–speed characteristic shown in Fig. 4-8(a). The motor-plus-load moment of inertia is .05 kg · m^2. Determine:

a) The general torque–speed equation.

b) The block diagram.

c) The transfer function.

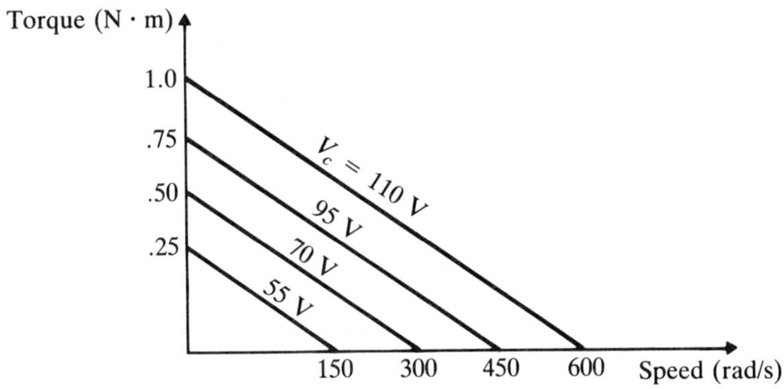

FIG. 4-8(a)

SOLUTION

a) Rewriting Eq. (4.33), we have

$$T = -m\omega + KV_c \qquad \text{(I)}$$

The slope of the torque–speed characteristic is

$$m = \frac{.25}{150} = .0017 \ \text{N} \cdot \text{m/rad/s}$$

The constant K cannot be established because the value of the control voltage is not given.

b) From the dynamics of the motor we have

$$T = J\frac{d\omega}{dt} \qquad \text{(II)}$$

Taking the transform of expressions (I) and (II), and neglecting initial conditions, we obtain

$$T(s) = -m\omega(s) + KV_c(s) \qquad \text{(III)}$$

$$T(s) = Js\omega(s) \qquad \text{(IV)}$$

Equations (III) and (IV) are represented in block diagram form in Fig. 4-8(b).

c) Equating Eqs. (III) and (IV), we obtain

$$-m\omega(s) + KV_c(s) = Js\omega(s)$$

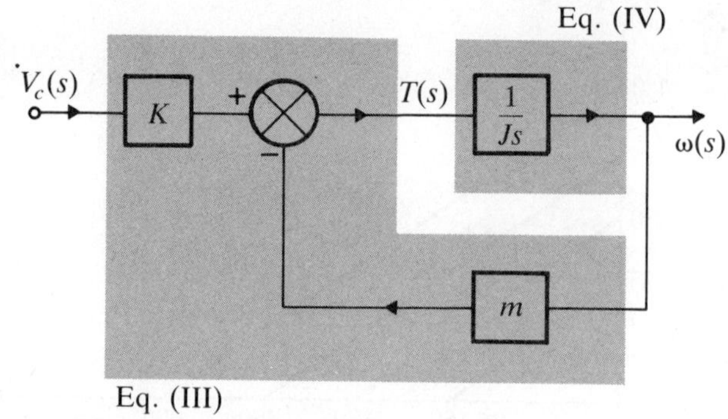

FIG. 4-8(b)

From the last relationship, we obtain

$$\frac{\omega(s)}{V_c(s)} = \frac{K}{m + Js}$$

Substituting the known parameters, and simplifying, we obtain

$$\frac{\omega(s)}{V_c(s)} = \frac{K}{.05\left(\frac{.0017}{.05} + s\right)} = \frac{20K}{.033 + s}$$

Exercise 4–3 When the control voltage of the two-phase motor described in Example 4-2 is a step function of 95 volts, determine:

a) The time constant of the motor's response

b) The speed of the motor as a function of time.

Answer (a) 30 s; (b) $\omega(t) = 450(1 - e^{-.033t})$ rad/s

4.2 Single-Phase Induction Motors

4.2.1 INTRODUCTION

Single-phase motors are used mainly in domestic applications for the operation of vacuum cleaners, washing machines, ventilating fans, compressors, etc. About 90% of all motors manufactured are of the single-phase type. Because their efficiency and power factor are low, their current rating may—depending on the kW rating of the motor—be relatively high.

Single-phase motors have low power rating and are generally noisy because the instantaneous power drawn from a single-phase system—the only kind available in an ordinary home—is pulsating in nature (it has a positive and negative value over a complete voltage cycle). In contrast, three-phase motors are less noisy because the instantaneous power drawn from a balanced three-phase system is constant. Nevertheless, more expensive single-phase motors—such as those used for residential room ventilators—incorporate circuits that minimize the negative component of the pulsating power.

The most commonly used types of motors that may be operated from a single-phase supply are

1. Induction motors, which are, internally, two-phase motors, at least during starting. This class of motors includes the so-called split-phase motor, the capacitor-start motor, the capacitor-run motor, and the capacitor-start–capacitor-run motor.
2. Shaded-pole motors.
3. Single-phase synchronous motors.
4. Series ac/dc motors.

A single-phase motor's theory of operation is based on the fact that a single stator winding produces two spacewise revolving fields* that rotate at synchronous speeds but in opposite directions. The development of these two rotating fields is discussed in the next section.

4.2.2 REVOLVING FIELDS

Consider Fig. 4-9. The mmf of the main winding is

$$\mathscr{F}_{\text{main}} = \mathscr{F}_{\text{m}}(\cos \omega t - \theta) \tag{4.39}$$

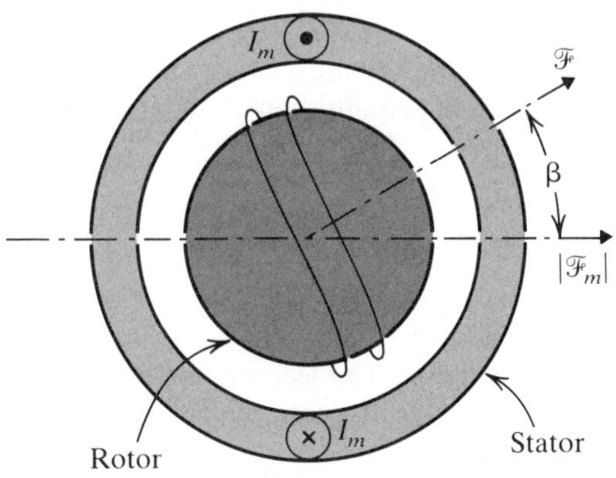

FIG. 4-9 Single-phase induction motor.

* The development of two fields in 1-ϕ induction motors further increases their noise in comparison to 3-ϕ induction motors in which a single field is produced.

where $\mathscr{F}_m$ is the maximum value of the mmf produced by the main winding, ω is the angular frequency of the stator voltage, and θ is the phase angle of the winding's impedance.

The field, at an angle of β degrees from the axis of the main winding's mmf, is

$$\mathscr{F} = \mathscr{F}_{\text{main}} \cos \beta \tag{4.40}$$

From Eqs. (4.39) and (4.40), we obtain

$$\mathscr{F} = \mathscr{F}_m \cos \beta \cos (\omega t - \theta) \tag{4.41}$$

From the above, by using the following trigonometric identities,

$$\cos (A \pm B) = \cos A \cos B \mp \sin A \sin B \tag{4.42}$$

we obtain (after rearranging and simplifying):

$$\mathscr{F} = \frac{\mathscr{F}_m}{2} [\cos (\omega t + \beta - \theta) + \cos (\beta - \omega t + \theta)] \tag{4.43}$$

Thus, a single-phase motor produces two *spacewise* oscillatory fields that rotate in opposite directions. The angular speed of the stator fields is designated with ω_s and is referred to as synchronous speed ($\omega_s = \omega$).

Rotor Stationary

When the rotor is stationary, the two revolving fields have equal amplitudes and, relative to the rotor, rotate at the same speed but in opposite directions. On account of this, no effective field is induced in the rotor windings, and thus the motor does not have a starting-torque capability. For this reason, some domestic fan motors will not start without an external "kicking." This kicking increases the torque in the direction of the external disturbance, and thus the equilibrium of the mmf's is not maintained. As a result, the motor starts to rotate.

Rotor Turning

When the rotor is turning, the relative angular velocities of the positive and negative fields, with respect to the rotor, are $s\omega_s$ and $(2 - s)\omega_s$, respectively. The two fields induce unequal voltages in the rotor winding. This leads to the development of torque, because the resulting mmf's try to align their magnetic axes.

4.2.3 EQUIVALENT CIRCUIT

A single-winding, single-phase motor can be considered as a two-winding two-phase motor with its auxiliary winding open-circuited. Then the motor's equivalent circuit is that of the two-phase motor, which is shown in Fig. 4-4. An alternative, but approximate, form of the equivalent circuit of a single-phase motor is shown in Fig. 4-10(a). The impedance $(R_1 + jX_1)$ represents the leakage impedance of the main winding. The magnetizing impedance (Z_m) and the components of the rotor impedance $(R_2 + jX_2)$ are shown as two equal parts connected in series. The justification for this form of equivalent circuit is as follows.

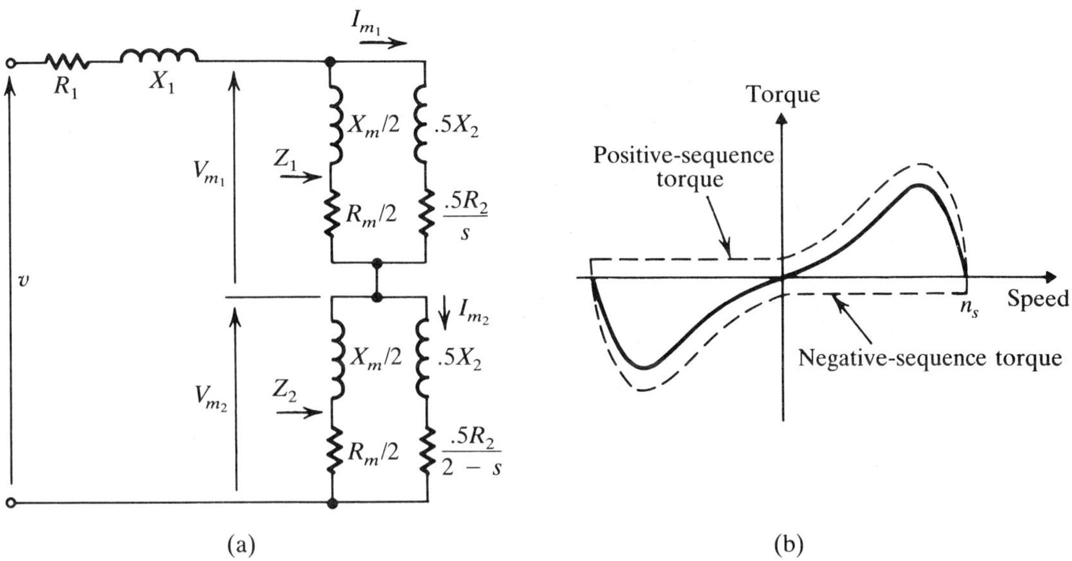

FIG. 4-10 Single-phase motor: **(a)** equivalent circuit, **(b)** torque–speed characteristic.

The positive- and negative-sequence components of the currents through the main winding are given by Eq. (4.20). Since $i_a = 0$, then

$$I_{m_1} = I_{m_2} = \frac{I_m}{2} \tag{4.44}$$

That is, the positive- and negative-sequence components of the current through the main winding are equal to each other. Because of this, the equivalent circuits for the positive and negative fields are connected in series, as shown in Fig. 4-10(a).

Disregarding the magnetizing impedance, the rotor's positive- and negative-sequence impedances are given by

$$Z_1 = .5X_2 + \frac{.5R_2}{s} \tag{4.45}$$

$$Z_2 = .5X_2 + \frac{.5R_2}{s - 2} \tag{4.46}$$

where s is the motor's operating slip. At standstill the positive-sequence impedance is equal to the negative-sequence impedance.

The positive- and negative-sequence components of the voltage across the rotor impedances are

$$V_{m_1} = I_{m_1} Z_1 \tag{4.47}$$

$$V_{m_2} = I_{m_2} Z_2 \tag{4.48}$$

From Eq. (4.19) and the above it is evident that, at standstill, there is no voltage induced in the auxiliary winding ($V_a = 0$). However, when the rotor is turning, Z_1 is not equal to Z_2 and thus there is a voltage induced in the auxiliary winding ($V_a \neq 0$).

For an accurate analysis, one should take into consideration the variation of the rotor's parameters with the operating slip.

4.2.4 TORQUE DEVELOPED

Since a single-phase motor's equivalent circuit is identical to that of a two-phase motor, the expressions for the torque developed, copper losses, and output power are also identical to those derived for two-phase motors. The net torque of the motor, and its positive- and negative-sequence components, are shown in Fig. 4-10(b).

Rewriting the equations for the motor's torque, we have

$$T = T_f - T_b \tag{4.49}$$

and

$$T = \frac{I_m^2}{\omega_s}\left(\frac{.5R_2}{s} - \frac{.5R_2}{2-s}\right) \text{ N} \cdot \text{m} \tag{4.50}$$

where I_m is the rms value of the main winding's current, and s is the motor's operating slip in per-unit.

EXAMPLE 4–3 A 120 V, 60 Hz, 4-pole, single-phase induction motor has the approximate equivalent circuit shown in Fig. 4-11. When the rotor rotates at 1764 r/min, estimate the torque it develops. Disregard the effects of the magnetizing impedance.

SOLUTION At 1764 r/min, the slip is

$$s = \frac{1800 - 1764}{1800} = .02 \text{ pu}$$

The impedance of the circuit is

$$Z = .15 + \frac{.1}{s} + \frac{.1}{2-s} + j1.2\ \Omega$$

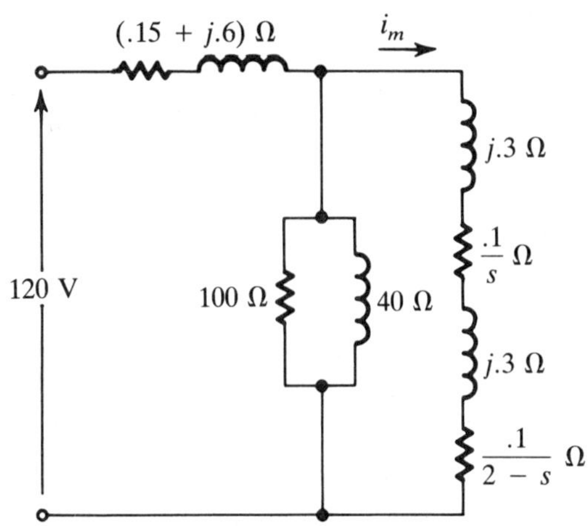

FIG. 4-11

Substituting the operating slip into the above, we obtain

$$Z = .15 + \frac{.1}{.02} + \frac{.1}{2 - .02} + j1.2$$
$$= 5.34\underline{/13^\circ}\ \Omega$$

The current is

$$I = \frac{V}{Z} = \frac{120\underline{/0^\circ}}{5.34\underline{/13^\circ}}$$
$$= 22.48\underline{/-13^\circ}\ \text{A}$$

The torque is calculated by substituting the known parameters into Eq. (4.50), as follows:

$$T = \frac{22.48^2}{1800\frac{2\pi}{60}}\left(\frac{.1}{.02} - \frac{.1}{2 - .02}\right)$$
$$= 13.41 - .135 = \underline{\underline{13.27\ \text{N} \cdot \text{m}}}$$

Exercise 4–4 For the motor of Example 4-3, estimate:

a) The voltage induced in the rotor windings and its positive- and negative-sequence components, all as seen from the stator.

b) The motor's copper losses.

Answer (a) $V_1 = 56.31\underline{/-9.6^\circ}$ V, $V_2 = 3.42\underline{/67.5^\circ}$ V, $V = 57.18\underline{/-6.2^\circ}$ V (b) 176.9 W

4.2.5 METHODS OF STARTING SINGLE-PHASE INDUCTION MOTORS

There are several methods of starting a single-phase induction motor. The motor's stator includes, in addition to the main winding, a starting, or auxiliary, winding. The motor's cost and its method of starting depend on the configuration of this auxiliary winding. The starting winding is always connected in parallel with the main winding. Because it has higher impedance than the main winding, it draws less current. The motor's starting torque depends on the impedance of its windings at standstill and on the effective turns ratio between the stator and rotor windings. Motors equipped with a starting winding are often referred to as "split-phase induction motors."

The various types of split-phase induction motors are described briefly below.

Auxiliary Winding, Permanently Connected

A split-phase motor with a permanently connected auxiliary winding is shown in Fig. 4-12(a). The auxiliary winding may—or may not—include a capacitor. The capacitor, depending on its relative size, will further increase the starting and full-load torques of the motor. A motor with a permanently connected auxiliary winding requires no maintenance and is used where a low starting torque is required, such as in the operation of a small ventilating fan. The torque–speed characteristic of this motor is shown in Fig. 4-15.

Phasor Diagram

The phasor diagram of a split-phase motor with a permanently connected auxiliary winding is shown in Fig. 4-12(b). The phasor diagram is constructed by following the sequence identified by the encircled numbers.

Phasors (1), (2), and (3)

The mmf of the auxiliary winding ($\mathscr{F}_a$) is arbitrarily drawn as shown. When this phasor is rotated by plus and minus 90°, it gives the phasors ($j\mathscr{F}_a$) and ($-j\mathscr{F}_a$), respectively.

Phasors (4), (5), and (6)

The mmf of the main winding ($\mathscr{F}_m$), relative to that of the auxiliary winding, is larger and more inductive. Its phasor is drawn as shown. Its positive-sequence component is found graphically by using Eq. (4.22).

Phasors (7) and (8)

The positive-sequence components of the main and auxiliary windings are equal in magnitude and in phase quadrature to each other. (Consider the mmf equivalent of Eq. (4.17).) The algebraic summation of the positive-sequence components yields the motor's net positive-sequence component ($\mathscr{F}_1$).

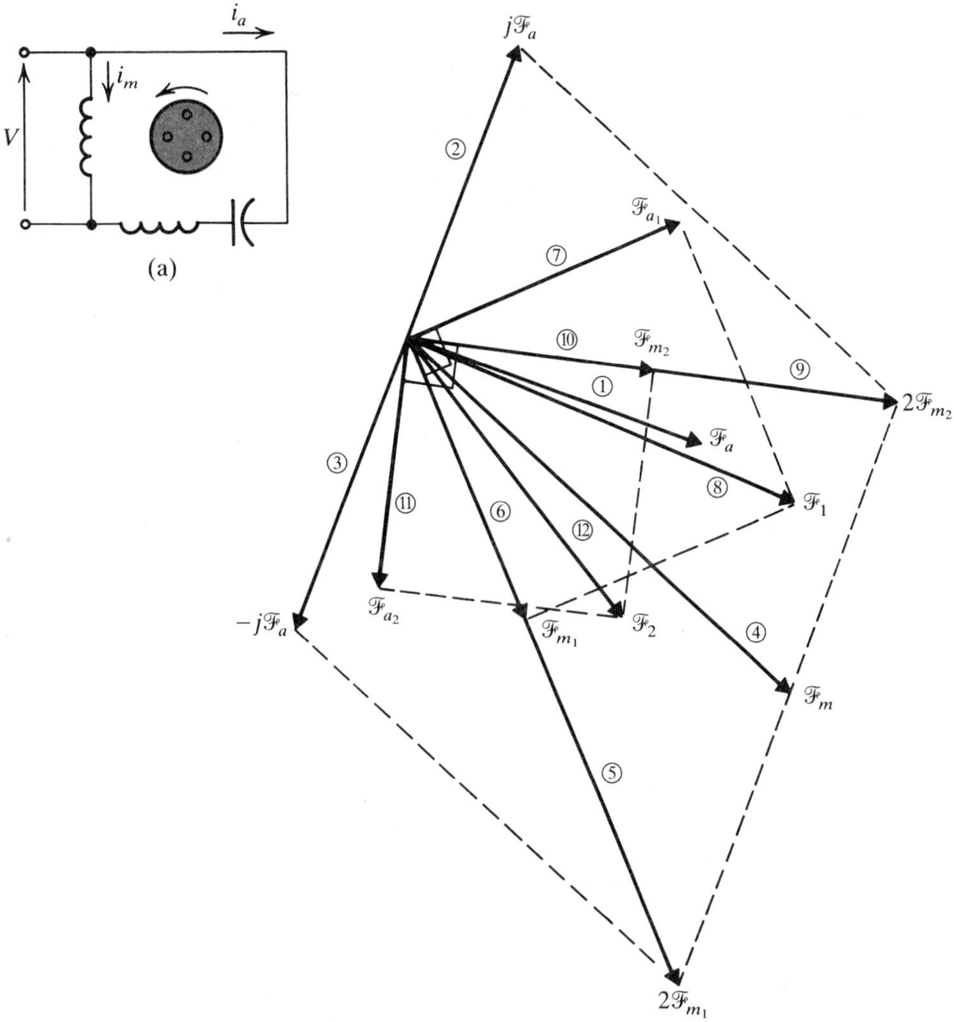

(b)

FIG. 4-12 Single-phase motor: **(a)** auxiliary winding, permanently connected, **(b)** phasor diagram.

Phasors (9) and (10)

The algebraic sum of the main winding's mmf and the phasor $j\mathcal{F}_a$ yields the negative-sequence component of the main winding's mmf (see Eq. (4.23)).

Phasors (11) and (12)

The negative-sequence component of the auxiliary winding's mmf lags the corresponding one of the main winding by 90° (see Eq. (4.18)). The algebraic sum of the negative-sequence components of the winding's mmf gives the motor's net negative-sequence component ($\mathscr{F}_2$) (see Eq. (4.25)).

Auxiliary Winding with Capacitor Start

The starting torque of the motor is substantially increased when a properly sized capacitor is connected in series with the auxiliary winding as shown in Fig. 4-13(a). The auxiliary winding is normally disconnected from the circuit by a centrifugal switch when the motor reaches about 75% of its rated speed. When the shaft of the motor is not accessible, as in hermetically enclosed compressors, then the contact of a voltage or current relay is used to disconnect the auxiliary winding.

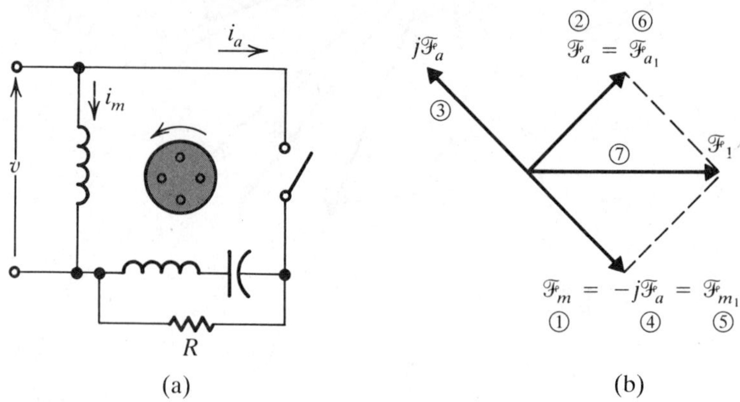

FIG. 4-13 Single-phase motor: **(a)** auxiliary winding with starting capacitor, **(b)** phasor diagram—ideal case.

Phasor Diagram

The phasor diagram of a capacitor-start motor in an ideal case is shown in Fig. 4-13(b). Here, the mmf's are arbitrarily assumed to be equal in magnitude and in space quadrature to each other. Their positive-sequence components can be obtained graphically (steps 5 and 6) by using Eq. (4.22) and the mmf equivalent of Eq. (4.17). The motor's net positive-sequence mmf is then obtained graphically (step 7) as shown.

As can be seen from the phasor diagram and Eq. (4.23), the negative-sequence component of the main winding's mmf is equal to zero. Under the stated assumptions, the motor develops only a positive-sequence mmf. As a result, it has maximum starting-torque capability.

Auxiliary Winding, Removed After Starting

A split-phase induction motor with an auxiliary winding that is removed after starting is shown in Fig. 4-14(a). The majority of motors, such as those used in ventilating fans and clothes washers/dryers, are of this type. The auxiliary winding can be removed

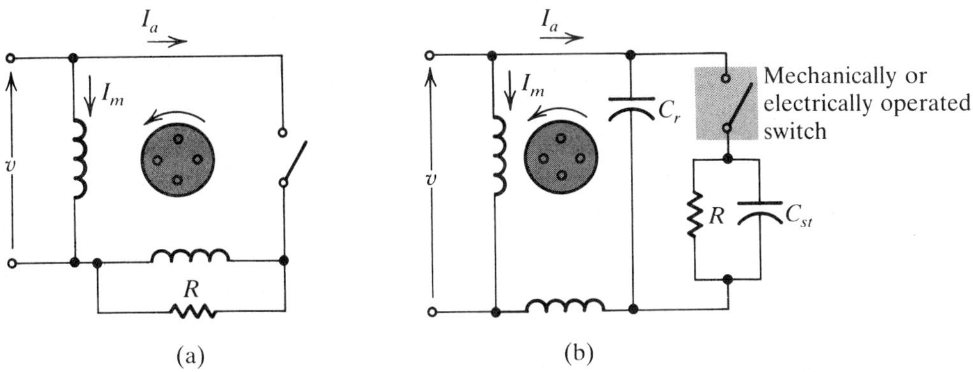

FIG. 4-14 Single-phase motor: **(a)** auxiliary winding, removed after starting, **(b)** with starting and running capacitors.

through a centrifugal switch or through the switching action of a relay. The auxiliary winding should be connected in parallel to a properly sized resistor that will dissipate the coils' stored energy following the current interruption. As can be seen in Fig. 4-15, the starting and running torques of this motor are smaller than those of a motor with a permanently connected auxiliary winding that includes a capacitor.

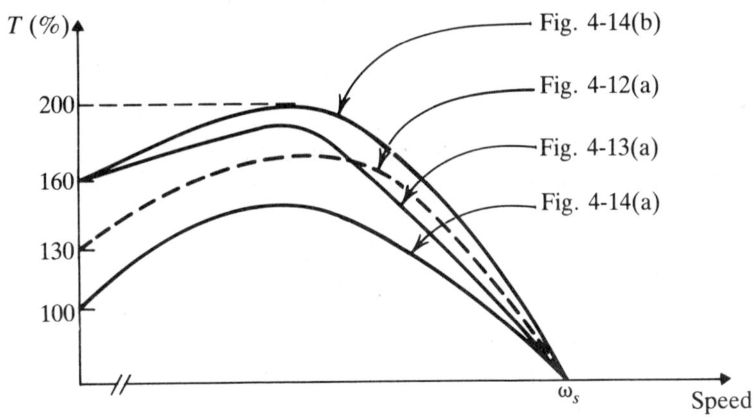

FIG. 4-15 Torque–speed characteristic of split-phase induction motors.

Auxiliary Winding with a Starting and a Running Capacitor

The highest starting and running torque of a single-phase induction motor is obtained when its split-phase winding is equipped, as shown in Fig. 4-14(b), with a starting and a running capacitor. For a 3 kW compressor motor, typical values for a starting and running capacitor are, respectively, 200 μF and 50 μF.

4.2.6 MAGNETIC FIELDS AT STARTING

As mentioned previously, the currents through the main and auxiliary windings produce—at standstill two magnetic fields that rotate synchronously in opposite directions. Designating the mmf's of the main and auxiliary windings $N_m I_m$ and $N_a I_a$, from Eq. (4.21) we have

$$\begin{bmatrix} \mathscr{F}_{m_1} \\ \mathscr{F}_{m_2} \end{bmatrix} = \frac{1}{2} \begin{bmatrix} 1 & -j \\ 1 & j \end{bmatrix} \begin{bmatrix} N_m I_m \\ N_a I_a \end{bmatrix} \tag{4.51}$$

where $\mathscr{F}_{m_1}$ and $\mathscr{F}_{m_2}$ represent, respectively, the positive- and the negative-sequence components of the motor's mmf's, and I_m and I_a are the rms values of the current through the main and auxiliary windings, respectively.

From the matrix equation we obtain:

$$\mathscr{F}_{m_1} = \frac{1}{2}(N_m I_m - jN_a I_a) \text{ ampere-turns} \tag{4.52}$$

and

$$\mathscr{F}_{m_2} = \frac{1}{2}(N_m I_m + jN_a I_a) \text{ ampere-turns} \tag{4.53}$$

From Eq. (4.53), the negative field is zero when the following condition is satisfied.

$$N_m I_m = -jN_a I_a \tag{4.54}$$

Replacing the currents by their corresponding voltage/impedance ratios (see Fig. 4-13(a) and assume $R = \infty$), we obtain

$$N_m \frac{V}{R_m + jX_m} = -jN_a \frac{V}{R_a + jX} \tag{4.55}$$

where V is the supply voltage and X is equal to the reactance of the auxiliary winding (X_a) minus the reactance of the capacitor (X_c). That is,

$$X = X_a - X_c \tag{4.56}$$

From Eq. (4.55) we have

$$N_m R_a + jN_m X = N_a X_m - jN_a R_m \tag{4.57}$$

Equating the real parts, we obtain

$$R_a = \frac{N_a}{N_m} X_m \tag{4.58}$$

Equating the imaginary parts, we obtain

$$X = -\frac{N_a}{N_m} R_m \tag{4.59}$$

or

$$X_a - X_c = -\frac{N_a}{N_m} R_m \tag{4.60}$$

From the above,

$$X_c = \frac{N_a}{N_m} R_m + X_a \tag{4.61}$$

Equations (4.58) and (4.61) can be used to design the stator and rotor windings in a way that optimizes the starting torque of the motor. In other words, when these equations are satisfied, a single rotating field is developed at starting.

EXAMPLE 4–4 A 120 V, 60 Hz, single-phase induction motor of the capacitor-start type has a main winding with 180 effective turns and an auxiliary starting winding with 250 effective turns. With the rotor stationary the input impedance of the main winding is $5 + j10$ ohms and that of the auxiliary winding is $13.89 + j15.50$ ohms. Determine at standstill:

a) The magnitude of the forward- and backward-rotating fields.

b) The size of the starting capacitor necessary to yield a single rotating magnetic field at starting.

SOLUTION

a) A schematic of the single-phase induction motor is shown in Fig. 4-16. The forward- and backward-rotating magnetic fields of the main winding are obtained from Eqs. (4.52) and (4.53), respectively.

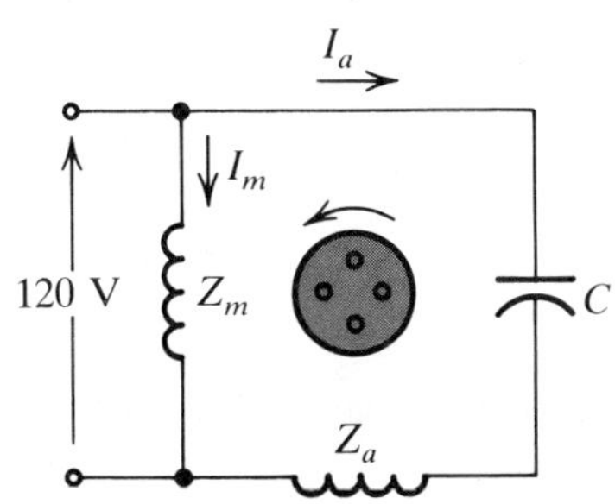

FIG. 4-16

$$\mathscr{F}_{m_1} = \frac{1}{2}(N_m I_m - jN_a I_a)$$
$$= \frac{1}{2}(120)\left(\frac{180}{5 + j10} - j\frac{250}{13.89 + j15.50}\right)$$
$$= 1349.10\underline{/-94.5^\circ} \text{ A rms}$$
$$\mathscr{F}_{m_2} = \frac{1}{2}(N_m I_m + jN_a I_a)$$
$$= \frac{1}{2}(120)\left(\frac{180}{5 + j10} + j\frac{250}{13.89 + j15.50}\right)$$
$$= 1041.7\underline{/-21.6^\circ} \text{ A rms}$$

The positive-sequence component of the auxiliary winding's mmf is

$$\mathscr{F}_{a_1} = j\mathscr{F}_{m_1} = 1349.10\underline{/-4.5^\circ} \text{ A rms}$$

and the negative-sequence component is

$$\mathscr{F}_{a_2} = -j\mathscr{F}_{m_2} = 1041.7\underline{/-111.6^\circ} \text{ A rms}$$

Thus, the net forward-rotating field is

$$\mathscr{F}_1 = \mathscr{F}_{a_1} + \mathscr{F}_{m_1} = 1341.1(1\underline{/-4.5^\circ} + 1\underline{/-94.5^\circ})$$
$$= \underline{\underline{1896.6\underline{/-49.5^\circ} \text{ A rms}}}$$

The net backward-rotating field is

$$\mathscr{F}_2 = \mathscr{F}_{a_2} + \mathscr{F}_{m_2} = 1041.7(1\underline{/-111.6^\circ} + 1\underline{/-21.6^\circ})$$
$$= \underline{\underline{1473.2\underline{/-66.6^\circ} \text{ A rms}}}$$

b) To have a single rotating field the backward-rotating mmf must be equal to zero. That is,

$$\frac{1}{2}(120)\left(\frac{180}{5 + j10} + j\frac{250}{Z_a}\right) = 0$$

The impedance of the auxiliary winding is

$$Z_a = 13.89 + j(15.5 - X_c)$$

where X_c is the reactance of the capacitor. From the last two relationships:

$$Z_a = -j\left(\frac{250(5 + j10)}{180}\right) = 13.89 + j(15.5 - X_c)$$

Solving for X_c, we obtain

$$X_c = 22.44\ \Omega$$

Thus,

$$C = \frac{1}{\omega X_c} = \frac{1}{2\pi(60)(22.44)}$$
$$= \underline{\underline{118.2\ \mu\text{F}}}$$

Alternatively, using Eq. (4.61),

$$X_c = X_a + \frac{N_a}{N_m} R_m = 15.5 + \frac{250}{180}(5) = 22.44\ \Omega$$

This reactance will give the same capacitance as that obtained previously.

Exercise 4–5

a) A 120 V, 60 Hz, 4-pole, single-phase induction motor operates at 1764 r/min. Determine the angular velocity of the rotating fields with respect to the rotor.

b) The net positive- and negative-sequence mmf's of a single-phase motor are as follows:

$$\mathscr{F}_1 = 100\underline{/-30^\circ}\ \text{A rms}$$
$$\mathscr{F}_2 = 80\underline{/-60^\circ}\ \text{A rms}$$

Find the mmf's of the main and auxiliary windings.

Answer (a) 36 r/min cw, 3564 r/min ccw
(b) $64.8\underline{/-34.1^\circ}$ A rms; $110.5\underline{/-48.7^\circ}$ A rms

---

Exercise 4–6 Refer to Fig. 4-14(b). Explain why the motor must not be restarted within a time interval that is substantially less than four times the time constant (RC) of the capacitor circuit.

Summary

The currents in the two stator windings of a split-phase motor are displaced from each other phasewise by either (1) making the resistance of one winding higher than that of the other or (2) connecting a capacitor *in series* with one of the windings. The resulting magnetic field may be resolved into two oppositely revolving components, one larger than the other, and this produces a net forward torque. To avoid overheating, the higher-resistance winding of the split-phase motor, or the winding with the capacitor in a capacitor-start motor, is disconnected when running speed is approached by either a relay's contact or a centrifugal switch. Once the rotor is turning, excitation of a single stator winding will yield forward torque. In the same class, insofar as operation while running is concerned, is the hand-started induction motor. This has only a single stator winding, but it will run in whichever direction it is started.

Shaded-Pole Motors

A shaded-pole, single phase induction motor has a squirrel-cage rotor and a salient-pole stator (see Fig. 4-17). Wound on the stator are the main multi-turn winding and the auxiliary single-turn winding (note that this differs from the distributed windings of split phase motors). The auxiliary winding is a short-circuited copper band commonly referred to as a shaded-pole winding.

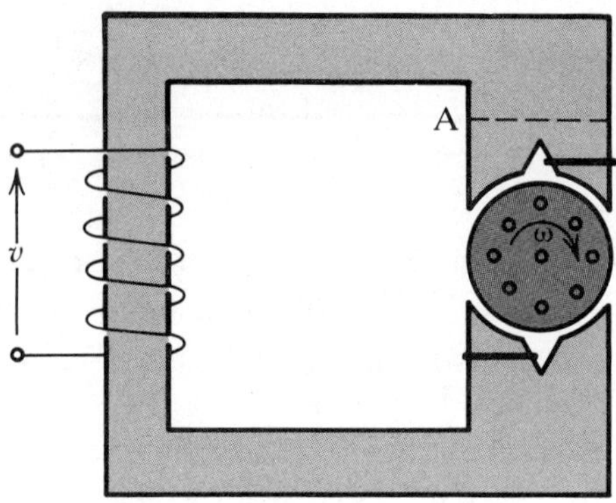

FIG. 4-17 Shaded-pole Motor

When the main winding is connected to the voltage supply, the flux through the shaded part of the pole lags the flux through the unshaded part because of the opposition to the flux that is generated by the induced currents in the auxiliary winding. As a result, the stator flux has two components, one of which lags the other. Each of these fluxes completes its loop through the stator poles, air gaps and rotor.

Revolving Field

The flux through the stator at section A-A (ϕ_A) is the vector sum of the fluxes through the shaded (ϕ_{sh}) and unshaded (ϕ_{un}) parts of the pole. Thus,

$$\phi_A = \phi_{un} + \phi_{sh} \tag{4.62a}$$

or

$$\phi_A = \phi_{un}(1\,\angle 0^\circ + K\,\angle -\theta) \tag{4.62b}$$

where: $K = \dfrac{\phi_{sh}}{\phi_{un}}$ and θ is the phase angle by which the flux through the shaded part of the pole lags the flux through the unshaded part of the pole. The parameters ϕ_{un}, K and θ are variables, at a particular point in the voltage waveform, have a unique value. As a result, the stator flux has two rectangular components that change as a function of time. This leads to the development of a kind of spacewise revolving field. The rotor is subjected to two magnetizing forces whose vector sum induces a rotor current and a corresponding rotor flux. The interaction of stator and rotor fluxes produces torque and rotation whose direction is from the unshaded to the shaded part of the pole.

Shaded-pole motors have a low power factor (.25 to .50) and efficiency (.25 to .40), but are very economical to manufacture. Their power rating is in the range of 1 mW to 200 W. Almost all motors rated at less than 50 W are of the shaded-pole type. Their starting torque is .30 to .85 pu and their maximum torque is 1.2 to 1.5 pu.

Usually, shaded-pole motors have 2, 4, or 6 poles. When supplied with integral gear systems, they can drive loads at speeds that are slower than one revolution per day. Shaded-pole motors are used in photocopy machines, humidifiers, fans, toys, phonograph turntables, etc.

Stepper Motors

Stepper motors* rotate in increments or steps that are determined by rotor construction and the type of stator excitation. The stator winding is often a multi-phase winding (sometimes up to four phases are used) and the excitation voltage is a series of current pulses (digital signal). The higher the number of stator phases, the smaller the increment of angular rotation per input pulse.

The rotor may be one of two types: permanent magnet or reluctance. A four-pole reluctance motor is shown in Fig. 4-18(a). The shaft of this kind of motor has one salient rotor configuration for each stator phase. That is, for a four-phase stator winding the rotor will have along its axis four distinct pole configurations.

When one of the stator windings is excited, the rotor will rotate in such a way that the path of least reluctance is established for the stator flux.

Reluctance motors can operate with a minimum rotation of 1.8°/pulse; that is, for a complete revolution 200 pulses are required. The characteristics of stepper motors [see Fig. 4-18(b)] are expressed in terms of the torque produced as a function of excitation pulses per second (pps).

The torque of the motor is constant up to a specific rate of excitation pulses. At a further increase in the rate of excitation pulses the torque is decreased either along the solid curve or along the dotted line characteristic.

In 4-18(b), the dotted lines represent motor operation without controlled acceleration and deceleration. This is referred to as error-free-start-stop (EFSS) operation. In this mode of operation, the higher the effective moment of inertia, the lower the motor's output torque. The solid line characteristic represents motor

• • •

* See reference 15 in Bibliography at the end of the book.

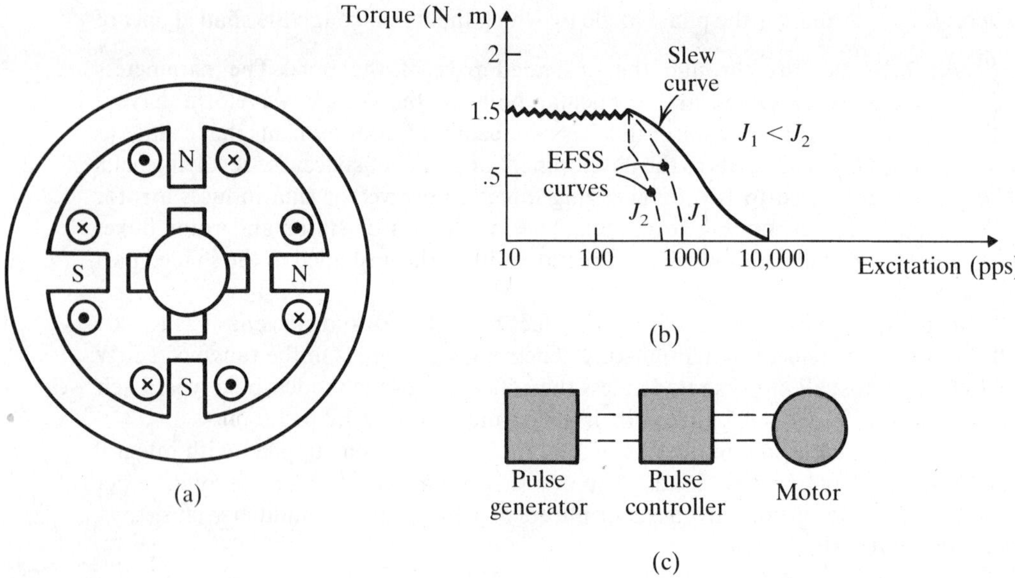

FIG. 4-18 Stepper motors **(a)** a four-pole reluctance motor, **(b)** torque: excitation characteristics, **(c)** control configuration.

operation with controlled acceleration and deceleration. The solid line characteristic is commonly known as the slew curve.

Permanent magnet stepping motors develop higher torque than reluctance motors, and rotate in larger steps per unit pulse excitation.

Control

Since the position of the rotor in a stepper motor depends on the number of pulses it receives, one can easily control the rotation of the shaft by controlling the number of pulses.

Fig. 4-18(c) shows the control configuration of a stepper motor. It is an open loop control where the generated pulses are applied to the motor's windings at a controlled rate through a programmable digital controller. The characteristics of a simple open loop control system are compatible to those of the closed loop design.

Stepper motors at steady state accelerate and decelerate with each pulse and consequently are not useful for applications that require smooth steady state speed, such as phonograph drives. Stepper motors are employed in the positioning of machine tools, computer disk drives, printers, typewriters, sewing machines, and others.

Exercise 4–7 A 5 A, 1.8°/pulse reluctance type stepper motor is to rotate 36° in .02 seconds. Determine:

a) The required number of excitation pulses.

b) The amplitude and width of each current pulse.

Answer (a) 20; (b) 10 A, .50 ms.

4.2.7 AC/DC SERIES MOTORS

Small single-phase motors that can operate with either dc or ac input voltage are called universal or series ac/dc motors. These motors are normally rated at less than 1 kW, and their speed is usually between 2000 and 12,000 r/min. They are generally used in hand tools and household appliances. Their operating principles are similar to those that govern the operation of dc series motors. The latter are discussed extensively in Chapter 6, and the student should read Section 6.1.7 before he attempts to understand the operation and analysis of series ac/dc motors.

The equivalent circuit of a series ac/dc motor and its torque–speed characteristics are shown in Figs. 4-19(a) and (b), respectively. The field current (I_f) is equal to the armature current (I_a) because the field and armature are connected in series. The voltage generated in the armature winding, V_g, is given by

$$V_g = K\phi\omega \tag{4.63a}$$

where K is a constant of the motor and ϕ is its effective field. The torque developed by the motor is proportional to the effective flux and to the current drawn from the line. Mathematically,

$$T = K\phi I_a \tag{4.63b}$$

The constant of proportionality is given by Eq. (6.14).

At a specified speed, the motor develops less torque for ac operation than it does for dc operation. This is due to magnetic losses and to the fact that, for ac operation, the armature current is reduced because of the inductive reactance of the motor's winding.

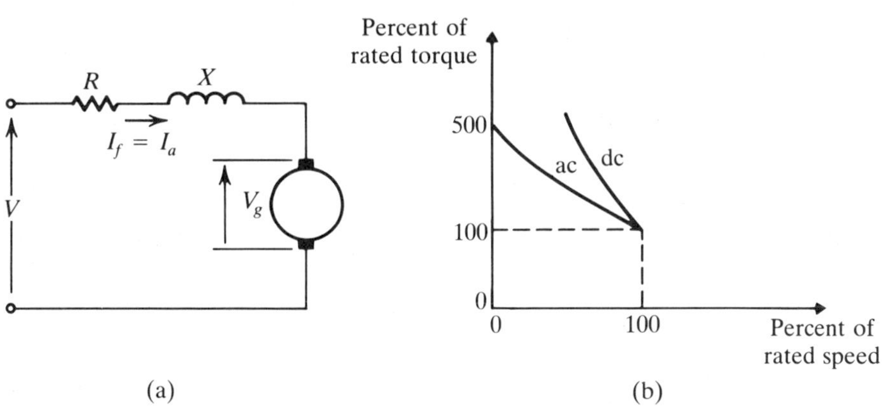

FIG. 4-19 Universal motor: **(a)** equivalent circuit, **(b)** typical torque–speed characteristics.

4.2.8 TACHOMETERS

All closed-loop speed-control systems are equipped with a device that measures the angular speed and produces an output signal directly proportional to the measured speed. The device used is a transducer (which changes angular speed to an electrical signal), which in its basic form is a simple single-phase generator. The three most commonly used types of single-phase generators are

1. The induction, or drag-cup, generator.
2. The permanent-magnet dc generator.
3. The permanent-magnet ac generator.

These generators are referred to as tachometers—that is, devices that measure speed.

AC and dc permanent-magnet generators measure the angular velocity of large rotating systems.

The induction, or drag-cup, generator is a small two-phase induction machine

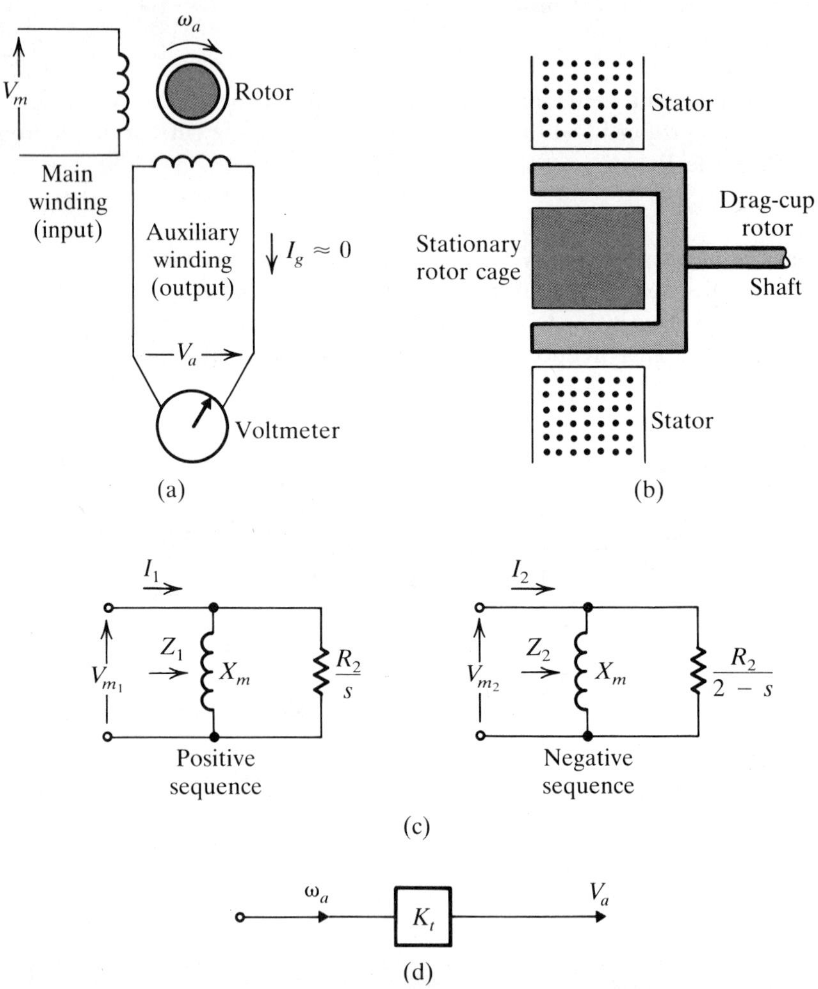

FIG. 4-20 Induction, or drag-cup, generator: **(a)** schematic representation, **(b)** rotor cross section, **(c)** approximate positive- and negative-sequence equivalent circuits, **(d)** block diagram.

with two windings in space quadrature. An elementary representation is shown in Fig. 4-20(a). Although the rotor can be of the squirrel-cage type, it is usually a thin copper cup which gives a minimum moment of inertia. The latter reduces the rotor's mechanical time constant. As a result, the time it takes for the generator to reach steady-state speed is minimized. The input and output windings are referred to as the main and auxiliary windings, respectively. The input winding is excited from a constant-voltage, constant-frequency ac source and constitutes the generator's field. When the generator's shaft is externally rotated, a voltage is induced in its output winding. From the basic principle of generator's operation, this voltage is given by

$$V_a = K_t\omega \tag{4.64}$$

where V_a is the output voltage of the generator, K_t is the tachometer constant, and ω is the angular speed of the generator's rotor.

Derivation of the Tachometer Constant

The tachometer constant depends on the parameters of the equivalent circuit and on the rated speed of the machine.

In terms of its positive- and negative-sequence components, the voltage of the auxiliary winding is

$$V_a = V_{a_1} + V_{a_2} \tag{4.65}$$

Also, from Eq. (4.19) we have

$$V_a = jV_{m_1} - jV_{m_2} \tag{4.66}$$

For small operating slips, the stator leakage impedance and rotor reactance are relatively small. Thus, they are not shown in the equivalent circuit.

From Fig. 4-20(c) we obtain

$$V_{m_1} = I_1 Z_1 \tag{4.67}$$

and

$$V_{m_2} = I_2 Z_2 \tag{4.68}$$

where Z_1 and Z_2 are the positive- and negative-sequence impedances, respectively. From the last two relationships we obtain

$$V_a = jI_1Z_1 - jI_2Z_2 \tag{4.69}$$

The current through the auxiliary winding is negligible ($I_a \approx 0$) because of the high impedance of the voltmeter. This results in equal positive- and negative-sequence currents. That is,

$$I_1 = I_2 \tag{4.70}$$

Also

$$V_m = V_{m_1} + V_{m_2} \tag{4.71}$$

$$= I_1 Z_1 + I_2 Z_2 \tag{4.72}$$

Thus,

$$V_m = I_1(Z_1 + Z_2) \tag{4.73}$$

From Eqs. (4.69), (4.70), and (4.73) we obtain

$$V_a = jI_1(Z_1 - Z_2) \tag{4.74}$$

and

$$V_a = j\frac{V_m}{(Z_1 + Z_2)}(Z_1 - Z_2) \tag{4.75}$$

From the approximate equivalent circuit we have

$$Z_1 = \frac{jX_m R_2}{R_2 + jX_m s}, \qquad Z_2 = \frac{jX_m R_2}{R_2 + jX_m(2 - s)} \tag{4.76}$$

Substituting Eq. (4.76) into Eq. (4.75) we obtain, after simplification:

$$V_a = jV_m \frac{X_m(1 - s)}{X_m - jR_2} \tag{4.77}$$

$$= \frac{jV_m X_m}{X_m - jR_2}\left(\frac{\omega_a}{\omega_s}\right) \tag{4.78a}$$

or

$$V_a = K_t \omega_a \tag{4.78b}$$

where

$$K_t = \frac{jV_m X_m}{\omega_s(X_m - jR_2)} \tag{4.79}$$

and ω_s is the speed of the synchronously rotating stator field in rad/s.

Thus the voltage across the output phase of the tachometer is proportional to the angular velocity of its rotor. The block diagram representation of a drag-cap generator is shown in Fig. 4-20(d). The constant of proportionality K_t can be calculated from Eq. (4.79).

EXAMPLE 4–5 A 120 V, 60 Hz, 4-pole induction generator has the approximate equivalent circuit shown in Fig. 4-21. Determine:

a) The constant of the tachometer.
b) The angular speed of the rotor when the output voltage is 100 volts.

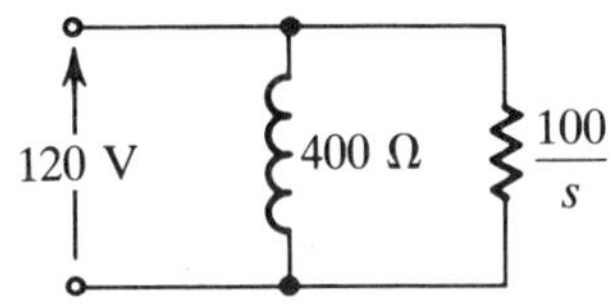

FIG. 4-21

SOLUTION

a) The constant of the tachometer is given by Eq. (4.79).

$$K_t = \left| \frac{V_m X_m}{\omega_s (X_m - jR_2)} \right| = \left| \frac{120(400)}{1800 \dfrac{2\pi}{60}(400 - j100)} \right| = \underline{\underline{.62 \text{ V/rad/s}}}$$

b) The angular speed of the rotor is

$$\omega_a = \frac{V_a}{K_t} = \frac{100}{.62} = \underline{\underline{161.91 \text{ rad/s}}}$$

Exercise 4–8 Explain the effects of the following on the tachometer constant:

a) Magnetic saturation.
b) Higher-than-nominal rotor winding temperatures.

4.3 Solid State Control

The speed of single-phase induction motors is seldom electronically controlled because the cost of the required solid state speed-control apparatus far exceeds the initial cost of the motor. However, in applications where precise speed control is of paramount importance, the motor's speed can be controlled through a single-phase frequency converter. A block diagram representation of a single-phase frequency converter is shown in Fig. 4-22.

The operation of single-phase frequency converters is similar to that of three-phase frequency converters, which is described in Section 3.6. That earlier description need not be repeated here; however, an exercise is presented whose results (Table 4.2) reveal the advantages of solid state control relative to other classical methods.

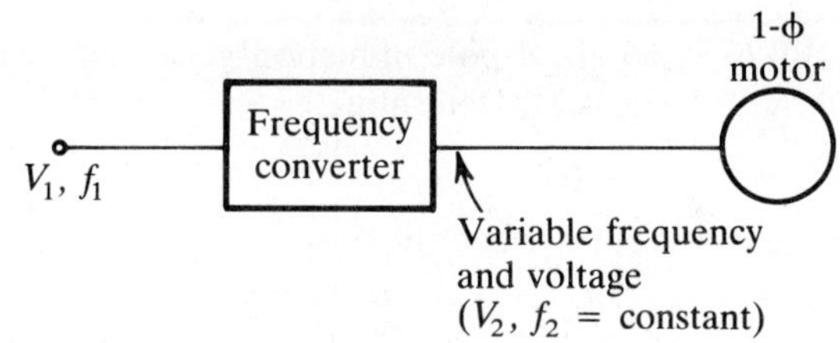

FIG. 4-22 Frequency converters.

Exercise 4–9 A 120 V, 60 Hz, 4-pole, single-phase induction motor has the equivalent circuit shown in Fig. 4-23. The motor drives a ventilating fan whose torque–speed characteristic is

$$T = 872 \times 10^{-6}\omega_a^2 \text{ N} \cdot \text{m}$$

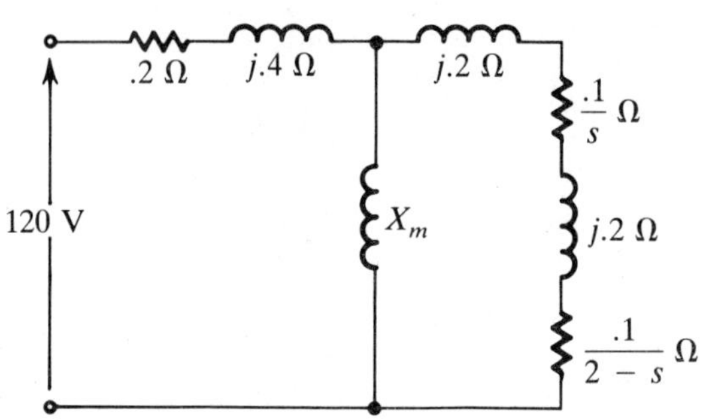

FIG. 4-23

In the following calculations, the effects of the magnetizing reactance and of the rotational losses are to be neglected.

1. Determine the speed of the motor.
2. The motor is equipped with a frequency converter which supplies voltage to the motor at 55 Hz. Determine:
   a) The voltage applied to the motor.
   b) The speed of the revolving fields.
   c) The actual speed of the motor.
   d) The output torque of the motor.
3. Instead of a frequency converter, an autotransformer is used that drives the motor at the same speed as that found in 2(c). Determine:
   a) The speed of the revolving field.
   b) The operating slip.
   c) The output voltage of the autotransformer.
   d) The output torque of the motor.

TABLE 4-2 Summary of Results of Exercise 4-9

Method of Speed Control	*Motor Parameters*				
	Actual Speed *r/min*	*Stator* Voltage *(V)*	*Stator* Frequency *(Hz)*	Speed of Revolving Fields *r/min*	Operating Slip *(pu)*
Frequency converter	1567.5	110	55	1650	.05
Auto-transformer	1567.5	105	60	1800	.13

4.4 Conclusion

The operation of two-phase induction motors is based on the fact that two balanced voltages (equal in magnitude and 90° out of phase with each other) produce a spacewise rotating field. The rotating field induces a voltage in the rotor windings, and thus a rotor flux is produced. The natural tendency of the rotor and stator fields to align their magnetic axes produces torque. The two-phase control motor is used in servomechanisms because of its linear torque–speed characteristic and its small electromechanical time constant.

Single-phase induction motors are used extensively in hand tools and in almost all home appliances. The power drawn from a single-phase voltage source produces two spacewise rotating electromagnetic fields. When the rotor is *stationary*, these two fields are equal in magnitude and opposite in direction. Because of this, single-phase motors have no starting-torque capability. However, by specially designing the stator windings (split-phase motors), one of the two fields can be reduced and the resultant field can develop the desired starting torque. It is theoretically possible to design the stator windings in a way that eliminates one of the rotating fields. This would result in maximum torque at starting. When the rotor is *rotating*, one of the revolving fields is greatly reduced. As a result, the motor develops nominal torque without any auxiliary circuit components.

The different nature of the fields at starting and running led to the development of two equivalent circuits from which the motor characteristics can be derived.

The tachometer is a special type of two-phase machine. On one of its stator windings it generates a voltage that is directly proportional to the speed of its rotor. Almost all speed-control systems incorporate a tachometer to measure and/or control speed.

The speed of a single-phase motor can be controlled smoothly through a solid state frequency converter or, in steps, through a voltage-reducing device (autotransformer, inductor, resistor, etc.).

Table 4-3 summarizes some of the essential mathematical relationships that govern the operation of single-phase and two-phase motors. Also included is manufacturer's data for a single-phase motor used in household clothes washers (see Table 4-4).

TABLE 4-3 Summary of Important Equations

Item	*Description*	*Remarks*
	Two-Phase Motors	
1	Rotating field $\mathscr{F} = \mathscr{F}_m \cos(\omega t + \beta - \theta)$	Eq. (4.7)
2	Speed of revolving field $n_s = 120 \dfrac{f}{p}$ r/min	Eq. (4.11)
	Unbalanced Operation	
3	Positive-sequence voltage $V_{m_1} = \dfrac{1}{2}(V_m - jV_a)$	Eq. (4.14)
4	$V_{a_1} = jV_{m_1}$	Eq. (4.17)
5	Positive-sequence slip $s = \dfrac{n_s - n_a}{n_s}$ pu	Eq. (4.26)
	Negative-sequence voltages	
6	$V_{m_2} = \dfrac{1}{2}(V_m + jV_a)$	Eq. (4.15)
7	$V_{a_2} = -jV_{m_2}$	Eq. (4.18)
8	Negative-sequence slip $s_b = 2 - s$ pu	Eq. (4.29)
	Torques	
9	Positive-sequence torque $T_1 = \dfrac{I_1^2 R_2}{\omega_s s}$	Eq. (4.30)
10	Negative-sequence torque $T_2 = \dfrac{I_2^2 R_2}{\omega_s(2 - s)}$	Eq. (4.31)
	Two-Phase Control Motor	
11	$KV_c(s) - T_L(s) = (Js + B + m)\omega(s)$	Eq. (4.38)
	Single-Phase Motors	
12	Forward-rotating field of main winding $\mathscr{F}_{m_1} = \dfrac{1}{2}(N_m I_m - jN_a I_a)$	Eq. (4.52)

(*Continues*)

TABLE 4-3 (*Continued*)

Item	*Description*	*Remarks*
	Single-Phase Motors	
13	Backward-rotating field of main winding $\mathscr{F}_{m2} = \frac{1}{2}(N_m I_m + jN_a I_a)$	Eq. (4.53)
	A single rotating field is produced when the parameters of the auxiliary winding are as follows:	
14	$R_a = X_m \frac{N_a}{N_m}$	Eq. (4.58)
15	$X_c = X_a + \frac{N_a}{N_m} R_m$	Eq. (4.61)
	Tachometer	
16	$V_a = K_t \omega_a$	Eq. (4.78b)
17	$K_t = j \frac{V_m X_m}{\omega_s (X_m - jR_2)}$	Eq. (4.79)

TABLE 4-4 Data for a split-phase, 120 V, 250 W, 1800 r/min, single-phase induction motor used in household clothes washers*

Parameter	*Winding* Main	*Winding* Auxiliary
Resistance in ohms	1.26	3.03
Reactance in ohms	Running condition	Locked-rotor condition
Positive sequence (X_1)	1.58	.69
Negative sequence (X_2)	1.12	.52
Zero sequence (X_0)	29.7	16.4
Effective turns ratio	50	65

* *Courtesy of General Electric Canada Inc.*

4.5 REVIEW QUESTIONS

1. What is meant by a balanced two-phase power supply?
2. For a two-phase motor, what are the frequencies of the positive- and negative-sequence components of the rotor current?
3. State some applications of two-phase motors.

4. If the voltage supply to a two-phase motor is unbalanced, excessive copper losses may damage the motor. Explain.

5. How can the equivalent-circuit parameters of a two-phase motor be measured in a laboratory?

6. Why are the characteristics of two-phase control motors approximately linear?

7. Why are the positive- and negative-sequence sections of the equivalent circuit of single-phase induction motors in series?

8. Why is the speed of some universal series motors more than 3600 r/min?

9. What is a tachometer, and how can its constant of proportionality be established?

4.6 PROBLEMS

4-1 The standard 120/240 V, single-phase voltage supply to residential homes is as shown in Fig. P4-1.

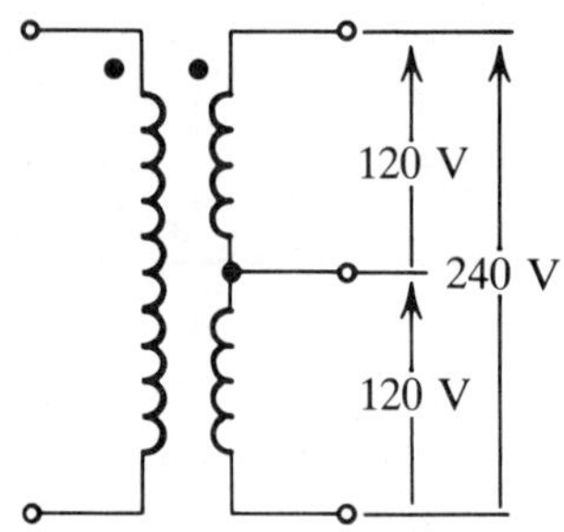

FIG. P4-1

a) Draw the instantaneous values and the phasor diagram of the 120 V supply.

b) Determine the symmetrical components of the supply voltages.

c) If a 120 V, two-phase motor is connected to such a voltage supply, will it operate?

4-2 Under balanced operating conditions, the starting torque of a 220 V, 60 Hz, 2 kW, two-phase motor is 2.5 per-unit. Determine the starting torque in per-unit when the following unbalanced voltages are applied to the motor:

$$V_m = 220\angle 0^\circ \text{ V}$$
$$V_a = 150\angle 45^\circ \text{ V}$$

4-3 The per-phase approximate equivalent circuit of a 110 V, 4-pole, two-phase induction motor is shown in Fig. P4-3. Rotational losses consume 4% of the power developed. The motor is supplied from a balanced voltage source of 110 V.

a) Determine:
   (1) The starting current and torque developed.
   (2) The output power and torque when the motor runs at 1740 r/min.

b) Repeat (a) if the motor is supplied from an unbalanced voltage supply of

$$V_m = 110\angle 0^\circ \text{ V}$$
$$V_a = 80\angle 60^\circ \text{ V}$$

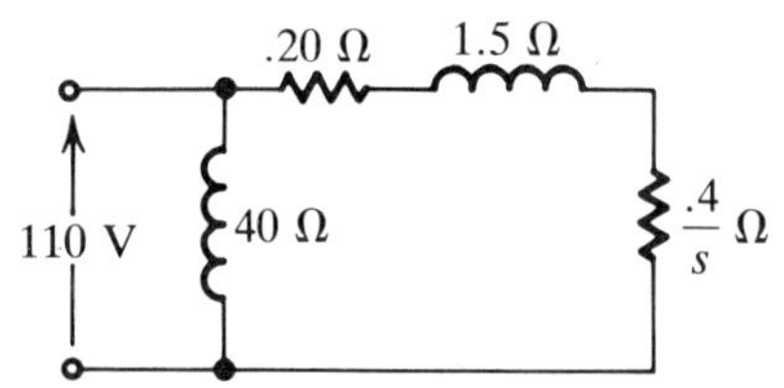

FIG. P4-3

4-4 A 60 Hz, 1.5 kW, 220 V, two-phase control motor has the torque–speed characteristic shown in Fig. P4-4. The load torque requirements are

$$T = .02\frac{d\omega}{dt} + .01\omega + .75 \text{ N} \cdot \text{m}$$

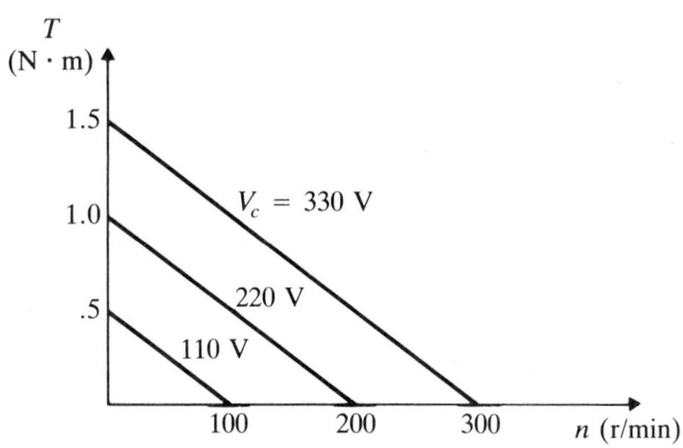

FIG. P4-4

Determine:

a) The general motor torque–speed equation.

b) The block diagram.

c) The transfer functions:

$$\frac{\omega_s}{T_L(s)}, \quad \frac{\omega(s)}{V_c(s)}$$

d) The steady-state speed when the control voltage is 330 V.

4-5 Figure P4-5 shows the equivalent circuit of a 60 Hz, 120 V, 2-pole single-phase induction motor. When the motor is running at a slip of 4%, neglecting the magnetizing current, determine:

a) The input current and power factor.

b) The forward torque.

c) The backward torque.

d) The torque developed.

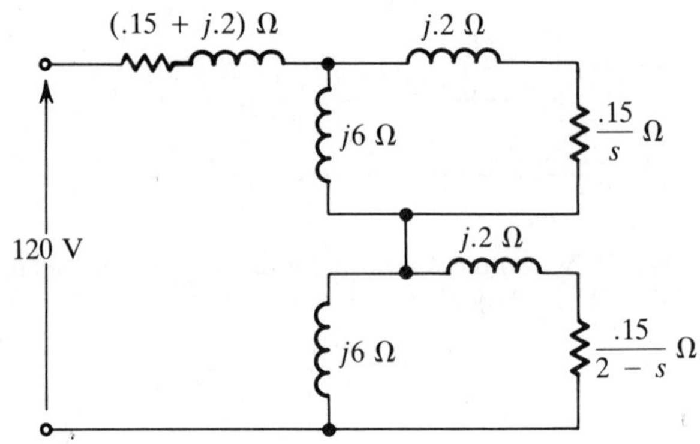

FIG. P4-5

4-6 A 120 V, 60 Hz, single-phase induction motor of the capacitor-start type has the following characteristics:

Parameter	Winding: Main	Winding: Auxiliary
Effective turns	200	250
Input impedance at standstill in ohms	$5 + j10$	$12.5 + j16$

Determine:

a) The amplitudes of the forward- and backward-rotating fields.

b) The size of the capacitor bank which, when connected in series with the auxiliary winding, will produce at starting only a forward-rotating field.

4-7 The approximate per-phase equivalent circuit of a small two-phase, 4-pole induction motor is shown in Fig. P4-7. With its main winding connected to a 120 V, 60 Hz source and its auxiliary winding connected across a voltmeter, the motor is used as a tachometer. Making reasonable assumptions, determine the speed of the motor when the voltmeter indicates 105 V rms.

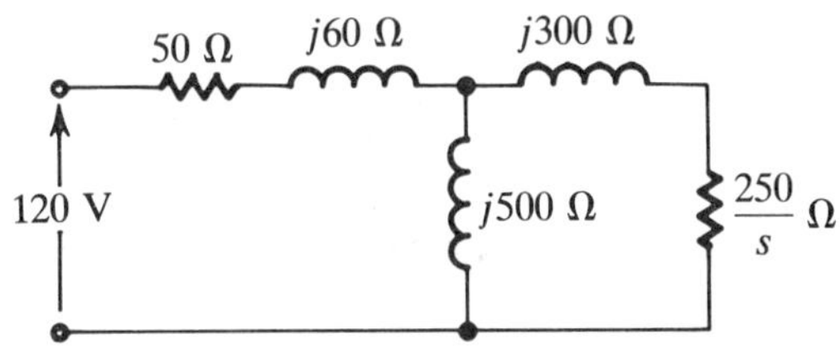

FIG. P4-7

4-8 For the shaded-pole motor shown in Fig. P4-8, prove the following:

a) When one of the switches (A or B) is closed, the direction of the rotation is from the unshaded part of the pole toward the shaded part of the pole.

b) The flux through the unshaded part of the pole is maximum at the instant the supply voltage is maximum.

c) The speed of the motor is maximum when the supply voltage is applied across the minimum number of winding turns.

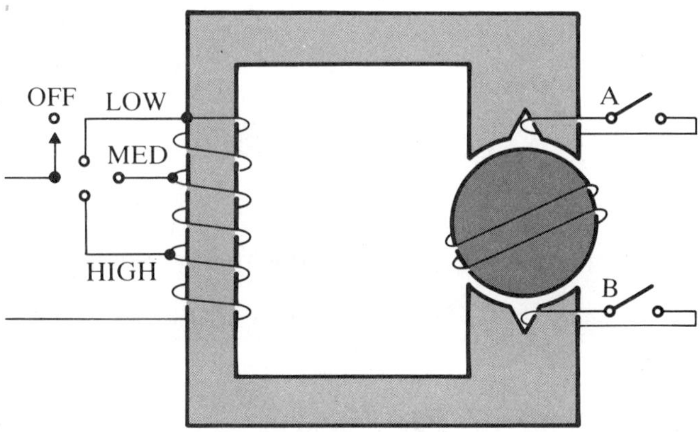

FIG. P4-8

4-9 Draw the torque–speed characteristics and explain the principle of operation of

a) Split-phase motors.

b) Shaded-pole motors.

c) Stepper motors.

FIVE

Synchronous Machines

5.0 Three-Phase Synchronous Machines

Three-phase synchronous motors are used to drive loads that operate at constant speed. Such loads include, for example, the grinding mills of the mining industry and the rolling mills of the paper and textile industries.

Synchronous motors operating at 1800 r/min and with ratings larger than 1000 kW have efficiencies up to 4% higher than those of equivalent three-phase squirrel-cage induction motors.

Synchronous generators produce, at their output terminals, a balanced three-phase voltage supply by transforming the mechanical energy transmitted to their rotors into electrical energy. The source of the mechanical power—commonly known as the prime mover—may be wind, hydropower, or steam.

Synchronous generators are used mainly by utility companies to generate their standard, three-phase, 60 Hz power. In fact, most ac power is generated through synchronous generators, the ratings of which sometimes exceed 1000 MVA. In the plants of consumers, smaller machines (up to 1 MVA) provide standby emergency power to critical loads in case the utility's power system fails. Remote locations use synchronous generators with ratings from 150 watts to several kilowatts.

Increasingly, one finds small synchronous generators—down to 20 kW—installed where a source of power can economically produce electricity. Such power may be supplied, for example, by small hydroelectric plants, the excess steam from chemical processes, the heat produced by factories burning dried cotton, and the like. Most automobiles are also equipped with three-phase generators.

Synchronous machines are classified in terms of rotor shape as *cylindrical rotor* or *salient rotor*. Machines of both types will be discussed in the sections that follow. The three-phase short circuits of the machines will be dealt with only briefly, for they are covered in more specialized publications.

The control schematics and the modern electronic control of synchronous machines are explained in the chapter on control schematics.

5.1 Three-Phase Cylindrical Rotor Machines: Motors

Three-phase synchronous motors are externally controlled, variable-power-factor machines. They run at synchronous speeds and, as shown in Fig. 5-1(a), require two

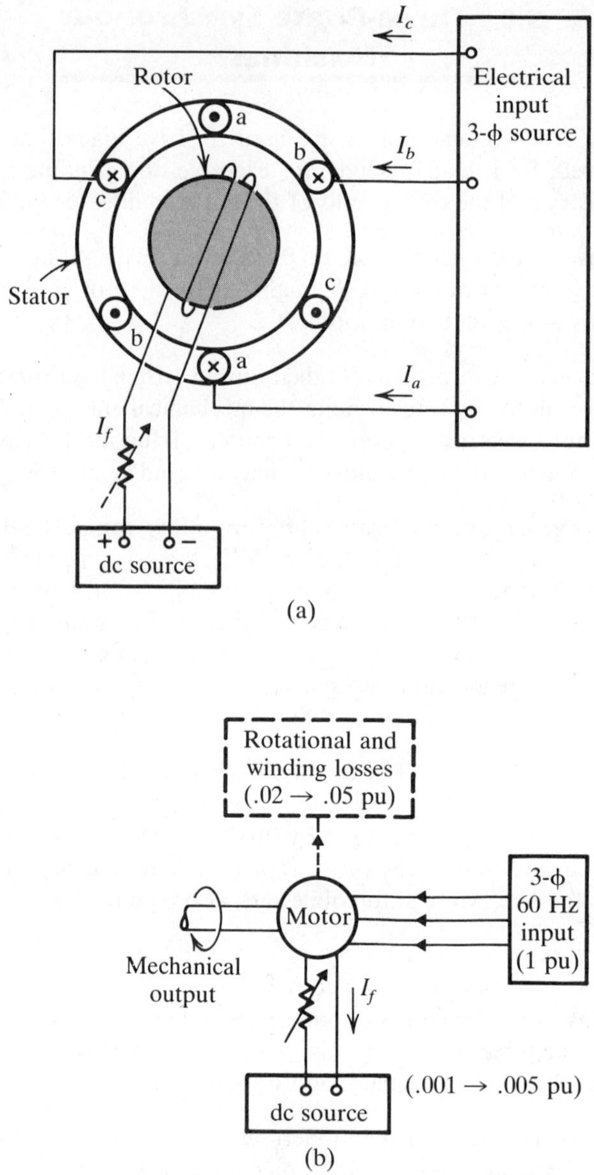

FIG. 5-1 Synchronous motor: **(a)** elementary representation of stator and rotor windings—cylindrical rotor, **(b)** direction of power flow and typical losses for a 100 kW machine.

sources of excitation:

1. A balanced three-phase ac source for the stator windings.
2. A dc source for the rotor windings.*

The structure and winding arrangement of the stator of a synchronous motor is identical to that of a three-phase induction motor. The rotor, however, is of the wound type and may be of either cylindrical or salient construction.

The dc current for the rotor windings may be obtained either from an external dc source through the slip rings of the rotor (classical machines) or by rectifying the ac voltage induced in a special set of rotor windings (modern or brushless machines). In the latter case the rotor carries—besides the dc *field winding*—a three-phase *ac winding* and a set of *solid state devices* (thyristors and diodes). These solid state devices are located between the ac and dc windings. When the rotor rotates, the mmf of an external dc supply induces a voltage in the ac winding.

Solid state devices rectify the induced voltage and control the timing of the application of the rectified voltage to the dc windings. A detailed description of the dc voltage source for the field windings of the motor is given in Section 7.3.3.

Brushless synchronous motors also include those synchronous machines that obtain their field excitation from an auxiliary generator, commonly known as an exciter. The exciter often is driven by the same prime mover as the synchronous machine.

The motor's mechanical output power comes from the transformation of the input ac power. The dc supply—although essential for the operation of the motor—transfers no power to the output. Figure 5-1(b) shows typical values of the input power and losses for a 100 kW motor.

The electrical input to the rotor windings of large synchronous machines is less than 1% of the total power. In efficiency calculations, therefore, it may be neglected. The field current essentially controls the reactive volt-ampere requirements of the stator circuits and has a pronounced effect on the motor's output torque.

In synchronous machines, the magnetic flux of the armature may aid or oppose the magnetic flux of the field current. This interaction, termed *armature reaction*, is also referred to—depending on its effects—as the magnetizing or demagnetizing effect of the armature current. Determination of the effects of the armature current is essential in analyzing and understanding synchronous machines.

The main advantage of synchronous motors is that their power factor ($\cos \theta$), input line current (I_L), and input reactive power (Q) can be easily controlled by properly adjusting the field current. In other words, synchronous motors—as seen from the network terminals—behave as either variable-shunt capacitors or variable-shunt inductors. This topic is discussed in detail in Section 5.1.9.

Synchronous motors provide variable torque for loads that require a constant speed. For ordinary applications, however, they are neither economical nor practical, because they operate at only one speed and require an ac source for stator excitation and a dc current for their rotor windings. Besides this, they have another crucial

. . .

*Permanent-magnet rotors may be used in small machines.

disadvantage: sudden or cyclic variations in the mechanical load torque may drive the motor out of synchronism. This may cause disturbances and lead to the shutdown of both the motor and the supplying electrical system.

One of the main engineering problems related to the steady-state operation of synchronous motors involves the calculation of the *field current* that corresponds to a specific armature current and power factor.

5.1.1 ROTOR

An elementary representation of a synchronous motor with a cylindrical rotor is shown in Fig. 5-1(a). In general, motors with rated speeds of 1800 or 3600 r/min have cylindrical rotors, while synchronous motors with lower rated speeds are designed, for economical reasons, with salient rotors.

Cylindrical-rotor machines, which are relatively simple to analyze, cease to operate when their field current is removed. Salient-pole machines—more difficult to analyze—can, without any excitation, develop a torque that may be up to 40% of rated torque.

FIG. 5-2 A 4-pole, 15 MW, 60 Hz synchronous motor. *Courtesy of General Electric Canada, Inc.*

The sections that follow will discuss the cylindrical-rotor motor's principle of operation, equivalent circuit, governing equations, phasor diagrams, and external characteristics, commonly known as "V curves." In the photographs of Fig. 5-2 to 5-5 some physical characteristics of synchronous motors are described in each figure.

5.1.2 ROTATING FIELDS

Synchronous machines have two synchronously rotating magnetic fields:

1. The dc field of the rotating rotor.
2. The stator or armature field.

FIG. 5-3 Stator of a 2.6 MW synchronous motor. *Courtesy of General Electric Canada, Inc.*

The stator field is identical to that produced by the armature of three-phase induction motors. For convenience its description is repeated here.
Refer to Fig. 5-6(a). The magnitude and speed of rotation of the stator field are given by

$$\mathscr{F}_s = \frac{3}{2}\mathscr{F}_1 \cos(\omega t - \beta) \text{ ampere-turns} \quad \textbf{(5.1)}$$

$$n_s = 120\frac{f}{p} \text{ r/min} \quad \textbf{(5.2)}$$

where:

$\mathscr{F}_s$ = is the rotating stator mmf produced by the three-phase armature currents

$\mathscr{F}_1$ = is the maximum value of the mmf produced by one phase only. For balanced armature currents and an equal number of turns per winding, we have

$$\mathscr{F}_1 = N_a I_{a_m} = N_b I_{b_m} = N_c I_{c_m} \text{ ampere-turns} \quad \textbf{(5.3)}$$

β = an arbitrary angle measured with respect to the horizontal line, as shown in Fig. 5-6(a)

ω = the angular speed of the stator currents

f = the frequency of the stator currents

n_s = the synchronous speed of the rotating field

p = the number of poles of the motor.

FIG. 5-4 Rotor of a 2.6 MW synchronous motor. *Courtesy of General Electric Canada, Inc.*

Equations (5.1) and (5.2) are derived in Section 3.1.3. The magnetic field of the armature is equivalent to the field of a synchronously rotating permanent magnet. Under normal operating conditions this field does not induce any voltage in the rotor winding, because the latter is rotating at the same speed.

As shown in Fig. 5-6(b), the field of the rotor is along the magnetic axis of the rotor. Since the rotor rotates at synchronous speeds, under normal operating conditions its field also travels at the same speed.

Under normal operating conditions, then, the stator and rotor fields travel at the same angular speed. This is referred to as the synchronous speed and is given by Eq. (5.2).

FIG. 5-5(a) Salient rotor synchronous motor: continuous-end rings. *Courtesy of General Electric Canada, Inc.*

FIG. 5-5(b) Salient rotor synchronous motor: discontinuous-end rings. *Courtesy of General Electric Canada, Inc.*

5.1.3 PRINCIPLE OF OPERATION

The motor's operation, or torque-producing capability, results from the natural tendency of two magnetic fields to align their magnetic axes. The fields under consideration are

1. The stator field (ϕ_s).
2. The rotor field (ϕ_f).

The flux that accompanies the synchronously rotating stator field (ϕ_s) completes its magnetic circuit by passing through the stator, air gaps, and rotor. The flux of the rotor field (ϕ_f) also passes through the rotor, the air gaps, and the structure of the stator. The vector sum of the armature and field flux, as shown in Fig. 5-6(b), gives the so-called resultant (ϕ_R) air-gap flux. Mathematically,

$$\vec{\phi}_R = \vec{\phi}_s + \vec{\phi}_f \tag{5.4}$$

The magnetic axis of the field flux tries to align itself with the magnetic axis of the stator

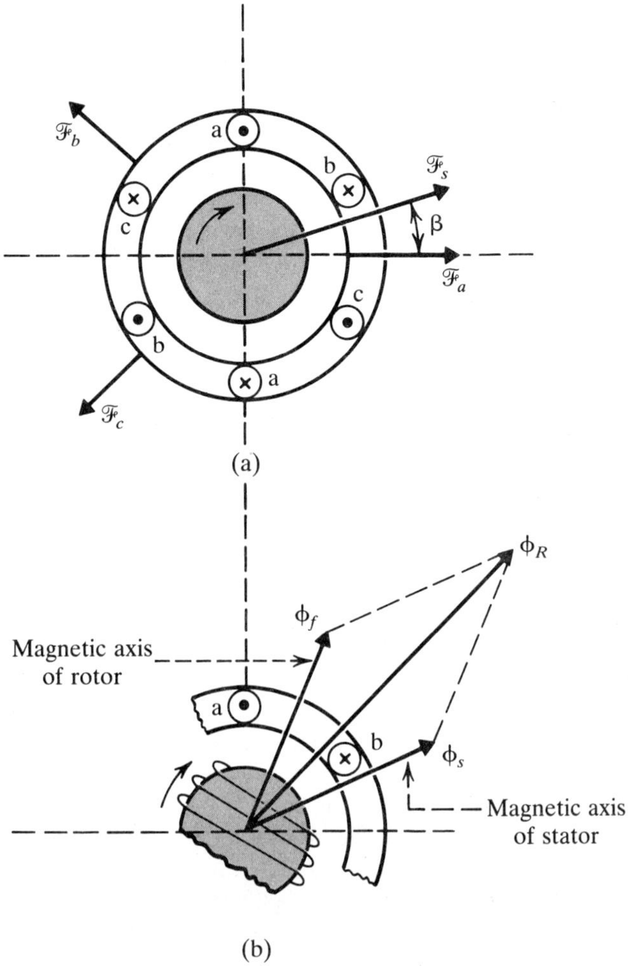

FIG. 5-6 Magnetic fields in a synchronous motor: **(a)** stator mmf's, **(b)** stator, rotor, and resultant magnetic flux.

flux. This process produces motor action. The mathematical expression for the torque developed was derived in the chapter on three-phase induction machines (see Section 3.1.2). This electromagnetic torque is in the direction of rotation of the stator field. When an additional load is applied to the shaft, the rotor or the magnetic axis of the field flux slips in space behind the magnetic axis of the resultant air gap flux. This results in a larger space angle; therefore, a larger torque is developed.

5.1.4 STARTING

Starting a three-phase synchronous motor is similar to starting a three-phase induction motor. Balanced voltages are applied to the stator windings, while the rotor winding is short-circuited or connected to a resistance. Synchronous motors are commonly equipped with damper windings, which are similar to the rotor windings of squirrel-cage induction motors. During any nonsynchronous-speed operation, damper windings contribute to the motor's output torque. The stator voltages produce a rotating magnetic field, which induces a voltage in the rotor winding. This voltage will circulate rotor current, which in turn will produce a magnetic field.

The interaction of these two fields produces motor action, which will bring the rotor up to a speed that is very close to synchronous. At this instant the rotor windings are switched to the normal dc supply, and the rotor pulls in synchronism with the armature rotating field. In practice, this switching often takes place at that particular instant when the flux of the field momentarily reduces the voltage already impressed in the stator. This results in high inrush currents in the armature windings, which are accompanied by a characteristic "thump."

The resistance (R_2) of the rotor circuit limits the rotor current and thus controls the starting torque of the motor. Generally speaking, the higher the resistance of the field windings, the higher the starting torque. Mathematically,

$$T_{st} \propto R_2 \tag{5.5}$$

The torque that the motor develops, at the instant the dc field is "switched in," is usually referred to as the pull-in torque. The pull-in torque must be larger than the load requirements at that speed; otherwise, the motor will not be able to synchronize.

At starting, synchronous motors have very low impedance. As a result, their starting current is five to six times greater than their rated current. Depending on the Thevenin's impedance of the supply network, this high current may cause a dip in the voltage delivered to the plant where the motor is operating. This voltage reduction will reduce the starting-torque capability of the motor ($T \propto V^2$) and may adversely affect other electrical loads, such as motors, lights, and magnetic contractors.

The accelerating time, the magnitude of the inrush current, and the torque are identical to those of three-phase induction machines.

Where circumstances do not permit high inrush currents, an autotransformer may be used to reduce the voltage applied to the motor. A more popular method is to design a motor with low starting torque and inrush current. Such a motor is coupled to the load through an air clutch. The motor starts unloaded, and after it is synchronized, the pressurized air—which is conducted through a properly sized opening along the inner section of the motor's shaft—engages or couples the air clutch to the load. The use of an air clutch not only permits the design of motors with lower starting torque and currents, but also develops pneumatic power that is sufficient to provide the high starting torque that some loads require.

The appendix to this chapter gives manufacturers' data for high- and low-starting-torque (soft or air-clutch starting) synchronous motors.

5.1.5 EQUIVALENT CIRCUITS

This section will develop and analyze the approximate equivalent circuit(s) of a synchronous motor. The flow of energy from the ac and dc sources and the electromagnetic coupling between their corresponding fields are shown in Fig. 5-7(a).

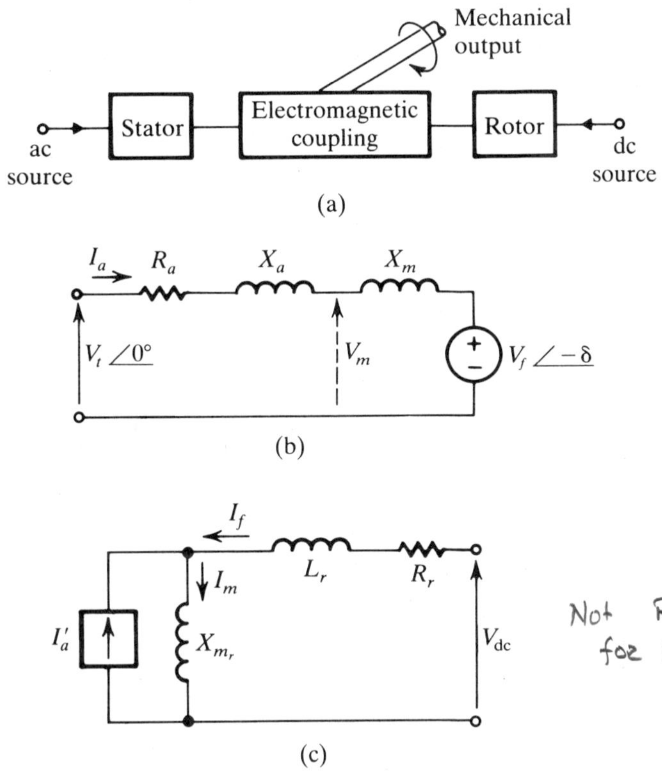

FIG. 5-7 Synchronous motor: **(a)** elementary presentation of energy flow, **(b)** approximate per-phase equivalent circuit as seen from the stator, **(c)** approximate equivalent circuit as seen from the rotor.

This electromagnetic coupling leads to the development of torque. From Ampere's law, the torque developed depends on the magnitude and relative position of the two interacting fields.

The approximate per-phase equivalent circuit of a motor, as seen from the stator (Fig. 5-7(b)), is derived by considering the effects of the two rotating fields on the stator. The resultant field magnetizes the stator (a process represented by X_m). The rotor field induces a voltage (V_f) in the stator windings whose phase angle is at $-\delta°$ to the stator voltage supply (V_t) because the synchronously rotating rotor field lags the stator field by the same angle. The leakage impedance of the stator windings is represented by $R_a + jX_a$.

The approximate equivalent circuit of a motor as seen from the rotor (Fig. 5-7(c)) is derived by considering the effects of the two rotating fields on the rotor. The resultant field magnetizes the rotor (a process represented by the reactance X_{m_r}), and the effects of the stator current are represented by a current source of I'_a amperes. The actual resistance and inductance of the rotor windings are as shown.

The stator and rotor equivalent circuits supplement each other. They give an in-depth view of the machine's conceptual design and behavior, enabling the application engineer to predict the motor's operation.

Equivalent-Circuit Parameters

The physical significance and nomenclature of the various parameters shown in the equivalent circuits are as follows.

Stator Parameters

V_t = *terminal voltage*. This is the per-phase ac stator voltage. Its magnitude depends on the voltage rating of the motor and on the type of connection (delta or star) of the armature windings. In phasor diagrams, or in actual calculations, this voltage is taken as reference.

V_m = *magnetizing voltage*. This is the voltage induced in the stator windings by the effective flux of the motor. This flux depends on the resultant mmf (vector sum of the stator and rotor's mmf's) and on the degree of magnetic saturation.

V_f = *excitation voltage*. This is the voltage induced in the armature windings by the dc rotating field. At the absence of saturation, the excitation voltage is equal to the open-circuit voltage that is measured across the armature terminals when the machine is driven as a generator. This voltage is sometimes referred to as the machines internal voltage or the open-circuit voltage. In motors, the excitation voltage—as explained in subsequent sections—always lags the supply or terminal voltage (V_t). Depending on the power factor, the excitation voltage may be larger or smaller than the terminal voltage. The excitation voltage is opposite in polarity to the terminal voltage and, depending on the motor's power factor, it may be smaller or larger than the supply voltage.

X_m = *magnetizing reactance* as seen from the stator. It represents the magnetization of the stator that results from the effective field of the armature. The magnetizing reactance depends on the characteristics of the magnetic material and the degree of saturation.

X_a = *stator leakage reactance.*

R_a = *stator resistance.* This is the effective or ac resistance of the armature windings—that is, the stator dc resistance adjusted to include magnetic losses and the effects of higher operating temperatures.

X_s = *synchronous reactance.* This is the sum of the stator leakage reactance and the magnetizing reactance. That is,

$$X_s = X_a + X_m \tag{5.6}$$

Z_s = *synchronous impedance.*

$$Z_s = R_a + jX_s \tag{5.7}$$

I_a = *armature current.* When the motor is star-connected, this current is equal to the line current.

Power Angle (δ)

The power angle, designated by the Greek letter delta (δ), is the phase angle between the excitation voltage (V_f) and the magnetizing voltage (V_m). It is also called the torque or the load angle. This angle is equal to the phase angle between the resultant air-gap flux (ϕ_R) and the field flux (ϕ_f).

The phase angle between the terminal voltage (V_t) and the excitation voltage (V_f) is approximately equal to the torque angle, because the leakage impedance is very small compared to the magnetizing reactance.

The torque angle for a synchronous motor is always negative (V_f lags V_t), regardless of operating power factor. The torque angle used for the steady-state analysis of the motors, or shown in the phasor diagrams, is expressed in electrical degrees. However, the torque angle used for the analysis of the machine's electromechanical transients is expressed in mechanical degrees. The relationship between mechanical (δ_m) and electrical degrees (δ) is derived in Section 3.1.3, and is rewritten here for convenience:

$$\delta_m = \frac{2}{p}\,\delta \tag{5.8}$$

where p is the number of poles of the machine.

Typical ranges of the parameters of synchronous motors are given in Table 5-1.

TABLE 5-1 Typical synchronous machine parameters

Parameter	*Minimum**	*Maximum**
X_a	.08	.30
X_m	.4	2.4
R_a	.002	.04

* *All values are in per-unit, based on the machine's rating.*

Rotor Parameters

V_{dc} is the dc voltage applied across the rotor windings. In all modern machines this voltage is obtained by rectifying the voltage induced in another set of rotor windings from a source that is magnetically coupled to the rotor but not electrically connected to it.

X_{m_r} is the machine's equivalent magnetizing reactance as seen from the rotor terminals.

R_r and L_r are the rotor resistance and self-inductance, respectively.

I_f is the dc current of the rotor winding.

I_m is the component of the field current that flows through the magnetizing reactance X_{m_r}.

I'_a is the component of the stator current that represents the armature reaction. This is the contribution of the armature current to the magnetization of the machine; it is expressed in equivalent field amperes. Depending on the power factor at which the motor draws power, the armature reaction may aid or oppose the actual field current. This is illustrated in the phasor diagrams of Fig. 5-10.

The ratio of the armature current (I_a) to its armature reaction component gives the effective turns ratio (N_e) between the stator and the rotor windings. Mathematically,

$$N_e = \frac{I_a}{I'_a} \tag{5.9}$$

The measurement of a machine's effective turns ratio and armature reaction is discussed in Section 5.2.

5.1.6 FIELD CURRENT

From the equivalent circuit of Fig. 5-7(c), the motor's magnetizing current (I_m) is given by

$$I_m = I_f + I'_a \tag{5.10}$$

where I_f is the actual field current and I'_a is the contribution of the armature current to the magnetization process. Rewriting the last relationship, we have

$$I_f = I_m - I'_a \tag{5.11}$$

I'_a is in phase with the armature current and is related to it by the effective turns ratio, as previously mentioned. The field current and its magnetizing component lag their corresponding voltages by 90°, as shown in Fig. 5-8.

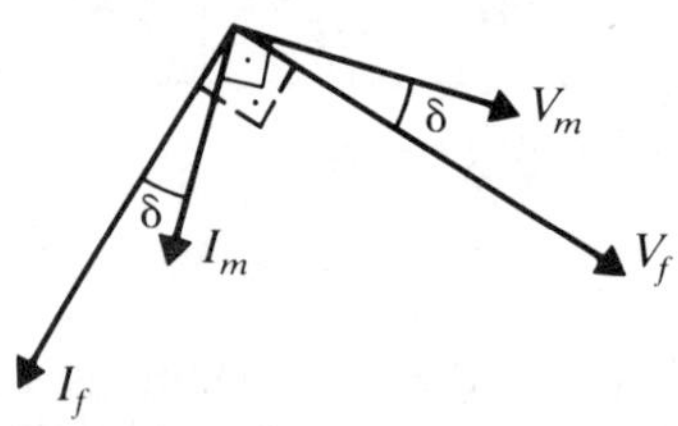

FIG. 5-8 Excitation and magnetization voltage phasors and their corresponding currents.

The field current and its magnetizing component can be obtained graphically by projecting their corresponding voltages on the machine's open-circuit characteristic (OCC). This is shown in Fig. 5-9.

The actual values of V_f and V_m at any particular motor operating condition can be found by applying KVL to the equivalent circuit of Fig. 5-7(b). The measurement

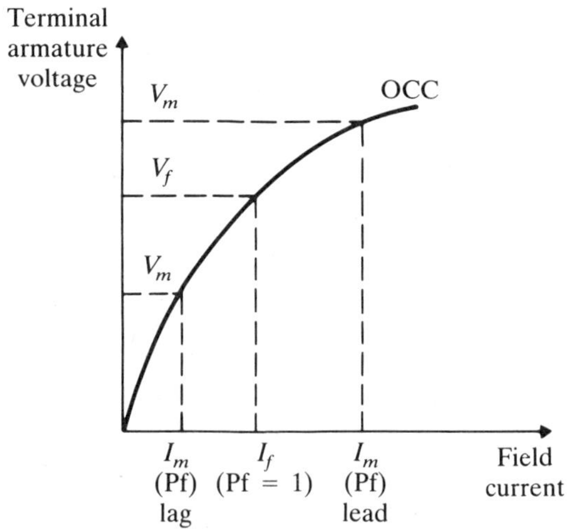

FIG. 5-9 The effects of the power factor on the magnetization voltage and on the magnetizing component of the armature current.

of the open-circuit characteristic, and further applications, are discussed in detail in Section 5.2.3.

Equation (5.10) may be written in terms of the interacting mmf's. That is, the effective magnetizing mmf is equal to the mmf of the field plus the mmf of the stator. Although this expression would be conceptually advantageous, it would be of limited usefulness because an additional unknown—the number of winding turns—is introduced.

5.1.7 PHASOR DIAGRAMS

A graphical representation, such as a phasor diagram, often simplifies the solution of a problem and enhances the understanding of the subject matter under consideration. Consider, for example, the following equation, which is obtained by applying KVL to the motor's equivalent circuit of Fig. 5-7(b):

$$V_f\underline{/-\delta} = V_t\underline{/0^\circ} - I_a(R_a + jX_s) \tag{5.12}$$

Usually the excitation voltage and its phase angle are unknown. One may mathematically determine the values of the unknowns, but the graphical solution is often simpler and has the additional advantage of giving a pictorial representation of the various terms and concepts under consideration. This section explains how to draw the phasor diagrams of Eqs. (5.11) and (5.12). These equations are central to the understanding of the motor's operation.

Figures 5-10(a), (b), and (c) show the phasor diagrams of a synchronous motor operating at unity, at leading, and at lagging power factors, respectively. The sequence

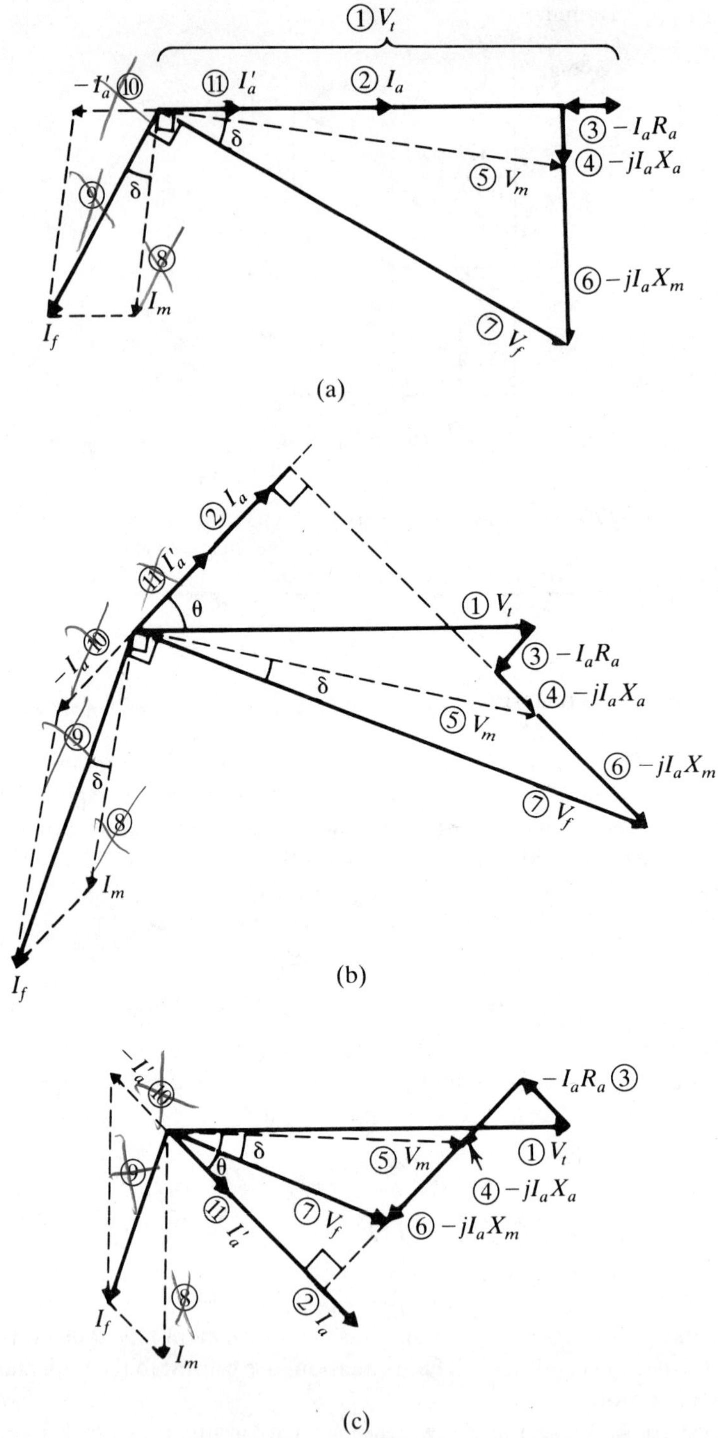

FIG. 5-10 Phasor diagrams for synchronous motors: **(a)** unity power factor, **(b)** leading power factor, **(c)** lagging power factor.

of steps required to complete the diagrams is outlined below and is identified on the phasor diagrams by the encircled numbers.

1. Draw the terminal voltage V_t to scale and take it as reference. Recall that for a star-connected motor:

$$V_t = \frac{V_{L-L}}{\sqrt{3}} \qquad \text{and} \qquad I_L = I_a$$

For a delta-connected motor:

$$V_t = V_{L-L} \qquad \text{and} \qquad I_L = \sqrt{3}\, I_a$$

Only the magnitudes of the various parameters are considered here. The subscript L indicates line quantities.

2. Calculate and then draw to scale the armature current I_a. The per-phase armature current is usually obtained from the rating and efficiency of the machine, as the following standard relationships indicate:

$$\text{input power} = \frac{\text{output power}}{\text{efficiency}} \tag{5.13}$$

Using mathematical symbols.

$$P_{\text{in}} = \frac{P_{\text{out}}}{\eta} \tag{5.14}$$

$$P_{\text{in}} = \sqrt{3}\, V_{L-L} I_L \cos\theta \tag{5.15}$$

where η and $\cos\theta$ represent the motor's efficiency and its power factor, respectively. Note that the output power is also called brake horsepower, name-plate power, or rated power.

3. From the tip of V_t draw the phasor I_aR_a. Because this represents the voltage drop in the stator resistance, it should be in phase with the armature current. However, owing to the negative sign in Eq. (5.12), it must be placed at an angle 180° from its normal direction.
4. From the tip of I_aR_a draw the phasor jI_aX_a. Because this represents the voltage drop in an inductor, it leads the current by 90°. However, the actual position of the phasor jI_aX_a is drawn at an angle 180° from its normal direction, owing to the negative sign of the voltage equation.
5. The tip of the phasor jI_aX_a locates the magnetization voltage V_m.
6. The phasor jI_aX_m is drawn at the extension of jI_aX_a.
7. The line drawn from the origin to the tip of the phasor jI_aX_m gives the excitation voltage V_f.
8. Having calculated V_m, obtain from the open-circuit characteristic the magnetizing component I_m of the field current.

9. Draw the field current I_f at 90° lagging V_f. The magnitude of this current is obtained from the motor's OCC by projecting on it the excitation voltage (see Fig. 5-9).
10. Obtain graphically the reverse of I'_a using Eq. (5.11).
11. Draw I'_a in phase with the armature current.

Comments

As can be seen from Fig. 5-10(c), when the motor operates at a lagging power factor, the field current is smaller than its magnetizing component because of the magnetizing effects of the armature current. Conversely, when the motor operates at a leading power factor, the field current is larger than its magnetizing component because of the demagnetizing effects of the armature current.

EXAMPLE 5–1 A 50 kW, 480 V, 60 Hz, 900 r/min, .8Pf leading, .93 efficient, star-connected synchronous motor has a synchronous impedance of .074 + j.48 ohms per phase. When the motor draws rated current, determine:

a) The excitation voltage.

b) The torque angle.

SOLUTION

a) The motor's rated armature current is

$$I_a = \frac{P_{\text{out}}}{\eta\sqrt{3}\,V_{L-L}\cos\theta}$$

$$= \frac{50{,}000}{.93\sqrt{3}(480)(.8)} = 80.83\,\underline{/36.9^\circ}\ \text{A}$$

Applying KVL to the equivalent circuit of the motor (Fig. 5-11(a)), we obtain

$$V_f\,\underline{/-\delta} = V_t/0^\circ - I_a(R_a + jX_s)$$

$$= \frac{480}{\sqrt{3}} - 80.83\,\underline{/36.9^\circ}(.074 + j.48)$$

$$= 295.62 - j34.63$$

$$= 297.64\,\underline{/-6.7^\circ}\text{V/phase}$$

and

$$V_f = 515.54\ \text{V, L–L}$$

b) The torque angle is

$$\underline{\underline{\delta = -6.7^\circ}}$$

The phasor diagram is shown in Fig. 5-11(b).

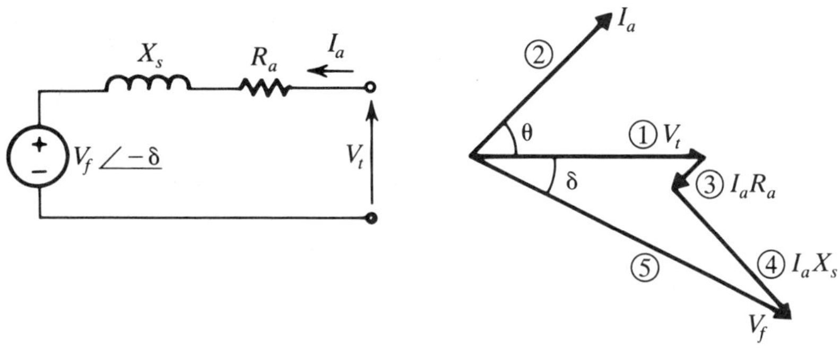

(a) (b)

FIG. 5-11

EXAMPLE 5–2 A 480 V, 3-ϕ, 60 Hz, 150 kW, .94 efficient synchronous motor has a per-phase synchronous impedance of .05 + j.75 ohms. The leakage reactance is .25 ohm per phase. The OCC in the operating region is given by

$$V_f = 10 + 20I_f$$

When the motor draws rated current at unity power factor, determine:

a) The approximate and exact values of the torque angle.
b) The magnetizing component of the field current.
c) The field current.
d) The armature reaction in equivalent field amperes.
e) The effective turns ratio.

SOLUTION

a) The rated current is

$$I_a = \frac{P_{\text{in}}}{\sqrt{3}\, V_{L-L} \cos\theta} = \frac{150{,}000}{.94}\left(\frac{1}{\sqrt{3}(480)(1)}\right) = 191.94\,\underline{/0^\circ}\ \text{A}$$

The magnetizing voltage is found by applying KVL to the equivalent circuit of Fig. 5-12:

$$\begin{aligned} V_m &= V_t - I_a(R_a + jX_a) \\ &= \frac{480}{\sqrt{3}}\,\underline{/0^\circ} - 191.94\,\underline{/0^\circ}(.05 + j.25) \\ &= 271.8\,\underline{/-10.2^\circ}\ \text{V/phase} \end{aligned}$$

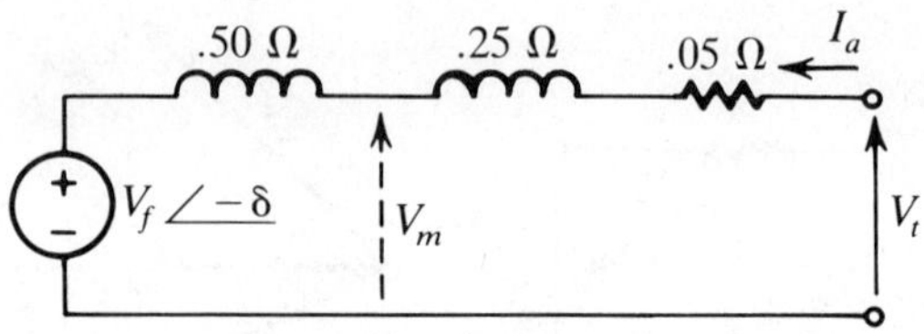

FIG. 5-12

Similarly, the excitation voltage is

$$V_f = \frac{480}{\sqrt{3}} \underline{/0^\circ} - 191.94\underline{/0^\circ}(.05 + j.75)$$
$$= 303.8\underline{/-28.3^\circ} \text{ V/phase}$$

Thus, the approximate and exact values of the torque angle in electrical degrees are

$$\delta_{\text{appr}} = \underline{\underline{-28.3^\circ}}$$
$$\delta_{\text{ex}} = -28.3^\circ - (-10.2^\circ) = \underline{\underline{-18.1^\circ}}$$

b) The magnitude of the magnetizing component of the field current can be found from the given open-circuit characteristic:

$$I_m = \frac{V_m - 10}{20} = \frac{271.8\sqrt{3} - 10}{20} = 23.04 \text{ A}$$

This current lags its corresponding voltage by 90°. Thus,

$$I_m = 23.04\underline{/-90^\circ - 10.2^\circ}$$
$$= \underline{\underline{23.04\underline{/-100.2^\circ} \text{ A}}}$$

c) Similarly, the magnitude of the field current is

$$I_f = \frac{303.8\sqrt{3} - 10}{20} = 25.81 \text{ A}$$

In phasor form:

$$I_f = 25.81\underline{/-90^\circ - 28.3^\circ}$$
$$= \underline{\underline{25.81\underline{/-118.3^\circ} \text{ A}}}$$

d) The armature reaction is found by using Eq. (5.11):

$$I'_a = I_m - I_f = 23.04\underline{/-100.2^\circ} - 25.81\underline{/-118.3^\circ}$$
$$= \underline{\underline{8.16\ \text{A}}}$$

e) The effective turns ratio is obtained from Eq. (5.9):

$$N_e = \frac{I_a}{I'_a} = \frac{191.94}{8.16} = \underline{\underline{23.51}}$$

Exercise 5–1 Prove the following:

a) The torque angle is equal to the phase angle between the field current and its magnetizing component.

b) The approximate value of the torque angle of a synchronous motor that operates at a leading power factor of θ degrees and has negligible armature resistance is given by

$$\delta = \arctan \frac{I_a X_s \cos \theta}{V_t + I_a X_s \sin \theta}$$

c) The per-unit copper losses in a motor are equal to the motor's per-phase armature resistance expressed in per-unit.

Exercise 5–2 A 375 kW, 2200 V, 60 Hz, .80 Pf lagging, .966 efficient, 900 r/min, star-connected motor has a synchronous impedance of $.015 + j.702$ pu. When the motor draws rated current at nominal power factor, determine:

a) The excitation voltage.

b) The approximate torque angle.

Answer (a) 1741.5 V, L–L; (b) -44.3°.

Exercise 5–3 A synchronous motor has negligible armature impedance. Under rated operating conditions its power factor is unity and its field current is 30 A. On open-circuit test, 26 A are required to give rated terminal voltage. Determine:

a) The armature reaction in equivalent field amperes.

b) The torque angle.

Answer (a) 14.97 A; (b) -29.9°.

5.1.8 POWER AND TORQUE DEVELOPED

In this section, general expressions will be derived for the complex, real, and reactive powers drawn by a synchronous motor that is connected to an infinite bus (see Fig. 5-13(a)). The system's per-phase equivalent circuit is shown in Fig. 5-13(b). The impedance (Z) is equal to the synchronous impedance of the motor plus the impedance of the transmission line between the motor and the infinite bus.

The complex power drawn from the infinite bus is

$$S = P + jQ = V_t I^* \tag{5.16}$$

where I^* is the conjugate of the current phasor and V_t is now the infinite-bus voltage. Using Ohm's law,

$$I^* = \left(\frac{V_t\angle 0^\circ - V_f\angle -\delta}{Z}\right)^* \tag{5.17}$$

By definition,

$$Z = |Z|\angle\theta \tag{5.18}$$

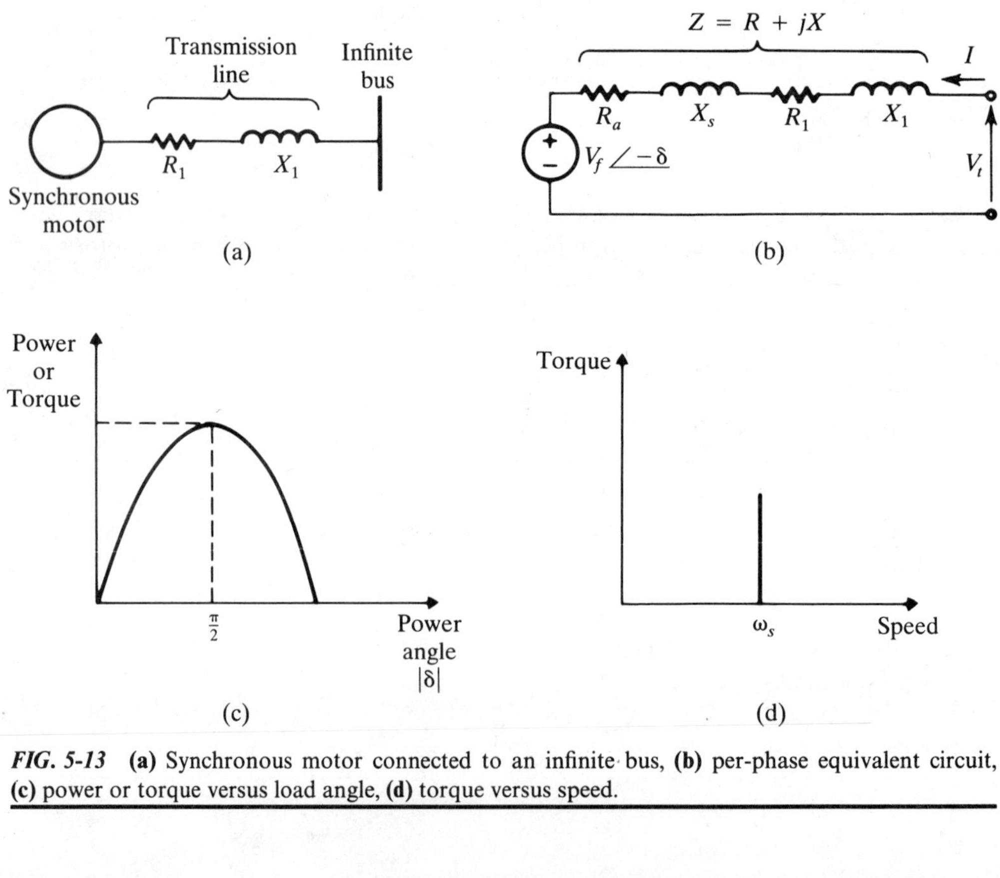

FIG. 5-13 **(a)** Synchronous motor connected to an infinite bus, **(b)** per-phase equivalent circuit, **(c)** power or torque versus load angle, **(d)** torque versus speed.

From Eqs. (5.16), (5.17) and (5.18) we obtain

$$S = \frac{V_t}{|Z|}(V_t\underline{/-\theta} - V_f\underline{/-\theta-\delta})^* \tag{5.19}$$

$$= \frac{V_t}{|Z|}[V_t(\cos\theta - j\sin\theta) - V_f(\cos(-\theta-\delta) + j\sin(-\theta-\delta))]^* \tag{5.20}$$

$$= \frac{V_t}{|Z|}[V_t\cos\theta - V_f\cos(-\theta-\delta) - j(V_t\sin\theta + V_f\sin(-\theta-\delta))]^* \tag{5.21}$$

Taking the conjugate of the expression within the brackets, we have:

$$S = \frac{V_t}{|Z|}[V_t\cos\theta - V_f\cos(-\theta-\delta) + j(V_t\sin\theta + V_f\sin(-\theta-\delta))] \tag{5.22}$$

The active power drawn from the infinite bus is given by the real component of the complex power. Thus,

$$P = \frac{V_t}{|Z|}[V_t\cos\theta - V_f\cos(-\theta-\delta)] \tag{5.23}$$

$$= \frac{V_t}{|Z|}(V_t\cos\theta - V_f(\cos\theta\cos\delta - \sin\theta\sin\delta) \tag{5.24}$$

Also, by definition,

$$\cos\theta = \frac{R}{|Z|} \quad \text{and} \quad \sin\theta = \frac{X}{|Z|}$$

Thus,

$$P = \frac{V_t}{|Z|^2}(V_tR - RV_f\cos\delta + XV_f\sin\delta) \text{ watts/phase} \tag{5.25}$$

The torque angle at which the motor draws maximum power is found by applying the calculus theory of minimum and maximum:

$$\frac{\partial P}{\partial\delta} = \frac{V_tV_f}{|Z|^2}(R\sin\delta + X\cos\delta) = 0 \tag{5.26}$$

From the above, the torque angle (δ_{m_1}) at which the motor develops maximum power is

$$\delta_{m_1} = -\arctan\frac{X}{R} \tag{5.27}$$

From Eq. (5.25), by neglecting the resistive component of the impedance and considering only magnitudes, we obtain

$$P = \frac{V_t V_f}{X} \sin \delta \text{ watts/phase} \tag{5.28}$$

The power input to the motor will be equal to the power developed. The latter provides the output power and the rotational losses.

By definition, the torque is

$$T = \frac{\text{power}}{\text{speed}} = \frac{V_t V_f}{\omega_s X} \sin \delta \ \text{N} \cdot \text{m/phase} \tag{5.29}$$

For constant excitation, the power is given by

$$P = K \sin \delta \tag{5.30}$$

and the torque by

$$T = K_1 \sin \delta \tag{5.31}$$

The constants K and K_1 can be obtained from Eqs. (5.28) and (5.29), respectively.

As shown in Fig. 5-13(c), the maximum torque of the motor occurs at a power angle of $\pi/2$ radians. Maximum torque is often referred to as breakdown torque or pull-out torque. Since the speed of the motor is constant, the torque-versus-speed characteristic will be as shown in Fig. 5-13(d).

The imaginary part of the complex power of Eq. (5.22) gives the reactive power drawn from the infinite bus. Thus,

$$Q = \frac{V_t}{|Z|} [V_t \sin \theta + V_f \sin(-\theta - \delta)] \tag{5.32}$$

and

$$Q = \frac{V_t}{|Z|} [V_t \sin \theta - V_f(\cos \theta \sin \delta + \sin \theta \cos \delta)] \tag{5.33}$$

Replacing $\sin \theta$ and $\cos \theta$ as before by their equivalent expressions, we get

$$Q = \frac{V_t}{|Z|^2} (XV_t - V_f R \sin \delta - XV_f \cos \delta) \ \text{VAR/phase} \tag{5.34}$$

The maximum reactive power drawn from the infinite bus will occur at a torque angle of

$$\delta_{m_2} = \arctan \frac{\text{R}}{\text{X}} \tag{5.35}$$

By neglecting the resistive component of the impedance in Eq. (5.34), we obtain

$$Q = \frac{V_t}{X}(V_t - V_f \cos \delta) \tag{5.36}$$

When the quantity within the parentheses of Eq. (5.36) is positive, the reactive power is also positive. Thus, the machine operates at a lagging power factor, and the motor is said to be underexcited. Converesly, when the quantity within the parentheses is negative, the motor operates at a leading power factor, and the motor is said to be overexcited.

When the parameters of Eqs. (5.25), (5.28), (5.34), and (5.36) are expressed in per-unit, then the derived formulae give the corresponding total output of the machine in per-unit. When a synchronous motor is used only to improve the power factor of a network, its torque angle is approximately zero. Thus, the reactive power drawn by the motor is

$$Q = \frac{V_t(V_t - V_f)}{X_s} \text{ VAR/phase} \tag{5.37}$$

Equation (5.37) makes it clear that, by varying the dc current, the reactive power of a synchronous motor can be varied smoothly in either direction. In comparison, the reactive power of a capacitor can change only in steps (discontinuously) and only in one direction.

Figure 5-14(a) shows a synchronous motor used to improve the power factor of a load. The phasor diagram of the pertinent parameters is shown in Fig. 5-14(b). The

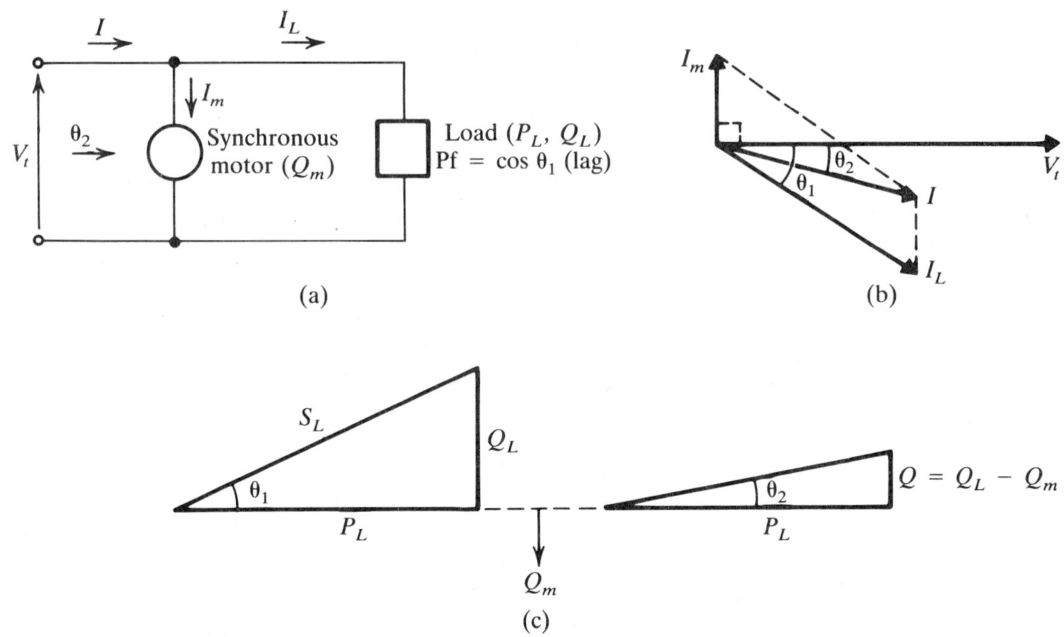

FIG. 5-14 Ideal synchronous motor and its effects on the overall power factor of a plant: **(a)** electrical network, **(b)** phasor diagram, **(c)** power triangles of the load, motor, and source.

power triangles of the load, motor, and source are shown in Fig. 5-14(c). Note that, in this case, a synchronous motor draws only reactive power.

Speed Control

The speed of a synchronous motor can be controlled either by changing the number of poles or by varying the frequency of the stator voltage. The frequency is varied through "frequency converters," and the number of poles is changed by reconnecting the stator and rotor windings into the desired pole configuration. The technique of frequency conversion is the same for both synchronous and three-phase induction machines, and so it need not be discussed again here. One exercise will be given, however, just to review the essentials.

5.1.9 EFFECTS OF FIELD CURRENT ON THE CHARACTERISTICS OF THE MOTOR

Previous sections discussed the effects of the field current on the motor's internal parameters—such as torque angle, excitation voltage, and the magnetization of the machine. This section will analyze the effects of the field current on the motor's external parameters. Specifically, qualitative proofs will be developed that give the power factor, the armature current, and the real and reactive powers as functions of the field current. The qualitative proofs are based on two assumptions: that the motor's output power remains constant, and that its stator resistance is negligible. These proofs will lead to the construction of the motor's external characteristics. One can easily predict these characteristics by first drawing a phasor diagram at unity power factor and observing the changes in the various parameters that accompany the increase or decrease of excitation voltage.

The curves of the power factor and of the armature current as a function of the field current resemble the letter V and are often referred to as the "Vee curves" of synchronous machines. From a practical point of view, these curves constitute one of the motor's most important characteristics.

Power Factor versus Field Current

The power-factor versus field-current characteristic of the motor is derived as follows:

1. The phasor diagram of a synchronous motor operating at *unity* power factor is shown in Fig. 5-15(a). The required field current for this operating condition corresponds to what is referred to as 100% excitation.
2. By increasing the field current above its normal value (Fig. 5-15(b)), the excitation voltage (V_{f_2}) will also be increased. From Eq. (5.28) it is clear that the torque angle δ_2 will be smaller than the torque angle δ_1. From the tip of the phasor V_t to the tip of the new excitation voltage V_{f_2} will be the phasor $I_{a_2}X_s$. This is so because of the following general voltage equation:

$$V_{f_2} = V_t - jI_{a_2}X_s \tag{5.38}$$

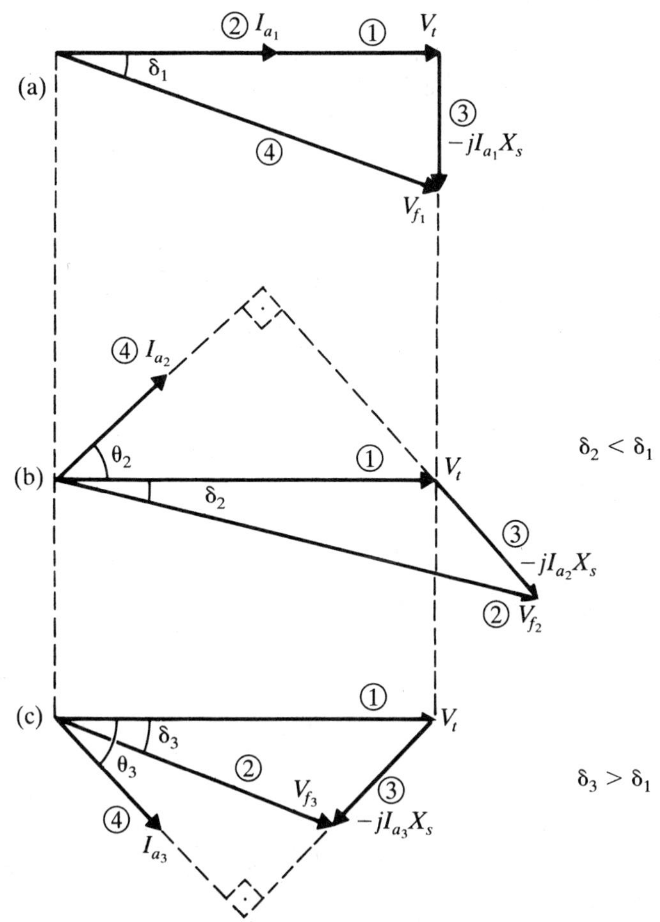

FIG. 15-15 Phasor diagrams for the qualitative derivation of Vee curves: **(a)** unity power factor (normal excitation), **(b)** leading power factor (overexcitation), **(c)** lagging power factor (underexcitation).

The extension of $I_{a_2}X_s$ (dotted line) must cross the line that represents the armature current at 90°. This is possible only when the armature current *leads* the terminal voltage, as shown in Fig. 5-15(b).

3. By decreasing the field current below its normal value (Fig. 5-15(c)), the excitation voltage (V_{f_3}) will also be decreased, and the new torque angle δ_3 will be larger than the torque angle δ_1. As before, from the tip of the phasor V_t to the tip of the new excitation voltage will be the phasor $I_{a_3}X_s$. The dotted extension of the phasor $I_{a_3}X_s$ must cross the line of the armature current at 90°. This is possible only when the armature current *lags* the terminal voltage, as shown in Fig. 5-15(c).

From the discussion, it should be clear that a change in the field current will be accompanied by a change in the power factor of the motor. Qualitatively speaking, the variation of the power factor as a function of the field current can be represented by a curve which resembles an inverted V, as shown in Fig. 5-16(a).

Magnitude of Armature Current versus Field Current

Neglecting armature resistance, the input power is equal to the power developed, which is assumed to remain constant. Mathematically,

$$V_t I_a \cos\theta = P_d \text{ watts/phase} \tag{5.39}$$

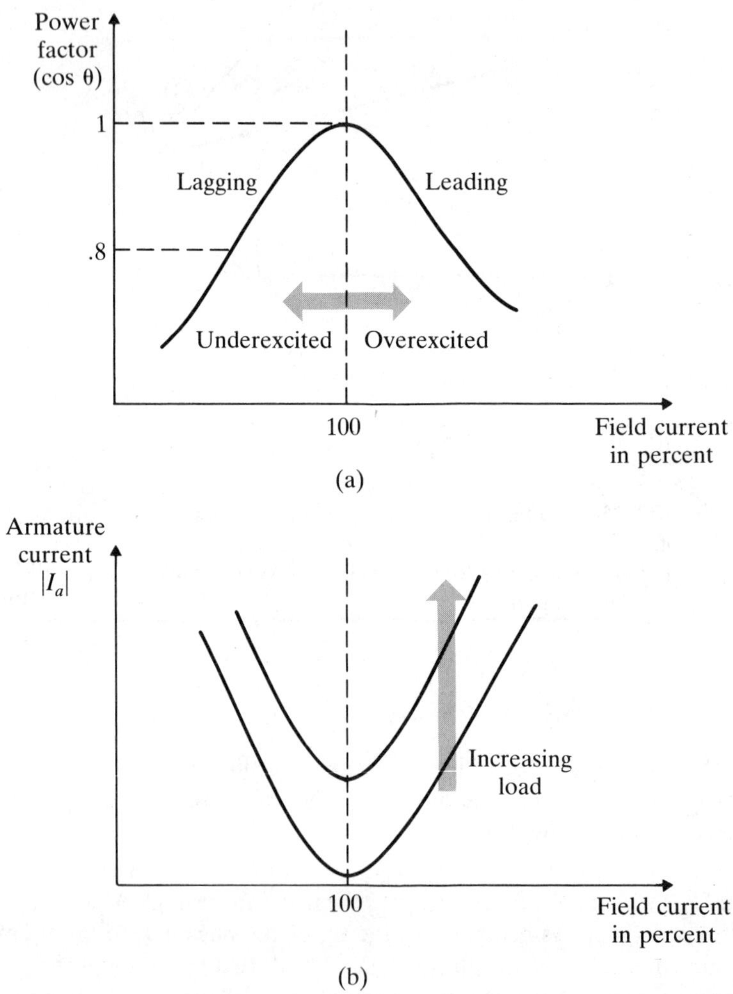

FIG. 5-16 Vee curves for a synchronous motor: **(a)** power factor versus field current, **(b)** magnitude of armature current versus field current.

from Eq. 5.39

$$I_a = \frac{K}{\cos\theta} \tag{5.40}$$

where the constant K is given by

$$K = \frac{P_d}{V_t} \tag{5.41}$$

Qualitatively speaking, the previously derived Vee curves of Fig. 5-16(a) indicate the following:

For variations of the field current up to 100% excitation,*

$$\cos\theta \propto I_f \tag{5.42}$$

For variations of the field current above 100% excitation,

$$\cos\theta \propto \frac{1}{I_f} \tag{5.43}$$

From Eqs. (5.40), (5.42), and (5.43) we obtain

$$\text{for excitation above 100\%:} \quad I_a \propto I_f \tag{5.44}$$

$$\text{for excitation below 100\%:} \quad I_a \propto \frac{1}{I_f} \tag{5.45}$$

The variation of the armature current as a function of field current may be represented by the curves shown in Fig. 5-16(b), whose shape resembles the letter V. It should be noted that armature current is minimum when the power factor is maximum. In other words, minimum armature current corresponds to 100% excitation or unity power factor.

Reactive Power versus Field Current

By definition, the reactive power drawn by a motor is

$$Q = V_t I_a \sin\theta \ \text{VAR/phase} \tag{5.46}$$

The Vee curves of Fig. 5-16(a) indicate that for excitation up to 100%:

$$\cos\theta \propto I_f$$

$$\therefore$$

* The proportionality symbols indicate the trend in the change of the variables but not true direct proportion or inverse proportion.

An increase in the cosine function will be accompanied by a decrease in the sine function. Thus,

$$\sin\theta \propto \frac{1}{I_f} \tag{5.47}$$

From Eqs. (5.45), (5.46), and (5.47) we get

$$Q \propto \frac{1}{I_f^2} \tag{5.48}$$

Similarly, for excitation above 100%, from Eqs. (5.43), (5.44), and (5.46) we obtain

$$Q \propto I_f^2 \tag{5.49}$$

From Eqs. (5.48) and (5.49) the reactive power versus the motor's field current can be drawn as shown in Fig. 5-17. Leading power factor corresponds to positive VAR and lagging power factor to negative VAR. At 100% field excitation the power factor is equal to unity and the reactive power is equal to zero.

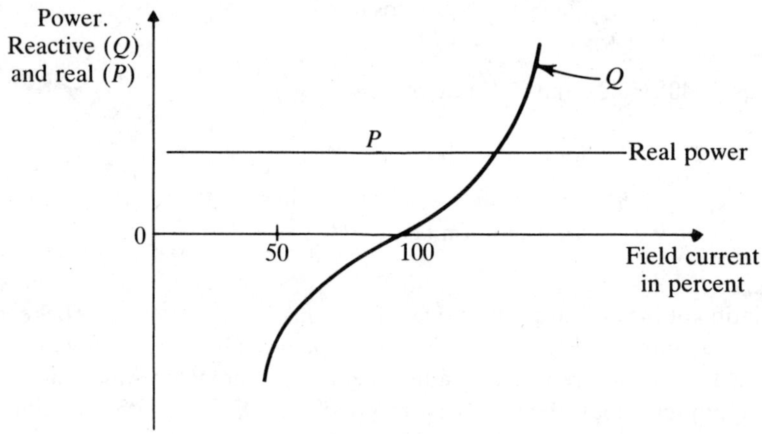

FIG. 5-17 A synchronous motor's real and reactive powers as a function of field current.

Real Power

The real power drawn by a synchronous motor is independent of the field current because the real component of the armature current ($I_a \cos\theta$) does not depend on the excitation. The variation of the real power as a function of field current is shown in Fig. 5-17. In an actual machine, however, the stator ($I_a^2 R_a$) losses, and hence the input power, will increase whenever the stator current increases as a result of changing the field current.

It should be emphasized that the curves derived in this section are approximate and are to be used for qualitative analysis only. Exact relationships can be derived from the magnetization characteristic of the particular motor and by the use of the appropriate equations discussed in this chapter.

EXAMPLE 5–3 Explain:

a) The fact that the torque angle of a synchronous motor always lags the terminal voltage.

b) The fact that a sudden increase in the torque requirements of a synchronous motor will momentarily decrease the synchronous speed of the rotor, and thus the torque angle will be enlarged.

c) The magnetizing and demagnetizing effects of armature current. Use phasor diagrams to illustrate the concepts.

SOLUTION

a) The output power of a synchronous motor is produced by the transformation of electrical input power. In other words, before the motor starts to rotate and produce an output, the electrical input (V_t) must first be applied; the rotation of the shaft will then follow (V_f). Since the torque angle is approximately equal to the phase angle between the terminal voltage and the excitation voltage, the torque angle is negative, or lags the applied voltage V_t. This is shown in Fig. 5-18. The excitation voltage is produced by the flux of the rotor, which rotates at the same angular speed as the shaft.

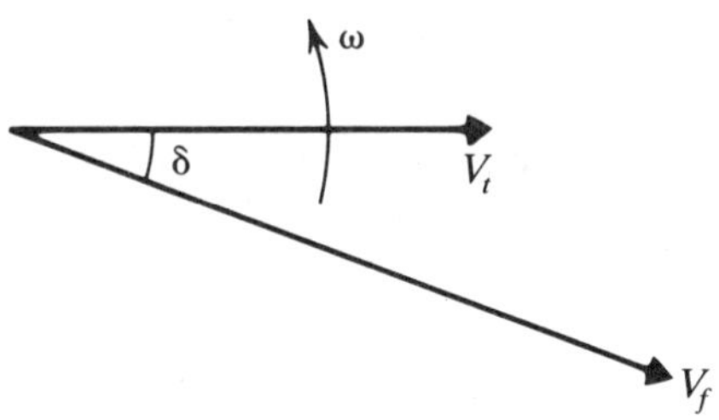

FIG. 5-18

b) As shown by the following equation, a sudden increase in the load torque requirements on a synchronous motor will be accompanied by an increase in the torque angle:

$$P = V_f \frac{V_t}{X_s} \sin \delta$$

Since the torque angle is always lagging, its enlargement will force the shaft to momentarily rotate at a lower speed relative to the stator field. Alternatively, owing to physical considerations, the loading of the shaft of the motor will momentarily reduce the rotor's angular speed and thus, as depicted in Fig. 5-18, the torque angle will become larger. This is a transient phenomenon. Once the power equilibrium is reestablished, the rotor will attain synchronous speed and the torque angle will settle at a magnitude that will satisfy the new power requirement.

c) Economic considerations dictate that, under normal operating conditions, the motor should operate at a point close to the knee of the magnetization curve. To achieve this level of magnetization a certain amount of field current is required. Since the field current has two components, its magnitude will depend on the relative magnitudes and positions of its components. Rewriting the equation for the field current, we have

$$I_f = I_m - I'_a$$

When the motor operates at a lagging power factor, as shown in Fig. 5-10(c), less field current is required than when it operates at a leading power factor, as shown in Fig. 5-10(b). At a lagging power factor, the flux of the armature current aids the magnetizing effect of the field current. This is referred to as the *magnetizing effects of the armature current.* At a leading power factor, on the other hand, the armature current produces flux that opposes the magnetization of the machine. This is referred to as the *demagnetizing effects of the armature current.*

EXAMPLE 5–4

1. Define power factor and give five advantages that result from its improvement. Why are synchronous condensers preferred over static capacitors?

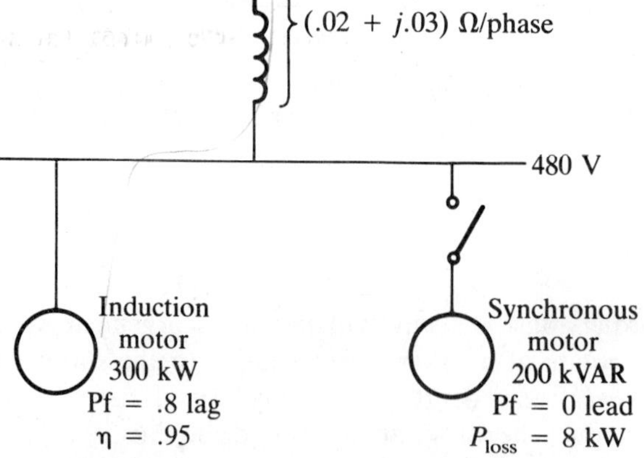

FIG. 5-19(a)

2. For the network shown in Fig. 5-19(a), calculate with and without the synchronous motor in the circuit:
   a) The losses in the feeder.
   b) The kVA delivered by the transformer.
   c) The losses in the transformer.
   d) The regulation.

SOLUTION

1. The power factor of a given circuit is equal to the cosine of the phase angle between the circuit voltage and current. The voltage is always taken as reference.

   The ordinary consumer of electrical energy is not interested in the phase angle between the voltage and the current applied to a motor. He is concerned not with power factor, but with the actual kW output of his motor. However, for constant voltage and output power, the current through the incoming feeders depends on the power factor, as shown by the following equation:

$$I = \frac{P_{\text{in}}}{\sqrt{3}\, V_{L-L} \cos\theta}$$

   For example, a reduction in the power factor from unity to .80 will result in a 25% increase in the current. Conversely, an increase (improvement) in the power factor will result in a reduction in the current, which will reduce the following:
   a) The losses in the upstream feeders.
   b) The required size of the upstream feeders.
   c) The required size of the transformer.
   d) The losses in the transformer.
   e) The voltage fluctuation across the motors and other electrical loads.

   Synchronous motors can provide output torque and leading or lagging kVAR to the network. Their operation is very smooth and their losses nominal. Capacitors can provide only kVAR lagging in steps, but they consume negligible power. For simplicity, it is assumed here that the motor does not provide any torque.

   Furthermore, the effects of harmonics and switching problems are more severe for capacitors than for synchronous motors. Nevertheless, it should be noted that a synchronous machine has maintenance requirements for its bearings, ventilation, etc., that static capacitors do not have.

2. *Without the synchronous motor in the circuit:*
   a) The line or feeder current is

$$I = \frac{300{,}000}{.95\sqrt{3}(.8)480} = \underline{474.79\,\angle -36.9^\circ}\ \text{A}$$

   Thus, the losses in the feeder (P_{lf}) are

$$P_{lf} = 3(.02)(474.79)^2 = \underline{\underline{13.53\ \text{kW}}}$$

b) The transformer should supply the power required by the load plus the power consumed by the feeder. From the circuit of Fig. 5-19(b), the voltage at the

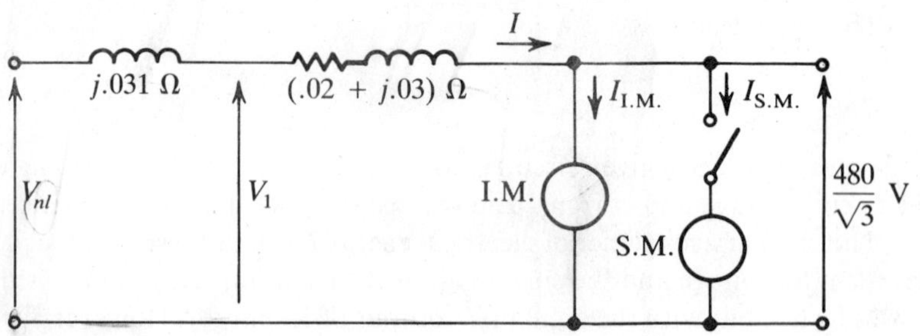

FIG. 5-19(b)

output terminals of the transformer is

$$V_1 = V_L + IZ = \frac{480}{\sqrt{3}}\underline{/0^\circ} + 474.79\underline{/-36.9^\circ}(.02 + j.03)$$
$$= 293.32\underline{/-1.1^\circ} \text{ V/phase, and } V_1 = 508.1 \text{ V L–L}$$

Thus, the apparent power delivered by the transformer is

$$S = \sqrt{3}\, V_{L-L} I_L$$
$$= (\sqrt{3}(508.1)(474.79) = \underline{\underline{417.8 \text{ kVA}}}$$

c) Assuming that the core loss is proportional to the square of the terminal voltage, the transformer's losses are

$$P_{l_{(t)}} = 3\left(\frac{508.1}{480}\right)^2 + 5\left(\frac{474.79}{601.4}\right)^2 = \underline{\underline{6.48 \text{ kW}}}$$

d) The no-load voltage at the secondary of the transformer is

$$V_{nl} = \frac{480}{\sqrt{3}}\underline{/0^\circ} + 474.79\underline{/-36.9^\circ}(.02 + j.03 + j.031)$$
$$= 302.61\underline{/3.3^\circ} \text{ V/phase}$$

and

$$V = \sqrt{3}(302.61) = 524.13 \text{ V, L–L}$$

$$R\% = \frac{524.13 - 480}{480}(100) = \underline{\underline{9.19\%}}$$

3. *With the synchronous motor in the circuit:*
   a) The current drawn by the synchronous condenser is found as follows:

$$Q = \sqrt{3}\, V_{L-L} I_L \sin\theta = 200 \text{ kVAR}$$
$$P = \sqrt{3}\, V_{L-L} I_L \cos\theta = 8 \text{ kW}$$

From the above, the current drawn by the synchronous motor is

$$I_{\text{S.M.}} = 240.76\,\underline{/87.7^\circ} \text{ A}$$

The current in the feeder is the sum of the currents drawn by the induction and synchronous motors. Thus,

$$I_f = 474.79\,\underline{/-36.9^\circ} + 240.76\,\underline{/87.7^\circ}$$
$$= 391.97\,\underline{/-6.5^\circ} \text{ A}$$

The new power factor is

$$\cos 6.5^\circ = .99 \text{ lag},$$

The losses in the feeder are

$$P_{lf} = 3(.02)(391.97)^2 = \underline{\underline{9.22}} \text{ kW}$$

   b) The voltage at the output of the transformer is

$$V_1 = \frac{480}{\sqrt{3}}\,\underline{/0^\circ} + 391.97\,\underline{/-6.5^\circ}(.02 + j.03)$$
$$= 286.15\,\underline{/-2.2^\circ} \text{ V/phase}$$

Then,

$$V_1 = \sqrt{3}(286.15) = 496.15 \text{ V, L-L}$$

The kVA delivered by the transformer is

$$S = \sqrt{3}(496.15)(391.97) = \underline{\underline{336.84 \text{ kVA}}}$$

   c) The transformer losses are

$$P_{l(t)} = 3\left(\frac{496.15}{480}\right)^2 + 5\left(\frac{391.97}{601.4}\right)^2 = \underline{\underline{5.33 \text{ kW}}}$$

   d) The no-load voltage at the substation is

$$V_{nl} = \frac{480\,\underline{/0^\circ}}{\sqrt{3}} + 391.97\,\underline{/-6.5^\circ}(.02 + j.03 + j.031)$$
$$= 288.53\,\underline{/4.55^\circ} \text{ V/phase}$$

Then

$$V_{nl} = \sqrt{3}(288.53) = 499.75 \text{ V, L-L}$$

The voltage regulation is

$$R\% = \frac{499.75 - 480}{480}(100) = \underline{\underline{4.1\%}}$$

For purposes of comparison, the results are summarized in Table 5-2.*

TABLE 5-2 Summary of results of Example 5–4

Synchronous Motor	*Power factor at the 480 V terminals*	*kVA delivered by the transformer*	*Losses in the transformer in kW*	*Losses in the feeder in kW*	*Current through the feeder in A*	*Voltage regulation in percent*
Disconnected	.8 lagging	417.8	6.48	13.53	474.79	9.19
Connected	.99 lagging	336.84	5.33	9.22	391.97	4.1

The advantages of higher-power-factor operation become very evident when one compares the tabulated results. The adverse effects of low-power-factor operation of large industrial customers on the upstream equipment cannot be tolerated.

. . .

* The student may evaluate some of the financial advantages of the high-power-factor operation by assuming the following:

- Cost of energy: \$1000/kW/year. Cost of transformers: \$50/kVA.
- Cost of a 3-conductor cable: \$.30/meter per ampere capacity.
- Length of cable: 200 meters.
- Monthly power-factor penalty:

$$\$10(\text{actual kW})\left(\frac{\text{Pf}_n}{\text{Pf}_a} - 1\right)$$

where

Pf_a = actual power factor
Pf_n = minimum permissible power factor, usually .90.

- The cost of a 480 V capacitor bank is about \$10/kVAR.

Exercise 5–4

a) A synchronous motor, when operated at .8 Pf leading, has an armature copper loss of 4 kW. *Estimate* the armature copper loss when the motor operates at unity power factor. Assume that the power drawn by the motor remains constant.

b) Derive Eq. (5.37).

Answer (a) 2.56 kW.

Exercise 5–5 The speed of a 480 V, 60 Hz, 10-pole, .94 efficient, unity-power-factor synchronous motor is controlled through a frequency converter. The synchronous reactance is 1.04 Ω/phase and the motor drives a 50 kW constant-power load. When the output voltage of the converter is 432 V, determine:

a) The speed of the motor.

b) The excitation voltage and the torque angle.

Answer (a) 648 r/min; (b) 256.5 V/phase, −8.6°.

5.2 Three-Phase Cylindrical Rotor Machines: Generators

5.2.1 EQUIVALENT-CIRCUIT AND PHASOR DIAGRAMS

This section deals with the principle of operation, governing equations, characteristics, and the measurement of the parameters of three-phase generators. As mentioned earlier, three-phase generators transform mechanical power into electrical power. The mechanical power of the prime mover rotates the shaft of the generator on which the dc field windings are installed. The speed of the prime mover is maintained at a constant level through electronically controlled speed regulators, commonly known as governors. The rotation of the dc flux cuts the windings of the armature and, because of the induction principle, a three-phase voltage is generated.

Three-phase generators are also called three-phase alternators. Structurally, they are similar to three-phase synchronous motors.

The direction of energy flow is shown in Fig. 5-20(a). The approximate per-phase equivalent circuit as seen from the stator and rotor terminals is shown in Figs. 5-20(b) and (c), respectively. The physical significance and nomenclature of each parameter is identical to that of three-phase synchronous motors.

The governing equations for alternators are

$$V_f \angle\delta^\circ = V_t \angle 0^\circ + I_a(R_a + jX_s) \tag{5.50}$$

$$I_f = I_m + I'_a \tag{5.51}$$

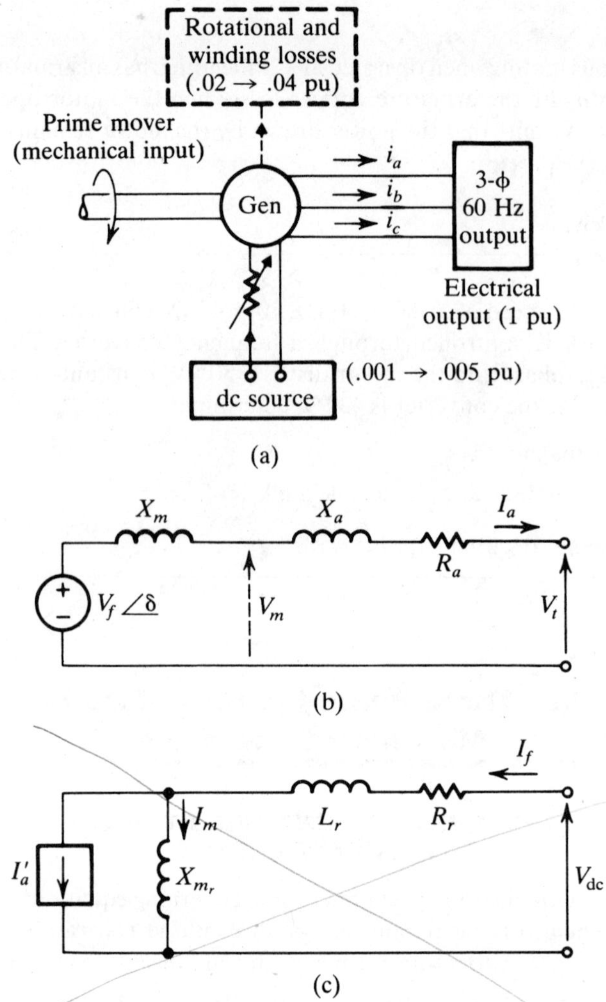

FIG. 5-20 Alternators: **(a)** direction of energy flow and typical losses for a 400 kVA alternator, **(b)** approximate equivalent circuit as seen from the stator, **(c)** partial equivalent circuit as seen from the rotor.

Note that these equations have a positive sign, whereas those for synchronous motors have a negative sign.

The torque angle of an alternator, owing to physical considerations, always leads the terminal voltage.

The Vee curves for alternators are similar to those for synchronous motors, except that the words "leading" and "lagging" should be interchanged. For example, an *underexcited* generator (leading Pf) will correspond to an *overexcited* (leading Pf) motor.

The power factor of a single alternator does not depend on its field current but on the impedance of the connected load. However, if an alternator operates in parallel with other generators, their individual power factors can be controlled by properly adjusting their corresponding field currents.

The phasor diagrams for unity and lagging power factor are drawn in Figs. 5-21(a) and (b), respectively. The order of construction is identified by the encircled numbers. The phasor diagrams of generators are easier to draw than those of motors because of the positive sign in the governing equations, Eqs. (5.50) and (5.51).

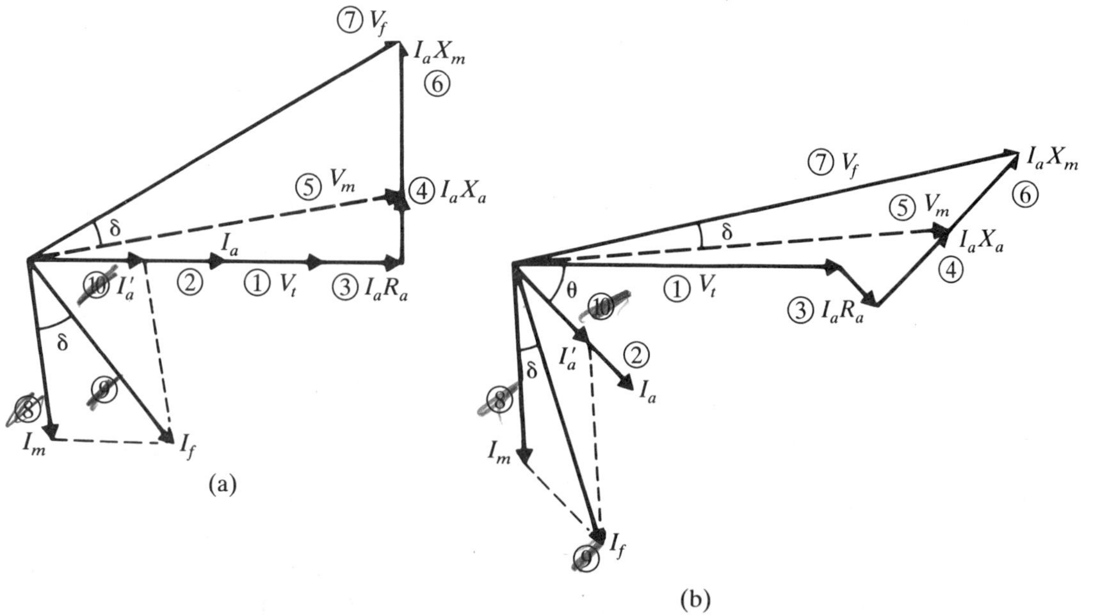

FIG. 5-21 Phasor diagrams for alternators: **(a)** unity power factor, **(b)** lagging power factor.

Notice that at lagging-power-factor operation a large field current is required in order to magnetize the stator and thus to enable the generator to provide rated terminal voltage. In other words, lagging-power-factor currents have a demagnetizing effect on the stator. For this reason normally a generator *can not* start when its load is an ac motor of about equal rating. At starting, an ac machine draws a highly inductive current ($\theta \approx 70°$) of a magnitude five to six times larger than rated value.

To select, thus, a standby generator, one has to carefully evaluate the nature of the connected load. It is normally the practice to consult with the manufacturers regarding the required size and *starting capability* of the unit under consideration. A general rule of thumb is that a generator can start an ac machine provided that the starting kVA of the motor is not more than twice the nominal kVA of the generator. When a properly sized capacitor bank is used, its leading-power-factor currents reduce the magnitude of the downstream inductive currents; thus the generator's starting capability is accordingly increased.

5.2.2 REGULATION OF ALTERNATORS

For practical application it is important to be able to determine the regulation of a three-phase synchronous generator. Mathematically, the regulation (R)—as explained in the chapter on transformers—is defined by

$$R = \left. \frac{|V_{nl}| - |V_{fl}|}{|V_{fl}|} \right|_{I_f = \text{constant}} \tag{5.52}$$

or

$$R = \left. \frac{|V_f| - |V_{fl}|}{|V_{fl}|} \right|_{I_f = \text{constant}} \tag{5.53}$$

where

V_{nl} = the voltage across the armature terminals at no-load. At the absence of saturation the no-load voltage is equal to the excitation or internal voltage (V_f). Under this assumption, this voltage can be calculated from Eq. (5.50) or can be obtained from the OCC, once the actual field current is known.

V_{fl} = the voltage across the armature terminals at full-load conditions. This is the rated voltage or the voltage specified on the name-plate of the alternator.

Owing to the effects of armature reaction, the regulation of an alternator often may vary by plus or minus 30%. This means that the output voltage of a generator rated at 1000 V will fluctuate between 700 and 1300 V, while the excitation remains constant and the armature current varies from its no-load to its full-load value.

The power factor of the load has a substantial effect on the generator's terminal voltage. Nevertheless, proper adjustments of the field current will result in constant output voltage or zero regulation, as shown in Fig. 5-22(a).

Figure 5-22(b) shows that, for a particular value of armature current, capacitive loads increase the terminal voltage while inductive loads decrease it. The reason is that lagging-power-factor armature currents have a demagnetizing effect on the machine, while leading-power-factor currents have a magnetizing effect.

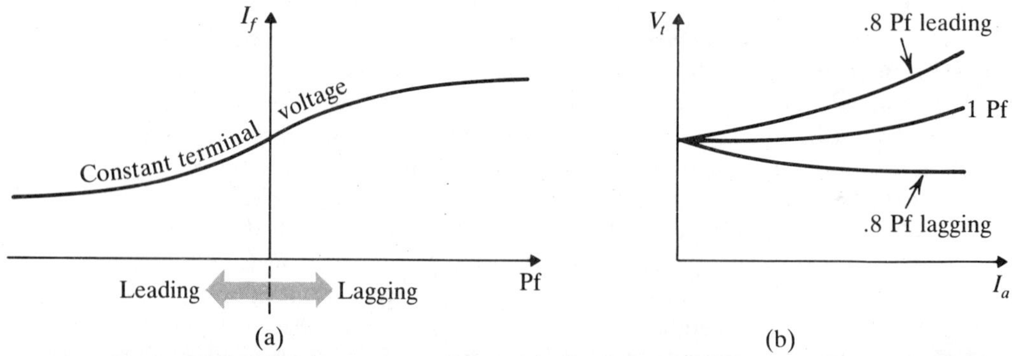

FIG. 5-22 Characteristics of alternators: **(a)** field current versus power factor for constant terminal voltage, **(b)** terminal voltage as a function of armature current for fixed excitation and variable power factor.

In practice, automatic regulators are used to adjust the field current in such a way as to maintain constant output voltage (zero regulation), independently of load condition.

The regulation of large three-phase alternators cannot be measured directly in the laboratory because of the difficulty in finding large enough loads to satisfy the required full-load condition. Theoretically, however, a generator's voltage regulation can be determined by the governing equations and from the data of its characteristics. These characteristics are discussed in the sections that follow.

5.2.3 CHARACTERISTICS OF ALTERNATORS

Open-Circuit Characteristic

The open-circuit characteristic (OCC) shows the relationship between the field current and the voltage across the open-circuited terminals of the armature windings. Laboratory data for this characteristic are obtained by driving the machine as a generator at rated synchronous speed (see Fig. 5-23(a)). When the armature and field

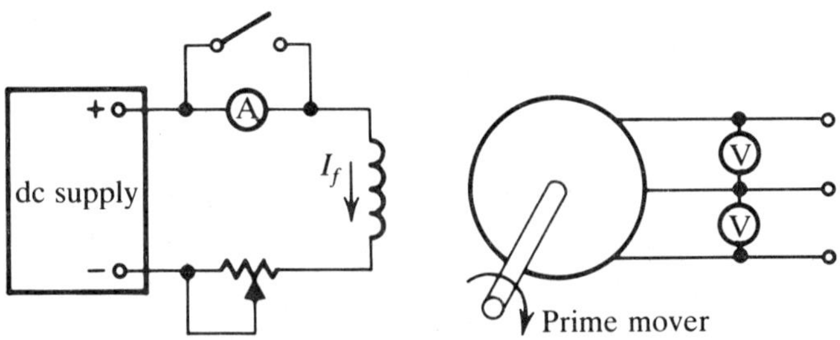

(a)

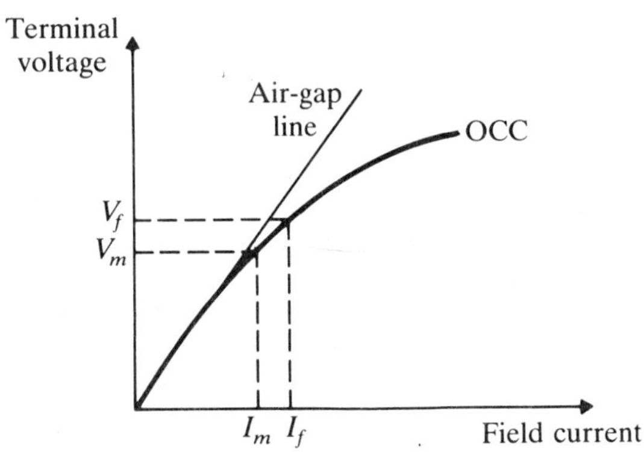

(b)

FIG. 5-23 Alternator: **(a)** schematic for the laboratory setup used to obtain data for the magnetization curve, **(b)** typical open-circuit characteristic.

currents are equal to zero, the power furnished by the prime mover constitutes the mechanical losses of the machine at no-load. A typical OCC is shown in Fig. 5-23(b).

The open-circuit characteristic can be used to obtain the magnetizing component (I_m) and/or the total field current (I_f) once the magnetizing voltage (V_m) and the excitation voltage (V_f) have been calculated from the machine's operating data and equivalent circuit. The open-circuit characteristic is also called the magnetization curve, or the no-load characteristic, or the saturation curve.

The line drawn tangentially to the linear portion of the open-circuit characteristic is called the air-gap line. This straight line is the magnetic characteristic of the machine when saturation is neglected.

Short-Circuit Characteristic

The short-circuit characteristic (SCC) shows the relationship between the field current and the armature current when the alternator terminals are shorted. The machine is driven as a generator, and the short-circuit armature current (I_a) is recorded while the field current (I_f) is gradually increased. The power delivered by the prime mover is equal to the rotational losses plus the stator copper losses.

A typical SCC, and the laboratory setup that may be used to obtain it, are shown in Figs. 5-24(a) and (b), respectively. Since the terminal voltage (V_t) is equal to zero, the

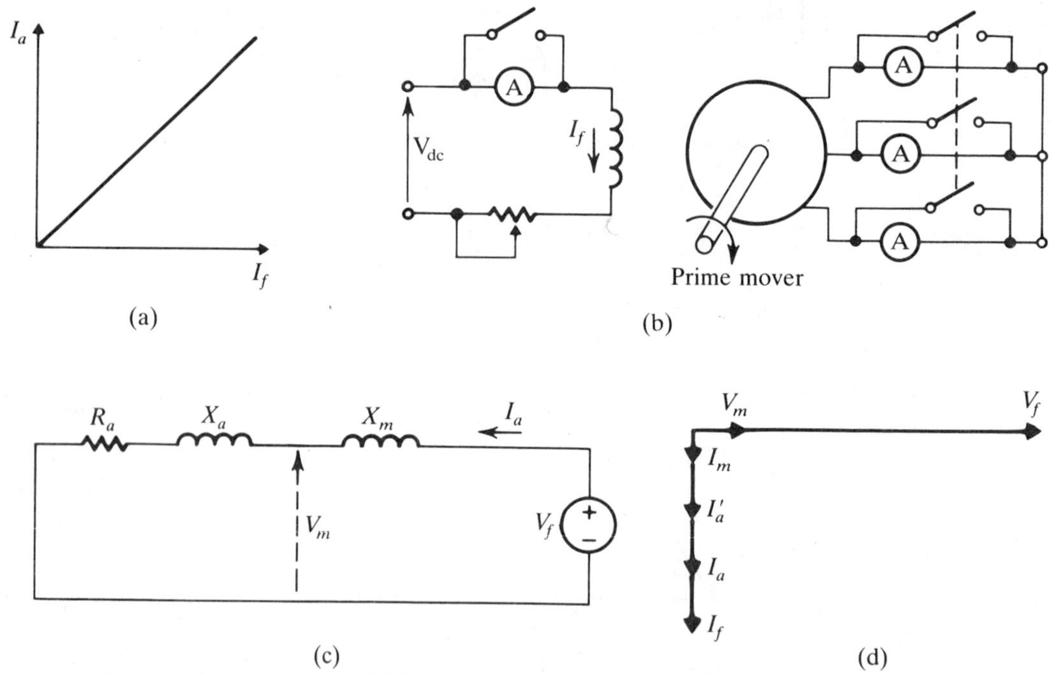

FIG. 5-24 **(a)** Typical short-circuit characteristic, **(b)** schematic for the laboratory setup that may be used to obtain data for the short-circuit characteristic, **(c)** equivalent circuit, **(d)** approximate phasor diagram ($R_a = 0$).

excitation voltage—as can be seen from the equivalent circuit of Fig. 5-24(c)—is given by

$$V_f = I_a(R_a + jX_s) \tag{5.54}$$

Usually, the armature resistance is negligible. Thus,

$$V_f \approx jI_aX_s \tag{5.55}$$

From the last relationship, it is evident that the armature current lags the excitation voltage by 90°. Rewriting (Eq. 5.51), we have

$$I_f = I_m + I'_a \tag{5.56}$$

I_m is very small because the magnetizing voltage is negligible. I_f also lags V_f by 90°. Therefore, the field current I_f and the armature reaction in equivalent field amperes (I'_a) are approximately in phase. This implies that a linear change in the armature current will be accompanied by a linear change in the field current. The phasor diagram for the short-circuit condition is shown in Fig. 5-24(d).

The short-circuit characteristic is also called the transformation-ratio curve, because its slope gives the approximate value of the effective turns ratio. Thus,

$$\text{slope of SCC} = \frac{I_a}{I_f} = \frac{I_a}{I_m + I'_a} \tag{5.57}$$

Since I_m is negligible,

$$\text{slope of SCC} \approx \frac{I_a}{I'_a} = N_e \tag{5.58}$$

(A more accurate value for this parameter can be obtained from the zero-power-factor characteristic that is discussed in the following section.)

By reducing the speed from its rated value, the excitation voltage (V_f) will be also reduced. From the generator principle (see Section 6.1)

$$V_f = K\phi\omega \tag{5.59}$$

where

K = a constant of the machine.
ϕ = the effective flux.
ω = the angular speed of rotation.

The impedance of the armature circuit, being largely inductive, will also be reduced accordingly:

$$Z_s \approx jX_s \tag{5.60}$$

and

$$Z_s \approx j\omega L_s \tag{5.61}$$

where L_s is the synchronous inductance of the machine. Thus, when the machine is short-circuited, the armature current is given by

$$I_a = \frac{V_f}{Z_s} \approx \frac{K\phi\omega}{j\omega L_s} = K_1 I_f \tag{5.62}$$

where K_1 is a constant of proportionality. This expression implies that, for a given field current and a moderate variation in the speed of the prime mover, the armature current is constant. A 25% change in the speed of the prime mover above or below its synchronous value does not appreciably affect the linearity of the short-circuit characteristic.

Zero-Power-Factor Lagging Characteristic

The zero-power-factor (ZPF) characteristic shows the relationship between the terminal voltage and the field current at constant rated armature current (I_a) and zero-load power factor. Lagging loads with a power factor up to .1 have negligible effect on the actual zero-power-factor curve.

A typical ZPF characteristic and a laboratory setup that may be used to obtain it are shown in Figs. 5-25(a) and (b), respectively. The load to the generator is a synchronous motor whose power factor and armature current are controlled through its field current. The terminal voltage is controlled by the generator's field current. The output power of the generator must be larger than that of the driven motor because of the latter's losses.

The readings of the instruments on the output terminals of the generator (voltmeters, ammeters, and wattmeters) are also used to calculate the power factor of the load. Instead of the two wattmeters (W_1 and W_2), one may use a polyphase wattmeter. The measurement of three-phase power with two wattmeters is explained in the Appendix (Section A.5).

5.2.4 MEASUREMENT OF PARAMETERS

Measurement of Synchronous Reactance X_s

The short-circuit characteristic, in conjunction with the magnetization curve, can be used to obtain the unsaturated ($X_{s_{un}}$) and saturated ($X_{s_{sat}}$) values of the synchronous reactance, as follows.

Unsaturated Value

Refer to Fig. 5-26. For any value of field current (I_{f_1}) there is a corresponding voltage on the air-gap line (V_{g_1}) and an armature current (I_{a_1}) on the short-circuit characteristic.

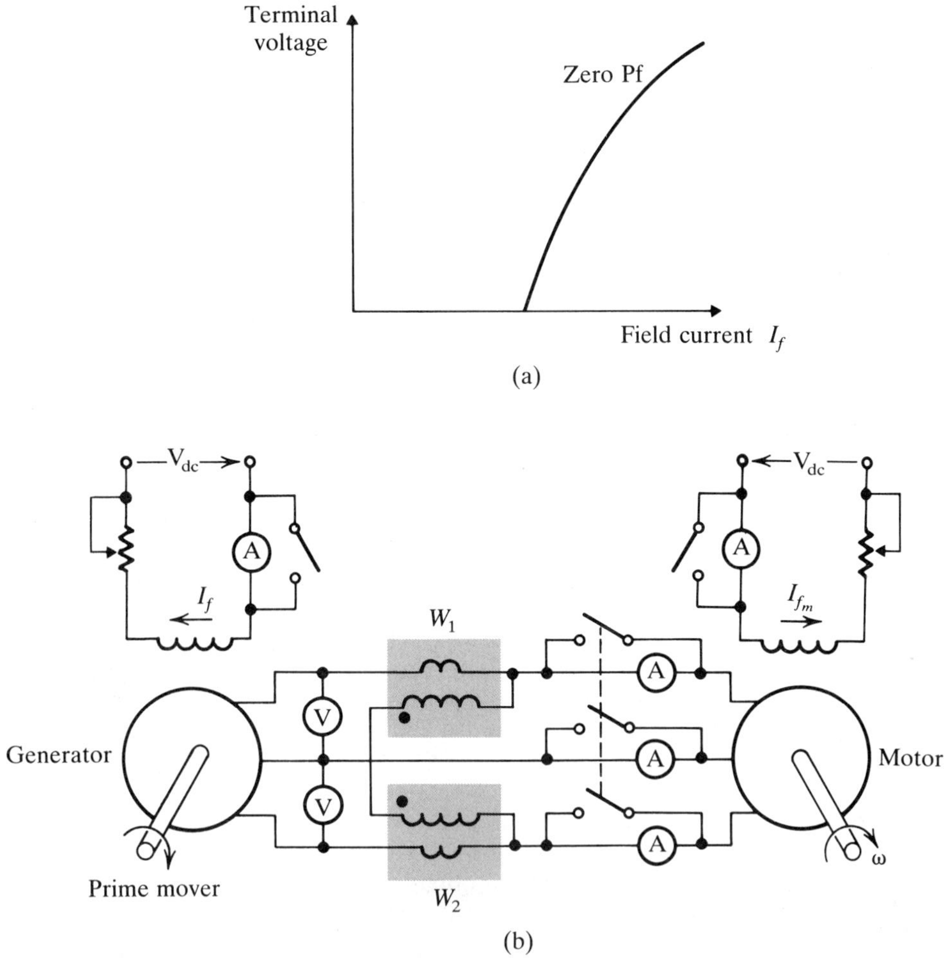

FIG. 5-25 **(a)** Typical lagging zero-power-factor characteristic, **(b)** schematic for a laboratory setup used to obtain data for the zero-power-factor curve.

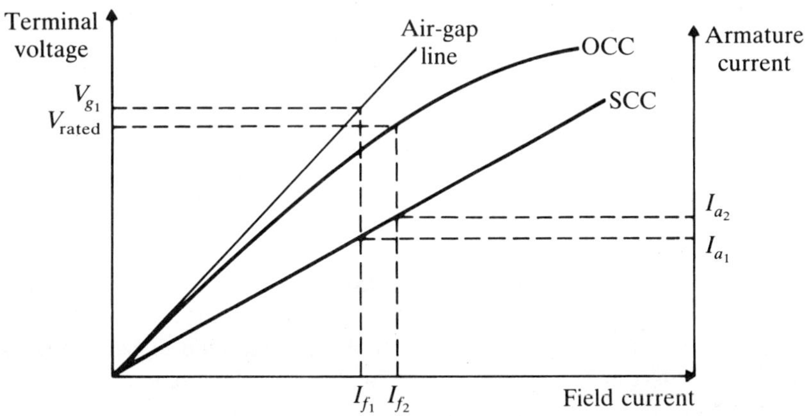

FIG. 5-26 Open- and short-circuit characteristics.

The unsaturated value of the synchronous reactance is given by

$$X_{S_{un}} = \left. \frac{\text{voltage obtained from the air-gap line}}{\text{armature current obtained from the SCC}} \right|_{\text{at any field excitation}}$$

or

$$X_{S_{un}} = \left. \frac{V_{g_1}}{I_{a_1}} \right|_{I_{f_1}} \tag{5.63}$$

Saturated Value

The ratio of the rated voltage (V_t) to the short-circuit current (I_{a_2}) at the same field conditions gives the approximate value of the saturated synchronous reactance ($X_{S_{sat}}$).

$$X_{S_{sat}} = \left. \frac{V_{\text{rated}}}{I_{a_2}} \right|_{I_{f_2}} \tag{5.64}$$

The per-unit value of the saturated synchronous reactance is the reciprocal of the short-circuit ratio (SCR). This parameter is defined as follows:

$$\text{SCR} = \frac{\text{field current obtained from OCC at rated terminal voltage}}{\text{field current obtained from SCC at rated armature current}} \tag{5.65}$$

The Potier Triangle: Measurement of X_a and I'_a

The zero-power-factor (ZPF) characteristic is used in conjunction with the open-circuit characteristic in order to obtain the Potier triangle. The Potier triangle (Fig. 5-27) is used to determine:

The stator leakage reactance (X_a).

The armature reaction (I'_a) in equivalent field amperes.

The following procedure is suggested in establishing the Potier triangle:

1. On the horizontal line at rated voltage draw AB:

$$AB = OM \tag{5.66}$$

2. From point A draw the line parallel to the air-gap line. This intersects the magnetization curve at point C. From this point draw the perpendicular to AB. This establishes point D.
3. The triangle BCD is called the Potier triangle.

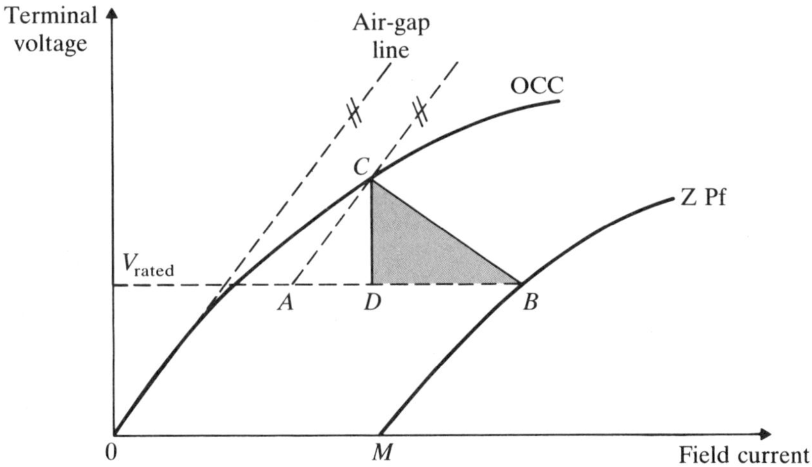

FIG. 5-27 Construction of the Potier triangle.

The base of this triangle gives the armature reaction in equivalent field amperes—that is, the I'_a. Thus,

$$BD = I'_a \tag{5.67}$$

The effective turns ratio (N_e) can be calculated from

$$N_e = \frac{I_a}{I'_a} \tag{5.68}$$

The other perpendicular side of the Potier triangle—that is, CD—gives the Potier reactance voltage drop. This is approximately equal to the voltage drop across the leakage reactance. Thus,

$$CD = I_a X_a \tag{5.69}$$

From this, the leakage reactance (X_a) can be calculated. Thus, the leakage reactance and the armature reaction are given, respectively, by the vertical and horizontal sides of the Potier triangle.

It should be noted that the origin of the ZPF characteristic corresponds to the value of field current required to circulate rated armature currents while the terminals of the machine are shorted. This particular point can also be obtained from the short-circuit characteristic.

Some Highlights of the Zero-Power-Factor Characteristic

Up to moderate saturation levels, the ZPF *lagging* characteristic can be obtained from the OCC by shifting the latter downward by an amount equal to $I_a X_a$, and horizontally to the right by an amount equal to I'_a, as shown in Fig. 5-28(a). Similarly, the ZPF *leading* curve can be obtained from the OCC by shifting it horizontally to the left by an amount equal to I'_a, and vertically upward by an amount equal to I_aX_a, as shown in Fig. 5-28(a). The leakage reactance (X_a) is assumed to remain constant. At high saturation levels this assumption is not valid.

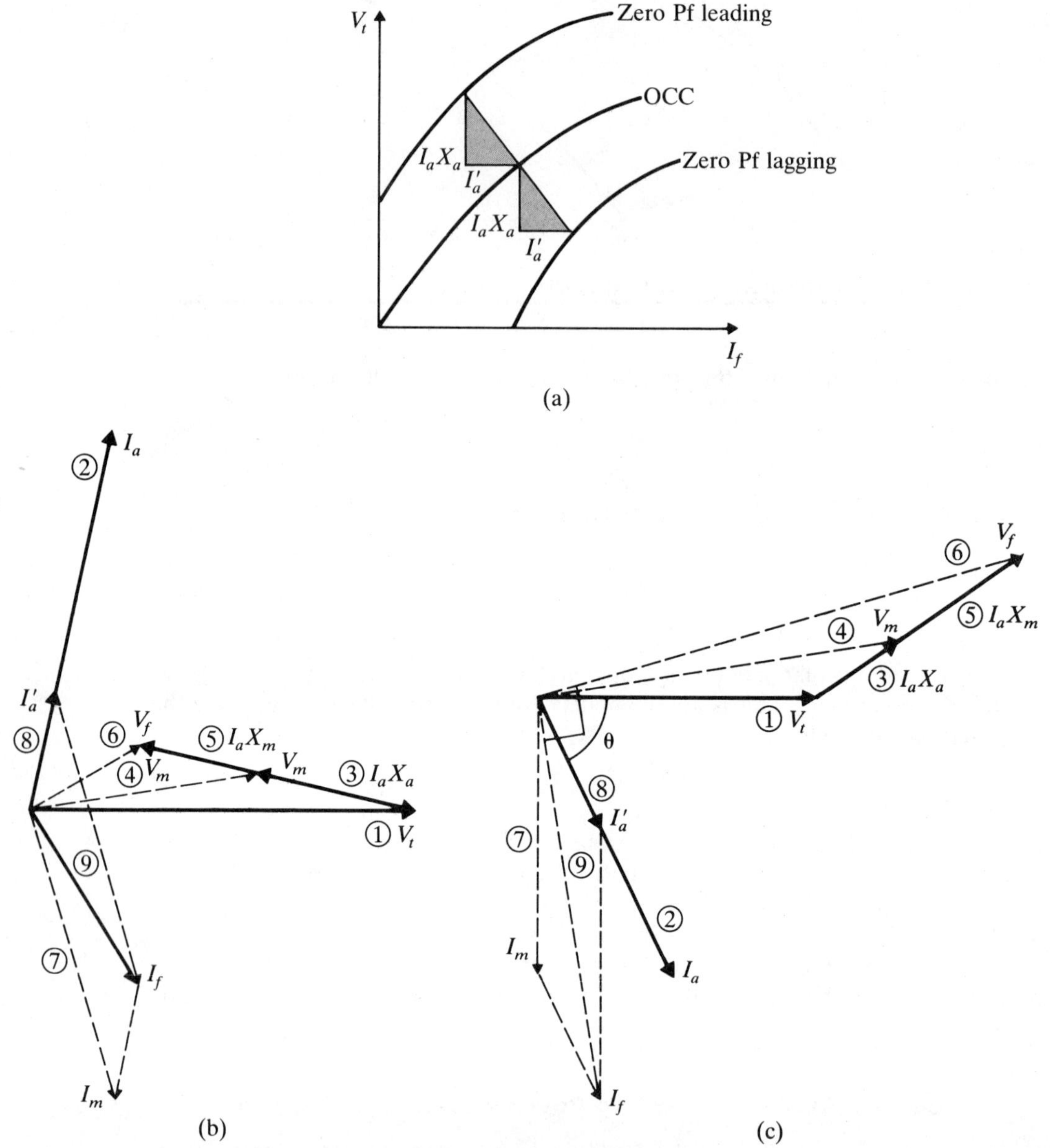

FIG. 5-28 Alternator: **(a)** zero-power-factor and open-circuit characteristics, **(b)** phasor diagram for near zero leading power factor, **(c)** phasor diagram for near zero lagging power factor.

When the power factor lags, more field current is required to magnetize the machine and to overcome the demagnetizing effects of the armature reaction. When the power factor leads, less field current is required, because the armature reaction aids the field in magnetizing the machine. At the origin of the zero-power-factor leading curve, the actual field current is zero, because the armature reaction is equal and opposite to the magnetizing component of the field current.

The phasor diagrams for ZPF leading and lagging loads are shown in Figs. 5-28(b) and (c), respectively.

EXAMPLE 5–5 The three-phase alternator shown in the one-line diagram of Fig. 5-29(a) has a per-phase synchronous impedance of $.03 + j.50$ per unit. The

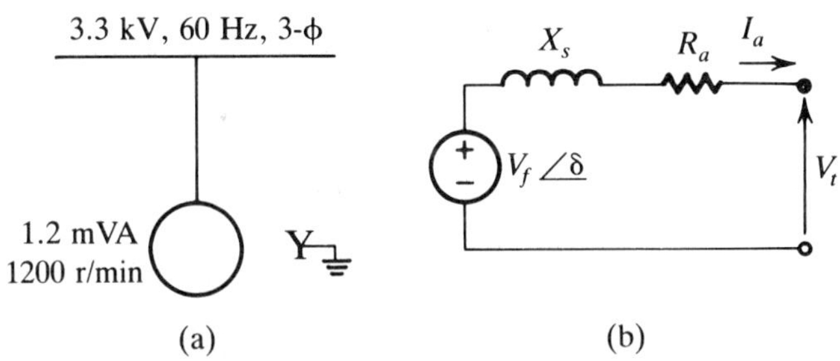

FIG. 5-29

magnetization characteristic at the operating region is given by

$$V_f = 70 + 55I_f$$

The generator delivers rated current to an infinite bus at .8 leading power factor. Determine:

a) The ohmic value of the synchronous impedance.
b) The excitation voltage.
c) The approximate torque angle in electrical and mechanical degrees.
d) The regulation.
e) The field current.

SOLUTION

a) The base parameters are given by the name-plate data of the generator:

$$V_b = 3.3 \text{ kV}, \qquad S_b = 1.2 \text{ MVA}$$

Then, by definition,

$$Z_b = \frac{(3.3)^2}{1.2} = 9.08 \ \Omega/\text{phase}$$

The ohmic value of the synchronous impedance is calculated as follows:

$$Z_{pu} = \frac{Z_{\text{actual}}}{Z_b}$$

$$Z_{\text{actual}} = 9.08(.03 + j.50)$$

$$= .27 + j4.54 = \underline{\underline{4.55\,\underline{/86.6^\circ}\ \Omega/\text{phase}}}$$

b) The magnitude of the rated current of the generator is

$$I = \frac{S}{\sqrt{3}\,V_{L-L}} = \frac{1200}{\sqrt{3}\,3.3} = 209.95\ \text{A}$$

The excitation voltage is calculated by applying KVL in Fig. 5-29(b).

$$V_f\,\underline{/\delta} = V_t + I_a Z_s$$

$$= \frac{3300}{\sqrt{3}}\,\underline{/0^\circ} + 209.95\,\underline{/36.9^\circ}(4.55\,\underline{/86.6^\circ})$$

$$= 1592.8\,\underline{/30^\circ}\ \text{V/phase}$$

and

$$V_f = \underline{\underline{2758.8\ \text{V, line-to-line}}}$$

c) The approximate value of the torque angle in electrical degrees is

$$\delta = \underline{\underline{30^\circ}}$$

In mechanical degrees its approximate value is

$$\delta_m = \frac{2\delta}{p} = \frac{2(30^\circ)}{6} = \underline{\underline{10^\circ}}$$

d) The regulation in percent is

$$R\% = \frac{2758.8 - 3300}{3300}(100) = \underline{\underline{-16.4\%}}$$

e) The field current is found from the given OC characteristic as follows:

$$I_f = \frac{2758.8 - 70}{55} = \underline{\underline{48.89\ \text{A}}}$$

EXAMPLE 5–6 Test results for a 3-ϕ, 60 Hz, 480 V, 400 kVA synchronous generator are as follows.

Open-circuit test:

Field current (A)	6	10	14	18	22	26
Armature voltage (V)	225	348	440	500	540	560

Short-circuit test: A field current of 8 A was required to produce rated armature current.

Zero-power-factor test:

$$I_a = \text{rated}$$
$$V_{L-L} = 500 \text{ V}$$
$$I_f = 30 \text{ A}$$

Assuming that the stator resistance is negligible, calculate the field current required to produce rated output voltage and current to a .90 Pf lagging load.

SOLUTION The characteristics of the generator are shown in Fig. 5-30.

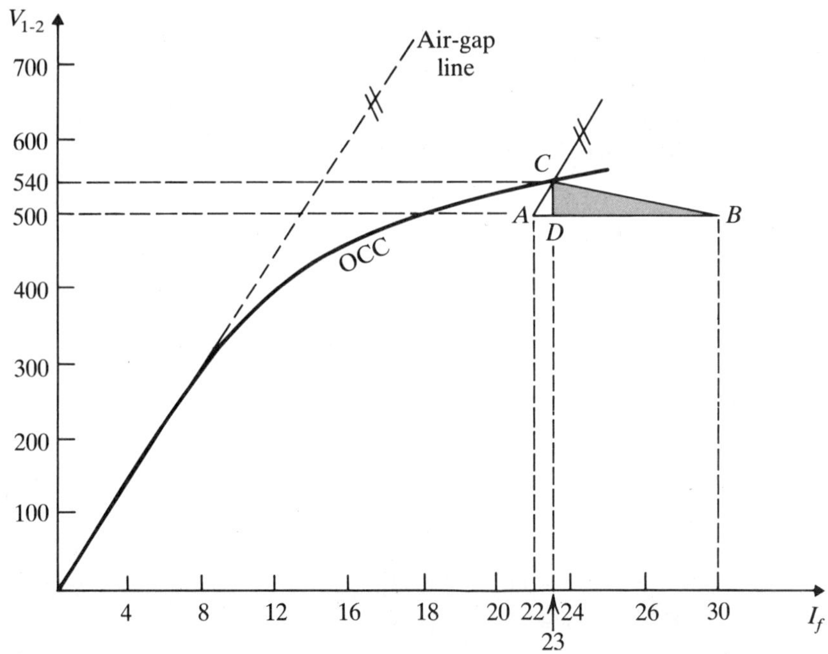

FIG. 5-30

From the diagram:

$$AB = 8 \text{ A}$$
$$AC \,/\!/\, \text{air-gap line}$$

From the Potier triangle:

$$CD = I_a X_a = 40 \text{ V, L–L}$$
$$DB = I'_a = 7 \text{ A}$$

The generator's rated current is

$$I_a = \frac{400}{\sqrt{3}(.480)} = 481.13 \text{ A}$$

In phasor form:

$$I_a = 481.13 \angle -25.8^\circ \text{ A}$$

The leakage reactance is

$$X_a = \frac{40/\sqrt{3}}{481.13} = .048 \ \Omega/\text{phase}$$

The magnetization voltage is

$$V_m \angle \beta = \frac{480}{\sqrt{3}} \angle 0^\circ + 481.13 \angle -25.8^\circ (j.048)$$

from which

$$V_m = 287.95 \angle 4.1^\circ \text{ V/phase}$$

and

$$V_m = 498.74 \text{ V, L–L}$$

From the OCC, the magnetizing component of the field current is 17.95 A. From Eq. (5.51), then, we obtain

$$I_f = 17.95 \angle -85.9^\circ + 7 \angle -25.8^\circ$$
$$= 22.3 \angle -70^\circ \text{ A}$$

Exercise 5–6 A 480 V, 50 kVA, 60 Hz, 4-pole synchronous generator delivers rated power at .8 Pf lagging. The synchronous impedance is $.2 + j1.4$ ohms per phase. Determine:

a) The excitation voltage.

b) The generator's regulation.

Answer (a) 593.38 V, L–L; (b) 23.62%.

5.3 Salient-Pole Synchronous Machines

5.3.1 INTRODUCTION

A two-pole salient-rotor synchronous motor is shown in Fig. 5-31. The length of the air gap between the rotor and the stator is not uniform. The axis along which the air gap is minimum is called the *direct axis,* and that along which the air gap is maximum is called the *quadrature axis.*

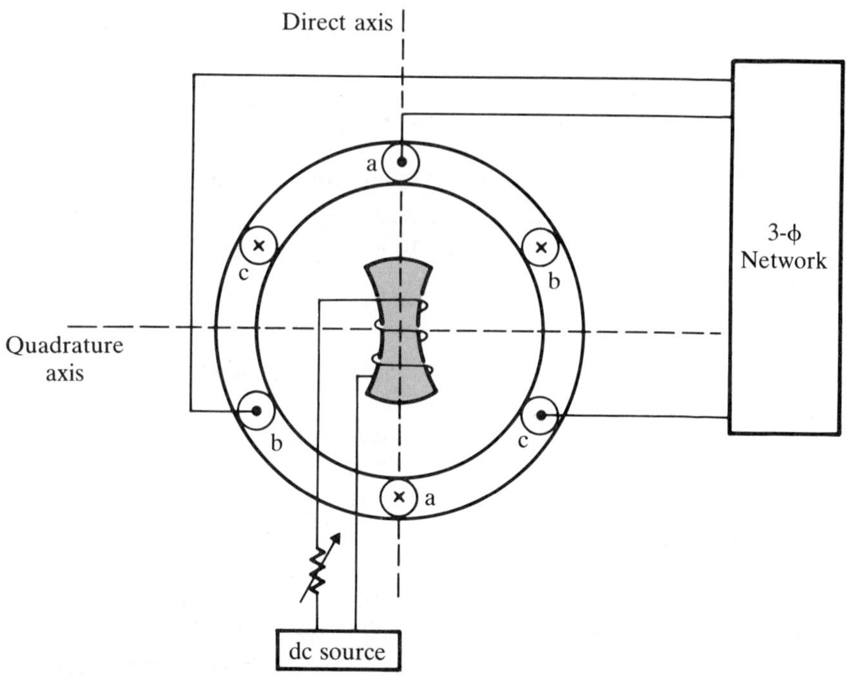

FIG. 5-31 An elementary representation of a two-pole salient-rotor synchronous machine.

Synchronous machines with salient-pole rotors usually have more than two poles and therefore rotate at relatively low synchronous speeds, a requirement of many industrial drives. They can develop up to 40% of their rated power without any excitation—zero field current—while cylindrical machines cease to function as motors or generators once the dc excitation is removed.

Saliency increases the ability of the machine to oppose the forces that tend to drive a motor out of synchronism. As a result, compared to cylindrical-rotor machines, salient machines are more suitable in applications where the motor might be subjected to occasional sudden torque variations.

In all modern synchronous motors there is no direct connection between the rotor windings and the field supply voltage source. The field current, as explained in Chapter 7, is obtained by rectifying the voltage induced in another set of rotor windings, commonly known as a "rotating armature."

The magnetizing flux through the machine's two distinct air gaps produces two different inductances. The inductance along the direct axis is designated as L_d, and that along the quadrature axis is designated as L_q. Since the inductance is an inverse function of the length of the air gap, the direct-axis (X_{m_d}) component of the magnetizing reactance is larger than its quadrature-axis (X_{m_q}) component. The ratio of these two reactances is about 1.5.

The effective reactance along either the direct (X_d) or the quadrature axis (X_q) is the sum of the armature leakage reactance (X_a) and its corresponding component of the magnetizing reactance. Mathematically,

$$X_d = X_{m_d} + X_a \tag{5.70}$$

$$X_q = X_{m_q} + X_a \tag{5.71}$$

where X_{m_d} and X_{m_q} are the direct- and quadrature-axis magnetizing reactances, respectively.

As mentioned in previous sections, the leakage reactance is, compared to the components of the magnetizing reactance, very small.

In the case of a cylindrical rotor:*

$$X_d = X_q = X_s \qquad \text{(synchronous reactance)} \tag{5.72}$$

To measure X_d and X_q, the excitation is removed from the machine and a voltage is applied across the stator terminals. Then the rotor is driven externally with a variable-speed dc motor. For different positions of the rotor magnetic axis relative to the stator axis, the maximum and minimum currents are recorded. The ratio of the applied voltage to the minimum and maximum currents gives, respectively, the direct-axis reactance and the quadrature-axis reactance. Under these test conditions the machine is not saturated, and thus the measured reactances represent nonsaturated values.

• • •

* Cylindrical-rotor machines have some saliency, which exists because the slots for the field windings are not uniform around the entire rotor.

Steady-State Analysis

The steady-state performance of salient-pole synchronous machines is satisfactorily predicted and analyzed by using Eq. (5.73). The positive signs are for a generator, the negative signs for a motor. This equation is derived from the De Blondel, or two-reaction, theory. This theory takes into consideration the effects of saliency but ignores the effects of magnetic saturation. Magnetic saturation is accounted for by the cylindrical-rotor theory, which ignores saliency. Under normal operating conditions, both theories yield slightly different but generally satisfactory results.

According to the De Blondel theory, and as shown by Eq. (5.74), the armature current is represented by two components: the direct-axis component (I_d) and the quadrature-axis component (I_q). The relationship of the armature current and its components is not given in phasor form because the direct-axis component may—depending on the power factor—lead or lag the quadrature-axis component for either a motor or a generator. However, the two components are 90° out of phase with each other, and the quadrature-axis component is always in phase with the excitation voltage. All other parameters of Eq. (5.73) are the same as those for cylindrical-rotor synchronous machines.

$$V_f \underline{/\pm\delta^\circ} = V_t \underline{/0^\circ} \pm I_a R_a \pm jI_q X_q \pm jI_d X_d \tag{5.73}$$

$$|I_a| = \sqrt{I_q^2 + I_d^2} \tag{5.74}$$

The development of the per-phase equivalent circuit of a salient-pole machine from which Eq. (5.73) could be derived is rather complex, and it is not presented in this text.

5.3.2 PHASOR DIAGRAMS

The steady-state analysis of salient-pole synchronous machines is simplified by using phasor diagrams—that is, the graphical representation of Eqs. (5.73) and (5.74). The following procedure is suggested in drawing the phasor diagrams for a motor.

From Eq. (5.73), write the governing voltage equation:

$$V_f \underline{/-\delta^\circ} = V_t \underline{/0^\circ} - I_a R_a - jI_q X_q - jI_d X_d \tag{5.75}$$

1. Draw to scale the terminal voltage (V_t) and take it as reference (see Fig. 5-32).
2. Draw the armature current at its given phase angle. In this case $\theta = 0^\circ$.
3. From the tip of the terminal voltage, draw the phasor $I_a R_a$. This is the voltage drop in the resistance (R_a) and should be in phase with the armature current (I_a). Here, however, $I_a R_a$ must be drawn at 180° to its normal direction because of the negative sign in the voltage equation.
4. Draw to scale a dotted line representing the phasor $I_a X_q$. This is the voltage drop in the reactance (X_q) and should lead the current (I_a) by 90°. Here again $I_a X_q$ must be drawn at 180° to its normal direction because of the minus sign in the voltage equation. The end of this phasor establishes the excitation line (V_f line) and therefore the torque angle. Although it is not in the equation, the term $I_a X_q$ is used

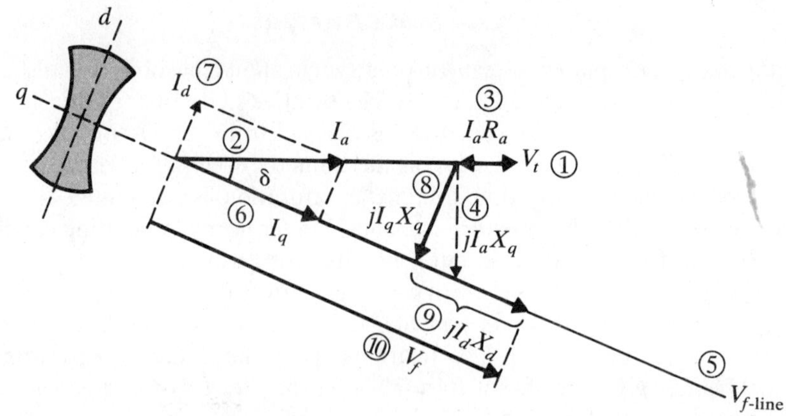

FIG. 5-32 Phasor diagram for a salient synchronous motor operating at unity power factor.

to locate the excitation or quadrature-axis line. For this reason, it is drawn dotted. The torque angle can also be obtained mathematically from the following relationship:

$$V_q \angle -\delta = V_t \angle 0^\circ - I_a R_a - jI_a X_q \tag{5.76}$$

where V_q represents the sum of the three phasors of Eq. (5.76). Its phase angle (δ) establishes the direction of the line on which the excitation voltage must lie.

5. Draw the quadrature or excitation line. This is designated as the V_f line.
6. Resolve the armature current into its rectangular components I_d and I_q. The quadrature-axis component (I_q) is always in phase with the excitation line.
7. The direct-axis current is 90° out of phase with the quadrature axis. In general, I_d may lead or lag I_q, depending on the machine's operating power factor.
8. Draw the phasor $I_q X_q$. This represents the voltage drop across the reactance (X_q) and should lead the current I_q by 90°. Again, because of the negative sign in the voltage equation, it must be drawn at 180° to its normal direction. The length of $I_q X_q$ does not need to be calculated, because it always starts at the tip of $I_a R_a$ and ends perpendicularly on the excitation line.
9. Finally, draw to scale the phasor $I_d X_d$. As this represents the voltage drop in the reactance (X_d), it should lead the current (I_d) by 90°. But on account of the equation's minus sign, it is drawn at 180° to its normal direction.
10. The excitation voltage (V_f) is drawn from the origin to the end of the tip of the phasor $I_d X_d$.

The procedure for drawing the phasor diagrams for salient alternators is similar. In fact, the phasor diagrams for a generator are simpler to construct because of the positive sign in the loop Eq. (5.73).

The phasor diagram of a salient synchronous generator operating at unity power factor is shown in Fig. 5-33.

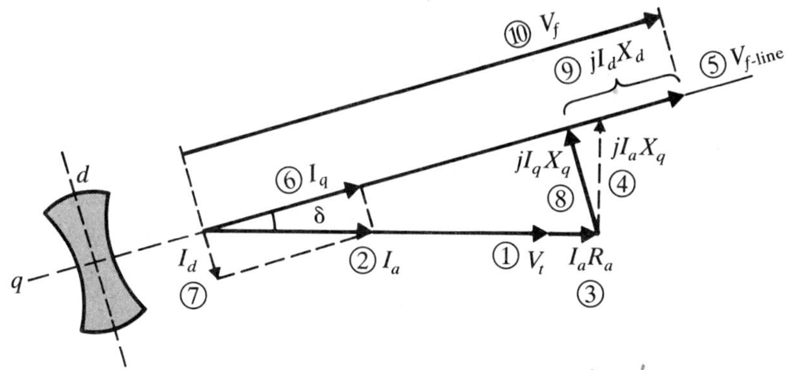

FIG. 5-33 Phasor diagram for a salient synchronous motor operating at unity power factor.

5.3.3 POWER DEVELOPED

To find the expression for the power developed by a salient synchronous motor, consider the phasor diagram shown in Fig. 5-34. Neglect armature resistance and assume a leading power factor. By definition,

$$P = V_t I_a \cos\theta \text{ watts/phase} \tag{5.77}$$

Say

$$\beta = \theta + \delta \tag{5.78}$$

Then,

$$P = V_t I_a \cos(\beta - \delta) \tag{5.79}$$

$$= V_t I_a \cos\beta \cos\delta + V_t I_a \sin\beta \sin\delta \tag{5.80}$$

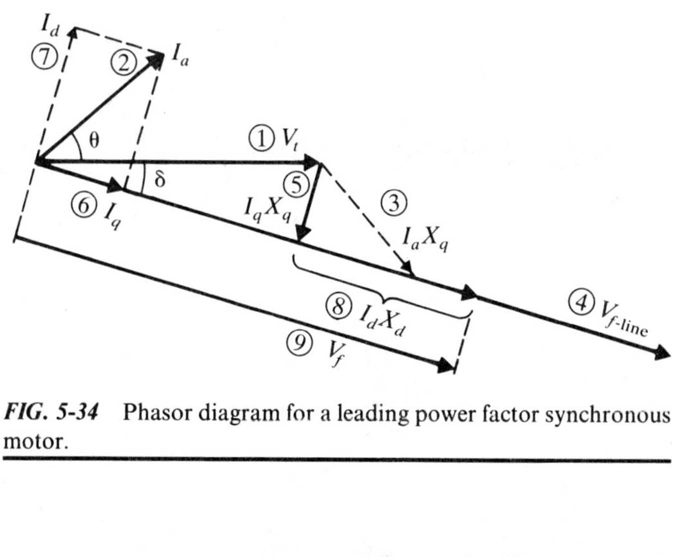

FIG. 5-34 Phasor diagram for a leading power factor synchronous motor.

From the phasor diagram,

$$I_q = I_a \cos \beta \tag{5.81}$$

$$I_d = I_a \sin \beta \tag{5.82}$$

Substituting, we get

$$P = V_t I_q \cos \delta + V_t I_d \sin \delta \tag{5.83}$$

From the phasor diagram, it is also evident that

$$I_d = \frac{V_f - V_t \cos \delta}{X_d} \tag{5.84}$$

and

$$I_q = \frac{V_t \sin \delta}{X_q} \tag{5.85}$$

Substituting Eqs. (5.85) and (5.84) into Eq. (5.83), we obtain

$$P = V_t \frac{V_t \sin \delta}{X_q} \cos \delta + \frac{V_t \sin \delta (V_f - V_t \cos \delta)}{X_d} \tag{5.86}$$

From the above,

$$P = \frac{V_t^2 (X_d - X_q)}{X_d X_q} \sin \delta \cos \delta + \frac{V_f V_t}{X_d} \sin \delta \tag{5.87}$$

using

$$\sin 2\delta = 2 \sin \delta \cos \delta \tag{5.88}$$

The expression in Eq. (5.87) becomes

$$P = \frac{V_f V_t}{X_d} \sin \delta + \frac{V_t^2 (X_d - X_q)}{2 X_d X_q} \sin 2\delta \text{ watts/phase} \tag{5.89}$$

The first term of Eq. (5.89) is identical to Eq. (5.28), which represents the power developed by a cylindrical synchronous machine. The second term represents the saliency and is referred to as the reluctance component. This component may account for up to 40% of a salient machine's output power. The power developed versus the torque angle is shown in Fig. 5-35.

By definition, the reactive power is

$$Q = V_t I_a \sin \theta \tag{5.90}$$

As before,

$$\theta = \beta - \delta \tag{5.91}$$

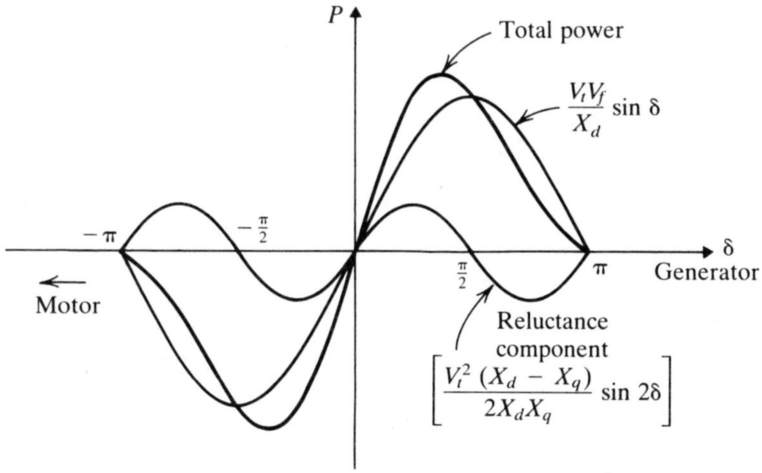

FIG. 5-35 Power versus torque angle.

Thus,

$$Q = V_t I_a \sin(\beta - \delta) \tag{5.92}$$

From Eqs. (5.84), (5.85), and (5.92), after expansion and simplification, we get

$$Q = \frac{V_t V_f}{X_d} \cos\delta - V_t^2 \left(\frac{\cos^2\delta}{X_d} + \frac{\sin^2\delta}{X_q}\right) \text{ VAR/phase} \tag{5.93}$$

When the parameters of the machine are expressed in per-unit, Eqs. (5.89) and (5.93) give the total input real and reactive power in per-unit.

5.3.4 TORQUE ANGLE FOR MAXIMUM POWER

When the motor is supplied from an infinite bus, and with a constant field current, then Eq. (5.89) can be written as

$$P = K_1 \sin\delta + K_2 \sin 2\delta \tag{5.94}$$

where the constants K_1 and K_2 are given, respectively, by

$$K_1 = \frac{V_f V_t}{X_d} \tag{5.95}$$

$$K_2 = \frac{V_t^2 (X_d - X_q)}{2X_d X_q} \tag{5.96}$$

In order to find the torque angle that corresponds to the maximum power, apply the calculus theory of maximum and minimum, as outlined below:

1. Find the derivative of the power with respect to the torque angle:

$$\frac{dP}{\partial \delta}$$

2. Set this derivative equal to zero:

$$\frac{dP}{d\delta} = 0$$

3. From the last expression, find the so-called critical values of δ.

Applying the above procedure, and after some mathematical manipulations, we find that the torque angle for maximum power is given by the following equation:

$$\delta = \arccos\left(-\frac{K_1}{8K_2} \pm \sqrt{\left(\frac{K_1}{8K_2}\right)^2 + \frac{1}{2}}\right) \tag{5.97}$$

The torque angle for maximum power is usually about 70°. The intermediate steps in the derivation of Eqs. (5.93) and (5.97) are left as an exercise (see Exercise 5–7(b)).

5.3.5 STIFFNESS OF SYNCHRONOUS MACHINES

The stiffness or toughness of a synchronous machine is a measure of its ability to resist and oppose the forces that tend to pull it out of synchronism. Mathematically, the stiffness is given by the partial derivative of the power developed with respect to the torque angle. In other words, the stiffness of a machine is given by the slope of the power-angle curve at the operating point under consideration. Maximum stiffness is referred to as synchronizing power.

Cylindrical Rotors

From Eq. (5.28), we can easily obtain the stiffness (S_T) of a cylindrical-rotor machine as follows:

$$S_T = \frac{dP}{d\delta} = 3\,\frac{V_f V_t}{X_s}\cos\delta \text{ watts/elect. radian} \tag{5.98}$$

where the factor of 3 accounts for the three phases of the machine.

The stiffness is maximum at the origin of the power-angle curve and minimum at the torque angle that corresponds to zero output power. In other words, at pull-out the stiffness of the machine is zero.

Salient Rotor

From Eq. (5.89), it can be shown that the stiffness of a motor with salient rotor is

$$S_T = \frac{dP}{d\delta} = 3\left(\frac{V_f V_t}{X_d}\cos\delta + \frac{V_t^2(X_d - X_q)}{X_d X_q}\cos 2\delta\right) \text{ watts/elect. radian} \qquad \textbf{(5.99)}$$

Although an increase of the field current does not affect the machine's output power, it does increase its toughness.

EXAMPLE 5–7 The synchronous motor shown in the one-line diagram of Fig. 5-36(a) draws rated current at a power factor of 0.95 leading. Its direct- and quadrature-axis reactances are .8 per-unit and .5 per-unit, respectively. Neglecting losses, determine:

1. a) The approximate torque angle in electrical and mechanical degrees.
   b) The excitation voltage.
   c) The maximum torque that the machine can develop when the excitation voltage, as determined in (b), is constant.
2. Repeat part 1 by using the cylindrical rotor theory.

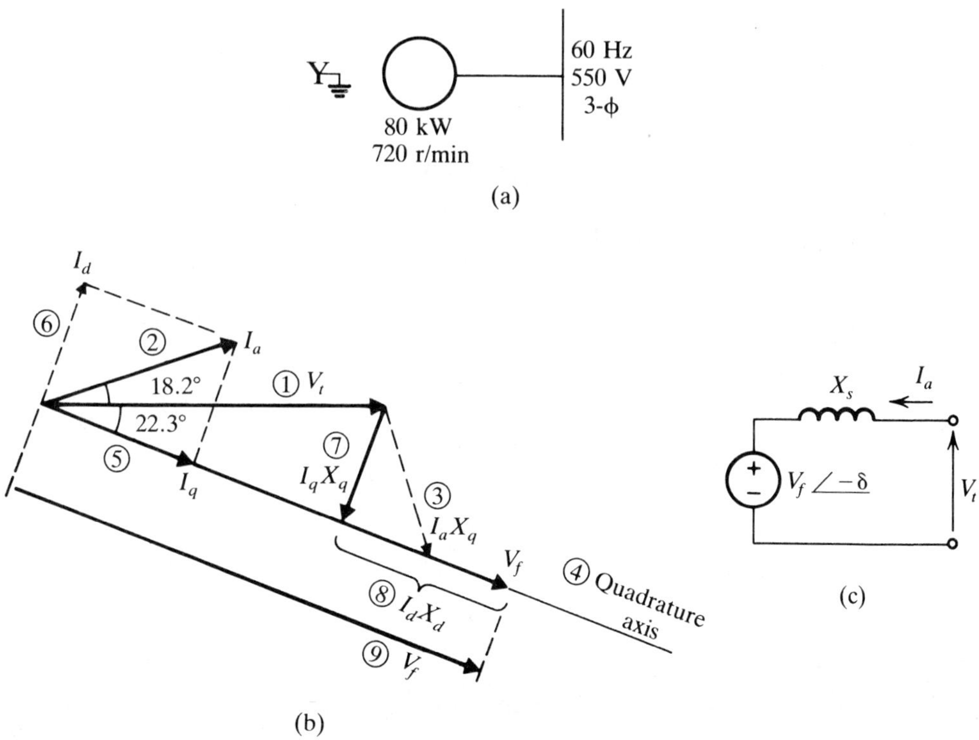

FIG. 5-36

SOLUTION

1. a) The torque angle is found by using Eq. (5.76).

$$V_q\underline{/-\delta} = V_t - jI_aX_q$$
$$V_q\underline{/-\delta} = 1.0\underline{/0^\circ} - j1.0\underline{/18.2^\circ}(.5) = 1.25\underline{/-22.3^\circ} \text{ pu}$$

Thus, the torque angle in electrical degrees is

$$\delta = \underline{\underline{-22.3^\circ}}$$

In mechanical degrees it is

$$\delta_m = \frac{2}{10}(22.3) = \underline{\underline{4.5^\circ}}$$

b) The phasor diagram, as shown in Fig. 5-36(b), is drawn by following the procedure outlined in Section 5.3.2. Then, from basic trigonometry:

$$I_q = I_a \cos(\theta + \delta)\underline{/-\delta}$$
$$= \cos(18.2^\circ + 22.3^\circ)\underline{/-22.3^\circ} = .76\underline{/-22.3^\circ} \text{ pu}$$

and

$$I_d = I_a \sin(18.2^\circ + 22.3^\circ)\underline{/90^\circ - 22.3^\circ}$$
$$= .65\underline{/67.7^\circ} \text{ pu}$$

Using Eq. (5.73), we obtain

$$V_f\underline{/-\delta} = 1.0\underline{/0^\circ} - j.76\underline{/-22.3^\circ}(.5) - j.65\underline{/67.7^\circ}(.8) = \underline{\underline{1.45\underline{/-22.3^\circ} \text{ pu}}}$$

c) The torque angle for maximum power is found from Eq. (5.97):

$$K_1 = \frac{1.45(1)}{.8} = 1.81$$
$$K_2 = \frac{(1.0)^2(.8 - .5)}{2(.8)(.5)} = .375$$
$$\delta = \arccos\left(-\frac{1.81}{8(.375)} \pm \sqrt{\left(\frac{1.81}{8(.375)}\right)^2 + \frac{1}{2}}\right)$$

from which

$$\delta = 70.9^\circ$$

The maximum torque is

$$T = \frac{\text{maximum power in per-unit}}{\text{speed per-unit}}$$

$$T = \frac{1.45(1)}{.8}\sin 70.9^\circ + \frac{(1.0)^2(.8 - .5)}{2(.8)(.5)}\sin 141.9^\circ = \underline{\underline{1.94}}\ \text{pu}$$

2. For a cylindrical rotor:

$$X_d = X_q = X_s = .8\ \text{pu}$$

a) Using the per-phase equivalent circuit of Fig. 5-36(c), from KVL we have

$$V_f\underline{/-\delta} = 1.0\underline{/0^\circ} - j1.0\underline{/+18.2^\circ}(.8) = 1.46\underline{/-31.3^\circ}\ \text{pu}$$

Thus,

$$\delta = \underline{\underline{-31.3^\circ}}$$

and

$$\delta_m = \frac{2}{10}(31.3) = \underline{\underline{6.3^\circ}}$$

b) The excitation voltage is

$$V_f = \underline{\underline{1.46\ \text{pu}}}$$

c) For maximum torque

$$\delta = 90^\circ$$

and

$$T = \left(\frac{1}{1.0}\right)\frac{1.46(1)}{.8} = \underline{\underline{1.83\ \text{pu}}}$$

For purposes of comparison, the results are summarized in Table 5-3.

TABLE 5-3 Summary of results of Example 5–7

Type of rotor	*Torque angle in electrical degrees*	*Excitation voltage in pu*	*Maximum torque in pu*
Salient	22.3	1.45	1.94
Cylindrical	31.3	1.46	1.83

Exercise 5–7

a) Draw the phasor diagram of a salient synchronous motor operating at unity power factor and show that the power angle is given by

$$\delta = \arctan \frac{I_a X_q}{V_t - I_a R_a}$$

b) Derive Eqs. (5.93) and (5.97).

c) Determine the torque angle for maximum power of a 480 V, 50 kW, .9 efficient, unity Pf, salient synchronous motor. The motor's direct- and quadrature-axis reactances are 3 Ω and 2 Ω, respectively.

Answer (c) 71°.

5.4 Three-Phase Faults

5.4.1 INTRODUCTION

This section discusses the basic principles and mathematical relationships that describe the stator and rotor currents produced when a synchronous machine is subjected to a three-phase short circuit. Knowledge of the short-circuit stator and rotor currents is critical: the safety of personnel and the protection of electrical equipment (motors, circuit breakers, fuses, etc.) depends on it.

A three-phase short circuit is often called a three-phase symmetrical fault.* This is the most critical fault that may occur on an *ungrounded generator*.

5.4.2 PRINCIPLE OF CONSTANT FLUX LINKAGE

Consider a coil that is connected to an ac supply of V volts. From KVL we have

$$v = iR + \frac{d}{dt}(Li) \tag{5.100}$$

or

$$v = iR + \frac{d\lambda}{dt} \tag{5.101}$$

. . .

* The type of fault depends on weather conditions, level of operating voltage, etc. About 70% are line-to-ground (L–G) faults; only 5% are three-phase faults. For a *solidly grounded* generator, typical values of short-circuit currents as a percentage of the 3-ϕ fault currents are: 3-ϕ fault: 100%; line-to-line fault: 86%; line-to-ground fault: 125%.

where R is the resistance of the coil and λ is its flux linkage. When the voltage supply is shorted, and neglecting the coil's resistance, we have

$$\frac{d\lambda}{dt} = 0 \tag{5.102}$$

That is, the flux linkage must remain constant when the coil is shorted.

A similar phenomenon occurs when the three-phase stator windings of a synchronous machine are shorted. However, the problem here is complicated by the interaction of stator and rotor fields. In general, at the inception of the fault the stator and rotor currents increase manyfold in order to satisfy the principle of constant flux linkage.

5.4.3 STATOR CURRENTS

Field observations and measurements indicate that the stator current (I_a) has an ac component (I_{ac}) and a dc component (I_{dc}). That is,

$$I_a = I_{ac} + I_{dc} \tag{5.103}$$

Typical ac and dc components in one phase of the stator current are shown in Fig. 5-37(a). The current waveforms of the other two phases are similar, but not identical. A description of these ac and dc current components follows.

AC Components of Stator Currents

Oscillographs of the armature current taken when an alternator is subjected to a three-phase short circuit indicate, as shown in Fig. 5-37(a), that the fault currents are characterized by three distinct periods: the subtransient, the transient, and the sustained fault period. The indicated duration of each period is approximate. At the inception of the fault, the rotor current follows the increase and decay of the stator currents, as shown in Fig. 5-37(b).

Subtransient Period

At the inception of a three-phase symmetrical fault, a rapid rise in the ac currents of the armature windings takes place. The almost purely inductive impedance of the circuit will make these currents highly demagnetizing. As a result, the currents in the field and damping windings will rapidly increase so that constant flux linkage is maintained.

The decay of the current in the damping and field windings will be accompanied by a corresponding decrease in the armature currents. The damping windings have a relatively short time constant (.01 to .04 s), so the currents induced in them will decay rapidly. The duration of the subtransient period is about .15 s (9 cycles).

The maximum rms value of the ac component (I'') during the subtransient period is given by

$$I'' = \frac{V_f''}{X_d''} \tag{5.104}$$

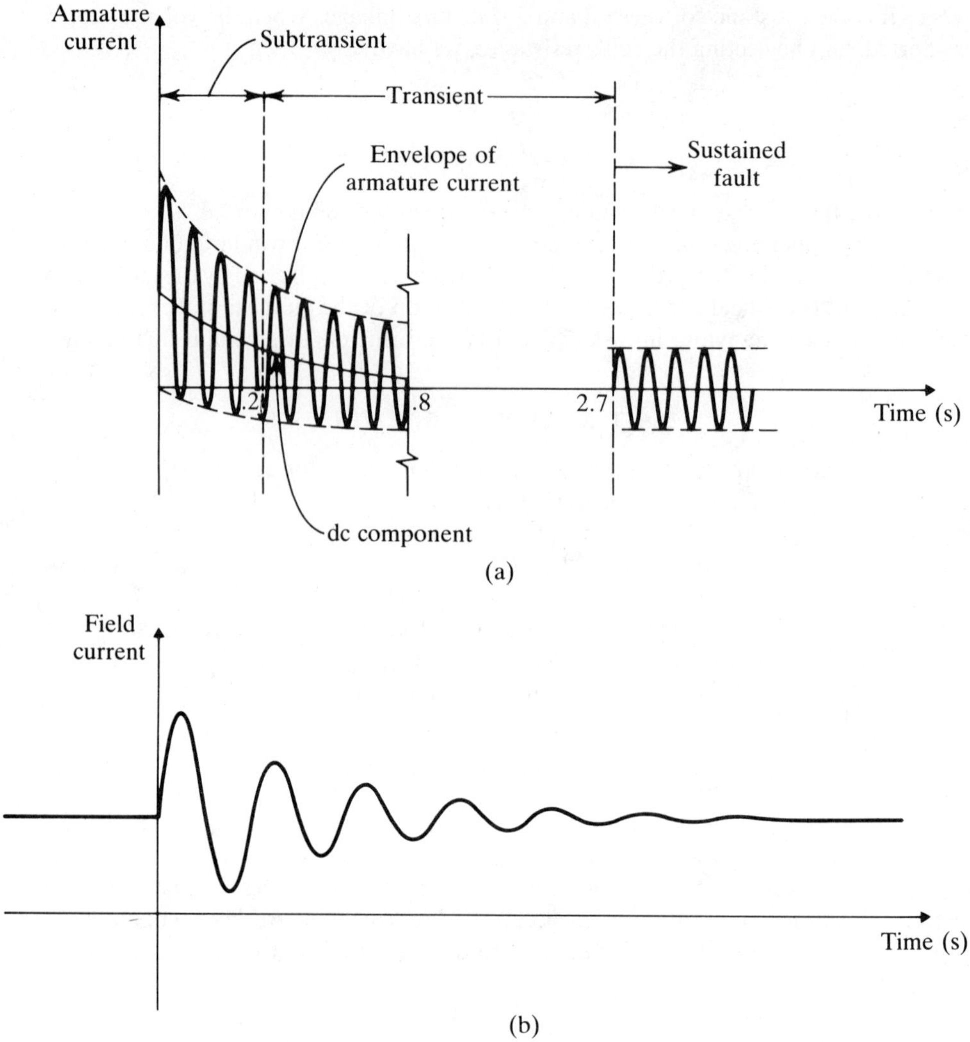

FIG. 5-37 Three-phase fault on an unloaded alternator: **(a)** armature current, **(b)** corresponding field current.

where

X''_d = the subtransient reactance, or the machine reactance during the subtransient period

V''_f = the subtransient voltage or the voltage behind the subtransient reactance X''_d. This voltage can be calculated from the following equation:

$$V''_f = V_t + jI_aX''_d \tag{5.105}$$

where V_t and I_a are the prefault armature voltage and the current of the generator, respectively. Typical values of a machine's subtransient reactance are shown in Table 5-4. The maximum rms value of the ac component given by Eq. (5.104) is referred to as the rms symmetrical value of the short-circuit current.

TABLE 5-4 Typical values of salient-pole synchronous machine constants*

Parameter	*Cylindrical rotor*	*Salient generators*	*Salient motors*
X''_d	.10	.22	.30
X'_d	.20	.40	.45
X_d	1.10	1.0	1.10
X_q	1.0	.65	.80
T'_{do}	4	5	2.5
T'_d	1.0	1.8	1.5
T''_d	.035	.035	.035
T_a	.16	.16	.16

** Reactances are in per-unit values based on the machines' ratings, and time constants are in seconds.*

Transient Period

The duration of the transient period can be up to approximately 5 s (300 cycles). The time constant of the field windings is relatively large (.25 to .50 s); therefore, the current induced in these windings decays rather slowly. This decay is accompanied by a corresponding decrease in the armature currents.

The maximum rms value of the ac component (I') during the transient period is given by

$$I' = \frac{V'_f}{X'_d} \tag{5.106}$$

where

X'_d = the direct-axis transient reactance. It varies between .2 to .45 per-unit.

V'_f = the transient excitation voltage or the voltage behind the transient reactance. This voltage can be calculated from the following KVL equation:

$$V'_f = V_t + I_a(jX'_d) \tag{5.107}$$

where V_t and I_a represent the prefault voltage and the current of the alternator, respectively.

Sustained Short-Circuit Period

The remaining period is the steady-state ac, or sustained short-circuit, condition. The rms value of the sustained short-circuit current is given by

$$I = \frac{V_f}{X_d} \quad \textbf{(5.108)}$$

where X_d is the direct-axis reactance, and V_f is the excitation voltage. This voltage is calculated from

$$V_f = V_t + jI_a X_d \quad \textbf{(5.109)}$$

The sustained short-circuit armature current given by Eq. (5.108) is normally less than the rated current of the alternator.

The descriptions of the subtransient, transient, and sustained fault periods indicate that the ac components of the stator currents are complex in waveform. They can be represented by a decaying sinusoidal waveform whose *approximate* rms value is given by Eq. (5.111).

The amplitude of the ac component at the origin will be dependent on the point of the voltage cycle (switching angle) at which the fault takes place. Neglecting second harmonics, the ac component of the stator current in phase A is given by the following approximate expression:

$$i_{ac} = I_m \sin(\omega t + \alpha) \quad \textbf{(5.110)}$$

where α is referred to as the switching angle.

The maximum rms value of I_m is given by the following *empirical* equation:

$$I_{ac} = \frac{V_f}{X_d} + \left(\frac{V'_f}{X'_d} - \frac{V_f}{X_d}\right) e^{-t/T'_d} + \left(\frac{V''_f}{X''_d} - \frac{V'_f}{X'_d}\right) e^{-t/T''_d} \text{ A (rms)} \quad \textbf{(5.111)*}$$

where

T''_d = is the subtransient time constant that varies between .03 to 0.04 s. Its actual value depends on the time constant of the damping windings.

T'_d = is the transient time constant. It varies from 1.0 to 1.2 s. This constant is given by

$$T'_d = T'_{do} \frac{X'_d}{X_d} \quad \textbf{(5.112)}$$

where T'_{do} is the direct-axis transient open-circuit time constant. It varies from 2.5 to 5 s and depends on the time constant of the field windings (L_f/R_f).

. . .

* By permission from A. E. Fitzgerald, C. Kingsley, and U. A. Kusko *Electric Machinery* (New York: McGraw-Hill, 1971).

If the three-phase short circuit takes place at a considerable distance from the generator terminals, the time constants and the reactances of Eq. (5.111) should be altered to include the impedance of the transmission line. Typical time-constant values for synchronous machines are given in Table 5-4.

DC Components of Stator Currents

As shown in Fig. 5-38, the dc or unidirectional component of the armature current is different in each phase and depends upon the point of the voltage cycle at which the fault occurs. The decrement or the rate of decay of the dc component is proportional to the ratio of the reactance to the resistance of the armature circuit.

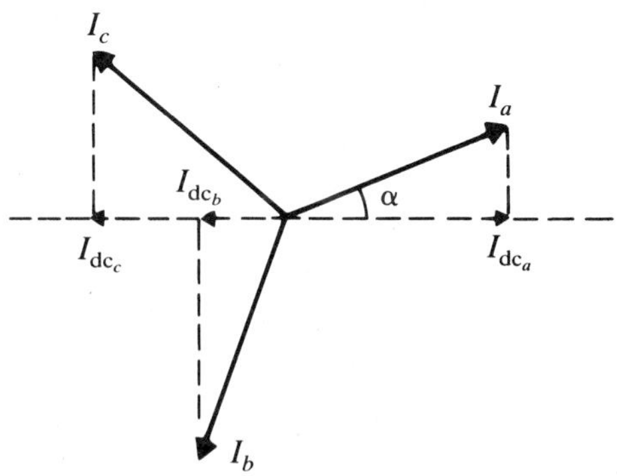

FIG. 5-38 Illustrating the calculation of the initial dc values of the components of the armature currents.

Mathematically, the dc component of the armature current is given by

$$I_{dc} = \sqrt{2}\,\frac{V_f''}{X_d''}(\cos\alpha)\,e^{-t/T_a} \qquad \textbf{(5.113)}$$

where

α = is the switching angle

T_a = is the armature short-circuit time constant.

$$T_a = \frac{L_2}{R_a} = \frac{\text{negative-sequence armature inductance}}{\text{stator winding resistance}} \qquad \textbf{(5.114)}$$

This time constant usually varies between .10 and .20 s.

The dc components (I_{dc_a}, I_{dc_b}, I_{dc_c}) of the three armature currents decay with the same time constant. As shown in Fig. 5-38, their initial values can be determined by finding the projection on the reference axis of the maximum instantaneous values of the ac stator currents.

From Eq. (5.113), the maximum value of the dc component is

$$I_{\text{dc}_{\max}} = \sqrt{2}\,\frac{V_f''}{X_d''} \tag{5.115}$$

In general, the total armature current is made up of three components, which are, in order of importance:

1. A fundamental frequency component.
2. A dc component.
3. A double frequency component. This second harmonic component is relatively negligible.

The fundamental frequency component is symmetrical* with respect to the time axis: its superposition on the dc component will give an asymmetrical waveform, which is the armature current. The degree of asymmetry depends upon the point of the voltage cycle at which the fault takes place.

Substituting into Eq. (5.103), Eqs. (5.104) and (5.115), we obtain the maximum rms value of the stator current:

$$I = \sqrt{I_{\text{ac}}^2 + I_{\text{dc}}^2} = \sqrt{\left(\frac{V_f''}{X_d''}\right)^2 + \left(\sqrt{2}\left(\frac{V_f''}{X_d''}\right)\right)^2} \tag{5.116}$$

from which:

$$I = \sqrt{3}\,\frac{V_f''}{X_d''} \text{ rms} \tag{5.117}$$

$$I = \sqrt{3} \text{ (symmetrical short-circuited current)} \tag{5.118}$$

In practice, equipment is rated in terms of the symmetrical short-circuit current. This is because it is difficult to know the switching angle and thus the dc component. Figure 5-39 shows actual armature short-circuit currents for a 500 kVA, 13.8 kV alternator.

It should be pointed out that the motors within a particular system also contribute to the short-circuit currents. The motor contribution depends on the internal impedance of the motor(s), the size and length of the cable from the motor to the point of fault, etc. It is customary to take the induction motor's contribution as being four times its rated current, unless otherwise specified.

• • •

* A *symmetrical* current is an alternating current that is balanced around the zero axis. An *asymmetrical* current is an alternating current that is not balanced around the zero axis. In other words, the positive and negative peaks of an asymmetrical waveform have different values.

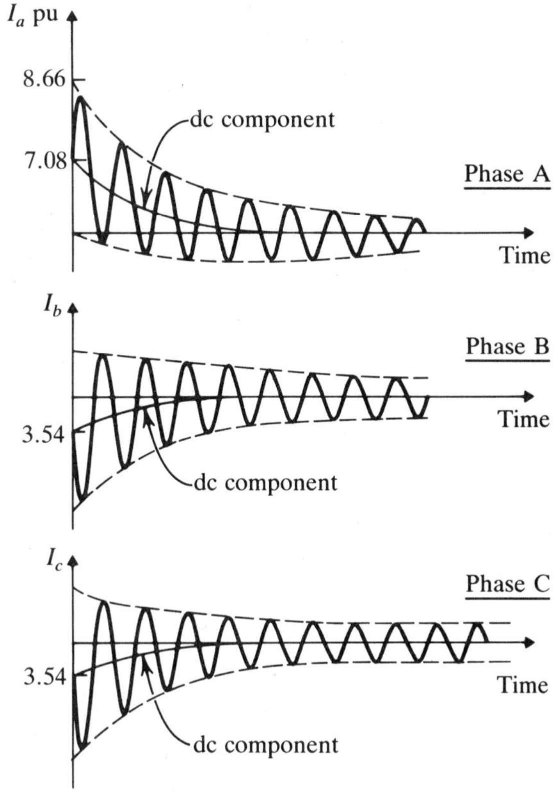

FIG. 5-39 Short-circuit currents resulting from a 3-ϕ short circuit on an unloaded 500 kVA, 13.8 kV synchronous generator. ($X_d'' = .2$ per unit, switching angle for phase A is 90°.)

5.4.4 FIELD CURRENT

The field current, like the stator current, consists of dc and ac components, as illustrated in Fig. 5-37(b). The ac component of the field current is comprised of fundamental and second-harmonic components. The second harmonic is usually negligible, and the fundamental-frequency component decays with the same time constant (T_a) as the dc component of the armature current.

The dc components decay with the same time constants—T_d' and T_d'', respectively—as the corresponding ac components of the stator current.

EXAMPLE 5–8 The three-phase alternator shown in the one-line diagram of Fig. 5-40(a) has a subtransient reactance of 20% and delivers rated current at .9 power factor lagging. Calculate:

a) The symmetrical short-circuit current that results from a 3-ϕ short on the terminals of the generator.

b) The maximum rms value of the short-circuit current.

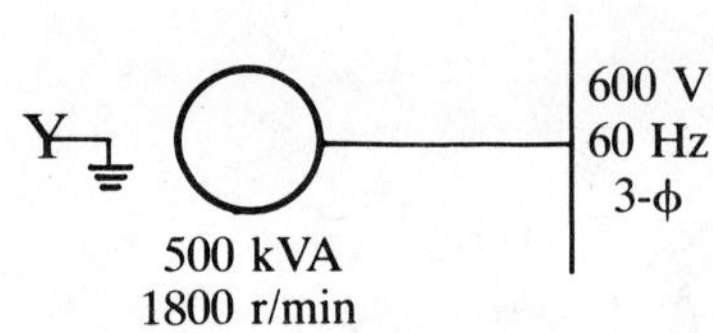

FIG. 5-40(a)

SOLUTION

a) The base quantities for the per-unit analysis are given by the name-plate data of the generator:

$$S_b = 500 \text{ kVA}$$
$$V_b = 600 \text{ V}$$

and

$$I_L = \frac{500}{\sqrt{3}(.6)} = 481.13 \text{ A}$$

The per-phase equivalent circuit of the generator is shown in Fig. 5-40(b). The voltage behind the subtransient reactance is calculated by substituting the known parameters into Eq. (5.105):

$$V_f''\underline{/\delta} = 1.0\underline{/0^\circ} + 1.0\underline{/-25.8^\circ}(j.2)$$
$$= 1.10\underline{/9.4^\circ} \text{ pu}$$

The symmetrical short-circuit current, or the maximum rms value of the ac component of the stator current, is obtained from Eq. (5.104):

$$I'' = \frac{V_f''}{X_d''} = \frac{1.10}{.2} = 5.51 \text{ pu} = 5.51(481.13) = \underline{\underline{2651.0 \text{ A}}}$$

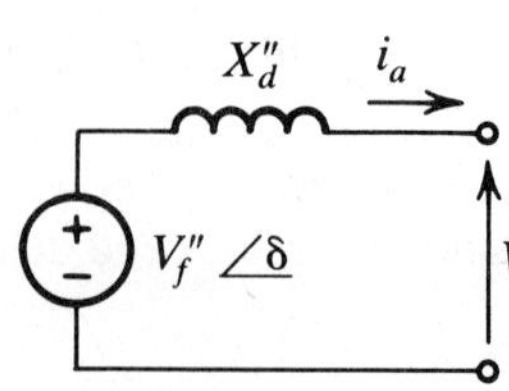

FIG. 5-40(b)

b) The maximum rms value of the stator short-circuit current is obtained from Eq. (5.117):

$$I_a = \sqrt{3}(5.51) = 9.54 \text{ pu}$$
$$I_a = \underline{\underline{4591.67 \text{ A rms, maximum}}}$$

Exercise 5–8 Explain:

a) The short-circuit armature currents of a generator have a demagnetizing effect.
b) The principle of constant flux linkage.
c) The sequence of events that occurs when a three-phase short circuit is suddenly applied to the terminals of an alternator.
d) The ac component of the current in the field windings decays with the same time constant (T_a) as the dc components of the armature currents.
e) The dc component of the currents in the field windings decays with the same time constants—(T'_d) and (T''_d)—as the ac component of the armature current.

Exercise 5–9 A 100 kW, 480 V, .90 efficient three-phase motor draws rated current at unity power factor. The synchronous impedance per phase is .04 + j.8 pu, and its subtransient reactance is .05 pu. Determine:

a) The symmetrical rms value of the short-circuit current that results when a three-phase fault occurs at the terminals of the motor.
b) The maximum rms value of the short-circuit current.

Answer (a) 2006.43 A; (b) 3475.23 A.

5.5 Conclusion

The operation of synchronous motors is based on the natural tendency of two fields to try to align their magnetic axes. The two fields—the stator and rotor fields—rotate at synchronous speeds.

The rotating stator field is produced by balanced three-phase currents flowing in properly distributed stator coils. The rotor field is produced by the direct current that flows through the synchronously rotating rotor windings.

The speed of a synchronous motor is directly proportional to the frequency of the stator voltage and inversely proportional to the number of poles. For a fixed-frequency source and number of poles, the speed is also fixed.

Because ordinary synchronous motors have these two crucial limitations—they operate at only one speed, and they require a dc voltage for their rotor windings and a balanced three-phase voltage for their stator windings—they have few applications.

At starting, synchronous motors function like squirrel-cage induction machines, and thus draw large inrush currents. As a result, the voltage drop in the impedance of the supply network may be considerable. The higher the inrush current, the greater the disturbance on the other electrical apparatus operating within the same distribution network.

High inrush currents also *reduce* the starting torque to levels that are sometimes too low to meet the requirements of a connected load. In this case, an engineer may select a soft-start motor. Such motors have relatively low starting torque and current, and they produce enough pneumatic power to provide the high starting-torque requirements of some loads.

These motors are coupled to high-starting-torque loads by an air clutch. The *motor* starts with the air clutch disengaged from the mechanical load. As soon as the *motor* is synchronized, air pressure—conducted to the air clutch through a properly sized opening inside the *motor's* shaft—couples the motor to the load. However, such a system has a high cost—about 15% of the cost of the synchronous *motor* itself.

The torque produced by a synchronous motor depends on its terminal voltage, its excitation voltage, and the phase angle between these two voltage phasors (commonly known as the power angle). Under normal operating conditions, the power angle is usually less than 30°.

The stator field may aid or oppose the rotor field. Once a machine is designed, this armature reaction depends on the operating power factor and on the stator current. As a result, a leading-power-factor motor requires more field current than a lagging-power-factor motor. Conversely, in order to produce a given terminal voltage, a synchronous generator requires much more field current when it operates at a lagging power factor than when it operates at a leading power factor. In either case, it is always useful to sketch and inspect a phasor diagram before arriving at any conclusions. As seen from the ac voltage supply, synchronous machines appear as variable capacitive or inductive impedances.

Vee curves give the variation of the power factor, and the armature current drawn by the motor, as a function of field current. In general, synchronous motors should not be designed for, or operated at, leading power factors, because their stator copper losses are increased. It is much more economical to operate at unity power factor and to purchase a capacitor bank to provide the required lagging kVAR power. There is, however, a disadvantage to reducing the power factor of a synchronous motor. Increasing the power factor from .8 leading to unity decreases the breakdown torque of the motor. Depending on load requirements, this may stall the motor. In short, each particular situation should be analyzed before any arbitrary adjustments are made.

Synchronous generators are used exclusively to generate three-phase power. Larger units—up to several hundred MVA's—are used by utility companies to generate commercially available three-phase power. Units with lower ratings—up to 1 MVA—are used in various plants to generate the standby power required to operate critical loads when the normal power source is temporarily interrupted. These generators are always driven at constant speeds by what are known as prime movers.

Theoretically, the terminal voltage of a generator—depending on the field current and power factor—varies over a wide range. In practice, however, owing to closed-loop voltage-regulator systems, it is kept constant at the desired level.

The various circuit parameters of a synchronous machine, when expressed in per-unit (as is the practice), are within a closed range and are available through a manufacturer's published data.

Salient-pole machines are stiffer than cylindrical-rotor machines—they can tolerate stronger load disturbances—and normally operate at speeds below 1800 r/min. Salient machines can be analyzed using the graphical or mathematical interpretation of the De Blondel equation. It is important to note here that salient machines can develop up to 40% of their rated torque without any excitation, while cylindrical-rotor machines cease to operate when the field current is removed.

When a three-phase synchronous machine is shorted, the resulting high stator currents create destructive forces. This must be taken into consideration when a machine is designed and installed. The short-circuit current contributed by a synchronous motor to faults remote from the motor but within the power grid of the plant contribute significantly to the overall short-circuit current. The short-circuit current contribution of a motor depends on its prefault operating condition and on its subtransient reactance.

Synchronous motors are not used (except when they are equipped with flywheels) to drive loads whose torque changes abruptly, or is cyclical in nature, because they can become unstable. Their stiffness, or capacity to oppose external disturbances, depends on their degree of saliency. Damping windings provide a higher starting torque and a stabilizing force each time the motor is forced to deviate from its synchronous speed.

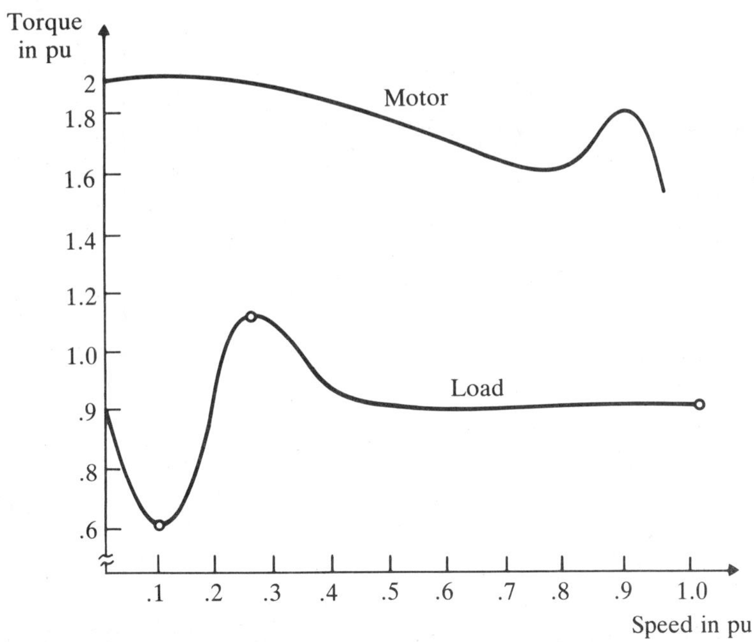

FIG. 5-41 Load and motor torque-speed characteristics. Motor: 2800 kW, 225 r/min, 60 Hz, 4000 V synchronous motor. Load: Ball grinding mill.

5.6 Review of Important Mathematical Relationships

Table 5-5 summarizes the main concepts of this chapter in their mathematical form.

TABLE 5-5 Review of important mathematical relationships

Item	*Description and Formula*	*Remarks*
	a) General	
1	Strength of the synchronously rotating field $$\mathscr{F}_s = \frac{3}{2}\mathscr{F}_1 \cos(\omega t - \beta)$$	Eq. (5.1)
2	Speed of synchronous machines $$n_s = 120\frac{f}{p} \text{ r/min}$$	Eq. (5.2)
	b) Cylindrical Rotor	
3	Synchronous impedance $$Z_s = R_a + jX_s$$	Eq. (5.7)
4	Relationship between mechanical and electrical radians (power angle) $$\delta_m = \frac{2}{p}\delta$$	Eq. (5.8)
5	Effective turns ratio $$N_e = \frac{I_a}{I'_a}$$	Eq. (5.9)
6	Field current $$I_f = I_m - I'_a$$	Eq. (5.11)
7	Excitation voltage $$V_f\angle{-\delta} = V_t\angle{0^\circ} - I_a(R_a + jX_s)$$	Eq. (5.12)
8	Power developed ($R_a = 0$) $$P = \frac{V_t V_f}{X}\sin\delta \text{ watts/phase}$$	Eq. (5.28)
9	Reactive power ($R_a = 0$) $$Q = \frac{V_t}{X}(V_t - V_f\cos\delta)$$	Eq. (5.36)

(*Continues*)

TABLE 5-5 (*Continued*)

Item	Description and Formula	Remarks
	c) Three-phase Generators	
10	Excitation voltage $V_f\angle\delta = V_t\angle 0^\circ + I_a(R_a + jX_s)$	Eq. (5.50)
11	Field current $I_f = I_m + I'_a$	Eq. (5.51)
12	Effective turns ratio $N_e = \text{slope of SCC} = \dfrac{I_a}{I'_a}$	Eq. (5.58)
13	Saturated value of synchronous reactance $X_{s_{\text{sat}}} = \left.\dfrac{V_{\text{rated}}}{I_{a_2}}\right\rvert_{I_{f_2}}$	Eq. (5.64)
	d) Salient-Pole Machines	
14	Excitation voltage $V_f\angle\pm\delta = V_t\angle 0^\circ \pm I_aR_a \pm jI_qX_q \pm jI_dX_d$ (positive sign for generator; negative sign for motor)	Eq. (5.73)
15	Power developed $P = \dfrac{V_fV_t}{X_d}\sin\delta + \dfrac{V_t^2(X_d - X_q)\sin 2\delta}{2X_dX_q}$ W/phase	Eq. (5.89)
	e) Three-phase Faults	
16	Symmetrical rms value of ac component $I'' = \dfrac{V''_f}{X''_d}$	Eq. (5.104)
17	Direct-current component $I_{\text{dc}} = \sqrt{2}\,\dfrac{V''_f}{X''_d}(\cos\alpha)e^{-t/T_a}$	Eq. (5.113)
18	Maximum rms value of stator current $I = \sqrt{3}\,\dfrac{V''_f}{X''_d}$	Eq. (5.117)

5.7 Manufacturer's Data

Tables 5-6, 5-7, and 5-8 present typical machine data, as furnished by manufacturers.

TABLE 5-6 Estimated characteristics for a three-phase, 514 r/min, 60 Hz, 4000 V, 3300 kW salient-pole synchronous motor—soft start—brushless exciter*

Item	Parameter	Power factor leading	
		.90	.80
1	Inrush current in percent	370	340
2	Torque in percent:		
	Starting	50	50
	Pull-in	65	62
	Pull-out	175	195
3	Efficiency in percent:		
	Full-load	95.2	94.8
	$\frac{1}{2}$ full-load	94.2	93.8
4	Reactances in per-unit:		
	X_d	2.2	2.46
	X'_d	.45	.55
	X''_d	.28	.32
5	Field excitation in kW	40	40

* *Courtesy of General Electric Canada, Inc.*

TABLE 5-7 Data for a three-phase, 225 r/min, 4000 V, .80 power factor leading, 2800 kW synchronous motor—hard start—brushless exciter*

Item	Parameter	Value	Remarks
1	Inrush current in percent	600	The starting characteristics of the motor and load (ball grinding mill) are shown in Fig. 5-41 (see p. 419).
2	Torque in percent:		
	Starting	2.2	
	Pull-in	1.4	
	Maximum	275	

(Continues)

TABLE 5-7 (*Continued*)

Item	*Parameter*	*Value*	*Remarks*
3	Reactances in per-unit:		
	X_d	1.1	
	X'_d	.29	
	X''_d (unsaturated value)	.126	
4	Efficiency at full-load	.95	
5	Exciter's amps at rated load	13.9	
6	Field resistance at 25°C in ohms	6.32	
7	Allowable stall time in seconds	5	
8	Accelerating time in seconds	4.5	

* *Courtesy of General Electric Canada, Inc.*

TABLE 5-8 Data for high- and low-starting-torque synchronous motors: 514 r/min, 60 Hz, .8 leading Pf, 4500 kW*

Item	*Description*	*Type of starting torque*	
		High	*Low*
1	Per-unit cost†	1.0	1.15
2	Inrush current in percent	6.5	3.75
3	Power factor at starting in percent	28	12
4	Efficiency at full-load	96	96.5
5	Torques in per-unit:		
	Starting	1.8	.4
	Pull-in	1.4	.7
	Pull-out	2	2
6	Torque angle at nominal operating conditions	28°	27°
7	Relative diameters	1	1.2

* *Courtesy of General Electric Canada, Inc.*
† *In 1988 the per-unit cost was about $400,000.*

5.8 REVIEW QUESTIONS

1. What are the advantages and disadvantages of synchronous motors?
2. Explain the magnetizing and demagnetizing effects of the armature current.
3. Draw and explain the Vee curves for a synchronous machine.
4. When starting a synchronous motor, a small resistance is placed in the rotor circuit; otherwise, the open-circuited windings may be damaged. Explain the reasons for this.
5. What conditions are necessary to produce a synchronously rotating magnetic field? How does this field compare with that produced by a permanent magnet rotating at the same speed?
6. The excitation voltage in a synchronous machine can be either larger or smaller than the terminal voltage, depending on the power factor. What are the reasons for this?
7. What is torque angle and on what external factors does it depend? How could it be observed in a laboratory?
8. What does the Potier triangle give? How would you measure it?
9. Define voltage regulation, and explain its significance and effects on the operation of electrical equipment.
10. What are the advantages and limitations of synchronous machines with salient rotors?
11. How does the impedance of a transmission line affect the power that can be transmitted between a generator and an infinite bus?

5.9 PROBLEMS

5-1 A three-phase, 60 Hz, 10-pole, 480 V, 100 kW synchronous motor has an overall efficiency of .92 and draws rated current at unity power factor. Its per-phase synchronous impedance is $.10 + j1.1$ ohms. Determine:

- **a)** The excitation voltage.
- **b)** The approximate value of the torque angle, in mechanical degrees.
- **c)** The stator electrical losses and the rotational losses.

5-2 A 300 kW, 4.16 kV, 600 r/min, 60 Hz, 3-ϕ synchronous motor has a synchronous impedance of $1 + j5$ ohms/phase. The motor delivers rated power and draws nominal current at a power factor of .9 leading. Its rotational losses are 12 kW. Determine:

- **a)** The armature current.
- **b)** The excitation voltage and the torque angle.
- **c)** The per-unit value of the synchronous impedance.

5-3 A 230 V, three-phase, 60 Hz, 8-pole, 37.3 kW star-connected synchronous motor has a synchronous reactance of .60 ohms per phase. The rotational losses are constant at 1800 watts and the stator resistance is negligible. Data for the open-circuit characteristic, at rated speed, is as follows:

Line-to-line voltage (V)	138	228	292	332	347
Field current (A)	2.0	4	6	8	10

Determine:

a) The field current when the motor draws rated current at a power factor of .80 leading.

b) The armature current when the load is halved and the excitation remains the same as in (a).

5-4 A 300 kW, 4-pole, 2.3 kV, 60 Hz, 3-ϕ synchronous motor has a synchronous impedance of .3 + j4.5 ohms/phase. Determine:

a) The torque angle at which the motor will draw maximum power from the infinite bus.

b) The torque angle at which the motor will draw maximum reactive power from the infinite bus.

5-5 A 750 kW, 1200 r/min, 2.2 kV, 60 Hz, 3-ϕ, .965 efficient synchronous motor has a per-phase synchronous reactance of 3.0 ohms and delivers power to a constant-torque mechanical load. The open-circuit characteristic at the operating region is given by

$$V_f = 254.37 I_f$$

When the motor draws rated current at unity power factor, determine:

a) **(1)** The excitation voltage.
(2) The torque angle.
(3) The field current.
(4) The magnitude of the armature current.
(5) The real, reactive, and complex power drawn by the motor.

b) If the excitation is changed by $\pm 25\%$ from its nominal value, while the load remains constant, repeat (a).

5-6 A 300 kW, 2-pole, 1.2 kV, 60 Hz, 3-ϕ synchronous motor has a leakage impedance of .1 + j1.0 ohms/phase. The effective turns ratio per pole is 20:1. The open-circuit characteristic is given by

$$V_f = 30 + 55 I_f$$

The motor receives electrical power at unity power factor. Its rotational losses are 17.05 kW. Determine:

a) The magnetization voltage.

b) The magnetizing component of the field current.

c) The armature reaction in equivalent field amperes.

d) The field current and the exact torque angle, in mechanical degrees.

e) The voltage induced in the armature windings due to the field current.

5-7 A 500 kVA, 1800 r/min, 4.16 kV, 60 Hz 3-ϕ, .94 efficient synchronous machine supplies rated current at .80 power factor lagging. Its synchronous impedance per phase is $1 + j12$ ohms. Determine:

a) The number of poles.

b) The per-unit value of the synchronous impedance.

c) The stator copper loss.

d) The electromechanical torque developed.

5-8 **a)** A 1.2 MVA, 1800 r/min, 6.6 kV, 60 Hz, 3-ϕ generator delivers rated current to an infinite bus at unity power factor. The synchronous reactance is 1.0 per-unit, based on its own rating. The open-circuit characteristic in the operating region is given by $V_f = 100 + 65I_f$. Determine:

(1) The excitation voltage.
(2) The torque angle.
(3) The armature current.
(4) The field current.
(5) The complex power delivered to the load.

b) Repeat part (a) if the field current is kept constant while the power supplied by the prime mover is changed by $\pm 25\%$.

5-9 A 750 kW, 1800 r/min, 6.3 kV, 60 Hz, 3-ϕ, .94 efficient synchronous motor when tested gave the following results:

Open-circuit test at rated speed:

Line voltage in volts	3500	5500	6500	7500	8000	8500
Field current in amperes	6	9.6	11.8	15	17.2	22

Short-circuit test at rated speed:

Line current	105 A
Field current	9 A

Zero-power-factor test:

Line current	105 A
Field current	20 A
Line voltage	5900 V

The motor delivers rated power and operates at a power factor of 0.80 leading. The resistance of the armature is 1.35 Ω/phase. Determine:

a) The Potier triangle.

b) The effective turns ratio.

c) The field current.

5-10 A 373 kW, 480 V, .8 Pf leading salient-pole synchronous motor has a direct-axis synchronous reactance of 1.0 per-unit and a quadrature-axis synchronous reactance of .65 per-unit. The rotational and stator copper losses are negligible. The field current is adjusted so that the motor draws minimum armature current while delivering rated power to a mechanical load. Under this operating condition, determine:

a) The power factor.

b) The power angle.

c) The excitation voltage.

d) The power developed. What percent of the total power is reluctance power?

5-11 A 5 MVA, 600 r/min, 6.6 kV, 60 Hz, 3-ϕ synchronous alternator delivers rated current at a power factor of .95 leading. Its direct-axis reactance and quadrature-axis reactance are 8.0 and 6.0 ohms, respectively. Determine:

a) The torque angle at full-load.

b) The regulation.

c) The reluctance torque in percent of the rated torque.

5-12 The symmetrical rms value of a generator's short-circuit currents is 2 kA. Determine the maximum dc components in each of the phase windings.

5-13 The three-phase synchronous generator shown in the one-line diagram of Fig. P5-13 delivers rated current at a power factor of .90 lagging. A solid three-phase fault occurs as shown. Neglecting the impedance of the transmission line and the current contribution of the system to the fault, determine:

a) The symmetrical three-phase fault current.

b) The maximum rms value of the armature current.

c) The dc components of the armature currents. Assume that the ac current is fully offset in phase A.

d) The rms value of the asymmetrical armature current as a function of time in phase A.

e) The rms value of the asymmetrical armature current 8 cycles after the fault inception in phase A.

f) The symmetrical three-phase fault current, when the three-phase short takes place at the end of the transmission line whose impedance is .02 + j.15 ohms/phase.

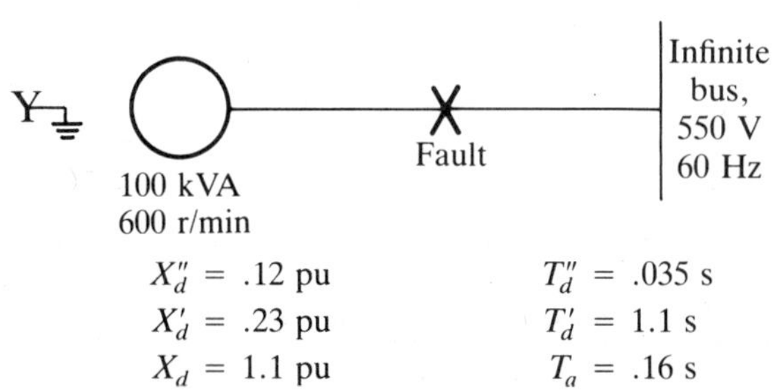

FIG. P5-13

SIX

DC Machines

6.0 Introduction

Direct-current motors are widely used in hoists, in steel mills, in cement plants, in the pulp and paper industry, and in mining operations. They are versatile and can be adjusted to meet tough job requirements, such as high starting torque, constant-power or constant-torque loads, quick acceleration and deceleration, and speed variations of up to $\pm 400\%$ of base value.

Although their torque–speed characteristics are easily adjustable, dc machines are not used for ordinary industrial applications because of the limitations, cost, and unavailability of dc voltage sources.

This chapter deals with the dc machine's principle of operation, steady-state operation, transient analysis, and modern techniques of speed control.

6.1 Steady-State Analysis

6.1.1 GENERAL

Like all the three-phase ac machines, direct-current machines have a stator and a rotor winding. The stator winding is referred to as the field winding and the rotor winding is referred to as the armature winding. Manufacturer's pictures of a dc motor and its stator and rotor windings are shown in Figs. 6-1, 6-2, and 6-3.

FIG. 6-1 A 375 kW dc motor. *Courtesy of General Electric Canada, Inc.*

FIG. 6-2 Stator of a 375 kW dc motor. *Courtesy of General Electric Canada, Inc.*

Field windings are wound around the poles of the stator and are supplied with dc current, which produces the main magnetic field of the machine. Depending on the machine type and rating, the stator structure may incorporate additional windings, such as the compensating, the commutating, and the auxiliary field windings. The functions and the relative physical location of these windings are described in Sections 6.1.5 and 6.1.7.

Armature windings are placed in the rotor slots, which are uniformly distributed around the rotor's periphery. The end connections of the armature windings terminate in the commutator segments (longitudinal bars of copper), which ride on the stationary brushes. The commutator segments are insulated from each other. In conjunction with the brushes, they rectify the ac current of the armature windings. The armature current of a dc motor is conducted from the external machine terminals to the brushes, then to the commutator segments, and from there to the armature conductors.

Depending on how the field winding is excited relative to the armature coil, dc machines are generally classified as shunt, series, or separately excited. As shown in Fig. 6-4, the field winding in a shunt machine is connected in parallel to the armature.

FIG. 6-3 Rotor of a 1200 kW, 600 V, 900 r/min dc motor. *Courtesy of General Electric Canada, Inc.*

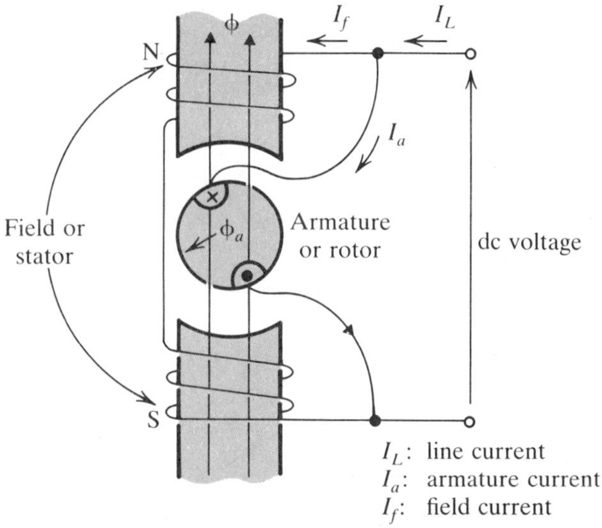

FIG. 6-4 An elementary representation of a 2-pole dc shunt motor. (The effect of the armature flux on the field flux is neglected.)

In series dc machines, the field winding is connected in series with the armature winding (see Fig. 6-14(a)). Depending on the location and relative magnetic orientation of the auxiliary field winding, dc shunt machines are further subdivided into other categories (see Section 6.1.7).

Regardless of the winding connection, as shown in Fig. 6-4, the current through the field windings (I_f) produces the field flux (ϕ_f), and the current through the armature windings (I_a) produces the armature flux (ϕ_a). The field flux combines with the armature flux to produce the machine's net effective flux (ϕ) per pole. The natural tendency of the stator and rotor fields to align their magnetic axes produces torque.

6.1.2 PRINCIPLES OF OPERATION

The equations that describe the generator-motor operation can be derived from the concept of the speed voltage (Eq. (1.100)) and Ampere's law (Eq. (1.164)) or by using the concepts associated with the energy stored in a coil and the definition of flux linkages. The latter method is used in this section because it gives a better understanding of the fundamentals.

Motor Principles

The operation of a dc motor is based on the tendency of a current-carrying coil, when placed in an external magnetic field, to rotate in such a way that the field it creates is parallel to, and in the same direction as, the external field. The general expression for the torque developed by a dc motor is derived as follows.

The armature winding shown in Fig. 6-5 has, for the sake of simplicity, only a single-loop coil. The instantaneous value of the torque developed (T_i) is determined by

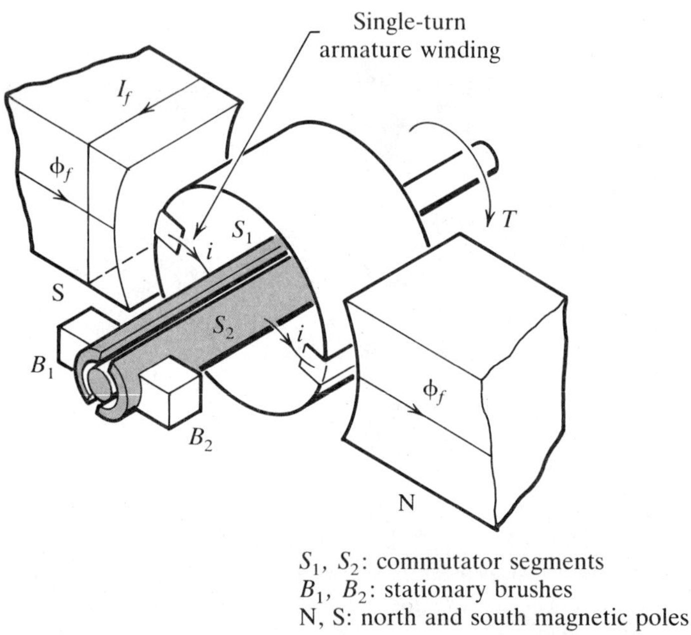

FIG. 6-5 An elementary representation of a dc machine.

the partial derivative of the energy stored (W_f) in the armature winding, with respect to its angle of rotation (θ). Mathematically expressed,

$$T_i = \frac{\partial}{\partial\theta} W_f \tag{6.1}$$

The incremental energy stored is given by

$$dW_f = v_g i \, dt \tag{6.2}$$

where v_g is the voltage induced in a single turn of the armature coil, and i is its current. From the above we obtain

$$T_i = \frac{v_g i \, dt}{\partial\theta} \tag{6.3}$$

Substituting for the voltage induced its equivalent expression in terms of the flux linkage (λ), we get

$$T_i = \frac{1}{\partial\theta}\left(i \frac{d\lambda}{dt} dt\right) \tag{6.4}$$

$$T_i = i \frac{d\lambda}{\partial\theta} \tag{6.5}$$

As the rotor turns dx of a revolution, the incremental change in the angle of rotation is

$$d\theta = 2\pi(dx) \text{ radians} \tag{6.6}$$

Each side of the conductor cuts, *only* once, through the regions of the north and south magnetic field. That is, the incremental change of the flux linkage for each side of the single-turn armature winding is

$$d\lambda = 2\phi(dx) \tag{6.7}$$

where dx, as before, is the incremental change in the rotation of the rotor. The change in the flux linkage for the entire length of the winding is

$$d\lambda = 4\phi(dx) \text{ for a two-pole machine} \tag{6.8}$$

$$= 2p\phi(dx) \text{ for a } p\text{-pole machine} \tag{6.9}$$

The current through the single winding is determined by the number of parallel paths of the armature winding and is related to the external armature current (I_a) by

$$i = \frac{I_a}{\beta} \tag{6.10}$$

where

β is the number of parallel paths of the armature winding

β = number of machine poles, for *lap-type** armature winding

$\beta = 2$, for *wave-type*† armature winding.

From Eqs. (6.5), (6.6), (6.9), and (6.10) the expression for the torque developed by a single conductor is

$$T_i = \frac{p}{\beta\pi}\phi I_a \ \text{N}\cdot\text{m/turn} \tag{6.11}$$

For N turns of the armature winding the total torque developed (T) by the motor is

$$T = \frac{Np}{\beta\pi}\phi I_a \ \text{N}\cdot\text{m} \tag{6.12}$$

or

$$T = K\phi I_a \ \text{N}\cdot\text{m} \tag{6.13}$$

where the constant of proportionality K is given by

$$K = \frac{Np}{\beta\pi} \tag{6.14}$$

Thus, the torque developed by a dc motor is directly proportional to the machine's effective field, to the armature current, to the number of poles, and to the number of armature conductors.

Generator Principles

The operation of generators is based on the induction principle, whereby voltage is induced in a conductor that is rotated through an external magnetic field. The polarity of the induced voltage, as per Lenz's law, is in such a direction as to produce a current whose mmf opposes the external magnetic field. The conductor under consideration is the coil of the rotor, the external magnetic field is provided by the stator coil, and the conductor's rotation is derived from a prime mover such as an induction motor, a steam or hydro turbine, etc.

. . .

* *Lap-type winding:* The starting and terminating ends of adjacent coils overlap each other's windings and are connected to adjacent commutator segments. The number of parallel paths is equal to the number of poles. This type of winding is used for low-voltage, high-current applications.

† *Wave-type winding:* The coils are connected in two parallel paths regardless of the number of poles. Most dc machines with ratings less than 75 kW are constructed using this type of winding.

The derivation of the general expression for the voltage generated is as follows. The voltage induced in a single-turn armature winding (Fig. 6-5) is given by the time rate of change of the armature coil's flux linkage. Considering only magnitudes, we have

$$v_g = \frac{d\lambda}{dt} \tag{6.15}$$

where v_g is the voltage generated and λ is the armature coil's flux linkage.

The change in the flux linkage of the armature winding that rotates through dx of a revolution is, as before,

$$d\lambda = 4\phi(dx) \text{ for a two-pole machine} \tag{6.16}$$

$$= 2p\phi(dx) \text{ for a } p\text{-pole machine} \tag{6.17}$$

The increment of time that the winding requires to travel through dx of a revolution is

$$dt = \frac{2\pi}{\omega}(dx) \text{ seconds} \tag{6.18}$$

where ω is the angular speed of the armature in radians per second. From Eqs. (6.15), (6.17), and (6.18) we obtain

$$v_g = \frac{p}{\pi}\phi\omega \text{ V/turn} \tag{6.19}$$

The total voltage generated in an armature winding depends on the number of coil turns connected in series. The number of coil turns connected in series is equal to the number of conductors divided by the number of parallel paths. That is,

$$\text{number of series conductors} = \frac{N}{\beta} \tag{6.20}$$

From Eqs. (6.19) and (6.20), the total voltage generated (V_g) in the armature conductors is

$$V_g = \frac{Np}{\beta\pi}\phi\omega \tag{6.21}$$

or

$$V_g = K\phi\omega \tag{6.22}$$

where the constant of proportionality K is the same as that given for the torque equation (see Eq. (6.14)).

The voltage generated in a single conductor, the machine's total voltage, and its rectification are further discussed in Section 6.1.6 and are shown in Fig. 6-10.

The voltage generated in the armature conductors is also referred to as the armature voltage V_a, or the countervoltage V_c, or the back emf V_b, or simply as the speed voltage. One must differentiate between the voltage generated in a machine and its terminal voltage. In this text the voltage generated will be designated as V_g and the terminal voltage as V_t.

From the preceding discussion it is evident that the principles of dc motor and generator operation are the same as those of polyphase ac machines. The essential difference between a motor and a generator is the direction of energy flow. A motor transforms electrical energy into mechanical form, while a generator transforms mechanical energy into electrical form.

6.1.3 POWER CONSIDERATIONS

The electromagnetic power developed (P_d) by a machine is equal to the product of the electromagnetic torque developed (T) times the speed (ω) of its rotor. In mathematical form,

$$P_d = T\omega \tag{6.23}$$

The torque is expressed in N · m, the speed in rad/s, and the power in watts. Applying the power-balance concept for a motor, we obtain

$$\begin{pmatrix}\text{input}\\ \text{electrical}\\ \text{power}\end{pmatrix} = \begin{pmatrix}\text{sum of various}\\ \text{winding copper}\\ \text{losses}\end{pmatrix} + \begin{pmatrix}\text{power}\\ \text{developed}\end{pmatrix}$$

Using mathematical symbols,

$$V_t I_L = I_f^2 R_f + I_a^2 R_a + P_d \tag{6.24}$$

where

V_t and I_L = the motor's supply (terminal) voltage and line current, respectively

$I_f^2 R_f$ = the copper losses of the field winding

$I_a^2 R_a$ = the copper losses of the armature winding.

The power developed provides the output power (P_{out}) plus the rotational losses ($P_{\text{r.l.}}$). Mathematically expressed,

$$P_d = P_{\text{r.l.}} + P_{\text{out}} \tag{6.25}$$

To a certain extent rotational losses depend on the speed of the motor, but throughout this chapter they will be assumed to remain constant. At nominal operating conditions, the output power of the motor is referred to as the full-load power, rated power, or name-plate power.

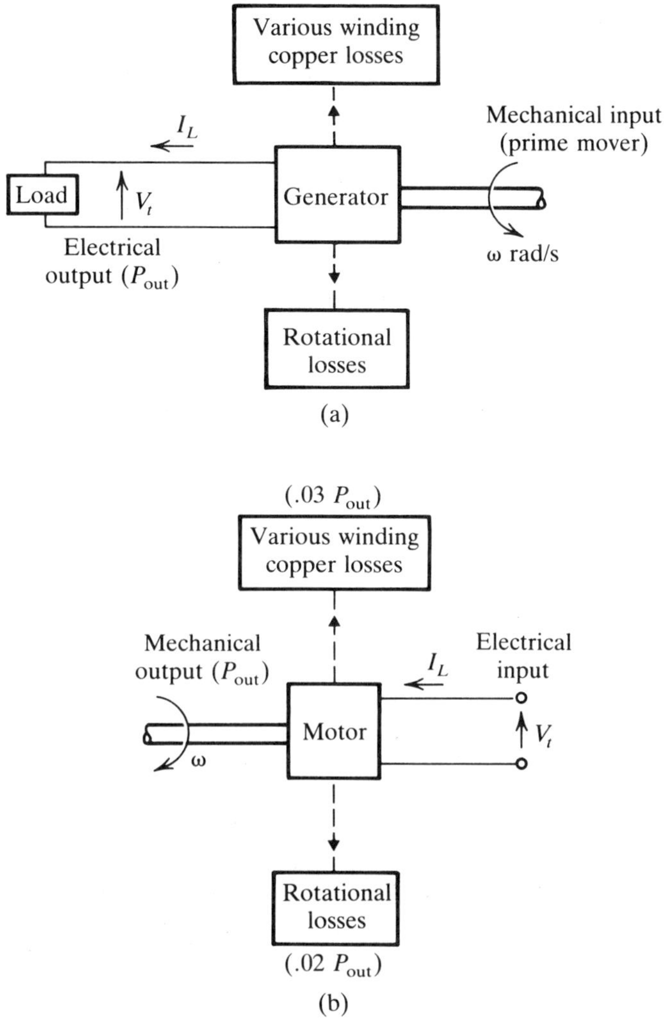

FIG. 6-6 DC machines—direction of power flow and typical losses for a 100 kW machine: **(a)** generator, **(b)** motor.

The various losses for a 100 kW dc machine, and the direction of energy flow, are shown in Fig. 6-6.

EXAMPLE 6–1 A 10 kW, 4-pole dc machine is magnetized at a flux level of .08 Wb and has a lap-type armature winding with 24 turns. When the motor delivers rated power at 1200 r/min, determine:

a) The voltage induced in the armature windings.

b) The armature current.

SOLUTION

a) The speed of the motor is

$$\omega = 1200 \frac{2\pi}{60} = 125.7 \text{ rad/s}$$

For a lap-type winding the number of parallel paths is equal to the number of poles. Thus,

$$\beta = 4$$

Substituting the given data into Eq. (6.21), we obtain

$$V_g = \frac{24(4)}{4\pi}(.08)(125.7) = \underline{\underline{76.8 \text{ V}}}$$

b) From Eq. (6.23) the torque is

$$T = \frac{\text{power}}{\text{speed}} = \frac{10{,}000}{125.7} = 79.6 \text{ N} \cdot \text{m}$$

From Eq. (6.12),

$$I_a = \frac{79.6}{24 \times 4(.08)}(4\pi) = \underline{\underline{130.21 \text{ A}}}$$

Exercise 6–1 A 5 kW, 2-pole dc motor has 200 W of rotational losses and draws an armature current of 30 A when it delivers rated power at 900 r/min. Determine:

a) The voltage generated in the armature windings.

b) The torque developed and the output torque of the motor.

Answer (a) 173.3 V; (b) 55.2 N · m, 53.1 N · m.

Exercise 6–2 A two-winding transformer has a primary and a secondary winding and two corresponding mmf's. The interaction of these two fields does not produce any torque. What are the reasons for this?

6.1.4 VOLTAGE AND TORQUE RELATIONSHIPS AS FUNCTIONS OF MUTUAL INDUCTANCE

General expressions for the voltage generated and the torque developed in a dc machine, as functions of the mutual inductance between the stator and rotor windings, will be derived in this section.

The voltage induced in the armature of a dc machine with constant field current is given by

$$v_g = -I_f \frac{d}{dt} L_{af(\theta)} \tag{6.26}$$

where $L_{af(\theta)}$ is the mutual inductance between the armature and field windings, as a function of the angle of rotation. The mutual inductance is nonsinusoidal in waveform. It has a fundamental term and harmonics of higher order. The negative sign gives the polarity of the induced voltage as per Lenz's law.

Rewriting the relationship of mechanical (θ_m) and electrical (θ) radians (see Eq. (3.15)), we have

$$\theta = \frac{p}{2}\theta_m \tag{6.27}$$

Differentiating with respect to time, we get

$$\frac{d\theta}{dt} = \frac{p}{2}\frac{d\theta_m}{dt} \tag{6.28}$$

From the above,

$$\frac{d\theta}{dt} = \frac{p}{2}\omega \tag{6.29}$$

where p is the number of poles and ω is the speed of the motor in rad/s.

Equation (6.26) can be rewritten as follows:

$$v_g = -I_f \frac{d\theta}{dt}\frac{dL_{af(\theta)}}{d\theta} \tag{6.30}$$

From the above and Eq. (6.29) we obtain

$$v_g = -I_f \frac{p}{2}\omega \frac{dL_{af(\theta)}}{d\theta} \tag{6.31}$$

Assuming that the mutual inductance has only a fundamental component, the maximum value of the induced voltage is

$$V_g = \frac{p}{2}\omega I_f L_{af_m} \tag{6.32}$$

where L_{af_m} is the maximum value of the fundamental component of the mutual inductance.

The equation of the torque is derived as follows:

$$T = \frac{\text{power}}{\text{speed}}$$

$$= \frac{\begin{pmatrix}\text{voltage induced} \\ \text{in the armature}\end{pmatrix} (\text{armature current})}{\text{speed}}$$

Thus

$$T = \frac{V_g I_a}{\omega} \tag{6.33}$$

From the above and from Eq. (6.32) we obtain

$$T = \frac{p}{2} I_f L_{afm} I_a \tag{6.34}$$

Comparing Eqs. (6.13) and (6.34), it becomes evident that the constant of proportionality of a given dc machine may also be expressed in terms of the mutual inductance between the field and armature windings. That is,

$$K = \frac{p}{2\phi} I_f L_{afm} \tag{6.35}$$

The maximum value of mutual inductance between the rotor and stator windings can be measured by one of the standard methods (see Chapter 1, Example 1–20) or can be evaluated indirectly from Eq. (6.32).

The voltage induced in the field windings due to armature current is negligible, because the structure of the machine drastically reduces the magnetic coupling of the stator windings to the flux of the armature current.

Exercise 6–3 A 5 kW, 1200 r/min, 4-pole dc machine has an armature current of 20 A and a field current of 2 A. Determine:

a) The amplitude of the fundamental component of the mutual inductance between the stator and rotor windings.
b) The voltage generated.

Answer (a) .497 H; (b) 250 V.

6.1.5 MAGNETIC SYSTEM, FLUX DISTRIBUTION, AND ARMATURE REACTION

This section discusses the dc machine's magnetic system, flux distribution, and the interaction of the stator and rotor fields, commonly known as armature reaction.

Magnetic System and Flux Distribution

Figure 6-7 shows an elementary magnetic system for a two-pole dc machine. The flux produced by the mmf of the field winding completes its loop by passing through the air gaps, stator, and rotor structures.

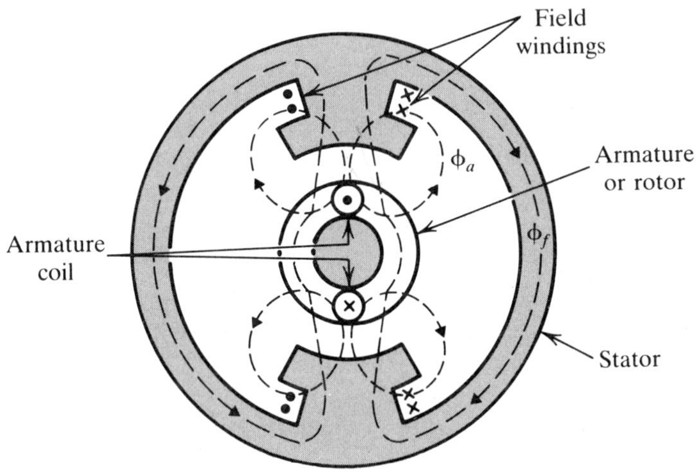

FIG. 6-7 An elementary magnetic system for a two-pole dc machine.

Applying Ohm's law for magnetic circuits, we have

$$\phi_f = \frac{N_f}{\mathscr{R}_f} I_f \tag{6.36}$$

where the subscript f indicates field parameters. The reluctance ($\mathscr{R}_f$) represents the opposition presented to the field flux by the intrinsic properties of the magnetic circuit components. This reluctance is determined by the sum of the magnetic resistances of the rotor, stator, and air-gap paths, through which the field flux completes its loop.

In general,

$$\mathscr{R} = \frac{l}{\mu A} = \frac{\text{length of closed magnetic path}}{(\text{permeability}) \times \begin{pmatrix}\text{cross-section area}\\ \text{perpendicular to the}\\ \text{flux}\end{pmatrix}} \tag{6.37}$$

Taking into consideration the three distinct parts of the field's magnetic circuit, we have

$$\mathscr{R}_f = \frac{l_r}{\mu_r A_r} + \frac{l_s}{\mu_s A_s} + \frac{l_g}{\mu_g A_g} \tag{6.38}$$

where the subscripts r, s, and g represent, respectively, the usual parameters of rotor, stator, and air gaps through which the field flux closes its magnetic circuit.

For operations along the linear section of the magnetic characteristic the permeability is constant. Thus from Eqs. (6.36) and (6.38) we obtain

$$\phi_f = K_f I_f \tag{6.39}$$

where K_f is a constant of proportionality that is proportional to the number of field winding turns and inversely proportional to the reluctance of the field's magnetic circuit.

Similarly, the flux of the armature coils (ϕ_a) is determined by

$$\phi_a = K_a I_a \tag{6.40}$$

where I_a is the armature current and K_a is a constant of proportionality which, as before, depends on the reluctance of the armature's magnetic circuit and on the number of turns of the armature winding.

It should be emphasized that Eqs. (6.39) and (6.40) are *not* applicable for machine operations along the nonlinear part of the magnetization characteristic.

Flux-Density Distribution

Figure 6-8 shows a simplified version of the flux-density distribution of a two-pole dc machine. The armature (B_a) and field (B_f) flux densities combine to produce the resultant (B_r) or effective flux density.

The field flux density under the north magnetic pole is designated as positive, while that under the south magnetic pole is negative. The relative polarity becomes obvious when one considers that in a N–S magnetic system the lines of force leave the north pole and enter the south pole.

The distribution of the armature flux density is of lower amplitude because of the larger reluctance path through which the armature flux completes its magnetic circuit. Notice also that for one complete revolution of the rotor, each side of the armature coil goes through a complete magnetic circuit. As a result, the armature coil will appear to travel two cycles for each magnetic field cycle.

The flux density distribution is zero under the stationary brushes because the latter are physically located at a point between the north and south magnetic poles where the field is zero.

The Concept of Armature Reaction

The flux of the armature current effectively *opposes* the flux of the field current and *distorts* the waveform of the field's flux-density space distribution. These two adverse effects of the armature flux are referred to as armature reaction. The qualitative analysis of the armature reaction is described below, while its quantitative analysis is given in Section 6.1.9.

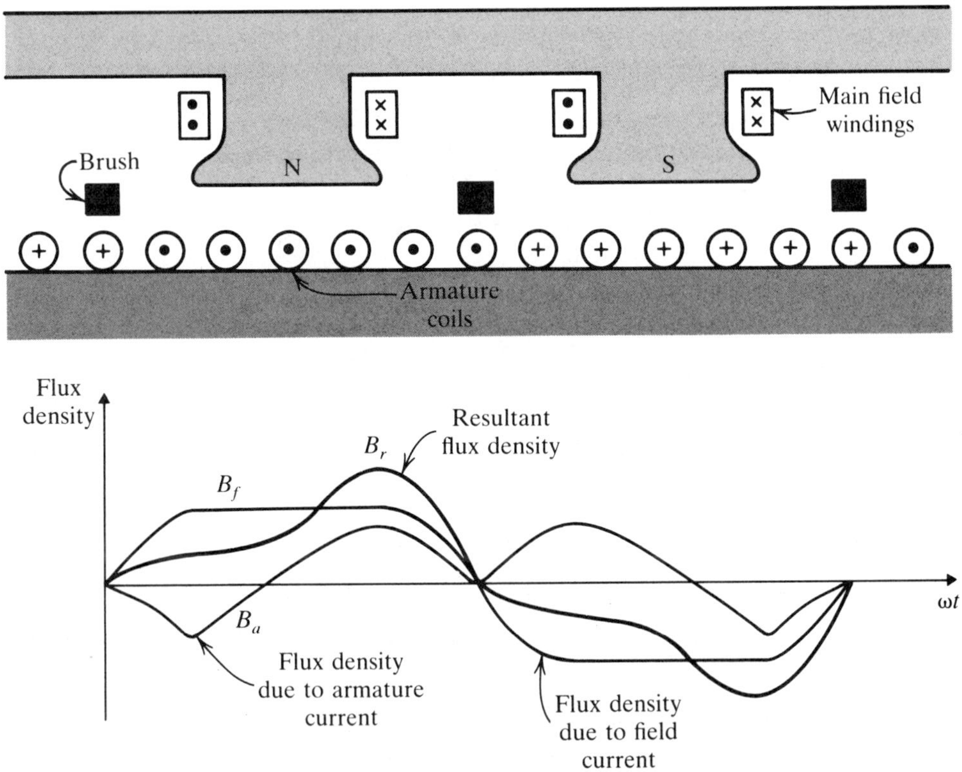

FIG. 6-8 Simplified view of a two-pole machine and the corresponding flux-density distributions. (Interpoles are not shown.)

Opposition to the Field

Depending on the rotor position, the flux of the armature *aids* the flux of the field on one side of the pole and *opposes* it on the other side. Aiding the field in one pole partly saturates the magnetic material, and thus the effective field is reduced. The overall effect of the armature mmf is, then, to distort and/or to oppose the flux of the field. This opposition, or armature reaction (AR), is usually less than about 5% of the field's flux at no-load.

With constant field current and rated armature current, an armature reaction of 4% means that

$$\phi_2 = .96\phi_1$$

where ϕ_1 is the machine's flux at negligible armature current ($I_{a_1} = 0$), and ϕ_2 is its effective flux at rated armature current (I_{a_2}). In other words, at full-load the effective magnetic field is 96% of the field at no-load.

The cross-magnetizing effect of the armature current may be minimized by placing, in longitudinal slots in the pole faces, a *compensating winding* (see Fig. 6-9). This

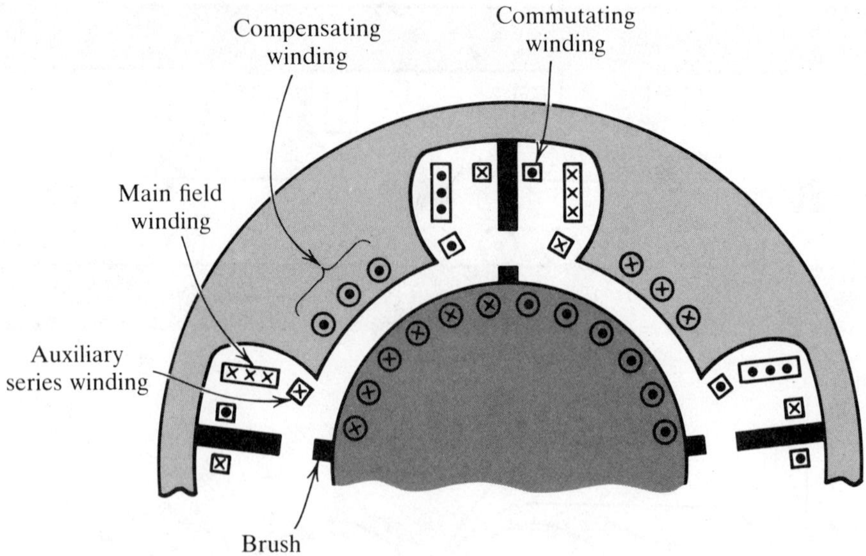

FIG. 6-9 DC machines. This figure illustrates the physical location and relative polarity of compensating and commutating windings.

winding is connected in series with the armature and hence carries the same current, but in the direction opposite to that of the adjacent armature conductors.

Distortion of the Field's Flux Distribution

As shown in Fig. 6-8, the flux of the armature coil distorts the field in the interpole region, particularly where the commutator segment rides under the positive or the negative polarity brush. This results in a nonlinear variation of the armature current, which can lead to severe sparking. In fact, one of the essential limitations of dc machines is the arcing and possible flashover that may accompany the reversal of the armature current, commonly known as commutation. (Commutation will be explained in the next section.) To cancel this adverse effect, a coil—the so-called *commutating winding*—is wound on the narrow poles of the interpole region and connected in series with the armature circuit. When properly designed, a commutating winding can minimize the distorting effect that the armature mmf has on the field flux distribution.

6.1.6 COMMUTATION

The simplified schematic of Fig. 6-10(a) shows that during one complete revolution of the shaft, each side of the armature conductor is rotated through two opposite magnetic fields (N and S). Thus the voltage, or the current, would be positive and negative or approximately sinusoidal in waveform, as shown in Fig. 6-10(b).

Properly reversing the end connections of the conductor causes a unidirectional or dc current to flow in the machine's external circuit. The current is reversed through the brushes, whose polarity is fixed by their physical location relative to the north and

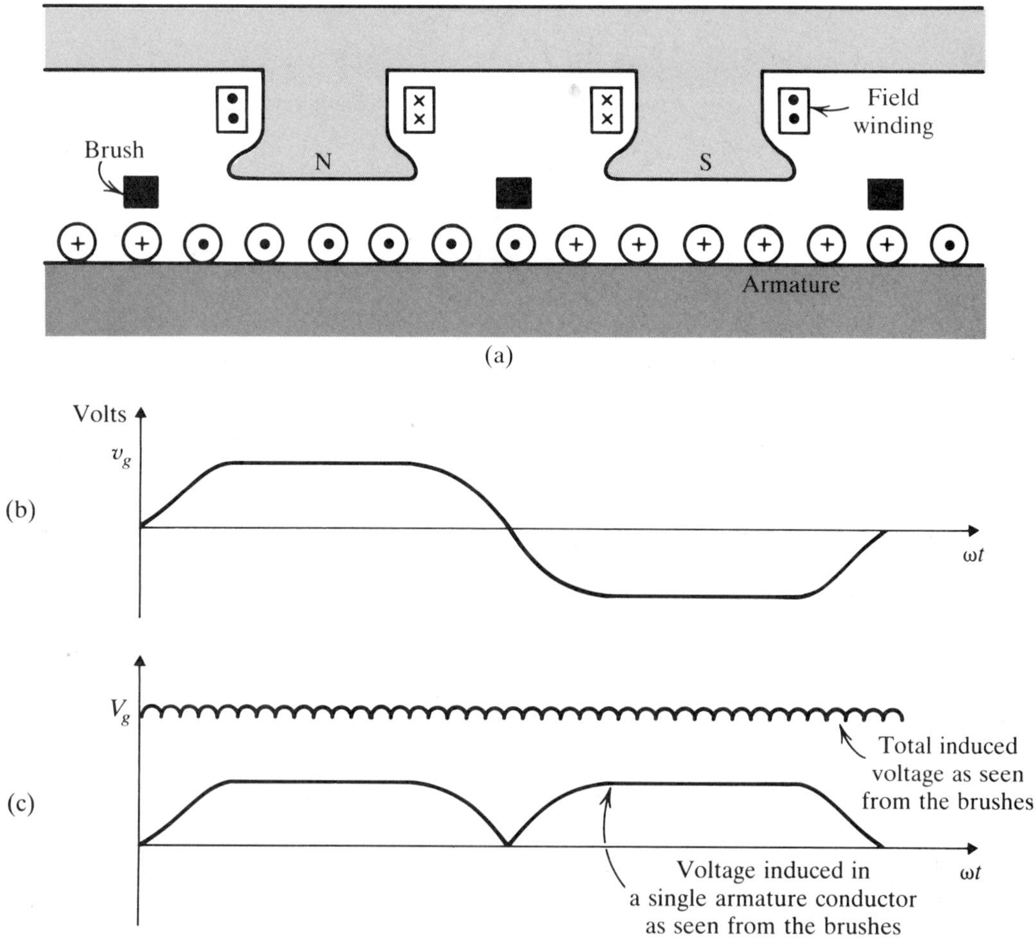

FIG. 6-10 **(a)** Simplified view of a two-pole machine, **(b)** voltage induced in a single armature conductor, **(c)** effect of the brushes on the induced voltage.

south poles of the field winding. The reversal of the armature current during half the period of its cycle is called *commutation.*

As the location of the brushes indicates, the start of the commutation takes place when the conductors are at the region of the weakest magnetic field (see Fig. 6-10(c)). Ideal or sparkless commutation occurs when the armature current is made to vary linearly during the commutation period. This occurs when the resistance of the shorting brush(es) is the predominant characteristic of the shorted armature conductor.

The instant the conductor leaves the north pole of the magnetic field and is about to enter the south pole, the brush, as shown in Fig. 6-11, short-circuits the conductor. As a result, sparking may take place. The degree of the arcing will depend on the magnitude of the armature current, on the voltage induced in the shorted conductor,

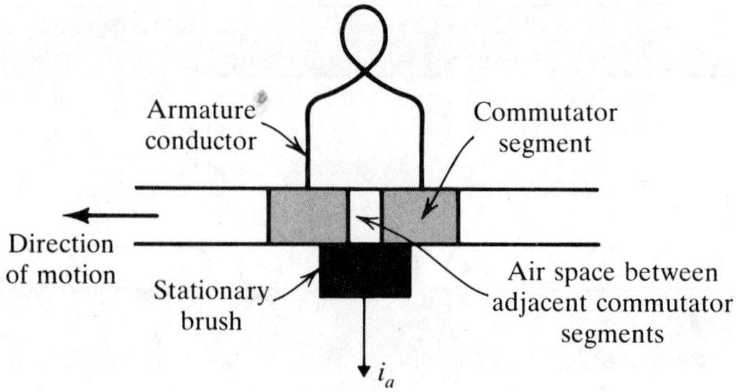

FIG. 6-11 Commutation, showing the location of brush and commutator segments when an armature conductor is shorted.

and on the resistance of the short-circuited path. Arcing is also increased by the distortion of the air-gap field that results from the armature reaction. The distortion of the air-gap field, as already discussed, can be completely neutralized by properly designing the commutating winding.

Excessive sparking may result in the breakdown of the air space between the commutator segments, and thus a flashover may engulf the entire periphery of the commutator. The flashover, as seen from the terminals of the motor, constitutes a short circuit to the external voltage source.

Mathematical Considerations

If one neglects the coil resistance, the voltage induced in an armature coil (V_{coil}) depends on its self-inductance, its mutual inductance, and the speed of the rotor. Mathematically,

$$V_{\text{coil}} = K\phi\omega - \frac{d\lambda_a}{dt} - \frac{d}{dt}(i(t)L(t)) \qquad \textbf{(6.41)}$$

where:

$K\phi\omega$ = the speed voltage or the voltage generated due to the induction principle. During the commutation period, this voltage changes because of variations in the magnitude and direction of the air-gap flux density. Its magnitude can be controlled by properly modifying the air-gap field distribution in the commutating region.

$\frac{d}{dt}(\lambda_a) = L_a \frac{di_{a(t)}}{dt}$ = the voltage drop caused by the coil's self-inductance and the rate of change of its current. During the commutation period this component of the voltage changes, because the current changes in magnitude and direction. By definition, this voltage is also given by the rate of change of the flux linkage of the coil undergoing commutation.

$\frac{d}{dt}(i(t)L(t))$ = the voltage drop due to the conductor's mutual ($L(t)$) inductance with respect to the adjacent conductors that carry an instantaneous current of $i(t)$ amperes. Here only one term is considered, but it is obvious that almost all coils are mutually coupled to the conductor that is undergoing commutation.

Before commutation takes place, these voltages are balanced by the external voltage. During commutation, however, the coil is short-circuited, and the sum of these voltages must be equal to zero. The generation or development of sparks results from nature's attempt to set up a voltage that satisfies KVL.

6.1.7 EQUIVALENT CIRCUITS AND EXTERNAL MACHINE CHARACTERISTICS

The equivalent circuit of a dc machine explicitly shows how the field winding is excited in relation to the armature winding. As such, an equivalent circuit reveals the type of machine (series, shunt, separately excited, etc.) it represents. An equivalent circuit may include the actual values of the winding impedances, so it can be used for transient and or steady-state analyses. In this section, however, only steady-state analyses are discussed. Consequently, the equivalent circuits include only the resistance of the armature (R_a) and field windings (R_f).

The resistance of the armature winding is relatively small, normally a fraction of an ohm. It includes the resistances of the armature conductors, the resistances of the compensating and commutating windings. The effects of the contact between the brush and the commutator segments is represented by a constant voltage drop (about 1 V).

The resistance of the field winding, when connected in series with the armature winding, is very small. When connected in parallel to the supply voltage, however, it is several hundred times larger than the resistance of the armature windings.

The external characteristics of dc machines can easily be derived from the basic motor-generator relationships and from the equations that can be obtained from the machine's equivalent circuit.

Shunt dc Motors

The equivalent circuit of a dc shunt motor is shown in Fig. 6-12(a). The voltage V_t represents the motor's dc supply or terminal voltage, and I_a represents the current of the armature windings. The armature current is determined by the supply voltage, the driven load, and the field current. The resistance of the field winding ($R_f > 100\ \Omega$) is much greater than the resistance of the armature coil ($R_a < .2\ \Omega$), and under normal operating conditions the field current is much smaller than the armature current.

Ordinarily, a dc shunt motor is equipped with a "starting box" that houses an external resistor whose value can be decreased in several steps. The purpose of the external resistor is to limit the armature current during starting (see Section 6.1.8).

When stopping a dc shunt motor, the armature circuit should be opened before, or at the same time as, the field circuit. If the field current is interrupted while the armature

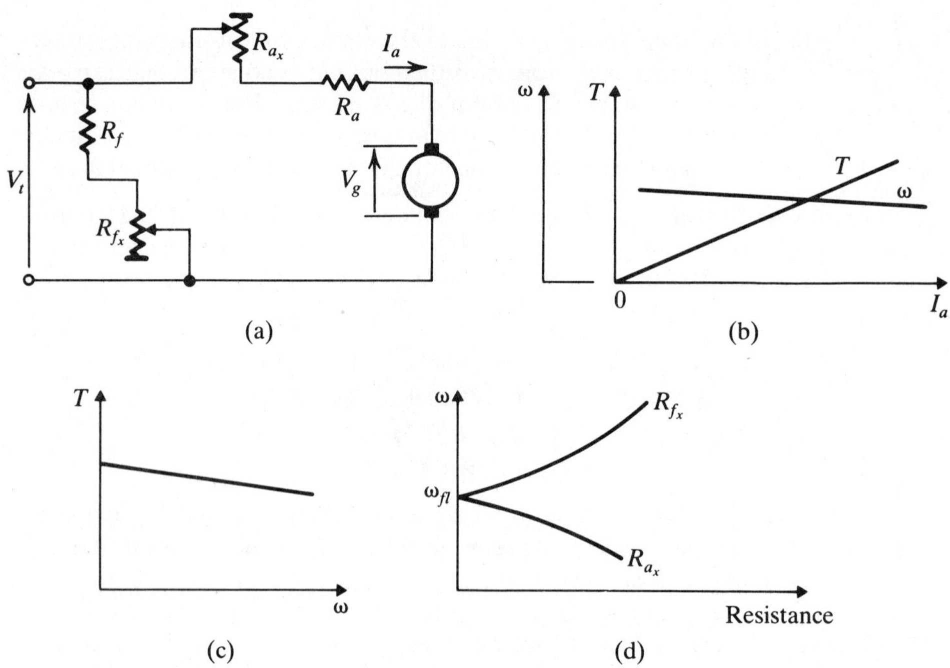

FIG. 6-12 Shunt dc motors: **(a)** equivalent circuit including the rheostat's resistances, **(b)** torque and speed versus armature current, **(c)** torque versus speed, **(d)** speed versus rheostat resistances.

circuit is energized, severe sparking on the commutator will result, and the motor's speed will increase to dangerously high levels.

For purpose of speed control, the motor may be equipped with variable external resistors in the armature (R_{a_x}) and field (R_{f_x}) circuits. The motor's external characteristics are derived as follows.

Torque versus Armature Current

The basic motor-torque equation, for a constant field current and negligible armature reaction, becomes

$$T = K_1 I_a \tag{6.42}$$

That is, the torque is directly proportional to the armature current. The torque–armature current characteristic is shown in Fig. 6-12(b).

Speed versus Armature Current

From the basic generator equation and KVL in the equivalent circuit ($R_{a_x} = 0$) we have

$$V_g = K\phi\omega \tag{6.43}$$

$$V_t = V_g + I_a R_a \tag{6.44}$$

From the above,

$$\omega = \frac{V_t}{K\phi} - \frac{R_a}{K\phi} I_a \tag{6.45}$$

For constant field current the coefficient of I_a is small and thus the speed of the motor does not appreciably change with moderate variations in the armature current. (See Fig. 6-12(b).)

Torque versus speed

From Eqs. (6.43) and (6.44), we obtain

$$V_t = I_a R_a + K\phi\omega \tag{6.46}$$

Also from the basic torque equation we have

$$T = K\phi I_a \tag{6.47}$$

From the last two equations we obtain

$$T = \frac{K\phi}{R_a} V_t - \frac{(K\phi)^2}{R_a} \omega \tag{6.48}$$

For a constant field current, the last equation is of the intercept-slope form, and because the slope is small, moderate variations in the speed do not appreciably change the torque. The torque–speed characteristic of the shunt motor is shown in Fig. 6-12(c).

Speed versus External Resistors

The speed of the shunt motor can be controlled through the external armature and field resistors.

Effects of Armature Rheostat (R_{ax})

Applying KVL in the equivalent circuit, we obtain

$$V_t = V_g + I_a(R_a + R_{ax}) \tag{6.49}$$

From Eqs. (6.43) and (6.49) we get

$$\omega = \frac{V_t - I_a R_a}{K\phi} - \frac{I_a}{K\phi} R_{ax} \tag{6.50}$$

For constant field and armature currents, the larger the external armature resistance, the smaller will be the motor's speed.

Equation (6.50) makes it evident that the external armature resistance can control the speed of the motor from zero up to its full-load value. This method of speed control is accompanied by high copper losses ($I_a^2R_{ax}$). For this reason, almost all modern speed-control apparatus uses solid state devices to reduce the terminal voltage and thus to control the speed of the motor. The speed of a shunt motor as a function of the external armature resistance is shown in Fig. 6-12(d).

Effects of Field Rheostat (R_{fx})

From the speed-voltage equation and Ohm's law, across the field winding we have

$$V_g = K_1 I_f \omega \tag{6.51}$$

$$I_f = \frac{V_t}{R_f + R_{fx}} \tag{6.52}$$

From Eqs. (6.44), (6.51) and (6.52) we obtain

$$\omega = \frac{(V_t - I_a R_a)(R_f + R_{fx})}{K_1 V_t} \tag{6.53}$$

Since $V_t \gg I_a R_a$, the last expression becomes

$$\omega \approx \frac{R_f + R_{fx}}{K_1} \tag{6.54}$$

Equation (6.54) shows that by increasing the external field resistance, the speed of the motor is also increased. The speed of the motor as a function of the field rheostat resistor is shown in Fig. 6-12(d).

The maximum attainable speed is limited by sparking in the brushes or by mechanical considerations of the centrifugal forces.

EXAMPLE 6–2 A 10 kW, 220 V shunt dc motor has field and armature resistances of 110 ohms and .20 ohms, respectively. At no-load the motor runs at 1200 r/min and has an armature current of 5.0 A. Determine the speed and electromagnetic torque when the motor draws 50 A from the supply. Assume an armature reaction that is:

a) negligible
b) 4%.

SOLUTION

a) The parameters marked with the subscript 1 will correspond to the no-load condition and those marked with subscript 2 will correspond to the load condition of 50 A. Applying KVL to Fig. 6-13 we obtain

$$\begin{aligned} V_{g_1} &= V_t - I_{a_1} R_a \\ &= 220 - .2(5) = 219 \text{ V} \end{aligned}$$

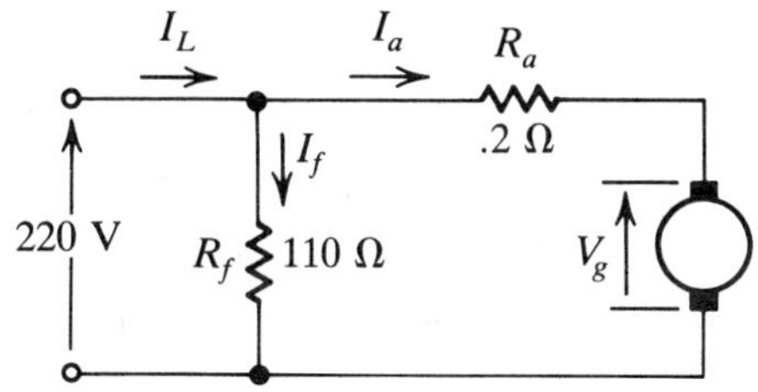

FIG. 6-13

When the motor draws 50 A, the armature current will be

$$I_{a_2} = I_L - I_f = 50 - \frac{220}{110} = 48 \text{ A}$$

and

$$V_{g_2} = 220 - .2(48) = 210.4 \text{ V}$$

For negligible armature reaction and constant field current,

$$\phi_1 = \phi_2$$

From the generator principle we have

$$\frac{V_{g_2}}{V_{g_1}} = \frac{\phi_2}{\phi_1}\frac{n_2}{n_1}$$

From the above,

$$n_2 = 1200\frac{210.4}{219} = \underline{\underline{1152.88 \text{ r/min}}}$$

Besides giving the rotational losses, the no-load data can also be used to establish the motor constant, as follows:

$$V_{g_1} = K\phi\omega_1$$

For constant field,

$$V_{g_1} = K_1\omega_1$$

Thus,

$$K_1 = \frac{219}{1200\dfrac{2\pi}{60}} = 1.74 \text{ V/rad/s}$$

The electromagnetic torque developed is

$$\begin{aligned} T &= K\phi I_a \\ &= K_1 I_a \\ &= 1.74 \times 48 = \underline{\underline{83.65\ \text{N}\cdot\text{m}}} \end{aligned}$$

Alternatively,

$$T = \frac{V_{g_2} I_a}{\omega} = \frac{210.4(48)}{1152.88\,\frac{2\pi}{60}} = \underline{\underline{83.65\ \text{N}\cdot\text{m}}}$$

b) For an armature reaction of 4%, we have

$$\phi_2 = .96\phi_1$$

From the generator principle we obtain

$$n_2 = \frac{V_{g_2}}{V_{g_1}} \frac{\phi_1}{\phi_2} n_1$$

Substituting the known parameters, we obtain

$$n_2 = 1200\,\frac{210.4}{219}\left(\frac{1}{.96}\right) = \underline{\underline{1200.91\ \text{r/min}}}$$

The torque is found as follows:

$$\begin{aligned} T &= K\phi_2 I_{a_2} \\ &= .96(K\phi_1 I_{a_2}) = .96(1.74 \times 48) \\ &= \underline{\underline{80.31\ \text{N}\cdot\text{m}}} \end{aligned}$$

For purposes of comparison, the results are summarized in the Table 6-1. Clearly, the armature reaction increases the speed and reduces the torque of the motor by a percentage about equal to that of the armature reaction.

TABLE 6-1 Summary of results of Example 6–2

Assumed armature reaction	*Armature current (A)*	*Speed r/min*	*Torque developed (N · m)*
0	48	1152.88	83.65
4%	48	1200.91	80.31

Series dc Motors

In a series dc motor, the field winding is connected in series with the armature winding (see Fig. 6-14(a)). For speed control, a variable resistor (R_{f_x}) is often connected in parallel to the field winding. The motor's external characteristics are derived as follows.

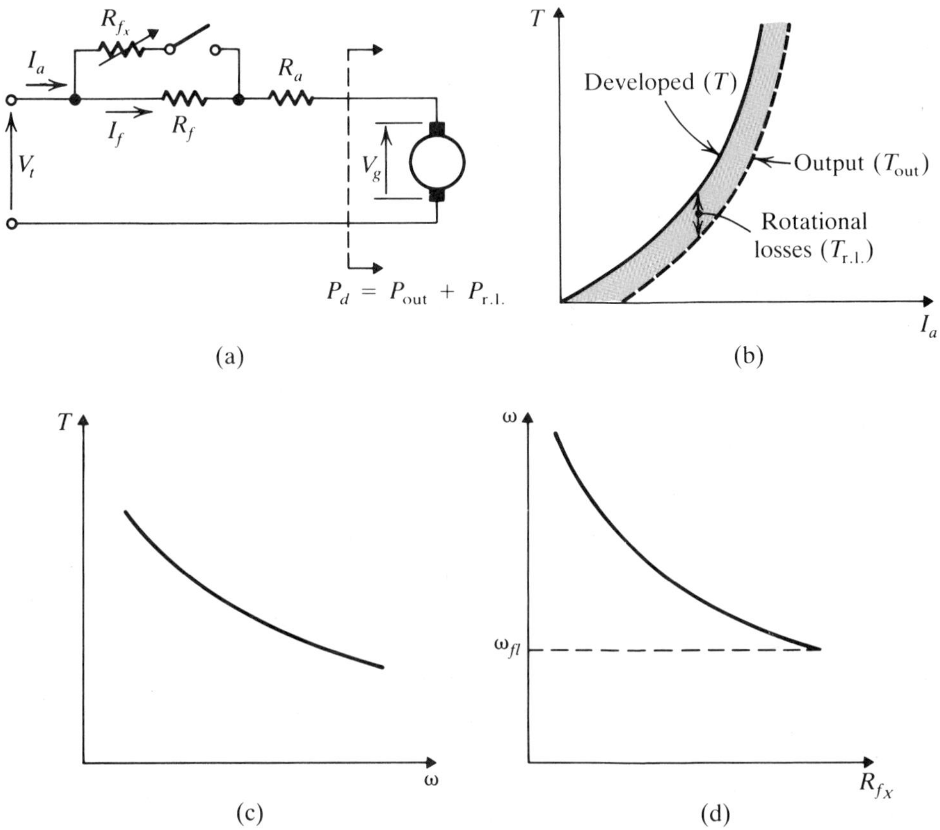

FIG. 6-14 Series dc motors: **(a)** equivalent circuit, **(b)** torque versus armature current, **(c)** torque versus speed, **(d)** speed versus external field resistance.

Torque versus Armature Current

In a series dc motor *without* an external field resistor, the field current is equal to the armature current. Hence, for operations along the linear portion of the magnetic characteristic, and in the absence of armature reaction, the machine's flux is directly proportional to the armature current. Rewriting the basic torque equations, we have

$$T = K\phi I_a \tag{6.55}$$

or

$$T = K_1 I_a^2 \tag{6.56}$$

In other words, the torque developed by a series dc motor is directly proportional to the square of the armature current. The torque-versus-armature-current characteristic is shown in Fig. 6-14(b). The actual curve, however, is displaced downward because of the machine's rotational losses.

Torque versus Speed

Applying KVL in the circuit of Fig. 6-14(a), we obtain

$$V_t = V_g + I_a(R_a + R_f) \tag{6.57}$$

Substituting for voltage generated its equivalent equation, we obtain

$$V_t = K_1 I_a \omega + I_a(R_a + R_f) \tag{6.58}$$

From Eqs. (6.56) and (6.58) we find

$$T = K_1 \left(\frac{V_t}{K_1 \omega + R_a + R_f} \right)^2 \tag{6.59}$$

Because the resistance of the armature and series field windings is relatively small, the equation becomes

$$T \approx \frac{V_t^2}{K_1 \omega^2} \tag{6.60}$$

The torque of a series motor, then, is inversely proportional to the square of the speed. As a result, series dc motors are suitable for mechanical loads that require high torque at low speeds and lower torque at higher speeds, so they are used in trains, trolleys, and the like.

The torque–speed characteristic is shown in Fig. 6-14(c).

Speed versus External Field Resistance (R_{f_x})

In practice the speed of a series motor is controlled through an external field resistor (R_{f_x}), which is connected in parallel to the field winding. The governing equation that describes the effect of the external field resistance on the speed of the motor is derived as follows.

By assuming operation along the linear part of the magnetization characteristic, the basic expression for the armature voltage becomes

$$V_g = K_1 I_f \omega \tag{6.61}$$

Neglecting the voltage drop across the small armature and field resistors, the voltage generated is equal to the terminal voltage. Thus:

$$V_t \approx K_1 I_f \omega \tag{6.62}$$

Considering the field resistors, from the current-divider concept we obtain

$$I_f = I_a \frac{R_{fx}}{R_f + R_{fx}} \tag{6.63}$$

From Eqs. (6.62) and (6.63) we get

$$\omega = \frac{V_t}{K_1 I_a R_{fx}} (R_f + R_{fx}) \tag{6.64}$$

or

$$\omega = \frac{V_t}{K_1 I_a} \left(1 + \frac{R_f}{R_{fx}}\right) \tag{6.65}$$

For a given armature current, then, the larger the external field resistance is, the smaller the speed of the motor will be. The corresponding motor characteristic is shown in Fig. 6-14(d).

EXAMPLE 6–3 A 300 V, 60 A, 1200 r/min series dc motor has an armature and a series field resistance, each .20 ohm. When a .10 ohm resistance is connected in parallel to the field winding, the motor's torque is doubled. Assuming negligible armature reaction and rotational losses, determine the motor's:

a) Armature current.
b) Speed.
c) Efficiency.

SOLUTION The given name-plate data establishes the constant of the motor. Using the equivalent circuit of Fig. 6-15(a), at 1200 r/min we have

$$V_{g1} = V_t - I_{a_1} R_{t_1}$$
$$= 300 - 60(.2 + .2) = 276 \text{ V}$$

As before, subscript 1 designates the motor's parameters at rated conditions and the subscript 2 the parameters that correspond to the new motor's operating condition.

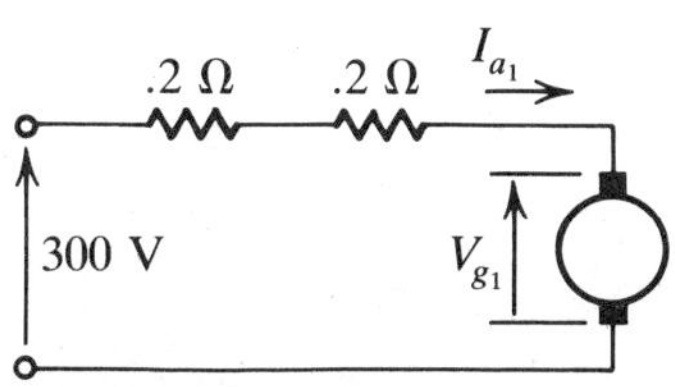

FIG. 6-15(a)

From the generated voltage equation we also have

$$V_{g_1} = K_1 I_{f_1} \omega$$

Solving for the constant of proportionality, we obtain

$$K_1 = \frac{276}{(60)1200\left(\dfrac{2\pi}{60}\right)}$$

$$= .0366 \text{ V/rad/s/field A}$$

a) The current through the .10 ohm resistor that is connected in parallel with the field winding (see Fig. 6-15(b)) does *not* contribute to the effective flux within the machine, because this resistance is physically connected outside the machine's structure.

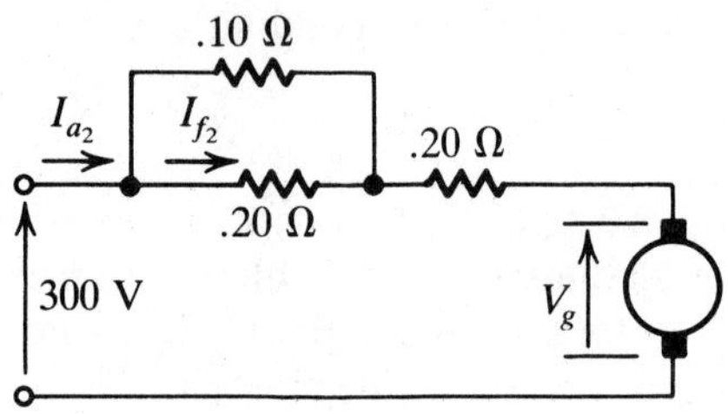

FIG. 6-15(b)

Using the current-divider concept, we obtain

$$I_{f_2} = I_{a_2}\left(\frac{.1}{.1 + .2}\right) = \frac{I_{a_2}}{3}$$

From the statement of the problem,

$$T_2 = 2T_1$$

or

$$K_1 I_{f_2} I_{a_2} = 2K_1 I_{f_1} I_{a_1}$$

From the above,

$$I_{a_2}^2\left(\frac{1}{3}\right) = 2I_{a_1}^2$$

Substituting and solving for the unknown, we obtain

$$I_{a_2} = 60\sqrt{6} = \underline{\underline{146.97 \text{ A}}}$$

b) The speed of the motor is determined as follows. From KVL, we have

$$V_{g2} = 300 - 146.97(.2 + .10 \,/\!/\, .20)$$
$$= 260.81 \text{ V}$$

and

$$\omega_2 = \frac{V_{g2}}{K_1 I_{f2}}$$
$$= \frac{260.81}{.0366\left(\frac{146.97}{3}\right)}$$
$$= 145.46 \text{ rad/s}$$

and

$$n_2 = \underline{\underline{1389 \text{ r/min}}}$$

c) The efficiency of the motor is

$$\eta = \frac{P_{out}}{P_{in}} = 1 - \frac{P_{loss}}{P_{in}}$$
$$= 1 - \frac{I_a^2(R_a + R_f)}{V_t I_a}$$
$$= 1 - \frac{(146.97)^2(.2 + .10 \,/\!/\, .20)}{300(146.97)} = .87$$

Alternatively,

$$\eta = \frac{V_{g2} I_{a2}}{V_t I_{a2}} = \frac{260.81}{300}$$
$$= \underline{\underline{.87}}$$

For purposes of comparison, the results are summarized in Table 6-2.

TABLE 6-2 Summary of results of Example 6–3

Load torque	*External field resistor (ohms)*	*Speed r/min*	*Line current (A)*
T_1	∞	1200	60
$2T_1$	.1	1389	146.97

Exercise 6–4 For a series dc motor, show that the speed as a function of armature current is given by

$$\omega_t = \frac{V_t}{K_1 I_a} - \frac{R_a + R_f}{K_1}$$

where K_1 is the motor's constant in V/A/rad/s.

Separately Excited dc Motors

In a separately excited dc motor, the field coil is supplied from a different voltage source than that of the armature coil, as shown in Fig. 6-16(a). The field circuit normally incorporates a rheostat through which the field current, and thus the motor's characteristics, can be externally controlled. This motor is mainly suitable for two types of loads: those that require constant torque for speed variations up to full-load speed, and those whose power requirements are constant for speed variations above nominal speed.

The power-versus-speed and torque-versus-speed characteristics of such loads are shown in Fig. 6-16(b) and 6-16(c), respectively. These diagrams also identify the method of a motor's speed control. The terminal voltage (V_t) control is used for loads that require constant torque for speed variations *up to* full-load, while the field current (I_f)

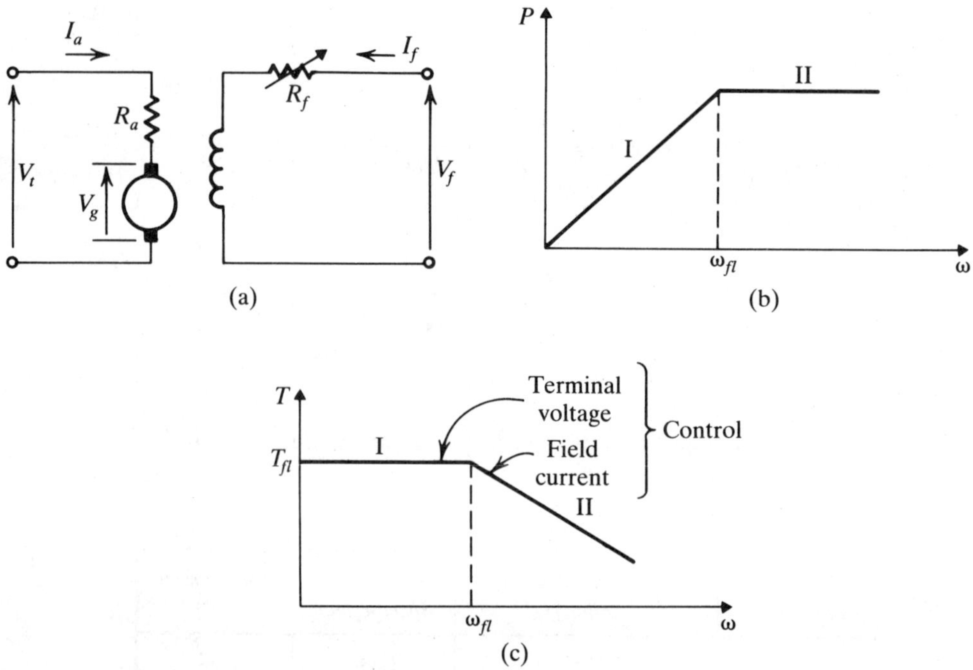

FIG. 6-16 Separately excited dc motor: **(a)** equivalent circuit, **(b)** power–speed characteristic of a load, **(c)** speed–torque characteristic of the load in (b).

control is used for constant power requirements for speed variations *above* the full-load speed. Both methods of speed control aim to supply the requirements of the mechanical load while not exceeding the nominal capability or rating of the machine.

The justification for each mode of speed control is derived by considering the following basic torque and armature-current relationships:

$$T = K\phi I_a \tag{6.66}$$

$$I_a = \frac{V_t - K\phi\omega}{R_a} \tag{6.67}$$

Terminal Voltage Control

From Eq. (6.66) for constant torque and armature current, the field current must also be constant. Then the armature current, as can be seen from Eq. (6.67), will be kept at its rated value, provided that the terminal voltage increases according to the increase in the speed.

Field Control

For loads that require constant power at speeds higher than rated value, the motor's terminal voltage is kept constant, and the increase of the speed is achieved by reducing the field current. The reduction of the field current for constant armature current results in a reduction of the torque and a corresponding increase in the speed. In practice, the speed control through the field current or through the terminal voltage is accomplished by using solid state speed-control apparatus (see Section 6.3).

EXAMPLE 6–4 A 220 V, 4 kW, 22 A, 1260 r/min, separately excited dc motor is driving a fan whose torque is proportional to the square of the motor's speed. The resistance of the armature winding is 0.50 ohm. Neglecting armature reaction, and assuming that the excitation remains constant:

a) Determine the terminal voltage that is required to lower the speed of the fan by 100 r/min.

b) Sketch the torque–speed characteristics of the motor and of the fan. Identify the operating points.

SOLUTION

a) The generated voltage at 1260 r/min is found by applying KVL in the armature circuit of Fig. 6-17(a).

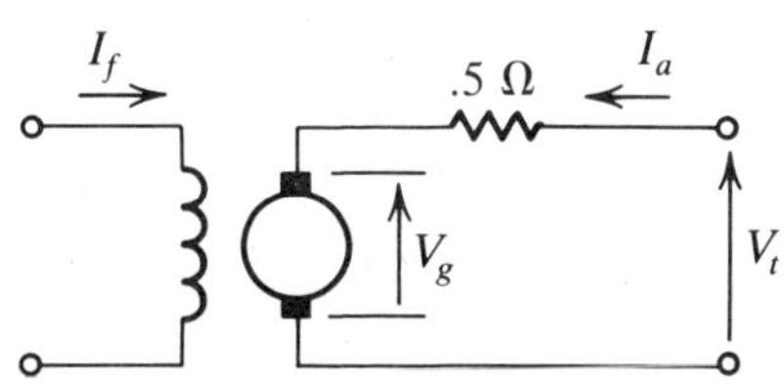

FIG. 6-17(a)

$$V_{g_1} = V_t - I_a R_a$$
$$= 220 - 22(.5) = 209 \text{ V}$$

From the generator principle, for constant field current we have

$$K_1 = \frac{V_{g_1}}{\omega} = \frac{209}{1260 \dfrac{2\pi}{60}}$$
$$= 1.58 \text{ V/rad/s}$$

From the motor principle and the given torque–speed relationship we obtain

$$\frac{T_1}{T_2} = \frac{I_{a_1}}{I_{a_2}} = \left(\frac{\omega_1}{\omega_2}\right)^2$$

Solving for I_{a_2}, and substituting the known parameters, we obtain

$$I_{a_2} = 22\left(\frac{1160}{1260}\right)^2 = 18.65 \text{ A}$$

The new generated voltage is

$$V_{g_2} = K_1\omega = 1.58\left(\frac{1160}{60}\right)2\pi = 192.41 \text{ V}$$

From KVL, the required terminal voltage is

$$V_t = 18.65(.5) + 192.41 = \underline{\underline{201.74 \text{ V}}}$$

b) The torque–speed characteristic of the fan is as shown in Fig. 6-17(b). The torque of the motor at a given speed must be equal to the torque requirement of the fan at the same speed.

The general expression for the torque–speed characteristic of the motor as a function of the terminal voltage is derived as follows:

$$T = K\phi I_a = K_1 I_a = K_1\left(\frac{V_t - V_g}{R_a}\right)$$
$$= K_1\frac{V_t}{R_a} - \frac{K_1}{R_a}(K\phi\omega)$$
$$= K_2 V_t - K_3\omega$$

The constants of proportionality, K_2 and K_3, have been introduced in order to shorten the algebra. The last expression is sketched in Fig. 6-17(b).

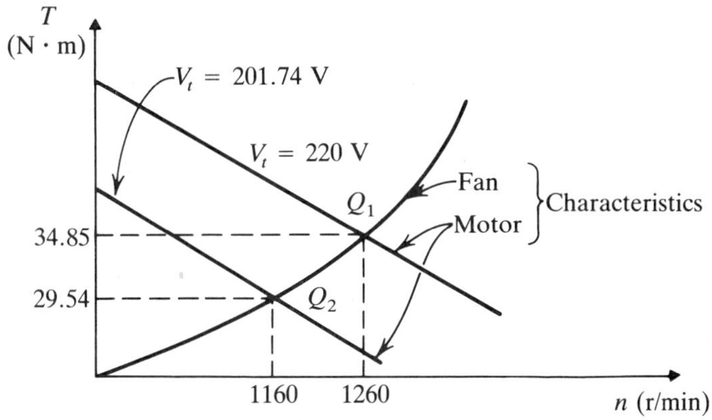

FIG. 6-17(b)

The torque at the given speeds is

$$T_1 = 1.58(22) = 34.85 \text{ N} \cdot \text{m}$$
$$T_2 = 1.58(18.65) = 29.54 \text{ N} \cdot \text{m}$$

The points Q_1 and Q_2, located at the intersection of the motor-load characteristics, represent the operating points.

Exercise 6–5 Redo Example 6–4 by assuming that the load's speed requirement is 1360 r/min instead of 1160 r/min. Will the motor be overheated?

Answer 238.41 V

Compound dc Machines

Shunt dc machines may be equipped with an *auxiliary field* winding. The function of this winding is to modify the machine's characteristics to make it more suitable for a particular load requirement.

The auxiliary and shunt field windings are wound around the main field poles. The auxiliary winding has very few turns. It carries either line or armature current, depending on its connection.

When equipped with an auxiliary field winding, shunt dc machines are called *compound* dc machines. Compound machines are further designated as cumulatively compounded or differentially compounded, depending on whether the auxiliary winding aids or opposes the main field winding.

When the auxiliary winding is connected *between* the terminal voltage and the junction of the armature and field circuits, the shunt machine is said to be connected in "short-shunt" compounded form. When the auxiliary field winding is connected *in*

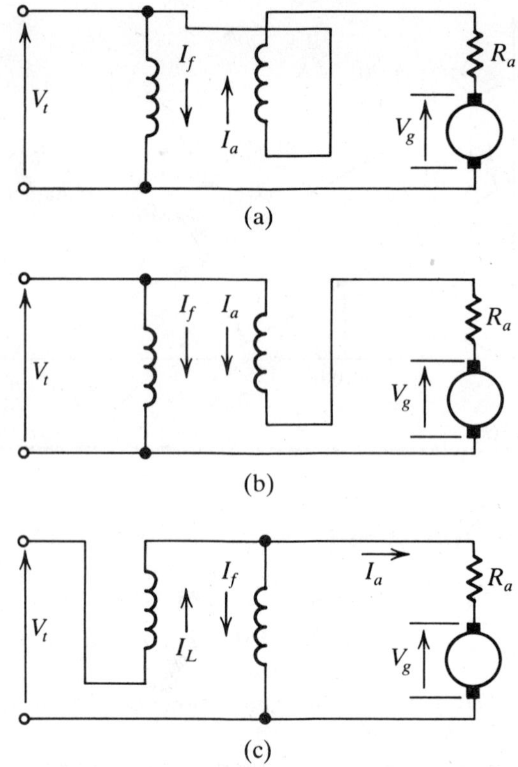

FIG. 6-18 Equivalent circuits of compound dc motors: **(a)** long-shunt, differentially compounded, **(b)** long-shunt, cumulatively compounded, **(c)** short-shunt, differentially compounded.

series with the armature resistance, the machine is said to be connected in "long-shunt" compounded form.

The equivalent circuits of the various compound dc motors are shown in Fig. 6-18. Normally, the resistances of the commutating and the compensating windings (that are part of most dc machines) are not shown separately but are included in the armature resistance.

As shown in Fig. 6-19, the torque–speed characteristic of a compound dc motor is essentially between that of the shunt and the series motor. Actually, a variation of the motor characteristic can be obtained by properly sizing and connecting the auxiliary winding.

Magnetic Circuit

Figure 6-20 shows the equivalent magnetic circuit per pole of a cumulatively compounded, long-shunt motor. The effective magnetic flux within the machine is essentially produced by the mmf ($N_f I_f$) of the main field.

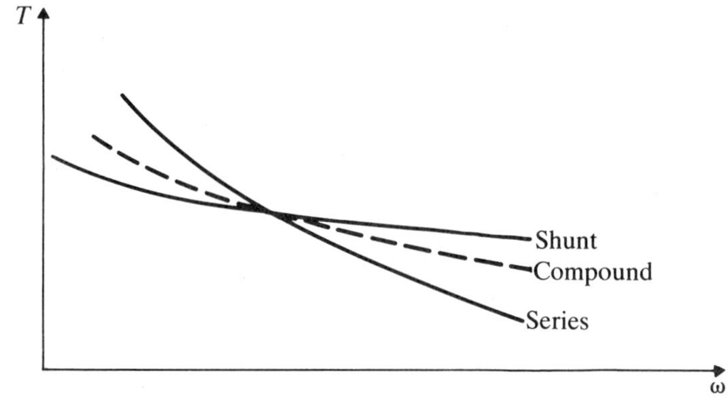

FIG. 6-19 Torque–speed characteristics.

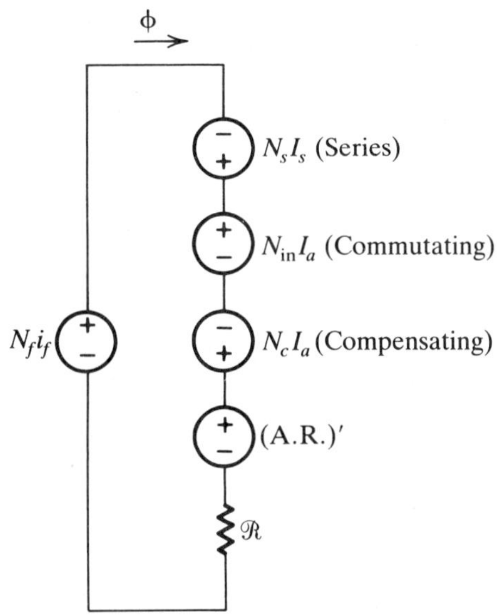

FIG. 6-20 Magnetic equivalent circuit of a long-shunt, cumulatively compounded motor.

The mmf contribution of the *auxiliary series winding* (N_sI_s) is shown as aiding the mmf of the main winding. However, the auxiliary series winding should be of opposite polarity if the machine is differentially compounded. In the long-shunt connection, the current through the auxiliary winding is equal to the armature current. In the short-shunt connection, the current through this winding is equal to the line current.

The armature reaction, in equivalent magnetic volts (A.R.)′ is of a polarity opposite to that of the main field. The compensating winding is designed in such a way that its mmf (N_cI_a) is equal and opposite to the armature reaction.

The mmf of the *commutating* or *interpole* winding ($N_{in}I_a$) has a qualititative effect on the flux density of the field winding, for it is specifically designed to minimize the distortion of the armature mmf on the main field flux density at the interpole region.

The effective reluctance $\mathcal{R}$ represents the sum of the reluctances of the air gaps, of the stator path, and of the path through the rotor of the machine. At relatively low flux levels the reluctance of the air gap predominates. At higher flux densities the reluctance of the constituent magnetic materials is, because of saturation, the controlling factor.

6.1.8 STARTING

The equivalent circuit of a shunt dc motor is shown in Fig. 6-21(a). At the instant of starting, the rotor in all dc motors is not revolving and thus the voltage generated is zero. As a result, the terminal voltage is applied across the relatively small armature resistance. Consequently, a very large current flows through the conductors of the armature. Mathematically, the armature's starting current ($I_{a_{st}}$) is

$$I_{a_{st}} = \frac{V_t}{R_a} \tag{6.68}$$

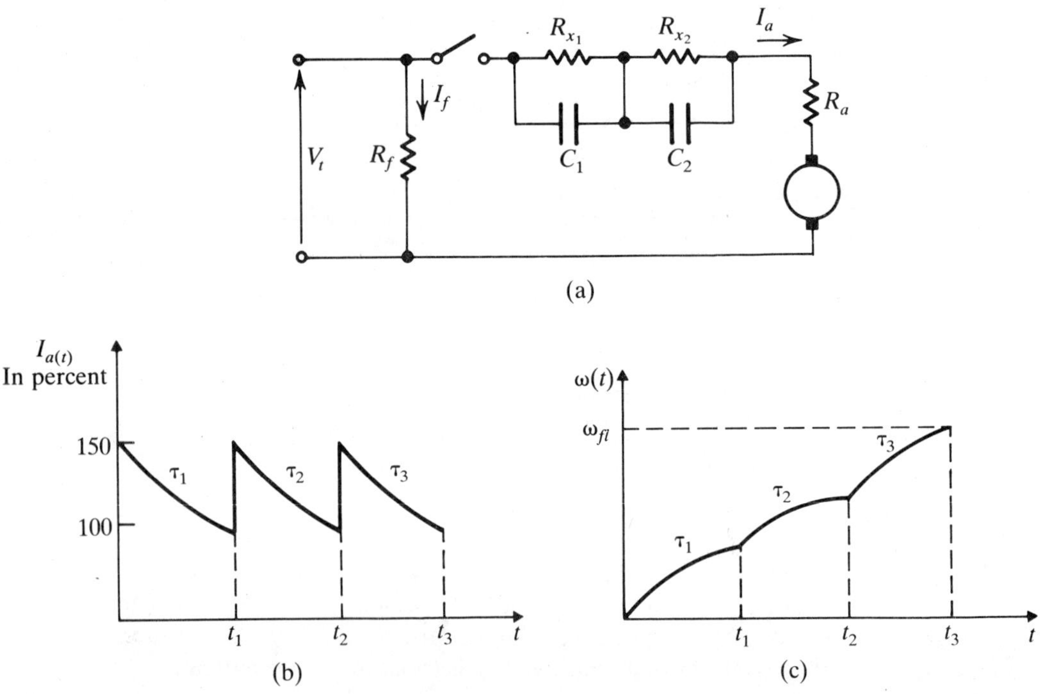

FIG. 6-21 Starting a dc shunt motor through step-by-step reduction of the external armature resistors: **(a)** equivalent circuit, **(b)** armature current versus time, **(c)** speed versus time.

High starting current is undesirable because it adversely effects the dc power supply, the commutation process, and the rate of heat loss within the machine.

In order to limit excessively large starting currents, dc shunt motors—like all other types of dc machines—are provided with external resistors that are connected in series with the armature circuit (see Fig. 6-21(a)). During starting, these resistors are shorted,* in steps, by the contactors (C_1 and C_2) until the armature current reaches its rated value. At the same time, the speed of the motor increases from zero up to its nominal value.

The variations in armature current and speed, as functions of time during the step-by-step reduction in the external resistors for a particular motor, are shown in Figs. 6-21(b) and (c), respectively.

At starting, both of the contactors (see Fig. 6-21(a)) are open. Therefore the armature current is limited to

$$I_{a_{st}} = \frac{V_t}{R_a + R_{x_1} + R_{x_2}} \tag{6.69}$$

As soon as the motor starts to rotate, the armature voltage increases. Consequently, the armature current is gradually reduced. From KVL we have

$$I_a = \frac{V_t - K_1\omega}{R} \tag{6.70}$$

where K_1 is constant of the machine, R is the total resistance of the armature circuit, and ω is the value of the motor's speed at the instant under consideration.

When the armature current reaches its rated value, the resistance R_{x_1} is shorted through the contactor C_1. At this instant, the speed of the motor cannot change instantaneously. Thus, the armature current must instantaneously increase in order to satisfy KVL. In Fig. 6-21(a) the resistors have been selected in such a way as to limit the armature current to 150% of its rated value.

For this variation of the armature current immediately before and after the removal of the first resistor (R_{x_1}) we have

$$I_{a_r} = \frac{V_t - K_1\omega_1}{R_a + R_{x_1} + R_{x_2}} \tag{6.71}$$

and

$$1.5\, I_{a_r} = \frac{V_t - K_1\omega_1}{R_a + R_{x_2}} \tag{6.72}$$

where I_{a_r} is the rated current of the armature windings and ω_1 is the speed of the motor at the instant when the resistor R_{x_1} is shorted. When the second external resistor is

* The operation of the contactors is explained in the chapter on control schematics.

shorted, the above equations can be rewritten with a new value for the speed of the motor. The resulting equations will yield the required values of the external resistors, and the speed of the motor, at the instant a resistor is removed.

At any instant after starting, the motor's armature current and speed, as functions of time, can be found by solving the governing differential equations. The solution of these equations gives the transient response of the dc machines, which is discussed in Section 6.2.

The time constants that control the decay of the armature current and the build-up of the speed (see Fig. 6-21(b)) are determined by the parameters of the motor plus load. Each time a resistor is removed, the corresponding time constant will, of course, increase.

6.1.9 OPEN-CIRCUIT CHARACTERISTICS AND DC GENERATORS

This section discusses the open-circuit characteristics of dc machines, their effective field current, and the various types of dc generators.

Open-Circuit Characteristic

The magnetization, or open-circuit characteristic (OCC), of a machine is obtained by driving the machine as a generator at constant speed. The terminal voltage, which is equal to the voltage generated (V_g), is recorded for various values of field current (I_f).

Figure 6-22(a) shows a laboratory set-up that may be used to obtain data to plot the open-circuit characteristic of a separately excited dc machine. The mechanical power input to the generator is equal to the sum of its mechanical losses and iron losses. The OCC of the machine is shown in Fig. 6-22(b).

At low flux densities the magnetization characteristic (MC) is linear, owing to the constant permeability of the air gap. At higher flux densities the MC becomes nonlinear, owing to the variations in the permeability of the magnetic material.

Alternatively, at relatively low flux densities the reluctance of the air gap predominates, and all input mmf is used to overcome the opposition to the flux that is presented by the air gap. As the flux density or the level of magnetization increases, the reluctance of the magnetic material becomes the controlling parameter of the magnetic circuit.

The machine's equivalent magnetic circuit per pole is shown in Fig. 6-22(c). The armature reaction (A.R.)′ is expressed in equivalent field ampere-turns.

The basic relationships of the first chapter show that the generated voltage (V_g) is proportional to the magnetic flux density (B), and also that the magnetic field intensity (H) is proportional to the field current (I_f). That is, the ordinate and the abscissa of the MC correspond to the magnetic field's flux density and flux intensity, respectively.

The flux of a coil is proportional to the coil's current and to the permeability of the magnetic material. That is:

$$\phi = K\mu I \tag{6.73}$$

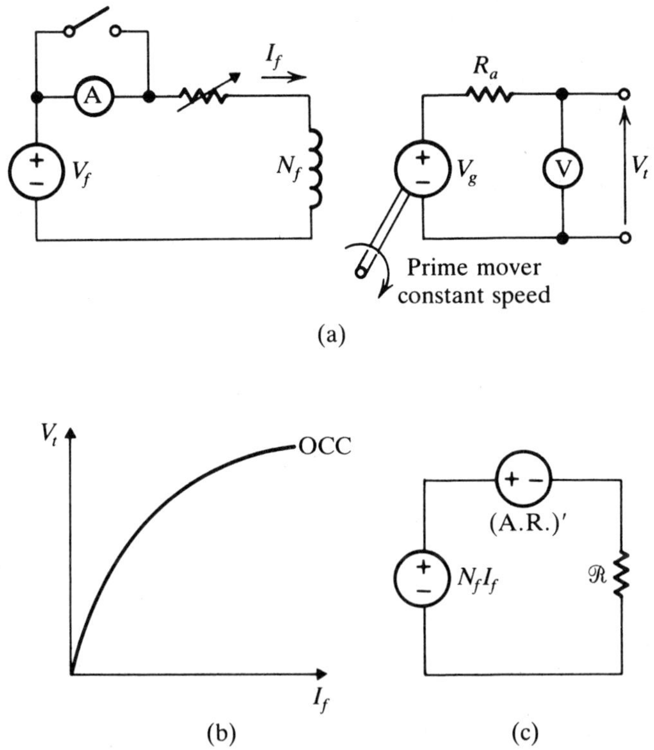

FIG. 6-22 Separately excited dc machine: **(a)** schematic for the laboratory set-up used to obtain data for the magnetization curve, **(b)** typical open-circuit characteristic, **(c)** equivalent magnetic circuit per pole. (At no-load the mmf of the armature reaction is zero.)

where K is a constant that is dependent upon the physical dimensions of the magnetic material and upon the winding's number of turns. As shown in Fig. 1-32, the permeability depends on the operating point or the level of magnetization.

From the generator principle, for two different speeds of operation ω_1 and ω_2 we have

$$\frac{V_{g1}}{V_{g2}} = \frac{\phi_1 \omega_1}{\phi_2 \omega_2} \tag{6.74}$$

When the effective flux ϕ_1 and ϕ_2 are the same, we obtain:

$$\frac{V_{g1}}{V_{g2}} = \frac{\omega_1}{\omega_2} \tag{6.75}$$

In Eq. (6.75) the generated voltages V_{g1} and V_{g2} must be calculated at the same effective field conditions. The magnetization curve can be used in conjunction with Eq. (6.75) to find the speed of the machine for a particular operating condition. Usually the generated voltage V_{g_1} is obtained from the OCC at the known speed ω_1 but at the same field condition as the generated voltage V_{g_2}. The latter corresponds to speed ω_2, which is normally obtained by applying KVL in the equivalent circuit of the machine under consideration. Clearly, then, the no-load characteristic can be used at full-load conditions, provided that the abscissa of the MC becomes the effective field current and its ordinate becomes the voltage generated at full-load conditions.

Effective Field Current

The general magnetic equivalent circuit per pole of a dc compound machine is shown in Fig. 6-23. Referring to Fig. 6-23(b), the effective flux within the machine is given by

$$\phi = \frac{N_f I_{\text{ef}}}{\mathscr{R}} \tag{6.76}$$

where $\mathscr{R}$ is the reluctance of the path through which the flux completes its loop, and $N_f I_{\text{ef}}$ is the effective potential that magnetizes the machine.

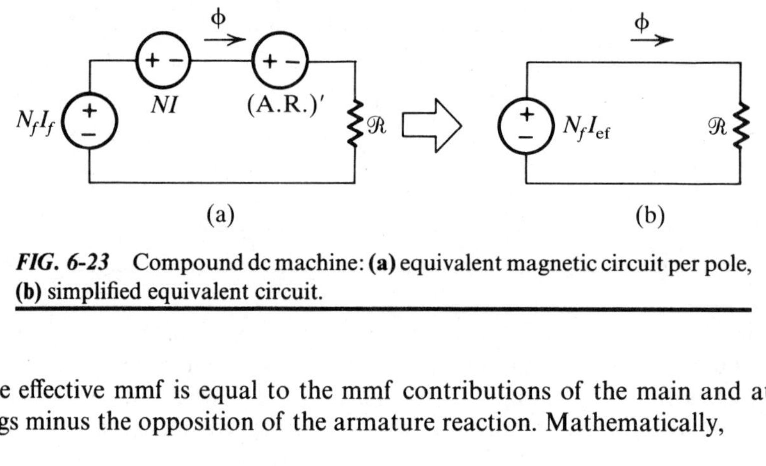

FIG. 6-23 Compound dc machine: **(a)** equivalent magnetic circuit per pole, **(b)** simplified equivalent circuit.

The effective mmf is equal to the mmf contributions of the main and auxiliary windings minus the opposition of the armature reaction. Mathematically,

$$N_f I_{f_{\text{ef}}} = N_f I_{f_a} \pm N_s I_s - (\text{A.R.})' \tag{6.77}$$

Dividing both sides by the turns of the main field winding N_f, we obtain

$$I_{f_{\text{ef}}} = I_{f_a} \pm \frac{N_s}{N_f} I_s - \text{A.R.} \tag{6.78}$$

where

$I_{f_{\text{ef}}}$ = the effective field current of the machine. Although this current cannot be seen in the machine's electrical equivalent circuit, it constitutes the abscissa of its OCC under load conditions.

I_{f_a} = the actual current through the main field winding of the machine. This current can easily be calculated from the given data and by using the equivalent circuit of the machine under consideration.

$N_s I_s$ = the magnetomotive force of the auxiliary field winding. This winding is a feature of compounded machines, and its flux may aid (cumulative connection) or oppose (differential connection) the flux of the main winding.

(A.R.)′ = the armature reaction in equivalent field ampere turns.

A.R. = the armature reaction in equivalent field amperes.

The cross-magnetizing mmf of the armature reaction is directly proportional to the armature current, but the resulting reduction in the effective, *voltage-generating flux* is approximately proportional to the *square of the armature current.*

The effects of the armature reaction on the OCC and on the full-load characteristics of a separately excited generator are shown in Fig. 6-24. In Fig. 6-24(b) the A.R.

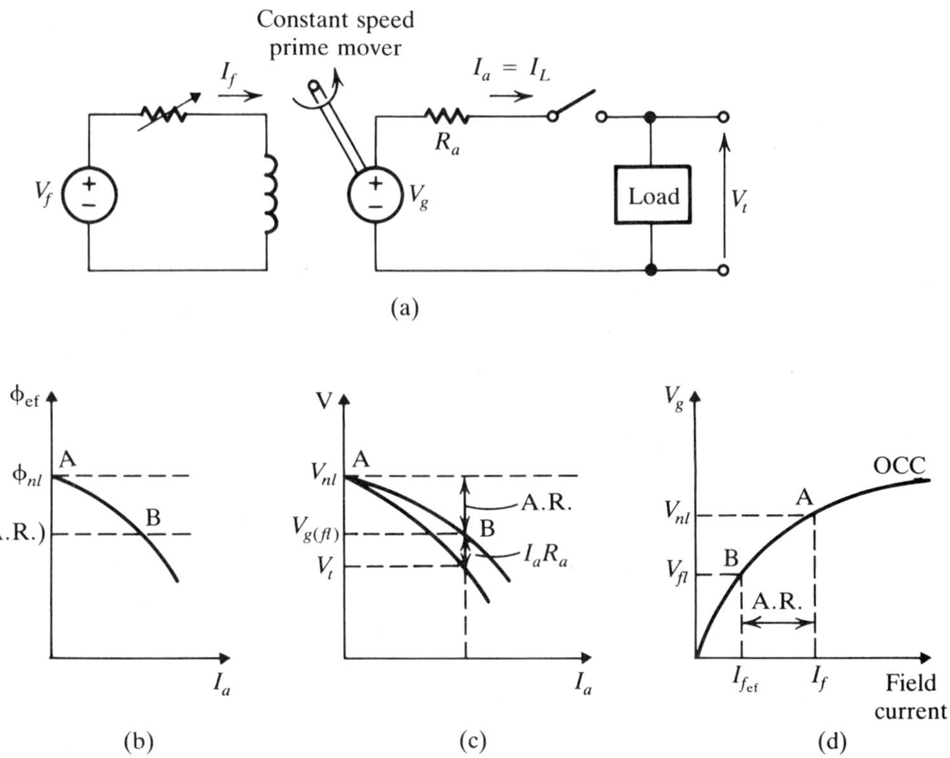

FIG. 6-24 Separately excited generator—an illustration of the three alternative units of armature reaction: **(a)** equivalent circuit, **(b)** effective flux versus load current (armature-reaction effects in percent of the flux at no-load), **(c)** generated and terminal voltages as functions of load current (armature-reaction effects in volts), **(d)** open-circuit characteristic (armature-reaction effects in terms of equivalent field amperes).

is expressed as a percentage of the flux at no-load. The letters A and B are used to identify the no-load and full-load conditions, respectively. In Fig. 6-24(c) the effects of the armature reaction are expressed in volts. This is seldom used in the analysis of problems because full-load terminal characteristics are unavailable. In Fig. 6-24(d) the armature reaction is expressed in equivalent field amperes. This is the most popular unit of measure of armature reaction. It should be noted that the open-circuit characteristic is used at no-load and at full-load conditions.

The effects, then, of the armature reaction can be expressed in terms of no-load flux, generated voltage, and/or field current.

EXAMPLE 6–5 A dc series motor is operated from a 450 V supply. The total armature and series field resistance is 0.40 ohm. Its magnetization characteristic, taken at 800 r/min, is

Volts	360	412	460	470
Amperes	35	45	55	65

The demagnetizing effect of the armature reaction, which may be assumed to be proportional to the square of the armature current, is such that for an armature current of 55 A the flux per pole is reduced by 10%. When the motor draws 65 A from the 450 V supply, determine:

a) The speed.

b) The electromagnetic torque developed.

c) The output power if the rotational losses consume 1500 W.

SOLUTION

a) The equivalent circuit of the motor is shown in Fig. 6-25(a). At 800 r/min and at 65 A from the given no-load characteristic, we obtain:

$$V_{g_1} = 470 \text{ V}$$

This voltage corresponds to a flux of ϕ_0 webers. Owing to the armature reaction,

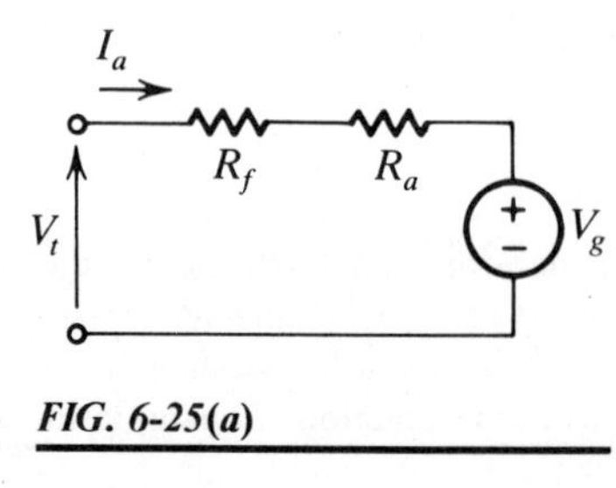

FIG. 6-25(a)

the effective flux ϕ_{ef} is

$$\phi_{ef} = \phi_0 - \left(\frac{10}{100}\right)\left(\frac{65}{55}\right)^2 \phi_0$$
$$= .86\phi_0$$

From the above, the actual voltage generated at full-load conditions, but at 800 r/min is

$$V_{g_1} = 470(.86) = 404.36 \text{ V}$$

The generated armature voltage at the unknown speed, but at the same flux level as V_{g_1}, is

$$V_{g_2} = V_t - I_a R$$
$$= 450 - 65(.4)$$
$$= 424 \text{ V}$$

Substituting the above data into Eq. (6.75), we obtain

$$n_2 = 424\left(\frac{800}{404.36}\right) = \underline{\underline{838.87 \text{ r/min}}}$$

b) The torque developed is

$$T = \frac{V_{g_2} I_a}{\omega_2} = \frac{424(65)}{838.87\left(\frac{2\pi}{60}\right)} = \underline{\underline{313.73 \text{ N} \cdot \text{m}}}$$

c) From the power-balance equation we have

$$P_d = P_{r_l} + P_o$$

from which

$$P_o = 424 \times 65 - 1500 = \underline{\underline{26.06 \text{ kW}}}$$

An Alternative—But Incorrect-Solution of Part (a)

The armature voltage at the unknown speed is, as before,

$$V_{g_2} = 424 \text{ V}$$

The effective value of the field current is

$$I_{ef} = (\text{actual field current}) - (\text{armature reaction})$$

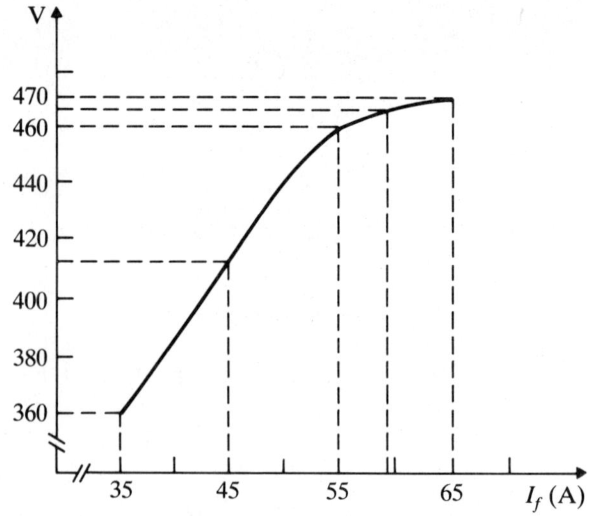

FIG. 6-25(b)

At 65 A,

$$I_{ef} = 65 - 65\left(\frac{10}{100}\right)\left(\frac{65}{55}\right)^2$$
$$= 59.92 \text{ A}$$

Projecting this current on the OCC (Fig. 6-25(b)), we obtain

$$V_{g_1} = 466 \text{ V}$$

and

$$\eta_1 = 424\left(\frac{800}{466}\right)$$
$$= 727.90 \text{ r/min}$$

which is lower than the speed already calculated by an amount which is equal to

$$\frac{838.87 - 727.9}{727.9}(100) = 15.25\%$$

In other words, it is a mistake to assume that the effective flux is dependent only on the field current. The change of permeability, and its effects on the voltage generated, cannot be assumed to be negligible.

Exercise 6–6 Given the following data for a dc motor that operates at saturation:

Condition	*Generated Voltage (V)*	*Effective Field (A)*	*Speed r/min*
1	213	1.4	900
2	208	1.2	?

a) Explain why the tabulated data, when substituted into Eq. (6.75), will not give the speed of the motor at operating condition 2.

b) What additional information is required to calculate the unknown speed?

Shunt Generators

Figures 6-26(a) and (b) show, respectively, the equivalent circuit and the OCC of a shunt-type generator. The shunt generator does not require any excitation for starting or operating. It is self-excited. This advantage results from the machine's residual magnetism—a characteristic of the magnetic material—which provides the required

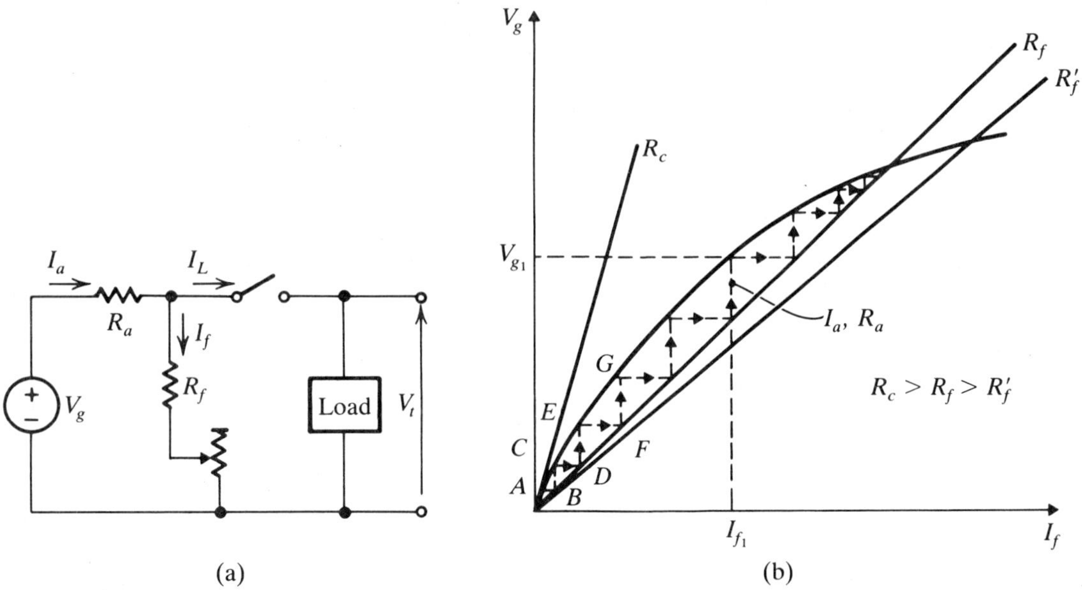

FIG. 6-26 Shunt generator: **(a)** equivalent circuit, **(b)** magnetization characteristic illustrating the build-up voltage for the no-load conditions.

field. As soon as the prime mover starts to turn the rotor, a voltage will be generated in the armature windings, because they are rotated through the field of the stator's residual magnetism.

Look at Fig. 6-26(b). When the armature starts to rotate, a small voltage will be generated, as indicated by point A. This voltage, impressed across the field circuit (point B), will cause a current to flow. If the polarity of the connection is correct, the mmf of this current will reinforce that of the residual magnetism; the strengthened magnetic field will then generate the voltage corresponding to point C, which in turn will increase the field current to the value corresponding to point D. The build-up process (*positive feedback*) will continue until the saturation curve intersects the field-resistance line. The slope of this line is equal to the total resistance of the field winding (resistance of the winding plus the portion of the external adjustable resistor that may be incorporated). The intersection of the magnetization curve and the field resistance line furnishes the operating point at no-load.

For a particular field current I_{f_1}, the generated voltage and the terminal voltage are given respectively by

$$V_{g_1} = I_{f_1}R_f + I_{a_1}R_a \tag{6.79}$$

$$V_{t_1} = I_{f_1}R_f \tag{6.80}$$

From the above it is evident that the vertical lines drawn from the OCC to the field resistance line represent the voltage drop across the armature resistance. The longest of these lines will give—indirectly—the maximum permissible value of armature current at rated speed.

By decreasing the field resistance (R'_f) below its rated value, the no-load voltage will increase slightly. By increasing the field resistance, the no-load voltage will decrease. The value of the field resistance, at which the build-up of the generated voltage cannot take place, is designated as the critical field resistance (R_c).

In general, the field resistance controls the starting, the terminal voltage, and the voltage regulation of the shunt generator.

Compound dc Generators

The equivalent circuits and the terminology of compound dc generators are similar to those of dc compound motors described in Section 6.1.7. The only differences between a compound dc generator and a compound dc motor are the energy flow and the external characteristics of the machine. The external characteristic of a motor is its torque–speed curve, while for a generator it is its terminal voltage versus its load-current curve.

Figure 6-27(a) shows an elementary representation of a machine's magnetic pole, on which the main and auxiliary windings are identified. The auxiliary winding is often referred to as the series winding to differentiate it from the shunt field winding.

The equivalent circuits of a long-shunt and a short-shunt compounded generator are shown in Figs. 6-27(b) and (c), respectively. The external characteristics of the compounded generators are shown in Fig. 6-27(d). It should be noted that the cumulatively compounded dc generators are further classified as overcompounded,

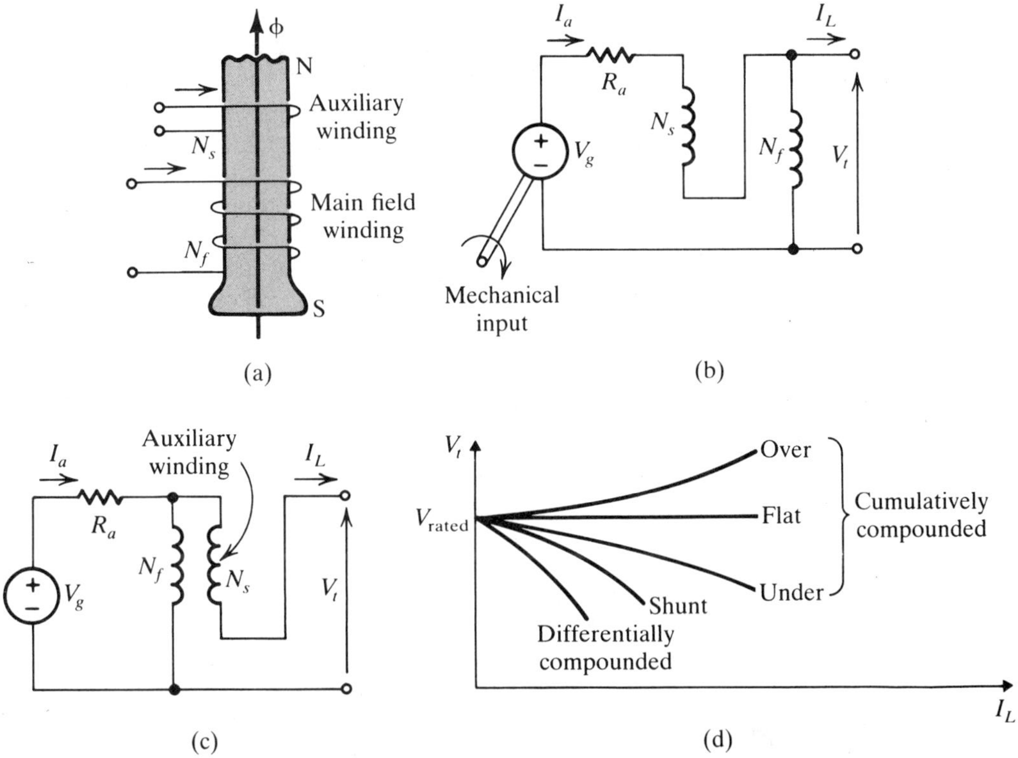

FIG. 6-27 Compound dc generators: **(a)** physical location of the main and auxiliary field windings, **(b)** long-shunt, **(c)** short-shunt, **(d)** characteristics of differentially and cumulatively compounded generators.

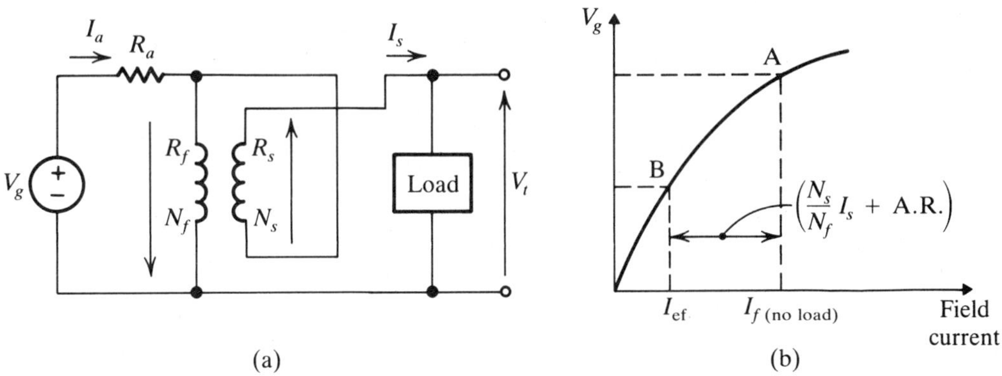

FIG. 6-28 Short-shunt differentially compounded generator: **(a)** equivalent circuit, **(b)** open-circuit characteristic.

flat-compounded or undercompounded, depending on whether the output voltage increases, remains constant, or decreases as the load current increases.

Figure 6-28(a) shows the equivalent circuit of a short-shunt differentially compounded dc generator. Figure 6-28(b) shows its OCC and the effects of the auxiliary winding and of the armature reaction on the magnetization of the machine.

EXAMPLE 6–6 A long-shunt, cumulatively compounded dc generator has 1800 shunt-field turns per pole and 12 series-field turns per pole. The armature resistance is 0.12 ohm, the series-field resistance is 0.04 ohm, and the shunt-field resistance is 210 ohms. Data for the open-circuit characteristic at 1700 r/min is as follows:

Volts	110	200	280	300
Amperes	.5	1.1	1.9	2.35

The generator supplies a load of 16 kW at 215 V when it is driven at 1700 r/min. Determine the speed at which the generator must be driven to supply a load of 23 kW at 270 V. Assume armature reaction is proportional to the armature current squared.

SOLUTION

The equivalent circuit of the long-shunt, cumulatively compounded dc generator is shown in Fig. 6-29(a). The series field winding, as the name of the generator implies, aids the shunt field winding. This is the reason the field currents flow in the same direction in the equivalent circuit. The OC characteristic is shown in Fig. 6-29(b). This characteristic can be used to obtain the no-load and full-load machine voltages and field currents only at 1700 r/min.

Load Condition: 16 kW at 215 V

The given data will be used to calculate the armature reaction as follows. Referring to Fig. 6-29(a), we have

$$\text{load current} = I_L = \frac{P}{V} = \frac{16{,}000}{215} = 74.42 \text{ A}$$

The current through the main field winding is

$$I_f = \frac{215}{210} = 1.02 \text{ A}$$

The armature current is

$$\begin{aligned} I_a &= I_L + I_f \\ &= 74.42 + 1.02 \\ &= 75.44 \text{ A} \end{aligned}$$

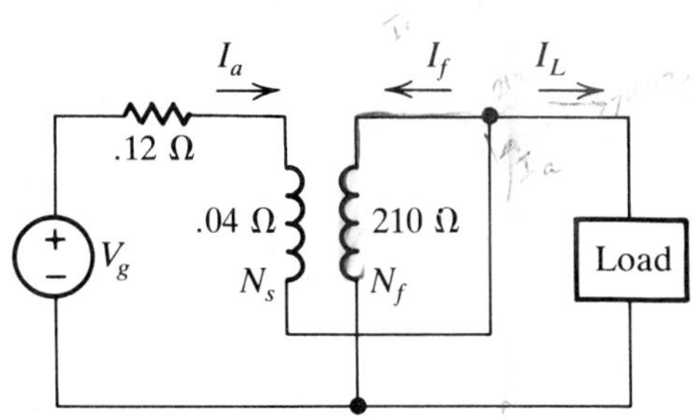

FIG. 6-29(a)

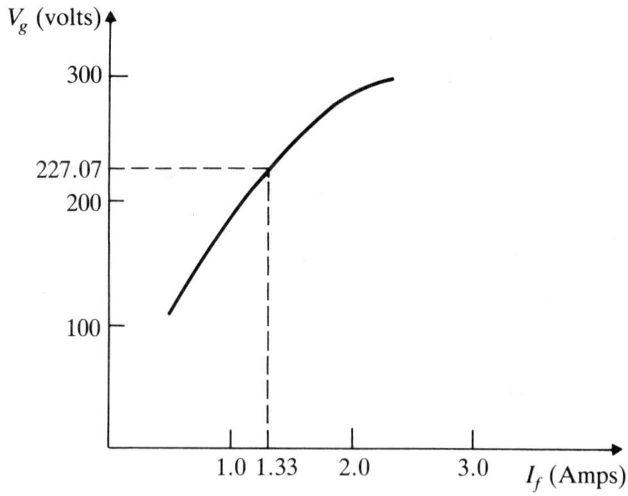

FIG. 6-29(b)

The generated voltage is

$$V_g = V_t + I_a(R_a + R_s) = 215 + 75.44(.12 + .04)$$
$$= 227.07 \text{ V}$$

Projecting this voltage on the given OCC, we obtain the effective field current:

$$I_{\text{ef}} = 1.33 \text{ A}$$

The armature reaction—at an armature current of 75.44 A—is found by using Eq. (6.78). Substituting the known data, we obtain

$$1.33 = 1.02 + \frac{12}{1800}(75.44) - \text{A.R.}$$

from which

$$\text{A.R.} = .197 \text{ equivalent field amperes}$$

Load Condition: 23 kW at 270 V

The load current at the unknown speed is

$$I_L = \frac{23{,}000}{270} = 85.18 \text{ A}$$

The current through the shunt field is

$$I_{f_a} = \frac{270}{210} = 1.28 \text{ A}$$

The armature current is

$$I_a = 85.18 + 1.28 = 86.47 \text{ A}$$

From the statement of the problem, the new armature reaction is

$$\begin{aligned} \text{A.R.} &= .197\left(\frac{86.47}{75.44}\right)^2 \\ &= .258 \text{ equivalent field amperes} \end{aligned}$$

Using Eq. (6.78), the new effective field current is

$$\begin{aligned} I_{\text{ef}} &= 1.28 + \frac{12}{1800}(86.47) - .258 \\ &= 1.60 \text{ A} \end{aligned}$$

Projecting this current on the given OCC gives the armature voltage at 1700 r/min. From the curve,

$$V_{g_1} = 255 \text{ V}$$

At the same field condition, but at the unknown speed, the generated voltage is obtained from the equivalent circuit as follows:

$$V_{g_2} = 270 + 86.47(.12 + .04) = 283.84 \text{ V}$$

The required speed is found by using Eq. (6.75). Substituting, we obtain

$$\frac{283.84}{255} = \frac{n_2}{1700}$$

from which

$$n_2 = \underline{\underline{1892.24 \text{ r/min}}}$$

In other words, the prime mover must rotate at 1892.24 r/min to meet the new load requirements.

For comparison purposes, the results are summarized in Table 6-3.

TABLE 6-3 Summary of results of Example 6–6

Condition	*Output* Power (kW)	Voltage (V)	Speed r/min	Shunt Field (A)	Effective Field (A)	Armature Reaction (A)
1	16	215	1700	1.02	1.33	.197
2	23	270	1892.24	1.28	1.6	.258

6.2 Transient Analysis

6.2.1 INTRODUCTION

The determination of machine variables as functions of time constitutes the transient analysis of machines. The knowledge of analytical expressions for speed and current is essential in determining the protection, control, and flexibility of machines.

Solving problems of transient nature requires a basic knowledge of differential equations or of Laplace transforms. In this section the Laplace transform technique will be used, because it is somewhat simpler and because it reveals—indirectly—the initial value, the final value, and the time constants of the time function under consideration.

Block diagrams, transfer functions, transformation of mechanical parameters and general machine equations will be discussed. All these concepts are necessary for a complete understanding of the problems concerning machine transients.

6.2.2 TIME-CONSTANT CONSIDERATIONS

The instantaneous values of a machine's torque, speed, and current have components of the form $e^{-t/\tau}$ that decrease exponentially. τ is a constant commonly known as time constant (TC). In general, the time constant of an exponentially decaying waveform is given by the value of t that makes the exponent equal to -1. For example, the time constant of the exponential function $e^{-.8t}$ is

$$TC = \frac{1}{.8} = 1.25 \text{ s}$$

For all practical purposes, it takes about four time constants for an exponentially decreasing waveform to reach a relatively negligible value ($e^{-4} = .0183$). That is, at the end of a period consisting of four times the time constant, the function has diminished to .0183 times its initial value. For example, the speed function

$$\omega(t) = 100(1 - e^{-5t}) \text{ rad/s}$$

will take about four time constants (4/5 = .8 s) to reach its final value of 100 rad/s. Thus, the time constant of an exponentially decreasing function reveals, by inspection, how long it will take for the function to reach its final value.

The time constant of a function can also be found from the function's Laplace transform. Consider the following Laplace transform:

$$\mathscr{L}(f(t)) = \mathscr{F}(s) = \frac{A}{s(1 + \tau_1 s)(1 + \tau_2 s)} \tag{6.81}$$

where $\mathscr{F}(s)$ is the Laplace transform of the function $f(t)$, A is a constant, and τ_1 and τ_2 are the time constants of the function.

Besides electrical time constants (L/R, CR), there are mechanical time constants (J/B) and electromechanical time constants, such as

$$\tau_m = \frac{R_a B + K^2}{R_a J} \text{ s} \tag{6.82}$$

where R_a, J, and B are the usual machine parameters, and K is the generator's or motor's constant, which depends on the type of machine and the data given.

Because of the physical significance of the time constants, the student must develop a feel for them and thus be able to predict, rather accurately, the time required for the variables of the machine (current, voltages, speed, etc.) to reach their final values after a given disturbance or when the machine starts from rest.

6.2.3 BLOCK DIAGRAMS AND TRANSFER FUNCTIONS

Mathematical relationships can be represented in a block diagram form. For example, the representation of Ohm's law:

$$i(s) = \frac{v(s)}{Z(s)} \tag{6.83}$$

in block diagram form is shown in Fig. 6-30. When multiplied by the input quantity, the parameters within the block will give the parameter shown in the output of the

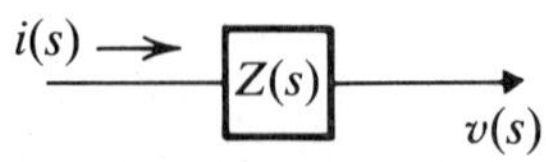

FIG. 6-30 Block diagram representation of Ohm's law.

block. Block diagrams simplify the formulation of a machine's transfer functions and the understanding of the machine's operation.

Transfer function is the ratio of the Laplace transform of the output to the Laplace transform of the input with all initial conditions equal to zero. The units of the transfer function depend on the units of the variables under consideration. In electric machines, the transfer functions are usually

$$\frac{\omega(s)}{\theta(s)}, \quad \frac{v(s)}{i(s)}, \quad \frac{\theta(s)}{v(s)}$$

and so on, where ω, v, i, and θ represent the machine's usual parameters.

The operations of addition and subtraction of equations are performed by the so-called summing devices.

In this section the transfer functions will be obtained from the machine's governing equations. However, when the block diagram representation of a machine is given, the desired transfer function can easily be found by using the "Mason gain" formula. The latter is discussed in most textbooks on automatic control.

An additional advantage of the block diagram representation is that it reveals to an experienced automatic control specialist some stability highlights of the system it represents.

6.2.4 TRANSFORMATION OF MECHANICAL PARAMETERS

Gears are used to transform the torque and the speed of a motor in the same way transformers are used to step voltages and currents up or down. This transformation is necessary to match the machine's output parameters to those required by the mechanical load.

The speed, the torque, the viscous coefficient of friction, and the moment of inertia of one gear are related to the corresponding parameters of the coupled gear as follows:

1. Referring to Fig. 6-31, and from basic considerations, we observe that the speed (ω) of one of the two coupled gears is inversely proportional to its number of teeth (N). That is,

$$\omega_1 \propto \frac{1}{N_1} \tag{6.84}$$

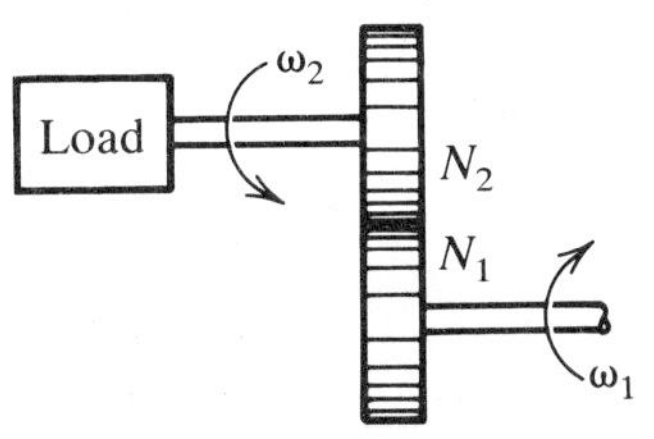

FIG. 6-31 Gears

and

$$\omega_2 \propto \frac{1}{N_2} \tag{6.85}$$

From the above,

$$\frac{\omega_2}{\omega_1} = \frac{N_1}{N_2} \tag{6.86}$$

2. From the power-balance equation (neglecting gear losses), we have

$$T_1\omega_1 = T_2\omega_2 \tag{6.87}$$

or

$$\frac{T_1}{T_2} = \frac{\omega_2}{\omega_1} \tag{6.88}$$

Substituting Eq. (6.86) in the above, we obtain

$$T_1 = T_2 \frac{N_1}{N_2} \tag{6.89}$$

3. The torque (T) of a gear that rotates at ω rad/s and drives a load which has a frictional viscous coefficient (B) N · m/rad/s is given by

$$T = B\omega \tag{6.90}$$

From the above (considering two gears) we obtain

$$\frac{T_1}{T_2} = \frac{B_1}{B_2}\left(\frac{\omega_1}{\omega_2}\right) \tag{6.91}$$

Substituting Eqs. (6.86) and (6.89) in the above, we find

$$B_1 = B_2\left(\frac{N_1}{N_2}\right)^2 \tag{6.92}$$

4. In a similar way, the moment of inertia of the load (J_2) (as seen from the driving gear) is

$$J_1 = J_2\left(\frac{N_1}{N_2}\right)^2 \tag{6.93}$$

In the above expressions, subscript 1 refers to the parameters of primary (driving) gear and subscript 2 to the secondary (driven) gear.

Speed and torque in either side of the gear are interrelated in the same way current and voltage are in either side of a transformer. Similarly, the moment of inertia and the coefficient of friction can be referred to either side of the gears in the same way the transformer's impedances are.

The essential concept here is that the power in either side of a lossless gear ($P = T_1\omega_1 = T_2\omega_2$) remains the same as the apparent power in either side of a lossless transformer ($S = V_1 I_1 = V_2 I_2$). The transformation property of gears is illustrated by the following example.

Assume a motor drives a 100 N · m mechanical load at 80 rad/s through a speed reducer of (1/2). The load's moment of inertia is 48 kg · m^2. Then speed, torque, and moment of inertia (as seen from the shaft of the motor) are as follows:

Speed:

$$\omega_1 = \omega_2 \left(\frac{N_2}{N_1}\right) = 80\left(\frac{2}{1}\right) = 160 \text{ rad/s}$$

Torque:

$$T_1 = T_2 \left(\frac{N_1}{N_2}\right) = 100\left(\frac{1}{2}\right) = 50 \text{ N} \cdot \text{m}$$

Moment of inertia:

$$J_1 = J_2 \left(\frac{N_1}{N_2}\right)^2 = 48\left(\frac{1}{2}\right)^2 = 12 \text{ kg} \cdot \text{m}^2$$

As a check, the power in either side of the gear is

$$P = T_1\omega_1 = T_2\omega_2 = 160(50) = 80(100) = \underline{\underline{8000 \text{ W}}}$$

6.2.5 GENERAL MACHINE EQUATIONS

The transient response of dc machines can easily be obtained from basic motor-generator equations and from voltage equations. The latter can be obtained from the equivalent circuit of the particular machine under consideration.

The differential equation that describes the torque developed by the motor is given by Eq. (3.49), which for convenience is rewritten here:

$$T = J\frac{d\omega}{dt} + B\omega + T_{\text{load}} \tag{6.94}$$

Neglecting initial condition and taking Laplace transforms, we have

$$T(s) = Js\omega(s) + B\omega(s) + T_L(s) \tag{6.95}$$

from which

$$\omega(s) = \frac{T(s) - T_L(s)}{Js + B} \tag{6.96}$$

Equation (6.96) in block diagram representation is shown in Fig. 6-32.

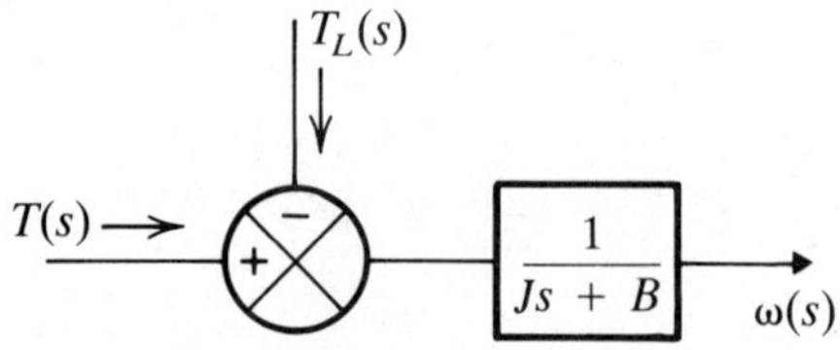

FIG. 6-32 The representation of Eq. (6.96) in block diagram form.

From the generator principle we have $V_g = K\phi\omega$. Taking Laplace and using the boundary conditions, we obtain

$$V_g(s) = K_1 I_f(s) \Big|_{\text{constant speed}} \tag{6.97}$$

$$V_g(s) = K_2 \omega(s) \Big|_{\text{constant field}} \tag{6.98}$$

Similarly, from the motor principle we have $T = K\phi I_a$. Then,

$$T(s) = K_2 I_a(s) \Big|_{\text{constant field}} \tag{6.99}$$

$$= K_3 I_f(s) \Big|_{\text{constant armature current}} \tag{6.100}$$

where the constants of proportionality K_1, K_2, and K_3 can easily be obtained from the basic relationships.

The underlying assumptions are that the machine operates along the linear part of its magnetization characteristic and that the armature reaction is negligible.

EXAMPLE 6–7 The separately excited dc motor shown in Fig. 6-33(a) has a constant excitation and drives a pure-inertia load. The moment of inertia of motor plus load is 2 kg · m^2. The motor constant is 2 N · m/rad/s. After the switch S is closed, determine:

a) The speed of the motor as a function of time.

b) The armature current as a function of time.

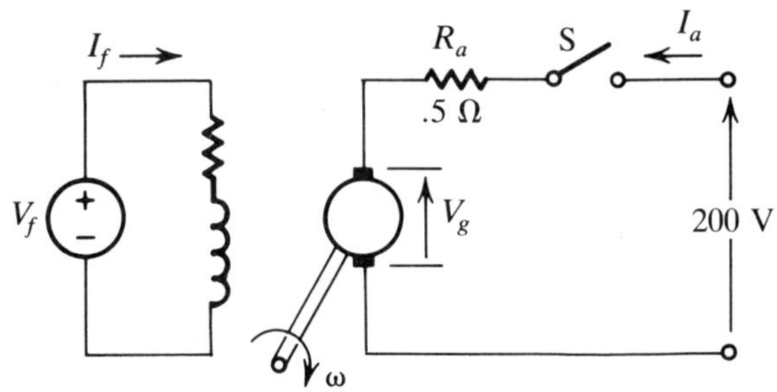

FIG. 6-33(a)

SOLUTION From the equivalent circuit of the motor (applying KVL), we have

$$V_t = I_a R_a + V_g$$

In Laplace,

$$V_t(s) = I_a(s) R_a + V_g(s) \tag{I}$$

From the motor principle,

$$T = K\phi I_a$$

Since the field current is constant, the expression for the torque becomes

$$T = K_1 I_a$$

In Laplace,

$$T(s) = K_1 I_a(s) \tag{II}$$

From Newton's law of rotating bodies we have

$$T = J\frac{d\omega}{dt}$$

Neglecting initial conditions and taking the Laplace transform, we obtain

$$T(s) = Js\omega(s) \tag{III}$$

From Eqs. (II) and (III) we find

$$I_a(s) = \frac{Js}{K_1}\omega(s) \tag{IV}$$

From the generator principle

$$V_g = K\phi\omega$$

or

$$V_g = K_1\omega$$

In Laplace transforms,

$$V_g(s) = K_1\omega(s) \qquad \text{(V)}$$

Equations (I), (II), (III), and (V) are shown in block diagram form in Fig. 6-33(b).

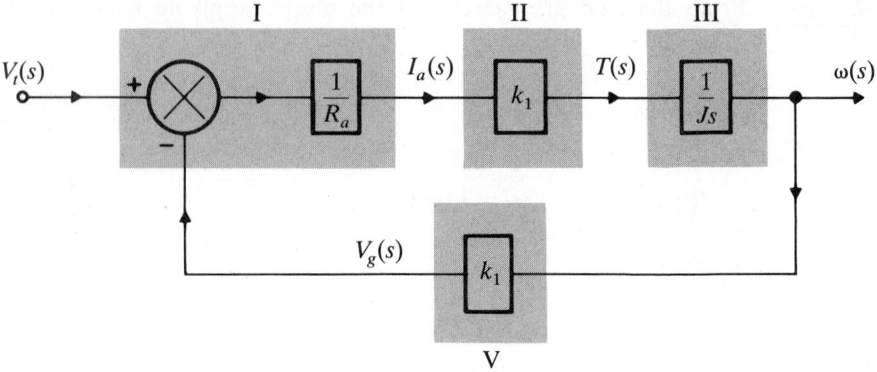

FIG. 6-33(b)

a) From Eqs. (I), (IV), and (V) we obtain

$$V_t(s) = R_a \frac{Js}{K_1}\,\omega(s) + K_1\omega(s)$$

From the above,

$$\omega(s) = \frac{K_1}{JR_a}\left(\frac{V_t(s)}{\tau + s}\right)$$

where $\tau = K_1^2/JR_a$. For a step input voltage, the equation for the speed becomes

$$\omega(s) = \left(\frac{K_1}{R_a J}\right)\frac{V_t}{s(\tau + s)}$$

Taking the inverse Laplace of the above and substituting the given values, we obtain

$$\omega(t) = \underline{\underline{100(1 - e^{-4t})\ \text{rad/s}}}$$

b) Similarly, from Eqs. (I), (IV), and (V) we obtain

$$V_t(s) = I_a(s)R_a + \frac{K_1^2}{Js} I_a(s)$$

From the above,

$$I_a(s) = \frac{V_t}{R_a(\tau + s)}$$

Taking the inverse transformation and substituting the given values, we obtain

$$I_a(t) = \underline{\underline{400e^{-4t} \text{ A}}}$$

As functions of time, the armature current and the speed are sketched as shown in Figs. 6-33(c) and (d), respectively. It should be noted that, in the absence of any rotational losses, the armature current will decay to zero. Thus no power will be drawn from the supply, although the motor will continue to rotate at 100 rad/s.

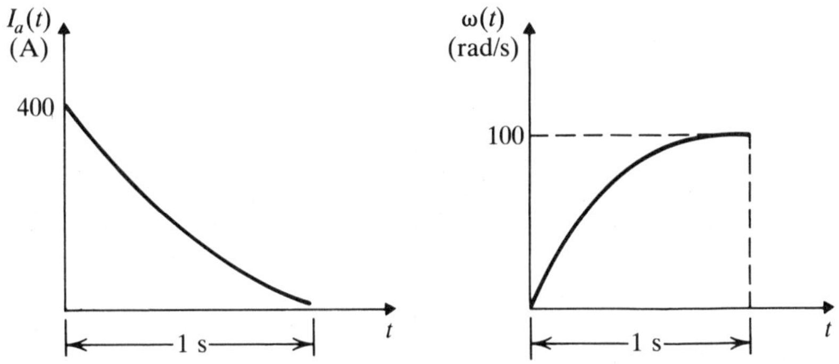

FIG. 6-33(c) and (d)

Exercise 6–7

a) Figure 6-34 shows a field winding of a dc machine and its discharge resistor R_1 connected across terminals A-B. The switch is suddenly opened after having been closed for a long time. This is actually the case when the input voltage to the field coil of a dc motor is switched OFF. Sketch the voltage V_{AB} for the following values of the discharge resistor:

(1) $R_1 = 100\ \Omega$
(2) $R_1 = 1000\ \Omega$

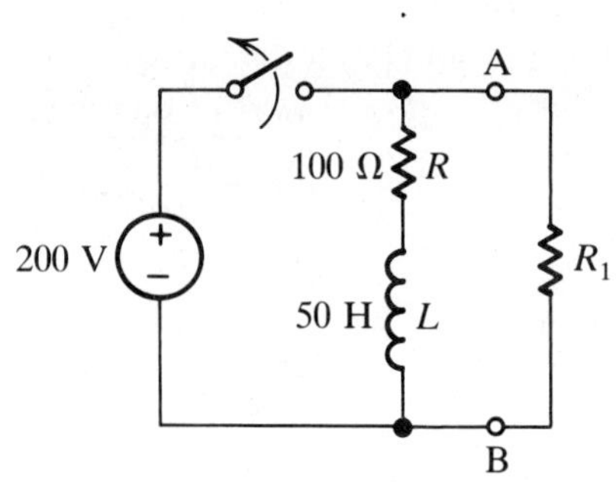

FIG. 6-34

b) Why must coils or field windings of dc shunt motors be connected across a resistor when their input voltage supply is switched OFF?

Answer (a) $200e^{-4t}$, $2000e^{-22t}$ V
(b) To harness the destructiveness of the energy stored in the magnetic field.

6.3 Modern Methods of Speed Control

6.3.1 RECTIFIERS

The most efficient and accurate method of controlling the starting current and the torque–speed characteristics of a dc motor is by using a variable armature and field voltage. The variable dc voltage is obtained through controlled rectifiers. Controlled rectifiers are used, as shown in Fig. 6-35(a), to convert single or polyphase ac voltage sources to dc controlled voltage supplies.

The controlling device for dc machines with low ratings is usually the transistor (Fig. 6-35(c)), while for higher machine ratings it is the thyristor* (Fig. 6-35(b)). The transistor, when used as a control switch, is discussed in Chapter 3. In this section, only the thyristor rectifying circuits will be analyzed. The dc voltage these circuits produce at their output terminals can be controlled smoothly by varying the thyristor's firing angle properly.

Controlled rectifiers, however, distort the power system's voltage waveforms, generate harmonics, and always result in an overall power factor of negative nature. The harmonics may be reduced or eliminated completely by using suitable filters, and the power factor can be improved by adding equipment that generates lagging (leading-power-factor) kVAR. Of course, this significantly increases the overall cost of the rectifying unit and decreases its reliability.

. . .

* The basic operation of transistors and thyristors is discussed in the Appendix.

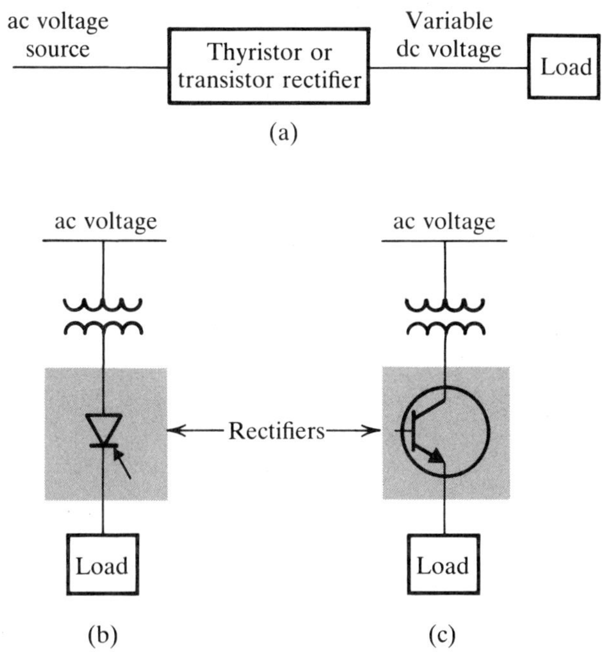

FIG. 6-35 AC-to-dc conversion: **(a)** block diagram representation, **(b)** one-line diagram representation, using thyristors, **(c)** one-line diagram representation, using transistors.

In general, a controlled rectifier system is comprised of an isolating or a voltage step-down power transformer, a thyristor bridge network, and various other control and protection apparatus.

In the next section the full-wave, single-phase, and three-phase controlled rectifiers will be discussed briefly. These rectifiers are also known, respectively, as two-pulse and six-pulse rectifiers because of the number of output voltage pulses they produce during one complete cycle of the input voltage.

6.3.2 SINGLE-PHASE, FULL-WAVE CONTROLLED RECTIFIERS

A single-phase, full-wave controlled rectifier is shown in Fig. 6-36(a). The resulting pertinent waveforms for a firing angle of less than 90 degrees are shown in Fig. 6-36(b).

The gating of the thyristors is synchronized to the supply voltage, and Q_1 and Q_2 are pulsed at a firing angle of α radians. The gating of Q_3 and Q_4 is initiated at a firing angle of $\alpha + \pi$ radians.

The load current is continuous: that is, current flows continuously over a complete cycle. As a result, one pair of thyristors is always conducting. Continuity results because thyristors Q_3 and Q_4 are turned ON before the current through Q_1 and Q_2 is reduced to zero. Continuity of current depends on the applied voltage, the thyristor's

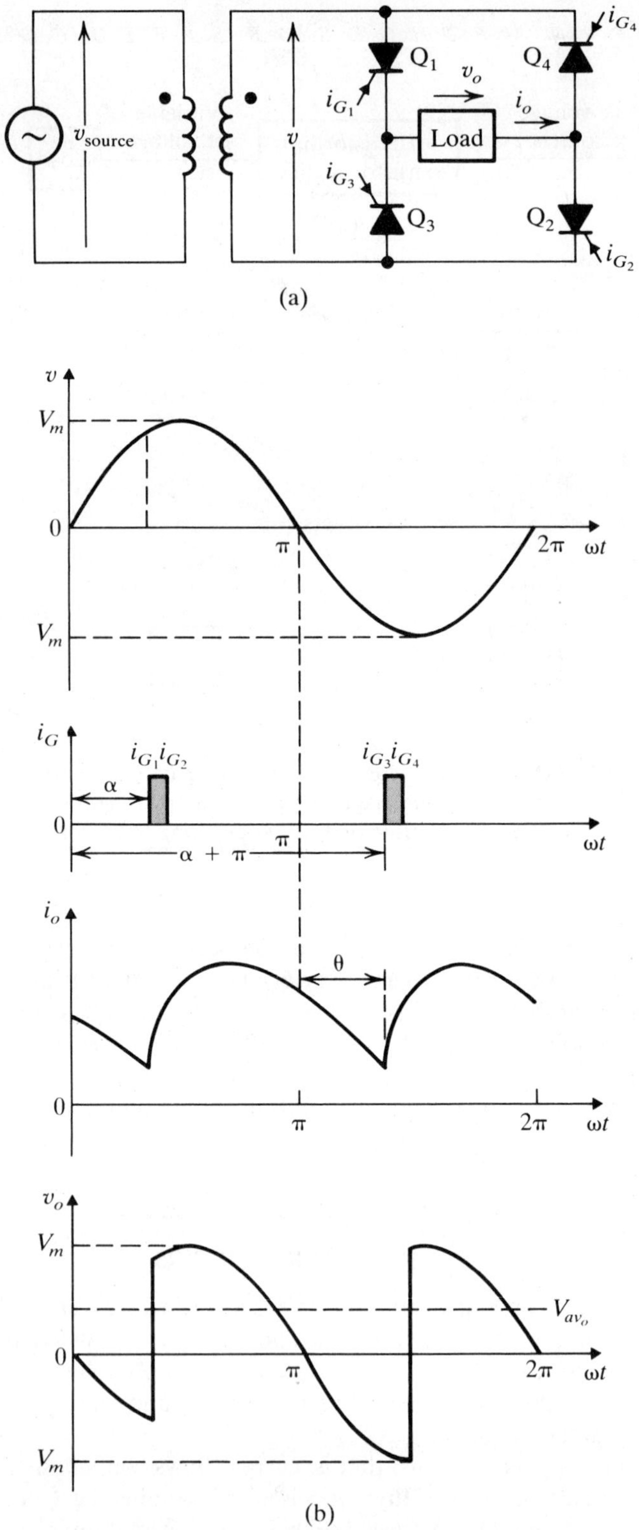

FIG. 6-36 Single-phase, full-wave controlled rectifier continuous load current: **(a)** circuit, **(b)** pertinent waveforms.

firing angle, and the impedance of the load. In an inductive impedance, the current reaches its zero value sometime after the applied voltage reaches its zero value. Thus, although the voltage increases negatively, the thyristor is not switched OFF until its current becomes zero.

Referring to Fig. 6-36(b), the average value of the output voltage is obtained as follows:

$$V_{av_o} = 2\left(\frac{1}{2\pi}\right)\int_{\alpha}^{\pi+\alpha} V_m \sin \omega t \, d\omega t \tag{6.101}$$

From which

$$V_{av_o} = 2\frac{V_m}{\pi}\cos\alpha \tag{6.102}$$

As can be seen from Eq. (6.102), the average value of the dc voltage can be varied smoothly by changing the firing angle of the thyristors accordingly. The average value of the output voltage, as a function of the angle of retard, is sketched in Fig. 6-37. At $\alpha = \pi/2$ radians, the dc value of the output voltage is equal to zero, and a further increase in the firing angle will result in negative output dc voltage.

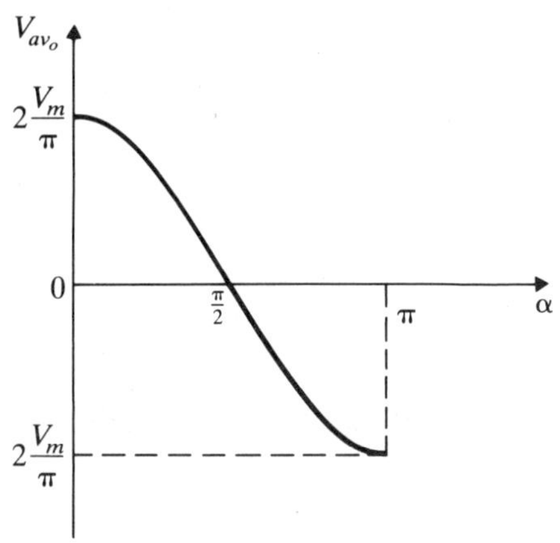

FIG. 6-37 Average value of output voltage versus angle of retard for a single-phase, full-wave controlled rectifier.

6.3.3 THREE-PHASE, FULL-WAVE CONTROLLED RECTIFIERS

A three-phase, full-wave controlled rectifier is shown in Fig. 6-38(a). This circuit is commonly known as a 3-ϕ bridge rectifier, a Graetz rectifier, or simply a six-pulse rectifier. Each of the output voltage pulses has a fundamental frequency, which is six times that of the supply voltage. The impedance Z_n of the R-L load to the nth harmonic current is

$$Z_n = R + jn\omega L \tag{6.103}$$

where $n = 6$, 12, etc., and ω is the angular rotation of the source voltage. Owing to increased impedances, the various harmonic components of the load current are much smaller than the corresponding ones produced by a two-pulse controlled rectifier.

The analysis of a six-pulse rectifier is simplified by considering the effects of the positive and negative line voltages, as shown in Fig. 6-38(b). During one complete cycle, the controlling range of each of these voltages is given by the interval through which the particular voltage has, relatively, the maximum positive value. For example, as can be seen in Fig. 6-38(b), the controlling range of the voltage v_{ab} starts at $\pi/3$ radians and goes up to $2\pi/3$ radians. The angle of retard for thyristor Q_1 is thus measured starting at $\pi/3$ radians. Therefore, for thyristor Q_1,

$$\alpha = \omega t - \frac{\pi}{3} \tag{6.104}$$

where ωt is, as shown in Fig. 6-38(b), the angle upon which the supply voltages depend.

The "turning ON" of the other thyristors is synchronized to the supply frequency, and their firing angles are spaced $\pi/3$ radians apart. The firing pulse of each thyristor is shown in Fig. 6-38(c). The end of conduction, as in all other circuits using thyristors, depends on the applied voltage and the load impedance. For rectifier operation, the angle of retard must be restricted to less than $\pi/2$ radians.

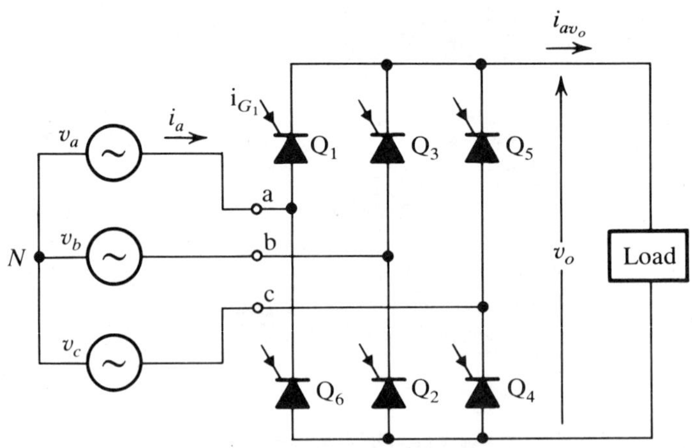

(a)

FIG. 6-38 Three-phase, full-wave rectifier: **(a)** circuit, **(b)** supply voltages, **(c)** thyristors' firing pulses, **(d)** output current waveform for resistive load and $\alpha = 0$, **(e)** current pulses through supply line "a".

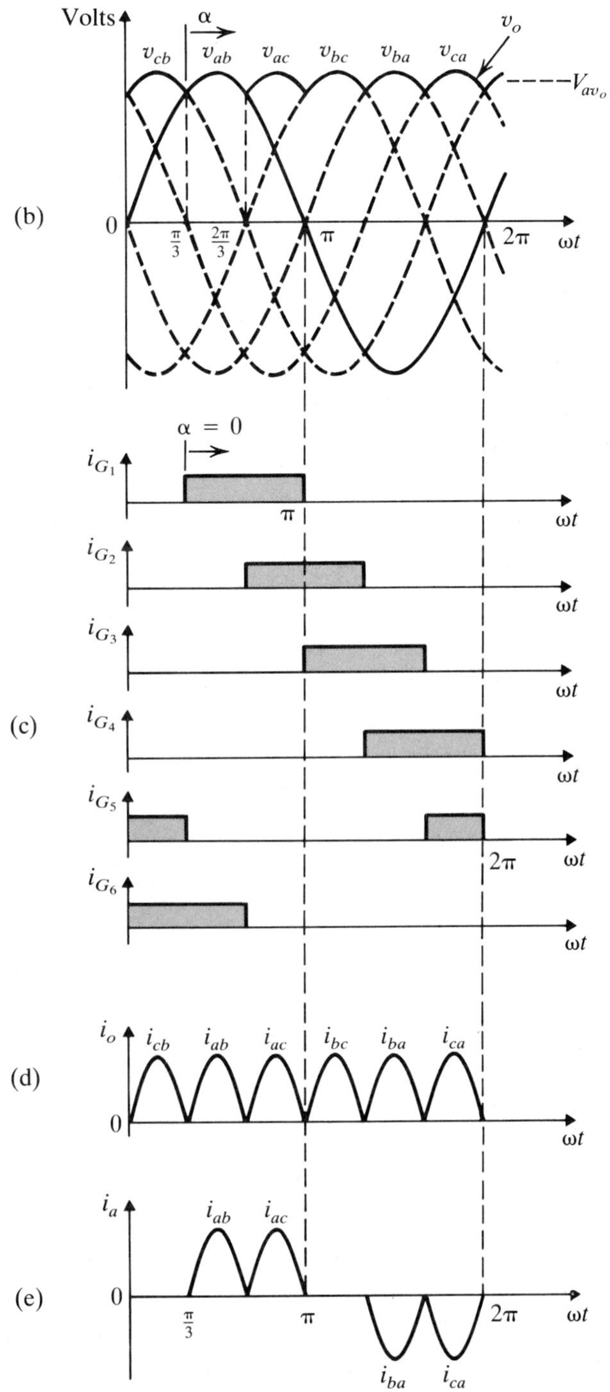

FIG. 6-38 *(Continued)*

Figure 6-38(d) shows the output current pulses for a firing angle of zero degrees. When one speaks of a firing angle for a particular thyristor, one refers to the time when the thyristor is gated relative to its corresponding line-to-line voltage.

Figure 6-38(e) shows the output current pulses that are conducted through supply line "a". Then in the six-pulse circuit, each of the supply lines carries four of the six output current pulses in an alternating symmetry and thus has no dc component. This permits the use of a 3-ϕ transformer to supply the voltage required by the bridge. Furthermore, the line-current waveform, as can be shown from Fourier series analysis, contains a fundamental component and various harmonics. The harmonics transport no real power to the load and distort the upstream current waveforms. This may lead to resonance problems and communication interference.

The type and duration of the output current pulses depend on the firing angle and the parameters of the load. The limits of integration for calculation of the various parameters depend on whether the output current is continuous or discontinuous.

For continuous load current, the dc value (V_{av_o}) of the output voltage is

$$V_{\text{av}_o} = \frac{6}{2\pi} \int_{(\pi/3)+\alpha}^{(2\pi/3)+\alpha} V_m \sin \omega t (d\omega t) = 3 \frac{V_m}{\pi} \cos \alpha \tag{6.105}$$

When the load to a six-pulse rectifier is a dc machine, Eq. (6.105) gives the motor's terminal voltage. Neglecting the resistance of the armature winding, from the generator principle and from KVL we have

$$V_g \approx 3 \frac{V_m}{\pi} \cos \alpha = K\phi\omega \tag{6.106}$$

Thus, the speed of the motor—as a function of the firing angle—is

$$\omega \approx \frac{3V_m}{K\phi\pi} \cos \alpha \tag{6.107}$$

That is, for constant field current the speed of the motor is directly proportional to the cosine of the firing angle.

From the above, then, it is evident that the thyristor's firing angle controls the motor's starting current and speed of operation.

Refer to Fig. 6-38(e). By assuming the load current is constant at I_{av_o}, we find the rms value of the supply line currents as follows:

$$I_{\text{rms}}^2 = \frac{2}{2\pi} \int_{\pi/3}^{\pi} I_{\text{av}_o}^2 \, d\omega t \tag{6.108}$$

$$= \frac{2}{3} I_{\text{av}_o}^2 \tag{6.109}$$

From Eq. (6.109),

$$I_{\text{rms}} = \sqrt{\frac{2}{3}}\, I_{\text{av}_o} \tag{6.110}$$

Equation (6.110) gives the rms value of the line current, not of its fundamental component.

The apparent power of the supply source is

$$|S| = 3V_{an}I_a = 3\frac{V_{L-L}}{\sqrt{3}}\sqrt{\frac{2}{3}}\, I_{\text{av}_o} = \sqrt{2}\, V_{L-L} I_{\text{av}_o} \tag{6.111}$$

where V_{L-L} is the rms value of the line-to-line supply voltage. Equation (6.111) can be used to size the isolating transformer, which supplies power to the rectifying bridge.

The apparent or total power factor ($\cos\theta_T$) at the input terminals of the bridge is found as follows:

$$\text{power} = \sqrt{3}\, V_{L-L} I_L \cos\theta_T \tag{6.112}$$

from which

$$\cos\theta_T = \frac{V_{\text{av}_o} I_{\text{av}_o}}{\sqrt{3}\, V_{L-L} I_L} \tag{6.113}$$

By substituting for V_{av_o} and I_L their equivalent expressions (given, respectively, by Eqs. (6.105) and (6.110)), we obtain

$$\cos\theta_T = \frac{3\dfrac{V_m}{\pi}\cos\alpha\, I_{\text{av}_o}}{\sqrt{3}\dfrac{V_m}{\sqrt{2}}\sqrt{\dfrac{2}{3}}\, I_{\text{av}_o}} \tag{6.114}$$

From the above,

$$\cos\theta_T = \frac{3}{\pi}\cos\alpha \tag{6.115}$$

The line current to the input terminals of the bridge has a fundamental component and various harmonics. Since the harmonics transport no average power to the load, the power measured with a standard power meter is

$$P = \sqrt{3}\, V_{L-L} I_1 \cos\theta \tag{6.116}$$

where I_1 is the rms value of the fundamental component of the line current and $\cos\theta$ is the actual power factor of the circuit. From Eqs. (6.112) and (6.116) we obtain

$$\cos\theta_T = g\cos\theta \tag{6.117}$$

where g is the ratio of the rms value of the fundamental component of the line current divided by the rms value of the current waveform.

EXAMPLE 6–8 The Δ-Y transformer shown in Fig. 6-39(a) supplies power to a six-pulse bridge rectifier. The line-to-line voltages across the secondary of the transformer are as shown in Fig. 6-39(b). The load current is constant and the transformer is ideal, with turns ratio equal to unity. For a firing angle of zero degrees, draw the waveforms of the current through:

a) The load.
b) The transformer windings.
c) The primary supply line (i_{L_1}).

SOLUTION

a) The waveform of the load current I_{av_o} is shown in Fig. 6-39(c). This current is continuous and of constant amplitude because of the load's inductance.
b) Each of the secondary transformer lines conducts during only two-thirds of the applied voltage's cycle. Thus, for continuous load current, each secondary transformer line conducts for ⅔π radians in the positive direction and ⅔π radians

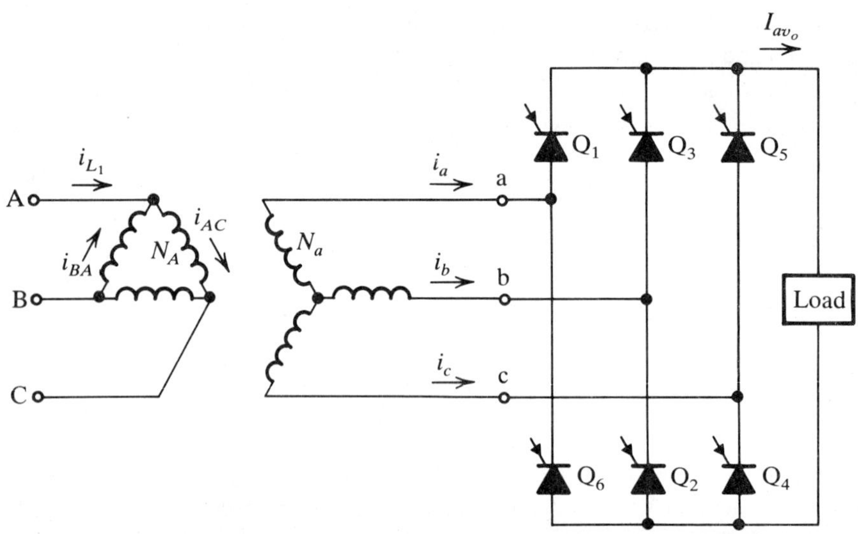

(a)

FIG. 6-39 Six-pulse rectifier. Waveforms of the currents through the load, the secondary windings, and a primary line of a Δ-Y transformer.

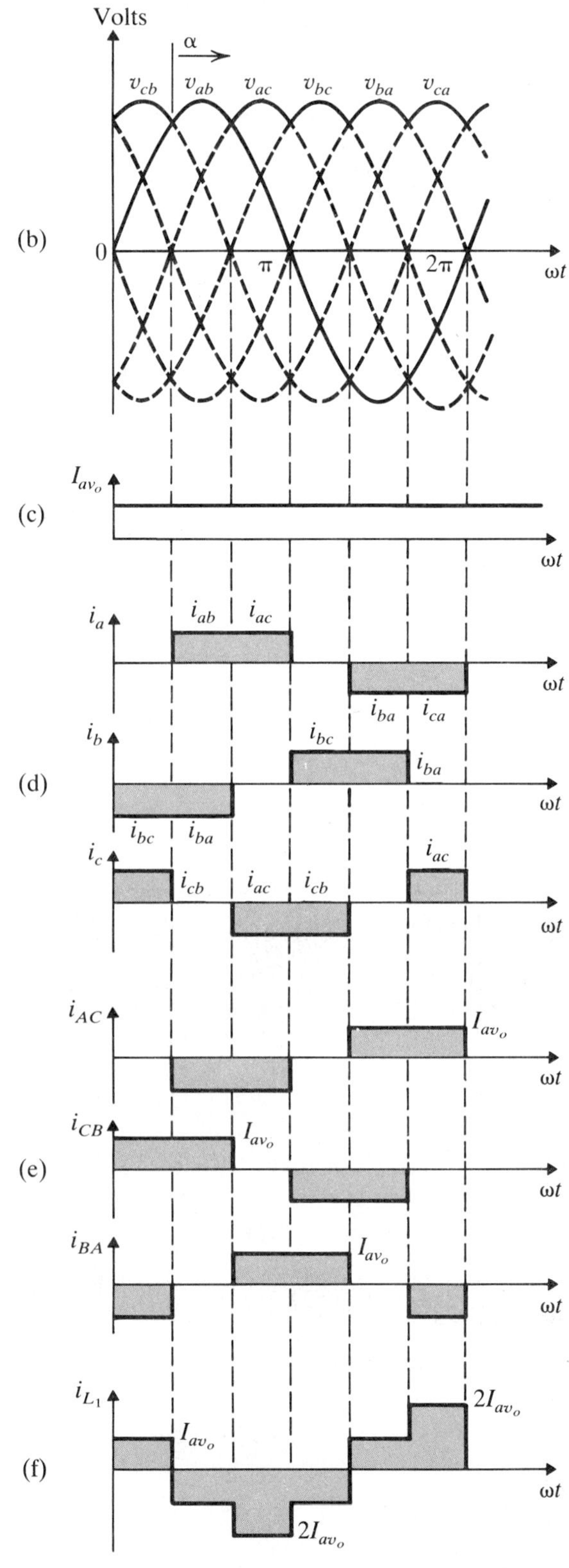

FIG. 6-39 *(Continued)*

in the negative direction. The particular conducting period of each secondary transformer line can be obtained by inspection of the given line-to-line voltage waveforms. For example, line "a" conducts in the positive direction when voltages v_{ab} and v_{ac} are the highest voltages in the circuit, and conducts in the negative direction when voltages v_{ba} and v_{ca} have the highest potential in the circuit. Therefore, line "a" conducts, as shown in Fig. 6-39(d), only during the time intervals between 60 to 180 degrees and 240 to 360 degrees. The currents in the other secondary lines (i_b and i_c) are drawn as shown in Fig. 6-39(d).

The transformer's primary winding currents are obtained from the secondary line currents by using KVL in magnetic circuits. That is,

$$N_A i_{AC} = N_a i_a$$

Since the numbers of primary and of secondary winding turns are equal, the magnitude of the winding currents must also be equal. However, these currents must be in opposite direction, as per Lenz's law.

In a similar manner, the currents through the other primary transformer windings can easily be determined. The resulting waveforms are shown in Fig. 6-39(e).

c) The primary line current is found by applying KCL in the delta-connected winding. Considering line "A", we have

$$i_{L_1} = i_{AC} - i_{BA}$$

From the above, the current through line "A" is obtained graphically by superimposing the negative of i_{BA} on the winding current i_{AC}. This is shown in Fig. 6-39(f).

EXAMPLE 6–9 A 20 kW, 220 V, 1200 r/min, .94 efficient dc motor is supplied through a six-pulse rectifier with a constant current. The line-to-line voltage across the secondary terminals of the transformer is 208 V. Determine:

a) The nominal firing angle of the thyristors.
b) The apparent power factor of the load, as seen from the secondary terminals of the rectifying transformer.
c) The line current toward the rectifying bridge.
d) The kVA rating of the rectifying transformer.

SOLUTION

a) The thyristor's firing angle is found from Eq. (6.105). Solving for α and substituting the given values, we obtain

$$\alpha = \arccos \frac{(220)\pi}{3\sqrt{2}\,208}$$

$$= \underline{\underline{38.45^\circ}}$$

b) From Eq. (6.115) we have

$$\cos\theta_T = \frac{3}{\pi}\cos\alpha$$

$$= \frac{3}{\pi}\cos 38.45$$

$$= \underline{\underline{.75 \text{ lagging}}}$$

c) The average value of the load current is

$$I_{av_o} = \frac{P}{V_{av_o}} = \frac{20}{.94(.220)}$$

$$= 96.71 \text{ A}$$

Thus, from Eq. (6.110) the rms value of the line current is

$$I = \sqrt{\frac{2}{3}}\, 96.71$$

$$= \underline{\underline{78.96 \text{ A}}}$$

d) The rms value of the load current is equal to its dc value. Thus, from Eq. (6.111), the kVA rating of the transformer is

$$|S| = \sqrt{2}(.208)(96.71) = \underline{\underline{28.45 \text{ kVA}}}$$

For practical purposes a 30 kVA, commercially available transformer should be used.

Exercise 6–8

a) Draw the output voltage and current waveforms of a single-phase, full-wave rectifier. Assume that the load current is continuous and $\alpha = 30°$.

b) A six-pulse, three-phase thyristorized rectifier supplies 100 A dc to a load. Determine the average and rms values of a thyristor's current.

Answer 33.33 A, 57.74 A.

Exercise 6–9 A separately excited, 100 kW, 480 V, .9 efficient dc motor has an armature resistance of .10 Ω. The motor receives its dc supply through a six-pulse rectifier as shown in Fig. 6-40. Determine:

a) The motor's current under nominal operating conditions.

b) The thyristor's firing angle under nominal operating conditions and at starting, if the motor's starting current is limited to 800 A.

Answer (a) 231.48 A; (b) 27.3°, 81.5°.

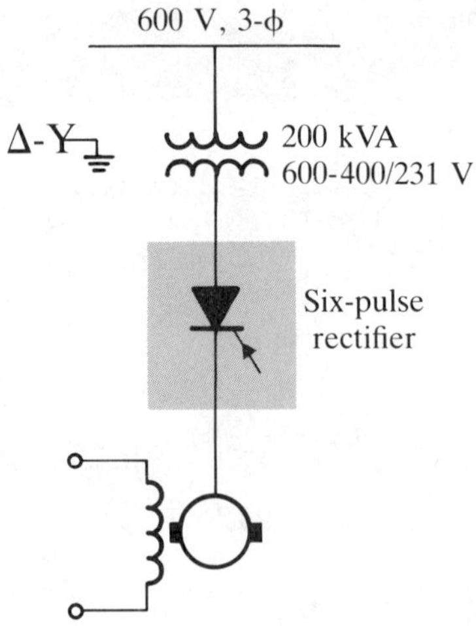

FIG. 6-40

6.4 Conclusion

The operation of a dc motor is based on the tendency of a current-carrying conductor—when placed in an external magnetic field—to rotate so that the field it creates is parallel to, and in the same direction as, the external field.

The operation of dc generators is based on the induction principle, according to which a voltage is induced in conductors that are rotated through an external magnetic field.

In their basic form dc machines have a field coil and an armature winding. The field coil is wound around the machine's stationary magnetic poles, while the armature winding is wound on its rotating shaft. The armature current is conducted from the external motor terminals to the stationary brushes, then to the rotating segments of the commutator, and from there to the armature conductors.

The rotated armature conductors cut, successively, the fields of the north and south magnetic poles. As a result, the generated armature current is of alternating waveform. However, as seen from the machine's external terminals, the armature current is of dc waveform, because it is rectified by the commutator-brush action. The process by which the armature current reverses direction during half the period of its cycle is called commutation.

During part of the rectification process each armature conductor is shorted through the brushes. As a result, sparking may take place. Sparking and flashover used to be the main limitations of dc machines. However, over recent years the design of the machines has been improved, and the commutation process is now sparkless. Nevertheless, the brushes require periodic replacement. In general, this increases the maintenance cost of dc machines.

The flux of the armature current effectively reduces the magnetic field and distorts it in the interpole region. This interaction between the flux of the field and armature current is commonly known as armature reaction. Armature reaction adversely affects the characteristics of the machine and the commutation process.

All large dc machines are equipped with *compensating* and *commutating windings.* These windings are connected in series with the armature coil. Their function is to cancel the *demagnetizing* and *field-distorting* effects of the armature current. The compensating winding is placed on the pole face of the machine and cancels the

TABLE 6-4 Summary of important equations

Item	*Description*	*Remarks*
1	The motor principle $T = K\phi I_a$	Eq. (6.13)
2	$K = \dfrac{Np}{\beta\pi}$	Eq. (6.14)
3	The generator principle $V_g = K\phi\omega$	Eq. (6.22)
4	Mutual coupling $V_g = \dfrac{p}{2}\omega I_f L_{af_m}$	Eq. (6.32)
5	$T = \dfrac{p}{2} I_f L_{af_m} I_a$	Eq. (6.34)
6	Effective field $I_{f_{ef}} = I_{f_a} \pm \dfrac{N_s}{N_f} I_s - AR$	Eq. (6.78)
	Two-pulse rectifier	
7	$V_{av_o} = 2\dfrac{V_m}{\pi}\cos\alpha$	Eq. (6.102)
	Six-pulse rectifier	
8	$V_{av} = 3\dfrac{V_m}{\pi}\cos\alpha$	Eq. (6.105)
9	$I_{rms} = \sqrt{\dfrac{2}{3}}\, I_{av_o}$	Eq. (6.110)
10	$\lvert S\rvert = \sqrt{2}\, V_{L-L} I_{av_o}$	Eq. (6.111)
11	$\cos\theta_T = \dfrac{3}{\pi}\cos\alpha$	Eq. (6.115)
12	$\cos\theta_T = g\cos\theta$	Eq. (6.117)

TABLE 6-5 Data for a 300 kW, 1150 r/min, .95 efficient, 1052 A, 300 V, 4-pole, separately excited dc motor.*

a) Resistances

Winding(s)	Resistance in ohms at 25°C
Armature	.005
Commutating and Compensating	.0045
Separate Field	2.07

b) Load characteristic

Load in Percent	*Armature*		*Field*		*Speed r/min*	*Commutation*
	Voltage (V)	*Current (A)*	*Voltage (V)*	*Current (A)*		
0	300	0	55	22.9	1175	Sparkless
25	300	270	55	22.9	1165	Sparkless
50	300	540	55	22.9	1160	Sparkless
75	300	800	55	22.9	1150	Sparkless
100	300	1052	55	22.9	1150	Sparkless
125	300	1350	55	22.9	1150	Sparkless
150	300	1600	55	22.9	1150	Sparkless

c) Temperature rise test

Air supply: 60 m^3/min

Time in Hours	*Load*	*Armature*		*Field*		*Speed r/min*
		Voltage	*Current*	*Voltage*	*Current*	
2	100%	300 V	1075 A	55 V	22.9 A	1150

	Armature				*Field Windings*			*Bearings*	
	Coil	*Core*	*Commut.*	*Frame*	*Main*	*Commut.*	*Comp.*	*Commutator Side*	*Opposite Side*
Temp., °C	48	45	55	18	35	33	32	20	15

d) Measurement of air gap

Main pole: 3.5 mm
Inter pole: 7.5 mm

* *Courtesy of Siemens Electric Limited.*

reduction of the field flux. The commutating coil is wound around the narrow poles in the interpole region and cancels the distorting effect on the field-flux distribution.

Besides the main field winding, compound dc machines have an auxiliary field winding. Depending on its connection, it may aid (cumulative connection) or oppose (differential connection) the flux of the main field. The function of the auxiliary winding is to modify the external characteristics of the machine.

The characteristics of dc machines can easily be adjusted through armature or field variable resistors or through variations of their dc voltage supply. The use of potentiometers results in higher energy losses. The use of variable dc voltage sources results in lower energy losses and in smoother torque–speed characteristics.

All modern dc drives are equipped with controlled rectifiers through which the motor characteristics are easily controlled. The six-pulse bridge rectifier is the most popular compromise between cost and desired characteristics. In general, controlled rectifiers are reliable and highly efficient, generate harmonics toward their ac voltage supply and a ripple voltage toward their dc output lines, and require KVAR compensation.

Owing to their efficiency, variable-speed dc drives have gradually displaced the more inefficient drives, such as eddy-current variable-speed drives, fluid couplings, the classical generator-motor set, and induction motors that drive fans whose flow is controlled through dampers.

Table 6-4 summarizes the essential relationships of dc machines and their control apparatus. Table 6-5 gives manufacturer's data for a 300 kW dc motor. Table 6-6 lists typical component losses of dc drives.

TABLE 6-6 Efficiency of dc drives at rated speed*

	Efficiency			
Rating kW	*Transformer*	*Control Circuit*	*Motor*	*Total*
15		.98	.92	.90
30		.99	.88	.87
150	.97	.99	.89	.85
225	.98	.99	.90	.87
750	.99	.99	.94	.92
1500	.99	.99	.94	.92

* *Courtesy of Siemens Electric Limited.*

6.5 REVIEW QUESTIONS

1. Describe armature voltage (V_a). Is this voltage ac or dc in waveform? Why is armature voltage also called voltage generated (V_g), back emf (V_b), countervoltage (V_c), or speed voltage?

2. Differentiate between speed voltage and transformer voltage.
3. What is armature reaction and what are its *two adverse effects* on the operation of dc machines?
4. Under what condition is the armature reaction proportional to the armature current squared?
5. What controls the shape of the magnetization characteristic?
6. What limits the operation of dc machines?
7. What are the *three adverse effects* of high starting current on dc machines?
8. Are the rotating parts of a dc machine the brushes or the commutator segments?
9. At what part of the flux-density distribution is the armature conductor located at the instant when commutation takes place? What is the reason for this?
10. How would you measure the mutual inductance between the armature and field windings of a separately excited motor?
11. Where are the series, the commutating, and the compensating windings located physically? What are their functions?
12. Draw the torque-versus-speed characteristics for series, shunt, and compound motors.
13. Draw the equivalent circuits of long-shunt and short-shunt dc compound machines. Consider differential and cumulative type connections.
14. Why are dc shunt motors normally equipped with a variable external resistor that is connected in series with the armature?
15. Define and give the mathematical relationships for the time constants of a dc motor and of a dc generator.
16. What is the power factor of a six-pulse thyristorized rectifier, as seen from its ac input terminals?
17. What is the average value of the dc voltage for a six-pulse thyristorized rectifier?

6.6 PROBLEMS

6-1 At 1800 r/min, a separately excited dc generator with constant excitation develops an open-circuit terminal voltage of 400 V. Determine the electromagnetic torque when the machine operates as a motor and draws an armature current of 60 A. Assume an armature reaction of 5%.

6-2 A 10 kW, 230 V, shunt dc motor has an efficiency of 88% at full-load. The resistances of the armature and shunt field are .12 and 230 ohms, respectively. At no-load, the motor rotates at 800 r/min and draws 4 A from the 230 V source.

- **a)** At full-load conditions, determine the armature current, the electromagnetic torque, and the speed of the motor (neglect armature reaction).
- **b)** Repeat (a), assuming an armature reaction of 5%.

6-3 A 200 V, 1200 r/min, 30 A, dc shunt motor drives a ventilating fan whose torque requirement is proportional to the speed squared. The resistances of shunt field and armature windings are 100 and .15 ohms, respectively. Neglecting rotational losses and armature reaction, determine:

a) The starting current of the motor.

b) The resistance that must be inserted in series with the armature to produce a motor speed of 1000 r/min.

c) The resistance of the field rheostat that must be connected in series with the field winding to yield a speed of 2000 r/min.

6-4 The generator-motor system shown in Fig. P6-4 is used as a variable-speed drive. In actual set-ups, the speed of the motor is controlled by varying the field of the generator. For this problem, however, assume that the fields of the motor and generator remain constant. The system's name-plate data and armature reaction are as shown on the diagram. The generator's prime mover is a 3-ϕ induction motor whose slip at no-load is negligible, and at full-load is 4%. Estimate:

a) The voltage regulation of the generator.

b) The speed regulation of the motor.

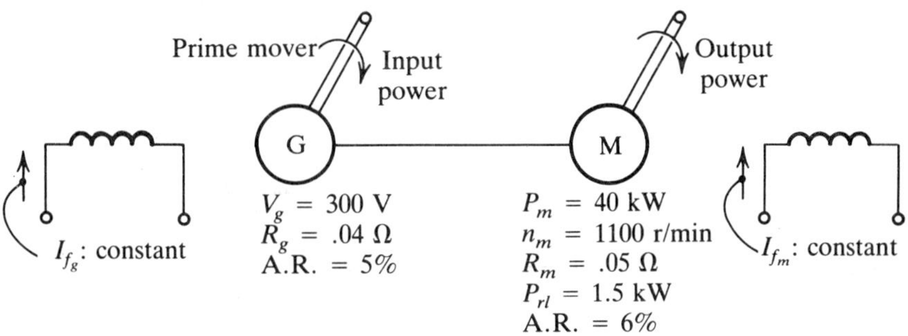

FIG. P6-4

6-5 The open-circuit characteristic of a 220 V dc series motor at 1800 r/min is approximated by

$$V_g = \frac{260 I_a}{25 + I_a}$$

The motor draws 90 A when it runs loaded at 1800 r/min. If the load conditions change and the motor draws 70 A, determine:

a) The speed.

b) The electromagnetic torque developed.

c) The useful output power if the mechanical losses within the motor consume a torque of 2 N · m.

d) The mathematical relationship that describes the open-circuit characteristic for the condition in (a).

6-6 A short-shunt compound generator has its auxiliary field winding connected in such a way that its magnetic field opposes that of the main field winding. The ratio of the effective number of turns per pole of the main field to auxiliary field winding is 175:1. The armature resistance is .10 ohm, the series-field resistance is .05 ohm, and the shunt-field resistance is 125 ohms. Data for the open-circuit characteristic at 1800 r/min are as follows:

Armature voltage (V):	50	100	200	260	300	318
Field current (A):	.21	.41	1.0	1.51	2.0	2.7

When driven at 1800 r/min the generator delivers 60 A to a load at 250 V. Determine the speed at which the generator must be driven in order to deliver 75 A at 300 V. Assume armature reaction is proportional to the armature current.

6-7 A 600 V, 7.46 kW, 1200 r/min series dc motor drives a constant torque-load of 50 N · m. The series and armature resistances are .10 and .20 ohm, respectively. In order to obtain variable speed, a resistance R_x is connected in parallel with the field winding. Neglecting armature reaction and rotational losses,

a) Show that the armature current as a function of R_x is given by

$$I_a = 11.48\sqrt{\frac{R_x + .1}{R_x}}$$

b) Show that the speed as a function of R_x is given by

$$\omega = 138.47\sqrt{\frac{R_x + .1}{R_x}} - \frac{.53}{R_x}(1.5R_x + .1)$$

c) Sketch the expressions found in (a) and (b).

6-8 In the Ward-Leonard speed control system shown in Fig. P6-8, the armature of the separately excited dc generator is connected directly to that of a separately excited dc motor. The generator is driven at constant speed, and its open-circuit characteristic at the operating region is described by

$$V_g = 5 + 10I_{f_g}$$

The motor has an efficiency of 95%. When the voltage at the terminals of the generator is 320 V, the motor runs at 1200 r/min and delivers 50 kW to the load.

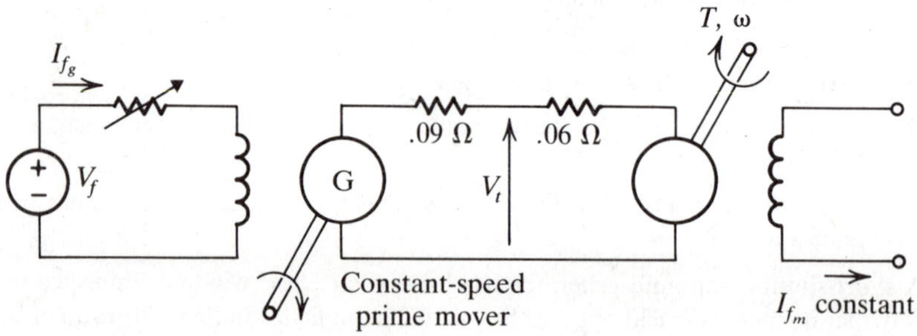

FIG. P6-8

a) For a constant torque-load requirement of 200 N · m, show that the field current of the generator is related to the speed of the motor by

$$I_{f_g} = .234\omega + .78$$

b) For a constant load-power (50 kW) requirement, show that the field current of the generator is related to the speed of the motor by

$$I_{f_g} = .234\omega + 1.97$$

c) Sketch the expressions derived in (a) and (b).

6-9 A six-pole, 1800 r/min separately excited dc generator has the following parameters.

Field voltage:	150 V
Field resistance:	50 Ω
Field inductance:	25 H
Generator constant K_1:	100 V/field amperes

The generator's terminals are open-circuited when the field voltage supply is switched on. Draw the control block diagram and determine, as functions of time:

a) The energy dissipated by the field winding.

b) The generator's terminal voltage.

6-10 A separately excited motor is to drive a load whose parameters are as shown in Fig. P6-10. The motor's constant is 2 V/rad/s. When the switch is closed, draw the control block diagram and determine:

a) The speed as a function of time.

b) The electromechanical time constant.

c) The armature current as a function of time.

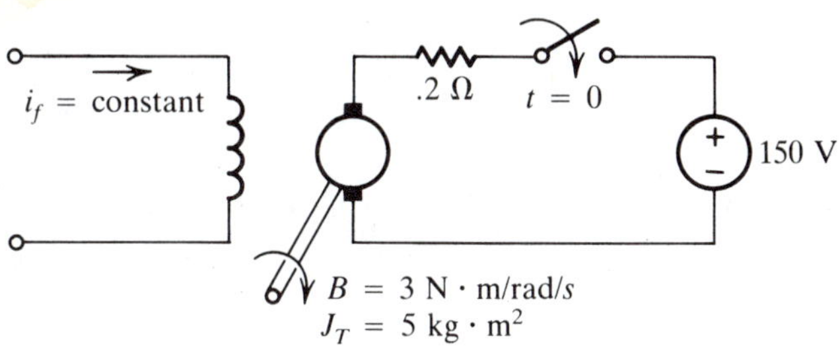

FIG. P6-10

6-11 A single-phase, full-wave rectifier operating at a firing angle of α degrees supplies power to a resistive load. Show that the apparent power factor of the load, as seen from the transformer's secondary terminals, is given by

$$\cos \theta_T = \sqrt{\frac{\pi - \alpha + \dfrac{\sin 2\alpha}{2}}{\pi}}$$

6-12 As shown in Fig. P6-12, a six-pulse rectifier supplies a 200 kW, 600 V, .93 efficient dc motor. The transformer is fed from a 4160 V substation. At nominal operating conditions the thyristor's firing angle is 36 degrees. Determine:

a) The transformer's turns ratio.

b) The kVA rating of the transformer.

c) The apparent power factor, as seen from the terminals of the substation.

Assume that the transformer is 96% efficient.

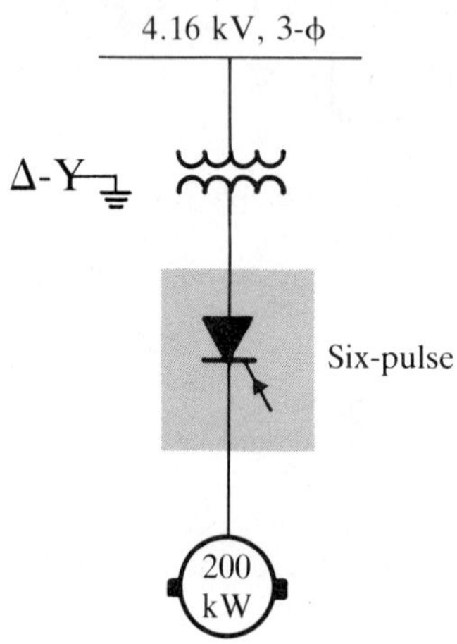

FIG. P6-12

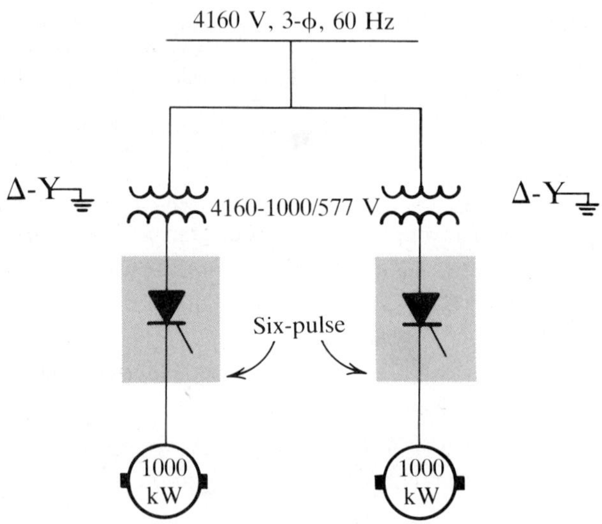

FIG. P6-13

6-13 Two dc drives, each 1000 kW, 800 V, .95 efficient, are connected, as shown in Fig. P6-13, to a 4.16 kV substation. Assuming that the motors draw rated current, determine:

a) The thyristor's firing angle.

b) The magnitude of the total current through the 4.16 kV feeder.

c) The apparent power factor, as seen from the secondary of one of the transformers.

SEVEN

Control Schematics

7.0 Introduction

The control schematic of a machine is an electrical drawing that uses conventional symbols to reveal and demonstrate the machine's starting, control, protection, deenergization, and safety features. Hence, the control schematic constitutes the interface between the engineer and the machine. As such, an understanding of control schematics, then, bridges the domains of theoretical and practical work in the field of electric machines.

Control schematics may be designed by using electromechanical components or solid state devices. The trend, of course, is toward solid state devices.

The design of a *good* control schematic requires a basic knowledge of the control components and of the principles of operation of the machine under consideration. The design of the *best* control schematic requires, additionally, some field experience and an up-to-date knowledge of the cost, availability, and limitations of the various control components.

The various functions and components of control schematics are represented on an engineering drawing by standard conventional symbols. Field specialists, however, do the actual hook-up of these devices using "wiring diagrams." A machine's *wiring diagram* shows the actual or physical interconnections between the various components. It does not identify the functional characteristics of each component.

The first section of this chapter explains the operation of the basic devices. Subsequent sections discuss the actual control schematics of single-phase motors, three-phase motors, and dc machines. The chapter ends with a detailed analysis of logic circuits and their applications in the computerized control of machines.

7.1 Basic Devices and Symbols

This section discusses relays, electromagnetic starters, pushbuttons, indicating lights, temperature sensors, and thermostats. The principle of operation and the symbol representation of these devices provides the basis for the understanding and the design of electrical control schematics.

7.1.1 ELECTROMAGNETIC RELAYS

As shown in Fig. 7-1(a), an electromagnetic relay in its basic form has a coil and a contact whose status change when the relay's coil is energized. The coil is wound around a stationary core, while the moving part of the contact is attached on a hinged iron bar called the armature. When the relay is energized, the armature is attracted to the electromagnet, and thus the contact of the relay changes status (Fig. 7-1(b)).

Physically, the coil and the contact are within the same enclosure. Electrically, however, the coil and the contact are generally connected to different circuits. As shown

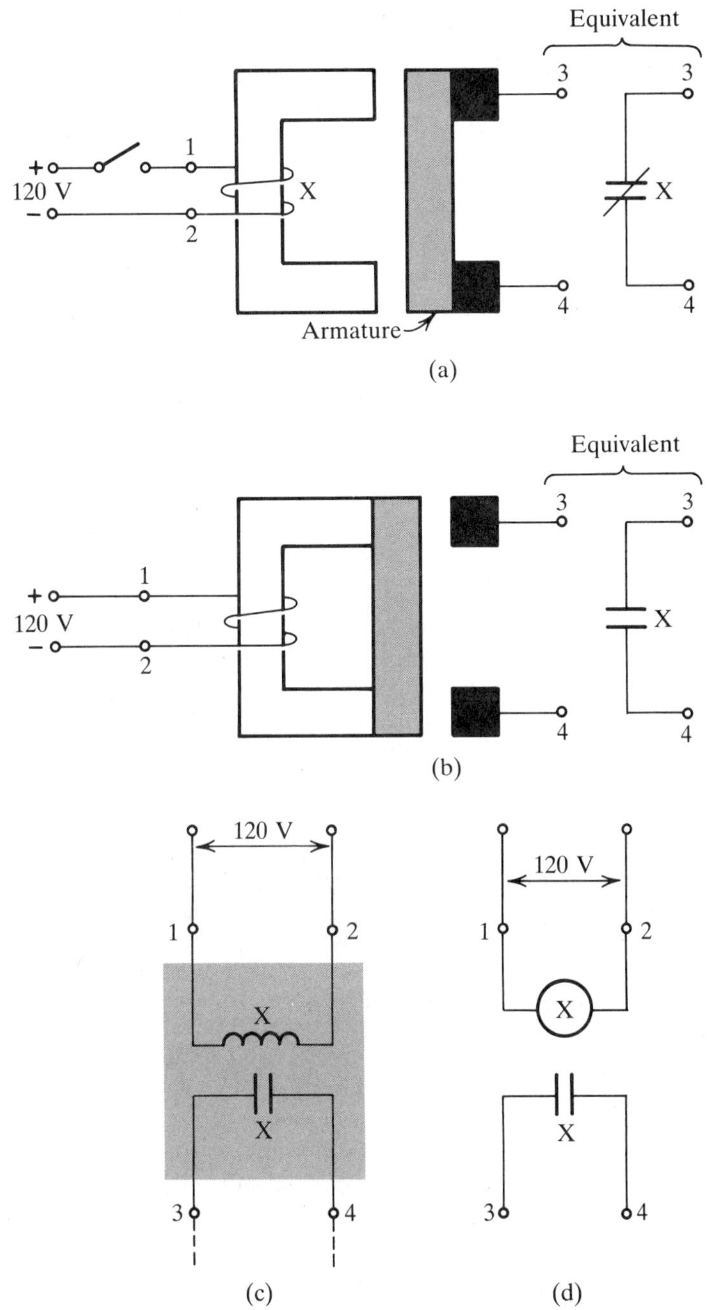

FIG. 7-1 Representation of an electromagnetic relay: **(a)** deenergized, NC contact, **(b)** energized, contact opened, **(c)** wiring diagram of (b), **(d)** control schematic of (b). The armature hinge is not shown.

in Fig. 7-1(c), the input terminals 1 and 2 of the relay receive the excitation signal, while the output terminals 3 and 4 are the external connections of the relay's contact.

Depending on the design, the coil of a given relay may require, for its energization, a certain voltage (voltage coil) or a certain current (current coil), or a combination of voltage and current. The energizing signal, where it is a voltage or a current, could be produced by either normal or abnormal machine operation, or by a sequence of events that accompany the operation of an electrical apparatus.

The energizing signal may be dc or ac; each has its advantages and disadvantages. Relays with ac coils cost less than those with dc coils but react slowly, consume more energy, and may malfunction owing to occasional transient fluctuations of the supply voltages. Although relays with dc coils are more expensive, they respond more quickly and are, within a certain voltage range, independent of the ac voltage fluctuations.

In a control schematic, the coil of a relay is represented by a circle in which is inscribed an arbitrary letter, as shown in Fig. 7-1(d). This letter is also used to designate the relay's contact. In this illustration the letter X appears, but very often the letter R is employed.

The contact(s) of the relay may be used to switch on an indicating light, to sound an alarm, to start a machine, or to remove a machine from its voltage source.

The contacts may be of the momentary or of the time-delay type, or a combination of both. When a relay's coil is energized, the momentary type of contacts change status instantaneously, while time-delay relays change the status of their contacts with some specific time delay. In a relay with a "time delay on energization," the beginning of the time delay starts immediately after its coil is energized. In a relay with a "time delay on deenergization," the time delay begins immediately after its coil is deenergized.

As shown in the control schematics, the status of the contacts of the various relays corresponds, by convention, to the deenergized status of the relays. The symbols and the general form of a device's contacts are further discussed in Section 7.1.4.

Electromechanical relays can be divided into two categories:

Relays for general purposes

Relays for the protection of equipment.

General-purpose relays are simple and economical and are used mainly for switching and annunciating functions. A simple general-purpose 120 V relay is shown in Fig. 7.1. The voltage rating of its coil is 120 V. However, the voltage and current ratings of its contact must be suitable for the circuit in which it will be connected.

Relays for the protection of equipment can have their current coils connected in series with the protected load, or the input signal to their coils can be obtained from instrument transformers that monitor the electrical condition of the load. Relays whose current coils are connected in series with the load are called *thermal overload relays.* Relays that receive their coil excitation through current or voltage transformers are generally known as *protective relays.* In either case the function of the relay is to protect the equipment, and therefore the assigned terminology is not functionally very meaningful. Because of their industrial importance, both of these types of relays are discussed extensively in the sections that follow.

7.1.2 ELECTROMAGNETIC STARTERS

As shown in Fig. 7-2, a single-phase electromagnetic starter—often referred to as a magnetic contactor—incorporates an electromagnetic coil (C), two main contacts (C_1), and an auxiliary contact (C_2). The coil and its contacts are usually identified by the same letter. This simplifies the understanding of multimachine control schematics. For the sake of clarity, however, in the introductory sections of this chapter the subscripts 1 and 2 will be used to differentiate between the main and auxiliary contacts.

Magnetic contactors are similar to relays. Their only electrical difference is that the main contacts of magnetic starters have higher voltage and current capacities than the contacts of the relays. The function of the magnetic contactor's assembly is to energize or deenergize a motor or any other electrical apparatus. When a suitable

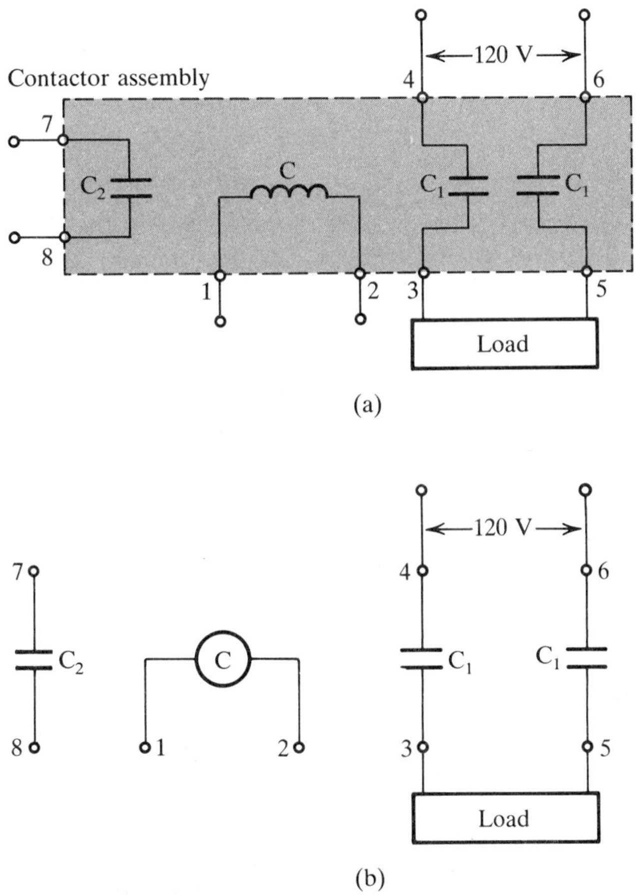

FIG. 7-2 Electromagnetic starter or contactor: **(a)** wiring diagram, **(b)** representation in a control schematic.

voltage is supplied to the coil (see Fig. 7-2(a), terminals 1 and 2) of the magnetic contactor, an electromagnetic force is developed which closes the normally open (NO) contacts C_1 and C_2. Closing the contacts C_1 connects the load to its voltage source.

For its energization, the coil of the starter (depending on its manufacturer) requires a minimum of about 80% of its rated voltage. Thus, the dipping of the supply voltage that usually accompanies the starting of ac machines may deenergize the electromagnetic starter.

The main contacts C_1 are connected in series with the controlled electrical load, so they must be capable of safely carrying and withstanding the currents that are produced by starting, operating, and deenergizing the load. The current capacities of the main contacts, on starting and on deenergizing the load, are referred to as the "making" and "breaking" capacity of the contacts, respectively. Typical ratings for three-phase electromagnetic starters are shown in Table 7-1.

TABLE 7-1 Typical ratings of electromagnetic starters*

Motor Rating: 480 V, 60 Hz, 3-ϕ	*Electromagnetic Starter*					
	Characteristics of the 120 V Coil			*Capacity of Main Contacts (A)*		
	Nominal Power (W)	*Inrush Current (A)*	*Continuous Current (A)*	*Make*	*Carry*	*Break*
1 kW	10	.41	.10	100	5	90
10 kW	13	.67	.14	400	50	300
100 kW	35	3	.37	1500	200	1200

* *Courtesy of Siemens Electric Limited*

As discussed in Section 7.1.6, the auxiliary contact C_2 is normally used to "seal-in" the start pushbutton.

The starter may be equipped, at an extra cost, with additional auxiliary contacts of either the momentary or time-delay type. The three-phase electromagnetic starter has three main contacts, one for each of the three power supply lines, as shown in Fig. 7-13.

7.1.3 THERMAL OVERLOAD RELAYS

Thermal overload relays are temperature-sensitive devices whose contacts close or open depending on the degree of heat produced by the current through the relay's coil. The coil of the relay, often referred to as its "heating element," is connected *in series* with the protected load. Since there is a definite time interval between the current flow and the heat produced by the heating element, thermal overload relays are used not for overcurrent* protection but for *overload* protection.

The heat generated on overload is proportional to the current squared. This heat is used either to melt an alloy, which permits a rachet wheel to turn and open a control contact, or to heat a bimetallic strip, causing it to bend and open a control contact.

As seen in Fig. 7-3(a), the thermal overload relay has, as a standard, a normally closed (NC) contact which is *manually reset* after current interruption. The more expensive thermal overload relays have an automatically reset NC contact and may also be equipped with a normally open (NO) contact for alarm or indication purposes. The control schematic representation of a thermal overload relay is as shown in Fig. 7-3(b).

The current level that can energize a thermal overload relay is externally adjustable by about $\pm 25\%$ of its base value. Typical time–current characteristic of a thermal overload relay is shown in Fig. 7-3(c). The thermal overload relays will trip in about 20 s when their coil's currrent is six times its rated value. When the load's current is 1.25 pu the thermal overload relay will trip in about 90 s. In some thermal overloads the temperature of the heating element at the instant of tripping is about 140°C.

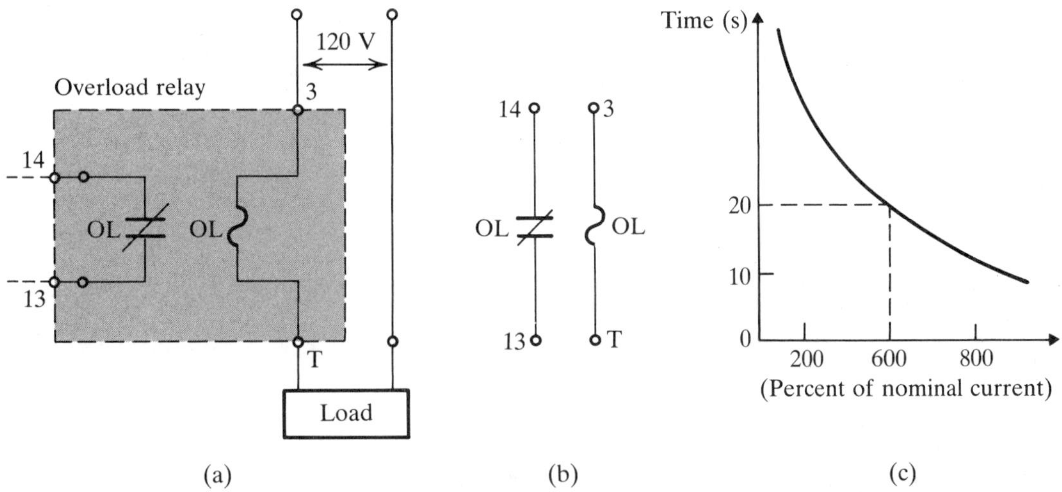

FIG. 7-3 Overload relay: **(a)** wiring diagram, **(b)** representation in a control schematic, **(c)** time–current characteristic.

* An overcurrent or short-circuit current is one that is more than six times the machine's nominal current. Overload is the current that is from one to six times the machine's rated current. Etymologically speaking, there is no clear-cut division between an overload and a short-circuit current. Generally, an overload current flows along the same path as the machine's full-load current, while an overcurrent flows partially along a different path.

Prolonged overloads deteriorate the winding's insulation, while uncontrolled overcurrents can damage all current-carrying and non-current-carrying parts of the machine.

The thermal overload protection of motors with rating smaller than 4 kW is normally an internal thermal protector that breaks the motor's circuit at a certain predetermined current level. After current interruption in such cases, one has to wait until the motor cools off before the thermal protector closes the opened circuit.

The thermal overload relay is usually attached to the magnetic starter; they form a unit. They can easily be separated, however.

7.1.4 ELECTRICAL CONTACTS

An electrical contact makes or breaks a circuit and therefore is used to energize or deenergize a load that is connected in series to it. A contact is either normally closed (NC) or normally open (NO). Generally speaking, a contact is the output of a device that senses a change in a variable and, at a predetermined level of the variable, produces a force that changes the status of its contacts. Relays, magnetic contactors, limit switches, and sensors are devices that sense a change in a variable and produce a change in the status of their contacts.

The driving force that changes the status of the contact could be manual, electromechanical, thermal, optic, or the like. In fact, many natural phenomena—changes in liquid levels (float switches) or pressure levels (pressure switches) or speed levels (speed switches)—can be used to change the status of a contact. When a driving force is applied, the contacts may change status instantaneously (momentary contacts) or after some time has elapsed (time-delay contacts). Figure 7-4 shows the standard symbols of some contacts that are used in the industry.

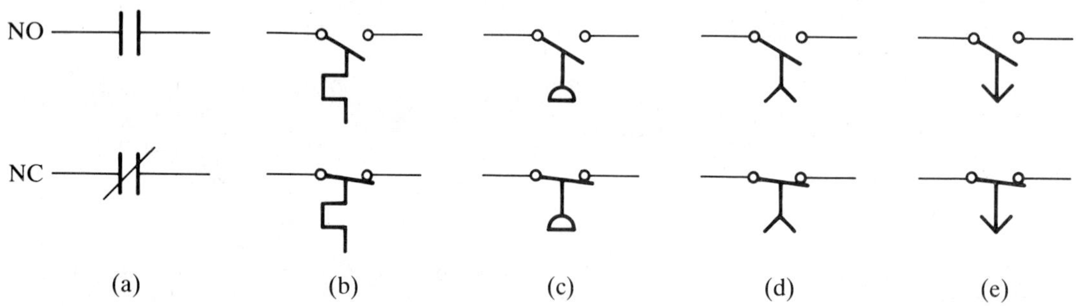

FIG. 7-4 Symbols of contacts: **(a)** instantaneous (relay or magnetic contactor), **(b)** temperature, **(c)** pressure, **(d)** time delay on energization, **(e)** time delay on deenergization.

The contacts of various devices have different configurations and are commonly known as contact forms. The single-pole single-throw (SPST) contact form (see Table 7-2) has only one contact, while the single-pole double-throw (SPDT) configuration has two or more contacts. The pole of the device is indicated with a bar terminating at a square. The pole's force, and thus its making or breaking capacity, is determined by a physical force upon which the design of the device is based. A device

TABLE 7-2 Contact forms

Design	*Sequence of Operation, Following the Application of the Force*	*Actual Configuration*	*Symbol*
SPST-NO	Make (1)	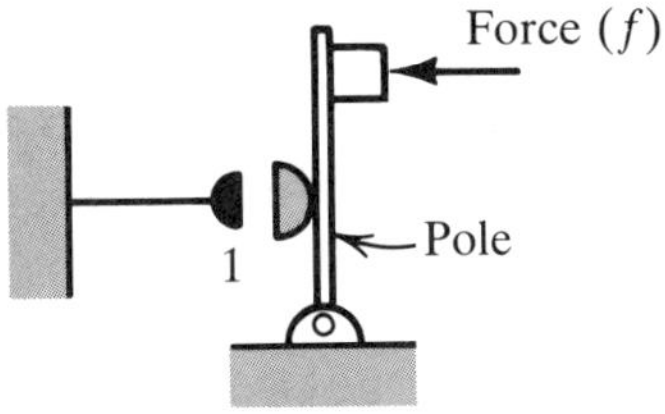	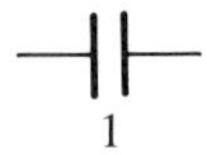
SPST-NC	Break (1)	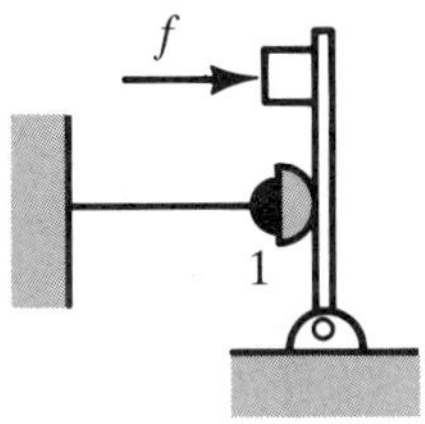	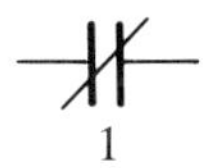
SPDT	Break (1) and then make (2)	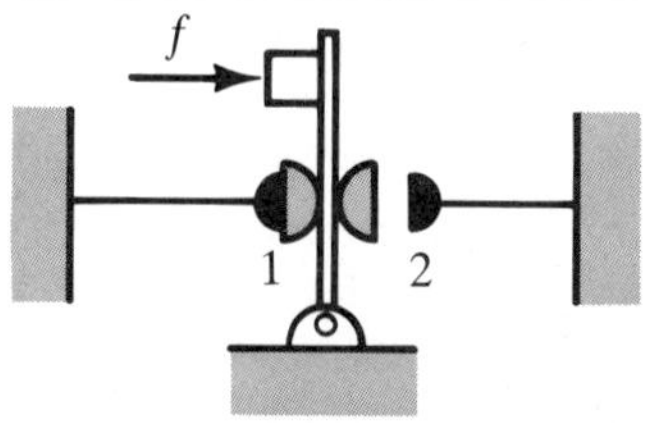	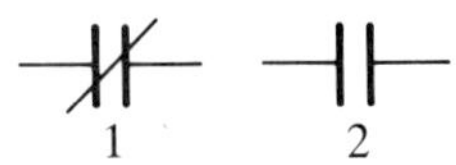
SPDT	Make (1) and then break (2)	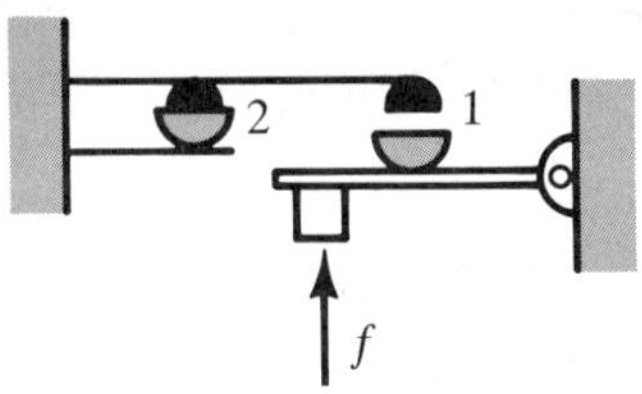	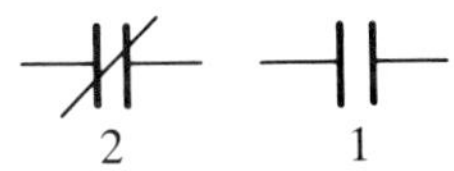

with a double-pole, double-throw (DPDT) configuration has, as the name implies, two poles and four contacts. Usually two of the contacts are NO and the other two NC. Devices with more than four contacts are more expensive and hence find limited use. In selecting a device with the SPDT contact form, it is important to understand the effects of the status of the contacts on the connected circuits during the transition period.

Since the function of a contact is to connect or disconnect a load from its supply voltage, its current and voltage specifications must equal or exceed those of the controlled load.

7.1.5 INDICATING LIGHTS

Indicating lights are usually of the incandescent type. As the name implies they reveal the status (ON or OFF) of an electrical apparatus. In more sophisticated systems the indicating lights may annunciate the cause of current interruption, such as line-to-ground fault, overload, overcurrent condition, etc. In solid state control systems the incandescent lights are, of course, replaced with light-emitting diodes (LED). The color of the lights for a particular indicating function is usually the choice of the customer.

In selecting indicating lights it is of paramount importance to know their voltage and resistance rating, because these parameters establish the minimum current rating of the contact that controls the switching of the lights.

A motor's fuses, circuit breakers, magnetic starters, and relays are all housed within a metallic cabinet—located on what is commonly known as a motor control center—on the front panel of which the indicating lights and the start-stop pushbuttons are located.

7.1.6 START-STOP PUSHBUTTONS

Start-stop pushbuttons are used to momentarily make or break an electric circuit, which in turn energizes or deenergizes—normally through an electromagnetic starter—an electrical load. In other words, ordinary start-stop pushbuttons are electrical contacts that change status manually. Their symbol is shown in Fig. 7-5(a).

Standard pushbuttons are spring operated. As soon as the depressing force is released, the pushbutton goes back to its original state. For this reason the "start" pushbutton must be connected in parallel to a NO contact—referred to as the "holding" or "maintaining" or "seal-in" contact—which becomes closed as a result of its coil's energization.

Since a "start" or "stop" pushbutton is used to make or break a circuit, it must be capable of withstanding the currents that flow through its circuit. These currents (make, break, and continuous) are determined by the load of the circuit, which might be light or heavy. Therefore pushbuttons are classified as standard or heavy duty. Their characteristics are shown in Table 7-3.

In an automated plant start-stop pushbuttons are replaced by equivalent contacts (using solid state devices) that are controlled by a signal originating in the main or auxiliary computer of the plant.

A manually operated device that *maintains* the circuit—as opposed to the pushbuttons, whose action momentarily makes or breaks the circuit—is the selector switch. The symbol for a selector switch is shown in Fig. 7-5(b).

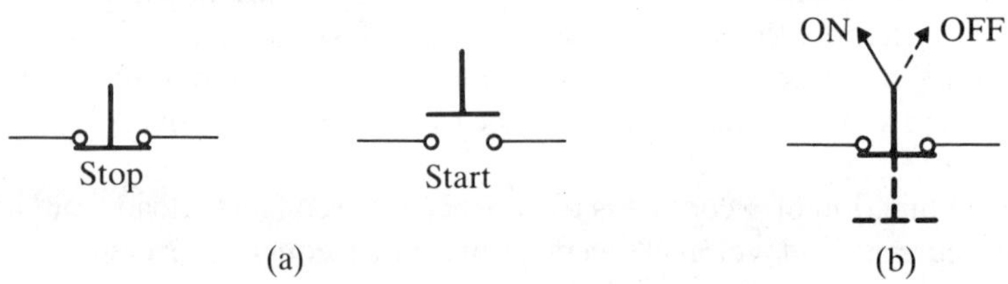

FIG. 7-5 Manually operated devices: **(a)** pushbuttons, **(b)** selector switch.

TABLE 7-3 Pushbutton characteristics

Type of Pushbutton	*Current Capacity in a*		
	Continuous	*Make*	*Break*
Standard	15	30	20
Heavy duty	20	40	30

7.1.7 INDUSTRIAL TIMERS

An industrial timer, following the energization of its coil, changes the status of its contacts cyclically. The time period during which the contacts could remain closed or open is usually externally adjustable, and the cycling period can vary from some milliseconds to several hundred hours. The number of the controlled contacts and the cycling period depend on the design of the timer. Figure 7-6(a) shows a timer's schematic. Figure 7-6(b) represents the status of its SPDT output contacts as a function of time.

Timers can be pneumatic, solid state, or motor-driven mechanisms. Solid state timers are normally incorporated within so-called programmable controllers (PC), which can be programmed to give to any number of contacts the desired cycling period.

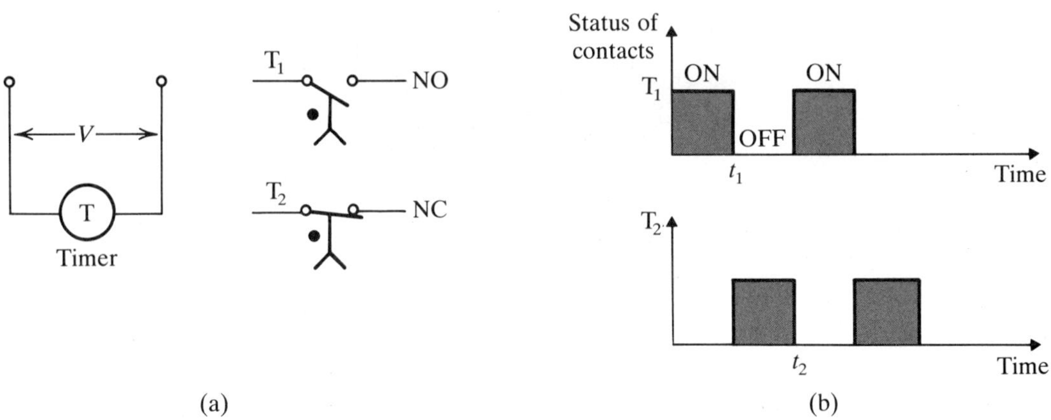

FIG. 7-6 Industrial timers: **(a)** symbol, **(b)** status of its output contacts as a function of time.

EXAMPLE 7–1 Referring to Fig. 7-7(a):

a) Explain its operation
b) Modify the control circuit so that the load could be energized or deenergized remotely.

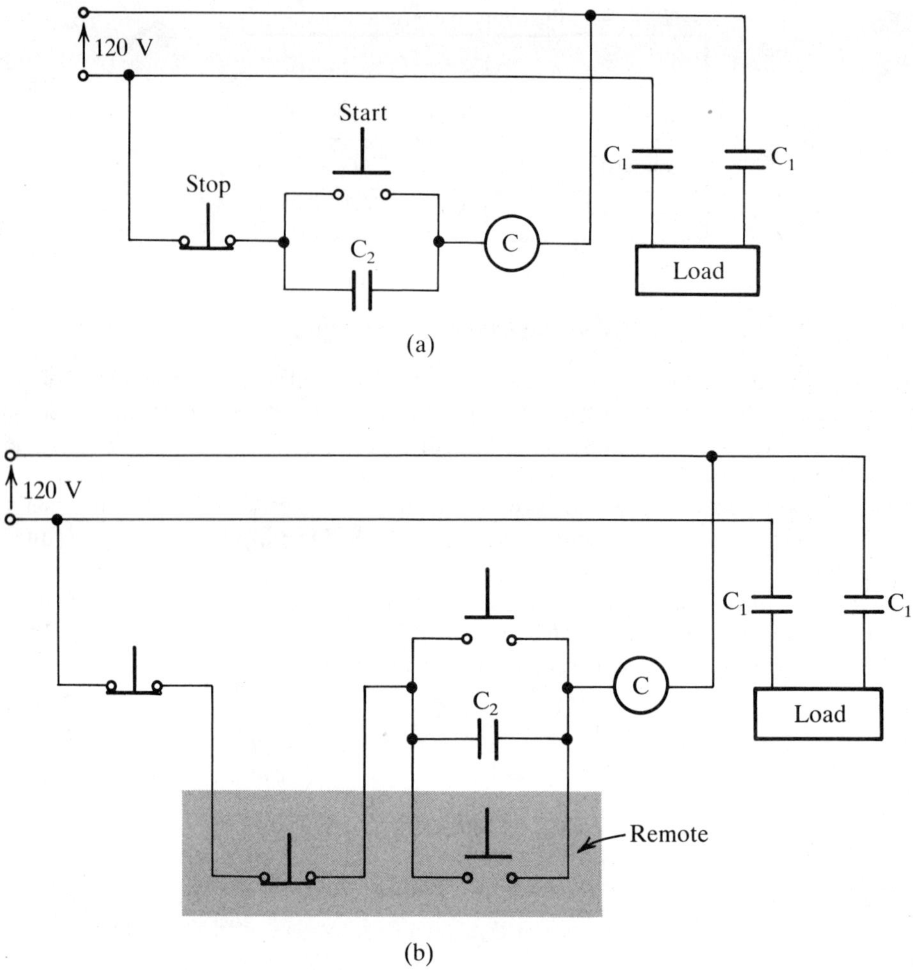

FIG. 7-7

SOLUTION

a) When the start pushbutton is depressed, the coil of the electromagnetic starter C is energized, and as a results its NO contacts close. The closing of the auxiliary contact C_2 maintains the 120 V supply to the coil, while the closing of the main contacts C_1 connects the load to its source voltage. When the stop pushbutton is depressed, the coil of the electromagnetic starter is deenergized. Its contacts open, and thus the load is removed from the voltage source.

b) The remote stop pushbutton is connected in series with the local stop pushbutton, as shown in Fig. 7-7(b), while the remote start pushbutton is connected in parallel to the local one. In either case the distance of the remote control station is limited by the impedance of the cable used. The higher the cable's impedance, the lower the voltage across the coil of the starter will be. This may adversely affect its operation.

Exercise 7–1 For ideal indoor comfort, such as can be found outdoors in the early mornings of some spring days, the following parameters should be at their optimum level:

1. Temperature (20°C)
2. Relative humidity (50%)
3. Rate of fresh air flow (.007 m^3/s/person)
4. Purity of air ($CO_2 < 600$ ppm).

When any of these conditions are not met, a switch is activated. Design a 120 V control schematic with indicating lights (red for undesirable condition, green for acceptable condition) that will identify the undesirable status of any of these parameters.

Exercise 7–2 A liquid storage tank is equipped with one inlet and two outlet pipes. The flow through the pipes is controlled by 120 V solenoid valves. The inlet valve is closed by the tank's high-level limit switch, while the outlet valves are operated by start-stop pushbuttons. The status of each valve is annunciated through standard green and red lights. Design the control schematic for the operation of the valves and of the lights.

7.1.8 TEMPERATURE SENSORS

A temperature sensor is a device that produces, at its output terminals, a signal that has some specific relationship to the sensor's ambient temperature. The output signal may be a variable resistance or simply an electrical contact (thermostat) that closes or opens, depending on the sensor's ambient temperature.

Temperature sensors of the variable-resistance type include thermistors and resistance temperature detectors (RTDs). These sensors have a specific resistance change in a given temperature range. Because thermistors and RTDs are the most widely used temperature sensors, a brief description of each will be useful.

Thermistors

Thermistors are nonlinear temperature devices whose resistance decreases exponentially with an increase in temperature. Thermistors are made from semiconductor materials that are doped with ceramics and metallic oxides. Their temperature characteristics are determined by their special atomic structures. In these materials, the atom's valance electrons form covalent bonds with the electrons of adjacent atoms. When the temperature increases, the thermal vibration of the electrons breaks some of these bonds, and thus some electrons are released. This changes the resistivity of the material and, consequently, its resistance. In the temperature range 0–100°C, a thermistor's resistance as a function of temperature is described by

$$R = R_0 e^{-K_1[(1/T_0)-(1/T)]} \tag{7.1}$$

where R and R_0 represent, respectively, the thermistor's resistances at temperatures T and T_0. The temperatures are expressed in degrees Kelvin (°K = 273 + °C). K_1 is a constant of the material and has a value which is in the range of 350° to 400°K. The thermistor's temperature response has a time constant of the order of a few

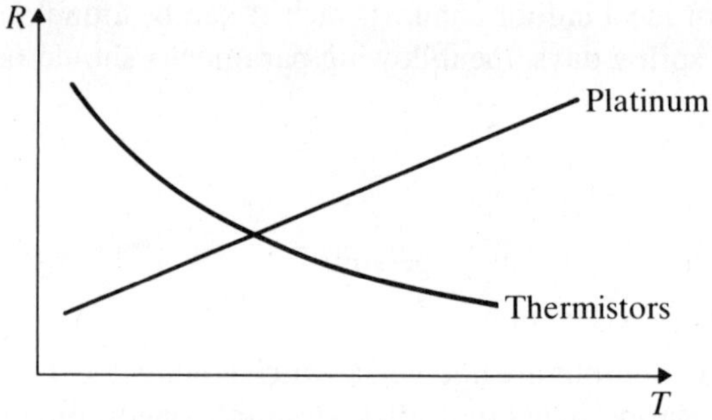

FIG. 7-8 Resistance-versus-temperature characteristics of platinum RTDs and thermistors.

milliseconds. The resistance–temperature characteristic of thermistors is shown in Fig. 7-8.

RTDS

The resistance of all metals changes as a result of change in temperature, and thus many materials could be used as temperature sensors. Platinum, however, has the most desirable characteristics (linear coefficient, high resistivity, acceptable thermal emf, resistance stability, broad temperature range, relatively low cost). For these reasons it is used extensively in devices that monitor and control the temperature of electric machines.

Almost all electric machines rated above 200 kW are equipped with two RTDs per phase. RTDs are permanently connected in the inaccessible parts of a machine's winding and furnish a signal to a monitor in a control cabinet. The monitor is designed to give a temperature indication and to provide an alarm and a trip signal when the temperature of the machine's winding exceeds a predetermined level. The resistance–temperature characteristic of platinum is shown in Fig. 7-8.

Thermostats

Thermostats are used to indicate and control ambient temperature. They constitute the main component of heating and cooling systems.

In its basic form a thermostat is a temperature sensor. It alters the status of its contacts each time the ambient temperature is above or below the thermostat's set point. Figure 7-9 shows the standard representation of a thermostat. A temperature-sensitive mercury column* within the thermostat—represented by dotted lines—will connect, depending on the set point and ambient temperature, the 24 V source to terminals X–W, in the case of heating demand, or to terminals X–Y in the case of cooling demand. Depending on the ambient temperature, in other words, a contact is made between terminals R and W or between terminals R and Y. The temperatures at which the thermostat will make these contacts are externally adjustable.

. . .

* Not all thermostats operate on the effects of temperature on a column of mercury.

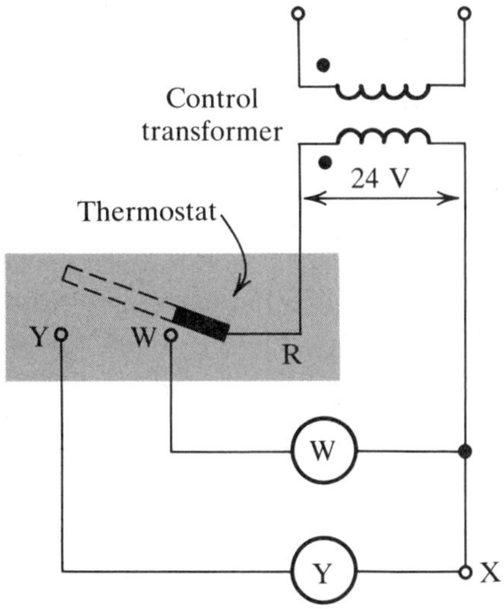

FIG. 7-9 Representation of a thermostat.

The energization of the terminals W or Y will energize their corresponding auxiliary relays, which in turn will control the operation of all those apparatuses that make up the heating or cooling system.

All modern thermostats make use of a microprocessor, which can be programmed to alter the status of a thermostat's contacts not only in response to changes in temperature but also in response to changes in time. In this way the efficiency of the heating and cooling system is optimized, which results in energy savings.

Operating Range and Differential Range

The set point of a sensor is the level at which the status of its contacts change as a result of a change in the monitored variable. For example, when a temperature sensor with a NO contact has a set point at 50°C, it means that its contact is open for all temperatures below 50°C and becomes closed for all temperatures above 50°C.

The set point of a sensor is usually adjustable within a given range, which is referred to as the sensor's "adjustable operating range."

In some sensors the status of its contacts will not alter when the monitored variable increases (additive differential range) or decreases (subtractive differential range) about the set point within a given range of the controlled variable change. For example, a temperature sensor with a NC contact, a set point of 30°C and an additive differential range of 10°C, means that on rising temperature its contact becomes closed for all temperatures less than 40°C. At temperatures above 40°C the contact opens and remains open until the temperature decreases to 30°C. At this temperature the contact closes and remains closed until the temperature is again increased to 40°C. The differential range characteristic of sensors stabilizes the operation of the controlled equipment, for it prevents them from continuous switching ON and OFF around the set point of the controlled variable.

7.1.9 MANUAL RESETS

A manual reset in a control schematic is a stop pushbutton which, in conjunction with an auxiliary relay, seals in a NO contact. This contact is usually derived from a field sensor (such as pressure, temperature, etc.), which, through auxiliary relays, shuts down a section or the entire process. When normal operating conditions are reestablished, the contact reverts back to its NO status, but the process remains interrupted because the manual reset had sealed in the control contact when the latter changed status. The operator must activate the reset pushbutton in order to remove the sealed-in circuit. Then the process is restarted. In this way the operator is made aware that a problem has developed; otherwise the process interruption might go unnoticed or unreported. The stop pushbutton can be connected in series or in parallel with the field contact. The function of a manual reset is demonstrated in the two examples that follow.

EXAMPLE 7–2 A NO contact of a temperature sensor is used to close a 120 V fail-safe solenoid valve (closed at the removal of the voltage supply) and to switch on a red (R) light. Design the control schematic:

a) Without manual reset
b) With manual reset.

SOLUTION

a) The control schematic without the manual reset is shown in Fig. 7-10(a). When the temperature contact is closed, the indicating light is switched ON and the power is removed from the solenoid valve. This closes the valve. When nominal temperatures are reestablished, the contact becomes NO again, the light is switched OFF, and the solenoid valve opens.
b) The control circuit with the manual reset is shown in Fig. 7-10(b). When the temperature contact closes, relay R_1 is energized and one of its contacts "seals in"

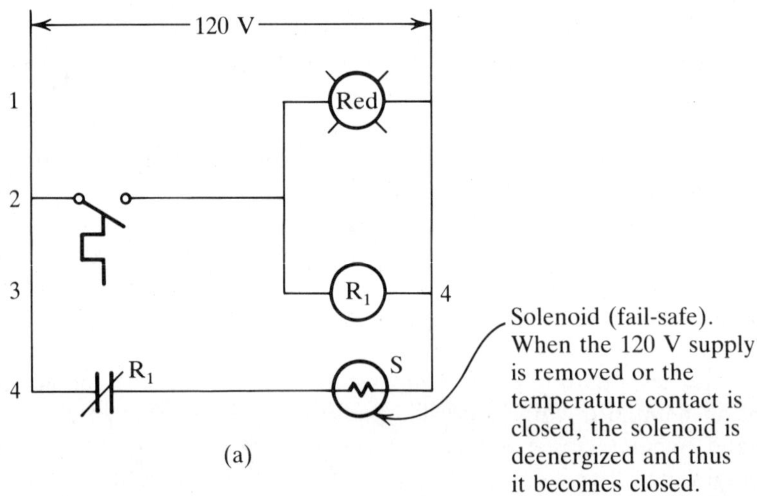

(a)

FIG. 7-10

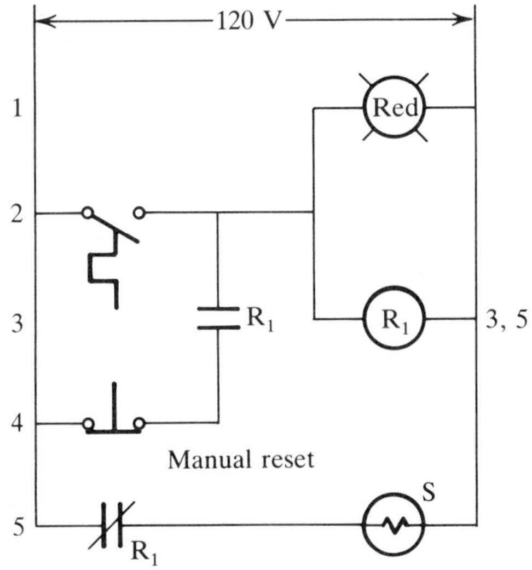

(b)

FIG. 7-10 (*Continued*)

its coil. The other contact becomes open, and as a result the solenoid is deenergized and thus becomes closed. When normal operating conditions are reestablished, the temperature sensor opens, but the solenoid remains closed until the manual reset is depressed.

The number(s) adjacent to the relay's coil in Fig. 7-10 identify where the relay's contacts are wired on the control schematic.

EXAMPLE 7–3 The sequence of operation of a process alarm light is as follows:

1. At a fault condition the alarm light starts flashing.
2. When the alarm reset pushbutton is depressed, while the fault condition is still present, the alarm light will remain ON steadily. If, however, the alarm condition is rectified, then the light should be switched OFF.

Design a control schematic to satisfy the above requirements.

SOLUTION The contact X in Fig. 7-11 is derived from a field transducer that monitors a particular process parameter. When a fault condition arises, this contact becomes closed; as a result, the ON–OFF timer T is energized. The contact of the timer switches the indicating light ON and OFF. When the alarm reset (AR) pushbutton is depressed, relay R_2 is energized, and as a result the pushbutton is "sealed in" and the supply voltage is steadily applied to the indicating light. Thus the ON–OFF function of the time-delay contact is bypassed.

When the process fault is rectified, contact X becomes open and the entire control circuit is deenergized.

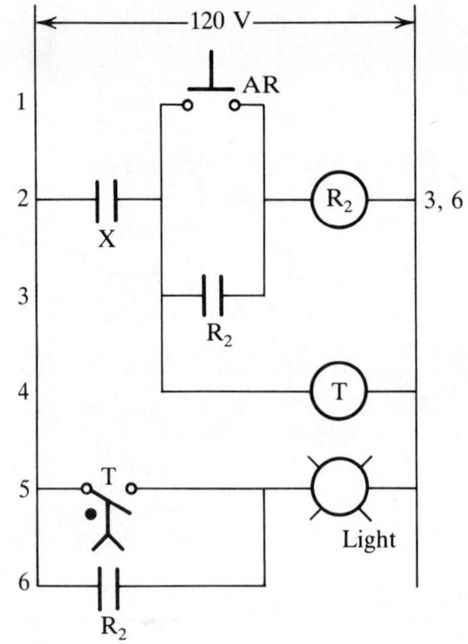

FIG. 7-11

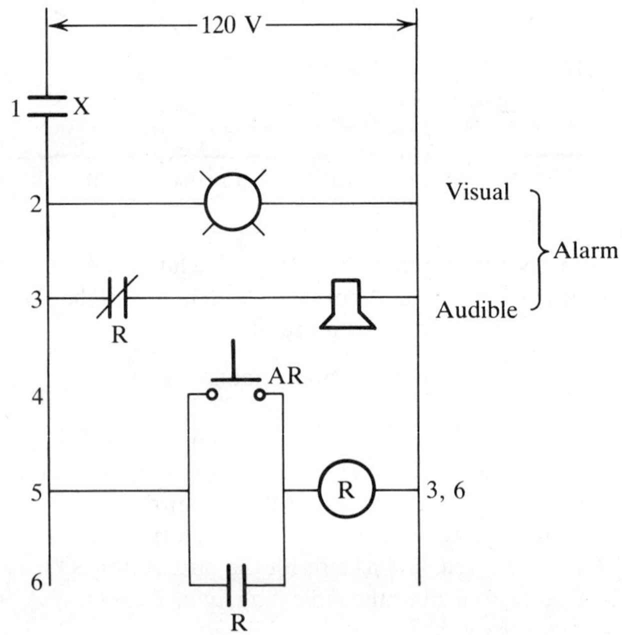

FIG. 7-12

Exercise 7–3 The NO contact X shown in Fig. 7-12 becomes closed at an undesirable level of a process's variable. Explain the operation of the circuit.

7.2 The Concept of Protection

7.2.1 GENERAL

Aside from the start-stop control components, the control schematics include additional devices that provide equipment protection and help to ensure the safety of personnel. Equipment protection involves the design, selection, and setting of those devices which detect an abnormal operating condition and as a result disconnect the faulted network or apparatus from the supply voltage.

An abnormal operating condition produces any one or a combination, of the following: overvoltages, overtemperatures, overloads, and overcurrents. An abnormal operating condition may arise from within the apparatus itself or may be caused by external factors, such as high transient voltages, unbalanced voltage supply, and high environmental temperatures.

Figure 7-13(a) shows the three-wire representation of a standard squirrel-cage, three-phase induction motor. The load-break disconnect and the contacts of the magnetic starter provide the motor's direct (manual) and indirect disconnecting means, respectively. The fuses and the overload devices provide, respectively, the motor's overcurrent and overload protection. Figure 7-13(b) shows the three-wire representation of a squirrel-cage, three-phase induction motor that has a circuit breaker for overcurrent protection. These devices are discussed in some detail in the next section.

An overview of the concept of protection is also given in Fig. 7-14. Figure 7-14(a) shows the one-line diagram of a 3-ϕ, squirrel-cage induction motor. On it are identified the functional characteristics of the various control and protection devices. In Fig. 7-14(b) are shown the general time-current characteristics of the protective devices and of the motor. The dotted line extensions of the characteristics of overload (curve d) and overcurrent (curve b) devices signify those parts of the curves that do not provide any effective protection.

A motor's operating and damage characteristics can be calculated from basic data or obtained from the motor's manufacturer. Those of the protective devices can be found in standard manufacturer's catalogues.

The first and most important step in motor protection studies is to obtain and sketch the motor's operating and damage characteristics on log-log paper. The protective devices must then be selected to ensure that their effective time–current characteristics lie between the motor's operating and damage characteristics. Protection devices must interrupt any undesirable current before the motor's damage condition is reached.

7.2.2 PROTECTIVE DEVICES

The equipment's protective devices include fuses, relays, and circuit breakers. Fuses, and most low-voltage circuit breakers, have fixed time–current characteristics. These characteristics indicate how long it will take a particular device to interrupt a given

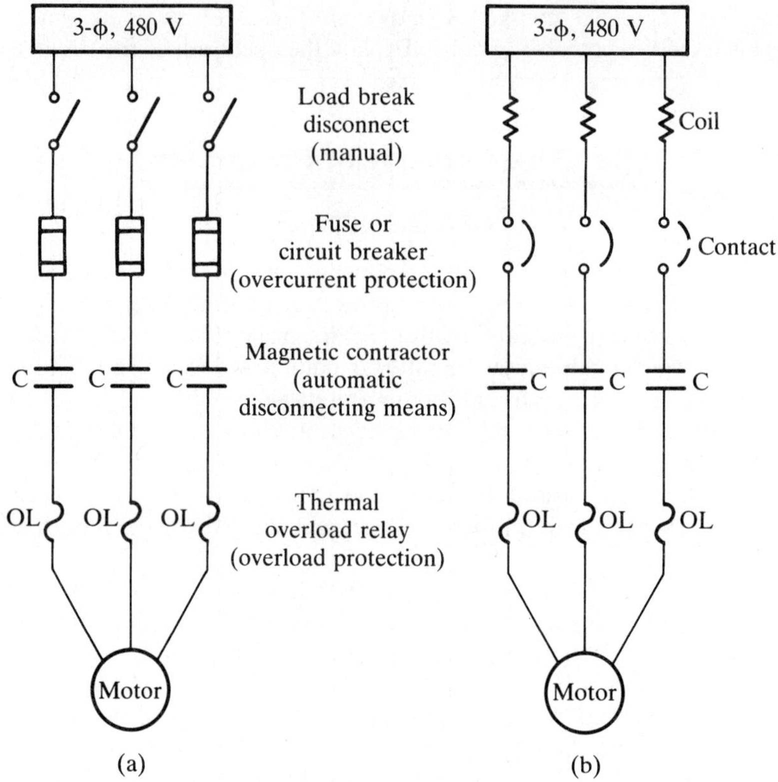

FIG. 7-13 Three-line representation of a 3-ϕ squirrel-cage induction motor: **(a)** fuses for overcurrent protection, **(b)** circuit breaker for overcurrent protection.

current. For purposes of illustration, consider the 15 A devices shown in Fig. 7-15. A current of 60 A will be interrupted by a *time-delay* fuse in 18 s, while the same current will be interrupted by the circuit breaker in 4 s. A fast-acting fuse will interrupt the same 60 A current in .6 s.

Protective devices must be capable of removing—disconnecting—the affected circuit from the supply when overcurrents flow. This is referred to as "interrupting capability." They also should be able to withstand the forces that will be produced when they close the circuit while a downstream fault condition exists. This is referred to as "making capacity" or "momentary capability."

Disconnects

Disconnects, or switches (shown in Fig. 7-13(a)), are normally operated manually and cannot interrupt the current on their own. There are two main types of interrupters: load-break switches and no-load disconnects. Load-break switches are capable of interrupting nominal load current. No-load disconnects can open up the circuit only

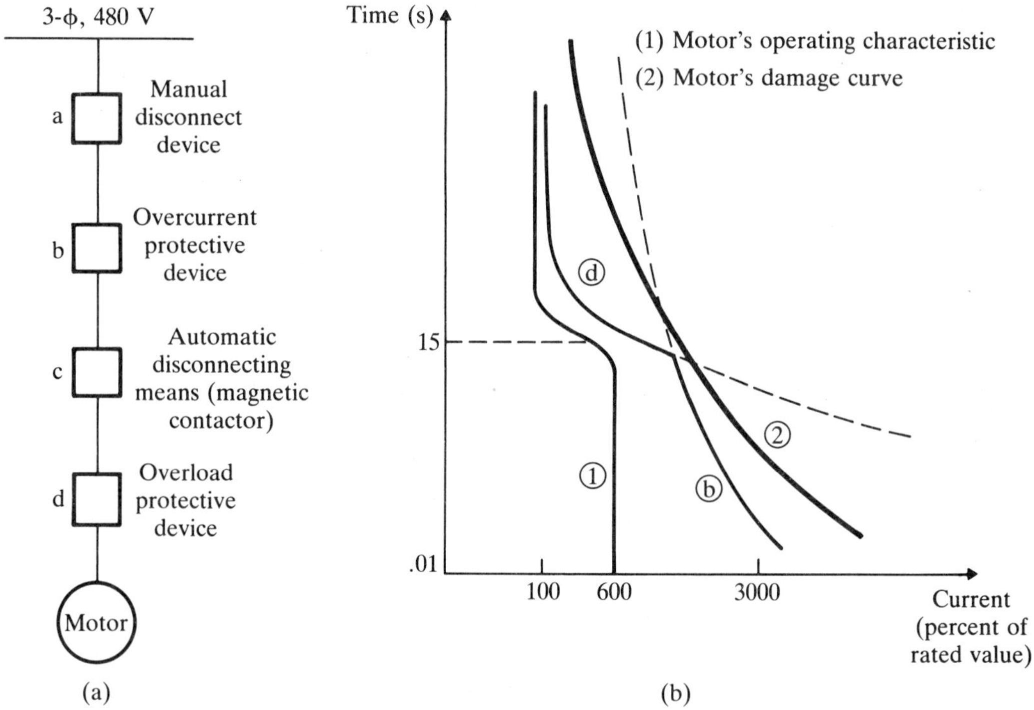

FIG. 7-14 The concept of protection: **(a)** one-line diagram of a 3-ϕ squirrel-cage induction motor, **(b)** time–current characteristics.

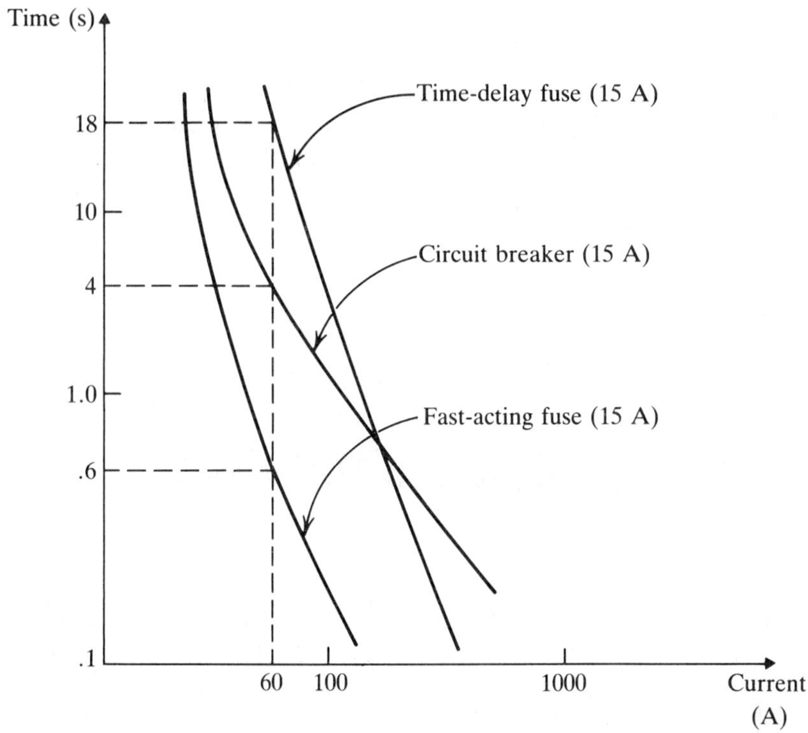

FIG. 7-15 Typical time–current characteristics of 15 A fuses and of a 15 A circuit breaker.

if the main downstream load is disconnected. All LV switches are of the load-break type while some of the MV switches are no-load interrupters. In the latter category belong the primary switches of some transformers that can interrupt only the no-load current of the transformer.

Fuses

The function of a properly selected fuse is to interrupt the current when it reaches an undesirable level. Low-voltage and low-current (below 60 A rating) fuses will interrupt 135% of their nominal current rating in 60 s. This is referred to as their fusing factor. A 50 A fuse, for example, will open up the circuit to which it is connected in 60 s when, during this time interval, the current through it is 67.5 A.

Considering the function they perform, fuses are quite economical. They are commercially available with different time–current characteristics. This is necessary because various types of equipment have different time–current start and damage characteristic curves.

Protective Relays

A protective relay is a device which detects and identifies a piece of equipment's abnormal operating condition. Upon detection and identification, the status of its output contacts changes. Their change of status is used too sound an alarm, or to initiate the tripping of a circuit breaker or a magnetic contactor.

Protective relays have adjustable time–current characteristics. The more expensive ones—the microprocessor types—are equipped with indicating meters that display the values of the parameters surveyed (such as temperature, current, unbalanced voltages, etc.). A protective relay normally is represented by a circle within which a number indicating its function is inscribed. The numbers 50 and 51, for example, indicate instantaneous and overload current protection, respectively. As per ANSI* recommendations (see Appendix F), the functions of a protective device and the corresponding identification numbers have been standardized.

Circuit Breakers

Ordinary low-voltage circuit breakers have a current coil which is connected in series with the NC contact (see Fig. 7-13(b)). The power to open the circuit breaker's contact originates from its coil each time the downstream or load current exceeds a predetermined value. The coil of the circuit breaker is usually not shown in one-line distribution drawings. The time–current characteristic of a 15 A, 480 V, 3-ϕ circuit breaker is shown in Fig. 7-15.

These low-voltage circuit breakers are commercially referred to as "molded-case circuit breakers" because they are constructed from some kind of molded material.

. . .

* ANSI: American National Standards Institute.

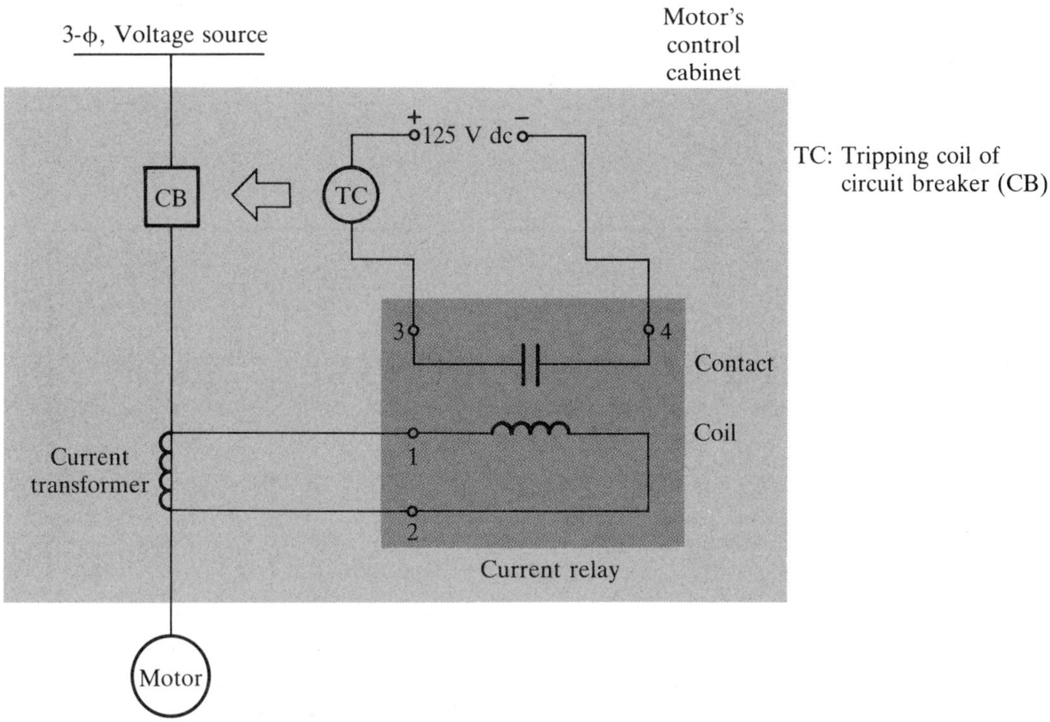

FIG. 7-16 Elementary representation of a relay and its tripping circuit.

The more expensive circuit breakers are tripped through relays and consequently can perform additional protective functions, such as line-to-ground or overcurrent protection. As illustrated in Fig. 7-16, when the current drawn by the motor increases to an undesirable level, the coil of the current relay is energized and its output contact closes. This in turn energizes the tripping coil (TC) of the circuit breaker. As a result, the motor is disconnected from its voltage supply.

The interruption of the current is within a chamber that is filled with a special oil, SF_6 gas, air, or vacuum. The commercial name of the circuit breaker reveals the quenching medium. For example, an air-circuit breaker indicates that the current interruption is in an air chamber.

EXAMPLE 7–4 Compare and contrast a contactor, a LV molded-case type of circuit breaker, and a thermal overload relay.

SOLUTION

A contactor provides a means of disconnecting but no protection. A circuit breaker provides a mean of disconnecting and overcurrent protection. An overload relay provides only thermal overload protection. Through the contactor it can switch OFF the equipment it protects.

Because the coil of the contactor is a voltage coil, it is connected in parallel to its supply voltage. The coils of the molded case circuit breakers and of the thermal overload relays, being current coils, are connected in series with the protected load.

For purposes of comparison the main characteristics of each device are summarized in Table 7-4.

TABLE 7-4 Comparison of protective devices

	Function			*Type of Coil*	
		Protection			
Device	*Disconnecting Means*	*Overload*	*Overcurrent*	*Voltage*	*Current*
Contactor	Yes	No	No	Yes	
LV circuit breaker	Yes	Yes, but often not effective	Yes		Yes
Thermal overload relay	Yes (indirectly)	Yes	Yes, but not effective		Yes

EXAMPLE 7–5 Figures 7-17(a) and (b) show, respectively, a one-line diagram of a motor and the time–current characteristics of the motor and its protective devices.

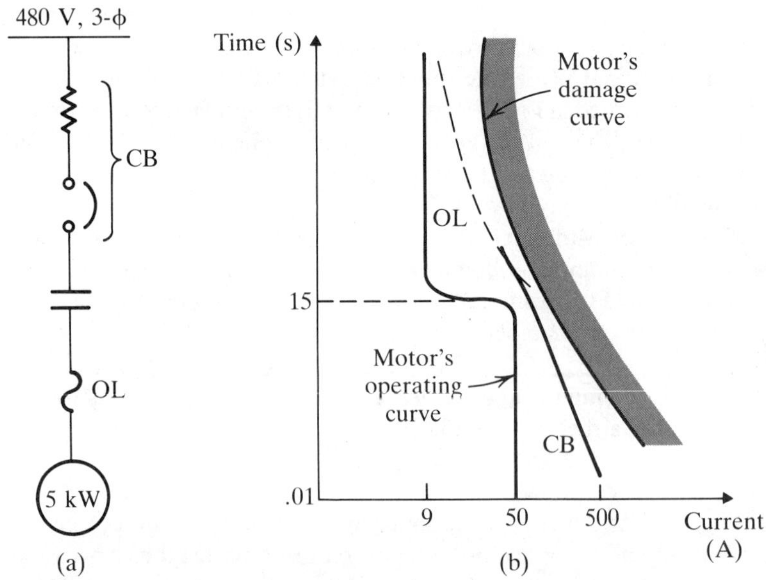

FIG. 7-17 Motor protection: **(a)** one-line diagram, **(b)** time–current characteristics of motor and protective devices.

a) Is the motor properly protected?

b) According to 1969 statistics, 19% of the fires in the United States were caused by motors. What were some of the reasons for the motors' damage and the subsequent fires?

SOLUTION

a) The motor is properly protected, because the time–current characteristics of the protective devices indicate that for any overload or overcurrent condition the motor will be switched OFF before its damage condition is reached. It should be noted that the given characteristics of the overload (OL) and circuit breaker are not extended over their entire range, because this would have made functional representation of the curves confusing. After all, the protective curves to the right of the motor's damage curve serve no purpose.

b) The damage to the motors and the subsequent fires could have been caused by any one, or a combination, of the following conditions;
   (1) The number of motor starts was higher than the maximum permissible.
   (2) The protective devices malfunctioned owing to aging, or poor maintenance, or some kind of manufacturing defect.
   (3) Poor motor maintenance resulted in an accumulation of dust that blocked the ventilation of the motor's windings. This is equivalent to shifting the damage curve of the motor to the left of its original position.
   (4) The overload relay's ambient temperature was much lower than the motor's.
   (5) The motor's ventilation system failed.

Exercise 7–4 Figure 7-18 shows a typical one-line diagram of a building's exterior lighting control. The photocell (PC) is normally an open contact which becomes closed at low illumination levels. The timer (T) requires 120 V for its operation and according to its setting can overide the function of the photocell. The contactor (C), when energized, switches ON the exterior lights. Draw the corresponding control schematic.

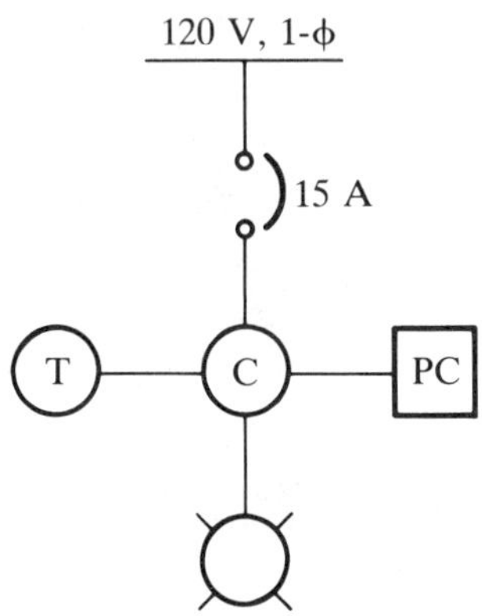

FIG. 7-18

7.3 Actual Control Schematics

Previous sections of the chapter discussed control devices and their conventional symbols. This section explains how these basic control devices are used to construct the control schematics of various types of single-phase induction motors, three-phase motors, and dc machines. The modern control of synchronous motors and dc machines also requires a knowledge of solid-state devices, which are explained in the Appendix of this book.

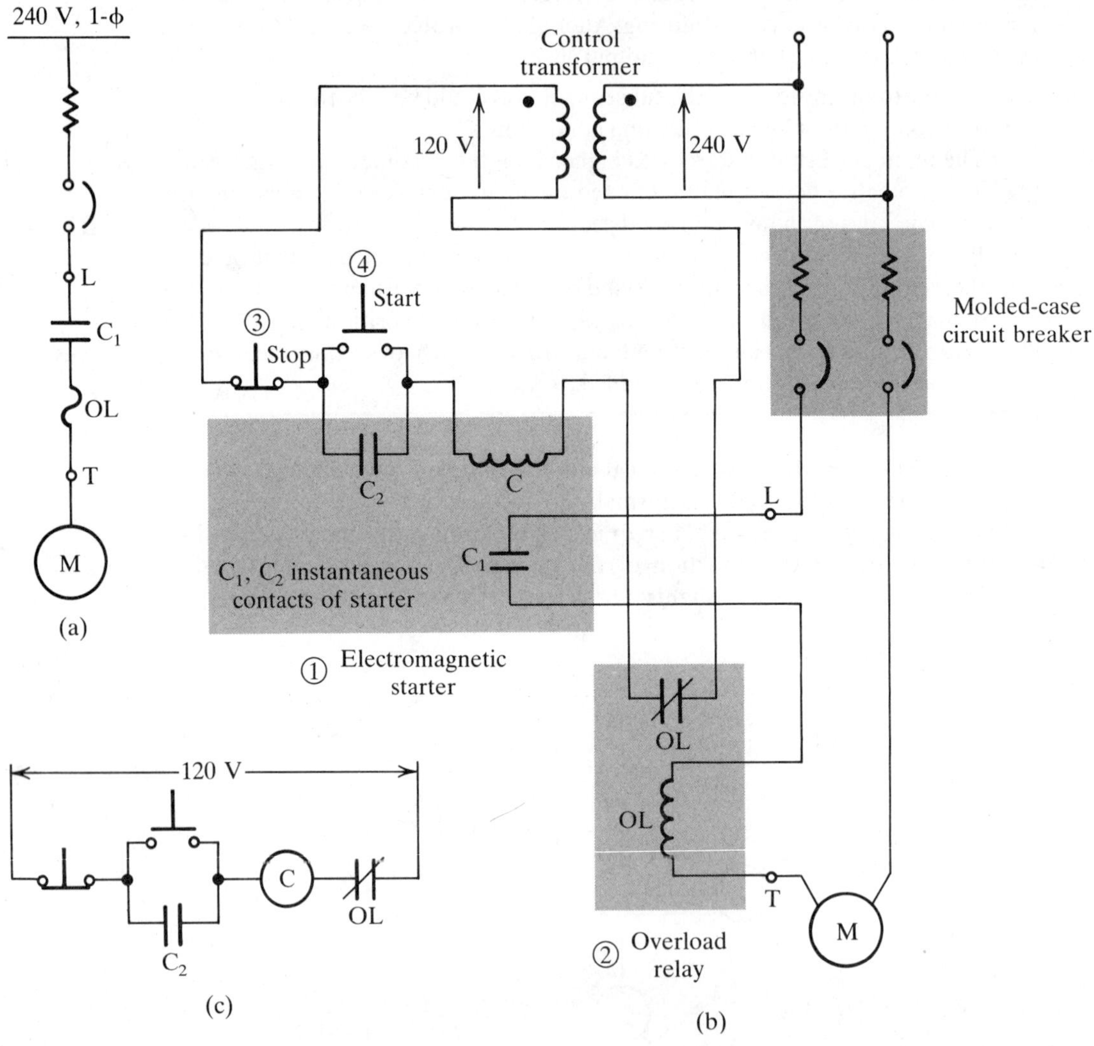

FIG. 7-19 Single-phase induction motor: **(a)** one-line diagram, **(b)** elementary wiring diagram, **(c)** control schematic.

7.3.1 SINGLE-PHASE INDUCTION MOTORS

There are various types of single-phase induction motors, and each has its own control schematic. The controls of the general-purpose motor, the capacitor-start/capacitor-run and the multispeed and speed-reversal motors are discussed here. The control schematic that brings a single-phase induction motor to a quick stop by the application of a dc voltage—commonly known as regenerative braking—is left as a student exercise (see Exercise 7-6).

General-Purpose Motor

The one-line diagram of a 240 V general-purpose single-phase induction motor is shown in Fig. 7-19(a). Figure 7-19(b) shows the motor's wiring diagram and identifies the various control devices. The motor's actual control schematic is shown in Fig. 7-19(c). Its sequence of operation is as follows:

1. When the start pushbutton is depressed, its momentary contact closes and remains closed while the button is depressed. Thus, 120 V are applied to the coil(C) of the electromagnetic starter.
2. After the coil is energized, its main (C_1) and auxiliary (C_2) contacts are closed.
3. The closing of the main contact completes the circuit of the motor to the 240 V supply voltage. Thus, the motor starts to accelerate toward its operating speed.
4. The closing of the auxiliary contact "maintains" the 120 V across the coil of the starter, although the force that depressed the start pushbutton is removed.
5. The motor can be stopped either by depressing the stop pushbutton or by opening the NC contact of the thermal overload relay.

Capacitor Start/Capacitor Run

Figure 7-20 shows a typical control schematic for a capacitor-start motor. Its sequence of operation is as follows:

1. When the start pushbutton is depressed, the coil of the contactor is energized. Its main contacts (C_1) connect the motor to its voltage supply, while the auxiliary contact (C_2) seals in the start pushbutton.
2. The application of the terminal voltage to the motor produces a starting torque and at the same time energizes the voltage relay (A). As soon as the motor starts, the voltage relay removes the starting capacitor from the circuit.

Features

1. The voltage relay remains energized during the motor's operation, and its voltage rating is equal to that of the motor's auxiliary winding. The voltage across the capacitors, however, is often larger than the supply voltage and must be carefully calculated and then experimentally verified.
2. The resistor across the starting capacitor protects the switching elements of the circuit from overvoltages.

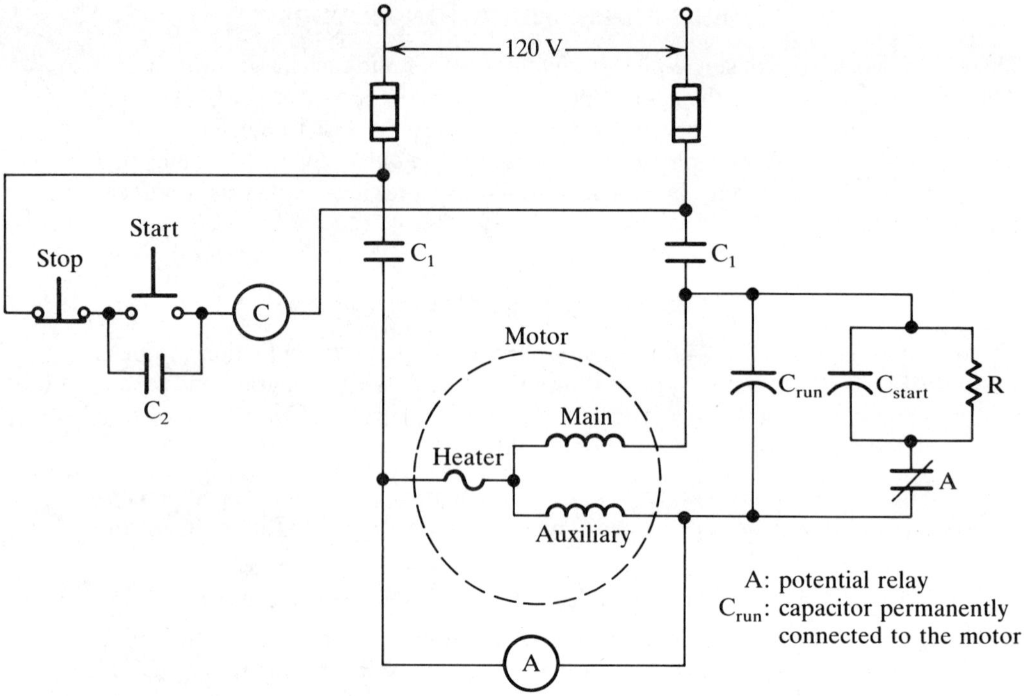

FIG. 7-20 Control schematic of a capacitor-start capacitor-run 1-ϕ induction motor.

3. The thermal overload protection (heater) of the motor, as shown in the circuit, is within the motor. Thus, when an overload causes the tripping of the motor, one has to wait several minutes until the motor's winding cools off. Then the termal protector closes its contact and the motor will start to operate. Its starting torque is reduced, owing to the already disconnected starting capacitor.
4. The use of a voltage relay for removing the starting capacitor is typical for hermetically enclosed motors such as those that drive the compressors of heating and air conditioning systems. In other applications a centrifugal switch is used that removes the capacitor following the starting of the motor.

Speed Reversal

In somc applications it is ncccssary to rcvcrsc thc dircction of motor rotation. A typical control schematic for a motor's speed reversal is shown in Fig. 7-21. The sequence of operation is as follows:

1. Before the motor starts to rotate, its main and auxiliary windings are disconnected from their supply voltages through the main contacts of the forward (F) and reverse (R) contactors.
2. When the *forward* pushbutton is depressed, the coil (F) is energized and its NO seal-in contact closes momentarily. Its time-delay-on-energization contacts close at a predetermined time (t_1). This connects the main and auxiliary motor windings to

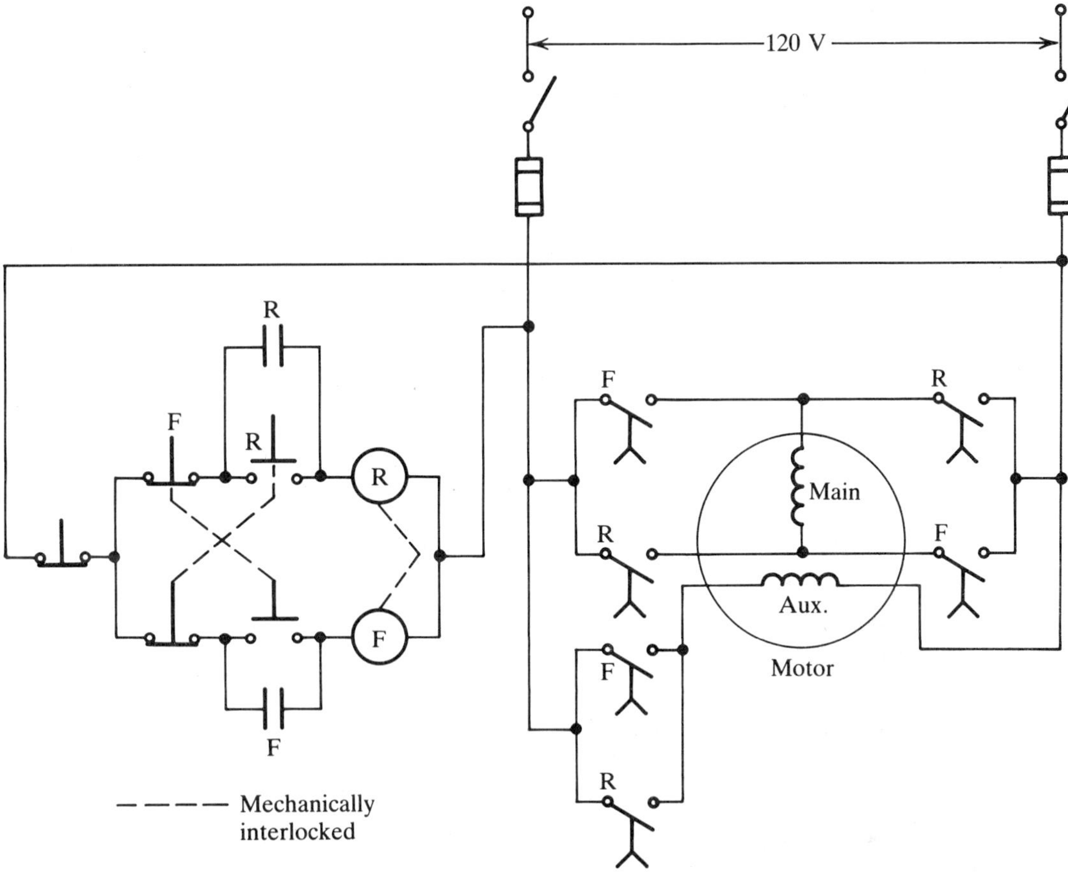

FIG. 7-21 Speed reversal of a single-phase induction motor.

their nominal voltage supply. The motor continues to run in the *forward* direction. The coil of the reverse starter remains deenergized.

3. When the *reverse* pushbutton (R) is depressed, the coil of the forward starter is deenergized, so its corresponding contacts open. Then the coil of the reverse starter is energized. This momentarily closes its seal-in contact. After a predetermined time delay (t_2), its time-delay-on-energization contacts close. This connects the main and auxiliary motor windings to their nominal ac supply. The current through the main winding flows in a direction opposite to that in which it flows when the forward pushbutton is depressed. As a result, the motor rotates in the *reverse* direction.

Features

The main features of this control schematic are as follows:

1. The coils of the forward and reverse starters cannot be simultaneously energized.
2. When the time delays t_1 and t_2 are properly selected the motor's windings cannot be damaged from the excess copper losses that accompany the motor's frequent speed reversals.

Multispeed Motors

The speed of a single-phase induction motor can be controlled by applying different voltages to its main winding. By energizing the low (L), medium (M), or high (H) speed contactors (see Fig. 7-22), three different voltages are seen by the motor's main winding. As a result, the motor can operate at three different speeds. Single-phase induction motors driving ventilating fans are supplied with multispeed winding, where the high and low winding connections will produce an operating speed of about ± 75 r/min in comparison with the speed that results from the medium-voltage winding connection.

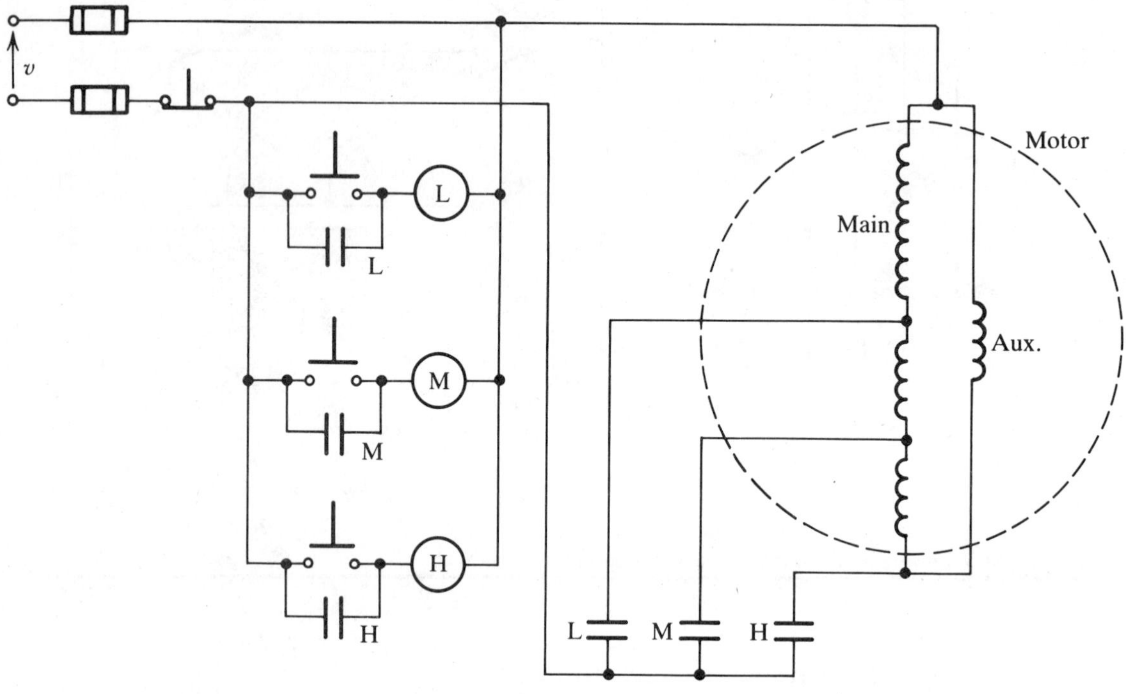

FIG. 7-22 Variable-speed, single-phase induction motor.

The speed of single-phase induction motors is more sensitive to variations in the supply voltage than is the speed of 3-ϕ induction machines. This is because the applied voltage in a single-phase induction motor produces two counteracting fields, while in a 3-ϕ induction motor the applied voltages produce only one stator field. Furthermore, as can be seen in Fig. 7-22, when the motor operates at medium or low speeds the mmfs produced by the bypassed section of the main winding are in a direction opposite to that produced by the rest of the winding. As a result, the motor's effective field is further reduced.

Multispeed single-phase motors are used to control air flow in all residential heating and cooling systems. By changing the speed (ω) of the motor, the rate of air flow

(Q') changes ($Q' \propto \omega$). Therefore, the air-flow requirements of different climates and seasons can be satisfied.

In an automated system the manual pushbuttons of Fig. 7-22 are replaced with contacts of relays whose coil's energization depends on ambient temperature.

EXAMPLE 7–6 Each of the four single-phase motors shown in the one-line diagram (Fig. 7-23(a)) is used to drive a ventilating fan that is part of a heating system in an industrial plant. The motors are internally protected (OL). To start the operation

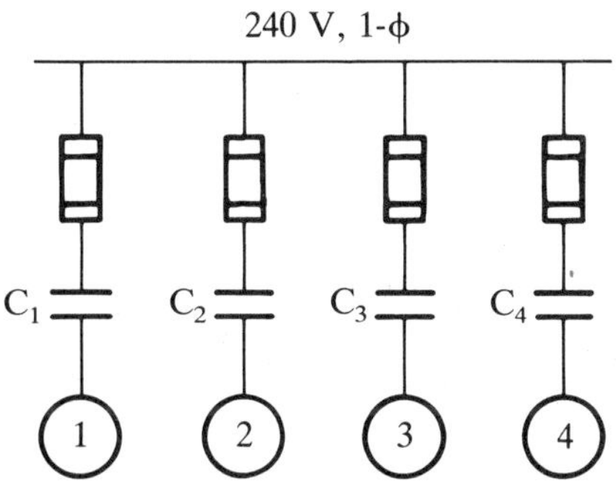

FIG. 7-23(a)

of the system two ambient temperature sensors are used. One temperature sensor has a NC contact for all temperatures below 66°F and the other has an NC contact for all temperatures below 72°F. The 24 V master control schematic must be constructed to satisfy the requirements as outlined in the logistics table. (*Hint:* Use relays and an industrial timer that has two contacts, one NC and the other NO, and whose status changes every 24 hours.)

Logistics Table

Condition	*Temperature (T)*	*Status of Motors* 1	2	3	4	*Remarks*
1	$T < 66°F$	ON	ON	ON	ON	
2	$66°F < T < 72°F$	ON	ON	OFF	OFF	50% of the time
3	$66°F < T < 72°F$	OFF	OFF	ON	ON	50% of the time
4	$T > 72°F$	OFF	OFF	OFF	OFF	

SOLUTION The required control schematic is shown in Fig. 7-23(b). The ambient temperature controls the status of the temperature sensors. The temperature sensors in turn control the energization of the relays. Through the relays, the entire system functions as per the requirements of the logistics table. Each of the auxiliary relays (R_1 and R_2) has two NO contacts. The timer has one NO and one NC contact.

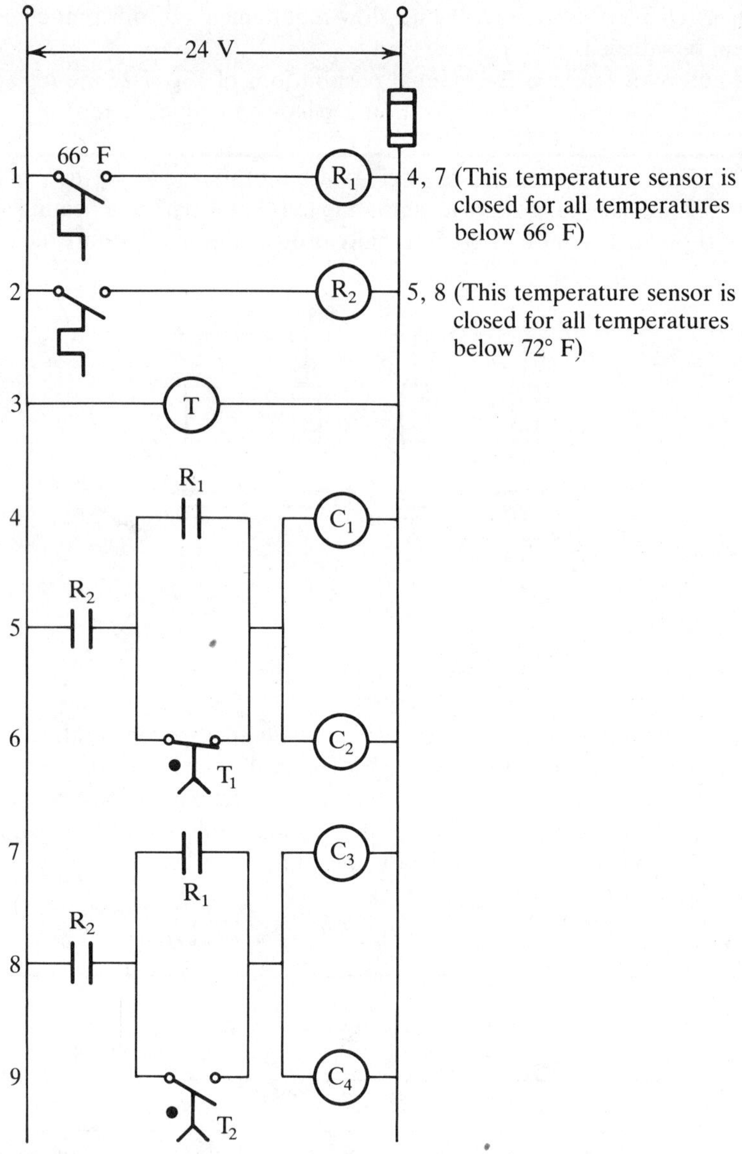

FIG. 7-23(b)

Condition 1, $T < 66°$

At temperatures below 66°F the contacts of the temperature sensors are closed and the relays are energized. This leads to the energization of the magnetic contactors and the subsequent starting and operation of all four motors. At temperatures below 66°F the timer does not perform any function because its contacts are bypassed or overruled by the contacts of the auxiliary relay R_1.

Conditions 2 and 3, 66° < T < 72°

At temperatures between 66°F and 72°F, the higher-temperature sensor maintains its NC contact, while the lower-temperature sensor opens its contact. This deenergizes relay R_1 and energizes relay R_2. However, the magnetic starters of motors 1 and 2, or 3 and 4, could be energized through the contacts of the timer. Since the timer changes the status of its contacts every 24 hours, the requirements of conditions 2 and 3 for the operation of only two motors—in the given temperature range—are satisfied.

Condition 4, T > 72°

At temperatures above 72°F the contact of the higher-temperature sensor is also open and thus the contactors of the magnetic starters are deenergized. As a result, all motors will be switched OFF.

Exercise 7–5

a) Modify the nominal control schematic of a single-phase motor (capacitor start/capacitor run) to include an indicating light that will reveal the status of the internal overload.

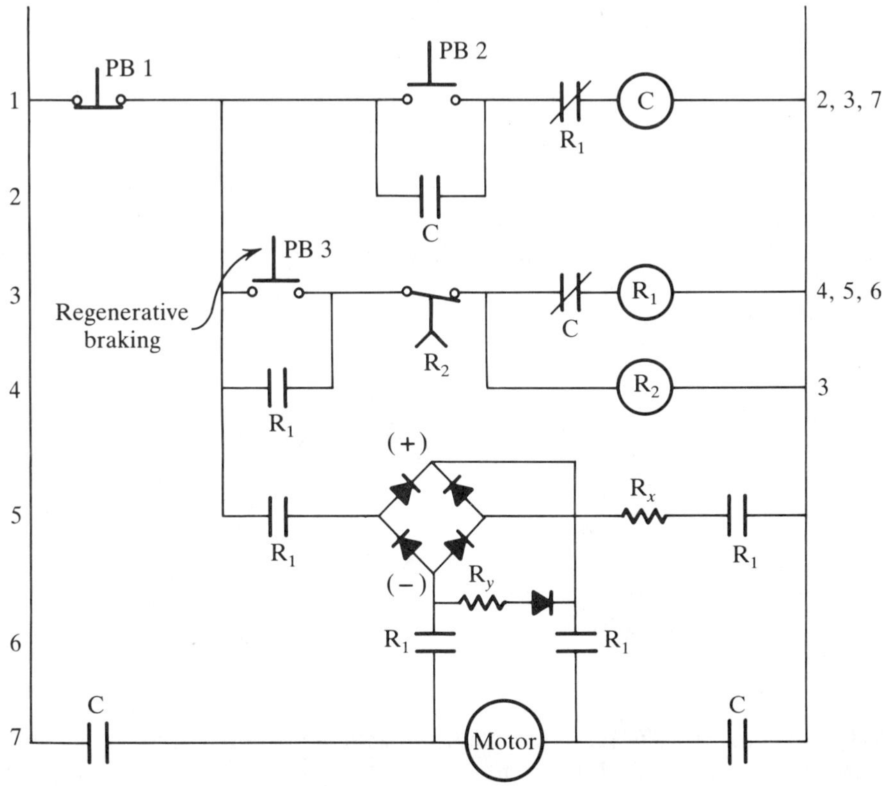

FIG. 7-24

b) Why are the voltages across the running capacitor and across the voltage relay much larger than the motor's supply voltage?

c) Why must manufacturers recall motors whose starting capacitor does not have a discharge resistor?

Exercise 7–6 The circuit shown in Fig. 7-24 is used for the ac operation of a single-phase induction machine. Applying a dc voltage to the stator winding will bring the operating motor to a quick stop. The latter is referred to as regenerative braking.

a) Describe the operation of the circuit.

b) What are the functions of the resistors R_x, R_y, and of the time delay on energization relay R_2?

7.3.2 CONTROL SCHEMATICS OF THREE-PHASE INDUCTION MOTORS

In the sections that follow the control schematics of LV squirrel-cage and wound rotor and of MV induction machines will be discussed. The control schematics of the MV motors, because of their higher power rating, are more elaborate and expensive.

Low-Voltage, Squirrel-Cage Rotor

An elementary control schematic used for the starting and the protection of 3-ϕ, LV induction machines is shown in Fig. 7-25. The sequence of operation and the essential features of this control schematic are as follows:

Sequence of Operation

1. When the start pushbutton is depressed, its momentary contact closes.
2. The coil (C) of the magnetic contactor is then energized, and as a result the main contacts (C_1) in the motor circuit plus the "seal-in" contact (C_2) are closed.
3. The motor starts to rotate.
4. The red (R) light is switched ON, indicating that the motor is operating.
5. When the momentary start pushbutton is released, its contact opens through the action of its own spring.
6. The "seal-in" contact maintains the circuit around the start pushbutton and the motor continues to rotate.
7. When the stop pushbutton is depressed, its contact opens.
8. The circuit to the coil (C) of the magnetic contactor is broken, the coil is deenergized and thus the main contacts in the motor circuit and the auxiliary seal-in contact are opened. The contact C_3 to the green light (G) is closed.
9. The motor stops.

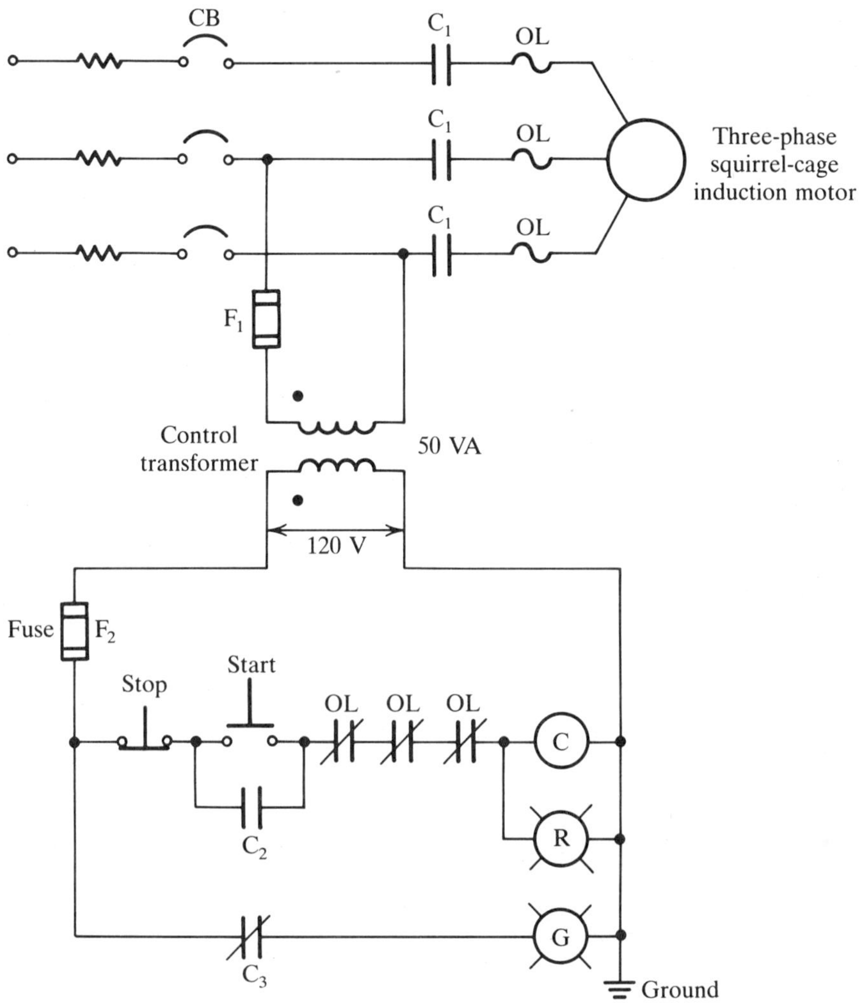

FIG. 7-25 Elementary control schematic for a three-phase, squirrel-cage induction motor 600 V or less.

Features

1. When the supply voltage is removed, the motor will stop and will *not* automatically restart after the control voltage is reapplied.
2. If the overload contacts are opened, the motor will stop. To restart the motor the overload relay must be manually or automatically reset (depending on its type) and the start pushbutton must be depressed.

Remarks

1. The volt-amperes capacity of the control transformer should equal or exceed the VA requirements of the devices connected to its secondary terminals.

2. In some applications, control schematics include additional control devices for intermittent motor operation (jogging). In such cases the motor starts and continues operation only when an instantaneous start pushbutton is maintained depressed. This is further explained in Example 7-12.

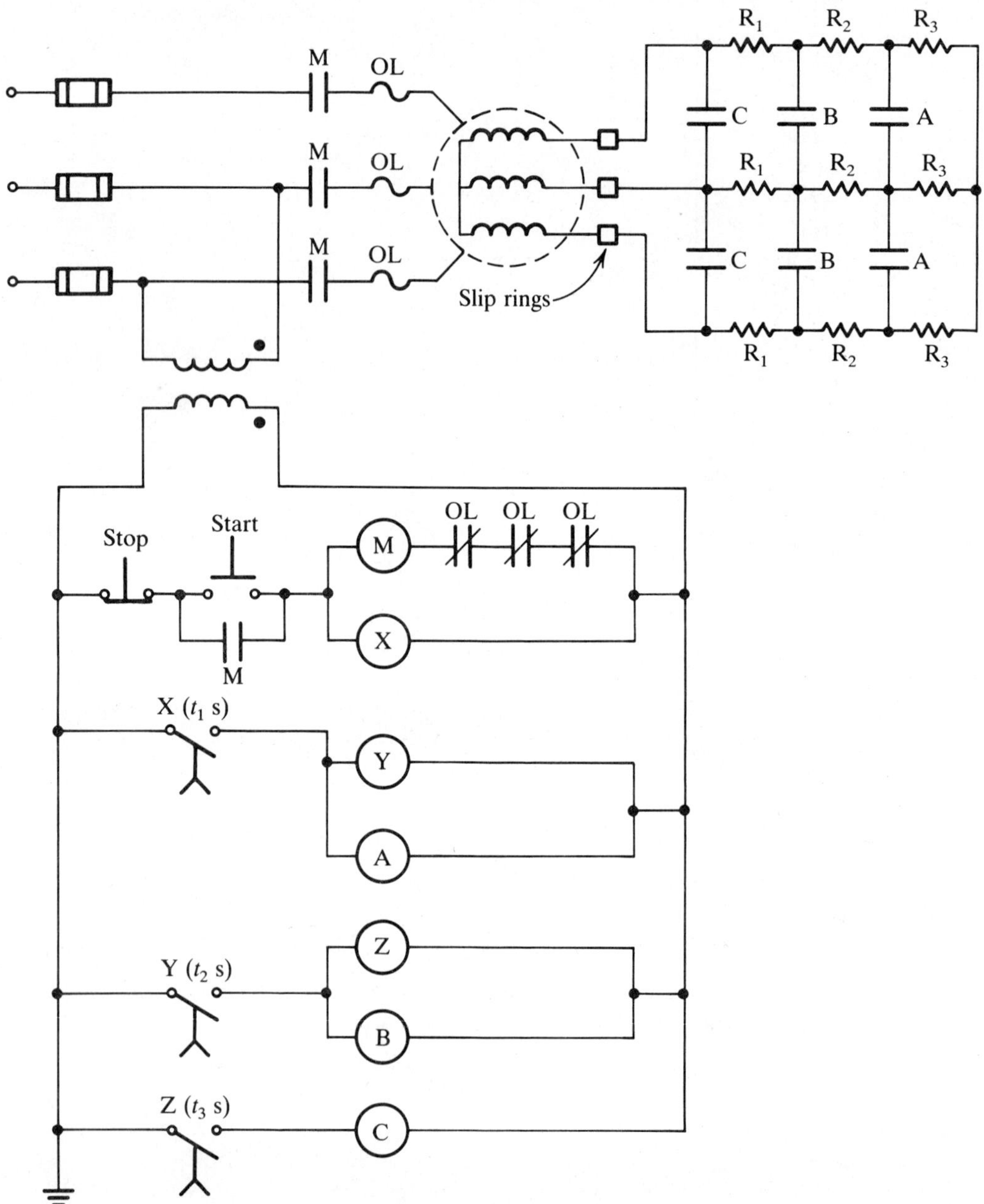

FIG. 7-26 Wound-rotor induction motor.

Low-Voltage, Wound-Rotor Motor

The control schematic of a three-phase, low-voltage, wound-rotor induction motor is shown in Fig. 7-26. The purpose of the rotor resistors is to reduce the motor's starting current and to increase its starting torque. Once the motor starts, the resistors are automatically shorted out in steps by the operation of the magnetic contactors (A, B, and C) and the time-delay relays. The time-delay relays are all of the "time-delay-on-energization" type. Briefly, the sequence of operation is as follows:

1. When the start pushbutton is depressed, the coils of the contactor M and of the time-delay relay X are energized. The motor then starts to accelerate. At starting, the per-phase rotor resistance is the sum of R_1, R_2, and R_3, and thus the starting current of the motor is accordingly reduced.
2. After t_1 s the contact of the time-delay relay X closes, energizing relay Y and contactor A. This removes the resistors R_3 from the rotor circuit.
3. After t_2 s the contact of the time-delay relay Y closes, energizing the relay Z and the contactor B. This removes the resistors R_2 from the rotor circuit. Similarly, after t_3 s the contact of the time-delay relay Z closes. The contactor C then removes the remaining resistors R_1 from the external rotor circuit. At about this time the motor operates at a steady-state condition.

EXAMPLE 7–7 A 550 kW, 480 V, 3-ϕ industrial plant is equipped with a stand-by 480 V, 500 kW, 3-ϕ diesel generator system. The plant's load includes two pumps of 50 kW each, which operate at different times.

a) Draw the one-line distribution diagram.
b) Design the control schematic, including the required interlocks, so that the diesel generator will deliver rated power and will supply only one of the two pumps that happened to be operating at the instant of power failure.

SOLUTION

a) The one-line diagram of the industrial plant is shown in Fig. 7-27(a). Under normal operating conditions the 500 kW diesel generator is OFF and the entire electrical load is supplied from the utility's power lines. When the utility's power is interrupted, the contacts of the transfer switch change status and the diesel generator starts to provide rated power. The voltage-sensing devices and auxiliary relays that are required to complete the transfer of the power sources are not shown.
b) The required control schematics are shown in Figs. 7-27(b) and (c). The main features of these circuits are as follows:
   (1) When one of the two pumps is chosen to be operated, its start pushbutton is depressed. As a result, the motor's magnetic contactor and the auxiliary relay (R_1 or R_2) in parallel to it are energized. The energization of the auxiliary relay disables the control schematic of the nonoperating motor.

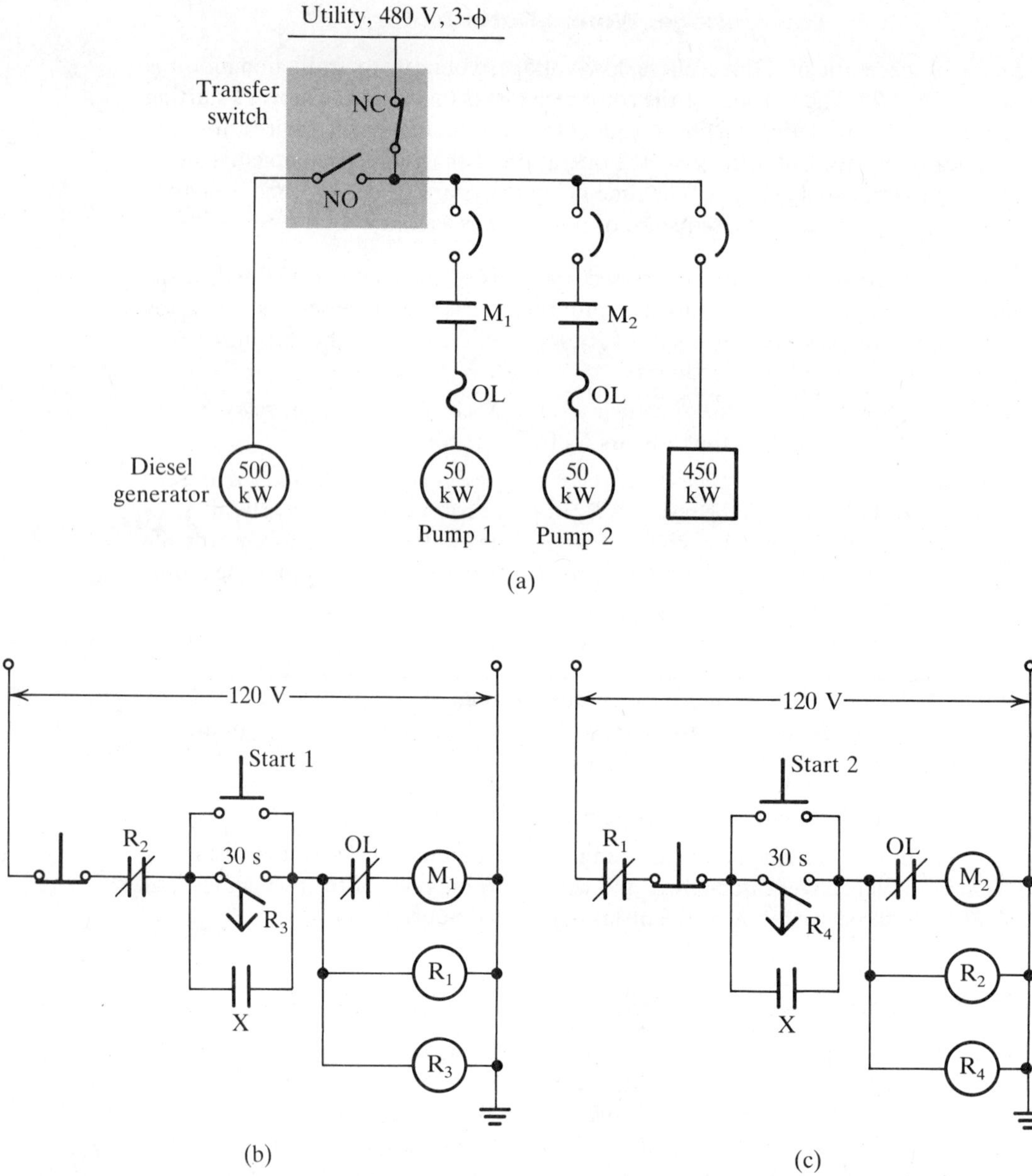

FIG. 7-27

(2) When energized, the time-delay relays (R_3 and R_4) close their contacts instantaneously, and upon deenergization keep their contacts closed for 30 s. Within this time delay the generator must be capable of being started.

(3) When the generator starts, the contacts X close, the start pushbuttons are sealed in, and so the previously operating pump continues its operation.

MV Induction Motors

As previously mentioned, three-phase induction motors of ratings larger than 200 kW are normally supplied, for economical reasons, at voltages between 2 and 15 kV. The metallic cabinet that houses the control and protection components of MV motors is referred to as the "medium-voltage starter." As shown in Fig. 7-28, the medium-voltage

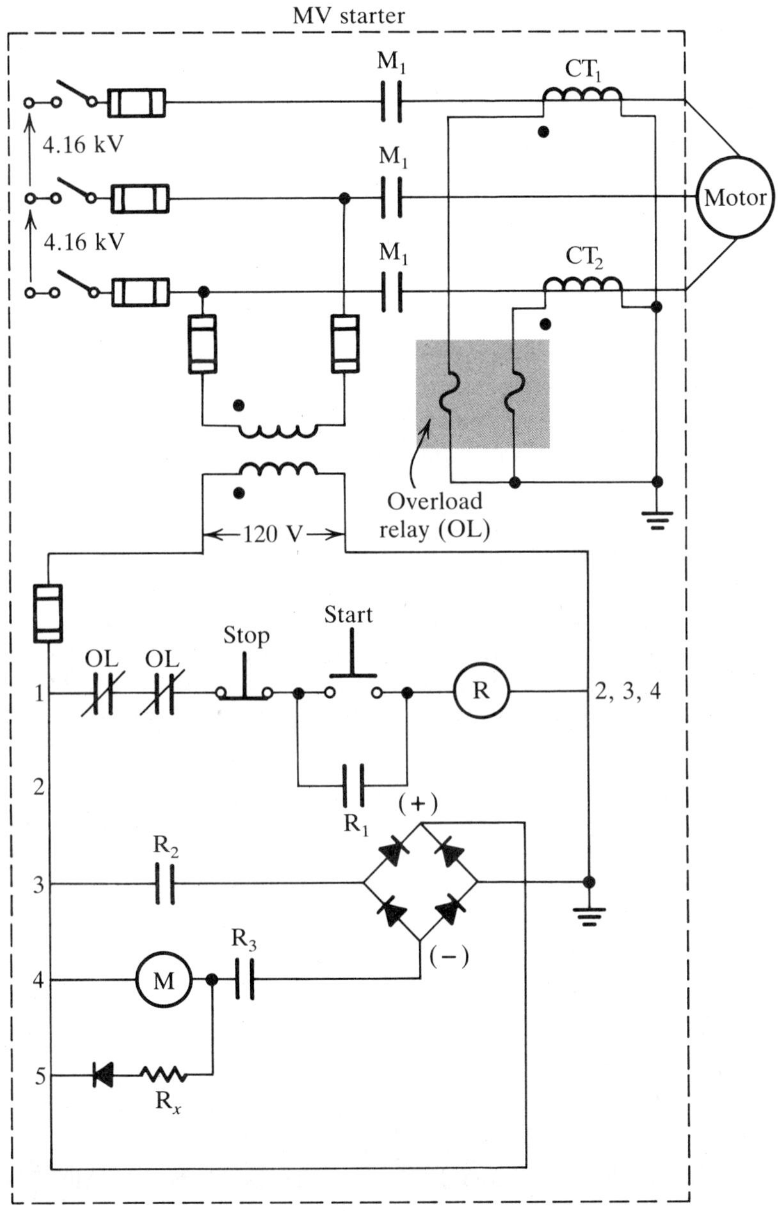

FIG. 7-28 Control schematic of a medium-voltage starter for a 4.16 kV three-phase induction motor.

starter incorporates power fuses, voltage and current transformers, overload relays, the rectifying bridge for the dc supply to the coil of the magnetic contactor, and so on. The sequence of operation of the control circuit is as follows:

1. When the start pushbutton is depressed, the auxiliary relay R is energized, and then its normally open contacts R_1, R_2, and R_3 close.
2. Upon energization of the auxiliary relay, the 120 V control voltage is applied to the full-wave diode bridge. The output dc voltage energizes the coil M of the magnetic contactor.
3. The energization of the coil of the magnetic contactor closes the NO main contacts M_1. Upon energization of the coil's contactor, then, the medium-voltage supply is connected to the motor's stator terminals and the motor starts to accelerate toward its full-load speed.

Features

1. The use of dc voltage for the operation of the contactor's coil results in faster coil operation and eliminates the undesirable capacitive and inductive effects of ac excitation.
2. Under normal operating conditions the diode-resistance network that is connected in parallel to the contactor's coil does not draw any current. When the motor is stopped, however, this network dissipates the energy stored in the contactor's magnetic field. DC coils which are frequently energized when not connected to such diode-resistance networks damage (carbonize) the in-series connected contacts of the other relays.
3. Fuses in the main power circuit provide short-circuit protection (50) and their maximum current rating is usually about 300% of the motor's full-load current. In actual cases the current rating of the fuse is selected by taking into consideration the damage and accelerating characteristics of the motor and the time–current curve of the overload relay.
4. The interruption of the motor's current—through the main contacts of the magnetic coil—takes place in a vacuum chamber. Originally the quenching medium was oil, which was subsequently replaced with air, but now almost all medium-voltage starters are of the vacuum type.
5. Overload current protection (51) is provided by a two-element overload relay. As shown in the diagram, the input current to the relay is supplied by the current transformers. A better overload protection would have used three overload relays. However, the monitoring and the protection of two motor windings, instead of all three, is acceptable.

EXAMPLE 7–8 Referring to the elementary control schematic of Fig. 7-29(a),

a) Why is an auxiliary relay X used to energize the coil of the magnetic contactor?
b) Modify the control schematic to include a mechanical door interlock by which the motor and the control components—except the control transformer—will be

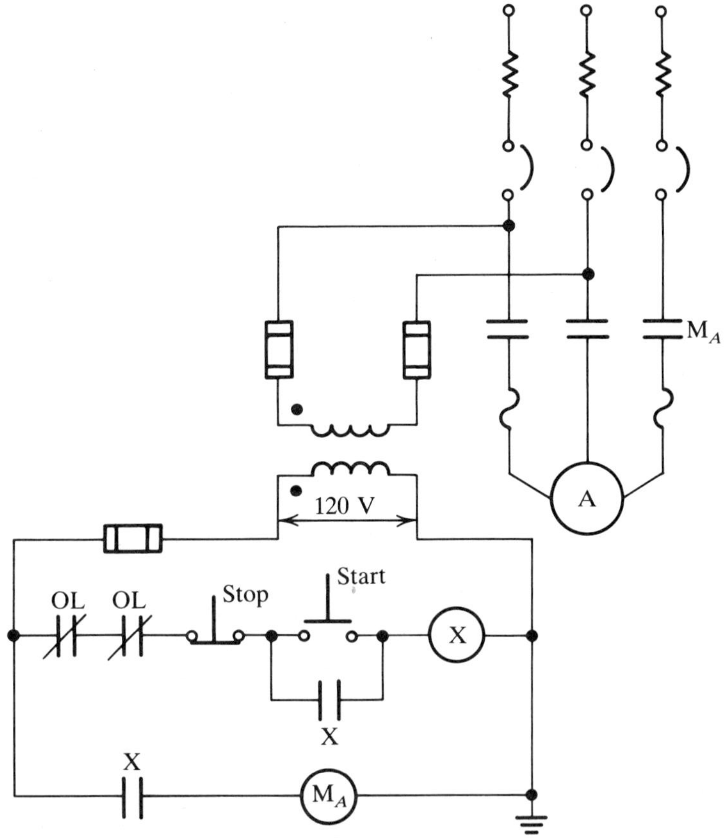

FIG. 7-29(a)

removed from their corresponding voltage supply each time the door of the control cabinet is opened.

c) Because of process requirements in an automated plant the starting and stopping of motors is done sequentially. That is, each motor starts or stops in relation to another motor within a specified time interval. Draw the elementary control diagrams of two such motors. Assume that the time interval in the sequential starting is 10 s.

SOLUTION

a) The main function of the auxiliary relay X is to provide low, continuous, make-and-break current for both the start-stop pushbuttons and the contacts of the overload relay. This results in a lower initial cost and in longer operating periods "free of troubleshooting." Conversely, when the coil of the magnetic contactor is connected in series with the start-stop pushbuttons, it draws much larger currents, and "heavy-duty" pushbuttons may be required as a result.

b) As shown in Fig. 7-29(b), a normally closed contact (identified as door interlock) is connected in series with the coil of the auxiliary relay. This contact is mechanically controlled by the position of the control cabinet's door. When the door is closed, the contact is closed, and when the door is opened, the mechanical interlock opens up the contact; thus the control circuit is deenergized and the motor stops. This is an important and absolutely necessary safety feature, especially when the cabinet houses the MV components of the starter.

c) One method of achieving sequential starting of a group of motors (A, B, or C, etc.) is to use, as shown in Fig. 7-29(b), a relay (Y) with a time delay on energization of 10 s. This relay is energized through the auxiliary relay of the first motor (A). The time-delay contact of this relay energizes the magnetic contactor of the second

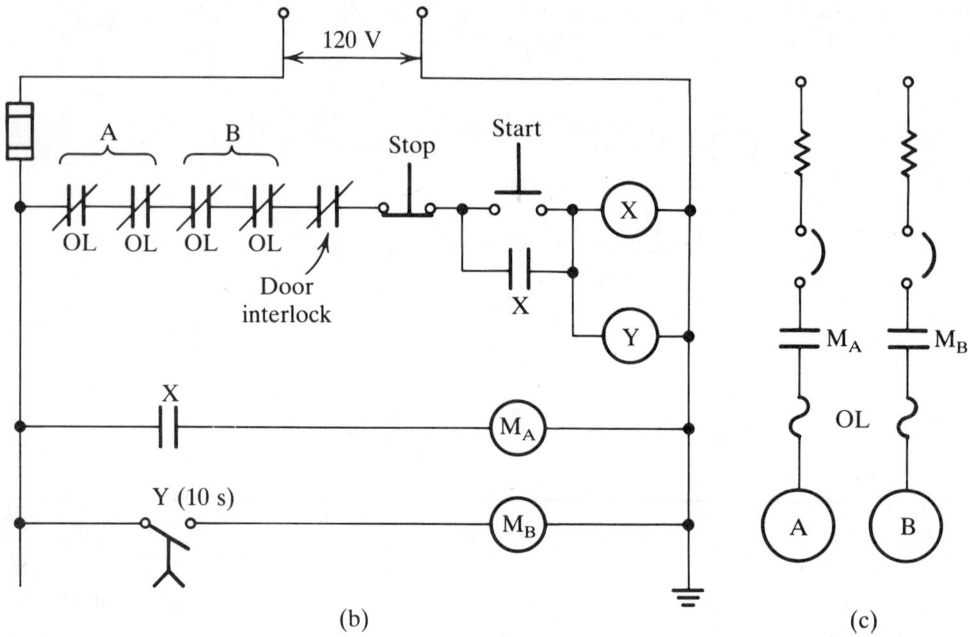

FIG. 7-29(b), (c)

motor (B). The time interval from the energization of the relay until the closing of its contact gives the sequential time between the starting of the two motors. When the start pushbutton is depressed, the auxiliary relays X and Y are energized. Relay X in turn energizes the contactor of motor A, while the time-delay relay Y energizes the contactor of motor B after 10 s. Nowadays, the status and the timing of the "seal-in" and time-delay contacts are controlled through programmable controllers or computers. The one-line diagram is shown in Fig. 7-29(c).

Exercise 7–7 Referring to Fig. 7-28:

a) Prove that the polarity of the full-wave rectifier is as shown.

b) When the motor stops, the energy stored in the conductor's coil is dissipated in 2 s. If the resistance of the coil plus that of R_x is equal to 50 Ω, estimate the minimum inductance of the coil.

Answer 25 H.

7.3.3 SYNCHRONOUS MOTORS

General

All modern synchronous motors are of the brushless type; that is, there is not any electrical connection between the external excitation system and the rotor's circuit. The dc current required to operate the motor is obtained by rectifying the voltage induced in another set of rotor windings, commonly referred to as coils of the rotating armature. As shown in Fig. 7-30(a), the rotating armature winding, the rectifying circuit, and the motor's field winding are all mounted on the motor's rotor.

The exciter's armature winding, or the so-called rotating transformer, produces, at its output, a three-phase ac voltage, which is rectified through the three-phase diode-bridge. As a result, under synchronous-speed operation only dc current flows through the motor's field winding. When one speaks about the measurement and control of the motor's dc field current, one is invariably referring not to the actual motor's field current, but to the exciter's dc current.

As demonstrated below, the operation of a brushless synchronous motor is based on the induction principle and on the operating characteristics of diodes and thyristors.

Principle of Operation

When ac voltage is applied to the motor's stator windings (Fig. 7-30(b)), a synchronously rotating field is generated that induces a voltage in the field windings (coil F_1F_2) of the rotor. The field produced by the rotor current attempts to align its magnetic axis to that of the stator field. As a result, the motor starts to accelerate. The frequency and magnitude of the alternating voltage induced in the rotor's windings depend on the effective turns ratio between the stator and rotor windings, and on the rotor's speed relative to the speed of the synchronously rotating stator field.

The frequency and the magnitude of the induced voltage are at a maximum when the rotor is stationary, and at zero when the rotor rotates with the same speed as the stator field. A detailed description of the relationships for a rotor's induced voltage and its frequency is provided in Chapter 3.

When terminal F_1 is positive with respect to F_2, the motor's field winding is shorted through the rectifier's diodes. When terminal F_2 is positive with respect to F_1, and of proper magnitude, the motor's field winding current is circulated through thyristors Q_1, Q_2 and resistor R_1. The gating pulse for the triggering or conduction of the thyristors is provided through the reverse-biased Zener diodes Z_1 and Z_2.

Up to a level of about 95% of the synchronous speed the voltage induced in the field windings *must be larger* than the threshold voltage (knee point) of the Zener diodes. Otherwise the Zener diodes will not contact.

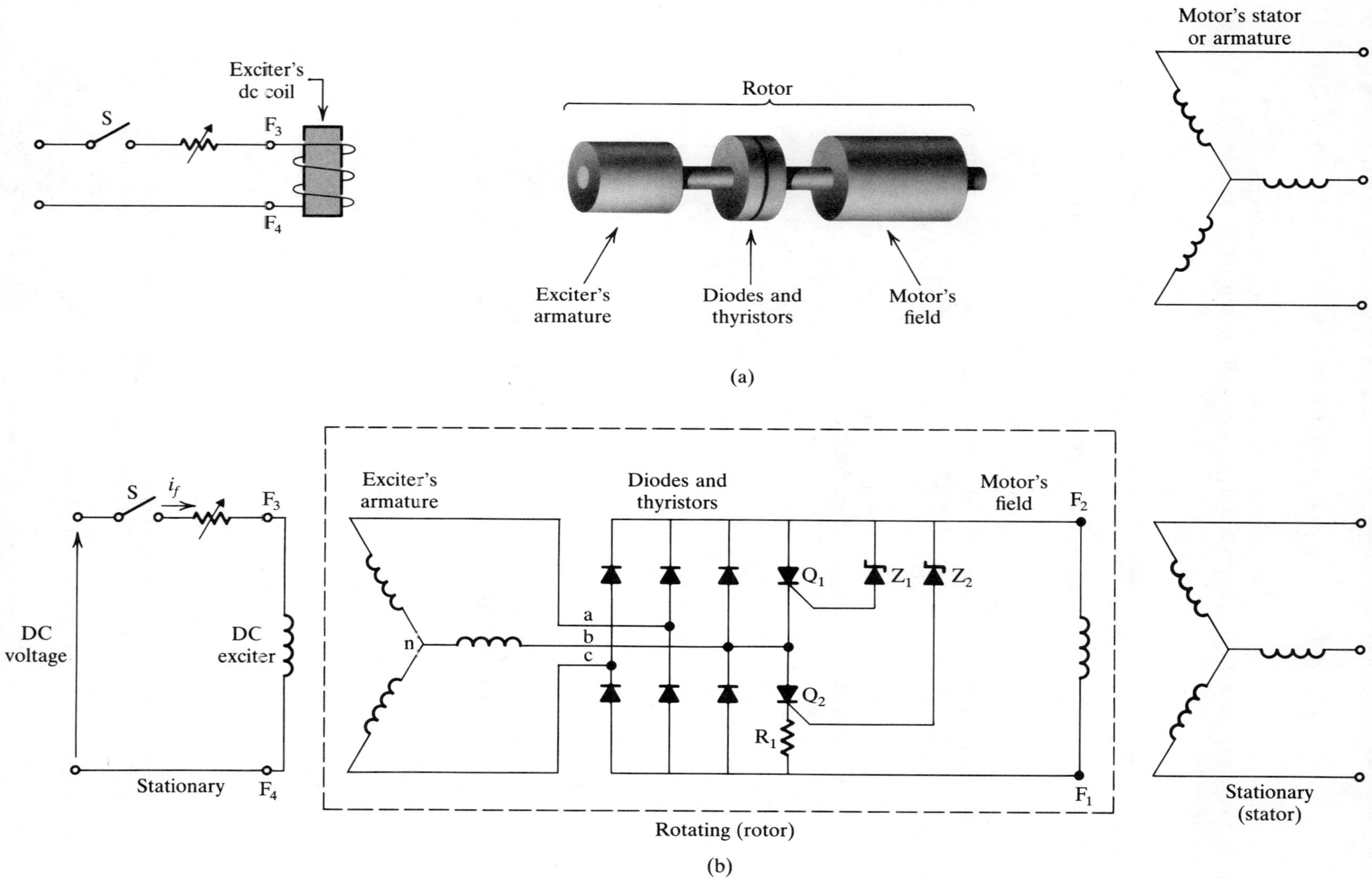

FIG. 7-30 Brushless synchronous motor: **(a)** physical representation, **(b)** circuit diagram. *Courtesy of General Electric Canada, Inc.*

When the rotor reaches about 95% of the synchronous speed, the control devices of the excitation circuit switch on the dc voltage supply. As a result, a voltage is generated in the rotating armature of the exciter. The voltage generated is proportional to the shaft's speed of rotation and to the effective field of the dc exciting coil. Mathematically,

$$V_{\mathrm{L-n}} = K_1 I_f \omega \tag{7.2}$$

where

$V_{\mathrm{L-n}}$ = the phase-to-neutral voltage induced in the rotating armature coils
I_f = the dc current of the exciter in A
ω = the rotor's angular speed of rotation in rad/s
K_1 = a constant of proportionality that depends on the effective number of winding turns between the rotating transformer and the excitation coil.

The average value of the voltage (V_{av}) at the output of the three-phase bridge rectifier is given by Eq. (3.112), which for convenience is rewritten here:

$$V_{\mathrm{av}} = \frac{3}{\pi} V_m \tag{7.3}$$

where V_m is the maximum line-to-line voltage at the output of the exciter's armature. The dc voltage given by Eq. (7.3) *must be smaller* than the threshold voltage of the Zener diodes; otherwise, the thyristors will be driven into conduction.

Starting and Protection

The control schematic for the starting and protection of a synchronous motor has to fulfill the following requirements:

1. The motor must be brought up to about 95% of its synchronous speed as an induction machine. During this period the dc rotor winding is disconnected from its nominal supply and is either shorted or connected across a discharge resistor.
2. At the end of this period the dc rotor source is connected to the motor's field winding.
3. The protective devices must be inoperative during this stabilizing or synchronizing period.
4. The circuit must incorporate protective devices that provide adequate motor protection and must include provisions for field-current adjustments.

Control Schematics of Brushless Synchronous Motors

A basic control schematic for the starting, operation, and control of a brushless synchronous motor is shown in Fig. 7-31. The sequence of operation is as follows:

1. By depressing the start pushbutton, the coil of the auxiliary relay X is energized. The "seal-in" contact maintains the voltage across its coil, and the other contact

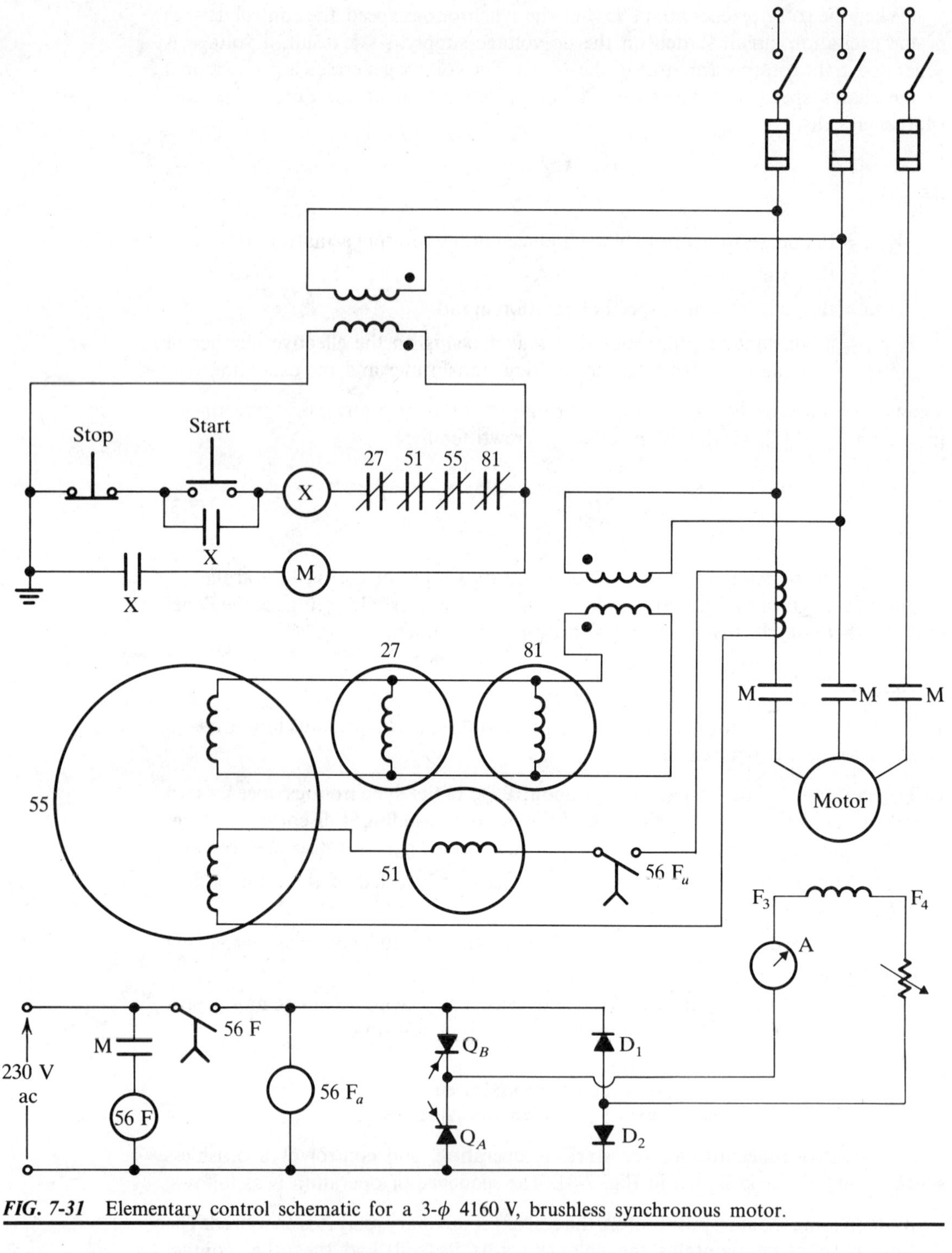

FIG. 7-31 Elementary control schematic for a 3-ϕ 4160 V, brushless synchronous motor.

energizes the motor's starter M. Thus the voltage source is connected to the motor's terminals. The motor starts to accelerate as a 3-ϕ induction motor.

2. Following the energization of the motor's contactor, the main field application relay (56F) is energized. This relay has an adjustable time-delay contact (5 to 20 s) on energization. The selected time is equal to the time that it takes for the motor to reach about 95% of its nominal speed.
3. At the end of the acceleration period—often determined in the field by trial and error—the contact of the main field application relay (56F) closes, the exciter's armature is energized, and consequently a dc voltage is applied to the motor's field winding. The rotor tries to pull into synchronism with the stator field (a natural tendency of fields to align their magnetic axes).
4. Closing the contact of relay (56F) also energizes the auxiliary field application relay ($56F_a$), which has an adjustable time delay of 2 to 10 s. This time interval is often referred to as the synchronization or stabilization period.
5. Following the energization of the relay ($56F_a$) and the closing of its contact, the pull-out or power-factor relay (55) is connected to its current supply.
6. The power-factor relay is energized (contact 55 opens) when the power factor of the motor is lower than the power-factor setting of the relay. The power-factor setting of some popular power-factor relays is adjustable from .85 lag to .8 lead. The tripping time, as per manufacturers' data, is adjustable from .5 to 2.5 s. At subsynchronous speeds, or at pull-out, or at low field current the power factor becomes drastically lagging and thus the relay initiates tripping and the motor is removed from its stator voltage. If this were not done, any of these conditions could damage the motor's field and/or stator windings.

Remarks

1. For high-speed performance the contactor of the motor is usually energized through a dc voltage supply (not shown in the schematic).
2. The input parameters of the power-factor relay are the motor's stator voltage and current. The energization signal for the undervoltage (27) and under/overfrequency (81) relay is the motor's stator voltage only.
3. The overload relay (51) provides the motor's overload protection. For the sake of simplicity the overload protection for the other stator lines is not shown. Line fuses provide the motor's overcurrent protection.
4. The exciter's dc current is adjustable through the firing angle of the thyristors Q_A and Q_B (not shown in the diagram) or through a rheostat. Where a variable resistor is used, the thyristors are replaced with diodes.
5. The motor's stator windings are usually equipped with a resistor (not shown in the control schematics) or a heater that protects them from condensation during the motor's nonoperating time.
6. The ammeter in the exciter's circuit indicates the dc current from which the operating power factor of the motor can be estimated (see the vee curves of Chapter 5).
7. When an electrical or mechanical disturbance drives the rotor out of synchronism,

the armature current and its power factor change according to the level of disturbance. Protective equipment (armature overcurrent relay and power-factor relay) will sense the change in these parameters and may, according to the settings, remove the electrical supply from the stator, bringing the motor to a complete stop. If the protective equipment does not remove the electrical supply from the stator, the motor's field winding will be shorted, as in the case of the accelerating period. This implies, of course, that the external disturbance forced the rotor to run at a speed that was outside the range of 95% to 100% of synchronous speed.

EXAMPLE 7–9 A 514 r/min, .8 Pf leading, 4160 V, 60 Hz, 500 kW brushless synchronous motor has an effective stator-to-rotor per-phase turns ratio equal to unity. Determine:

a) The effective value of the voltage induced in the motor's field just prior to the switching ON of the dc supply. Assume this to take place at 95% of synchronous speed.

b) The dc voltage across the motor's field winding at nominal operating conditions, given that the voltage constant between the dc exciter and its armature is 2 V/rad/s/phase.

c) How the setting of the power-factor relay controls the minimum breakdown torque developed by the motor.

SOLUTION

a) The voltage induced in the rotor's winding (V_r) is proportional to the stator voltage (V_s), the slip, and the effective turns ratio between the stator and the rotor. Mathematically,

$$V_r = sV_s \frac{N_r}{N_s} \tag{I}$$

where N_r/N_s is the ratio of rotor to stator number of turns per phase. The slip s is given by

$$s = \frac{\omega_s - \omega_a}{\omega_s} = \frac{1 - .95}{1} = .05 \text{ pu}$$

Substituting into equation (I), we obtain

$$V_r = .05 \left(\frac{4160}{\sqrt{3}}\right)(1) = 120.09 \text{ V}$$

b) The line-to-line voltage across the motor's rotating armature is

$$V_{L-L} = K\omega = 2(514)\frac{2\pi}{60}\sqrt{3} = 186.46 \text{ V, L–L}$$

The dc voltage across the motor's field windings is determined by Eq. (7.3):

$$V_{av} = \frac{3}{\pi}(186.46)\sqrt{2} = \underline{\underline{251.81 \text{ V}}}$$

c) The breakdown torque of the motor, as can be seen from Eq. (5.29), depends on the motor's field current. In turn the field current controls the power factor of the motor. Therefore, the setting of the power-factor relay controls the minimum permissible value of the field current and the minimum value of the breakdown torque.

Exercise 7–8

a) Referring to Fig. 7-30(b), what are the functions of:
   (1) The diodes
   (2) The stationary dc coil
   (3) The rotating exciter's armature.
b) Referring to Fig. 7-31, what are the functions of:
   (1) The power-factor relay (55)
   (2) The undervoltage relay (27)
   (3) The frequency relay (81).

7.3.4 CONTROL SCHEMATICS OF dc MACHINES

The starting and control of dc machines is implemented either electronically or through a network of resistors. The network of resistors is controlled through standard electromagnetic contactors and relays, which provide speed changes in steps and are accompanied by high resistor copper losses. The control schematic of a shunt dc motor using a network of resistors and magnetic contactors is given as a student exercise (see Problem 7-10).

Electronic controls are relatively expensive but provide efficient and continuous operation over a wide range of speeds. Electronic control is the predominant method of speed control for all dc motors. A simple version of it is discussed here.

Electronic Control of dc Motors

Figures 7-32(a) and (b) show, respectively, the one-line diagram and the control schematic of a separately excited dc motor. The main components of the one-line diagram function as follows.

The main power transformer isolates the electronic devices from dangerous upstream transients and, if necessary, provides a suitable voltage for the operation of the six-pulse rectifier.

The control transformer furnishes the required 120 V to all control circuits. Fuses provide the overcurrent protection, and device 51 provides current overload protection to the six-pulse rectifying bridge and the motor.

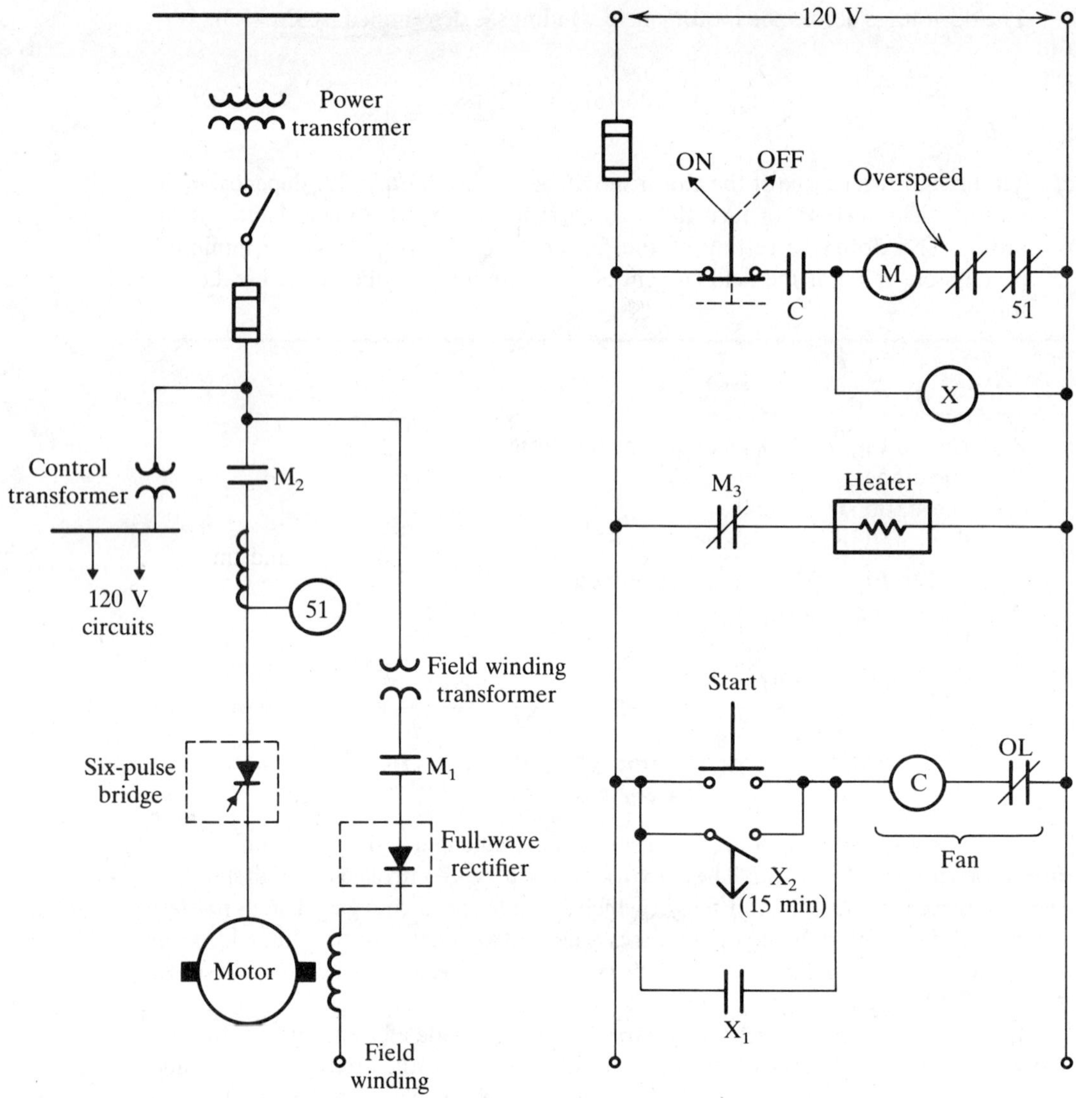

(a) (b)

FIG. 7-32 Separately excited dc motor: **(a)** one-line diagram representation, **(b)** elementary control schematic.

For large dc motors the switching element of the six-pulse rectifying bridge is the thyristor, while for dc motors of lower power ratings (up to several hundred kW) it is the transistor. In either case a controlled dc voltage proportional to the motor's speed is available at the output terminals of the bridge.

The transformer to the field winding provides the necessary voltage and power to the motor's field winding. The contacts M_1 and M_2 serve as remotely controlled ON–OFF switches.

The dc motor and its controls are referred to as a "dc drive." All the components of a dc drive are generally supplied by the same manufacturer. As a result, the control schematic of these drives incorporates not only the motor's protection but also the protection of the drive's components, such as the power transformer, the thyristorized bridge, and the motor's field. For reasons of simplicity, however, the control schematic shown in Fig. 7-32(b) incorporates only the basic motor protection and control.

Description of Operation

Refer to the elementary dc drive shown in Fig. 7-32. The motor's sequence of operation is as follows:

1. When the start pushbutton is depressed, the coil of the ventilating fan's contactor C is energized and consequently its contact is closed.
2. This results in the energization of the motor's starter coil M and of the auxiliary relay X. Relay X has two contacts, one being of the instantaneous type (X_1) and the other of the time-delay-on-deenergization type (X_2).
3. The energization of the relay X closes its contacts instantaneously, and thus the control voltage for the fan's operation is maintained.
4. The energization of the motor's starter coil M closes its main contacts M_2 and M_1, and thus nominal voltages are applied to the input terminals of the six-pulse and full-wave bridge networks.
5. The gate signal is then applied to the thyristors of the six-pulse rectifier—by means not shown in the diagram—and depending on the magnitude of the firing angle (α) a dc voltage (V_{av}) is generated across the motor's armature terminals. For a six-pulse controlled rectifier this voltage is given by Eq. (6.105), which is rewritten here for convenience:

$$V_{av} = \frac{3}{\pi} V_m \cos \alpha \tag{7.4}$$

   where V_m is the maximum value of the line-to-line voltage on the bridge's input terminals.

   At the motor's starting, the armature current (I_a) is given by

$$I_a = \frac{V_{av}}{R_a} \tag{7.5}$$

   where R_a is the resistance of the armature circuit.

   From Eqs. (7.4) and (7.5) it is evident that the motor's starting current can be limited by a proper selection of the thyristors' firing angle.
6. At the thyristors' firing, the motor starts to accelerate toward its desired speed.
7. By turning the selector switch to the OFF position, the coil of the magnetic contactor is deenergized, and thus the voltage supply is removed from the terminals of the armature and field winding. The fan will cool the motor for another 15 min because of the time delay on the deenergization contact (X_2).

8. As shown in the elementary control diagram, the motor can also be disconnected from the source voltage for any of the following reasons:
   a) Motor and/or fan overload.
   b) Motor overspeed. The motor overspeed contact is controlled from the output of a tachogenerator, which is not shown in the diagram.

Operational Features

1. The power transformer steps down the ac voltage and dampens the effects of the high transient currents on the control components of the dc drive. When the transformer voltage ratio is equal to 1, an inductor can be used to reduce the possible destructiveness of the transients that originate in the upstream network.
2. Protection from overspeed (an undesired motor and load operating condition) can be furnished either through the field circuit or through the tachogenerator. The speed (ω) of the motor is given by

$$\omega = \frac{V_{\text{av}} - I_a R_a}{K\phi} \tag{7.6}$$

   where ϕ is the effective flux of the field and K is the motor's constant. By limiting the minimum value of the field current, overspeed protection can be obtained.
3. The motor will not operate when the fan is OFF. The heater protects the motor's windings from condensation.
4. The motor is further protected by the built-in protection of the thyristorized circuit. The bridge network incorporates either *current-limiting fuses* or special *high-speed circuit breakers*, and thus for any downstream faults the bridge and the motor will be removed from the source voltage. The voltage and current capabilities of the thyristors are much higher than those of the motor, but it is common practice to select the built-in control protection according to the motor's rating. This provides an additional safeguard.

EXAMPLE 7–10 In the separately excited dc motor shown in Fig. 7-32(a), the voltage is supplied by an isolating transformer of voltage ratio 480/480 V line-to-line. The motor's constant is 5 V/rad/s. The permissible range of the firing angle of the thyristors is 30 → 90 degrees, and the motor's field current is constant. At what speed will the overspeed device stop the motor?

SOLUTION Neglecting the motor's armature resistance, we have

$$V_t \approx K\omega$$

Also,

$$V_t = V_{\text{av}}$$

From the above and Eq. (7.4) we obtain

$$V_t = 3\frac{V_m}{\pi}\cos\alpha$$

Maximum speed corresponds to maximum terminal voltage, which for this example is produced at a firing angle of 30°. Thus, from the above,

$$\omega = 3\frac{V_m}{\pi K}\cos\alpha = \frac{3\sqrt{2}(480)}{\pi(5)}\cos 30° = \underline{\underline{112.28\ \text{rad/s}}}$$

Exercise 7–9 Referring to Fig. 7-32:

a) Why is a selector switch used instead of a momentary stop pushbutton?

b) Is the motor protected from undervoltages and overcurrents?

c) Is any time delay required for the sequential energization of the motor's field and the armature winding? Explain.

d) What is the function of the 15 min time delay on deenergization contact X_2?

7.4 Solid-State Controls

This section discusses the basic building blocks of logic diagrams (gates and memory devices). A detailed explanation is given of how these basic building blocks are used to design logic diagrams (equivalent to electromechanical control schematics).

7.4.1 GATES

Gates are solid state devices that are constructed by using transistor(s) or transistors and diodes. The input and output voltage signals of the gates can be of high or low voltage level. The high voltage level for the majority of gates is about 5 V, while the low voltage level is about 0 V. In actual circuits the high-level voltage signal is within the range of 4 to 5 V, while the low-level signal is within the range of .2 to 1 V. The actual and ideal voltage levels of a gate are shown in Figs. 7-33(a) and (b), respectively.

The high-voltage signal is usually referred to as a logic level of 1, and the low-voltage signal as a logic level of 0. The logic levels of 1 and 0 are the only digits of the binary system, upon which the design and operation of calculators, microprocessors, and computers are based. The high and low logic levels are also referred to as "true" and "false" or "yes" and "no" logic levels. The input–output relationships of a gate's signals may be written in Boolean algebra notation or in a tabular form that is referred to as the gate's truth table.

There are six gates: NOT, OR, NOR, Exclusive OR, AND, and NAND. The symbols that the various standards use for the gates are summarized in Table 7-14. In this text only the symbols of the MIL (military) standards will be used.

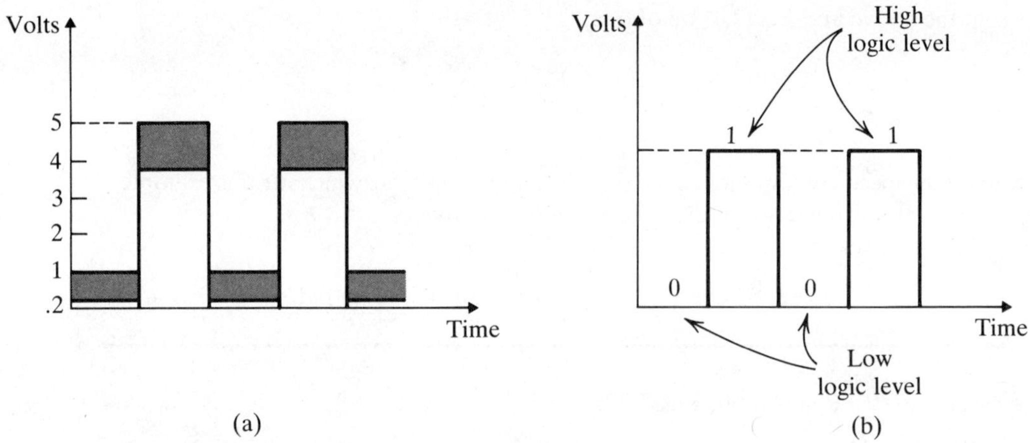

FIG. 7-33 Input–output voltage levels of gates: **(a)** actual, **(b)** ideal.

All the gates, except the NOT gate, have two or more inputs. For each gate we are concerned with its symbol, input–output relationship in Boolean algebra notation, truth table, and electromechanical equivalent diagram.

NOT Gate

The NOT gate has, as shown in Fig. 7-34(a), only one input and is usually referred to as an inverter. The output of a NOT circuit takes on the 1 state if, and only if, the input does not take the 1 state. The governing equation of a NOT gate is

$$Y = \bar{A} \tag{7.7}$$

This equation is read: "*Y* is the complement of *A*"; or "*Y* equals NOT *A*"; or "*Y* is the opposite of *A*"; or "*Y* is the negation of *A*." The truth table of the NOT gate is shown in Table 7-5.

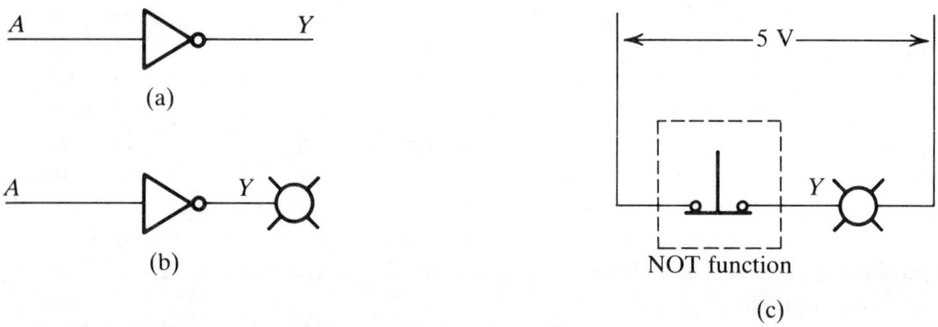

FIG. 7-34 NOT gate: **(a)** symbol, **(b)** inverter operating as a light switch, **(c)** equivalent electromagnetic control schematic of (b).

TABLE 7-5 Truth table of NOT gate

Input A	*Output* Y
0	1
1	0

Figure 7-34(b) shows an inverter operating as a light switch. When no signal is applied to A, the output of the inverter is at high logic level and thus the light is ON. When a high logic signal is applied to A, the output of the inverter is low and thus the light is switched OFF. The inverter of the logic diagrams is equivalent to a normally closed contact or to the stop pushbutton of the electromechanical control schematics.

As shown in Fig. 7-34(c), the stop pushbutton in a nonactivated state (no input or application of external force) provides an output or lets electric current go through it. When an external force is applied to it, however, the current is interrupted and no signal is transmitted to the output device, as long as it remains depressed.

OR Gate

The symbol of the OR gate is shown in Fig. 7-35(a). The output of an OR gate assumes the 1 state if one or more of the inputs assumes the 1 state. The truth table of the OR gate is given in Table 7-6. The Boolean algebra equation of an OR gate is given by

$$Y = A + B + \cdots + N \tag{7.8}$$

This equation is read: "Y equals A or B or N."

If $A = B$, then $Y = A + A = A$. Also, if $B = \bar{A}$, then $Y = A + \bar{A} = 1$.

The above relationships are referred to as Boolean algebra identities.

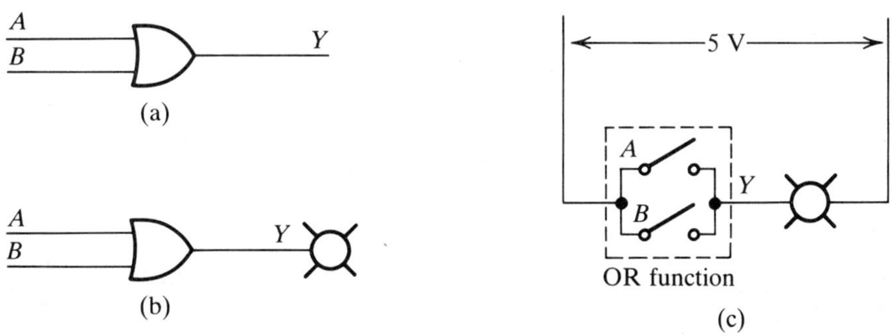

FIG. 7-35 OR gate: **(a)** symbol, **(b)** operating as a light switch, **(c)** equivalent electromagnetic control schematic of (b).

TABLE 7-6 Truth table of OR gate

Input		Output
A	B	Y
0	0	0
1	0	1
0	1	1
1	1	1

Figure 7-35(b) shows an OR gate operating as a light switch. Its equivalent control schematic is shown in Fig. 7-35(c). When one or both of the input signals is at high logic level (the corresponding manual switch is closed), the light will be switched ON.

NOR Gate

A NOR gate's function, is opposite that of an OR gate. The logic-circuit symbol of a NOR gate is shown in Fig. 7-36(a). The output of a NOR gate assumes the 1 state if and only if both of the inputs are at low logic level. The truth table of the NOR gate is given in Table 7-7.

The governing equation of a NOR gate is

$$Y = \overline{A + B} \tag{7.9}$$

This is read: "Y is the complement of the logic summation of A or B."

Figure 7-36(b) shows a NOR gate operating as a light switch. Its equivalent control schematic is shown in Fig. 7-36(c). When one or both of the input signals is at high logic level (the corresponding manual switch is open), the light will be switched OFF.

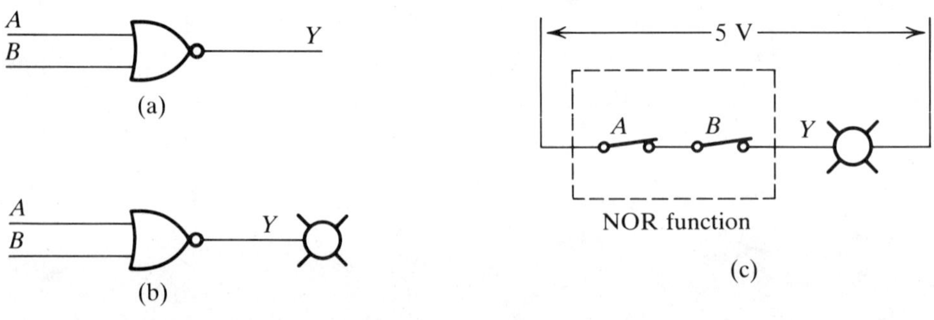

FIG. 7-36 NOR gate: **(a)** symbol, **(b)** operating as a light switch, **(c)** equivalent electromechanical control schematic of (b).

TABLE 7-7 Truth table of NOR gate

Inputs		Output
A	*B*	*Y*
0	0	1
0	1	0
1	0	0
1	1	0

A NOR gate can be constructed from an OR gate cascaded with a NOT gate, as shown in Fig. 7-37.

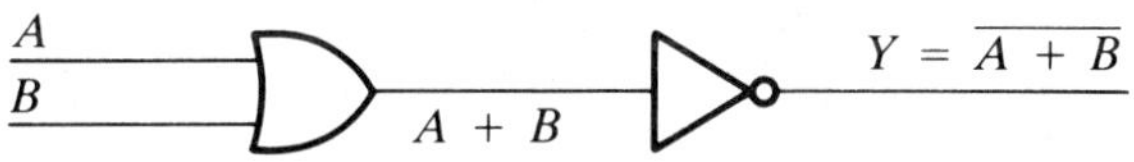

FIG. 7-37 Equivalent NOR gate.

AND Gate

The logic-diagram representation of an AND gate is shown in Fig. 7-38(a). The output of an AND gate assumes the 1 state if and only if all the inputs assume the 1 state. The equation describing an AND gate is

$$Y = AB \cdots N \tag{7.10}$$

This equation is read: "Y equals A and B and $\cdots$ N."

The truth table of the AND circuit is given in Table 7-8.

TABLE 7-8 Truth table of AND gate

Inputs		Output
A	*B*	*Y*
0	0	0
1	0	0
0	1	0
1	1	1

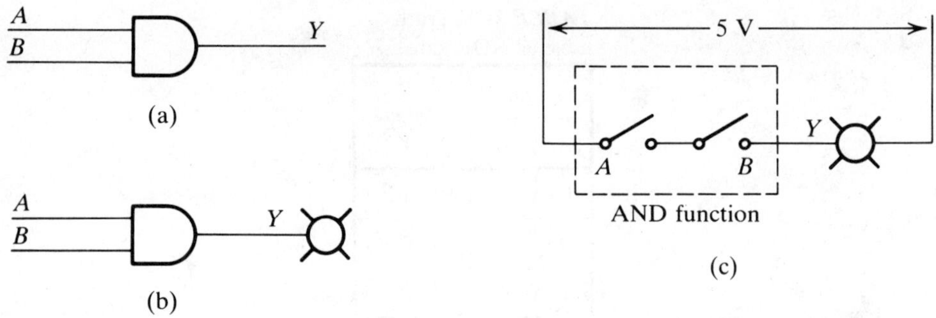

FIG. 7-38 AND gate: **(a)** symbol, **(b)** operating as a light switch, **(c)** equivalent control schematic of (b).

From the truth table, if A is equal to B, then $Y = AA = A$, and if $B = \bar{A}$, then $Y = A\bar{A} = 0$

Figure 7-38(b) shows an AND gate operating as a light switch. Its equivalent control schematic is shown in Fig. 7-38(c). When a signal of high logic level is applied to A and B, the light will be switched ON.

Exercise 7–10 Show that

$$\overline{AB} \neq \bar{A}\bar{B}$$

(*Hint:* Construct the truth table.)

NAND Gate

The symbol of a NAND gate is shown in Fig. 7-39(a). The NAND gate's function, as the name implies, is opposite that of an AND gate. The truth table of the NAND circuit is

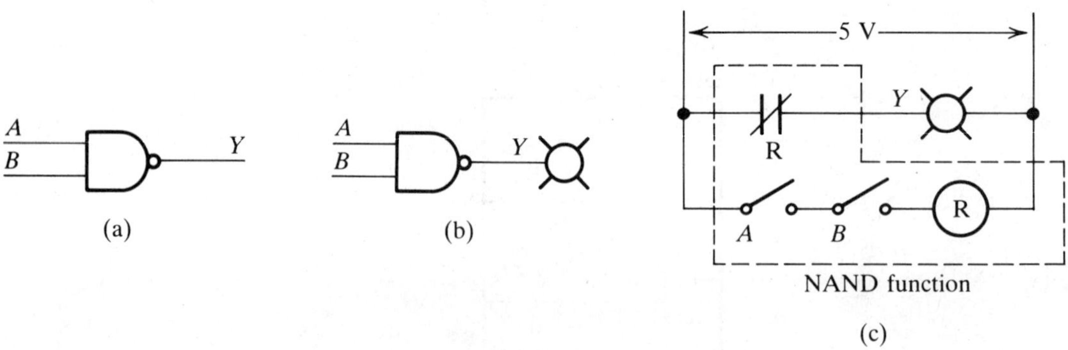

FIG. 7-39 NAND gate: **(a)** symbol, **(b)** operating as a light switch, **(c)** equivalent control schematic of (b).

TABLE 7-9 Truth table of NAND gate

Inputs		Output
A	*B*	*Y*
0	0	1
0	1	1
1	0	1
1	1	0

given in Table 7-9. The governing equation of a NAND gate is

$$Y = \overline{AB} \tag{7.11}$$

This equation is read: "*Y* is equal to the complement of *A* and *B*."

A NAND gate operating as a light switch is shown in Fig. 7-39(b). Its equivalent control schematic is shown in Fig. 7-39(c). When *A* and *B* are both at high logic level (switches *A* and *B* are closed), the light is switched OFF. For any other combination of input signals the light will be ON because relay R is not energized.

Exercise 7–11 The circuit shown in Fig. 7-40 is used to switch the light from two different locations (*A* and *B*). The positions of each switch are identified as high (logic 1) and low (logic 0).

a) Write the truth table of the circuit.
b) Draw the equivalent logic diagram.

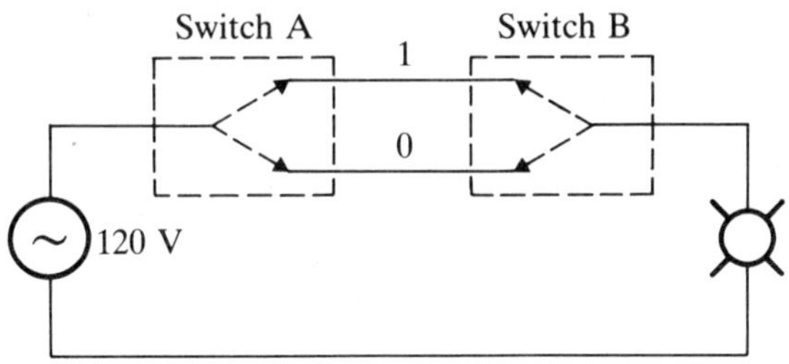

FIG. 7-40

Exclusive OR Gate

The logic diagram of an Exclusive OR gate is shown in Fig. 7-41(a) and its truth table is given in Table 7-10. The output of the Exclusive OR gate assumes the 1 state if

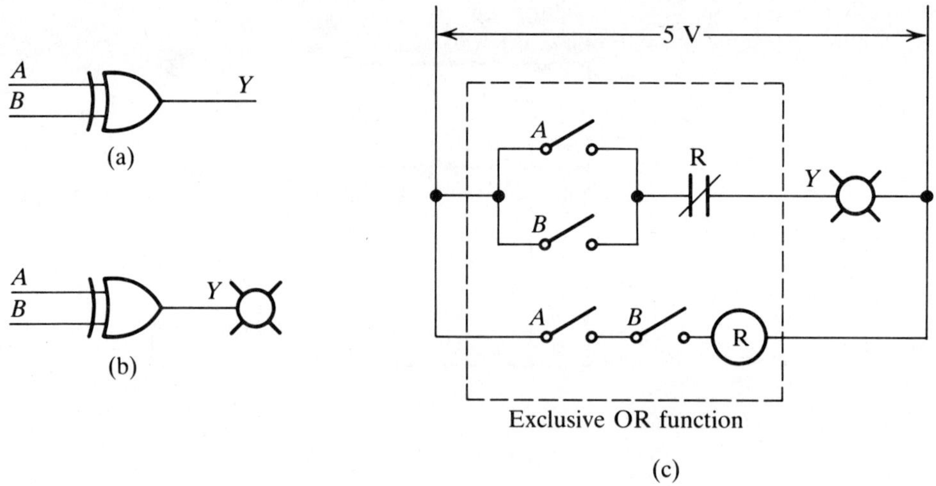

FIG. 7-41 Exclusive OR gate: **(a)** symbol, **(b)** operating as a light switch, **(c)** control schematic equivalent to (b).

TABLE 7-10 Truth table of Exclusive OR gate

Inputs		*Output*
A	*B*	*Y*
0	0	0
1	0	1
0	1	1
1	1	0

either *A* or *B* assumes the 1 state. The governing equation of an Exclusive OR gate is

$$Y = A\bar{B} + B\bar{A} \tag{7.12}$$

Figure 7-41(b) shows an Exclusive OR gate operating as a light switch. Its equivalent control schematic is shown in Fig. 7-41(c). When either of the input signals is at high logic level, the light will be switched ON. When both of the inputs are at high logic level, the relay R is energized and its NC contact opens, switching the light OFF.

The symbol, the equivalent control schematic, the truth table, and the governing equation of each gate are summarized in Table 7-11.

TABLE 7-11 Summary of logic gates

Name	*Symbol*	*Equivalent Control Schematic*	*Truth Table*	*Governing Equation*
NOT	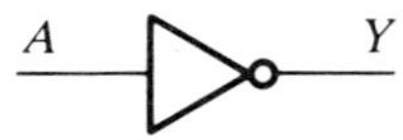	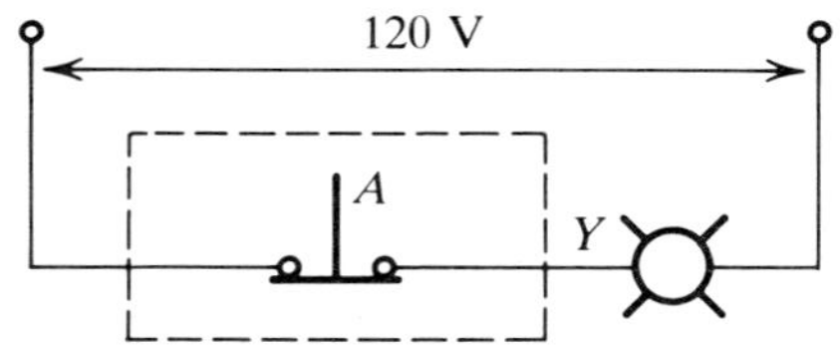 NOT function	A Y 0 1 1 0	$Y = \bar{A}$
OR	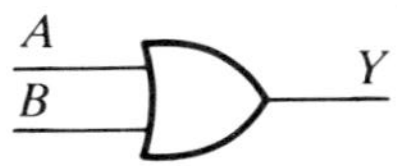	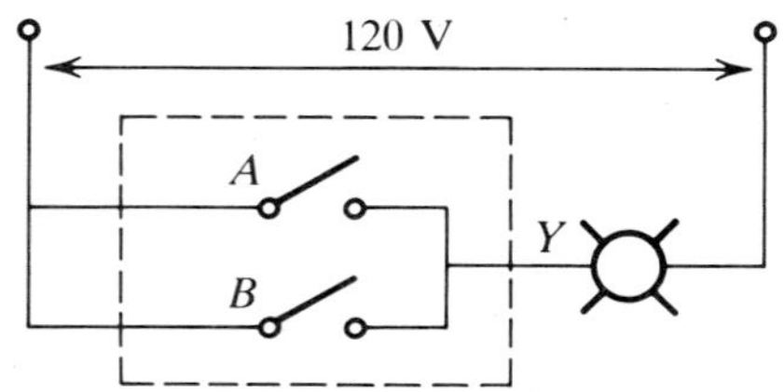 OR function	A B Y 0 0 0 1 0 1 0 1 1 1 1 1	$Y = A + B$
NOR	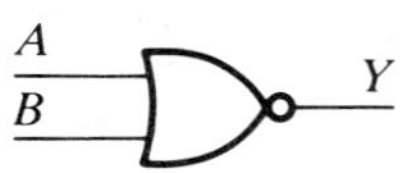	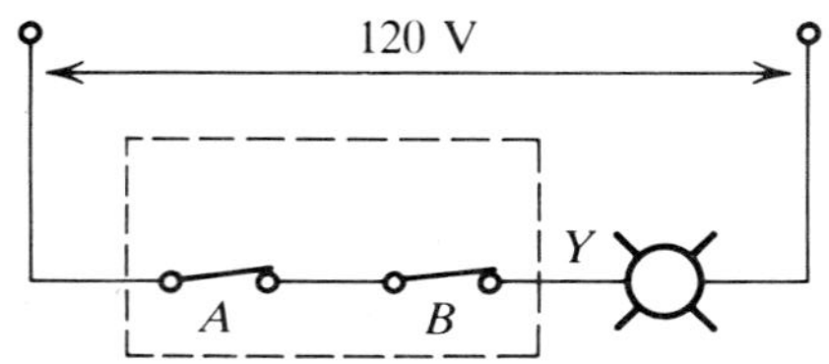 NOR function	A B Y 0 0 1 1 0 0 0 1 0 1 1 0	$Y = \overline{A + B}$

(Continued)

TABLE 7-11 (*Continued*)

Name	*Symbol*	*Equivalent Control Schematic*	*Truth Table*	*Governing Equation*
Exclusive OR	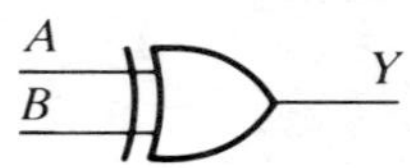	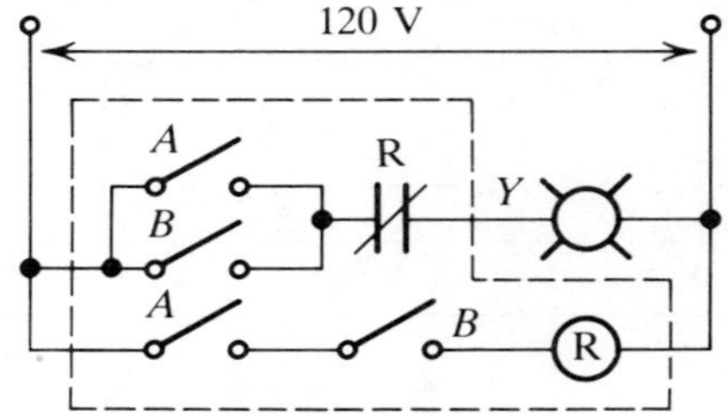 Exclusive OR function	*A B Y* 0 0 0 1 0 1 0 1 1 1 1 0	$Y = A\bar{B} + \bar{A}B$
AND	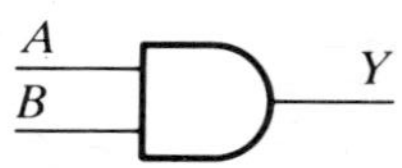	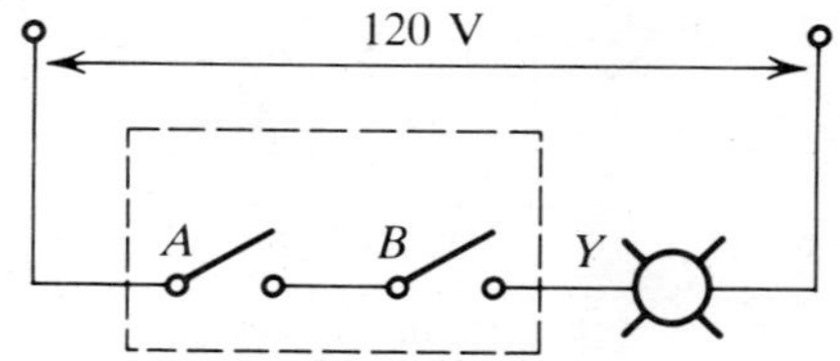 AND function	*A B Y* 0 0 0 1 0 0 0 1 0 1 1 1	$Y = AB$
NAND	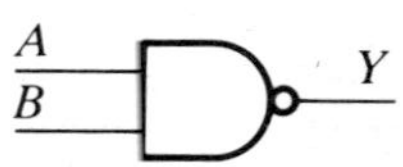	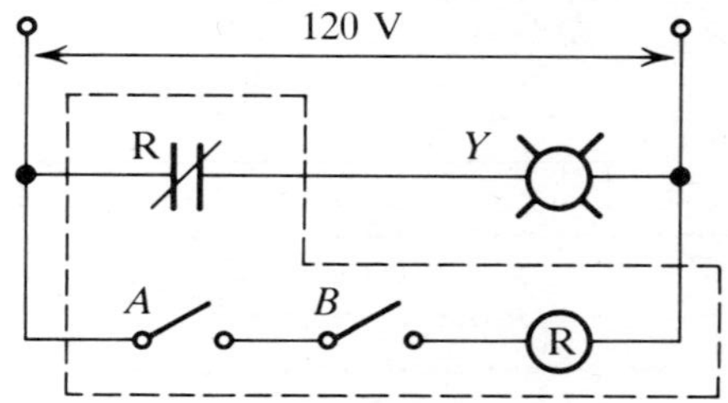 NAND function	*A B Y* 0 0 1 1 0 1 0 1 1 1 1 0	$Y = \overline{AB}$

EXAMPLE 7–11 The one-line diagram of a three-phase, 480 V load and its control schematic are shown, respectively, in Figs. 7-42(a) and (b). Draw the equivalent logic diagram of the control schematic.

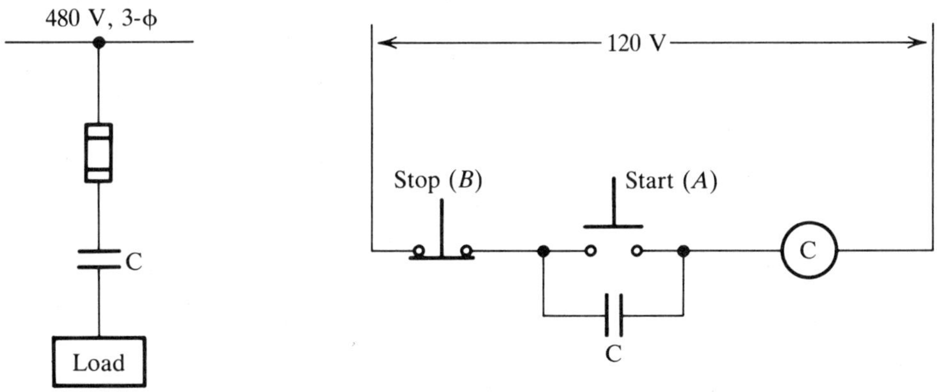

FIG. 7-42(a), (b)

SOLUTION The equivalent logic diagram of Fig. 7-42(b) is shown in Fig. 7-42(c). The nonactivated stop pushbutton is at a logic level of 0.

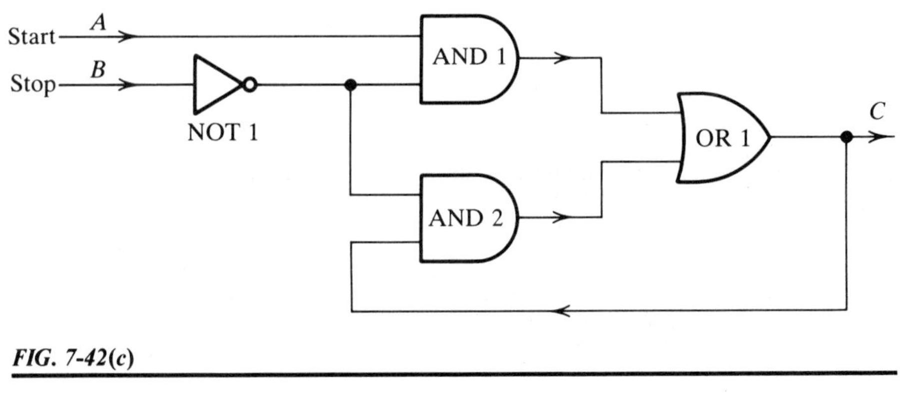

FIG. 7-42(c)

Starting

When a high logic signal is applied to A, the output of AND gate 1 is at high logic level, and thus the contactor's coil is energized and remains energized through the feedback signal and AND gate 2.

Stopping

The motor can be stopped by depressing the stop pushbutton (B is shifted to a logic level of 1). In this case the output of the two AND gates is at a logic level of 0. Thus no voltage is applied to the contactor's coil and the 480 V are removed from the load.

EXAMPLE 7–12 The logic diagram shown in Fig. 7-43(a) is the standard circuit used for nominal and intermittent (JOG) motor operation. In the latter case, when the pushbutton J is held depressed, the motor rotates.

a) Write the output–input relationships in terms of Boolean algebra notation.
b) Draw the equivalent electromechanical control schematic. (*Hint:* Use one auxiliary relay.)

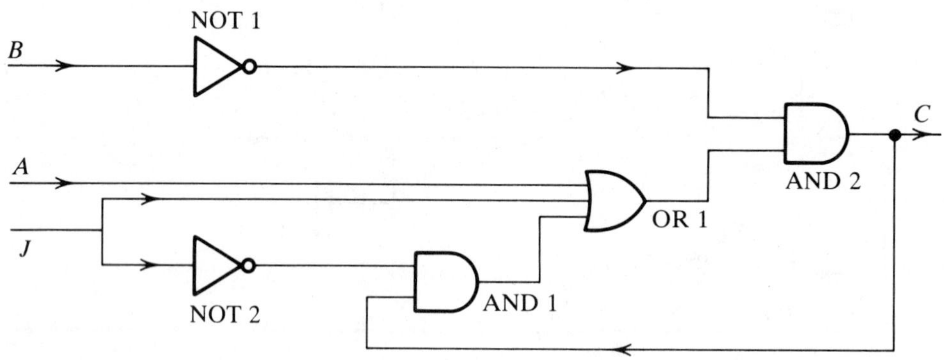

FIG. 7-43(a)

SOLUTION

a) From the characteristics of the gates we obtain

$$C = \bar{B}(A + J + \bar{J}C)$$

b) The equivalent control schematic is shown in Fig. 7-43(b).

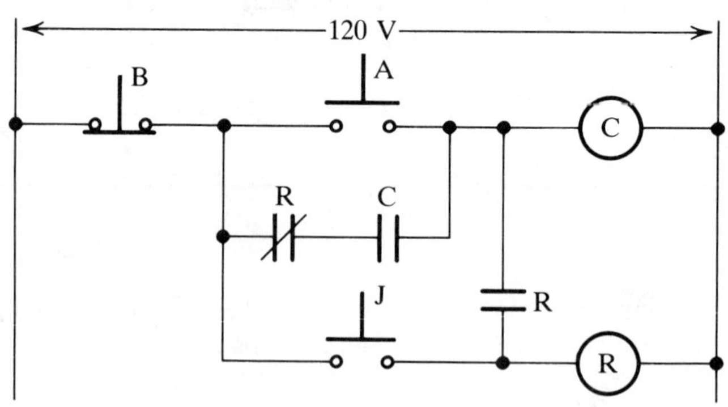

FIG. 7-43(b)

(1) When the momentary start pushbutton A is depressed, the coil C of the motor's magnetic contactor is energized. Pushbutton A is sealed in, and the motor's supply voltage (not shown in the diagram) is delivered to the motor's terminals.

(2) By removing the 120 V supply or by depressing the stop pushbutton B, the contactor's coil is deenergized, and the motor comes to a full stop.
(3) For intermittent motor operation, the motor can operate only when the start pushbutton J is held depressed.

Exercise 7–12 Prove the following De Morgan's theorems.

a) The negation of a logical product is equivalent to the logical sum of the negations (complements) of the variables comprising the logical sum. In Boolean algebra notation,

$$\overline{A \cdot B} = \bar{A} + \bar{B}$$

b) The complement of a logical sum is equivalent to the logical product of the negations of the variables making up the logical sum. In Boolean algebra notation,

$$\overline{A + B} = \bar{A} \cdot \bar{B}$$

(In the above equations the lefthand side can be implemented with two gates, while the righthand side can be implemented with three gates.)

Exercise 7–13 For the logic diagram shown in Fig. 7-44:

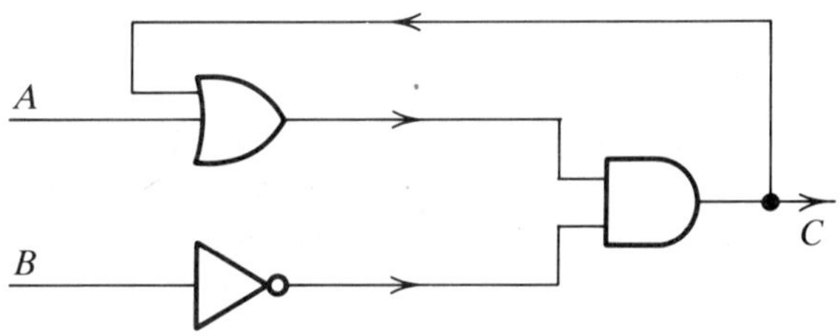

FIG. 7-44

a) Draw its equivalent control schematic.
b) Write its input–output relationship in terms of Boolean algebra notation and compare it to that of Fig. 7-42(c).

7.4.2 MEMORY DEVICES

Memory devices are capable of memorizing* digital data permanently or temporarily. That is, they can memorize bits of 1 or 0, which they deliver at their output terminals

• • •

* Programmable pocket calculators may have a memory of up to 1000 bits, while a personal calculator has about one million bits of memory. It has been estimated that the average human brain has the equivalent of 10^{14} bits of memory.

when they receive the proper combination of input signals. The fundamental unit of memory is the latch. A latch can hold, store, or memorize one bit.

When a latch's operation is synchronized to a clock signal, it is referred to as a flip-flop. In this text only latches are discussed briefly. Understanding computerized motor controls requires a basic knowledge of latches.

Latches

The general symbol of a latch is shown in Fig. 7-45. A latch has two outputs (designated as Q and $\bar{Q}$), one of which is always the inverse of the other.

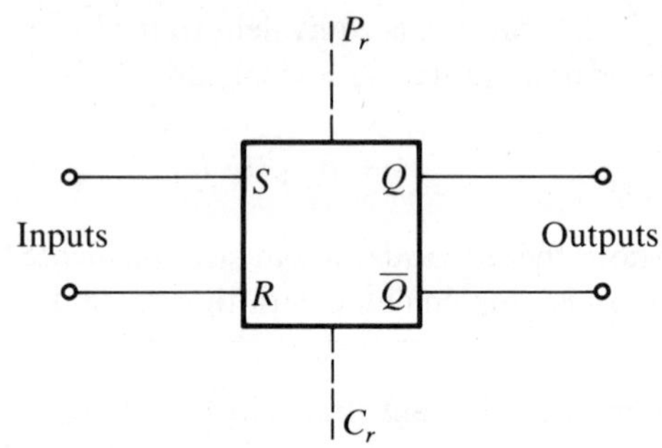

FIG. 7-45 Symbol of a latch.

The status of a latch refers to the binary level of Q. When Q is at a logic level of 1, the latch is said to be at high level; when Q is at a logic level of 0, the latch is said to be at low level. In the absence of input signals the latch is stable; that is, its outputs do not change or drift with the passage of time. The input data signals set (S) and reset (R) are used to control the status of the latch.

A latch is set or initialized through the external signals, preset (P_r) and clear (C_r). These signals are asynchronous and operate independent of the other inputs. They are shown by dotted lines in Fig. 7-45. When the latch is set at high logic level, $C_r = 1$ and $P_r = 0$. When it is at low logic level, $C_r = 0$ and $P_r = 1$. After the latch is set as desired, the initializing signals are kept at high logic level; that is, $P_r = C_r = 1$.

A latch can be made from two NOT and two NAND gates, as shown in Fig. 7-46. The outputs (Q and $\bar{Q}$) are given by

$$Q = \overline{\bar{S} P_r \bar{Q}} \qquad \textbf{(7.13)}$$

$$\bar{Q} = \overline{\bar{R} C_r Q} \qquad \textbf{(7.14)}$$

From these equations and the results of Exercise 7.12(a) we obtain

$$Q = S + \bar{P}_r + Q \qquad \textbf{(7.15)}$$

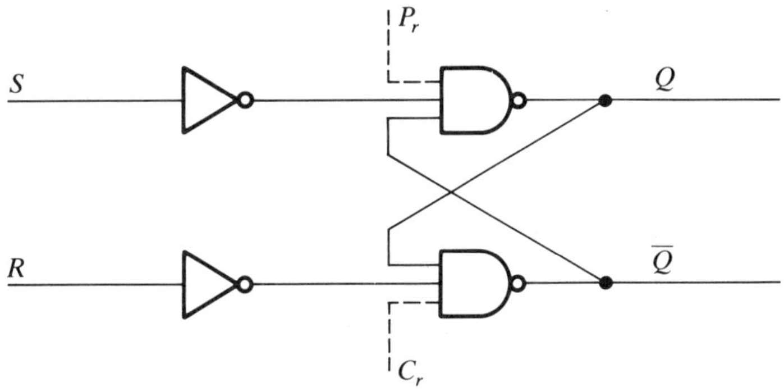

FIG. 7-46 A NOT–NAND latch.

and

$$\bar{Q} = R + \bar{C}_r + \bar{Q} \tag{7.16}$$

The truth table of the NOT–NAND latch is given in Table 7-12.

TABLE 7-12 Truth table of a NOT–NAND latch

S	R	P_r	C_r	Q
0	0	0	1	1
		1	0	0
1	0	1	1	1
0	1	1	1	0
1	1	1	1	Prohibited

Notes

1. When $S = R = 0$, the status of the latch depends on the logic level of the preset and clear signals. Then, to find out what was already stored in the latch, one has to just set the input signals equal to zero. This constitutes the memory capacity of the latch.
2. From Eqs. (7.15) and (7.16) it is evident that the condition $S = R = 1$ is prohibited, because it results in $Q = \bar{Q} = 1$, which by construction of the latch is impossible.

EXAMPLE 7–13 A motor is protected from overloads through a relay whose coil is connected in series with the motor. The motor's overload current will energize the coil of the relay, and as a result its NC contact will disconnect the motor from its supply voltage. The motor's control circuit incorporates the standard stop-start pushbuttons and the coil of the contact. Draw:

a) The logic diagram
b) The equivalent control schematic of (a) and the motor's one-line diagram.

SOLUTION

a) The logic diagram is shown in Fig. 7-47(a).

The motor is idle. The latch is set at $Q = 0$. That is,

$$\left.\begin{aligned} S &= 0 \\ R &= 0 \end{aligned}\right\} \quad Q = 0$$

The motor starts. When the start pushbutton is activated, then from the basic operation of the gates we obtain

$$\left.\begin{aligned} S &= 1 \\ R &= 0 \end{aligned}\right\} \quad Q = 1$$

This leads to the energization of the coil of the motor's contactor. As a result, the motor is connected to its voltage supply.

The motor operates. When the force that activated the start pushbutton is

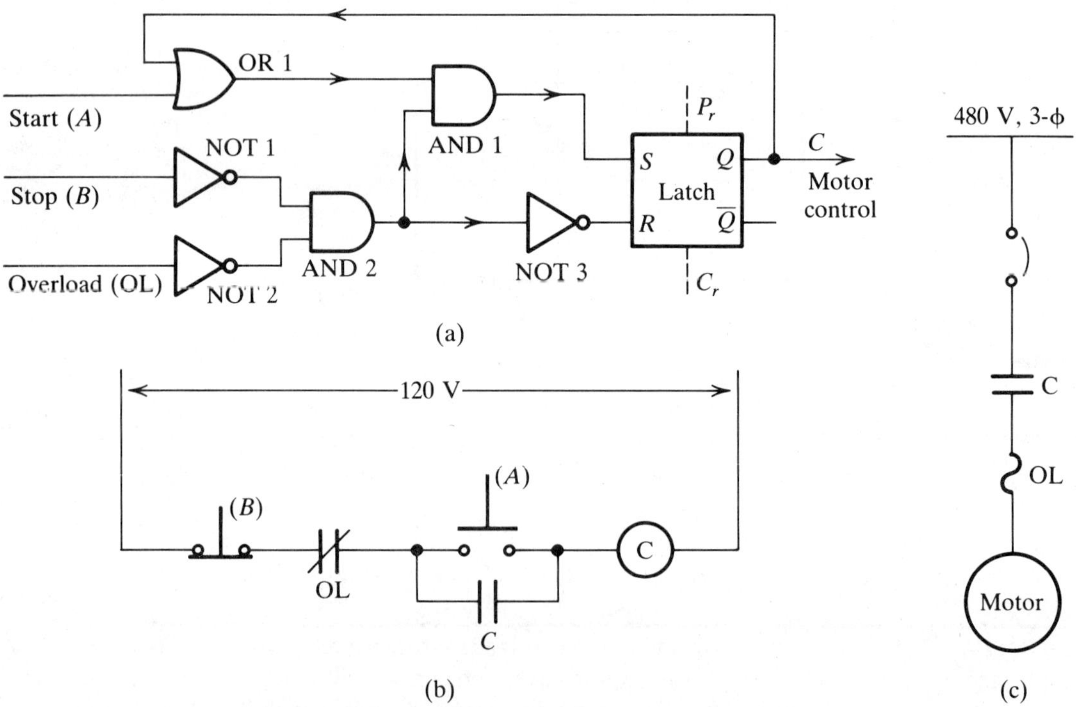

FIG. 7-47 Example 7-13: **(a)** logic diagram, **(b)** control schematic, **(c)** one-line diagram.

removed, the status of the latch ($Q = 1$) does not change, because the feedback signal maintains $S = 1$. The motor continues its operation.

The motor stops. On an undesirable condition (overload or an activation of the stop pushbutton) the output of the AND gates 1 and 2 becomes 0 and thus

$$S = 0$$
$$R = 1$$

The output of the latch is then

$$Q = 0$$

The motor stops and will not start till the undesirable condition is removed.

b) The motor's control schematic and one-line diagram are shown, respectively, in Figs. 7-47(b) and (c).

Exercise 7–14 Referring to Fig. 7-48, design the logic diagram of the black box so that the following truth table is satisfied. (*Hint:* A feedback signal may be used.)

Input		*Output*
A	*B*	*Q*
0	1	0
0	0	1
0	1	1
1	1	0

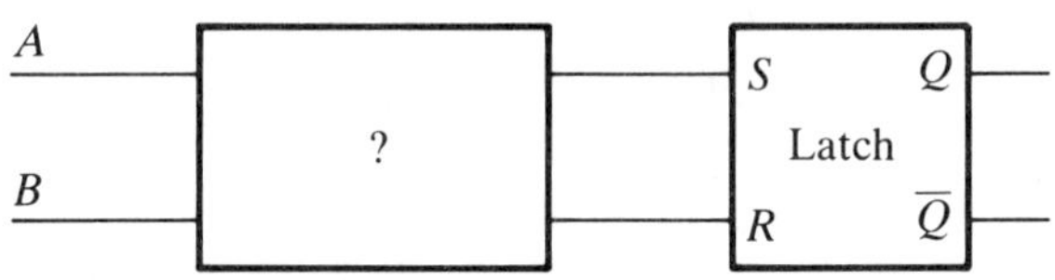

FIG. 7-48

7.4.3 PROGRAMMABLE CONTROLLERS

The programmable controller (PC) has, as shown in Fig. 7-49(a), three distinct sections: the central processing unit (CPU), the input module, and the output module. The input module constitutes the interface between CPU and the incoming field signals, while the output module constitutes the interface between the CPU and the outgoing field signals. The CPU is easily programmed to function as a desired control schematic with many features, such as ordinary relays, timers, and flip-flops.

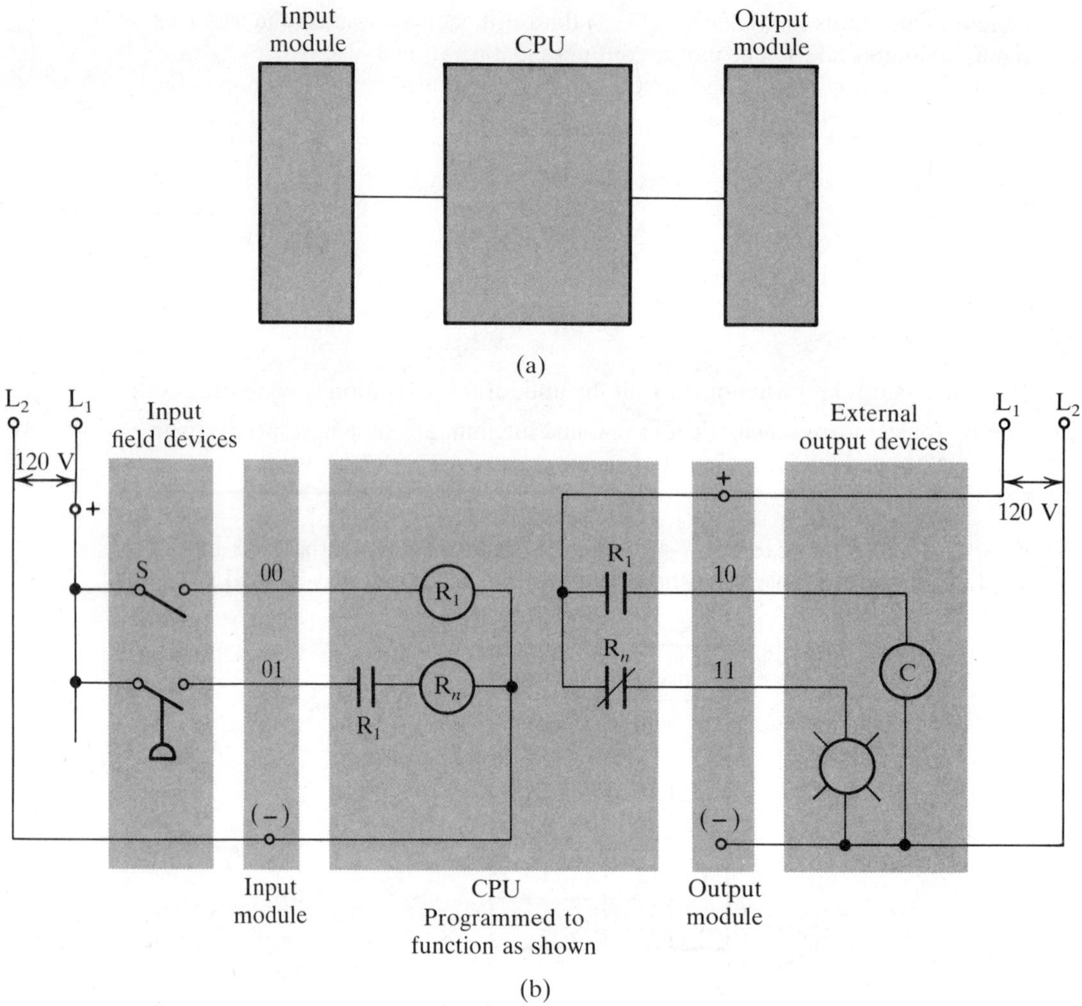

FIG. 7-49 Programmable controller: **(a)** block diagram, **(b)** hook-up.

A typical hook-up of a PC is shown in Fig. 7-49(b). When the field limit switch S closes, the coil of relay R_1 is energized, and as a result the coil of the contactor C is energized. If the pressure switch is also closed, the annunciating light is switched OFF. The relays R_1 and R_n and their contacts are set up in the CPU section by following the manufacturer's programming instructions.

Owing to its appearance, the programmed control schematic is often referred to as a ladder diagram.

7.5 Conclusion

The control schematics of a plant constitute the interface between the engineer and the plant's electrical equipment. Using conventional symbols, they reveal the modc of operation and protection of each electrical apparatus. Understanding and interpreting control schematics is simple, but their overall design requires an in-depth knowledge of the equipment's operation and protection. The latter is a specialty in itself.

Conventionally, control schematics show start and stop pushbuttons, magnetic contactors, relays, fuses, circuit breakers, voltage and current transformers, and many other auxiliary control devices whose complexity and worth depend on the relative importance and cost of the controlled equipment.

By far the most widely used control devices are magnetic contactors.

Thermal overload relays are included in almost all motors rated below 200 kW. Fuses or circuit breakers generally provide short-circuit protection. The more expensive circuit breakers are tripped through protective relays, and thus they can provide additional protection from ground faults, unbalanced voltages, and so on.

Each type of motor, depending on its application, has a particular control schematic. In most cases the control schematics are standard and can easily be found in manufacturers' catalogues. However, since each process has its own peculiarities, the design engineer must use the most economical and effective control alternative.

The control schematics of single-phase induction motors are discussed in Section 7.3.1. A "start capacitor" is used whenever it is desirable to increase the motor's starting-torque capability. This capacitor must always be connected in parallel to a discharge resistor, which dissipates the energy stored in the capacitor's electric field at the instant the capacitor is disconnected from the circuit. The regenerative braking of a single-phase motor is used to minimize the stopping time of the motor. This is accomplished by applying an unregulated dc voltage to the motor's winding.

The controls for a squirrel-cage motor are quite simple. Those of the wound motor are rather more complex, because their design requires the selection of the rotor's external resistors, which depend on the moment of inertia of the load and motor.

The control schematics of three-phase synchronous motors are unique in that they require a thorough knowledge of the motor's operation and a knowledge of special types of protective devices. The control schematics of single-phase and three-phase induction motors are quite simple and have been in use for a long time, while the control schematics of synchronous machines and dc drives are relatively new. These machines are started and controlled through solid state devices which, since the 70s, have completely replaced their electromechanical counterparts.

The basic components of solid state control devices are the gates and the memory devices. These solid state components are the basic building blocks of computers and programmable controllers. The latter can easily be programmed to function as either simple or complex control networks. Their potential is unlimited in the field of motor control.

Control schematics have evolved considerably. Electromagnetic devices (*electrical*) were replaced by solid state devices (*electronic*), which in turn have been supplanted by *microprocessor* controls. The basic difference between electronic and microprocessor controls is that the latter can easily be programmed to provide not only switching functions but continuous control of all parameters of a given process as well.

The various symbols of the electromechanical devices and of the logic circuits are shown in Tables 7-13 and 7-14, respectively.

TABLE 7-13 Summary of symbols for control schematics

Symbol	Description	Notes
a. 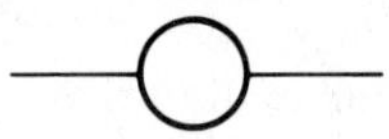	Voltage coil of a relay, electromagnetic starter, timer, or the like.	
b. 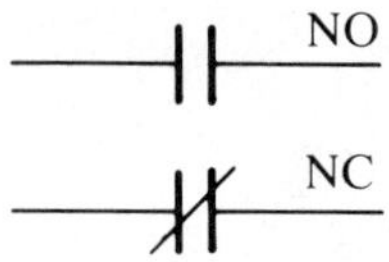	Instantaneous contacts: normally open (NO) and normally closed (NC)	
c. 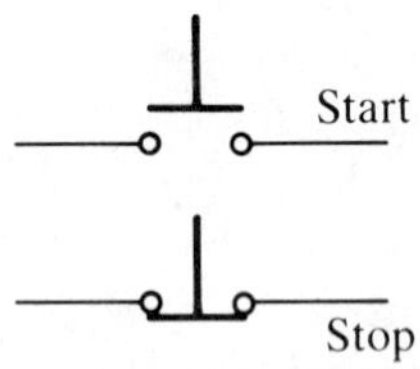	Momentary start and stop pushbuttons	
d. 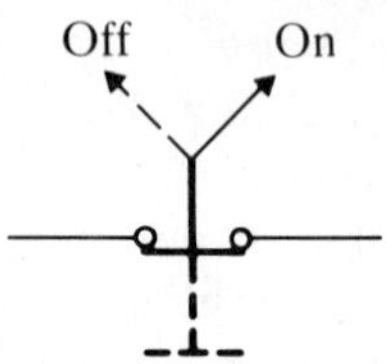	Selector switch	
e. 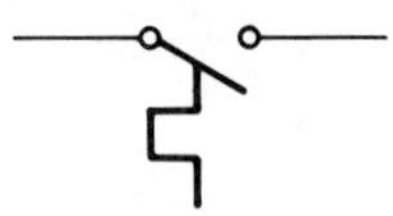	Contact of a temperature sensor	
f. 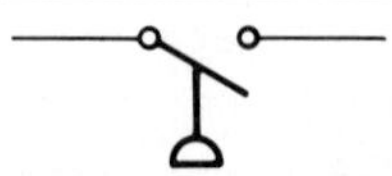	Contact of a pressure sensor	
g. 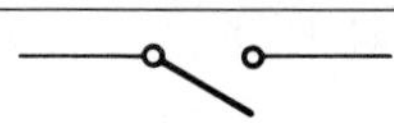	Limit switch	
h. 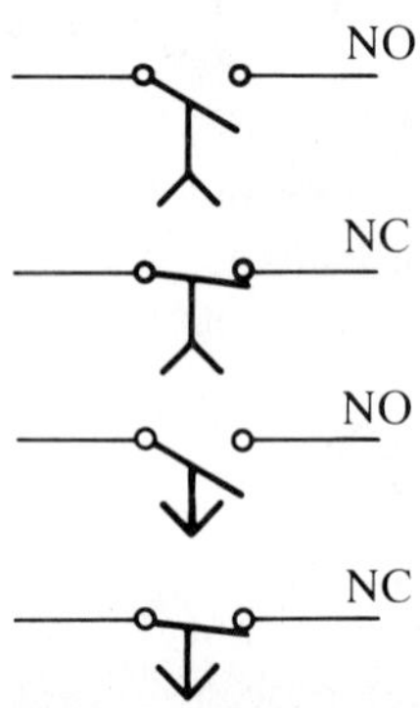	Time contacts. Contact action is retarded when the coil is: Energized Deenergized	When a dot is placed adjacent to a time-delay contact, it implies a cyclic variation of its status while the coil is maintained energized.

(Continued)

TABLE 7-13 *(Continued)*

Symbol	Description	Notes
i. 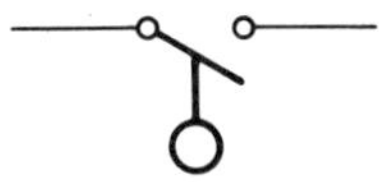	Float switch	
j. 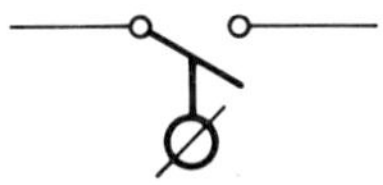	Contact of a relative-humidity sensor	
k. 	Disconnect	
l. 	Fuse	
m. 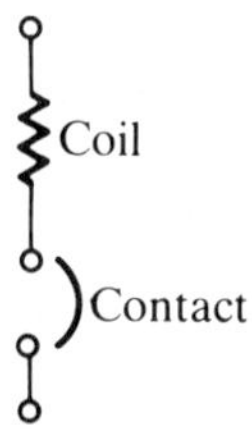	Molded-case circuit breaker	
n. 	Circuit breaker, operated through relays	
o. 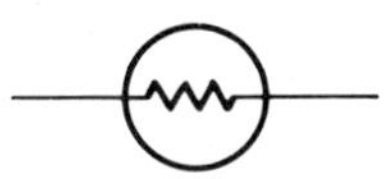	Solenoid	
p. 	Pump	
q. 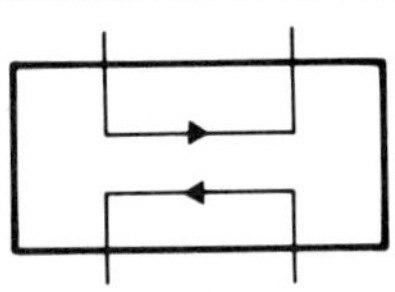	Heat exchanger	

TABLE 7-14 Symbol conversion list

	MIL	*Metric*
NOT	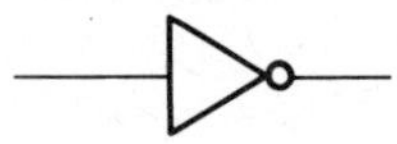	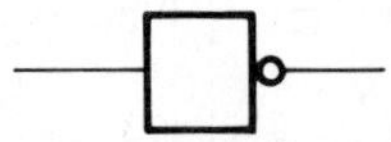
AND	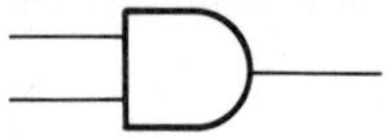	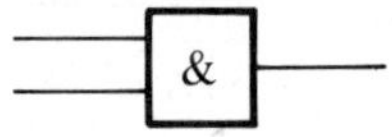
OR	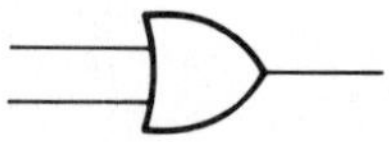	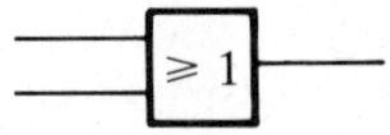
NAND	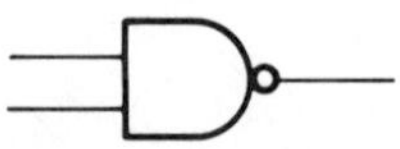	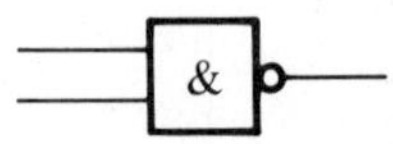
NOR	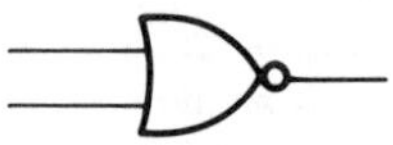	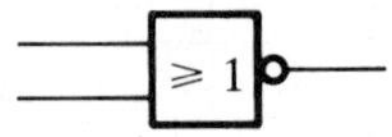
Timer		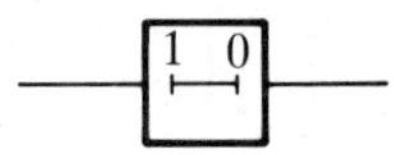

7.6 REVIEW QUESTIONS

1. What is the difference between a control schematic of a motor and its wiring diagram?
2. Draw a single-pole, single-throw and a single-pole, double-throw switch.
3. Explain what a thermal overload relay is.
4. Explain the differential range of a sensor.
5. Differentiate between a thermal overload relay and a fuse or low-voltage circuit breaker.
6. Draw a typical thermostat and explain the significance of the terminals, X, R, Y, and W.
7. What is a "seal-in" contact?
8. Explain the concept of regenerative braking as applied to ac machines.
9. Why does a change in the volts/turn ratio have a more drastic effect on the speed of a 1-ϕ induction motor than on a 3-ϕ ac machine?
10. What is the difference between an overload and an overcurrent condition?
11. What do the protection numbers 50 and 51 signify?
12. Draw the starting and damage characteristics of a motor and of the motor's protective devices on the same time—current base.

13. Differentiate between LV and MV levels.
14. Describe the starting of a brushless synchronous motor.
15. What is the function of the ventilating fan shown in the control schematic of a dc motor?
16. Explain the function(s) of a gate and of a latch.
17. How many latches are required to represent all the possible combinations of the alphanumeric characters?

7.7 PROBLEMS

7-1 Draw a 24 V control schematic that includes two pushbuttons and a NO pressure switch, which upon activation illuminates a light and provides an audible alarm. One of the push butttons is used to silence the audible alarm, and the other is used to determine if the indicating light is burned out.

7-2 Draw a 120 V control schematic that incorporates a manual reset and a NO temperature contact, which upon activation energizes a fail-safe solenoid valve and an annunciator light.

7-3 Figures P7-3(a) and (b) show, respectively, the control schematic and the one-line diagram of a 3-ϕ, 480 V motor.

a) Modify the control schematic so that the motor can be started and stopped from a remote control station.

b) Explain why it is improper to wire the contacts of the overload relay as shown. (*Hint:* Consider what would happen to the motor if the overload contacts opened while the coil of the magnetic contactor was accidentally grounded.)

c) If the pickup value of the starter's coil is 80% of its nominal voltage, what is the maximum permissible length of the wire (size #14, $R = .0085\ \Omega/\text{m}$) that connects the remote interlock to the control circuit? The coil's starting current is $2\angle{-70^\circ}$ A

Answer 2723.48 m.

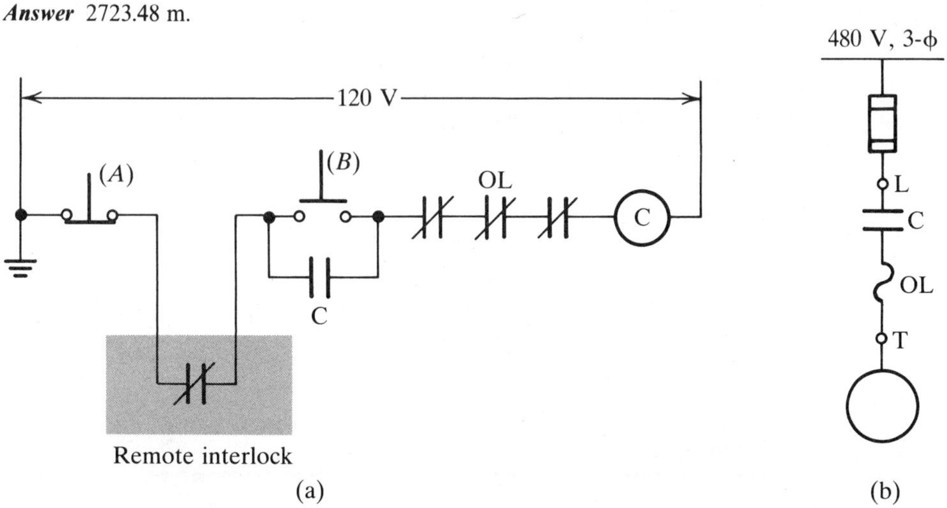

FIG. P7-3

7-4 Referring to the control schematic of Fig. P7-3:

a) If the supply voltage is removed and then reapplied, will the motor start automatically? Explain.

b) When the motor's current increases above its rated value, the contacts of the thermal overload relay will be opened and thus the motor will stop. Will depressing the start pushbutton start the motor? Explain.

c) The number one danger to a motor is its frequent starts. Modify the control schematic so that the plant's operators or electricians will not be able to restart the motor within 10 min after it has been stopped.

7-5 A pump is used to circulate and pressurize a compressor's lubricating oil. The oil flow system is equipped with a pressure switch, which will shut down the compressor–pump system each time the pressure of the lubricating oil is reduced to an undesirable level. The drive, however, can start and continue to operate at low pressures for 40 s. If, in the meantime, the pump is unable to maintain the oil's pressure at nominal operating level, the entire system is switched OFF. Draw a control schematic that satisfies these requirements.

7-6 Referring to the basic circuit diagram of a brushless synchronous motor (Fig. 7-30(b)).

a) Identify the path of the current through which the motor's field current flows during the accelerating period, and explain the reasons for the thyristor's conduction.

b) Estimate the maximum permissible line-to-line output voltage of the rotating armature if the Zener's nominal voltage is 16 V.

c) Explain why the resistor R_1 increases the motor's starting torque.

7-7 Draw a control schematic for a brushless synchronous motor that includes a dc contactor for motor control, a winding heater, a ventilating fan, three overload relays, an overtemperature bearing protection, and a contact that is derived from a rotor-vibration detector.

7-8 The diagram shown in Fig. P7-8 is used to control the operation of a motor. The motor is equipped with an internal thermal overload, and the start-stop pushbuttons are of the instantaneous type.

a) Determine the elements of the logic circuit and draw the motor's equivalent control schematic and one-line diagram.

b) Why are the latch and the motor's contactor interfaced through an auxiliary relay?

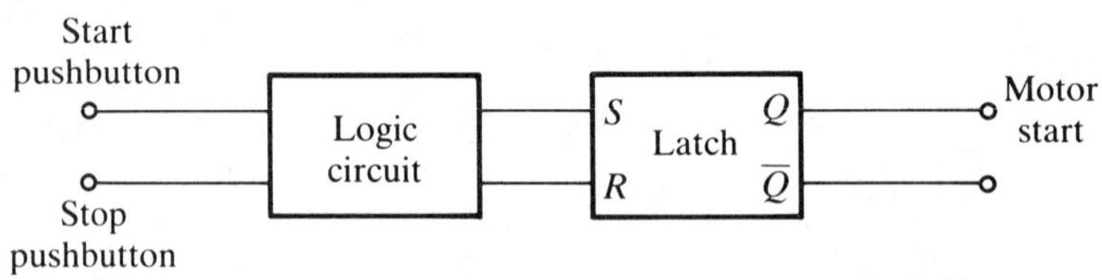

FIG. P7-8

7-9 Draw the basic control schematic for a dc drive that incorporates relays whose contacts will annunciate the following:

a) The drive is ready to run (the power is ON).

b) The drive is running.

c) The drive has stopped because of motor's overheating.

7-10 Figure P7-10 shows a typical control schematic for a shunt dc motor.

a) Explain the circuit's operation and draw a typical current-versus-time motor characteristic.

b) Describe the function(s) of resistors R_x, R_a, and R_f.

c) Why are two different fuses (FU_1 and FU_2) used?

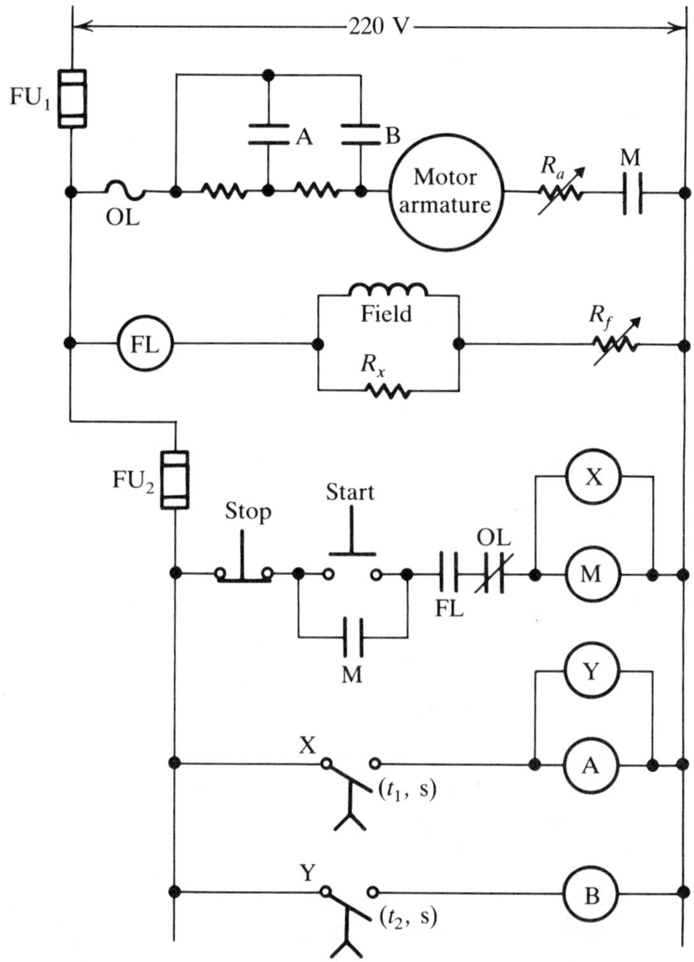

FIG. P7-10

7-11 Two pressure sensors each with a NO contact, are used to control the operation of two pumps, M_1 and M_2. The characteristics of the sensors are as follows:

Sensor	*Pressure Pick-Up in Psi*	*Differential Range in Psi*
P_1	30	10
P_2	35	10

The operation of the pumps as a function of pressure is as shown in Fig. P7-14. Design the electromechanical and logic equivalent control schematic to satisfy the system requirements.

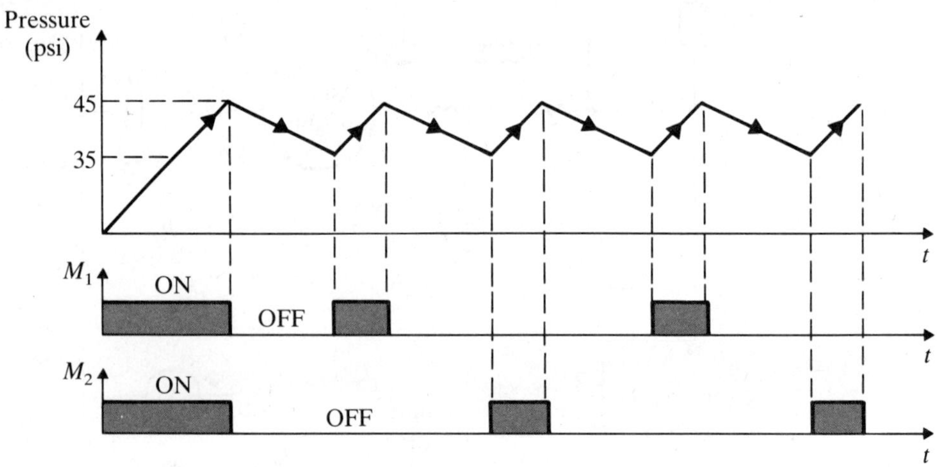

FIG. P7-11

7-12 The logic diagram of Fig. P7-12 represents a section of a process alarm circuit. Under normal operating conditions the field contact C is closed. The alarm reset (AR) is a switch that, when activated, produces a high logic level. Following the rectification of an undesirable field condition, the AR and the latch are reset to their nominal status.

a) Describe the circuit's operation.

b) Draw the equivalent control schematic.

7-13 The circuit of Fig. P7-13 is part of the elementary logic diagram that is used for the computerized control of the motors in a multimachine plant. The NO contact C is controlled from the coil of the motor's starting contactor. The latch is initialized (by means not shown in the diagram) each time the motor comes to a full stop. Describe the motor's sequence of operation and control.

7-14 Figure P7-14 shows, in a simplified version, the chronological events and the operators' actions that led to the Chernobyl nuclear accident. The switches are manually operated and each relay's coil, when energized, implements through a circuit (not shown in the diagram) the function as described adjacent to the coil. The accident, of course, took place because the plant was producing power while the operators performed a series of steps that were permissible when and only when the reactor's output power was negligible.

a) Modify the control schematic to incorporate an interlock that would not permit the operators to do what they did while the reactor was apparently still producing power. The interlock must be a contact whose status is controlled by the temperature of the reactor.

b) Draw the logic diagram equivalent to (a).

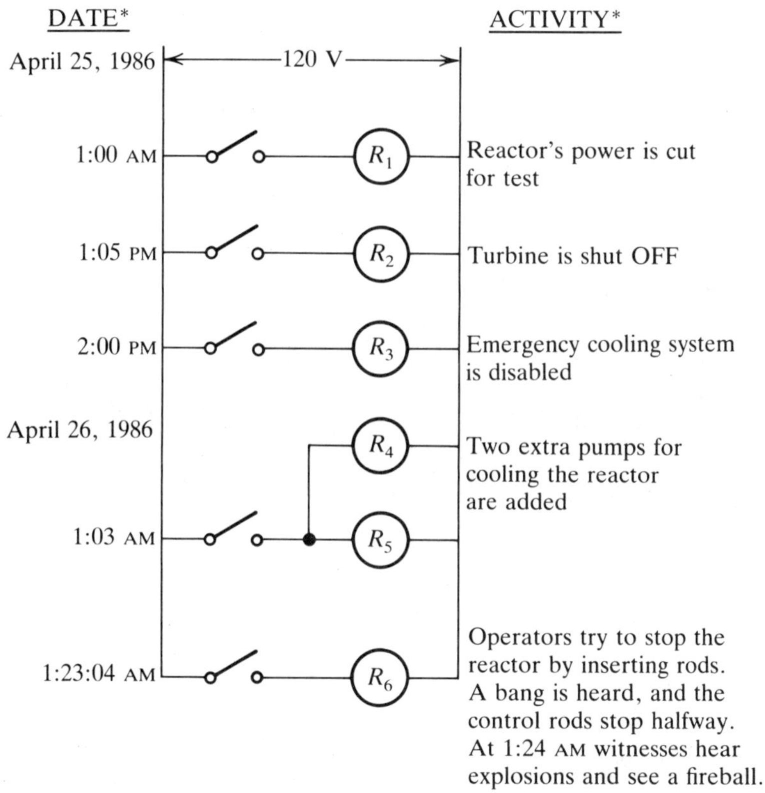

FIG. P7-14

* Courtesy of the New York Times.

7-15 Figure P7-15 shows a typical section of a process flow sheet for rock crushing in a cement plant. The system requirements are as follows:

1. Each motor should have a local JOG/stop pushbutton.

2. Motors should start in a sequence that is the reverse of the material flow. The time interval between the sequential starting of motors is about 15 s.

3. The operation of each motor's light should be as follows:
When a motor is not running, its light will be OFF.
When a motor is running, its light will be ON.
When an overload condition occurs, the motor's light will be flashing. By depressing the alarm reset, the light will be switched OFF.

a) Design a *control schematic* that will satisfy the system's requirements.

b) Draw the equivalent logic diagram of (a).

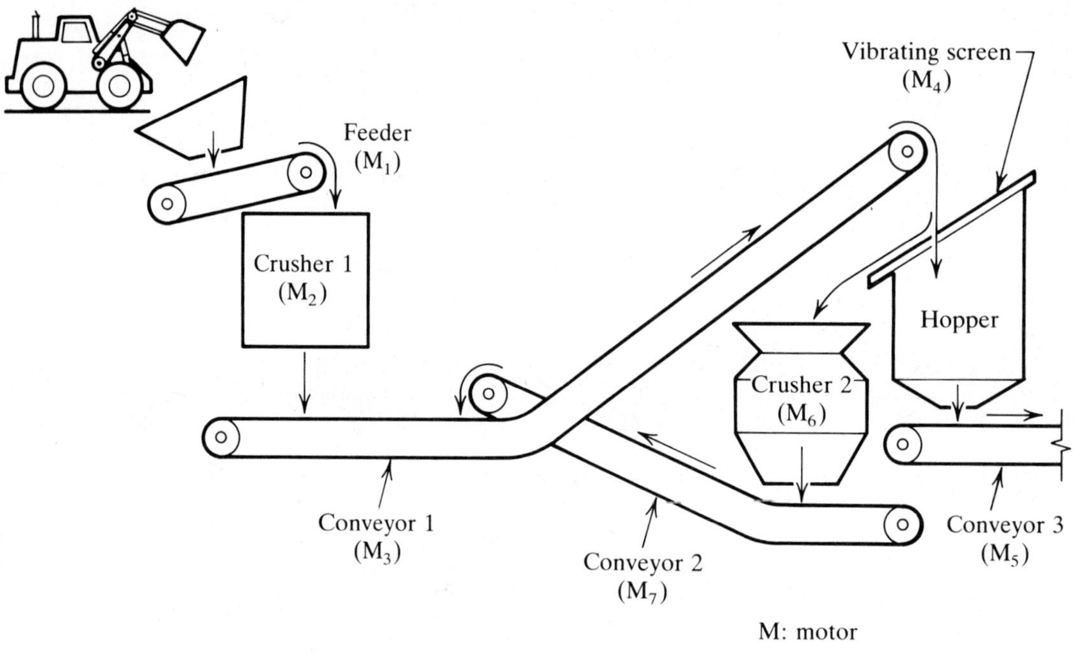

FIG. P7-15 Rock-crushing process flow sheet.

7-16 Figure P7-16 shows the essential components of a ground-source heat-pump system that is used for cooling–heating and for supplying the hot-water demand of a household. Heat is extracted from the ground (heating season) or injected to the ground (cooling season) through a closed loop of underground plastic pipe that carries brine liquid (antifreeze). The starting of the hot-water pump is through a temperature sensor located on the hot-water tank. This sensor has a contact that is NC for all temperatures below 110°F and an additive differential range of 2°F.

a) Draw the one-line diagram of the electrically operated devices and design the control schematic that will satisfy the requirements of the logistics table and also the following: The compressor must start 30 s after the brine pump, and on hot-water demand the water's temperature increases until the refrigerant's high-pressure switch (set at the equivalent of 180°F) is activated.

b) Draw the equivalent logic diagram of (a).

Logistic Table

			Motors				
Season	*Demand*	*Thermostat's Output Signal at*	*Brine Pump 1*	*Compressor 2*	*Reversing Valve 3*	*Fan 4*	*Hot-Water Pump 5*
Heating	Heating	Y, G	ON	ON	OFF	ON	OFF
	Hot water	G	ON	ON	OFF	OFF	ON
	Heating and hot water	Y, G	ON	ON	OFF	OFF	ON
Cooling	Cooling	Y, G, 0	ON	ON	ON	ON	OFF
	Hot water	G, 0	ON	ON	OFF	OFF	ON
	Cooling and hot water	Y, G, 0	OFF	ON	ON	ON	ON

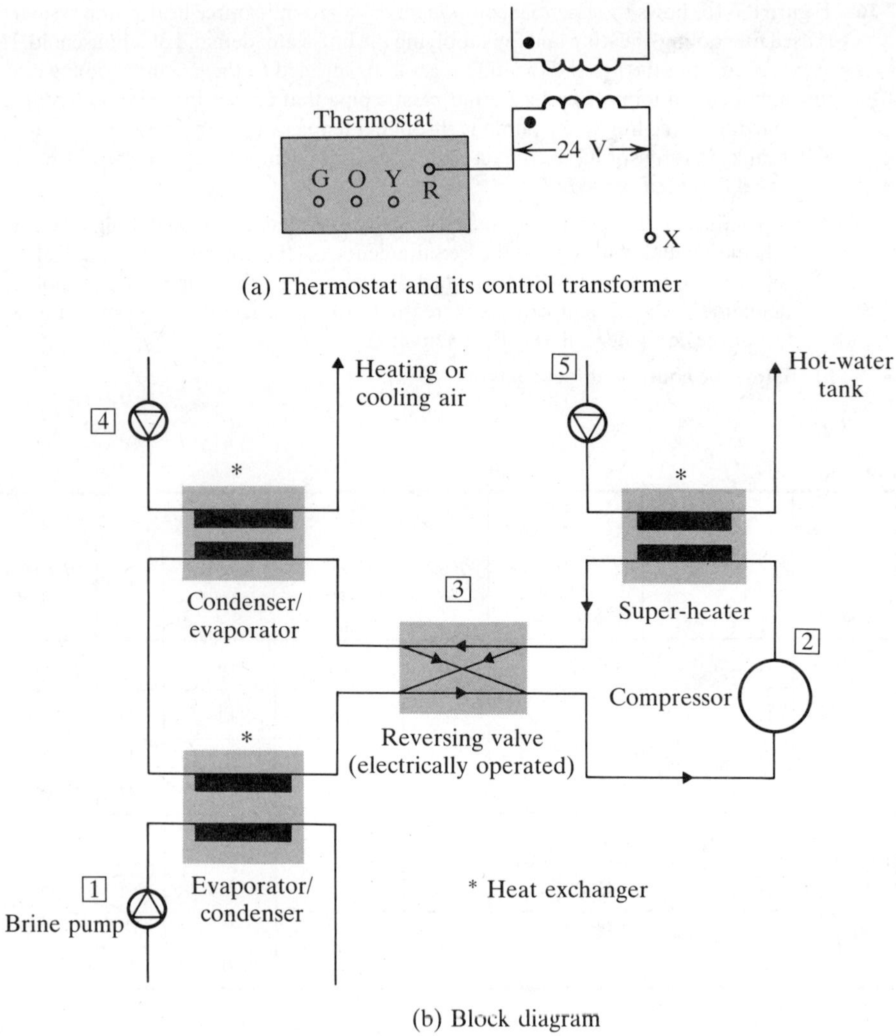

FIG. P7-16 Ground-source heat pump.

7-17 The control schematic shown in Fig. P7-17 is used to operate a 480 V motor at three different speeds: slow forward (SF), 900 r/min, fast forward (FF), 1800 r/min, and slow reverse (SR), 900 r/min. The time-delay relays (TD_1, TD_2, and TD_3) are of the "OFF-delay" type, and each has an adjustable time range of 20 to 200 s. The motor's internal winding connections (T_1, T_2, T_3, T_4, T_5, and T_6) are not shown. Explain the operation of the circuit, and describe how to estimate the setting of the time-delay relays.

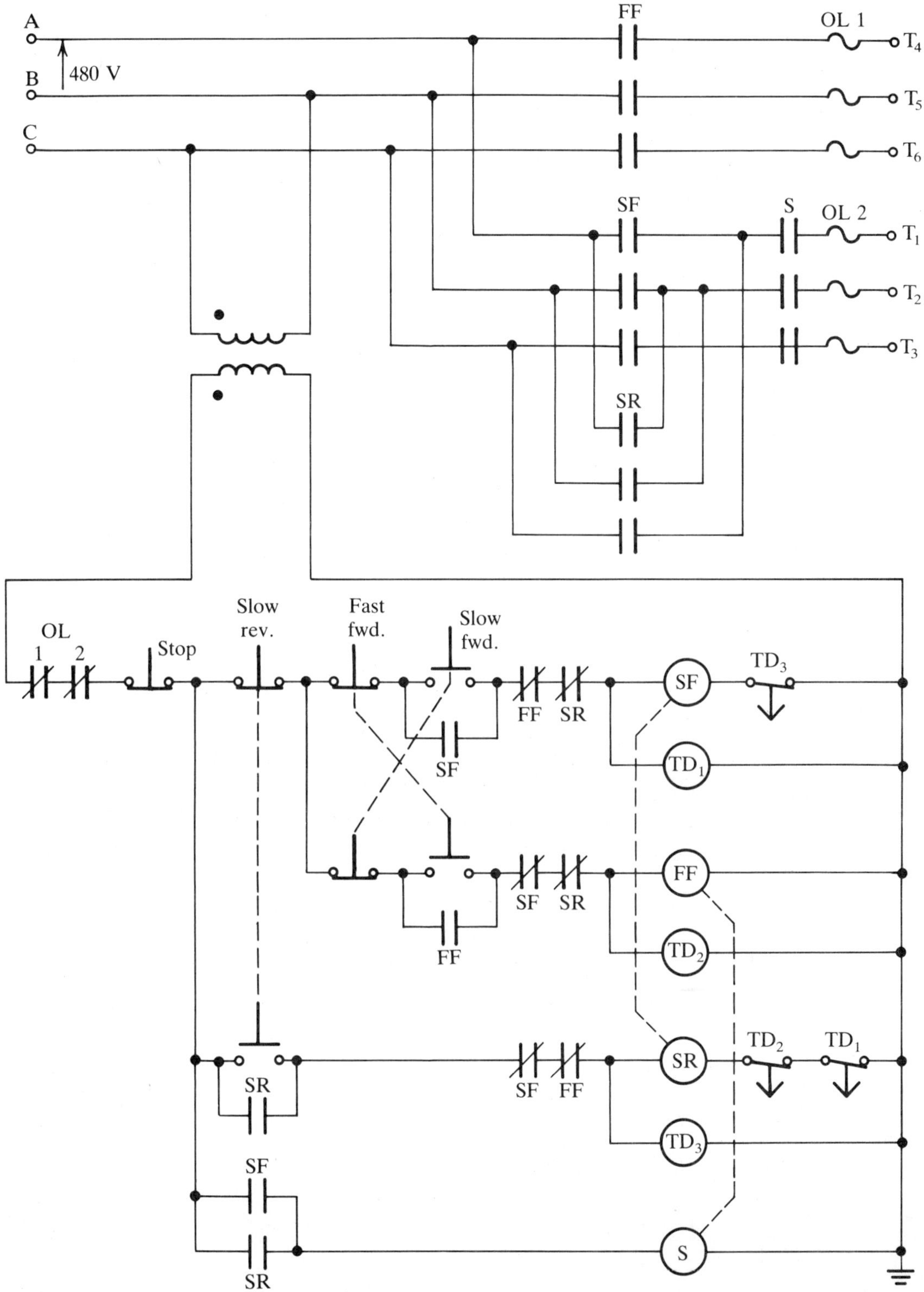
A
B
C
480 V
FF
OL 1
T_4
T_5
T_6
SF
S
OL 2
T_1
T_2
T_3
SR
OL
1
2
Stop
Slow rev.
Fast fwd.
Slow fwd.
FF
SR
SF
SF
TD$_3$
TD$_1$
SF
SR
FF
FF
TD$_2$
SR
SF
FF
SR
TD$_2$
TD$_1$
TD$_3$
SF
SR
S

FIG. P7-17

Appendixes

Introduction

The Appendixes include discussions of three-phase systems, per-unit values, Laplace transforms, solid state devices, and the formulas that are necessary for the economic evaluation of equipment. Many of these topics are covered in detail in other courses, for which the student may have already earned credit. Nevertheless, they are briefly discussed here in order to better prepare the student for this course and to further enhance his or her understanding of electric machines.

The Appendixes also include tables that give information about units and protective devices.

A THREE-PHASE SYSTEMS

A three-phase generator and its connected load are shown in Fig. A-1(a). The advantages of 3-ϕ systems over single-phase ac or dc transmission systems have mainly to do with economics and flexibility. The information given below is essential to an understanding of 3-ϕ systems.

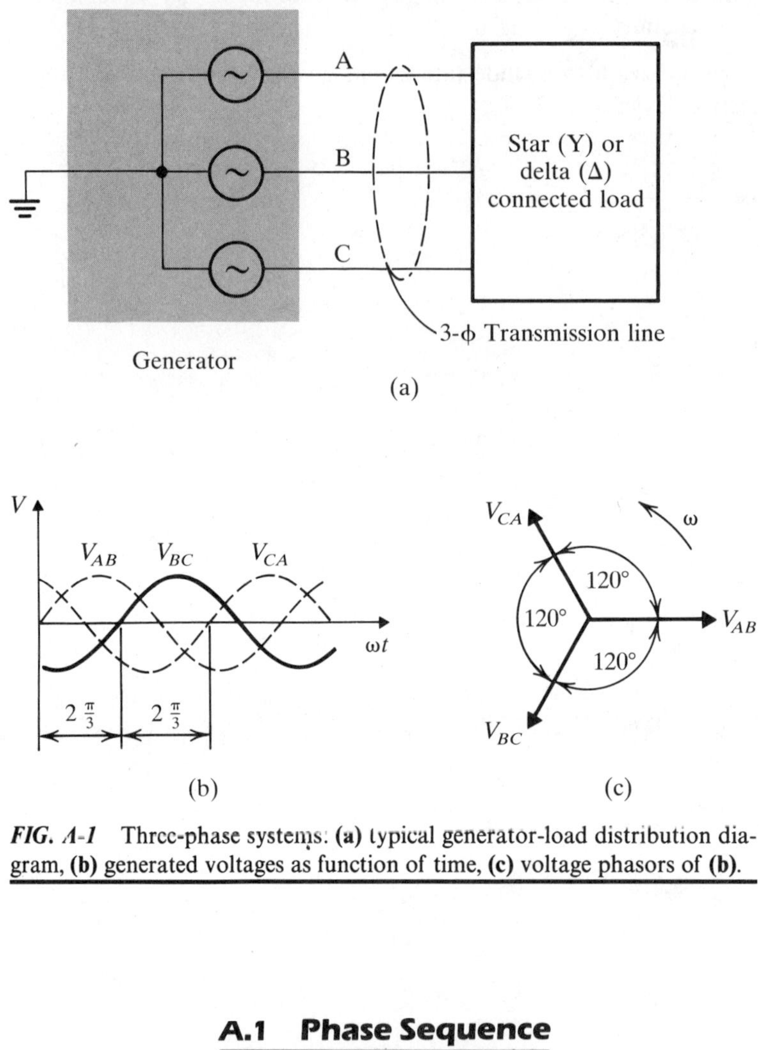

FIG. A-1 Three-phase systems: **(a)** typical generator-load distribution diagram, **(b)** generated voltages as function of time, **(c)** voltage phasors of **(b)**.

A.1 Phase Sequence

The three line-to-line voltages (V_{AB}, V_{BC}, and V_{CA}) produced at the terminals of a generator are normally balanced; each voltage phasor is equal in magnitude to the others but displaced 120° from the others in phase. The three balanced line-to-line voltages can be represented either by their instantaneous values or by their rotating phasors, as shown in Figs. A-1(b) and (c), respectively.

The angular speed of rotation of the voltage phasors for a 60 Hz system is

$$\omega = 2\pi f = 2\pi(60) = 377 \text{ rad/s} \tag{A.1a}$$

In electrical engineering practice, the direction of rotation of phasors in the complex plane is counterclockwise.

The phase sequence of a given system of voltages can be determined by considering its instantaneous waveforms or the equivalent phasor diagrams. As shown in Fig. A-1(b), the positive maximum value of the voltages occurs first in phase A and then, successively, in phases B and C. In such cases it is said that the order of rotation of the three-phase voltages is ABC. If the positive maximum value occurs first in phase A and then in phases C and B, the order of rotation of the three-phase voltages is said to be ACB. The order of rotation is often referred to as the *phase sequence*.

Two possible phase sequences, and their directions of rotation, are shown in Fig. A-2. For phase sequence ABC or ACB, the phase voltages rotate in the same counterclockwise direction; when applied to a three-phase motor, however, the phase sequence ABC produces a rotating field in the direction opposite to that produced by the phase sequence ACB. Under normal operating conditions the motor rotates, and produces torque, in the same direction as the predominant rotating field. On account of this, when three-phase motors are installed or new plants are commissioned, the phase sequence of the supply voltages is measured and checked against the requirements of the equipment before any three-phase field interconnections take place.

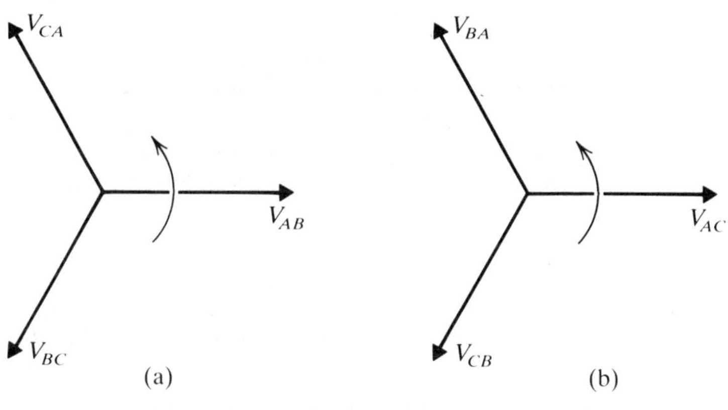

FIG. A-2 Phase sequences: **(a)** ABC, **(b)** ACB.

For a mathematical analysis of the balanced operation of motors and other three-phase symmetrical loads the phase sequence of the applied voltages is not very important. In such cases we customarily do the analysis per phase, and which voltage follows the other does not come into consideration. However, when the loads or the applied voltages are unbalanced, the phase sequence is very important, because the solutions of the various problems are based entirely on the phase sequence of the applied voltages.

VOLTAGE UNBALANCE*

The voltage unbalance is calculated as follows:

$$\text{voltage unbalance} = \frac{V_{\text{max}} - V_{\text{av}}}{V_{\text{av}}}(100) \quad \textbf{(A.1b)}$$

where V_{max} and V_{av} are the maximum and average rms values of the line-to-line voltages, respectively. For example, if the rms values of the line-to-line voltages in a 3-ϕ, 208 V power distribution system are 199 V, 209 V, and 213 V, then

$$V_{\text{max}} = 213 \text{ V}, \qquad V_{\text{av}} = \frac{199 + 209 + 213}{3} = 207 \text{ V}$$

and

$$\text{voltage unbalance} = \frac{213 - 207}{207}(100) = \underline{\underline{2.9\%}}$$

Some disadvantages of unbalanced voltages are overheating of motors, generation of harmonics, vibrations, and increased copper losses. A general rule is that the percentage *increase* in a motor's copper losses is equal to the voltage unbalance.

A.2 Star-Delta Connected Systems

Three-phase generators and their loads can be connected in either a star (Y) or a delta (Δ) configuration. The essential characteristics of star- and delta-connected systems are discussed in the sections that follow.

STAR-CONNECTED SYSTEM†

Three-phase, star-connected systems are very popular because of the advantages of earthing the neutral point and of supplying power to line-to-neutral loads. Three-phase, 4-wire distribution systems—common to most plants—require a star-connected supply.

A star-connected load is shown in Fig. A-3(a). By inspection, the line current (I_L) is equal to phase current (I_P). That is,

$$I_L = I_P \quad \textbf{(A.2)}$$

. . .

* According to recent published surveys of 83 utilities, 43% allowed voltage unbalance in excess of 3%, while 30% allowed voltage unbalance of 5% or higher.

† Star-connected systems are sometimes referred to as *wye* (Y) or *tee* (T) connected systems.

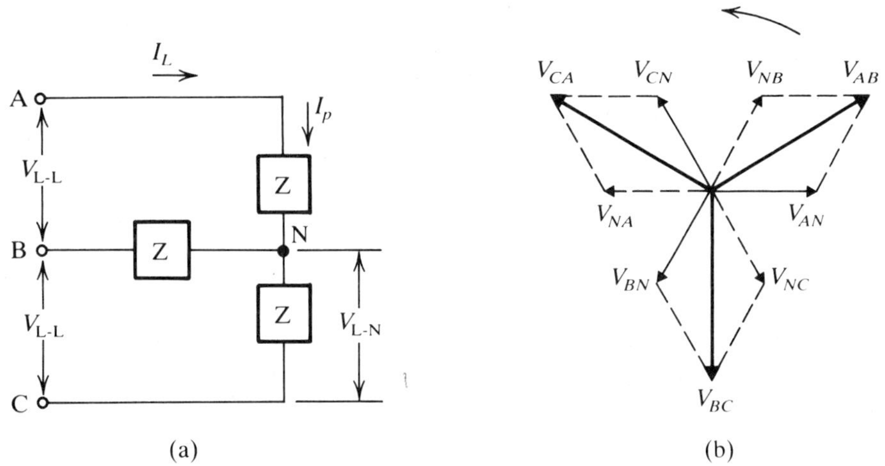

FIG. A-3 **(a)** A star-connected load, **(b)** phasor diagrams of line-to-line and line-to-neutral voltages for phase sequence ABC.

The line-to-line voltage is larger than its corresponding line-to-neutral voltage by a factor equal to $\sqrt{3}$, and leads it by 30°. Mathematically,

$$V_{\text{L-L}} = \sqrt{3}\, V_{\text{L-N}} \angle 30^\circ \qquad \textbf{(A.3)}$$

Equation (A.3) can easily be verified by first drawing the three-balanced line-to-neutral voltages and then using KVL ($V_{AB} = V_{AN} + V_{NB}$).

Each line-to-line voltage, as seen in Fig. A-3(b), has two line-to-neutral components, *one* that leads the line-to-line voltage by 30°, the other that lags it by 30°. Then the question arises which phase voltage is represented by Eq. (A.3).

The phasor diagram of phase sequence ABC makes it clear that V_{AB} leads V_{AN}, V_{BC} leads V_{BN}, and V_{CA} leads V_{CN}. If the phase sequence is ACB, then V_{AC} will lead V_{AN} by 30°, and so on.

DELTA-CONNECTED SYSTEMS

A delta-connected load and its phasor diagram are shown in Figs. A-4(a) and (b), respectively. The line-to-line voltages are equal to the phase voltages, while the line currents are larger than the phase currents by a factor equal to $\sqrt{3}$. Mathematically,

$$V_{\text{L-L}} = V_{\text{phase}} \qquad \textbf{(A.4)}$$

$$I_L = \sqrt{3}\, I_P \angle -30^\circ \qquad \textbf{(A.5)}$$

The main advantage of delta-connected loads (such as can be found in some metal-processing plants) is that a single line-to-ground fault does not necessarily result in interruption of power, as is the case in a star-connected load. A delta connection is sometimes referred to as a *pi* (π) connection.

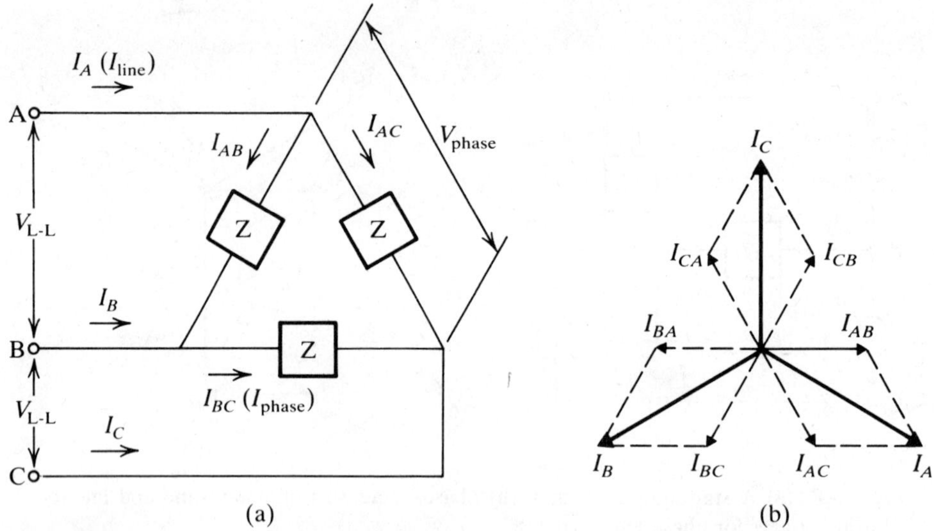

FIG. A-4 **(a)** A delta-connected load, **(b)** phasor diagram of line and phase currents (phase sequence ABC).

DELTA-STAR TRANSFORMATION

A three-phase delta-connected impedance (see Fig. A-5) can be changed to an equivalent star-connected impedance by using the following equation:

$$Z_{y_1} = \frac{Z_{\Delta_1} Z_{\Delta_2}}{Z_{\Delta_1} + Z_{\Delta_2} + Z_{\Delta_3}} \qquad \textbf{(A.6)}$$

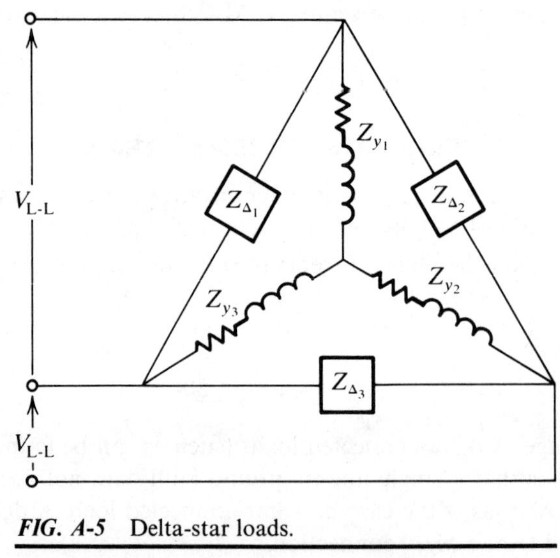

FIG. A-5 Delta-star loads.

where Z_{Δ_1} and Z_{y_1} are, in general, the delta- and star-connected impedances, respectively. This equation is derived from basic principles. One can easily remember it by noticing that each star-connected load is equal to the product of the two adjacent delta-connected impedances, divided by the sum of the three impedances. When the loads are balanced, from Eq. (A.6) we obtain

$$Z_y = \frac{Z_\Delta}{3} \tag{A.7}$$

In case of ideal capacitors, Eq. (A.7) gives

$$C_y = 3C_\Delta \tag{A.8}$$

where C_y and C_Δ are, respectively, the star and its equivalent delta-connected capacitor.

The equations above are very useful for changing a delta-connected load to an equivalent star-connected load. This leads to a simplified per-phase analysis.

For a given star-connected load, each of the equivalent delta-connected impedances is given by the ratio of the sum of the products of the y-impedances taken two at a time, and then divided by the opposite y-impedance. Considering Z_{Δ_1}, we have

$$Z_{\Delta_1} = \frac{Z_{y_1}Z_{y_2} + Z_{y_1}Z_{y_3} + Z_{y_2}Z_{y_3}}{Z_{y_2}} \tag{A.9}$$

Z_{Δ_2} and Z_{Δ_3} are calculated in a similar manner.

A.3 Instantaneous Power

In a balanced three-phase system the total instantaneous power is constant, while in a single-phase system—as explained in Section 1.1.8—the instantaneous power changes or is said to be pulsating.

The general expression for the instantaneous three-phase power is derived as follows. Consider the balanced three-phase system shown in Fig. A-6(a). The instan-

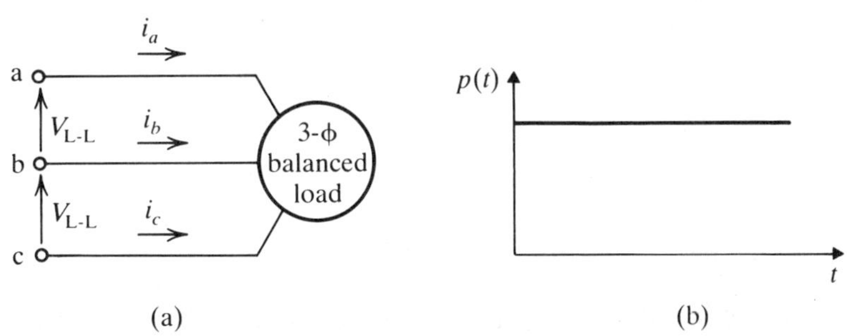

FIG. A-6 Three-phase load: **(a)** distribution diagram, **(b)** total instantaneous power.

taneous power ($p(t)$) delivered to the load is equal to the sum of the instantaneous powers (p_a, p_b, and p_c) drawn by the load. That is,

$$p(t) = p_a + p_b + p_c$$

From basic definitions, this expression can be written as follows:

$$p(t) = v_a i_a + v_b i_b + v_c i_c \tag{A.10}$$

where

v_a, v_b, v_c = the instantaneous values of the phase-to-neutral voltages
i_a, i_b, i_c = the instantaneous values of the phase currents.

The phase voltages for a phase sequence ABC are

$$\begin{aligned} v_a &= V_m \cos \omega t \\ v_b &= V_m \cos (\omega t - 120°) \\ v_c &= V_m \cos (\omega t + 120°) \end{aligned} \tag{A.11}$$

where V_m is the maximum value of the phase-to-neutral voltage. The corresponding phase currents for a load impedance of $|Z|\underline{/\theta}$ ohms per phase are

$$\begin{aligned} i_a &= \frac{V_m}{|Z|} \cos (\omega t - \theta) \\ i_b &= \frac{V_m}{|Z|} \cos (\omega t - 120° - \theta) \\ i_c &= \frac{V_m}{|Z|} \cos (\omega t + 120° - \theta) \end{aligned} \tag{A.12}$$

Substituting Eqs. (A.11) and (A.12) into Eq. (A.10), we obtain

$$\begin{aligned} p(t) = \frac{V_m^2}{|Z|} [&\cos \omega t \cos (\omega t - \theta) + \cos (\omega t - 120°) \cos (\omega t - 120° - \theta) \\ &+ \cos (\omega t + 120°) \cos (\omega t + 120° - \theta)] \end{aligned} \tag{A.13}$$

Using the identity $\cos x \cos y = ½(\cos (x + y) + \cos (x - y))$, we get

$$\begin{aligned} p(t) = \frac{V_m^2}{2|Z|} [&\cos (\omega t + \omega t - \theta) + \cos (\omega t - (\omega t - \theta)) \\ &+ \cos ((\omega t - 120°) + (\omega t - 120° - \theta)) + \cos ((\omega t - 120° - (\omega t - 120 - \theta)) \\ &+ \cos ((\omega t + 120° + (\omega t + 120° - \theta)) + \cos ((\omega t + 120° - (\omega t + 120° - \theta))] \end{aligned} \tag{A.14}$$

Summing up the angles within the parentheses, we obtain

$$p(t) = \frac{V_m^2}{2|Z|}[\underline{\cos(2\omega t - \theta)} + \cos\theta + \underline{\cos(2\omega t - 240° - \theta)} + \cos\theta + \underline{\cos(2\omega t + 240° - \theta)} + \cos\theta] \tag{A.15}$$

The three underlined terms are equal in magnitude and are displaced from each other by 120°. Thus, their sum is equal to zero. Therefore,

$$p(t) = \frac{V_m^2}{2|Z|} 3\cos\theta \tag{A.16}$$

or

$$p(t) = \frac{3V_m I_m}{2}\cos\theta = 3\frac{V_m}{\sqrt{2}}\frac{I_m}{\sqrt{2}}\cos\theta \tag{A.17}$$

Using equivalent rms values, we obtain

$$p(t) = 3V_{\text{L-N}} I_P \cos\theta \tag{A.18}$$

where $V_{\text{L-N}}$ and I_P are the rms values of the line-to-neutral voltage and phase current, respectively.

For a star-connected load, Eq. (A.18) becomes

$$p(t) = 3\frac{V_{\text{L-L}}}{\sqrt{3}} I_L \cos\theta \tag{A.19}$$

from which

$$p(t) = \sqrt{3}\, V_{\text{L-L}}\, I_L \cos\theta \tag{A.20}$$

For a delta-connected load, Eq. (A.18) becomes

$$p(t) = 3V_{\text{L-L}}\frac{I_L}{\sqrt{3}}\cos\theta \tag{A.21}$$

from which

$$p(t) = \sqrt{3}\, V_{\text{L-L}} I_L \cos\theta \tag{A.22}$$

Thus, the total instantaneous power drawn by a three-phase balanced load is constant, regardless of whether a star or a delta type of load connection is used.

Instantaneous power as a function of time is shown in Fig. A-6(b).

A.4 Real, Reactive and Complex Power

REAL POWER

The real or average power (P) drawn by a three-phase balanced load is

$$P = \frac{1}{T}\int_0^T p(t)\,dt \qquad \textbf{(A.23)}$$

where $p(t)$ is the instantaneous power and T is its period of oscillation. Substituting Eq. (A.22) for $p(t)$ after integration, we obtain

$$P = (\sqrt{3}\,V_{\mathrm{L-L}}I_L \cos\theta)\frac{1}{T}\,t\Big|_0^T$$

from which

$$P = \sqrt{3}\,V_{\mathrm{L-L}}I_L \cos\theta \qquad \textbf{(A.24)}$$

Equation (A.24) gives the average power consumed by a three-phase load, which is often referred to as the *power consumed* or the *active power*. Comparing Eqs. (A.22) and (A.24), it is evident that the instantaneous power drawn by a three-phase balanced load is equal to the average power consumed by the load.

The power consumed by a three-phase balanced load can also be determined by

$$P = 3I_p^2R \qquad \textbf{(A.25)}$$

where R is the per-phase resistance of the three-phase load. The power reading that appears in a utility's invoice is called "power demand." Power demand is the maximum of the average power recorded over any 15-minute interval of the billing period. (The time interval depends on the utility and could be 30 minutes or even 60 minutes.) Furthermore, the actual power demand is adjusted to include a power-factor penalty if applicable. For example, if a utility's minimum power-factor requirement is .90, while the maximum average consumption over a 15-minute interval is 150 kW at a power factor of .75, then the charged power demand is

$$150\left(\frac{.9}{.75}\right) = 180\ \text{kW}$$

Usually the cost of power demand is in the range of 10% to 25% of the total electricity cost.

REACTIVE POWER

By definition, the reactive power drawn by a three-phase load is

$$Q = 3V_{\mathrm{L-N}}I_P \sin\theta \qquad \textbf{(A.26)}$$

For a star-connected load, Eq. (A.26) becomes

$$Q = 3\frac{V_{\text{L-L}}}{\sqrt{3}} I_L \sin\theta \tag{A.27}$$

from which

$$Q = \sqrt{3}\, V_{\text{L-L}} I_L \sin\theta \tag{A.28}$$

Similarly, for a delta-connected load Eq. (A.26) becomes

$$Q = 3V_{\text{L-L}}\frac{I_L}{\sqrt{3}} \sin\theta$$

from which

$$Q = \sqrt{3}\, V_{\text{L-L}} I_L \sin\theta \tag{A.29}$$

COMPLEX POWER

By definition, the complex power drawn by a three-phase load is

$$S = 3V_{\text{L-N}} I_P^* \tag{A.30}$$

where I_P^* is the conjugate of I_P. For a star-connected system, Eq. (A.30) becomes

$$S = 3\frac{V_{\text{L-L}}}{\sqrt{3}} I_L^* \tag{A.31}$$

from which

$$S = \sqrt{3}\, V_{\text{L-L}} I_L^* \tag{A.32}$$

Similarly, for a delta-connected system, Eq. (A.30) becomes

$$S = 3V_{\text{L-L}}\frac{I_L^*}{\sqrt{3}} \tag{A.33}$$

from which

$$S = \sqrt{3}\, V_{\text{L-L}} I_L^* \tag{A.34}$$

Thus the real, reactive, and complex powers drawn by a balanced three-phase load are independent of the type of load connection (delta or star). For the steady-state analysis of a balanced three-phase system, it is advantageous to draw the so-called power

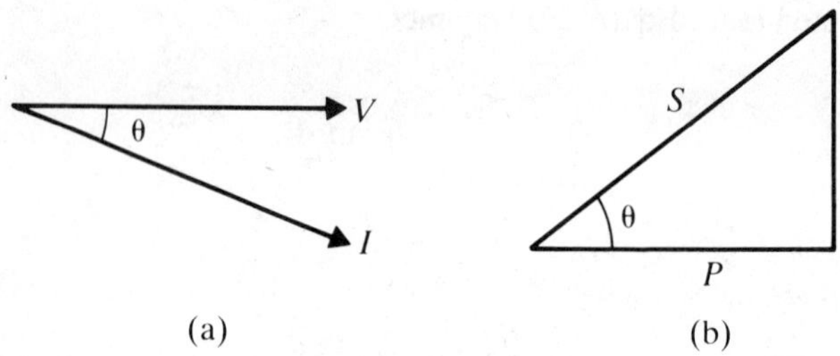

FIG. A-7 **(a)** Voltage and current phasors, **(b)** power triangle.

triangles of the individual loads and then to obtain the system's power triangle by adding algebraically the real and reactive power components of the individual loads. A typical voltage-current phasor and the power triangle of an inductive load are shown in Figs. A-7(a) and (b), respectively.

It should be noted that a lagging-power-factor load will result in a positive complex power.

A.5 Measurement of Power and Power Factor

POWER

The average power in a balanced three-phase system can be measured with a polyphase meter or with two wattmeters, as shown in Fig. A-8. For a three-phase system,

$$P = \sqrt{3}\, V_{L-L} I_L \cos\theta \tag{A.35}$$

or

$$P = W_1 + W_2 \tag{A.36}$$

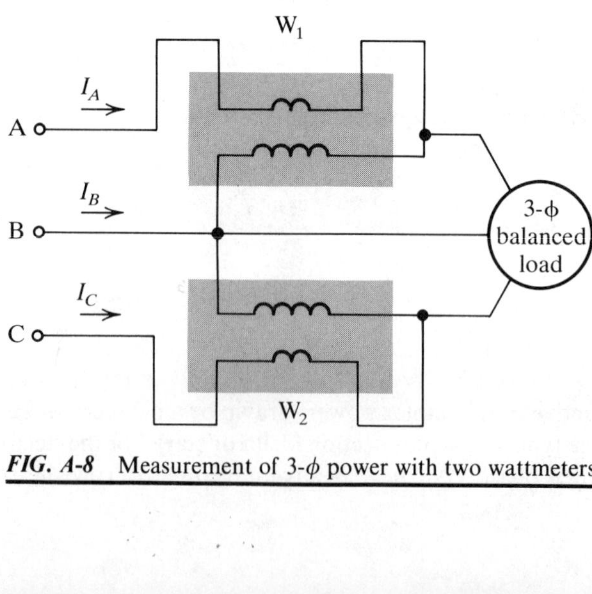

FIG. A-8 Measurement of 3-ϕ power with two wattmeters.

where the wattmeter readings W_1 and W_2 are given by

$$W_1 = V_{L-L} I_A \cos(\theta + 30°) \tag{A.37}$$

and

$$W_2 = V_{L-L} I_C \cos(\theta - 30°) \tag{A.38}$$

Equations (A.37) and (A.38) can easily be derived from the phasor diagram of a three-phase system supplying power to a load whose per-phase impedance has a phase angle of $\theta°$. By adding up Eqs. (A.37) and (A.38), it can be seen that the sum of the wattmeter readings gives the average power consumed by a balanced three-phase load. This is demonstrated in Example A-1.

POWER FACTOR

Using Eqs. (A.37) and (A.38), it can be shown that the per-phase power factor of the load is given by

$$\cos\theta = \frac{1}{\sqrt{1 + 3\left(\dfrac{1-a}{1+a}\right)^2}} \tag{A.39}$$

where $a = W_1/W_2$.

In many installations a two-element kWh meter is used. On its front panel this meter has two visible rotating disks, which derive their rotation from the power indication of their internal wattmeters (W_1 and W_2). By measuring the time (t_1 and t_2) that is required for the disks to complete one revolution, and by knowing the corresponding constants of the meter (so many kWh per revolution), the actual kW indication of each element can be calculated. If the wattmeter readings are known, then from Eq. (A.39) the instantaneous power factor of the load can be calculated. This technique—compared to the reading of standard switchboard Pf meters—is by far the most accurate method of measuring the power factor of a three-phase load.

A.6 Three-Phase, 5-Wire Distribution System

Figure A-9 shows the industrially popular three-phase, 5-wire distribution system. Between a line and the neutral, or between two lines, the single-phase loads are connected. Under normal operating conditions the neutral conductor carries the resultant current of all loads connected between lines a, b, c, and the neutral. The fifth wire or cable is the ground conductor of the system. In a well-balanced system and under normal operating conditions the ground conductor carries no current but is sized to withstand corrosion and accidental ground-fault currents. In drawing the

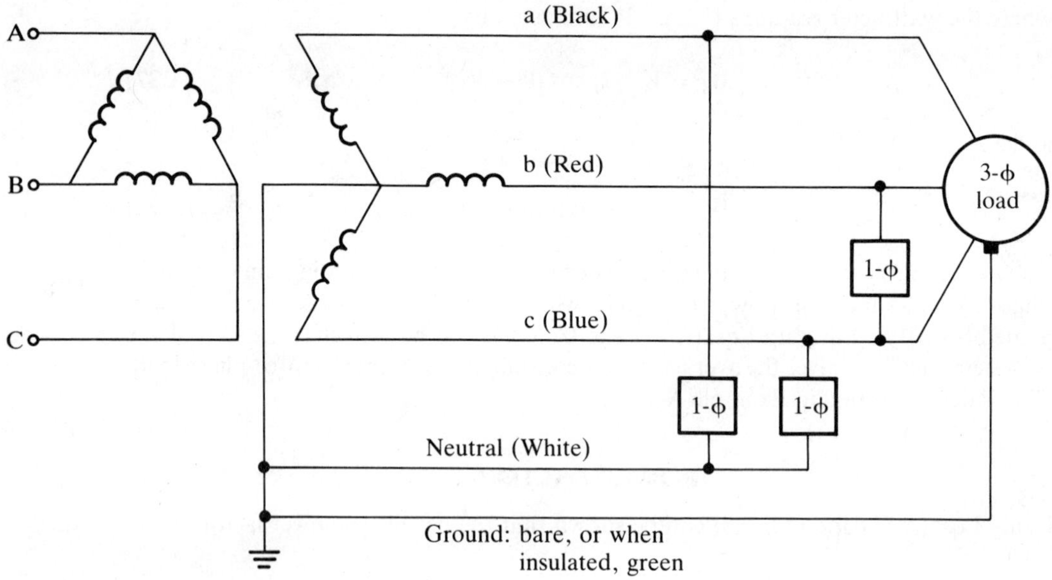

FIG. A-9 Three-phase, 5-wire distribution system.

phasor diagrams of single- and three-phase loads of a 5-wire distribution system, the line-to-neutral voltage is usually taken as reference. Often in practice the three-phase, 5-wire distribution system is referred to as three-phase, 4-wire, because the ground wire, which is used for safety reasons, is conductively joined to the distribution system at only one point, and (ignoring possible inductive and capacitive couplings) does not influence the voltage and currents within the distribution system.

EXAMPLE A–1 Figure A-10(a) shows a 480 V generator supplying power to a delta-connected load whose impedance is $10\underline{/30^\circ}\ \Omega$ per phase. For a phase sequence ABC draw the phasor diagram of the currents and determine:

a) The phase and line currents.

b) The wattmeter readings and total power consumed by the load.

SOLUTION

a) For an assumed phase sequence ABC, the generated voltages are drawn as shown in Fig. A-10(b). The encircled numbers indicate the order or the sequence of steps that are used in drawing the individual phasors. Disregarding the current through the voltage coil of the wattmeters, and by applying KCL, we obtain

$$I_{a_1a} = I_{ab} + I_{ac} = \frac{V_{ab}}{Z_{ab}} + \frac{V_{ac}}{Z_{ac}}$$

The phase angle of the voltages V_{ab} and V_{ac} are obtained from the phasor diagram.

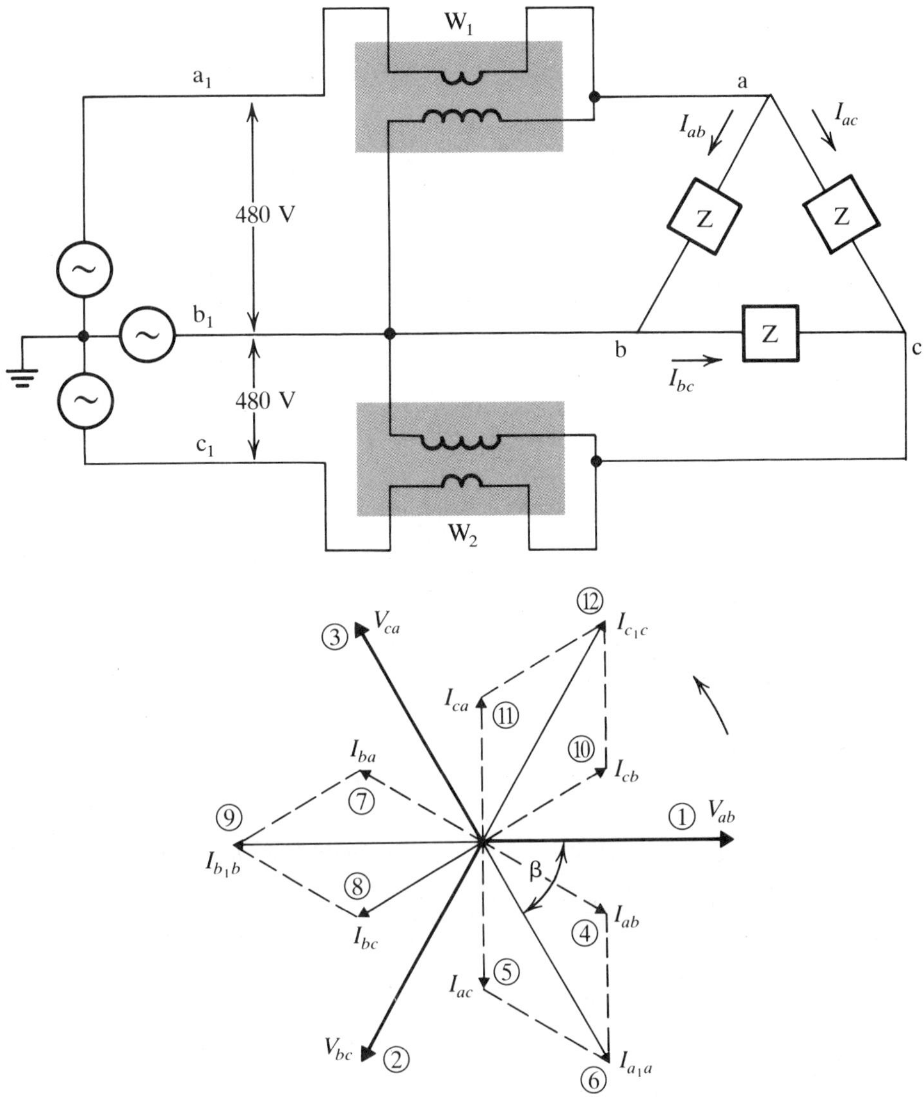

FIG. A-10(a),(b)

Substituting, we get

$$I_{a_1a} = \frac{480\angle 0^\circ}{10\angle 30^\circ} + \frac{480\angle -60^\circ}{10\angle 30^\circ} = \underline{\underline{\sqrt{3}(48)\angle -60^\circ \text{ A}}}$$

Alternatively, by using Eq. (A.5), we have

$$I_{a_1a} = \sqrt{3}\, I_{ab}\angle -30^\circ = \sqrt{3}\left(\frac{480}{10\angle 30^\circ}\right)\angle -30^\circ = \underline{\underline{\sqrt{3}(48)\angle -60^\circ \text{ A}}}$$

Similarly,

$$\begin{aligned} I_{b_1b} &= I_{ba} + I_{bc} \\ &= \frac{V_{ba}}{Z_{ba}} + \frac{V_{bc}}{Z_{bc}} = \frac{480\,\underline{/-180^\circ}}{10\,\underline{/30^\circ}} + \frac{480\,\underline{/-120^\circ}}{10\,\underline{/30^\circ}} \\ &= \underline{\underline{48\sqrt{3}\,\underline{/180^\circ}\ \text{A}}} \end{aligned}$$

and

$$\begin{aligned} I_{c_1c} &= I_{ca} + I_{cb} = \frac{V_{ca}}{Z_{ca}} + \frac{V_{cb}}{Z_{cb}} \\ &= \frac{480\,\underline{/120^\circ}}{10\,\underline{/30^\circ}} + \frac{480\,\underline{/60^\circ}}{10\,\underline{/30^\circ}} \\ &= \underline{\underline{48\sqrt{3}\,\underline{/60^\circ}\ \text{A}}} \end{aligned}$$

From Eq. (A.5) and the above results, we conclude that the magnitude of each phase current is equal to 48 A.

b) By definition, the wattmeter reading is equal to the product of the voltage across its voltage coil, times the current through its current coil, times the cosine of the angle (β) between the voltage and the current. Considering W_1, we have

$$W_1 = V_{ab} I_{a_1a} \cos \beta_1 \qquad \text{(I)}$$

As seen from Fig. A-10(b), the magnitude of the phase angle between V_{ab} and I_{a_1a} is 60°. Thus,

$$\beta_1 = -60^\circ$$

Substituting into equation (I), we obtain

$$W_1 = (480)(48\sqrt{3}) \cos 60^\circ = \underline{\underline{19.95\ \text{kW}}}$$

Similarly,

$$W_2 = V_{cb} I_{c_1c} \cos \beta_2$$

The phase angle between V_{cb} and I_{c_1c} is zero degrees. Thus,

$$W_2 = 480(48\sqrt{3}) \cos 0^\circ = \underline{\underline{39.91\ \text{kW}}}$$

(1) The total power consumed by the load is

$$P = W_1 + W_2 = 19.95 + 39.91 = \underline{\underline{59.86\ \text{kW}}}$$

(2) Alternatively, for a balanced load,

$$P = \sqrt{3}\, V_{L-L} I_L \cos\theta = \sqrt{3}(480)(48.0\sqrt{3}) \cos 30 = \underline{\underline{59.86\ \text{kW}}}$$

(3) Or, substituting into Eq. (A.25), we obtain

$$P = 3(48)^2(10 \cos 30^\circ) = \underline{\underline{59.86\ \text{kW}}}$$

EXAMPLE A–2 A balanced, star-connected load of $10\angle 26^\circ\ \Omega$ per-phase impedance is connected to a 480 V supply, as shown in Fig. A-11. Determine:

a) The line-to-neutral voltage V_{an}.
b) The phase current I_{an}.
c) The line current I_a.
d) The power consumed by the load.

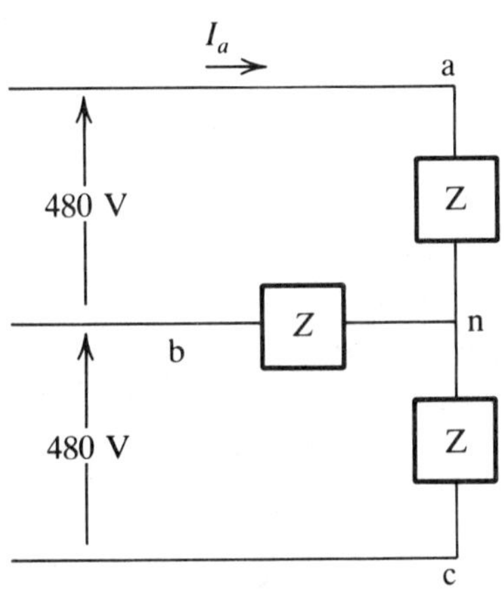

FIG. A-11

SOLUTION

a) For a phase sequence ABC, the magnitude of the line-to-neutral voltage (V_{an}) is

$$V_{an} = \frac{|V_{ab}|}{\sqrt{3}} = \frac{480}{\sqrt{3}} = \underline{\underline{277.1\ \text{V/phase}}}$$

b) The phase current I_{an} is

$$I_{an} = \frac{V_{an}}{Z} = \frac{277.1}{10\angle 26^\circ} = \underline{\underline{27.71\angle -26^\circ\ \text{A}}}$$

c) As seen in Fig. A-11, the line and phase currents are equal. Therefore,

$$I_a = \underline{\underline{27.71\angle -26^\circ \text{ A}}}$$

d) The power consumed is determined by substituting the known data into Eq. (A-24). Thus,

$$\begin{aligned} P &= \sqrt{3}(480)(27.71)\cos 26^\circ \\ &= \underline{\underline{20.71 \text{ kW}}} \end{aligned}$$

Alternatively, using Eq. (A.25), we obtain

$$P = 3(27.71)^2(10\cos 26^\circ) = \underline{\underline{20.71 \text{ kW}}}$$

EXAMPLE A–3 A 480 V, star-connected, three-phase ac generator supplies power to a delta-connected load through a transmission line. The per-phase impedances of the load and of the transmission line are $(16 + j20)\ \Omega$ and $(1 + j2)\ \Omega$, respectively. Determine:

a) The line current.

b) The magnitude of the line-to-line voltage across the load.

c) The efficiency of transmission.

SOLUTION The given network is sketched as shown in Fig. A-12(a). Using Δ-Y transformation, the circuit is redrawn as shown in Fig. A-12(b). Since the per-phase impedances of the transmission line are equal, and because the load is balanced, we can do the analysis per phase.

a) The per-phase equivalent circuit is shown in Fig. A-12(c). Using Ohm's law, we obtain

$$I_{a_1a} = \frac{V_{a_1n}}{Z_1 + Z_2} = \frac{\dfrac{480\angle 0^\circ}{\sqrt{3}}}{1 + j2 + \dfrac{16 + j20}{3}} = 25.82\angle -53.8^\circ \text{ A}$$

Here the line-to-neutral voltage is taken as reference.

b) The per-phase voltage across the load is

$$\begin{aligned} V_{an} &= \frac{480\angle 0^\circ}{\sqrt{3}} - IZ_1 \\ &= \frac{480\angle 0^\circ}{\sqrt{3}} - 25.82\angle -53.80^\circ(1 + j2) \\ &= 220.41\angle -2.5^\circ \text{ V/phase} \end{aligned}$$

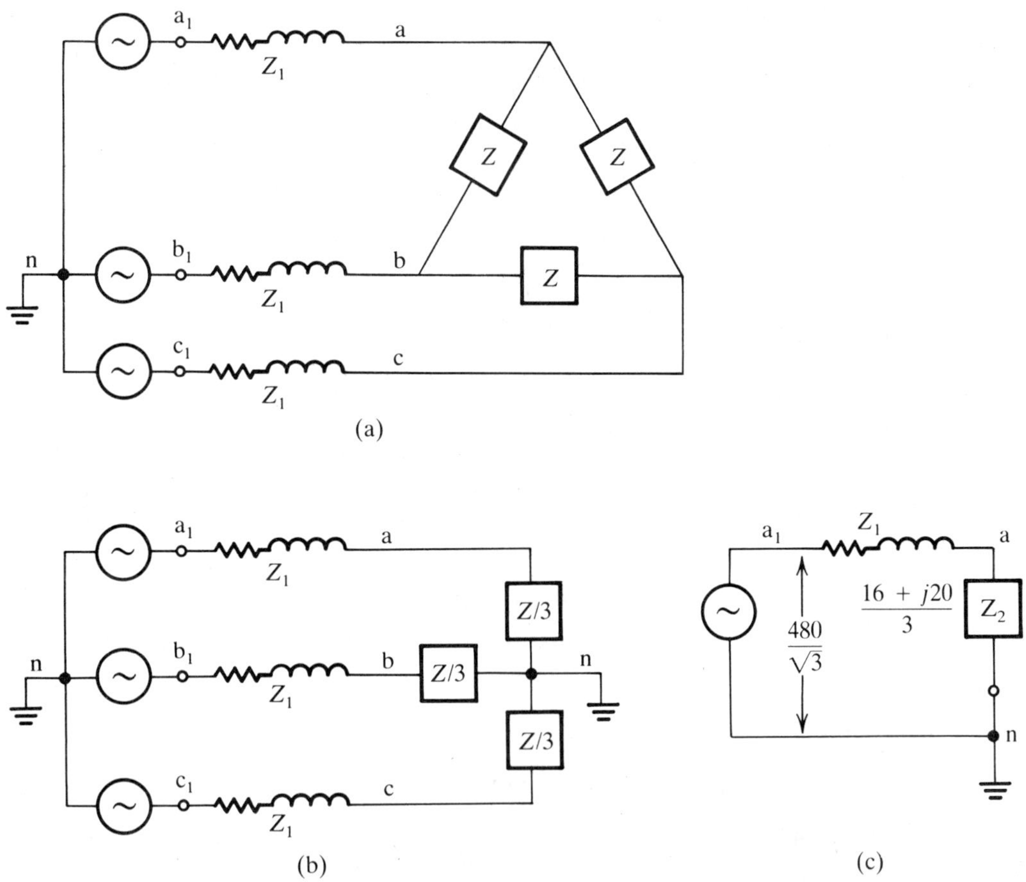

***FIG. A-12* (a) (b) (c)**

The magnitude of the line-to-line voltage is

$$V_{ab} = \sqrt{3}(220.41) = 381.77 \text{ V, L–L}$$

Alternatively,

$$V_{an} = IZ_2 = 25.82\underline{/-53.8^\circ}\left(\frac{16 + j20}{3}\right)$$

$$= 220.41\underline{/-2.5^\circ} \text{ V/phase}$$

and

$$V_{ab} = \underline{\underline{381.77 \text{ V, L–L}}}$$

c) The efficiency of transmission is

$$\eta = \frac{\text{power delivered}}{\text{power generated}}$$

$$= \frac{3I^2R_{\text{load}}}{3I^2(R_{\text{load}} + R_{\text{transmission line}})}$$

$$= \frac{R_L}{R_L + R_{\text{T.L.}}} = \frac{\frac{16}{3}}{\frac{16}{3} + 1}$$

$$= \underline{\underline{84\%}}$$

EXAMPLE A–4 A 3-ϕ, 60 Hz, 480 V source delivers power to a balanced, star-connected load of 100 kW and .8 power factor lagging. Determine:

a) The magnitude of the line current and the power triangle of the load.
b) The size of the Δ-connected capacitor which, when connected across the load, will improve the power factor to .95 lagging.

SOLUTION

a) Referring to Fig. A-13(a), and using Eq. (A.24), we obtain

$$I_L = \frac{100 \times 10^3}{\sqrt{3}(480)(.8)}$$

$$= \underline{\underline{150.35 \text{ A}}}$$

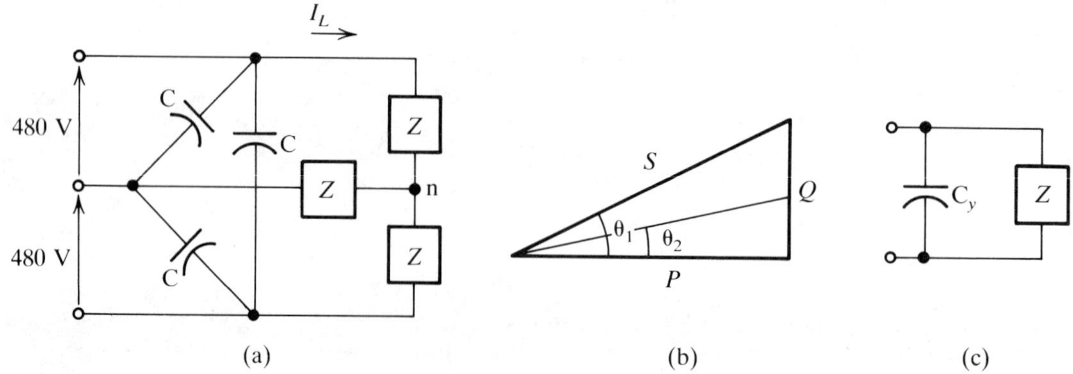

FIG. A-13

The reactive power drawn by the load is

$$Q = P \tan \theta$$
$$Q = 100 \times 10^3 \tan 36.87^\circ$$
$$= \underline{\underline{75 \text{ kVAR}}}$$

The complex power is

$$S = \frac{P}{\cos \theta} = \frac{100}{.8}$$
$$= \underline{\underline{125 \angle 36.87^\circ \text{ kVA}}}$$

The power triangle, with and without the capacitor bank in the circuit, is shown in Fig. A-13(b). The per-phase equivalent circuit is shown in Fig. A-13(c).

b) The per-phase star-connected capacitance that is required to improve the power factor from $\cos \theta_1$ to $\cos \theta_2$ (see Fig. A-13(b)) is found as follows. The reactive power of a capacitor is

$$Q = VI \sin \theta_c$$
$$= V \left| \frac{V}{Z} \right| \sin 90^\circ$$
$$= V^2 \omega C$$

Thus,

$$C = \frac{Q}{V^2 \omega}$$

Also, from the geometry of Fig. A-13(b),

$$Q = (P/\text{phase})(\tan \theta_1 - \tan \theta_2)$$

From the above,

$$C = \frac{P/\text{phase}}{V^2 \omega}(\tan \theta_1 - \tan \theta_2) \text{ F/phase}$$

where

$$\theta_1 = \arccos .8 = 36.9^\circ$$
$$\theta_2 = \arccos .95 = 18.2^\circ$$
$$V = \frac{480}{\sqrt{3}}$$

Substituting, we obtain

$$C = \frac{100 \times 10^3}{3\left(\frac{480}{\sqrt{3}}\right)^2 (377)} (\tan 36.9° - \tan 18.2°)$$

$$= 485.1 \ \mu\text{F/phase}$$

Since the commercially available capacitor bank is delta-connected, we use Eq. (A.8) to obtain

$$C_\Delta = \frac{C_Y}{3} = \frac{485.1}{3} = \underline{\underline{161.7 \ \mu\text{F/phase}}}$$

Usually a capacitor bank is specified in terms of its voltage and kVAR rating. In this case, $V = 480$ V, L–L and $Q = 100 (\tan 36.9° - \tan 18.2°) = 42.13$ KVAR.

EXAMPLE A–5 The two-element energy meter shown in Fig. A-14 is used to measure the energy consumed by a 3-ϕ, 3w load. The constant of the meter for each rotating wheel is 5 kWh/revolution. At a given instant the rotating wheels require 30 and 50 s, respectively, to complete one revolution. Calculate the power factor of the load at that instant.

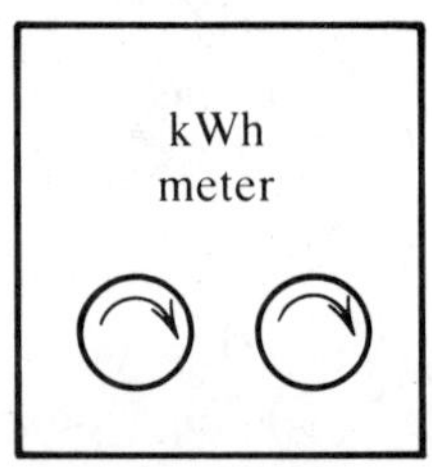

FIG. A-14

SOLUTION Designating the time that each wheel takes to complete one revolution as t_1 and t_2, the ratio of the wattmeter's reading is

$$\alpha = \frac{W_1}{W_2} = \frac{\frac{(\text{energy})_1}{t_1}}{\frac{(\text{energy})_2}{t_2}} = \frac{\frac{5}{30}}{\frac{5}{50}} = 1.67$$

Substituting into Eq. (A.39), we obtain

$$\cos\theta = \frac{1}{\sqrt{1 + 3\left(\frac{-.67}{2.67}\right)^2}} = \underline{\underline{.92}}$$

B PER-UNIT SYSTEM

The per-unit system is a set of mathematical relationships that transform the standard parameters of voltage, current, power, impedance, etc. into their equivalent per-unit values. The use of equivalent per-unit values greatly simplifies the understanding and the solution of many engineering problems.

B.1 Calculation of Per-Unit Values

By definition, the per-unit value of a parameter is given by

$$\begin{array}{c}\text{per-unit value}\\ \text{of a parameter}\end{array} = \frac{\text{actual value of the parameter}}{\text{base value for this parameter}} \tag{B.1}$$

For example, if the actual value of an impedance is 2 Ω and its base value is 5 Ω, then the per-unit value of the impedance is

$$Z_{pu} = \frac{Z_{\text{actual}}}{Z_{\text{base}}} = \frac{2}{5} = .4 \text{ pu}$$

The per-unit value of a quantity (X) is equal to the quantity expressed in percent, divided by 100. That is,

$$X_{pu} = \frac{X\%}{100}$$

The base values of the various parameters are selected as follows.

BASE POWER (S_b)

For a single machine, in performing the per-unit computations of real, reactive, and complex power, the base power is always taken as the magnitude of the machine's complex power. The magnitude of the complex power is referred to as the *apparent power*. Mathematically,

$$S_b = (VA)_{\text{base}} \tag{B.2}$$

For example, if a load draws $1000\,\underline{/30}$ VA, the per-unit values of the complex, real, and reactive powers are calculated as follows. By definition,

$$S_b = 1000 \text{ VA}$$

The per-unit value of complex power,

$$S = \frac{\text{actual value}}{\text{base value}} = \frac{S_a}{S_b}$$

$$= \frac{1000\,\underline{/30^\circ}}{1000} = 1\,\underline{/30^\circ} \text{ pu}$$

The per-unit value of real power,

$$P = \frac{\text{actual value}}{\text{base value}} = \frac{P_a}{S_b}$$

$$= \frac{1000 \cos 30^\circ}{1000} = .87 \text{ pu}$$

The per-unit value of the reactive power,

$$Q = \frac{\text{actual value}}{\text{base value}} = \frac{Q_a}{S_b}$$

$$= j\frac{1000 \sin 30^\circ}{1000} = j.5 \text{ pu}$$

Thus, the power drawn by the load in per-unit is

$$S = .87 + j.5 = 1.0\underline{/30^\circ} \text{ pu}$$

The base power of a transformer or generator is its rated output VA. The base power of a motor is its input apparent power. In the case of transformers and generators their name-plate data furnishes their output apparent powers, while a motor's input volt-amperes can be calculated from its output power, overall efficiency, and input power factor as follows:

$$S_b = \frac{\text{output in watts}}{(\text{efficiency})(\text{power factor})} \tag{B.3}$$

For a three-phase system, the apparent power is

$$S_b = \sqrt{3}\, V_{L-L} I_L = \frac{P_{\text{out}}}{\eta \cos \theta} \tag{B.4}$$

Very often an educated guess or assumption has to be made regarding the overall efficiency and input power factor of a motor. In the case of a system that comprises transformers, cables, and machines, the base power is arbitrarily chosen and remains *constant* for each component of the system. In other words, once the base power is selected, it does not change throughout the system.

BASE VOLTAGE (V_b)

The base voltage for a motor or generator is its name-plate voltage. For a transformer, the base voltage is different on either side of the transformer and hence will be designated—depending on the winding under consideration—as

$$V_{b_H} \quad \text{or} \quad V_{b_L}$$

where V_{b_H} is the base voltage of the high-voltage winding and V_{b_L} is the base voltage of the low-voltage winding. Thus the base voltage in a multicomponent system will change each time one advances from one side of the transformer to the other.

BASE CURRENT (I_b)

The base current is normally the rated or full-load current of the machine. For single-phase loads,

$$I_b = \frac{S_b}{V_b} \tag{B.5}$$

For a three-phase system,

$$I_b = \frac{S_b}{\sqrt{3}\, V_{b\text{L-L}}} \tag{B.6}$$

The base current on either side of the transformer is different.

BASE IMPEDANCE (Z_b)

For per-unit computations of impedance, resistance, and reactance the base impedance is used. This is given by the ratio of base voltage to base current. That is,

$$Z_b = \frac{V_{b\text{L-L}}/\sqrt{3}}{I_b} \tag{B.7}$$

or

$$Z_b = \frac{V_{b\text{L-L}}/\sqrt{3}}{S_b/\sqrt{3}\, V_{b\text{L-L}}} \tag{B.8}$$

From the above,

$$Z_b = \frac{V^2_{b\text{L-L}}}{S_b}\ \Omega/\text{phase} \tag{B.9}$$

If the base voltage of a three-phase machine or transformer is taken as the line-to-neutral voltage, then the base power should be the per-phase apparent power. If, however, the line-to-line voltage is used, then the base power must be the total or three-phase apparent power of the machine under consideration.

As seen from the high- and low-voltage windings, the base impedance for a transformer is different. However, the per-unit values of the impedances are the same, regardless of which side is chosen as the reference.

TORQUE AND SPEED

The base values of a machine's torque and speed are, respectively, its full-load torque and speed.

MULTICOMPONENT SYSTEMS

In a multicomponent system comprising transformers, cables, and machines, a convenient base power and voltage are chosen. As already mentioned, the base power remains the same for each component of the system, while *the base voltage and the base current change* for either side of the transformer. Very often, in the one-line diagram of a power distribution system, the per-unit impedances of the individual machines and transformers are given as calculated from their own base parameters. The calculations are simplified if all per-unit impedances are expressed on the same base.

To change the per-unit value of an impedance from one base (old) to a new base, the following relationship is used:

$$Z_{pu_{\text{new}}} = Z_{pu_{\text{old}}} \left(\frac{V_{b\text{-old}}}{V_{b\text{-new}}}\right)^2 \left(\frac{S_{b\text{-new}}}{S_{b\text{-old}}}\right) \tag{B.10}$$

For example, if on a base of 480 V and 1000 kVA an impedance is .2 per-unit, then at a new base of 1200 V and 1500 kVA the per-unit value of the impedance is

$$\begin{aligned} Z_{\text{new}} &= .2\left(\frac{480}{1200}\right)^2\left(\frac{1500}{1000}\right) \\ &= \underline{\underline{.048 \text{ pu}}} \end{aligned}$$

B.2 Advantages of Per-Unit Values

The per-unit system is very popular in the power field of electrical engineering because of the following advantages:

1. The per-unit value of *impedance, voltage,* and *current* of a transformer are the same, regardless of whether they are referred to the low- or high-voltage side. This is a tremendous advantage, because it helps to bypass all the confusions and misunderstandings that result when the transformer's parameters are referred to either the high- or low-voltage windings.
2. The per-unit system makes engineering calculations easier because the rated voltage and current for a single machine or a transformer, when expressed in per-unit on its own machine rating, are each equal to 1.0.
3. The per-unit values of the parameters of a machine vary within a small range for all types and ratings of machines produced by different manufacturers. This gives the designer and the application engineer a better understanding of the machine, and a

tool with which similar units may easily be compared. For example, the ohmic values of the reactance of alternators could vary from .1 to 20 ohms; the same reactance, expressed in per-unit, would be in the range of .10 to 1.1 per-unit.

4. The per-unit value of the resistance of a machine furnishes almost at a glance its electrical losses in percent of its rated power. For example, a transformer operating under rated conditions at unity power factor with a winding resistance of .01 per-unit has a copper loss of (1%):

$$I^2R = (1.0)^2(.01) = .01 \text{ pu}$$

This is very useful to a power systems engineer, because he or she can estimate and locate the quantity of the various copper losses simply by looking at the one-line impedance diagram.

5. The per-unit values of the impedance of a single- or simple multicomponent power system readily provide the short-circuit current at different points on the network. For example, a transformer with a per-unit impedance of 5% would result—without taking into consideration the motor's short-circuit current contribution—in a short-circuit current (I_{sc}) of

$$I_{sc} = \frac{1}{Z_{pu}} = \frac{1}{.05} = 20 \text{ pu}$$

Knowing the short-circuit current is very important in the selection and coordination of protective devices.

6. The per-unit system simplifies the analysis of problems that include star-delta (Y-Δ) types of winding connections. The factor of $\sqrt{3}$ is not used for the per-unit analysis. For example, consider the expression for the power:

$$P = VI \cos \theta$$

When the voltage and the current are expressed in per-unit, this relationship gives the total power in per-unit, regardless of delta or star winding connection.

7. Per-unit parameters simplify the simulation of simple or complex power-system problems on analog and digital computers. Such simulations are important for transient and steady-state analyses.

EXAMPLE B–1 For the three-phase transformer shown in the one-line diagram of Fig. B-1, determine the per-unit values of voltages, currents and impedances, as seen from the high- and low-voltage windings, when the transformer operates at rated condition.

SOLUTION

Low-Voltage Winding.

The base parameters are

$$V_{b_L} = 480 \text{ V}, \qquad S_{b_L} = 100 \text{ kVA}$$

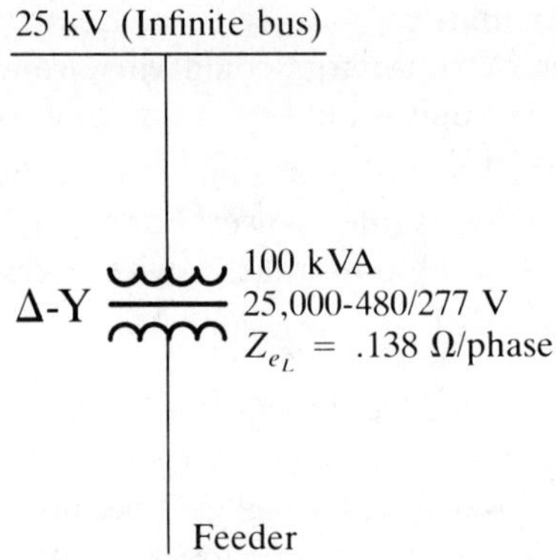

FIG. B-1

and

$$I_{b_L} = \frac{S_{b_L}}{\sqrt{3}\, V_{b_{L-L}}} = \frac{100 \times 10^3}{\sqrt{3}\, 480} = 120.28 \text{ A}$$

From Eq. (B.9),

$$Z_b = \frac{(V_{b_{L-L}})^2}{S_b} = \frac{(480)^2}{100 \times 10^3} = 2.3 \ \Omega/\text{phase}$$

Thus, the per-unit value of the impedance is

$$Z_{pu} = \frac{Z_{\text{actual}}}{Z_{\text{base}}} = \frac{.138}{2.3} = \underline{\underline{.06 \text{ pu}}}$$

The per-unit value of the current is

$$I_{pu} = \frac{I_{a_L}}{I_{b_L}} = \frac{120.28}{120.28} = \underline{\underline{1.0 \text{ pu}}}$$

The per-unit value of the voltage is

$$V_{pu} = \frac{V_{a_L}}{V_{b_L}} = \frac{480}{480} = \underline{\underline{1.0 \text{ pu}}}$$

High-Voltage Winding.

The base parameters are

$$V_{b_H} = 25 \text{ kV}, \qquad S_b = 100 \text{ kVA}$$

Thus,

$$Z_{b_H} = \frac{(V_{b_{L-L}})^2}{S_b} = \frac{(25)^2}{.10} = 6250.0 \ \Omega/\text{phase}$$

$$I_{b_H} = \frac{100}{\sqrt{3}\,25} = 2.31 \ \text{A}$$

The ohmic value of the winding impedance referred to the high-voltage delta-connected winding is

$$Z_H = Z_L\left(\frac{N_1}{N_2}\right)^2$$

$$= .138\left(\frac{25000}{480/\sqrt{3}}\right)^2 = 1123.1 \ \Omega/\text{phase}$$

Changing this impedance to an equivalent star, we obtain

$$Z_y = \frac{Z_\Delta}{3} = \frac{1123.1}{3} = 374.35 \ \Omega/\text{phase}$$

Thus, the per-unit values are

$$Z_{pu} = \frac{374.35}{6250} = \underline{\underline{.06 \ \text{pu}}}$$

The per-unit value of the voltage is

$$V_{pu} = \frac{\text{actual value}}{\text{base value}} = \frac{25 \times 10^3}{25 \times 10^3} = \underline{\underline{1.0 \ \text{pu}}}$$

Similarly, the per-unit value of the current is

$$I_{pu} = \frac{2.31}{2.31} = 1.0 \ \text{pu}$$

Thus, the per-unit values of the impedance, voltage, and current are the same for the high- and low-voltage sides of the transformer.

EXAMPLE B–2 For the power system shown in the one-line diagram of Fig. B-2(a) determine:

a) The per-unit value of all reactances on the base of the generator rating.
b) The magnitude of the short-circuit current at points *A* and *B*. (Neglect the motors' short-circuit current contributions.)

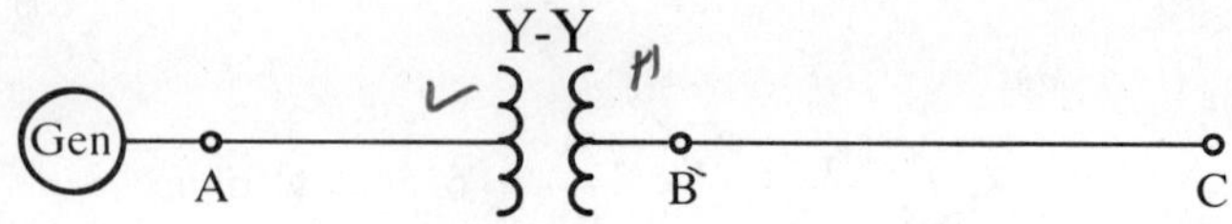

50 MVA
$X_d'' = .95\ \Omega$/phase
13.8 kV, 60 Hz

20 MVA
13.8-69 kV
$X_{e_H} = 17.85\ \Omega$/phase

Cable
$X_c = 41.4\ \Omega$/phase

FIG. B-2(a)

SOLUTION

a) The base values are given by the generator rating:

$$S_b = 50 \text{ MVA}, \qquad V_{b_{L-L}} = 13.8 \text{ kV}$$

$$I_b = \frac{50 \times 10^3}{\sqrt{3}(13.8)} = 2091.85 \text{ A}$$

The base impedance at 13.8 kV is

$$Z_b = \frac{(13.8)^2}{50} = 3.81 \ \Omega/\text{phase}$$

Generator. The per-unit value of the generator's reactance is

$$X_d = \frac{.95}{3.81} = \underline{\underline{.25 \text{ pu}}}$$

Transformer. Referring the transformer's reactance to the low-voltage winding, we have

$$X_{e_L} = X_H \left(\frac{N_1}{N_2}\right)^2$$

$$= 17.85 \left(\frac{13.8}{69}\right)^2$$

$$= \underline{\underline{.714 \ \Omega/\text{phase}}}$$

In per-unit,

$$X_e = \frac{.714}{3.81} = \underline{\underline{.187 \text{ pu}}}$$

Cable. Referring the cable's reactance to the low-voltage winding of the transformer, we have

$$X_{c_L} = 41.4 \left(\frac{13.8}{69}\right)^2 = 1.656 \ \Omega/\text{phase}$$

In per-unit,

$$X_c = \frac{1.656}{3.81} = \underline{\underline{.435 \text{ pu}}}$$

The per-phase equivalent circuit, using per-unit values, is shown in Fig. B-2(b).

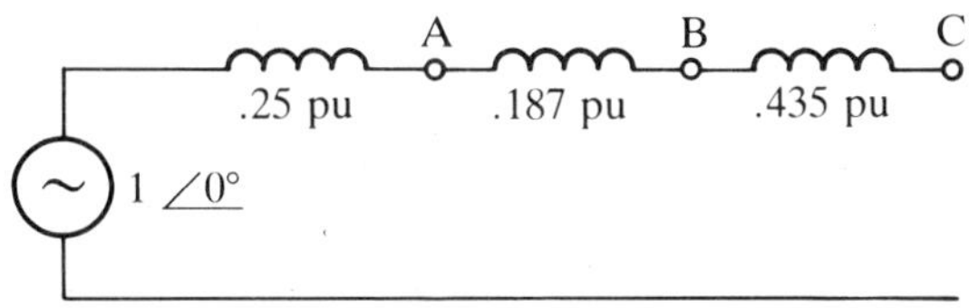

FIG. B-2(b)

b) The magnitude of the short-circuit current at *A* is:

Using Ohmic values:

$$I = \left|\frac{V_{\text{L-N}}}{Z}\right| = \frac{13{,}800/\sqrt{3}}{.95} = \underline{\underline{8.39 \text{ kA}}}$$

Using per-unit values:

$$I = \left|\frac{V}{Z}\right| = \frac{1.0}{.25} = 4.0 \text{ pu}$$

$$= 4(2091.85) = \underline{\underline{8.39 \text{ kA}}}$$

The magnitude of the short-circuit current at *B* is:

Using Ohmic values:

$$I = \frac{V_{\text{L-N}}}{Z} = \frac{13{,}800/\sqrt{3}}{(.95) + .714} = \underline{\underline{4.79 \text{ kA}}}$$

Using per-unit values:

$$I = \left|\frac{V}{Z}\right| = \frac{1.0}{.25 + .187}$$

$$= 2.29 \text{ pu}$$

$$= 2.29(2091.85)$$

$$= \underline{\underline{4.79 \text{ kA}}}$$

It should be noted that when we speak of a short circuit we usually mean a solid, three-phase short circuit.

C LAPLACE TRANSFORMS

The transient response of a system, whether that of a simple circuit or an electric machine, can be obtained by solving the system's differential equations. The solution of a differential equation can be determined by using either the so-called classical method or the technique of Laplace transforms. The Laplace method uses the tabulated Laplace transforms.

The Laplace transform $\mathbf{F}(s)$ of any function $f(t)$ is found by the following equation:

$$\mathbf{F}(s) = \mathscr{L}(f(t)) = \int_0^\infty f(t)e^{-st}\,dt \qquad \textbf{(C.1)}$$

where s is a complex frequency variable defined as

$$s = \sigma + j\omega \qquad \textbf{(C.2)}$$

where ω is the angular frequency and σ is a real number.

EXAMPLE C–1 Find the Laplace transform of e^{-at}.

SOLUTION

From Eq. (C.1) we obtain

$$\begin{aligned}\mathscr{L}(e^{-at}) &= \int_0^\infty e^{-at}e^{-st}\,dt \\ &= -\left.\frac{e^{-t(s+a)}}{s+a}\right|_0^\infty \\ &= \frac{1}{s+a}\end{aligned}$$

LAPLACE TABLES

Transformation tables give the Laplace transforms of many time functions. These tables are derived from Eq. (C.1) and the properties of the frequency-domain functions. The tables can be used not only to find the Laplace transforms of the tabulated time functions, but also to find the time functions (inverse transformation) from their corresponding Laplace transforms.

Table C-1 includes the Laplace transforms of some functions that are used for the transient analysis of simple machine problems. By using Laplace transforms, a machine's transient response can easily be determined by the following procedure:

1. Write the machine's equations in their differential form.
2. Find the Laplace transform of the differential equation and bring it to the standard form that is available in the tables.
3. Take the inverse transform of (2).

The solution of differential equations by using Laplace transforms is demonstrated by the following example.

TABLE C-1 Laplace Tables

No.	$f(t)u(t)$*	$\mathbf{F}(s)$
1	k	$\frac{k}{s}$
2	e^{-at}	$\frac{1}{s+a}$
3	$\frac{1}{a}(1-e^{-at})$	$\frac{1}{s(s+a)}$
4	$L\frac{di}{dt}$	$LsI(s) \pm LI_0$†
5	$\frac{1}{C}\int i\,dt$	$\frac{I(s)}{sC} \pm \frac{V_{co}}{s}$†
6	t	$\frac{1}{s^2}$

* *The unit-step function* u(t) *insures that the time function does not exist at negative time* (t).

† *Where* I_0 *and* V_{co} *are, respectively, the initial inductor current and the initial capacitor voltage.*

EXAMPLE C–2 Refer to the circuit in Fig. C-1(a). After the switch is closed, find the current $i(t)$.

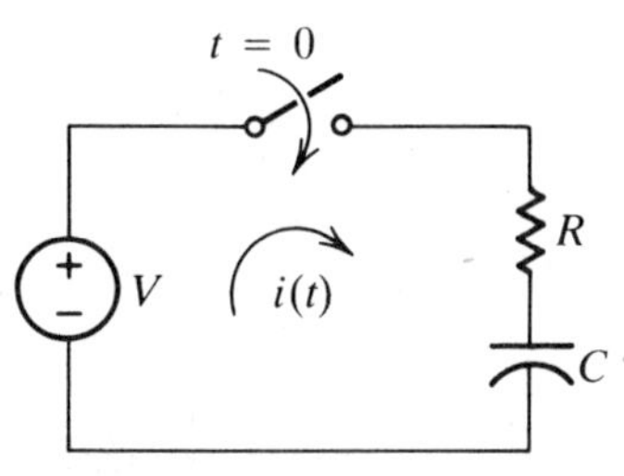

FIG. C-1(a)

SOLUTION

1. From KVL we have

$$V = Ri(t) + \frac{1}{C}\int i(t)\,dt$$

2. Taking the Laplace transform of the above, we obtain

$$\frac{V}{s} = I(s)R + \frac{I(s)}{sC}$$

Solving for $I(s)$, and after simplification, we obtain

$$I(s) = \frac{V}{R\left(\dfrac{1}{CR} + s\right)}$$

3. Taking the inverse Laplace transform (No. 2 of Table C-1) of the last expression, we find the current as a function of time:

$$i(t) = \frac{V}{R} e^{-t/RC}$$

The current waveform is sketched in Fig. C-1(b).

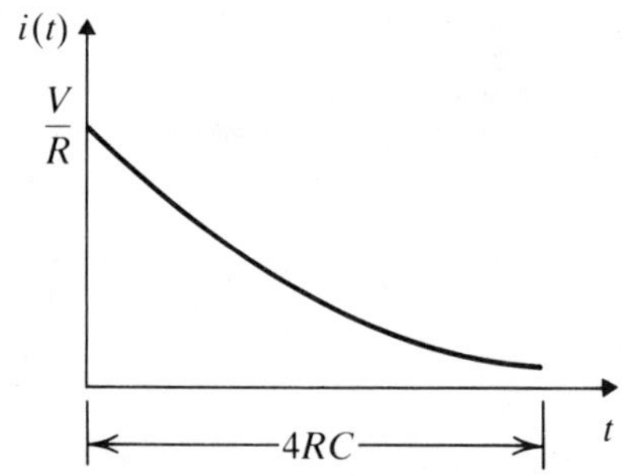

FIG. C-1(**b**)

INITIAL- AND FINAL-VALUE THEOREMS

When the Laplace transform $\mathbf{F}(s)$ of a function $f(t)$ is known, then the value of the function at $t = 0$ and at $t = \infty$ can be found as follows:

$$f(t)\Big|_{t=0} = \lim_{s\to\infty} s\mathbf{F}(s) \tag{C.3}$$

$$f(t)\Big|_{t=\infty} = \lim_{s\to 0} s\mathbf{F}(s) \tag{C.4}$$

Equation (C.2) and (C.3) are known, respectively, as the initial- and the final-value theorems. One can easily remember these equations by noting that when $t \to 0$, $s \to \infty$ and vice versa. From Eqs. (C.3) and (C.4), it can be said that the time domain begins where the frequency domain ends and vice versa. These equations are often used to evaluate a machine's variable at starting ($t = 0$) and at steady state ($t = \infty$) from its equivalent Laplace transform. The latter can be easily established from basic principles. Furthermore, from Eq. (C.3) and the definition of a coil's reactance, it can be seen why a coil presents an infinite impedance to the propagation of upstream electrical impulses or spikes.

EXAMPLE C–2 The Laplace transform of a motor's speed is given by

$$\omega(s) = \frac{125}{s(s+2)} \text{ rad/s}$$

Determine the speed of the motor at $t = 0$ and at $t = \infty$.

SOLUTION

$$\omega(0) = \lim_{s \to \infty} s \frac{125}{s(s+2)} = 0$$

$$\omega(\infty) = \lim_{s \to 0} s \frac{125}{s(s+2)} = \underline{\underline{62.5 \text{ rad/s}}}$$

D SOLID STATE DEVICES

This section discusses the operation of diodes, transistors, and thyristors. These devices act as switches (ON or OFF) to the flow of current; as such, they constitute the basic building blocks of all solid state speed-control apparatus.

D.1 Diodes

Diodes are made from *pn*-type semiconductor material and act as electronic switches because they permit the current to flow in only one direction.

In Fig. D-1(a), the *p*-type semiconductor material has an excess of positive charges, and the *n*-type material has an excess of negative charges. The excess, or so-called majority carriers (charges), are shown encircled in order to differentiate them from the minority carriers or charges.

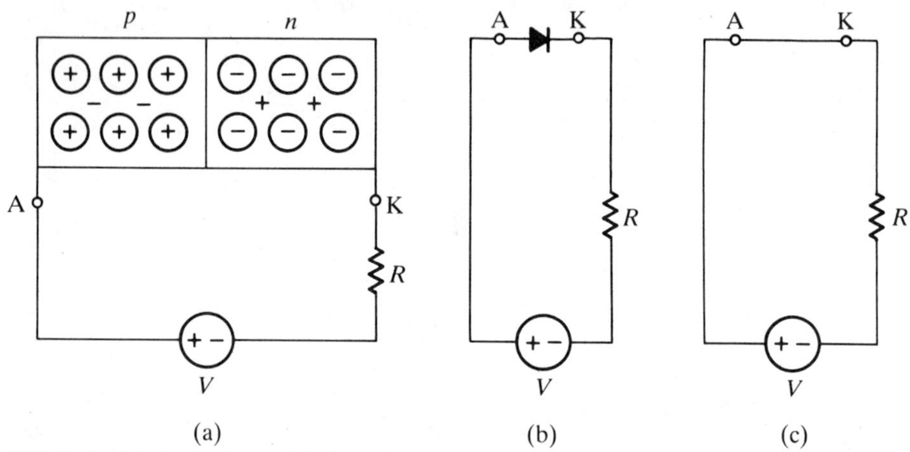

FIG. D-1 Diode: **(a)** forward-biased, **(b)** electric circuit representation, **(c)** equivalent circuit of (b).

The terminal of the diode that is connected to *p*-type material is the anode (A), and the terminal that is connected to *n*-type material is the cathode (K).

In an ideal diode, when *p*-type material is connected to the positive terminal of the battery, current will flow, and the diode is said to be "forward biased" or short-circuited. The operation of the diode can easily be understood by considering the equivalent circuits shown in Figs. D-1(b) and (c). As per Ohm's law, the current flowing in the circuit depends on the polarity of the voltage source and the resistance of the circuit.

When *p*-type material is connected to the battery's negative terminal (Fig. D-2(a)), no current will flow, and the diode is said to be "reverse biased" or open-circuited. The equivalent circuit representation of Fig. D-2(a) is shown in Fig. D-2(b). The switching function of the diode is illustrated in Fig. D-2(c).

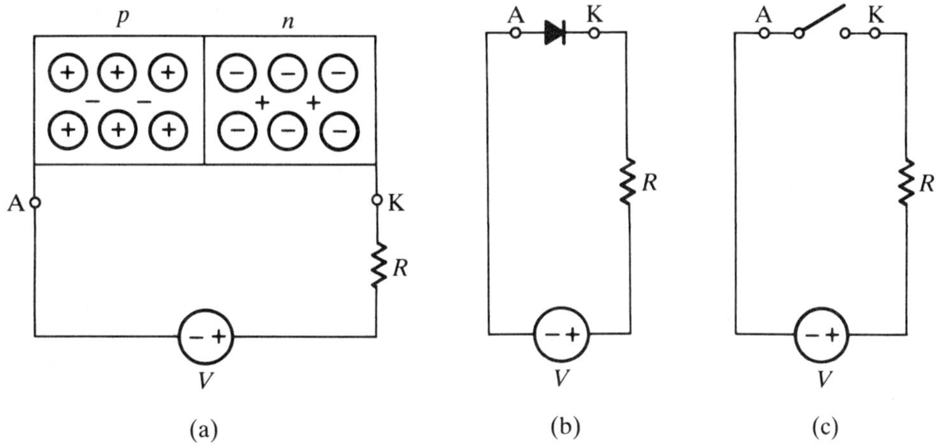

FIG. D-2 Diode: **(a)** reverse-biased, **(b)** electric circuit representation, **(c)** equivalent circuit of (b).

The actual voltage–current characteristic of a diode is shown in Fig. D-3. When the diode is forward biased, a potential of about .6 V must be applied before conduction starts. This voltage is referred to as the diode's threshold voltage. When the diode is biased in the reverse direction, about 4 μA will flow. However, when the reverse-biased potential increases above the diode's maximum voltage capability (breakdown voltage), the diode will fail or lose its unidirectional current-flow characteristic. The maximum voltage a diode can withstand in the reverse direction is called "peak inverse voltage" (PIV).

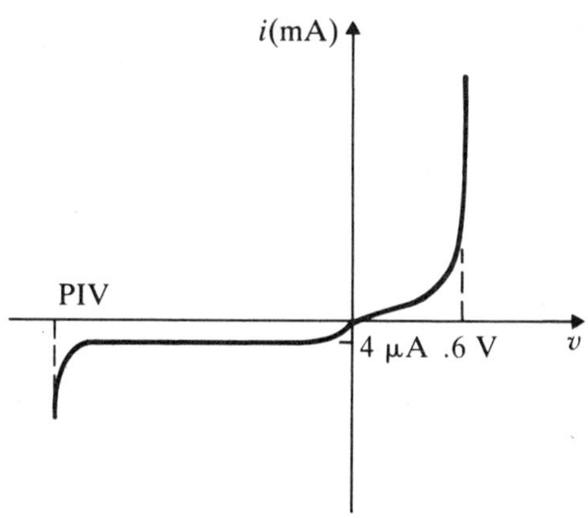

FIG. D-3 Voltage–current characteristic of a diode.

Figure D-4 illustrates the effects of an ideal diode on the output voltage waveform. When the input voltage is positive, the diode is forward biased. Thus, the output

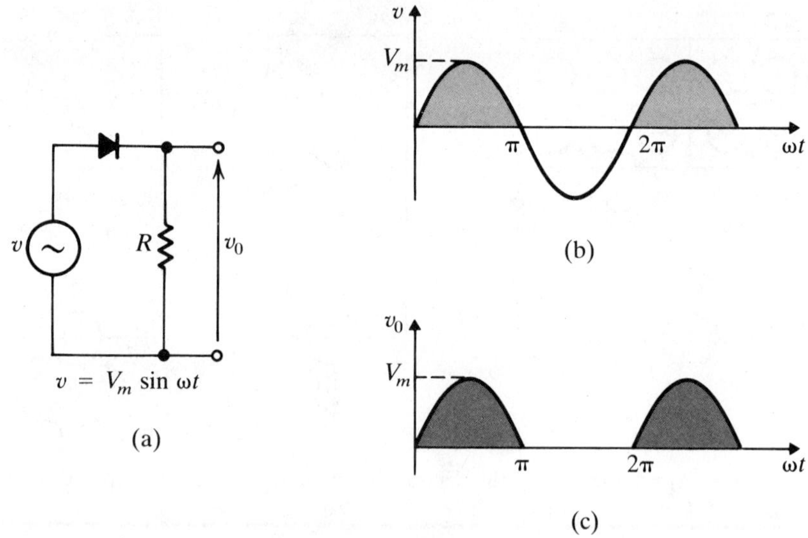

FIG. D-4 Diode: **(a)** circuit, **(b)** input voltage, **(c)** output voltage.

voltage is equal to the input voltage. When the input voltage is negative, the diode is reverse biased and no current flows. As a result, the output voltage is zero.

EXAMPLE D-1 For the circuit in Fig. D-4(a), determine:

a) An expression for the rms value of the output voltage.
b) The diode's minimum peak inverse voltage rating for a safety factor of 200%.

SOLUTION

a) By definition, the rms value V_o of the output voltage is

$$\begin{aligned} V_o^2 &= \frac{1}{T}\int_o^T v_o^2\,dt \\ &= \frac{1}{2\pi}\int_o^{\pi} (V_m \sin \omega t)^2\,d\omega t = \frac{V_m^2}{4\pi}\int_o^{\pi} (1 - \cos 2\omega t)\,d\omega t \\ &= \frac{V_m^2}{4\pi}\left[\omega t\Big|_o^{\pi} - \frac{1}{2}\sin 2\omega t\Big|_o^{\pi}\right] = \frac{V_m^2}{4} \end{aligned}$$

Thus,

$$V_o = \frac{V_m}{2}$$

b) Under nominal operating conditions the diode is subjected to a maximum reverse voltage of V_m V. For a safety factor of 200%, the diode's peak inverse voltage (PIV) capability should be

$$\text{PIV} = \underline{\underline{3V_m}}$$

ZENER DIODES

A Zener diode is a special type of semiconductor diode whose symbol representation and ideal characteristic are shown, respectively, in Figs. D-5(a) and (b). The outstanding characteristic of a Zener diode is that when it is reverse biased (current flows from cathode to anode), the voltage drop across the diode is constant. This voltage is designated as V_z and is commonly known as the Zener voltage, or the diode's threshold voltage. Each class of Zener diodes has its own Zener voltage. The particular Zener voltage is obtained by adding a specific amount of impurity to the *pn* material.

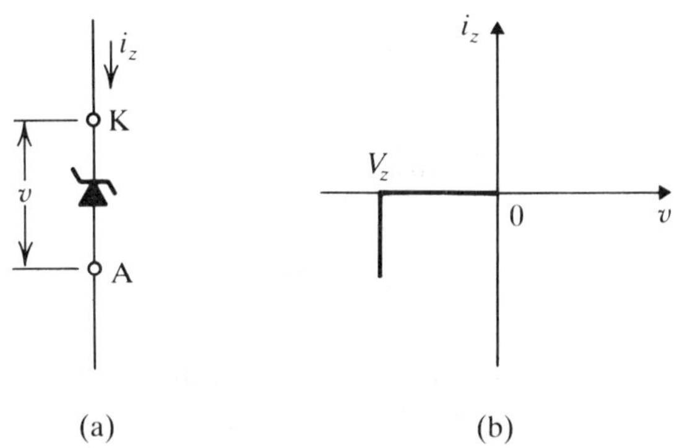

FIG. D-5 Zener diode: **(a)** symbol, **(b)** characteristic.

In Fig. D-6(a), the circuit's current i_z depends on the diode's mode of biasing. When the diode is forward biased,

$$i_z = \frac{v}{R} \qquad \text{(terminal A is at a higher potential than terminal K)}$$

When the diode is reverse biased,

$$i_z = \frac{v - V_z}{R}, \qquad \text{provided that } v > V_z$$

The biasing of the diode depends, of course, on the polarity of the applied potential. When the input voltage is a sinusoidal waveform, the output voltage and current waveforms are as shown in Fig. D-6(b).

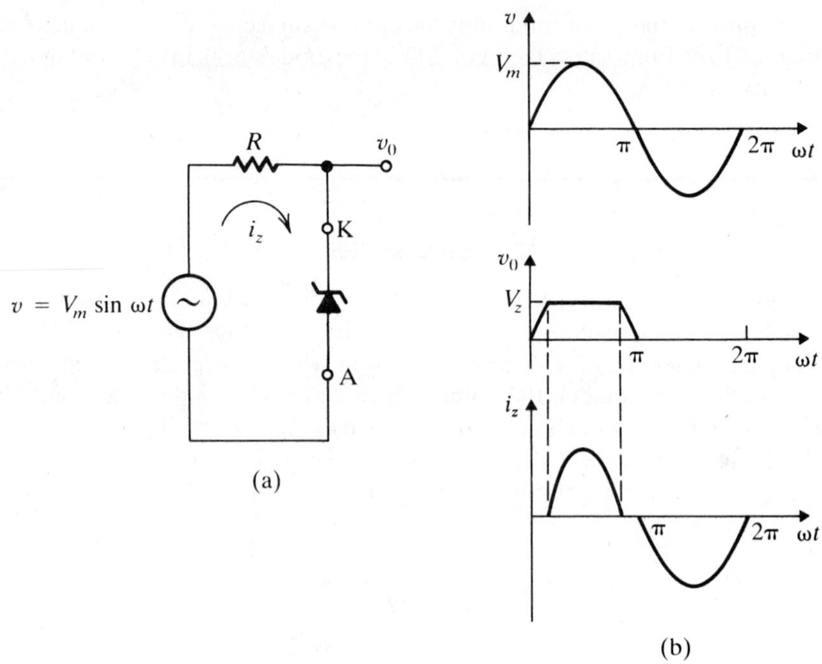

FIG. D-6 Zener diode: **(a)** basic circuit, **(b)** pertinent waveforms.

D.2 Transistors

Transistors are made from *pnp* or *npn* semiconductor material. As shown in Fig. D-7, they have three terminals: the base (B), the collector (C), and the emitter (E). Depending on the base current (I_B) and on the voltage from collector to emitter (V_{CE}), transistors may operate as amplifiers or as controlled switches. The discussion here will be restricted to the switching function of transistors. Figure D-8(a) shows the collector current (I_C) versus the voltage from collector to emitter (V_{CE}) as a function of the base current (I_B). (The collector current is in amperes, while the base current is in milliamperes.) These characteristics are the graphical representation of the mathematical principles or equations that interrelate the transistor's parameters.

The ratio of the collector current to the base current, represented by h_{fe},* is called the transistor's current gain. For dc operation,

$$h_{fe} = \frac{I_C}{I_B}$$

The current gain of the transistor ranges from 10 to 150.

• • •

* The subscripts of this parameter identify a *npn* transistor in which the input signal is connected to the base of the transistor and the output voltage is obtained from collector-to-ground.

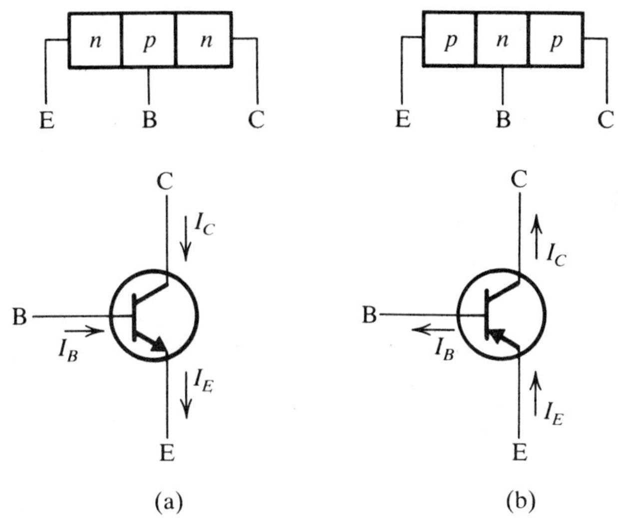

FIG. D-7 Transistor's polarity and symbol: **(a)** *npn*, **(b)** *pnp*.

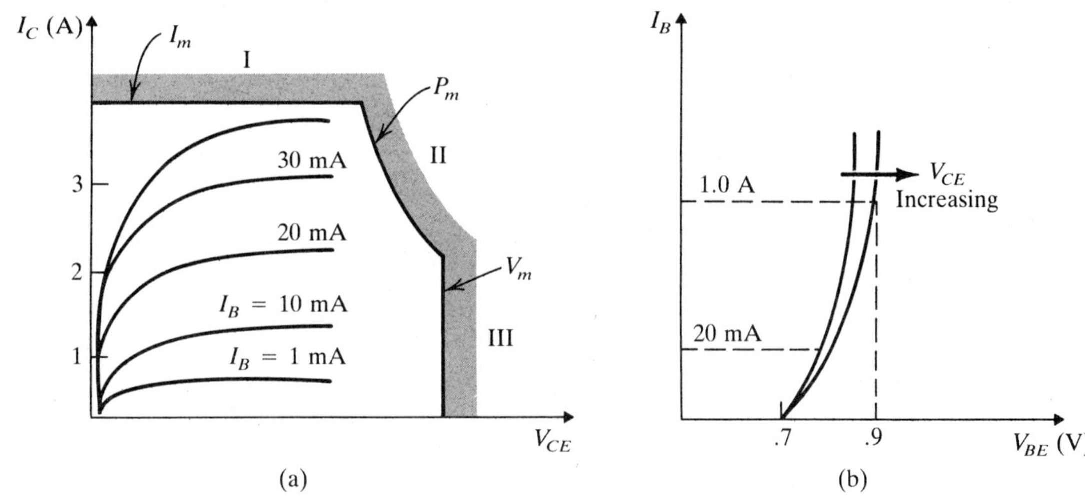

FIG. D-8 Transistor's characteristics (*npn*): **(a)** I_C versus V_{CE}, **(b)** I_B versus V_{BE}.

The power dissipated within a transistor, at any particular operating condition, is given by the product of the actual collector current (I_C) times the voltage from collector to emitter (V_{CE}); that is,

$$P_{\text{loss}} = V_{CE} I_C \qquad \textbf{(D.1)}$$

Under normal operating conditions the power dissipated must be less than or equal to the power rating (P_r) of the transistor. Since for a given transistor the power rating is constant, the maximum permissible power dissipation for a transistor is a curve of an equilateral hyperbola (the product of two variables is equal to a constant).

Depending on the particular device, the transistor's temperature, collector current, voltage (V_{CE}), and power consumption must be below a certain level. Otherwise the device will be permanently damaged. The limiting values are shown in Fig. D-8(a).

Figure D-8(b) shows the base current versus the voltage from base to emitter (V_{BE}) as a function of the voltage from collector to emitter. It should be noted that if the base-to-emitter voltage is increased slightly above .7 V,* the base current—and subsequently the collector current—will increase substantially. This is the process by which the transistor is driven into saturation ($V_{CE} \approx 0$). When the base-to-emitter voltage is reduced below .7 V, the base current becomes zero and the transistor acts as an open circuit ($I_C = 0$). Figure D-9(a) shows a typical transistorized circuit. The output voltage of the transistor is usually obtained across the collector-to-emitter terminals.

The circuit represented in Fig. D-9(b) is an alternative to (a). Electrically, the two circuits are identical. Figure D-9(c) is obtained by applying Thevenin's theorem to the circuit in Figure D-9(b). From basic circuit theory, Thevenin's equivalent resistance (R_{th}) and voltage (V_{th}) are given by the following relationships:

$$R_{th} = R_B = \frac{R_1 R_2}{R_1 + R_2}$$

$$V_{th} = V_{BB} = R_2 \frac{V_{CC}}{R_1 + R_2}$$

Applying KVL in the collector circuit of Fig. D-9(c), we obtain

$$V_{CC} - I_C R_L - V_{CE} = 0$$

from which

$$I_C = -\frac{V_{CE}}{R_L} + \frac{V_{CC}}{R_L} \tag{D.2}$$

where V_{CC} is the collector supply and V_{CE} is the voltage from collector to emitter. Equation (D.2) is of the slope-intercept form and is referred to as the dc load-line equation. The intersection of the load line and the curve that represents the actual base current gives the operating parameters of the transistor; that is, it furnishes the base current, the collector current, and the voltage from collector to emitter. These parameters, for a given operating condition, are identified by the subscript q as follows:

$$I_{B_q}, \qquad I_{C_q} \quad \text{and} \quad V_{CE_q}$$

* This is the nominal conduction voltage for silicon-type semiconductor material.

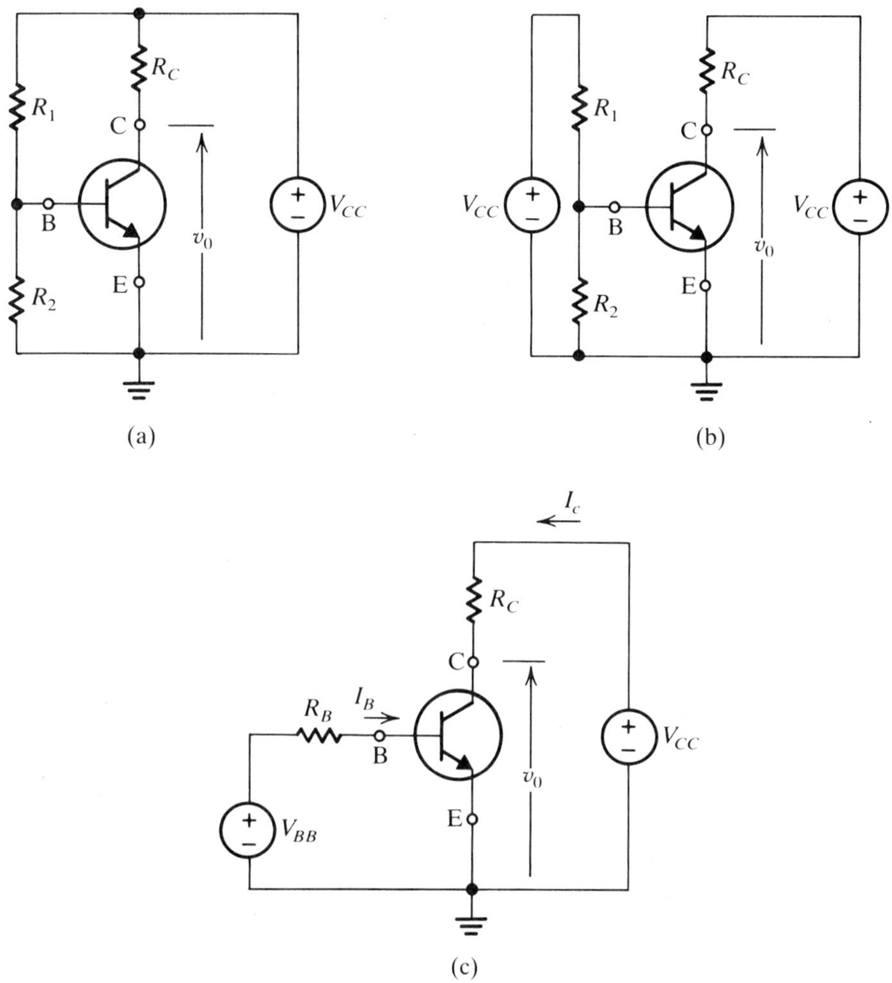

FIG. D-9 DC analysis of transistors: **(a)** typical circuit, **(b)** equivalent circuit of (a), **(c)** Thevenin's equivalent of (b).

When the transistor in Fig. D-10 operates as a switch, the base current varies from zero to a value of I_{B_n}. The resulting dc operating points are identified by Q_2 and Q_1. The regions identified by the numerals I and II represent, respectively, the saturation and cut-off regions. Saturation and cut-off correspond, respectively, to short-circuiting and open-circuiting the transistor. When the transistor is short-circuited, the ac output voltage (v_o) is equal to zero, and when the transistor is cut off, the output voltage is equal to the collector's dc supply V_{CC}.

In Fig. D-11(a), switch S is operated electronically (by means not shown in the diagram) at a rate that controls the flow of base current and, as a result, the switching function of the transistor. For the base current to flow, it must be of sufficient magnitude to saturate the transistor. Thus, for all practical purposes, the output

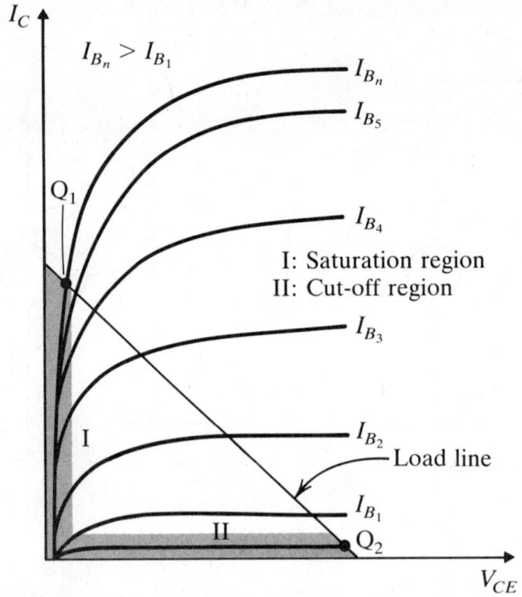

FIG. D-10 Operating points of a transistor.

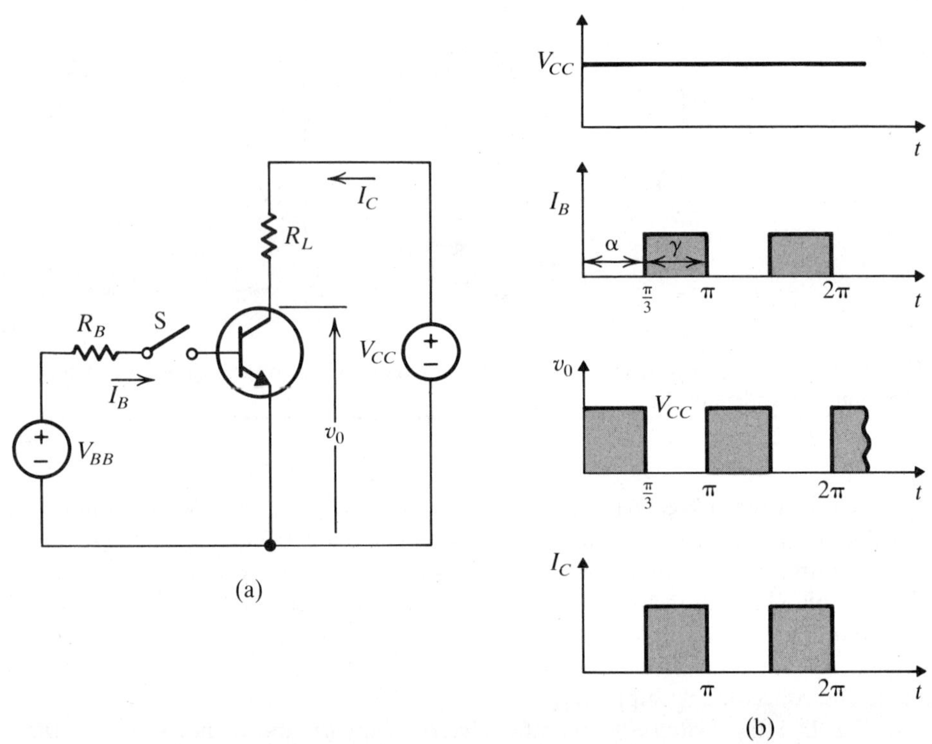

FIG. D-11 The transistor as a switch—dc supply: **(a)** circuit, **(b)** input–output waveforms.

voltage is equal to zero. The time interval during which the base current flows defines the transistor's conducting period, identified by the angle γ. The time interval during which the base current is switched OFF defines the transistor's nonconducting period, identified by the angle α. This angle is referred to as the firing, or triggering, angle of the transistor.

Figure D-11(b) shows the output voltage waveform that results from a firing angle of $\pi/3$ radians and a conduction angle of $2(\pi/3)$ radians.

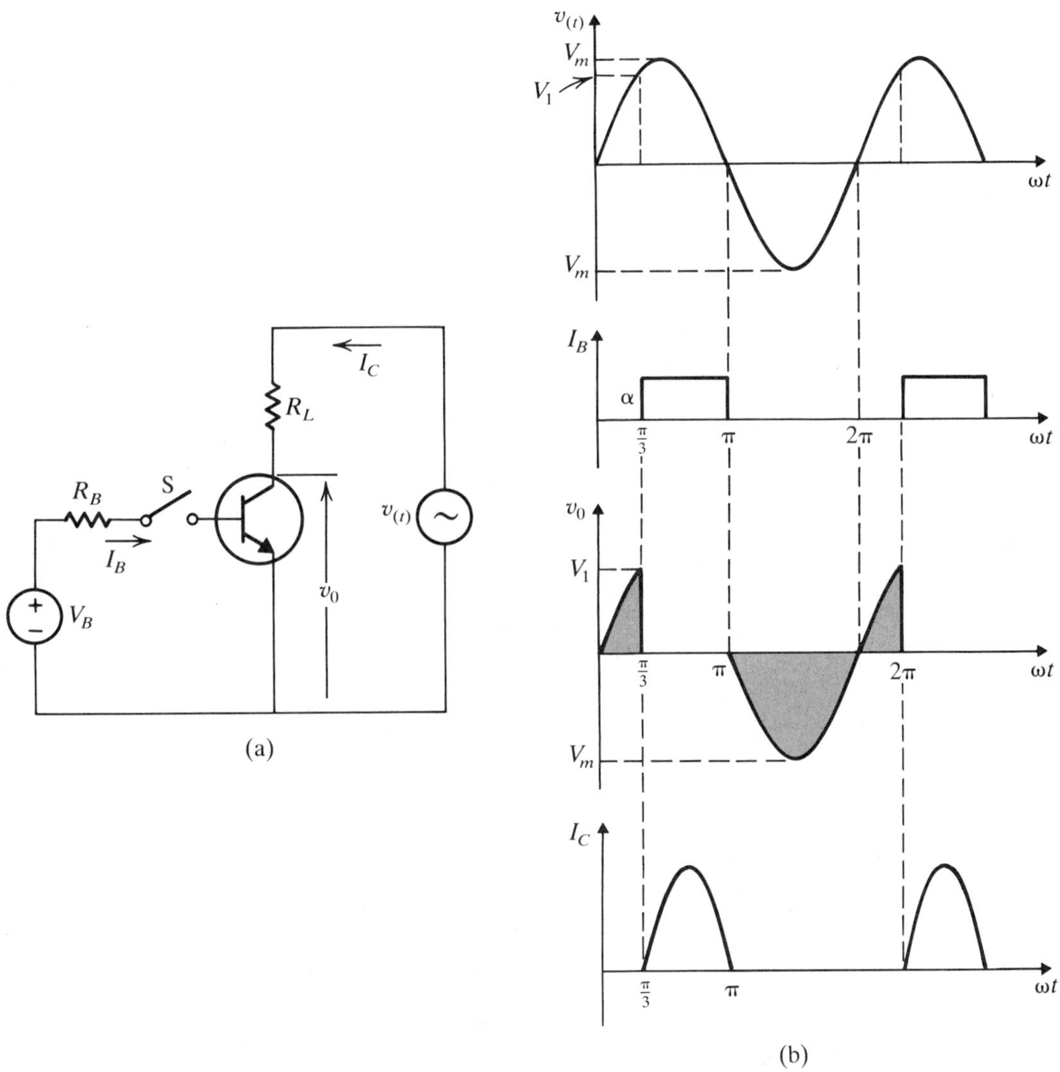

FIG. D-12 The transistor as a switch—ac supply: **(a)** circuit, **(b)** input–output waveforms.

It should be noted that when the polarity of the ac supply is negative, the transistor is switched OFF—regardless of base current. This is because the negative voltage reverse-biases the emitter junction and no current can flow.

EXAMPLE D–2 For the circuit shown in Fig. D-12(a):

a) Derive an expression for the dc component of the output voltage as a function of the supply voltage and the transistor's firing angle.

b) Evaluate the dc component of the output voltage for

$$v = 100 \sin \omega t \text{ V} \quad \text{and} \quad \alpha = \frac{\pi}{3} \text{ radians}$$

SOLUTION

a) When the transistor does not conduct (is switched OFF), the voltage from collector to emitter (v_o) is equal to the supply voltage. Since the transistor's nonconducting period is given by the firing angle (α), then from basic considerations we have

$$V_{av_o} = \frac{1}{2\pi} \int_o^{2\pi} v \, dt$$

$$V_{av_o} = \frac{1}{2\pi} \left[\int_o^{\alpha} V_m \sin \omega t \, d\omega t + \int_{\pi}^{2\pi} V_m \sin \omega t \, d\omega t \right]$$

The limits of integration can easily be established from Fig. D-12(b). From the above we obtain

$$V_{av_o} = -\frac{V_m}{2\pi} \left[\cos \omega t \Big|_o^{\alpha} + \cos \omega t \Big|_{\pi}^{2\pi} \right]$$

$$= -\frac{V_m}{2\pi} (1 + \cos \alpha)$$

b) Substituting, we obtain

$$V_{av_o} = -\frac{100}{2\pi} (1 + \cos 60^\circ) = -23.87 \text{ V}$$

D.3 Thyristors

Thyristors are three-terminal, solid state devices that can electronically control currents up to 50 kA and voltages up to 5 kV. They constitute the main control element of all high-power, solid state controls.

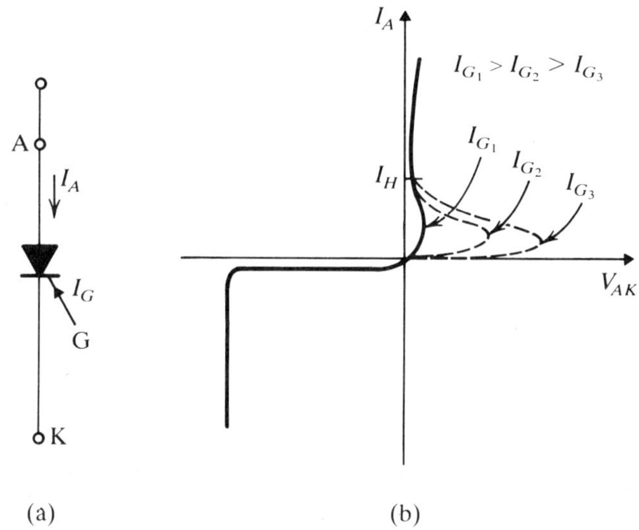

FIG. D-13 Thyristor: **(a)** symbol, **(b)** *i-v* characteristic.

The thyristor's three terminals, as shown in Fig. D-13(a), are identified as anode (A), cathode (K), and gate (G). These terminals correspond to the collector (C), emitter (E), and base (B) terminals of transistors. A thyristor's typical current–voltage characteristic is shown in Fig. D-13(b). The current–voltage characteristic depends on the operating temperature, applied voltage, and gate current.

As control switches, thyristors operate either at the saturation (short-circuited) or cut-off (open-circuited) region. Their switching, or firing, requires a proper anode voltage supply V_{AA} (Fig. D-14(a)) and a suitable *gate current*. The latter is normally a pulse whose amplitude is less than 500 mA and whose duration is less than 500 μs.

Once switched ON, thyristors with positive supply voltage V_{AA} continue to operate (conduct) regardless of gate current. Transistors are switched OFF once their base current is reduced below a nominal value, while thyristors continue to conduct even after their gate-current pulse is reduced to zero. Thyristors are switched OFF when their anode-to-cathode current is reduced below a specific minimum value, the so-called threshold current (I_H). The reduction of the current below the threshold limit depends on the load impedance and the applied positive voltage. The thyristor will remain in the OFF state provided that the positive supply voltage is not reapplied within the time interval given by the thyristor's "turn-OFF" time. The "turn-OFF" time is the time required for the reduction of the concentration and mobility of the charge carriers that resulted from the operating temperature. The "turn-OFF" time is usually less than 100 μs.

When the voltage supply is of alternating form, at a certain point in the voltage cycle the conduction current will be reduced to less than the threshold limit. Thus the thyristor will be switched OFF. It will not restart conduction until proper gate pulse and suitable anode voltage are reapplied.

When commutation of a thyristor (the process the thyristor goes through before its current is reduced to zero) is initiated by the alternating nature of the driving voltage,

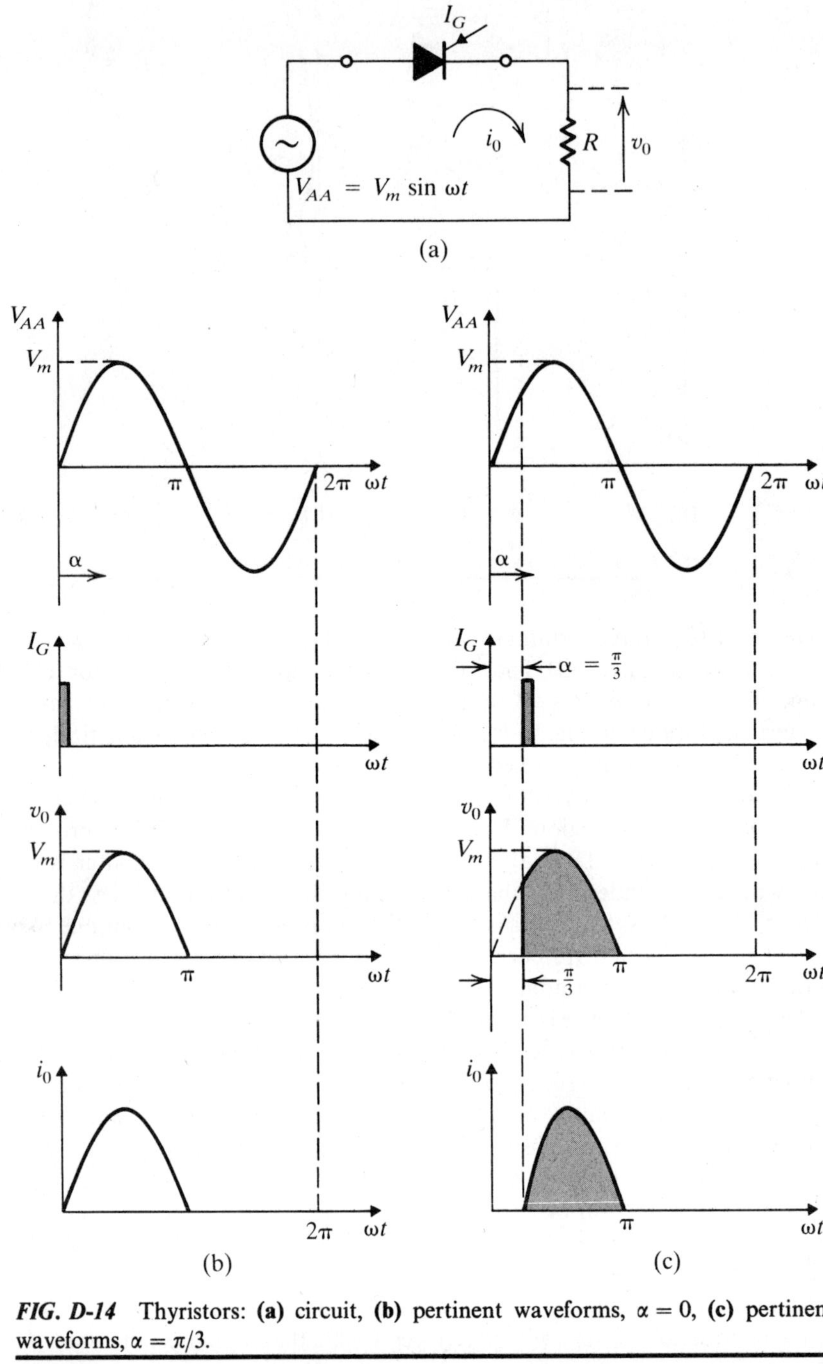

FIG. D-14 Thyristors: **(a)** circuit, **(b)** pertinent waveforms, $\alpha = 0$, **(c)** pertinent waveforms, $\alpha = \pi/3$.

the thyristor is said to be phase, ac, or line commutated. In some circuits, such as those used to convert dc sources to ac supplies (inverters), commutation is more difficult to accomplish. Since the driving, or input, voltage is dc, the switching OFF of the thyristor (reduction in the driving voltage) requires properly designed external or load circuits. In such cases the thyristor is said to be externally or load commutated.

The instant at which the current pulse is introduced in the gate circuit, relative to the anode-to-cathode voltage supply, defines the firing angle* (α) of the thyristor. Figure D-14 shows pertinent voltage waveforms that result from a simple half-wave controlled rectifier for a firing angle of 0° and of 60°. It should be noted that when the thyristor's anode-to-cathode potential becomes negative, conduction is discontinued, and thus the voltage across the load is zero.

EXAMPLE D–3 For the circuit shown in Fig. D-15(a), sketch the voltage across the thyristor for a firing angle of 30°.

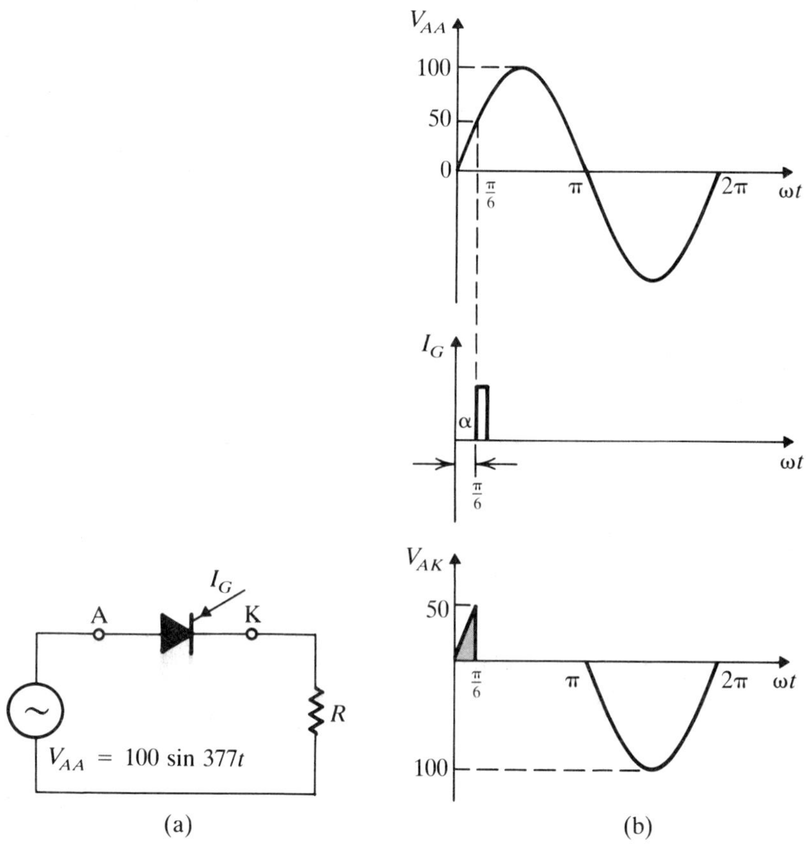

FIG. D-15(a) and (b)

• • •

* This angle is also referred to as the switching angle, the triggering angle, or the angle of retard.

SOLUTION The input voltage waveform and the voltage across the thyristor are shown in Fig. D-15(b). The reference point for the measurement of the thyristor's firing angle is the origin of the supply voltage. Thus, when the gate pulse is applied at $\pi/6$ radians, the input voltage is $100 \sin 377t \,|_{\pi/6} = 50$ V. It should be noted that when the thyristor is switched OFF, the thyristor voltage (V_{AK}) is equal to the supply voltage.

E BASIC ECONOMIC CONSIDERATIONS

The following formulae are very useful for the economic analysis of various subjects. These formulae are based on simple interest compounded once per year.

E.1 "Future Worth" of a One-Time Investment

The future worth of a one-time investment is given by

$$F = P(1 + r)^n \tag{E.1}$$

where

F = the future worth

P = the present worth of money deposited

r = the rate of interest compounded once per year

n = the number of years.

E.2 "Future Worth" of Equal Annual Installments

The future worth of equal annual installments is given by

$$F = A\left[\frac{(1 + r)^n - 1}{r}\right] \tag{E.2}$$

where

A represents the amount of equal annual installments

F, r, and n have the same meaning as in Eq. (E.1).

E.3 "Present Worth" of Equal Annual Installments

The present worth of equal annual installments is given by

$$P = A\left[\frac{(1 + r)^n - 1}{r(1 + r)^n}\right] \tag{E.3}$$

where

P is the present worth of the equal annual installments
A, r, and n have the same meaning as in Eq. (E.2).

E.4 "Present Worth" of Uniform Incremental Increases in the Annual Deposits

The present worth that results from uniform annual deposit increases is given by

$$P = A\left[\frac{(1 + r_x)^n - 1}{r_x(1 + r_x)^n}\right] \tag{E.4}$$

where

$$r_x = \frac{1 + r}{1 + r_e} - 1 \tag{E.5}$$

where r_e is the uniform annual percentage increase of the parameter under consideration. The other parameters are the same as in the previous expressions. Equations (E.4) and (E.5) are very useful in evaluating the present worth of equipment energy consumption while the latter's cost increases at a constant annual rate.

EXAMPLE E–1 One thousand dollars is deposited in a bank at a simple interest rate of 10% for 5 years. Find its worth at the end of the fifth year.

SOLUTION Substituting the given data into Eq. (E.1), we obtain

$$F = 1000(1 + .1)^5 = \underline{\underline{\$1,610.50}}$$

EXAMPLE E–2 At the beginning of each year, for 10 years, a businessman makes a bank deposit of \$4000. The annual interest rate is 12%. Find:

a) The worth of these deposits at the end of 10 years.
b) The equivalent present worth of these deposits.

SOLUTION

a) Substituting the given data into Eq. (E.2), we obtain

$$F = 4000\left(\frac{1}{.12}\right)[(1 + .12)^{10} - 1]$$
$$= \underline{\underline{\$70,195}}$$

b) The present worth is found by substituting the given data into Eq. (E.3).

$$P = 4000 \frac{(1 + .12)^{10} - 1}{.12(1 + .12)^{10}} = \underline{\underline{\$22{,}601}}$$

EXAMPLE E–3 A 25-year mortgage of \$50,000 is obtained at a 12% annual interest rate. Find:

a) The monthly installments.

b) The actual money paid by the borrower over the 25-year period.

c) The effective worth of the mortgage at the end of 25 years.

SOLUTION

a) From Eq. (E.3), the equivalent annual payments are

$$A = \frac{50{,}000(.12)(1 + .12)^{25}}{(1 + .12)^{25} - 1} = \$6{,}375/\text{year}$$

Thus the monthly payment is

$$\frac{\$6{,}375}{12} = \underline{\underline{\$531.25/\text{month}}}$$

b) The actual money paid by the borrower over the 25-year period is

$$12 \times 25 \times 531.25 = \underline{\underline{\$159{,}375}}$$

c) The effective worth of the mortgage at the end of 25 years is found from Eq. (E.1).

$$F = 50{,}000(1 + .12)^{25} = \underline{\underline{\$850{,}003}}$$

This is the money you would receive if you deposited \$50,000 in the bank and left it there for 25 years at an annual interest rate of 12%.

EXAMPLE E–4 Find the present-worth cost of 1 kW of power that is continuously consumed over a 5-year period, assuming that the energy cost is 6¢/kWh and increases annually at 6%. The nominal interest rate is 10% per year.

SOLUTION From Eq. (E.5),

$$r_x = \frac{1.1}{1.06} - 1 = .038$$

and

$$A = (.06)(365 \times 24) = \$525.6/\text{kW}$$

Substituting into Eq. (E.4), we obtain

$$P = 525.6\left[\frac{(1.038)^5 - 1}{.038(1.038)^5}\right] = \underline{\underline{\$2{,}354.84/\text{kW}}}$$

PROBLEMS

1. **a)** The electrical rates of a power company in North America are as follows:

Demand charge:	$5.00 per kW
Energy charge:	4 cents per kWh
Reactive power charge:	.35 cents per kVAh

Calculate the total cost per month for a plant that operates continuously and draws 8 MVA at a power factor of .8 lagging.

b) Why should power utilities compensate consumers whose plants operate at leading power factors?

2. Refer to the accompanying partial invoice of an industrial plant. The total cost includes charges of energy and power demand. The minimum power factor requirement is 95%.

a) Determine the actual power demand for each month and the cost of electricity per kW and per kWh for the months of August and September.

b) Determine the penalty due to the plant's poor power factor.

c) Estimate the magnitude (kVAR) and the cost ($10/kVAR) of the capacitor bank that is required in order to improve the power factor to the minimum required level.

d) Show that the utilization factor for the month of August is 41.1%.

Partial Invoice of an Industrial Plant

Month	*Billing Period (days)*	*Peak Demand Charged in kW*	*Energy Consumption in kWh*	*Utilization Factor (U.F.)* in Percent*	*Actual Power Factor in Percent*	*Total Cost*
July	30	6110	1,795,200	44	88.1	$ 69,282
Aug.	32	6110	1,790,400	41.1	88.1	$ 71,659
Sept.	28	5928	1,464,000	40.5	86.2	$ 59,845
Oct.	31	6293	1,588,000	34.2	94.2	$ 67,921
Nov.	28	6840	1,977,600	43.2	94.7	$ 91,030
Dec.	32	8064	2,784,000	45	97.7	$138,864

$$*\ U.F. = \frac{\text{Actual energy consumed}}{\text{Maximum possible energy consumption}}$$

3. A three-phase 208 V, 60 Hz power line has two loads connected to it. The first is delta-connected and draws 25 kW at .70 power factor lagging. The second is star-connected and draws 6.25 kVA at .80 power factor leading. What are the total line current and the combined power factor?

4. A balanced star-connected load having a phase impedance of $(8 + j6)\ \Omega$ is connected to a three-phase supply of 480 V. Determine:
   - **a)** The line-to-neutral voltage.
   - **b)** The phase current.
   - **c)** The line current.
   - **d)** The power factor of the load.
   - **e)** The power consumed by the load.
   - **f)** The equivalent delta-connected load.

5. The per-phase impedance of a 480 V delta-connected load is $4 + j3\ \Omega$. The load is supplied through a transmission line whose per-phase impedance is $.1 + j.3\ \Omega$. Determine the efficiency of transmission and the source voltage.

6. For the 480 V, 3-ϕ, 60 Hz system shown in Fig. 1:
   - **a)** Calculate the phase and the line currents, and draw a phasor diagram showing all phase currents, line currents, and line voltages.
   - **b)** Calculate the reading of each wattmeter and the total power.
   - **c)** Repeat (b) if $Z = 20 + j60\ \Omega$.

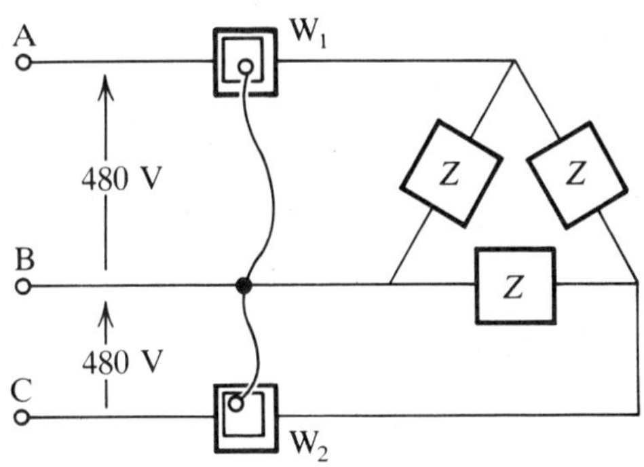

Phase sequence: ABC
$Z = (17.32 + j10)\ \Omega$

FIG. 1

7. The phase sequence of the voltages applied to the three-phase network of Fig. 2 is ABC. Determine the magnitude of the line currents by using any three of the following methods:
   - **a)** The concept of the power triangle.
   - **b)** The per-phase analysis using ohmic values.
   - **c)** The per-unit values.
   - **d)** KCL, without changing the delta load to its equivalent star.

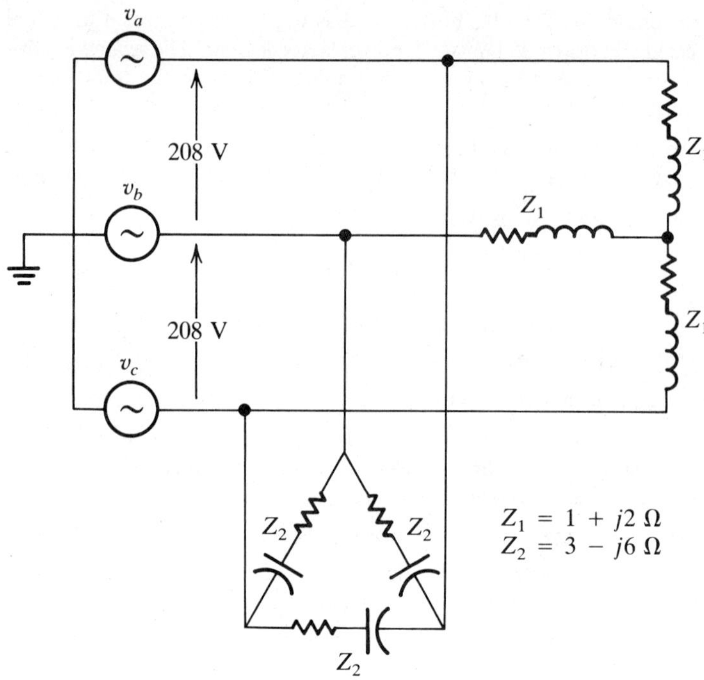

FIG. 2

8. The circuit shown in Fig. 3 is used to identify the phase sequence of the line voltages. Prove that for a phase sequence ABC the voltmeter's indication is less than the line-to-line voltage, while for a phase sequence ACB the voltmeter's indication is larger than the line-to-line voltage. (*Hint:* The voltmeter's internal impedance is relatively very high.)

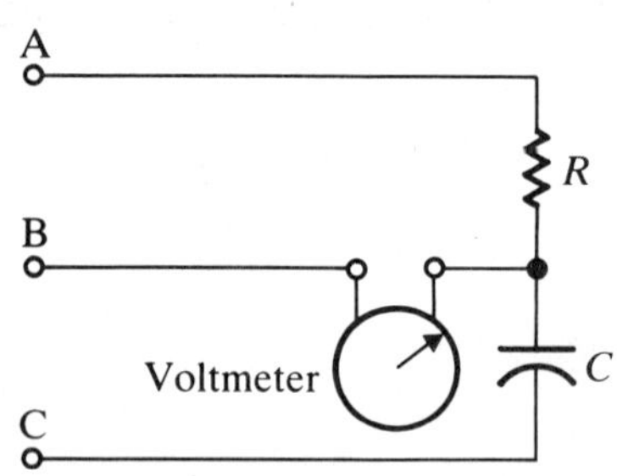

FIG. 3

9. A 3-ϕ, 480 V, 500 kW, .9 Pf lagging, .95 efficient load receives rated current through a transmission line whose per-phase impedance is $.01 + j.04\ \Omega$. Determine:
   a) The voltage across the load.
   b) The capacitor bank that when connected across the load will maintain its voltage at 480 V.

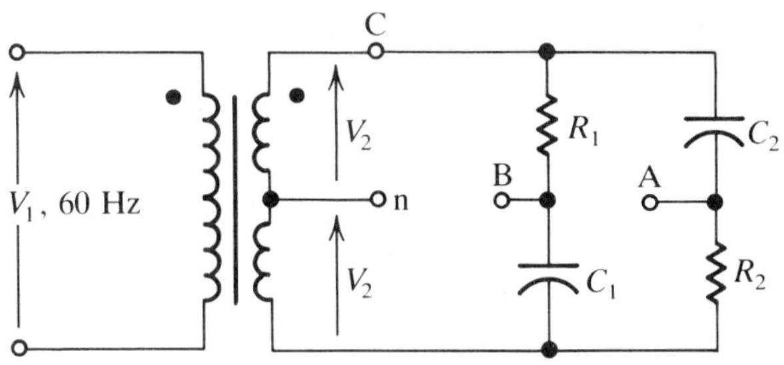

FIG. 4

10. The circuit shown in Fig. 4 is used to produce balanced three-phase output voltages (V_{nA}, V_{nB}, V_{nC}) from a single-phase supply.

a) Show that the steady-state output voltages are

$$V_{nC} = V_2$$

$$V_{nB} = V_2\left(\frac{j\omega\tau_1 - 1}{1 + j\omega\tau_1}\right) = V_2\underline{/180^\circ - 2\theta_1}$$

$$V_{nA} = V_2\left(\frac{1 - j\omega\tau_2}{1 + j\omega\tau_2}\right) = V_2\underline{/-2\theta_2}$$

where

$$\tau_1 = C_1R_1 \qquad \theta_1 = \arctan \tau_1\omega$$

and

$$\tau_2 = C_2R_2, \qquad \theta_2 = \arctan \tau_2\omega$$
$$\omega = 377 \text{ rad/s}$$

b) Determine the time constants τ_1 and τ_2 so that the output voltages are at 120° to each other.

c) If $R_1 = R_2 = 50$ kΩ, find C_1 and C_2.

d) If the phase sequence is ABC, what color should each line wire be?

11. A 3-ϕ, 4160-480/277 V, 1500/2000 kVA transformer has a leakage impedance of .01 + j.0575 pu. Determine:

a) The ohmic values of the leakage impedance.

b) The short-circuit MVA when the transformer's secondary is shorted while the off-load voltage taps are at 1.05 pu.

12. For the power system shown in the one-line diagram of Fig. 5:

a) Determine the per-unit value of each reactance on the base of its own machine rating.

b) Determine the per-unit value of each impedance on the base of the generator rating.

c) Draw the one-line impedance diagram with values obtained in (b) above. Indicate clearly the bases of voltages, currents, and impedances.

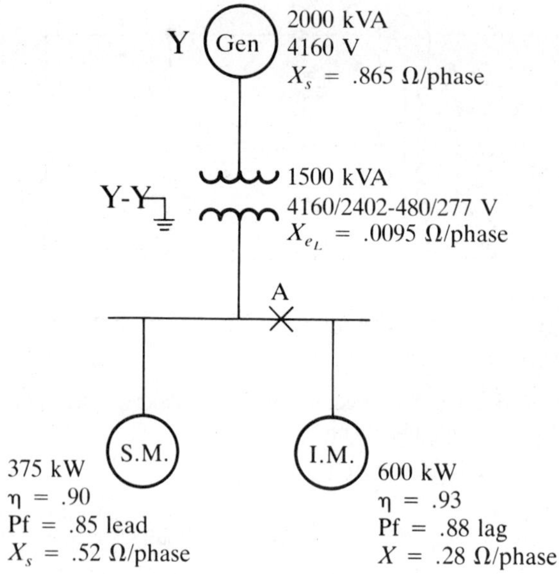

FIG. 5

d) Draw the per-phase equivalent circuit and calculate the line-to-line voltage on bus A. Assume that the motors draw rated currents.

13. For the circuit shown in Fig. 6, sketch the output voltage (v_o) and determine the power produced by the 12 V source. The input voltage is a periodic function with a period of 2 s.

$$v_i = 20t, \text{ V}, \qquad 0 \leqslant t \leqslant 1 \text{ second}$$
$$v_i = 20 \text{ V}, \qquad 1 \leqslant t \leqslant 2 \text{ seconds}$$

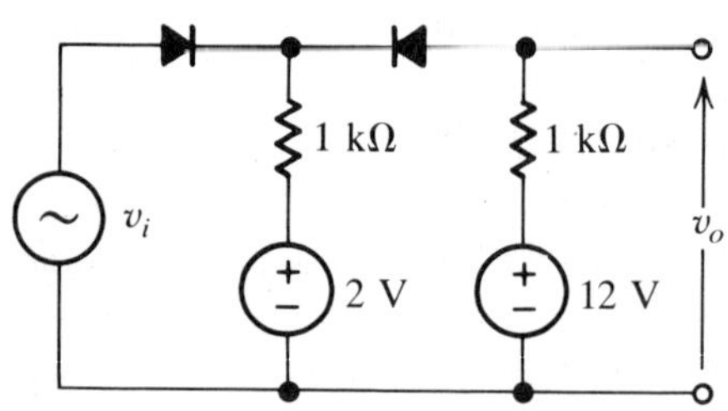

FIG. 6

14. The circuit shown in Fig. 7 is used to supply rated voltage to the field coil of a separately excited dc motor. The rectifier does not draw any appreciable current from the ac source.

a) Sketch the input voltage to the rectifier, assuming that $V_z = 100$ V.

b) What are the functions of the Zener diodes?

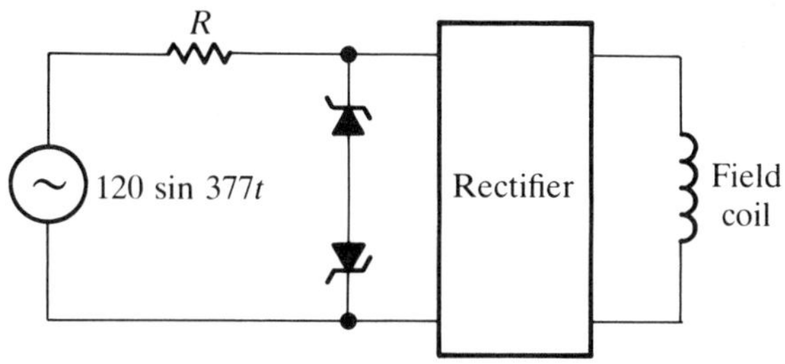

FIG. 7

15. The control circuit of Fig. 8 does not draw any appreciable current from the ac source. If the threshold voltage of the Zener diodes is $V_z = 120$ V.

   a) Sketch the voltage to the control circuit and the current through the diodes.

   b) Determine the minimum power rating of the diodes.

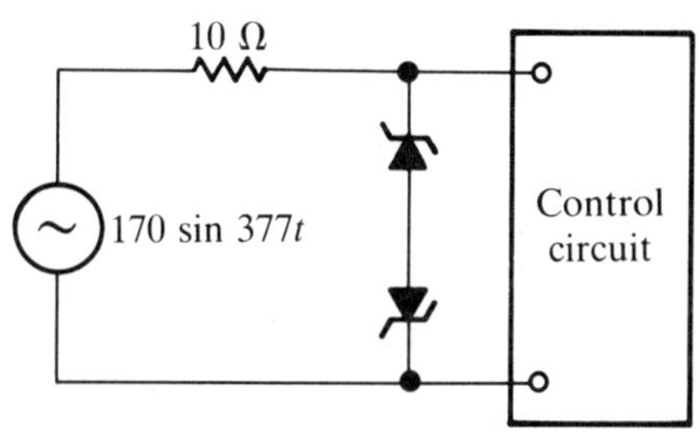

FIG. 8

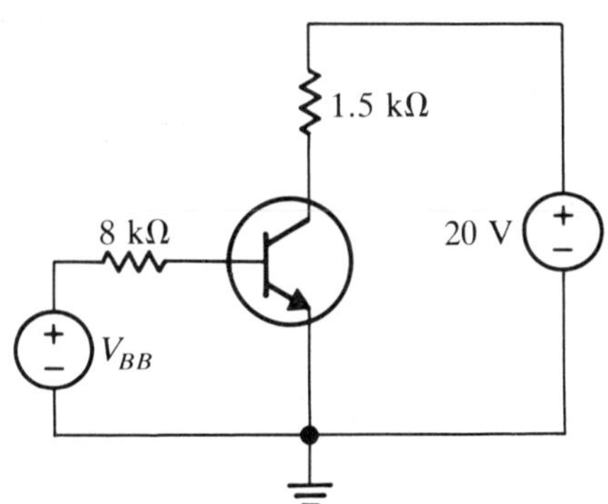

FIG. 9

16. For the circuit shown in Fig. 9, determine the base voltage V_{BB} needed to saturate the transistor, given that $(V_{CE})_{sat} = .2$ V, $(V_{BE})_{sat} = .8$ V, and $h_{fe} = 60$.

17. The circuit shown in Fig. 10 is used to energize (apply voltage) or deenergize the relay X whose resistance is $R\ \Omega$. When the switch (S) is closed, the transistors are saturated

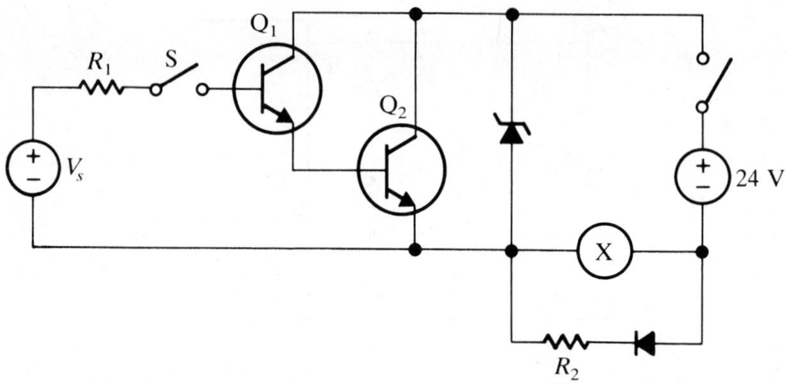

FIG. 10

($V_{CE_2} = 2$ V, $h_{fe} = 10$); when the switch is open, the transistors are switched OFF. The characteristics of the Zener diode are

$$V_z = 20 \text{ V}, \qquad I_{z_{max}} = 100 \text{ mA}$$

Making reasonable assumptions, determine:

a) The resistance of the relay.

b) The maximum current through the collector of transistor Q_1.

c) The functions of the diode–resistor network and of the Zener diode.

18. The power drawn by the 25 Ω resistive load shown in Fig. 11 is controlled through a thyristor that has the following characteristics:

V_{GK} (V)	I_G (mA)	I_H (mA)	Forward Voltage Drop V_{AK} (V)
.8	20	150	1.5

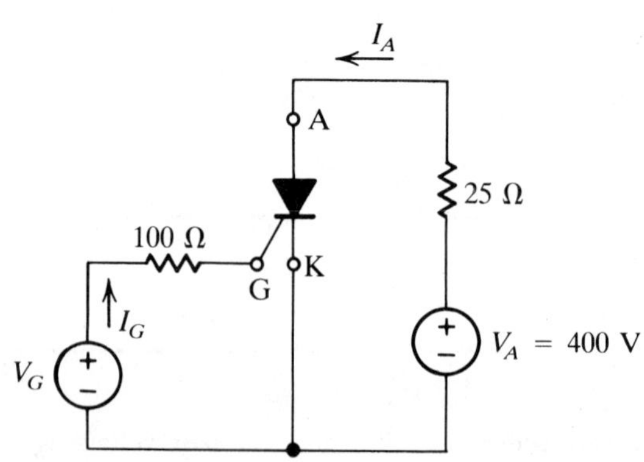

FIG. 11

Determine:

a) The gate supply voltage V_G.

b) The load current.

c) The rate of heat dissipation in the thyristor.

d) The minimum value of the supply voltage V_A below which conduction will cease.

19. The thyristor whose rating was given in Problem 18 is used to control the power to a resistive load as shown in Fig. 12. Neglecting the forward voltage drop:

a) Sketch the output current waveform, and determine the power consumed by the load and the effective power factor as seen by the 220 V source.

b) Repeat (a) if $v_g = 5 \sin(377t - 60°)$.

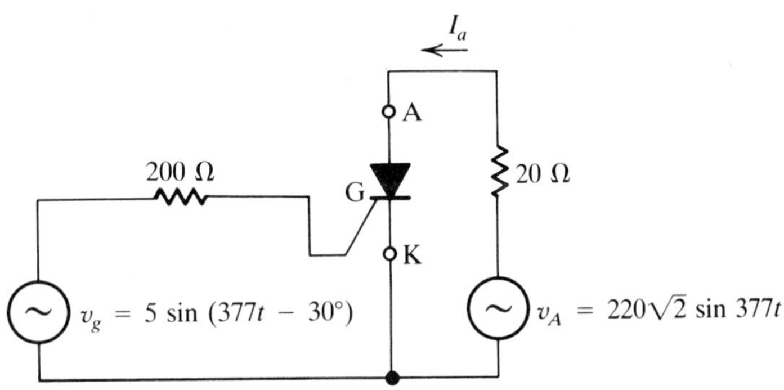

FIG. 12

F TABLES

F.1 Constants and Conversion Factors

CONSTANTS

Permeability of free space:	$\mu_o = 4\pi \times 10^{-7}$ H/m
Permittivity of free space:	$\varepsilon_o = 8.854 \times 10^{-12}$ F/m
Velocity of light:	$C = 2.998 \times 10^8$ m/s

CONVERSION FACTORS

Energy	1 Joule (J)	= 1 watt-second
		= .7376 foot-pound (ft-lb)
		= 10^7 ergs
		= 9.48×10^{-4} British Thermal Units (BTU)
Length:	1 m	= 3.281 ft
		= 39.37 in.
Mass:	1 kg	= 2.205 lb_m
		= .0685 slug
Force:	1 newton (N)	= .225 lb_f
		= 7.23 poundals
		= 10,000 dynes
		= 102 grams
Torque:	1 newton · meter (N · m)	= .738 ft · lb
		= 141.7 oz · in
Power:	1 watt (W)	= 1.341×10^{-3} hp
Moment of inertia:	1 kg · m^2	= 23.7 lb · ft^2
		= .732 slug
Magnetic flux:	1 weber (Wb)	= 10^8 maxwells (lines)
Magnetic flux density:	1 tesla (T)	= 1 Wb/m^2
		= 10,000 gauss
		= 64.5 $kilolines/m^2$

Magnetizing force: $\dfrac{1 \text{ ampere-turn}}{\text{meter}} = \dfrac{.0254 \text{ ampere-turn}}{\text{in.}}$

F.2 ANSI Standard Device Function Numbers

Table F-1 gives a partial list of *American National Standards Institute* (*ANSI*) standard device function numbers.

TABLE F-1 ANSI standard device function numbers (partial list)

Device Number	*Definition and Function*
26	Apparatus thermal device functions when the temperature of the shunt field or the amortisseur winding of a machine, or that of a load-limiting or load-shifting resistor or of a liquid or other medium, exceeds a predetermined value; or if the temperature of the protected apparatus, such as a power rectifier, or of any medium decreases below a predetermined value.
27	Undervoltage relay is a device that functions on a given value of undervoltage.
32	Directional power relay is one that functions on a desired value of power flow in a given direction, or upon reverse power resulting from arcback in the anode or cathode circuits of a power rectifier.
38	Machine's bearing temperature protection.
49	Machine, or transformer, thermal relay is a device that functions when the temperature of an ac machine armature, or of the armature or other load-carrying winding or element of a dc machine, or converter or power rectifier or power transformer (including a power rectifier transformer), exceeds a predetermined value.
50	Instantaneous overcurrent or rate-of-rise relay is a device that functions instantaneously on an excessive value of current, or on an excessive rate of current rise, thus indicating a fault in the apparatus or circuit being protected.
51	AC time overcurrent relay is a device with either a definite or inverse time characteristic that functions when the current in an ac circuit exceeds a predetermined value.
52	AC circuit breaker is a device used to close and interrupt an ac power circuit under normal conditions or to interrupt this circuit under fault or emergency conditions.
55	Power-factor relay is a device that operates when the power factor in an ac circuit becomes above or below a predetermined value.
56	Field application relay is a device that automatically controls the application of the field excitation to an ac motor at some predetermined point in the slip cycle.
59	Overvoltage relay is a device that functions on a given value of overvoltage.

(*Continued*)

TABLE F-1 (*Continued*)

Device Number	*Definition and Function*
63	Liquid or gas pressure, level, or flow relay is a device that operates on given values of liquid or gas pressure, flow, or level, or on a given rate of change of these values.
81	Frequency relay is a device that functions on a predetermined value of frequency—either under or over or on normal system frequency—or rate of change of frequency.
86	Locking-out relay is an electrically operated hand or electrically reset device that functions to shut down and hold an equipment out of service on the occurrence of abnormal conditions.
87	Differential protective relay is a protective device that functions on a percentage or phase angle or other quantitative difference of two currents or of some other electrical quantities.

G BIBLIOGRAPHY

Circuits

1 Boylstad, R. *Electronic Devices and Circuit Theory.* Columbus, OH: Charles Merrill, 1982.

2 Nilsson, J. W. *Electric Circuits*, 2d ed. Reading, MA: Addison-Wesley, 1985.

Electric Machines

3 Bartholomew, D. *Electrical Measurements and Instrumentation*, Boston: Allyn and Bacon, 1963.

4 Del Toro, V. *Electric Machines and Power Systems.* Englewood Cliffs, NJ: Prentice-Hall, 1985.

5 Fitzgerald, A. E., Kingsley, C., and Umans, S. *Electric Machinery*, 4th ed. New York: McGraw-Hill, 1983.

6 General Electric Company, *Permanent Magnet Manual.* Edmore, MI: Magnetic Materials Business Section.

7 McPherson, G. *An Introduction to Electric Machines and Transformers.* New York: John Wiley and Sons, Inc., 1981.

8 Match, L. W., and Morgan, J. D. *Electromagnetic and Electromechanical Machines*, 3d ed. New York: Harper & Row, 1986.

9 O'Kelly, D., and Simmons, S. *Introduction to Generalized Electrical Machine Theory.* New York: McGraw-Hill, 1968.

10 Schulz, R., and Smith, A. *Introduction to Electric Power Engineering.* New York: Harper & Row, 1985.

11 Slemon, G. R., and Straughen, A. *Electric Machines.* Reading, MA: Addison-Wesley, 1980.

Controls

12 Harwood, P. B., *Control of Electric Motors.* New York: John Wiley, 1952.

13 Kosow. I. L., *Control of Electric Machines.* Englewood Cliffs, NJ: Prentice-Hall, 1973.

Power Systems

14 Stevenson, W. *Elements of Power System Analysis.* New York: McGraw-Hill, 1982.

Step Motors

15 Bodine Electric Company: Handbook, Chicago, Ill.

ANSWERS TO PROBLEMS

Chapter 1

1-1 **a)** 133%; **b)** 5 s. **1-2** **a)** 5.77 V; **b)** 0 V; **c)** 8.33 W.

1-3 73.52 V. **1-4** $I_1 = 10.3$ A, $I_2 = .37$ A, $I_3 = 1.1$ A. **1-5** 13.6 V.

1-8 **b)** 1.45 mV/m, rms. **1-9** $-UB[L_1 + L_2 + L_3(\cos\theta_1 + \sin\theta_1)]\tilde{a}_y$ V.

1-10 **a)** 3023.9; **b)** 63.7 μWb (max); **c)** .57 T (rms); **d)** 150.9 A/m;
e) 90.1 A.

1-11 .75 mWb. **1-12** 1.88 H. **1-13** $\dfrac{-\mu_0 b^2 i_1 i_2}{\pi a(a+b)}\tilde{a}_x$

1-15 **b)** **(1)** .34 H; **(2)** .67 J; **(3)** -59.57 N; **(4)** .22 T.

1-16 **a)** 1.68 H; **b)** 83.96 J; **c)** 339.6 J.

1-17 **a)** $-2L_{12}I_m^2 \sin 2\beta \sin^2(\omega t - \theta)$ N · m. **b)** $\dfrac{L_{12}I_m^2}{2}\sin 2\theta$ N · m.

Chapter 2

2-1 **b)** 222.93 V. **2-2** **b)** 2509.8 V; **c)** 4.57%, .949; **d)** $.05\underline{/50.2^\circ}$ pu.

2-3 **a)** 300.96 A; **b)** .98/.97; **c)** .90 leading.

2-4 **a)** 1551.6 W, 387 W; **b)** 322.5 W.

2-5 **a)** 240 V; **b)** 2.12%, $-.7$%, 4.2%.

2-6 **a)** .013 pu, 4.07%; **b)** 1.05 pu.

2-7 **a)** $I_{AB} = 4$ A; **b)** $I_{AB} = I_{CB} = 4$ A; **c)** $I_a = 167.1$ A; $I_{AB} = I_{BC} = I_{CA} = 11.1$ A.

2-8 **a)** .016 + j.026 pu, .043 + j.087 pu; **b)** 55.2 kW; **c)** 27.4 kV.

2-9 **a)** 90.7 kVA; **b)** $I_a = 70.77$ A, $I_A = 4.68$ A,
$I_b = 95.2$ A, $I_B = 6.3$ A,
$I_c = 68$ A, $I_C = 4.5$ A.

2-10 **a)** **(1)** 277.1 A, $554.3\underline{/120^\circ}$ A, $277.1\underline{/-60^\circ}$ A;
(2) 0 V; **(3)** fuse in line c will blow;
b) **(1)** 277.1 A, $277.1\underline{/-120^\circ}$ A, $277.1\underline{/120^\circ}$ A; **(2)** 277.1 V;
c) **(1)** $i_c = \infty$ A, **(2)** 0 V, **(3)** fuse in line c will blow;
d) **(1)** 277.1 A; **(2)** 277.1 V.

2-11 **a)** 25 kVA; **b)** 20 kVA.

2-12 **a)** 680 kVA, 70.6%; **b)** .99; **c)** 10.75 kA.

2-13 **a)** 4350.8 V; **b)** 787.8 A, 433 A; **c)** 655 kVA, 360 kVA;
d) 180.2 A; **e)** 3.3 kA.

2-14 **(1)** 441.9 V, 4 A, 1202.8 A; **(2)** 0 V, 35.6 A, 10.7 kA; **(3)** 481.1 V.

2-15 **a)** $54,259, $54,516; **b)** $44,928, $47,129; **c)** $48,660, $49,928.

Chapter 3

3-1 **a)** 3.33%; **b)** 4; **c)** $I_L = 11.8$ A.

3-3 **a)** **(1)** 1200 r/min **(2)** .05 pu; **(3)** 309.84 A;

(4) 64.82 A, .98 Pf, 200.6 N · m, .90; **(5)** .05 + j.07 pu, .05 + j.12 pu;
(6) .2425 pu, 213.2 A; **(7)** 447.3 N · m; **b)** 312.8 A.

3-4 **a)** .05 pu; **b)** .20 pu; **c)** .8 pu; **d)** 2.10 pu.

3-5 **a)** $R_1 = .08\ \Omega$, $R_2 = .13\ \Omega$, **b)** 226.4 N · m.
$X_1 = .17\ \Omega$, $X_2 = .25\ \Omega$,
$R_m = 2.47\ \Omega$, $X_m = 20.67\ \Omega$;

3-6 **a)** .382 N · m/rad/s; **b)** .155 Ω/phase; **c)** 1674 r/min.

3-8 **a)** $I_A = 199.1$ A, $I_B = 166.1$ A, $I_C = 129.2$ A, $I_n = 75.9$ A;
b) $I_A = 102.8$ A, $I_B = 94.3$ A, $I_C = 76.5$ A, $I_n = 75.9$ A.

3-9 **a)** 230 kVAR; **b)** .99 lag.

3-10 **a)** .993 pu; **b)** 1793 kVAR; **c)** 1.9 pu.

3-11 **a)** 16.8 A; **b)** 115.1 N · m.

3-12 **a)** .81; **b)** 1.71 pu; **c)** 266.5 kVAR.

3-13 **a)** 1140 r/min, **b)** 2.2 s; **c)** 1.34 s. **3-14** **a)** 3.05 s; **b)** 114.1 rad/s.

3-15 **a)** 78.96 kW-s, **b)** 3.24 s, 236.87 kW-s; **c)** 4.70 s, 315.83 kW-s;
d) 1.46 Ω, **e)** 1.58.

3-16 **b)** 59.4 N · m. **3-19** $11,019, $6086.

Chapter 4

4-2 1.2 pu. **4-3** **a)** **(1)** 70.65 A, 19.68 N · m; **(2)** 1783.6 W, 9.79 N · m;
b) **(1)** 70.65 A, 12.39 N · m; **(2)** 1116 W, 6.1 N · m.

4-4 **a)** $T = -.048\omega + KV_c$; **c)** $\dfrac{\omega(s)}{T_L(s)} = \dfrac{-50}{2.9 + s}, \dfrac{\omega(s)}{V_c(s)} = \dfrac{50K}{2.9 + s}$; **d)** 124 r/min.

4-5 **a)** 29.84 A, .99 lagging; **b)** 8.86 N · m; **c)** .18 N · m; **d)** 8.70 N · m.

4-6 **a)** 1663.4 and 2006 A; **b)** 119.22 μF. **4-7** 184.4 rad/s.

Chapter 5

5-1 **a)** 520 V, L–L; **b)** 5.7°; **c)** 5.13 kW, 3.57 kW.

5-2 **a)** $49.23\angle 25.8°$; **b)** 4289.8 V, L–L, −5.6°; **c)** .02 + j.10 pu.

5-3 **a)** 7.5 A; **b)** 99.5 A. **5-4** **a)** −86.2°; **b)** −3.8°.

5-5 **a)** **(1)** 2441.97 V, L–L;
(2) −25.7°;
(3) 9.6 A;
(4) 203.96 A;
(5) 777.2 kW, 777.2 kVA, 0 kVAR;

b) $1.25I_f$: **(1)** 3052.46 V, L–L;
(2) −20.3°;
(3) 12 A;
(4) 240.5 A;
(5) 777.2 kW, 916.6 kVA, 485.9 kVAR;

$\underline{\underline{.75I_f}}$: **(1)** 1831.5 V, L–L;
(2) −35.4°;
(3) 7.2 A;
(4) 245.1 A;
(5) 777.2 kW, 934 kVA, 518 kVAR.

5-6 **a)** 1203.7 V, L–L; **b)** $21.3\angle{-103^\circ}$ A; **c)** 7.8 A;
d) $24.3\angle{-121.2^\circ}$ A, −18.2°; **e)** 1367.2 V, L–L.

5-7 **a)** 4 poles; **b)** $.03 + j.35$ pu; **c)** 14.45 kW; **d)** 2257.5 N · m.

5-8

Questions	Parameter	Operating at Rated Conditions, Unity Pf	Constant Field Excitation: Increase in the Input Power of the Prime Mover by 25%	Constant Field Excitation: Decrease in the Input Power of the Prime Mover by 25%
1	Excitation voltage in pu	$\sqrt{2}$	$\sqrt{2}$	$\sqrt{2}$
2	Torque angle	45°	62.1°	32°
3	Armature current in pu	$1.0\angle{0^\circ}$	1.30	.78
4	Field current in A	142	142	142
5	Real power in pu	1.0	1.25	.75
	Reactive power in pu	0	.34	.20

5-9 **a)** 1100 V, 7 A; **b)** 15; **c)** 18.5 A.

5-10 **a)** 1.0; **b)** −27.5°; **c)** 1.26 pu; **d)** 298.4 kW, 27.5%.

5-11 **a)** 39.8°; **b)** 62%; **c)** .18 pu.

5-12 **a)** 2.83 kA; **b)** 1.41 kA; **c)** 1.41 kA.

5-13 **a)** 8.82 pu; **b)** 15.27 pu; **c)** 12.47 pu, 6.23 pu;
d) $1.62 + 3.25e^{-t/1.1} + 3.95e^{-t/.035}$ pu; **e)** 4.58 pu; **f)** 6.42 pu.

Chapter 6

6-1 120.96 N · m.

6-2 **a)** 48.41 A, 132.7 N · m, 781.10 r/min; **b)** 126.10 N · m, 822.1 r/min.

6-3 **a)** 1335.33 A; **b)** 1.74 Ω; **c)** 82.62 Ω. **6-4** **a)** 11.7%; **b)** 7.4%.

6-5 **a)** 1946.3 r/min; **b)** 71.1 N · m; **c)** 14.1 kW; **d)** $281.13\left(\frac{i_a}{25 + i_a}\right)$

6-6 2000 r/min. **6-9** **a)** $450(t - .75 + e^{-2t} - .25e^{-4t})$ J; **b)** $300(1 - e^{-2t})$ V.

6-10 **a)** $65.22(1 - e^{-4.6t})$ rad/s; **b)** .22 s; **c)** $97.83 + 652.2e^{-4.6t}$ A.

6-12 **a)** 7.58; **b)** 278.4 kVA; **c)** .74 lag.

6-13 **a)** 53.67°; **b)** 516.5 A; **c)** .57 lag.

Appendix

1 **a)** $228,416;

2 **a)** $1.34/kW, $.035/kWh; **b)** $2027; **c)** 1395 kVAR.

3 .81 lagging, 102.86 A.

4 **a)** 277.13 V/phase; **b)** $27.71\angle -36.9^\circ$ A; **c)** 27.71 A; **d)** .8 lagging; **e)** 18.43 kW; **f)** $30\angle 36.9^\circ\ \Omega$.

5 .93, 557.3 V, L–L.

6 **a)** 24 A, 41.57 A; **b)** 9.98 kW, 19.95 kW, 29.93 kW; **c)** −1.27 kW, 4.72 kW, 3.45 kW.

7 48 A. **9** **a)** 446.9 V, L–L, **b)** 415 kVAR. **10** **c)** .092 μf, .03 μf.

11 **a)** $.12 + j.66\ \Omega$/phase; $.0015 + j.009\ \Omega$/phase; **b)** 27 MVA.

12 **d)** 473.7 V, L–L. **13** 14.25 mW. **15** 99.2 W. **16** 2.56 V.

17 **a)** 40 Ω; **b)** 45.8 mA.

18 **a)** 2.8 V; **b)** 15.94 A; **c)** 23.93 W; **d)** 5.25 V.

19 **a)** 423.8 W, .42 lagging; **b)** 118.6 W, .22 lagging.

INDEX

D

E

J

K

L

M

N

O

P

Q

R

S

T

U

V

W

Z